The Routledge Handbook of Henri Lefebvre, The City and Urban Society

The Routledge Handbook of Henri Lefebvre, The City and Urban Society is the first edited book to focus on Lefebvre's urban theories and ideas from a global perspective, making use of recent theoretical and empirical developments, with contributions from eminent as well as emergent global scholars.

The book provides international comparison of Lefebvrian research and theoretical conjecture and aims; to engage with and critique Lefebvre's ideas in the context of contemporary urban, social and environmental upheavals; to use Lefebvre's spatial triad as a research tool as well as a point of departure for the adoption of ideas such as differential space; to reassess Lefebvre's ideas in relation to nature and global environmental sustainability; and to highlight how a Lefebvrian approach might assist in mobilising resistance to the excesses of globalised neoliberal urbanism. The volume draws inspiration from Lefebvre's key texts (*The Production of Space*; *Critique of Everyday Life*; and *The Urban Revolution*) and includes a comprehensive introduction and concluding chapter by the editors. The conclusions highlight implications in relation to increasing spatial inequalities; increasing diversity of needs including those of migrants; more authoritarian approaches; and asymmetries of access to urban space. Above all, the book illustrates the continuing relevance of Lefebvre's ideas for contemporary urban issues and shows – via global case studies – how resistance to spatial domination by powerful interests might be achieved.

The Handbook helps the reader navigate the complex terrain of spatial research inspired by Lefebvre. In particular the Handbook focuses on: the series of struggles globally for the 'right to the city' and the collision of debates around the urban age, 'cityism' and planetary urbanisation. It will be a guide for graduate and advanced undergraduate teaching, and a key reference for academics in the fields of Human Geography, Sociology, Political Science, Applied Philosophy, Planning, Urban Theory and Urban Studies. Practitioners and activists in the field will also find the book of relevance.

Michael E. Leary-Owhin has an international reputation in the fields of urban planning and regeneration. He has over 30 years' experience in the field and has practised in the public and private sectors, recently giving expert witness evidence at a major urban regeneration public inquiry in the UK.

John P. McCarthy is Associate Professor in Urban Studies in The Urban Institute, School of Energy, Geoscience, Infrastructure and Society at Heriot-Watt University. He worked as a planning practitioner in the public sector in London in the 1980s, and has worked in academia at the University of Dundee and Heriot-Watt University.

The Routledge Handbook of Henri Lefebvre, The City and Urban Society

Edited by Michael E. Leary-Owhin and John P. McCarthy

LONDON AND NEW YORK

First published 2020
by Routledge
4 Park Square, Milton Park, Abingdon, Oxon OX14 4RN
605 Third Avenue, New York, NY 10017

First issued in paperback 2023

Routledge is an imprint of the Taylor & Francis Group, an informa business

British Library Cataloguing-in-Publication Data
A catalogue record for this book is available from the British Library

Library of Congress Cataloging-in-Publication Data
Names: Leary-Owhin, Michael Edema, editor.
Title: The Routledge handbook of Henri Lefebvre, the city and urban society / edited by Michael E. Leary-Owhin and John P. McCarthy.
Description: New York: Routledge, 2020.
Identifiers: LCCN 2019029003 (print) | ISBN 9781138290051 (Hardback) | ISBN 9781315266589 (eBook)
Subjects: LCSH: Urbanization–History. | Cities and towns–History. | City planning–Environmental aspects. | Lefebvre, Henri, 1901–1991.
Classification: LCC HT111 .R669 2020 (print) | LCC HT111 (ebook) | DDC 307.7609–dc23
LC record available at https://lccn.loc.gov/2019029003
LC ebook record available at https://lccn.loc.gov/2019029004

ISBN: 978-1-03-256994-9 (pbk)
ISBN: 978-1-138-29005-1 (hbk)
ISBN: 978-1-315-26658-9 (ebk)

DOI: 10.4324/9781315266589

Typeset in Bembo
by Deanta Global Publishing Services, Chennai, India

Publisher's Note
The publisher has gone to great lengths to ensure the quality of this reprint but points out that some imperfections in the original copies may be apparent.

Contents

Editors

Michael E. Leary-Owhin has an international reputation in the fields of urban planning and regeneration. He has over 30 years' experience in the field and has practised in the university, public and private sectors, recently giving expert witness evidence at a major urban regeneration public inquiry in the UK. He has carried out funded research and published widely. He sits on the board of several academic journals and is the author of numerous peer-reviewed journal articles, the co-editor of *The Routledge Companion to Urban Regeneration* and the author of the (2016) research monograph, *Exploring the Production of Urban Space*. He regularly chairs sessions and presents papers at major international conferences. He was until recently the course director for the MA Planning Policy and Practice at London South Bank University. He is now an independent academic, researcher and planning consultant.

John P. McCarthy is Associate Professor in Urban Studies in The Urban Institute, School of Energy, Geoscience, Infrastructure and Society at Heriot-Watt University. He worked as a planning practitioner in the public sector in London in the 1980s and has worked in academia at the University of Dundee and Heriot-Watt University. He has published books, book chapters and journal articles in a variety of journals and presented at many conferences, including as keynote speaker. He has also carried out funded research for a range of funding bodies, including research councils and government. His current interests include tourism- and leisure-based regeneration as well as culture-led regeneration.

Contributors

Nick Bailey
Emeritus Professor, Faculty of Architecture and the Built Environment
University of Westminster
London, UK

Bülent Batuman
Associate Professor, Department of Urban Design and Landscape Architecture
Bilkent University
Ankara, Turkey

Iain Borden
Professor of Architecture and Urban Culture, Bartlett School of Architecture
University College
London, England

Chris Butler
Lecturer, Griffith Law School
Griffith University
Nathan, Brisbane, Queensland, Australia

Rebio Diaz Cardona
Assistant Professor
LaGuardia Community College-CUNY
New York, USA

Carl Cassegård
Associate Professor of Sociology
University of Gothenburg
Sweden

Damiano Cerrone
Project Researcher, School of Architecture
Tampere University of Technology
Tampere, Finland

Nathaniel Coleman
Reader, School of Architecture, Planning and Landscape
Newcastle University
Newcastle upon Tyne, UK

Hans-Peter Meier Dallach
Professor
CulturProspectiv/World Drives Association
Zurich, Switzerland

Ian Ellison
Partner
3edges Workplace Ltd.
Northallerton, UK

Gülçin Erdi
CNRS Senior Researcher, Center for Research on Cities, Environment, Territories and Society
University of Tours François Rabelais
Tours, France

Pierre Filion
Professor, School of Planning
University of Waterloo
Canada

Derek R. Ford
Assistant Professor of Education Studies
DePauw University
Greencastle, IN

Miguel Torres García
Independent researcher
Seville, Spain

Michael Granzow
Mitacs Research Intern, Faculty of Arts
University of Alberta
Canada

Oded Haas
Vanier Scholar, Faculty of Environmental Studies
York University
Toronto, Canada

Steve Hanson
Associate Lecturer
Manchester Metropolitan University
UK

Gunter Heinickel
ScienceScapes
Berlin, Germany

Chris Hesketh
Senior Lecturer in International Political Economy
Oxford Brookes University
UK

Marie Huchzermeyer
Professor, School of Architecture and Planning
University of the Witwatersrand
Johannesburg, South Africa

Alasdair J.H. Jones
Assistant Professor, Department of Methodology
London School of Economics
UK

Nick Jones
Lecturer in Film, Television and Digital Culture, Department of Theatre, Film and Television
University of York
UK

Silvia Jorge
Faculty of Architecture
University of Lisbon
Portugal

Siew-Peng Lee
Visiting Research Fellow, Department of Anthropology
Brunel University
UK

Panu Lehtovuori
Professor of Planning Theory, School of Architecture
Tampere University of Technology
Tampere, Finland

Ho Hon Leung
Professor, Department of Sociology
State University of New York College
Oneonta, NY

Saara Liinamaa
Assistant Professor, Department of Sociology and Anthropology
University of Guelph
ON, Canada

Ran Liu
Associate Professor, College of Resource Environment and Tourism
Capital Normal University
Beijing, China

Anna Ludovici
PhD student, Institute of Geography and Spatial Planning
University of Lisbon

Eveliina Lyytinen
Postdoctoral Researcher (Academy of Finland)
Migration Institute of Finland
Eerikinkatu, Finland

Jorge Malheiros
Associate Professor, Institute of Geography and Spatial Planning
University of Lisbon

Luca P. Marescotti
Emeritus Professor, Dipartimento di Architettura
Politecnico di Milano
Italy

Gregory Marinic
Associate Professor of Design, School of Design
Syracuse University
New York, USA

Rishika Mukhopadhyay
Doctoral Researcher, Department of Geography
University of Exeter
UK

Sana Murrani
Lecturer in Architecture and Associate Head of School for Graduate Affairs
School of Art, Design and Architecture
University of Plymouth
UK

Mee Kam Ng
Professor, Department of Geography and Resource Management
The Chinese University of Hong Kong
China

Oscar Olivier-Didier
Senior Urban Designer, Borough of the Bronx New York City, Department of City Planning and Adjunct Faculty, Visual Arts Department
Fordham University,
New York, USA

Daniel Paiva
Researcher, Centre for Geographical Studies
Institute of Geography and Spatial Planning
University of Lisbon
Portugal

Margarida Queirós
Assistant Professor, Institute of Geography and Spatial Planning
Centre of Geographical Studies
University of Lisbon

Mark Rainey
Centre for Cultural Studies, Goldsmiths
University of London
UK

Claire Revol
Lecturer in Philosophy and Planning, Institut de Géographie Alpine
Grenoble-Alpes University
Grenoble, France

Ashraf M. Salama
Professor, Department of Architecture
University of Strathclyde
Glasgow, Scotland

Nicholas A. Scott
Assistant Professor of Sociology, Department of Sociology and Anthropology
Simon Fraser University
Burnaby, BC, Canada

Hilary J. Shaw
Director and Senior Research Consultant
Shaw Food Solutions
Newport Shropshire, UK

Julia J.A. Shaw
Professor of Law and Social Justice, Faculty of Business and Law
De Montfort University
Leicester, UK

Rob Shields
Henry Marshall Tory Chair and Professor, Faculty of Arts
University of Alberta
Canada

Marcelo Lopes de Souza
Professor of Geography, Department of Geography
Federal University of Rio de Janeiro
Rio de Janeiro, Brazil

Michael Spacek
PhD candidate
Carleton University
Ottawa, Canada

Jani Tartia
Grant Researcher and Doctoral Student, School of Architecture
Tampere University of Technology
Tampere, Finland

Matthew Thompson
Postdoctoral Research Fellow, Heseltine Institute for Public Policy and Practice
University of Liverpool
UK

Charalampos Tsavdarolgou
Marie Curie Fellow, Department of Human Geography, Planning and International Development Studies
University of Amsterdam

Sílvia Leiria Viegas
Architect and Postdoctoral Researcher, Centre for Social Studies
University of Coimbra
Portugal

Florian Wiedmann
Lecturer in Architecture
Frankfurt University of Applied Sciences
Frankfurt am Main
Germany

Tai-Chee Wong
Professor, Faculty of Humanities and Social Sciences
Southern University College
Jalan Skudai, Johor Bahru, Malaysia

Esther H.K. Yung
Assistant Professor, Department of Building and Real Estate
Hong Kong Polytechnic University
China

Introduction: 'Urban' ideas for two centuries

Michael E. Leary-Owhin and John P. McCarthy

> As for the reproduction of knowledge, it incorporates not only the reproduction of social relations (through the relation between teacher and student), but that of ideologies, mixed with concepts and theories, that take the form of topics, citations, whether concealed or revealed, 'research', evaluations, redundancies, mixed with information, reductions that are more or less contradicted and so on. A certain relation between knowledge and non-knowledge, which ideology holds together, is also transmitted – especially when it concerns Marxism and the city.
>
> *(Lefebvre 2016: 143)*

> Instead of the spectre of communism, it is the spectre of the urban, the shadow of the city that now haunts Europe. What this means is that urban questions are extremely important, and cannot be ignored in studies of quite diverse topics. There are questions about the political economy of the urban – rent and the distribution of living space – and questions about the politics of the urban – questions of centre-periphery relations, marginalization, ghettoization, segmentation, the organization of its space and the power relations that course through it.
>
> *(Elden 2004: 143)*

Context and rationale

Did Henri Lefebvre obliterate the city as a theoretical and analytical construct in 1970? Yes, and no. Yes, because in the book *The Urban Revolution* he did force us to *reconsider* the idea of the city as a physical entity that at its boundary gives way to the rural and the countryside. No, because Lefebvre did not deny the bounded physical city as conceived in the 19th century with its sprawling tentacles of economic and power relations that *encompassed* the countryside and the globe. Manchester, England, the world's first modern industrial and globalised trading city, was at the forefront of rethinking the essence of the city. In recognition of this rethinking of the city – its 'globalisation' – Manchester's 1842 coat-of-arms depicted a globe and an ocean-going sailing ship. The ceiling of the Great Hall of the city's 19th century neo-Gothic Town Hall portrays the flags of many cities and countries around the world – the city's global trading partners and source of wealth. Rather than a physically bounded entity, the city began to be reconceptualised as an unbounded assemblage of economic and social relationships – at the centre of globalised capitalism. In similar ways, London was reconceptualised as the globalised centre of Empire.

Debate about the essence of cities had been rumbling along for centuries. William Shakespeare stressed the city as a collection of social interactions, when he had the Plebeians say in the play *Coriolanus*, the people are the city. Georg Simmel considered the 19th-century modern industrial city of impersonal commercial relationships, to be a state of mind of the blasé metropolitan. And Robert Park of the 1920s Chicago School went further, arguing that the city is not merely a physical form but also the site of social relations: institutions, customs, traditions and organised attitudes and sentiments. Lefebvre's reconsideration of the city necessitated of course an appreciation of the role of the capitalist mode of production in the formulation and reproduction of the city and the ensuing urbanisation. Neoliberalism is the current productive moment; rethinking the city on a planetary scale is still happening and is one of the themes of this Handbook.

Lefebvre's influence on our understandings and study of the city and the urban has been immense. Citations of his work are superabundant. His formulation of the notions of the production of space, planetary urbanisation and the right to the city elevate him to the level of one of the most important philosophers of the 20th century. This book is proposed at a propitious moment for three reasons. First, the recent explosion of Lefebvrian-inspired production of space research. Second, the implosion of right to the city (RTC) struggles concentrated in cities worldwide. Third, the recent collision of debates around the championing or deriding of: the urban age, 'cityism' and planetary urbanisation and the diffusion of capitalism across the globe. Each of these developments has complicated the already existing complex terrain of Lefebvrian literatures regarding cities, urbanisation and the production of space. From the intermittent drip drip drip of the 1990s, urban empirical research based on Lefebvre's ideas has become a worldwide torrent. The range of applied topics and geographical locales has also multiplied at a bewildering rate. Despite or because of this there remains no agreement about how to interpret Lefebvre's production of space ideas or even the fundamentals of the spatial triad. A somewhat disparate body of literature has emerged rather than one with a significant degree of cohesion.

The title for the book is on the one hand a straightforward, unproblematic signal as to content, on the other it is meant to be a provocative intervention. 'The city' as an analytical category continues, with the advent of Lefebvrian urbanisation, to provoke controversy and debate. We contend that as a material and lived social reality it is unlikely to wither away any time soon. The city remains crucial 'virtually and materially, for progressive politics' (Merrifield 2013: xii). Clearly, for Lefebvre from the 1970s, the urban transcends, industrialisation and the bounded physical city and the distinction between city and rural becomes problematic. The urban as a focus for analysis consists of: space, everyday life (including work *and* leisure) and neo-capitalist economic and social relations. Lefebvrian urbanisation implies increasing alienation but also the capacity for leisure and festival. Similarly, 'urban society' is something that can be grasped immediately but is also replete with contested understandings. For Lefebvre, urban society offers a critique of planetary urbanisation and everyday life under the abstract space of globalised neo-capitalism, but simultaneously, it offers a tantalizing glimpse of a post-capitalist 'utopian' society and the achievement of the possible-impossible and the dialectical relationship between these two moments.

Over the years, Lefebvrian-inspired urban space research has fragmented theoretically, empirically, thematically and geographically. What may be called the traditional Lefebvrian empirical research approach in the broad production of urban space field tends to focus on counter-posing official representations of space with somehow more authentic, quotidian spaces of representation. Privileging spaces of representation leads to invigorating research but simultaneously is rather at odds with Lefebvre's insistence that the dialectical interaction of the three moments of the spatial triad together produce urban space. This being so, the nature of the urban space produced would become problematised in positive ways if the focus was to shift away from spaces of representation. Lived space, everyday space and abstract space also figure predominantly in the literatures and, though at times highly insightful, these foci tend to elide other powerful spatial ideas that Lefebvre championed after what might be called his urban turn in the 1960s.

Being a neo-Marxist, Lefebvre was suspicious of the state's role in neo-capitalism. These suspicions were crystallised in the 1960s when the French state went into partnership with multinational business to build the new town of Mourenx (see Leary-Owhin in Part 6). He is explicit about the dangers of the alliance of neoliberal governance, private property development and planning in the Preface to the 1986 edition of *The Production of Space* (TPOS) (Lefebvre in Elden et al. 2003: 207). And he is wary about the implications of neoliberalism. Since the 1980s neoliberalism through globalisation has become even more significant in the production of urban space and resisting some of the calls for the right to the city. Metropolitan centres and world cities in the Global North have seen a concentration of power and resources and the formation of new urban elites who are skilled at operating a 'neoliberal development model' (Schmid 2011: 53).

Planetary urbanisation facilitates a tightening grip of globalised neoliberal capitalism even at a time when genuine resistance to its more unsavoury consequences is being documented. The unbounded city of planetary urbanisation calls attention to a variety of environmental crises, and given that Lefebvre's first research interests were rooted in the rural rather than the urban, it is surprising that little attention has been given to his ideas regarding the transition from absolute space of 'nature' to the abstract space of exchange value neo-capitalism and back again. Planetary urbanisation has several key aspects: the globalised spread of capitalism, globalised capitalism's harmful impacts on nature (or what we would now call the environment), globalised capitalism's harmful impacts on everyday life and the potential to transcend capitalist alienation and exploitation in the achievement of not so much a post-industrial or post-consumerist but an urban society. In planetary terms, Lefebvre demurred from applying his production of space ideas universally, but in fact this has happened, as documented in this book.

In city districts of the Global North and South where lives and communities are disrupted and displaced by insensitive private redevelopment and state-led urban regeneration schemes, rights to the city are being asserted or repulsed in all manner of ad hoc and contradictory ways. Activists draw on Lefebvre's RTC ideas whether consciously or not, and a stream of literature stresses the importance of city centrality rather than an explosion of urban fragments into an amorphous urbanisation. More fundamentally, the reopening of the 'what is a city?' question, allied to some strong critiques of the emerging planetary urbanisation thesis, offers the prospect of fascinating theoretical development. However, there is also a danger that the cityism-urbanisation debate allows theorists to talk past each other, rather than deploy Lefebvre's spatial ideas to unravel and understand some of the major issues facing the planet especially in cities, whether conceptualised as bounded, unbounded or both.

Book aims

Lefebvre's urban society concept requires therefore an intense engagement with: urbanisation, the production of space, everyday life and state implicated neo-capitalism. Stuart Elden reminds us, in the second epigraph to this Introduction, that urban questions cannot be ignored, including, we maintain, questions about the production of city and urban space.

With this emerging context in mind we propose five ambitious aims for the book:

1. To provide an accessible forum across the Global North and South (accepting that these terms are neither definitive or unproblematic) in the hybrid field of city-urbanisation, for

the international comparison of Lefebvrian research and theoretical conjecture that gives new impetus for a fresh take on Lefebvre's ideas in the challenging neoliberalised global context of the early 21st century.

2. To critique, rethink and extend Lefebvre's spatial ideas in the context of early 21st-century contingencies, we endeavour to engage afresh with Lefebvre's theoretical musings from the vantage point of various current discourses resulting from recent urban, social and environmental upheavals across the planet.
3. To use the spatial triad both as a research tool *and* a point of departure for the adoption of other key but neglected Lefebvrian ideas, especially: differential space, absolute space, counter-projects, spatial coalitions and a historical approach to the exploration of the production of urban space.
4. To reassess what a Lefebvrian approach to issues of nature and environmental sustainability might mean, bearing in mind the widespread paradox of relentlessly expanding yet simultaneously shrinking cities in the Global North and expanding megacities in the Global South.
5. To highlight emerging theoretical developments and potentials and explain how a Lefebvrian approach might be developed to further mobilise RTC local resistance by activists and politicians, to the harmful excesses of globalised neoliberal urbanism.

The book draws its primary inspiration from the ideas, themes and arguments developed in the Anglophone editions of three of Lefebvre's most celebrated texts regarding cities and urban society: *The Production of Space* (1991), *Critique of Everyday Life* Volume III (2008[1981]) and *The Urban Revolution* (2003). It draws also on three other texts: *Right to the City* (1996, *Rhythmanalysis* (co-authored with his last wife, Catherine Régulier-Lefebvre and published posthumously in 1992) and *Introduction to Modernity* (1995 [1962]). This does not of course preclude the incorporation of concepts and approaches derived from Lefebvre's other major texts. That said, we acknowledge that our book is also limited in scope. Lefebvre was a polymath with a vast array of publications covering: philosophy, Marxism, communism, nationalism, modernity, rural sociology, architecture, fascism, existentialism, literary criticism and European history. In a volume of this size we could not hope to include the whole range of the Lefebvre canon but focus instead on the city and the urban. In 2016 the English translation of *Marxist Thought and the City* appeared (Lefebvre 2016). In it, Lefebvre interrogates the works of Marx and Engels first-hand to explicate their understandings of the importance of cities for capitalism and neo-capitalism. Although it now has obvious relevance for this volume it was published too late to be incorporated substantially.

Production of Lefebvrian literature

A Lefebvrian inspired literature has emerged that has in the last two decades become massive, multi-disciplinary and worldwide. We make some sense of it by identifying four 'waves'. Our waves draw on but are distinct from the three waves of Kipfer et al. (2008: 3). They indicate our thematic approach to the literature rather than a chronological or mutually exclusive classification and exist in parallel. Similarly, an author may well appear in more than one wave. Lefebvre wrote from a 'Western' and European perspective, albeit that he was well-travelled and his work is certainly inflected with an internationalist perspective and appreciation. So it is no surprise that his ideas were not adopted in the Global South until the second millennium, but such authors are now well represented in the four waves. The notable exception is the adoption legally of the right to the city in Brazil in the 1980s. It is evident that Lefebvre's provocative and challenging ideas, expressed often in a repetitive and at times convoluted writing style, were interpreted first in the Anglophone world by what we call a first wave of urban theorists who engaged with the literature in the original French. It will become obvious that the range of authors, countries, issues and research approaches sketched out is more extensive than is possible in this Handbook, but it does serve to locate the volume in the global literature.

Lefebvre came to the attention of the Anglophone world in five ways. First, by Anglophone academics reading his Marxist political work in the original French and writing about it in English (for instance Litchtheim 1966). Second, the early English translations of his books (e.g. *The Explosion: Marxism and the French Upheaval* Lefebvre 1969; *Dialectical Materialism* Lefebvre 1968). Third, Anglophone academics reading his city and urban work in the original French but writing about it in English (Harvey 1973 and Castells 1977). Fourth, the easier availability of English translations of his major works, especially TPOS after 1991. Fifth, the explosion of Lefebvrian scholarship after 2000, due in part to the contributions of Ed Soja and Rob Shields.

David Harvey first brought Lefebvre's urban ideas to the attention of the academic Anglophone world in the fields of human geography and a loosely defined urban studies, in the conclusions to his 1973 book *Social Justice and the City*. However, this is disputed (see Gottdiener below). This consideration of Lefebvre's urban work was, Harvey (1973: 302) admits readily, based mainly on his last-minute reading in French of *inter alia* Lefebvre's book *La Revolution Urbaine* (2003/1970), not translated into English until 2003 as *The Urban Revolution*. Readers will find the foreword to this version by Neil Smith most illuminating. Harvey was followed by Manuel Castells, whose severe critique (in French in 1972), *The Urban Question*, appeared in English in 1977 (discussed below). The likes of: Pickvance (1976), Saunders (1981), Anderson (1983), Ross (1983), Smith (1984), Gottdiener (1985), Soja (1989, 1996, 2010), Katznelson (1992), Merrifield (1993), Fernandes (2007) and Stanek (2011) enrich the first wave. These writers tended to engage with Lefebvre theoretically. And while they had their own particular interpretations of Lefebvre's key ideas, they tended not to use the production of space as a framework for empirical research. Of course, many scholars worldwide read the original French versions of Lefebvre's works, including some of the authors in this book.

A second wave of Anglophone writers takes a more biographical approach (see Harvey 1991; Shields 1999; Soja 1996; Elden 2004; Entrikin and Berdoulay 2005; Merrifield 2006; and Stanek 2011). Each author brings a different perspective to their task, for example: Shields and Merrifield tend to stress Lefebvre the man and sometimes quirky human being, Elden seeks comprehensiveness, while Stanek tends to focus on Lefebvre's empirical research and his interest in the practicalities of architecture and urban planning. Their ability and willingness to engage with Lefebvre in the original French certainly assists their endeavours.

It was the publication of the English version of TPOS in 1991 that sparked a third wave of Lefebvrian research and literature. It is regarded by many as the culmination of all his urban (and rural) work (see Elden, in Lefebvre 2016: viii). These researchers, primarily across the Global North, constitute the largest wave and did use a Lefebvrian theoretical framework for empirical research with varying locations and themes, notably: London, Allen and Pryke (1994); Glasgow, Los Angeles-Amsterdam, Soja (1996); racism, McCann (1999); Baltimore, Harvey (2000); bias of academic urbanism, Gottdiener (2000); Los Angeles, Dear (2000); California, Borden (2001); Coventry Hubbard et al. (2003); New Mexico, Carp (2008); Finland, Lehtovuori (2010); New York State, Buser (2012); Cleveland, Zingale et al. (2014); Czech Republic, Mulíček et al. (2016); Italy D'Ascoli (2018); and Pittsburgh, Neumann (2018).

We can see that this wave of empirical research and further contemplation stimulated theoretical engagements that seek to go beyond Lefebvre (Merrifield 2011; Leary-Owhin 2016) and render his approach and ideas relevant for a host of researchers and for contemporary issues, for example: the modern French novel and neoliberalism, Willging (2019); urban rights,

Harmon (2019); abstract space, Dimendberg (1998); utopia, Harvey (2000), Gardiner (2004) and Coleman (2014); Calvinist worship, MacDonald (2002); RTC and politics of the inhabitant, Purcell (2002, 2014), Mitchell (2003) and Harvey (2012); politics of nature, Janzen (2002); differential space, Groth and Corijn (2005); the senses, Degen (2008); counter-space, Leary 2009; collaborative planning, Andres (2013); revolutionary romanticism, Grindon (2013); planetary urbanisation, Merrifield (2014) and Buckley and Strauss (2016); spatial triad, Gustavsson and Ingemar (2016); RTC-differential space, Leary-Owhin (2016); rhythmanalysis, Edensor and Larsen (2017); conflict-urbanism, Oldenburg (2018); and spontaneous short-lived protests, Sun and Huang (2018).

Global South academics came later to Lefebvrian research and tend to cluster in the second wave, including empirical research in: Jakarta, Simone (2019); India, Goswami (2004) and Parikh (2019); Brazil, Monte-Mór (2005) and Friendly (2013); Mexico, Wilson (2011) and Hesketh (2013); northern Africa, Karplus and Meir (2013); Delhi, Bhattacharyya (2015); Yan'an, China, He and Lin (2015); Global South, Huchzermeyer (2015) and De Melo (2017); sub-Saharan Africa, Chirozva (2017); Lagos, Agbiboa (2017); and Congo/London, Garbin and Millington (2018). In addition, we see a thematic focus in the work of Manzi et al. (2018), neoliberal RTC; and Patel (2018), urban studies. That said, we do not seek to present an oversimplified Global North–South divide since: these imprecise categories remain moot, writers may originate in the North but research about and in the South and vice versa, and in part this reflects the globalisation of the emerging Lefebvrian academy as is evident in this Handbook.

The most recent fourth wave of Lefebvre scholarship takes the form of edited collections that tend to focus on theoretical development *and* empirical research dealing with: difference and everyday life (Goonewardena et al. 2008), urban revolution (Stanek et al. 2014) and planetary urbanisation process and outcome (Brenner and Schmid 2015). It is rather as one might expect for edited collections, a rather eclectic wave. These collections, like the present volume, provide a reference source for certain aspects of Lefebvre's work and through the careful and comprehensive discussion and of use citations, assist the reader in locating the papers in their Lefebvrian, historical and contemporary contexts. They also show Lefebvre being 'brought up to date' for the early 21st century, and this volume seeks to continue that emerging tradition.

Some key ideas and a surprise

Urban space is understood as both outcome and process. A Humanist neo-Marxist, rather than orthodox Marxist, Lefebvre departed company with Karl Marx regarding the significance of urban space (see Lefebvre's *The Urban Revolution* 2003). Although the recent publication in English of *Marxist Thought and City* (Lefebvre 2016) demonstrates that Marx was not silent on the significance of cities for social change. Lefebvre stressed the importance of urban space and its production through dialectical relationships for the maintenance of state-regulated and -implicated neo-capitalist society, whereas Marx of course stressed the importance of land, capital accumulation and dialectical struggle between the proletariat and bourgeoisie supported by a complicit state. For Lefebvre (1991), although class politics are important, the focus is on the relative power of those who use their positions of power to produce and reproduce urban space.

A defining feature of Lefebvre's theories is the importance of power relationships and the linkages between the private sector and the state, for the reproduction of neo-capitalist society. We would add that over recent decades the importance of civil society groups has also become apparent. Urban space cannot therefore be produced with out the formation of coalitions of interest, sometimes disparate coalitions. It is strange that Lefebvre's thoughts on the importance of coalitions in the production of space have been largely overlooked in the literature.

The production of space, based on the notion of the spatial triad, is one of the key themes running through this whole volume. Contradiction and conflict were key themes underlying Lefebvre's approach to understanding cities and the urban. Dialectical contradiction and struggle underpin Lefebvre's analysis of urban space and his ideas about societal change. Lefebvre gave prominence to the spatial triad in striving to understand neo-capitalism in the 20th century. Although suffused with complexity and some divergence of interpretation, the spatial triad does have an intuitive simplicity, and our approach sees its elements as follows (based on Leary-Owhin 2016: 13–15):

- **Spatial practice** has three major elements: 1) the physical, material city and its routine maintenance; 2) major urban redevelopment in the context of existing neo-capitalist and state power structures; 3) routines of daily life that conform with official representations of space. It is space directly perceptible through the senses – perceived space.
- **Representations of space**: rational, intellectualised, official conceptions of urban areas for analytical, administrative and property development purposes. They are produced by technocrats: architects, engineers, urbanists and planners but also artists with a scientific bent. They are the dominant representations and may be in the form of the written word, for example in city-wide zoning plans and strategy documents, or quasi-scientific visual representations of various kinds such as maps, master plans and design guides – conceived space.
- **Spaces of representation**: have two major elements: 1) urban everyday space as directly understood and lived by inhabitants and users in ways informed not so much by representations of space as by associated cultural memories, images and symbols imbued with cultural meaning; 2) emotional, artistic interpretations of city space by poets, writers and painters and other artists. These kinds of space overlay physical space and value places in ways that run counter to the dominant representations of space – imaginative and lived space.

Surprisingly perhaps, although the term 'spaces of representation' is used widely, it does not appear in Lefebvre (1991). Nicholson-Smith translated '*les spaces de représentation*' as 'representational spaces' Shields (1999: 161). The term spaces of representation is used instead and first appeared in English in Frank Bryant's translation of Lefebvre's *The Survival of Capitalism* (Lefebvre 1976: 26), originally published the year before as TPOS. Interestingly, it is here that Lefebvre first develops the idea of the spatial triad, referring to spaces of representation and representations of space, 'mental (abstract representations of space)' and 'social (real spaces – spaces of representation)' respectively. For Shields (1999: 161), the term spaces of representation is preferable because the Nicholson-Smith translation makes the triad 'more difficult to comprehend'). Interestingly, we can discern within the three moments of the spatial triad rhythms within spatial practice and lived space; we can also detect the importance of everyday understandings and practices in lived space.

In addition to the triad, three other Lefebvrian spatial concepts are important for understanding cities and the urban; they are:

- **Abstract space**; the urban spaces of state-regulated neo-capital characterised by private ownership, restricted access, restricted performance, commodified exchange value and the tendency to homogenisation. Abstract space is created as capitalist economic relations colonise the absolute space of nature in processes of urbanisation.
- **Differential space**; privileges inclusiveness and use value rather than the exchange value of abstract space, it may be long-lasting or transitory space and arises from the inherent vulnerabilities of abstract space. At some times Lefebvre imbues it with an immediate

'here-and-now' quality. It is also associated with a more utopian post-capitalist phase of urban society.

- **Spatial coalition**; social groupings of local or non-local inhabitants, the locus of social struggle, may include reactionaries, progressives, liberals or radicals and even revolutionaries, often forming unlikely alliances, that come together to promote a counter-project (or counter-plan) to produce a counter-space (sites within cities), in opposition to ones promoted by officialdom through representations of space and spatial practice.

The elements of Lefebvre's spatial triad exist in a state of dialectical interaction. This involves class contradiction but also the tendency of neo-capitalism to produce exchange value abstract space in opposition to spatial coalitions that tend towards the production of use value counter-space and differential space.

For Lefebvre, the production of a new space can never be brought about by any one particular social group and must necessarily result from coalitions based on relationships between diverse groups which may include: 'reactionaries', 'liberals', 'democrats' and 'radicals' (Lefebvre 1991: 380–1). It should be no surprise therefore when space-related issues spur collaboration between quite different kinds of interests and actors. The empirical research to be presented later certainly points to the importance of a variety of coalitions that were vital for the production of urban space. It becomes apparent that Lefebvre understood 'urban social space' to be composed of diverse structures 'reminiscent of flaky *mille-feuille* pastry than of the homogeneous and isotropic space' (Lefebvre 1991: 86, emphasis in original). So an appreciation of social and spatial diversity is a key theme in the production of space.

Producing contradiction and controversy

Continuous contradiction is how Lefebvre understood the world of state-regulated and supported capitalism, or neo-capitalism for short. His long life was full of conflict and contradiction, starting with the contradictory religious beliefs of his parents and extending to conflicts with other academics, the Situationists, the Nazis, the French Communist Party and various aspects of authority. Given his lifelong adherence to principles of dialectical materialism, the exploration of contradiction, as a method of understanding the world, Lefebvre would no doubt be pleased that his work continues to provoke controversy. Castells' well known, first-hand and direct structuralist contradiction of Lefebvre's Marxism as expressed in *The Urban Revolution* and views on urban space as an agent of change and the possibility of an urban revolution, first published in *The Urban Question* (1972), was just one of the first critiques of Lefebvre's urban work. Castells was one of Lefebvre's students and then a colleague at Nanterre University, and it may be the ultimate compliment that some of Lefebvre's most important work, especially in *The Survival of Capitalism* (first published in 1973, reprinted in 1976), TPOS and *Rhythmanalysis* seems to offer a response to Castells' critiques. Castells also criticised Lefebvre's production of space theory, partly because it was bad theory based on a corruption of Marx, but also because it lacked an empirical basis (in Stanek 2011: vii); something disputed by Elden (2004: 143). Subsequently, Castells later moved away from Marxism and retracted some of his criticisms (Stanek 2011: 52), and in fact applied Lefebvre's concept of urban revolution in the book *The City and the Grassroots* (1983). We would add that Castells' urban social movements concept may well have been inspired by Lefebvre's notions of spatial coalitions, counter-projects and counter-space.

Castells' critique remains important because he established a kind of dialectic in his approach. First Lefebvre's work is presented, then the contradiction with Castells' ideas is presented. What Castells also does is offer profuse praise. So he famously affirmed that Lefebvre was probably the greatest philosopher on cities we have had (in Elden 2004: 142). Although this may be faint praise since he drew a distinction between researchers, who know about the material and social world, and philosophers, who do not. What, then, a resolution? No. Rather than Hegelian synthesis or resolution, the contradictions produce continuing debate. We see this pattern repeated by other critics such as Harvey, who also demurred from Lefebvrian Marxism and empirical credentials regarding capitalism, but offered lavish praise. A familiar complaint is that Lefebvre's written work is often poorly structured, vague, difficult to follow, with meanings that are not elucidated fully. Even those generally sympathetic hold his view (for instance Stewart 1995).

Taking Schmid (2008: 29) as an example, he maintains that the three moments central to the theory of the production of space 'exist in a state of uncertainty'. The implication is that the meaning of the three constituents of the spatial triad is not clear. Shields and Soja in particular share Schmid's concern, avowing that his ideas regarding the production of space are confused (quoted in Elden 2004: 37). Lefebvre seems responsible for this since he introduced the elements of the triad in TPOS as approximations, then modified them in theoretical excursions throughout the book. Apparently, this allowed the reception of the theory to produce 'near-total confusion of opinion about these three dimensions' (29). It is reasonable for Schmid to reason that 'The meaning of the three dimensions becomes clear only in the overall context of the theory' (29). But his claim that the meaning can only be constructed via an engagement with 'Lefebvre's *entire work*' (29, emphasis added) is at best dubious.

Three things strike us about production of space theory that may be applied to Lefebvre's other major urban ideas and derive from his literary style. First, the main concepts are not so much uncertain or confused as complex and multi-faceted. They are often short-hand for an assemblage of facts, propositions and deliberate provocations. Second, his key concepts tend to be elaborated at length, drawing on his decades of scholarship and research. Third, Castells' comment that Lefebvre 'had a genius for intuiting what was really happening. Almost like an artist' (in Elden 2004: 142) is instructive. In his urban phase, Lefebvre was working on a catalogue of original theoretical and empirical insights or reformulations of the familiar. His work was 'mature-exploratory'. That being so, it is unreasonable to expect his ideas to be fully formed and pitch-perfect. They are rough, rather than polished diamonds. Lefebvre dared us to think differently about things we thought we knew well.

Schmid offers criticism of secondary sources such as in the work of Shields, Soja and Elden. And we share Schmid's view that the elements of the triad must not be treated as separate entities but should be understood as contributing to the contradictory dialectical process that produces space. In particular, we note that Lefebvre does not privilege one or other of the elements, and we do not see, for example, heroic spaces of representation as somehow 'better' than official representations of space, they are just different. We assert that to his credit Schmid's astute conclusions about the spatial triad underlying TPOS can be derived from a close and unprejudiced reading of 400-plus pages of TPOS. Despite the near total confusion, Schmid accepts that some 'promising analyses do exist', including his own (43). We believe that the chapters in this Handbook augment that corpus.

Nowadays, the acceptance and ubiquity of Lefebvre's ideas belies the fact that his urban work was largely ignored in France, Europe and North America after publication in the late 1960s and early 70s. Harvey's intervention in 1973 produced little or no ripple effect. What did was the 'appropriation' of Lefebvre's urban writing by postmodern Marxist geographers such as Soja in the 1980s and 90s. This appropriation, as it is called, has met with rancour in some quarters, including Elden (2001: 809), who declares, 'Postmodernism often seems to be an intellectual 'pick 'n' mix'; it sees 'disparate thinkers are lumped together' and thrown into 'a great melting pot' that is 'seasoned with invective, half-baked and served up to the various disciplines'.

Gottdiener continues this tradition. He castigates Harvey and those who assert he first brought Lefebvre's urban ideas to Anglophone attention. Gottdiener (2018: 266–7) is concerned deeply by the 'disconcerting way his work has been disseminated, absorbed, and commented on' in Anglophone academia. He is adamant that there is a 'disciplinary conspiracy of silence' because Lefebvrian geographers, such as Merrifield, cite sociologists so infrequently. Unfortunately, although provocative, Gottdiener's main argument regarding Harvey falls away when he says that the British sociologist Pickvance (1976) first brought French critical urbanists like Lefebvre to an Anglophone audience. Harvey of course did this in 1973.

Perhaps the most severe critique is that by Unwin (2000) – a geographer. He starts by stressing that there are fundamental problems with Lefebvre's conceptualisation of space. We do not dwell on these here, suffice to say that Unwin finds Lefebvre's discussion of space largely unhelpful, but instead concentrate on Unwin's comments on the paucity of useful methodological insights in Lefebvre's research. Unwin (2000: 23) is circumspect regarding the implications 'for our empirical research practice', finding little practical methodological merit in Lefebvre's work. Unwin asks 'what are the implications of this [Lefebvre's ideas and research approach] for our empirical research practice?' and, in studying the production of space, what should we study? He finds little help from Lefebvre and is worried that this has serious consequences for future research. He offers a familiar critique of Lefebvre's literary style and his 'project' – researching the production of space:

> The complexity of Lefebvre's arguments is closely related to the elusiveness with which he develops them. Reading *The production of space* can be compared to walking across quicksand, or trying to find the end of a rainbow. No sooner does one think that one has understood what he is trying to say, then he shifts his position, so that what was once thought to be acceptable is now shown to be problematic. At the heart of Lefebvre's project there is thus an intention to make complex the taken-for-granted, and to force the reader to question her or his own understandings of space. As well as being elusive, though, there is a tension within Lefebvre's work, because this very character of being contradictory, and lacking certainty, to some extent runs counter to his own certainty that space is actually produced.
>
> *(Unwin 2000: 14)*

In summary, he pronounces that Lefebvre (plus Harvey and Soja) wastes space and his work cannot contribute to radical politics partly because the implications for empirical research practice are bleak. So it appears Lefebvrian-inspired empirical research either cannot proceed or cannot inform policymakers and campaigners. Researchers around the globe disagree with Unwin increasingly since 2000. See Leary-Owhin (2018) regarding urban regeneration. A recent survey of Lefebvrian-inspired empirical research finds that:

> In Anglo-America – but not only there – Lefebvre is now taken in various directions. This trend is welcome to the extent that it enriches theory, research, and strategy while proposing often much-needed critiques and correctives of Lefebvre's work. It is also Lefebvrean in the sense of being open to a plurality of struggles and theoretical currents. We have also suggested, however, that in contemporary debates, sustained points of contact should be maintained to Lefebvre's open, integral, and differential marxism and the dialectical urbanism that helped shape it.
>
> *(Kipfer et al. 2012: 127)*

This Handbook was conceived in light of the proliferation of Lefebvrian scholarship, especially in academic journals, not all of which are easily accessible. In particular, right to the city research has seen not just an academic and institutional proliferation but has 'emerged as a demand by an impressive array of movements for example in Brazil, German-speaking Europe, the USA and the Americas' (Ibid.), and many other countries as shown above and in some of the chapters that follow. Interestingly too, right to the city emerged as the most popular topic for this Handbook requiring two parts. And it is to the structure of the book that we now turn.

Structure of the book

There is no obvious way to thematise and structure a large book of this kind, where the ambition must be to offer a comprehensive but coherent text. Our thematic threads and the six parts of the book reflect our reading of Lefebvre over the last 15 years, our interests, backgrounds and contacts within academia. We did not fetter the authors with our ideas and preferences but did encourage them to propose chapters that could have a range of approaches and styles. Neither did we try to impose this or that definitional or interpretational approach to Lefebvre's contested ideas. No doubt the inevitable variation in the book reflects to some extent what Lefebvre hints at in the epigraph to this introduction, as academics reevaluate and rework their own thinking in their particular contexts. We sought deliberately to include chapters related to the countries and issues of the Global South. Topics range from: the organic modes of building adaptation, politics of language, social anthropology, the 'concrete abstract', the right to the 'centre', writing in the right to the city; to: the production of air, neoliberal urbanism, refugees' struggles, aesthetics of spatial justice and 'co-induced in-between spaces'. Notably, too, the geographical settings range across the globe from Barcelona, Detroit and Tokyo to Lisbon, New York and Maputo. The 47 substantive chapters are structured into six parts from authors around the world, giving a global reach covering countries of the Global North and South. Each part has a brief introduction and rationale written by the editors. The main introductory chapter provides a context and interpretive framework for the chapters that follow.

There is a range of substantive chapter types, with a balance between theoretical, critical review or empirical research chapters, or those that focus on discussions of new and potential future developments in the field. The volume includes contributions from prominent established scholars and from emerging researchers and academics across various disciplines. Chapter authors were briefed to write for a knowledgeable readership that has a good grasp of the basic concepts and issues and has engaged thoughtfully with some of the key secondary literature. All substantive chapters are specially commissioned contributions. Chapters are long enough to achieve convincing depth of discussion and argument and synthesis but short enough to be accessible and readable in one sitting. Chapter types vary from those that focus directly on Lefebvre's writings to those that concentrate on theory, empirical research, secondary sources or state of the art reviews. Since Castells' *The Urban Question*, Lefebvre has had his critics and chapters that offer new lines of critique were welcomed and are included in the book. The structure and organisation of the book is set out below, as are the rationales for the six parts.

A comprehensive book of this size, dealing as it does with complex ideas and wide-ranging literatures, requires a substantial introduction, and the main introduction is specified below. Readers are encouraged to read this chapter first. The principal introduction serves two main roles in the book. It provides the rationale and five aims for the book and it outlines each of the six parts. It argues that with the global consolidation of neoliberalism, growth of megacities in the Global South and environmental crises, the perspective of a Lefebvrian lens is more apposite now than ever before. The book's five aims are explained, giving particular attention as to why a Lefebvrian approach to space must engage critically with the divergent notions of the city and what is called urban society. We provide an overview of recent literature within which to situate the Handbook. What this chapter and indeed the whole book do not attempt is cover the entire Lefebvre canon (even in English), nor the whole range of responses to it and derivations from it. Of necessity, there is a focus on key texts and certain major aspects of cities and urban society, and these are explained in this chapter. Here the editors provide a summary of the book's essential features, structure and the key attributes of each part.

Chapter One has a biographical element and provides a more detailed engagement with the development of Lefebvre's ideas about space, cities and the urban. The chapter provides critical insights into Lefebvre's life journey from the rural to the urban, albeit he never quite considered himself a fully metropolitan Parisian. A frequent source of confusion for many budding Lefebvre and city researchers is the manner in which Lefebvre switches seamlessly from abstractions and metaphors such as the everyday and the spatial triad, to their understandings in concrete reality. This has led the likes of David Harvey, Andy Merrifield, Ed Soja and Kristin Ross to muse about the fundamental nature of Lefebvre's spatial ideas. This chapter starts to unpack key Lefebvrian notions regarding the production of space under neoliberalised neo-capitalism. Given the contested nature of Lefebvre's spatial and other concepts, this chapter maps out some of the interpretative terrain. A major intention is to help ensure that those readers who have only a passing knowledge of Lefebvrian scholarship are prepared for the subsequent substantive chapters. Obviously, Lefebvre did not produce his major works in isolation from his forebears and contemporaries, and this chapter points to some of his intellectual relationships with other key thinkers on space and related themes. The production of space concept has inspired multi-faceted, multi-disciplinary areas of policy and practice, and we locate these in the context of Lefebvre's key writings. The chapter, along with the main introduction, locates the book within the framework of the classic (English language) Lefebvrian texts and emerging literatures. It provides pointers to the importance of key debates both for the globalised context and for the production of urban space. We argue that the diverse global ubiquity of Lefebvrian research and scholarship needs to be contextualised within real constraints and contestations, while accepting that there are opportunities for continued developments. Below we provide a brief summary rationale for each part of the book.

Part 1: Globalised neoliberal urbanism: hegemony and opposition

There are many publications offering interpretations of Lefebvre's main texts, some aimed at the aficionados and others at newcomers. Part 1 assists reader orientation by placing two of Lefebvre's key ideas, bottom-up struggle and neoliberalism, in several important contexts. Lefebvre addressed the problematic of the big state, big private corporation nexus, or neoliberalism, in a range of his writing. This first part serves two major roles. First, it provides an overarching context for the rest of the book by placing ideas, policy and practice regarding the production of urban space in the context of the global phenomenon of neoliberalism. Second, it provides a roadmap showing how Lefebvrian ideas and scholarship have been taken up in response to perceived neoliberal threats to and opportunities for cities and urban space. What Part 1 does not provide are chapters consisting of extracts from Lefebvre's own major texts in English, as these are now readily available in a variety of formats. Part 1 highlights the hegemonic but vulnerable nature of neoliberalism, while also explaining how Lefebvre's spatial ideas point to the potential for significant opposition that may take advantage of inherent cracks in the neoliberal façade.

Part 2: Rethinking the spatial triad and rhythmanalysis

The spatial triad and its role in the production of space is without doubt Lefebvre's most popular theoretical framework for empirical urban research, dating from at least Fyfe (1996).

It is an innovative and powerful tool for understanding and interrogating apparently inert container-like urban space. Nonetheless, as with most Lefebvrian concepts, it is susceptible to a variety of interpretations, and some argue, misinterpretations. Each element of the triad is simultaneously intuitively simple and easily grasped but also, on closer inspection, assumes increasing complexity and ambiguity. Spatial practice, for example, includes routine ways of life that are qualitatively different from the everyday life of spaces of representation. In turn, everyday life is woven by Lefebvre into ideas regarding rhythmanalysis. Strangely, perhaps, few have tried to graphically represent Lefebvre's ideas apart from e.g. Gregory (1994: 401), with his 'Eye of Power', and Leary-Owhin's (2016: 15) visual reconceptualisation of the triad. This is even more surprising when one considers the large number of geographers who have turned to Lefebvrian research. We return to reconsider this issue in the main conclusions.

Rhythmanalysis adds the dimension of time and events to Lefebvre's long-term analysis of space. Along with the production of space and the critique of everyday life, it is often seen as his greatest achievement, and being his last major work, it did encompass much of his thinking about the urban. This part endeavours to stand back from the spatial triad by providing some radical rethinking and reinterpretation. While not seeking to denigrate previous research that counter-poses representations of space and spaces of representation, Part 2 highlights the range of approaches to Lefebvre's core ideas underpinning the spatial triad and rhythmanalysis. It also demonstrates new ways in which the spatial triad and rhythmanalysis can provide intellectual inspiration for production of space research.

Part 3: Representing and contesting urban space

In this part the book focuses on Lefebvre's deceptively simple idea of spatial representations and the manner in which contestation of space is key to much of Lefebvrian thinking about the urban. Though apparently simple, Lefebvre was adamant that official representations of space can be treacherous because they often serve to mystify social relations and the wielding of power. Lefebvre presents increasingly sophisticated understandings of representations of space, including visual representations under conditions of neo-capitalism, or what may also be called neo-liberal urbanism. Representations of space are problematised in three main ways. First, through examination of the idea that they can serve to exclude certain voices, communities and ways of seeing. Second, by exploring how representations of space are contested within officialdom. Third, by investigating how urban space is moulded through processes of contestation where differing groups, including civil society groups, offer alternative representations of space. From the traditional paper map, plan, perspective and axonometric, this part allows the elaboration of how new technologies, big data, the internet and social media are impacting on the production of space in so-called smart cities and elsewhere.

Part 4: Planetary urbanisation and 'nature'

In the late 1990s, Neil Smith (1998) commented that Lefebvre's rethinking of nature was rather poor. That said, Lefebvre argued controversially that a rather passive nature is progressively destroyed by abstract space. Since then ideas about nature and environmentalism have taken on global importance as ideas of the Anthropocene are debated. Lefebvre's contributions regarding nature and environment are perhaps some of the least explored and developed avenues of his writing and theorising. They are, however, related to ideas about planetary urbanisation and the production of urban abstract space. This part brings together the latest theoretical and empirical developments that seek to employ Lefebvre's ideas to further our understanding of the city, the urban and urbanisation and nature in the early 21st century. Actual planetary urbanisation over recent decades in the Global North and South seems to be confirming as a concrete phenomenon – what Lefebvre described in the early 1970s as something virtual and impending. Whether this means we can abandon the idea of the city, the rural and the wilderness remain, however, a moot point. An overriding contention here is that seeing the city through a Lefebvrian lens allows us to move beyond the reignited debate as to whether cities are best understood primarily as bounded physical sites or complex economic and social processes and power relations. This part explores the implications when such processes are unleashed globally via the tendency of neoliberal urbanism to commodify and appropriate space globally.

Part 5: Rethinking the right to the city

Perhaps even more than Lefebvre's ideas regarding the production of space and the spatial triad, his right to the city (RTC) propositions seem to have captured the imagination worldwide of academics, politicians and activists in the last decade. Notably, the incorporation of the RTC into official national and city policy legislation e.g. in Brazil from the 1980s and into official United Nations policy, Habitat III has caused controversy. Some argue this is the antithesis of Lefebvre's RTC project that foresaw the state withering away and workers themselves, not the state, guaranteeing the RTC. Citizens would create self-managed RTC through autogestion and self-management, involving a withdrawal of the state from some of its familiar roles. The contested understandings of RTC provide ample space for heated debate and a range of strongly expressed views. Advocates seize on RTC as a vital set of governance principles and hopes for urban citizenry, but critics bemoan its vagueness and the usurping of RTC by conservative interests and a privileged neoliberal urban elite. Rather than simply a dogmatic reaffirmation of what is after all a highly malleable concept, this part allows a more critical exploration of what RTC might mean, whether it has outlived its usefulness and why it continues to fascinate in the early 21st century.

Part 6: Right to the city, differential space and urban utopias

Lefebvre was an optimist, many say a Utopian, when it came to his view of the long-term future for cities and urban space. Nowhere is this more apparent than in his ideas about the RTC and the related concept of differential space. Both these multi-faceted ideas have been springboards for research and street-level activism. A key feature of Part 6 is the manner in which differential space and the RTC are regarded as being intimately entwined. Rather like the spatial triad itself, this part argues that the RTC and differential space are metaphorical but also concern the exercise of real rights in the concrete public spaces of cities worldwide. There is a tension between self-managed versus official RTC. The United Nations deploys RTC ideas in various policy statements and it is *also* a rallying cry for many city-based 'Occupy' movements and a variety of urban campaigns in the Global North and South. Global human rights discourses add greater salience to RTC and differential space ideas. For Lefebvre, differential space has the potential to subvert abstract space in the long term, in ways sympathetic to everyday use value. In the short term there are myriad opportunities to produce locally important inclusive democratic urban space, especially in cities. Part 6 offers the chance to consider the extent to which concepts such as RTC are a key component of utopia. Lefebvre's political utopianism was of course dialectical, and centred on the method of transduction. This means he advocated detecting the possible within the apparently impossible constraints of neoliberal neo-capitalism. In this final part we should bear in mind how researchers have risen to the challenge of being Lefebvrian.

Routledge Handbooks bring together the world's leading scholars and emerging academic talent to provide a cutting-edge overview of current research in the field. They provide an authoritative guide to theory and method, the key sub-disciplines and the primary debates. This Handbook is at the weighty end of the publishing spectrum, and few readers would want to consume it sequentially and continuously like a novel. But like a novel, there are twists and turns and the unexpected. It is comprehensive, but we do not claim it is exhaustive or definitive. *The Routledge Handbook of Henri Lefebvre: The City and Urban Society* is an *oeuvre*, a work in progress. We are concerned of necessity in this book with the production of knowledge, something we return to in the main conclusions. There are inevitably issues or even geographies that some may feel needed their own part in the book. We did consider a part on the everyday but decided against it since it is a recurring refrain in the Handbook. Ultimately, the book's structure reflects our backgrounds as editors and of course the chapters that reached us. The Handbook is offered as a critical friend rather than a source of correction or discipline. Guided by the introductions (and conclusions), we hope readers will enjoy the experience of exploration and discovery the Handbook offers.

Acknowledgements

Thanks are due first to the chapter authors for the time and intellectual effort they devoted to producing the compelling high-quality chapters in this book. Michael Leary-Owhin is grateful for the enthusiasm with which Routledge's Andrew Mould supported his idea for an edited book about Lefebvre and the city. He is also grateful for the *hospitalité* received when visiting Mme Catherine Régulier-Lefebvre and Léa Régulier-Lefebvre in the Lefebvre family home in Navarrenx, Béarn. He will endeavour to 'Voyageur Lefebvrien en Béarn'. We would like to thank Egle Zigaite (Routledge) for expert editorial support. In the course of the book project we received and acknowledge encouragement and assistance from Stuart Elden, Andy Merrifield and Vincent Berdoulay. Finally, our appreciation goes out to the three anonymous referees for their perceptive, constructive criticism of the book proposal.

References

Agbiboa, D.E. (2017) 'God's time is the best: the fascination with unknown time in urban transport in Lagos', in: Baumbach, S., Henningsen, L. and Oschema, K. (eds) *The Fascination with Unknown Time*, Basingstoke: Palgrave Macmillan.

Allen, J. and Pryke, M. (1994) 'The production of service space', *Environment and Planning D: Society and Space*, 12:4 453–475.

Anderson, P. (1983) *In the Tracks of Historical Materialism*, London: Verso.

Andres, L. (2013) 'Differential spaces, power hierarchy and collaborative planning: a critique of the role of temporary uses in shaping and making places', *Urban Studies*, 50:4 759–775.

Bhattacharyya, R. (2015) 'Understanding the spatialities of sexual assault against Indian women in India', *Gender, Place & Culture*, 22:9 1340–1356.

Borden, I.M. (2001) *Skateboarding, Space and the City: Architecture and the Body*, Oxford: Berg.

Brenner, N. (ed.) (2014) *Implosions/Explosions: Towards a Study of Planetary Urbanization*, Berlin: Jovis.

Brenner, N. and Schmid, C. (2015) 'Towards a new epistemology of the urban?', *City*, 19:2–3 151–182.

Buckley, M. and Strauss, K. (2016) 'With, against and beyond Lefebvre: planetary urbanization and epistemic plurality', *Environment and Planning D: Society and Space*, 34:4 617–636.

Burgel, G., Burgel, G. and Dezes, M.G. (1987) 'An interview with Henri Lefebvre', *Environment and Planning D: Society and Space*, 5 27–38.

Buser, M. (2012) 'The production of space in metropolitan regions: a Lefebvrian analysis of governance and spatial change', *Planning Theory*, 11:3 279–298.

Butler, C. (2012) *Henri Lefebvre: Spatial Politics, Everyday Life and the Right to the City*, London: Routledge.

Carp, J. (2008) '"Ground-truthing" representations of social space: using Lefebvre's conceptual triad', *Journal of Planning Education and Research*, 28:2 129–142.

Castells, M. (1977) *The Urban Question. A Marxist Approach*, London: Edward Arnold.

Chirozva, C. (2017) 'Community agency and entrepreneurship in ecotourism planning and development in the Great Limpopo transfrontier conservation area', in: Backman, K. and Munanura, I.E. (eds) *Ecotourism in Sub-Saharan Africa: Thirty Years of Practice*, London: Routledge.

Coleman, N. (2014) *Lefebvre for Architects*, London: Routledge.

D'Ascoli, A. (2018) *Public Space: Henri Lefebvre and Beyond*. Milan: Mimesis International.

De Melo, E.O. (2017) 'Lefebvre and the periphery: an interview with Professor Marie Huchzermeyer', *International Journal of Urban Sustainable Development*, 9:3 365–370.

Dear, M. (2000) *The Postmodern Urban Condition*. Oxford: Blackwell.

Degen, M. (2008) *Sensing Cities: Regenerating Public Life in Barcelona and Manchester*, London: Routledge.

Dimendberg, E. (2004) *Film Noir and the Spaces of Modernity*, Harvard: Harvard University Press.

Dimendberg, E. (1998) 'Henri Lefebvre and abstract space', in: Light, A. and Smith, J.M. (eds) *The Production of Public Space*, Oxford: Rowman and Littlefield.

Edensor, T. and Jonas Larsen, J. (2017) 'Rhythmanalysing marathon running: a drama of rhythms', *Environment and Planning A*, 50:3 730 746.

Elden, S. (2001) 'Politics, philosophy, geography: Henri Lefebvre in recent Anglo-American scholarship', *Antipode*, 33:5 809–825

Elden, S. (ed.) (2004) *Understanding Henri Lefebvre: Theory and the Possible*, London: Continuum.

Elden, S., Lebas, E. and Kofman, E. (eds) (2003) *Henri Lefebvre: Key Writings*, London: Continuum.

Entrikin, N. and Berdoulay, V. (2005) 'The Pyrenees as place: Lefebvre as guide', *Progress in Human Geography*, 29:2 129–147.

Erdi-Lelandais, Gülçin (ed.) (2014) *Understanding the City: Henri Lefebvre and Urban Studies*, Cambridge: Cambridge Scholars Publishing.

Essex, S. and Brayshay, M. (2007) 'Vision, vested interest and pragmatism: who re-made Britain's blitzed cities?', *Planning Perspectives*, 22:4 417–441.

Fernandes, E. (2007) 'Constructing the "right to the city" in Brazil', *Social & Legal Studies* 2007, 16:2 201–219.

Friendly, A. (2013) 'The right to the city: theory and practice in Brazil', *Planning Theory & Practice*, 14:2 158–179.

Fyfe, N. (1996) 'Contested visions of a modern city: planning and poetry in post war Glasgow', *Environment and Planning A*, 28:3 387–403.

Gardiner, M. (2004) 'Everyday utopianism: Lefebvre and his critics', *Cultural Studies*, 18:2–3 228–254.

Garbin, D. and Millington, G. (2018) '"Central London under siege": Diaspora, "race" and the right to the (global) city', *The Sociological Review*, 66:1 138–154.

Goonewardena, K., Kipfer, S., Milgrom, R. and Schmid, C. (eds) (2008) *Space Difference and Everyday Life: Reading Henri Lefebvre*, London: Taylor Francis.

Goswami, M. (2004) *Producing India: From Colonial Economy to National Space*, Chicago: Chicago University Press.

Gottdiener, M. (2000) 'Lefebvre and the bias of academic urbanism: what can we learn from the "new" urban analysis?', *City*, 4:1 93–100

Gottdiener, M. (1993) 'A Marx for our time: Henri Lefebvre and the production of space', *Sociological Theory*, 11:1 129–134.

Gottdiener, M. (1985) *The Social Production of Urban Space*, Austin: University of Texas.

Gottdiener, M. (2018) 'Who owns Lefebvre? The forgotten sociological contribution to the new urban sociology', *Contemporary Sociology*, 47:3 266–271.

Gregory, D. (1994) *Geographical Imaginations*. Oxford: Blackwell.

Grindon, G. (2013) Revolutionary romanticism: Henri Lefebvre's revolution-as-festival, *Third Text*, 27:2, 208–220.

Groth, J. and Corijn, E. (2005) 'Reclaiming urbanity: indeterminate spaces, informal actors and urban agenda setting', *Urban Studies*, 42:3 503–526.

Gustavsson, E. and Ingemar Elander, E. (2016) 'Sustainability potential of a redevelopment initiative in Swedish public housing: the ambiguous role of residents' participation and place identity', *Progress in Planning*, 103 1–25.

Harmon, J. (2019) 'The right to exist: homelessness and the paradox of leisure', *Leisure Studies*, first published online at doi: 10.1080/02614367.2019.1640775.

Harvey, D. (1973) *Social Justice and the City*, London: Edward Arnold.

Harvey, D. (1991) 'Afterword', in: Lefebvre, H. (ed.) *The Production of Space*, Oxford: Blackwell.

Harvey, D. (2012) *Rebel Cities: From the Right to the City to the Urban Revolution*, New York: Verso Books.

Harvey, D. (2000) *Spaces of Hope*, Oxford: Blackwell.

He, S., and Lin, G. C. (2015) 'Producing and consuming China's new urban space: State, market and society', *Urban Studies*, 52:15 2757–2773.

Hesketh, C. (2013). 'The clash of spatializations: Geopolitics and class struggles in southern Mexico', *Latin American Perspectives*, 40:4 70–87.

Hubbard, P., Faire, L. and Lilley, K.D. (2003) 'Contesting the modern city: reconstruction and everyday life in post-war Coventry', *Planning Perspectives*, 18:4 377–397.

Huchzermeyer, M. (2015) *Reading Henri Lefebvre from the 'global south': The legal dimension of his right to the city*, UHURU Seminar Series, Rhodes University.

Jameson, F. (1992) *Postmodernism: Or, the Cultural Logic of Late Capitalism*, London: Verso Books.

Janzen, R. (2002) 'Reconsidering the politics of nature: Henri Lefebvre and the Production of Space', *Capitalism Nature Socialism*, 13:2 96–116.

Karplus, Y. and Meir, A. (2013) 'The production of space: a neglected perspective in pastoral research', *Environment and Planning D: Society and Space*, 31:1 23–42.

Katznelson, I. (1992) *Marxism and the City*. Oxford: Clarendon Press.

Kipfer, S., Goonewardena, K., Schmid, C. and Milgrom, R. (2008) On the production of Henri Lefebvre, in: Goonewardena, K., Kipfer, S., Milgrom, R. and Schmid, C. (eds.) *Space Difference and Everyday Life: Reading Henri Lefebvre*, London: Taylor & Francis.

Kipfer, S., Saberi, P. and Wieditz, T. (2012) 'Henri Lefebvre: Debates and controversies', *Progress in Human Geography*, 37:1 115–134.

Leary, M. E. (2009) 'The production of space through a shrine and vendetta in Manchester: Lefebvre's spatial triad and the regeneration of a place renamed Castlefield', *Planning Theory & Practice*, 10:2 189–212.

Leary-Owhin, M.E. (2016) *Exploring the Production of Urban Space: Differential Space in Three Post-Industrial Cities*, Bristol: Policy Press.

Leary-Owhin, M.E. (2018) *Proof of Evidence for the Aylesbury Estate Regeneration CPO Public Inquiry*, London: Southwark (full text at academia.edu/MichaelLearyOwhin).

Lefebvre, H. (1968 [1940]) *Dialectical Materialism*, London: Jonathan Cape.

Lefebvre, H. (1969 [1968]) *The Explosion. Marxism and The French Upheaval*, New York: Monthly Review Press.

Lefebvre, H. (1976/1973) *The Survival of Capitalism: Reproduction of the Relations of Production*, London: Allison & Busby.

Lefebvre, H. (1991/1974) *The Production of Space*, Oxford: Blackwell.

Lefebvre, H. (1996/1968) 'Right to the city', in: Kofman, E. and Lebas, E. (eds) *Writings on Cities*, Cambridge, MA: Blackwell.

Lefebvre, H. (2003/1970) *The Urban Revolution*, Minneapolis, MN: University of Minnesota Press.

Lefebvre, H. and Regulier-Lefebvre, C. (2004/1992) *Rhythmanalysis: Space, Time and Everyday Life*, London: Continuum.

Lefebvre, H. (2008/1981) *Critique of Everyday Life Volume Three: From Modernity to Modernism (Towards a Metaphilosophy of Daily Life)*, London: Verso.

Lefebvre, H. (2016/1972) *Marxist Thought and the City*, Minneapolis, MN: University of Minnesota Press.

Lehtovuori, P. (2010) *Experience and Conflict: The Production of Urban Space*, Farnham: Ashgate.

Lichtheim, G. (1966) *Marxism in Modern France*, New York and London: Columbia University Press.

Light, A. and Smith, J.M. (eds) (1998) *The Production of Public Space*, Lanham, MD: Rowman & Littlefield.

Limonad, E. and Monte-Mór, R. L. (2005) 'Beyond the right to the city: Between the rural and the urban', *Les Cahiers du Développement Urbain Durable*, 103–115.

MacDonald, F. (2002) 'Towards a spatial theory of worship: some observations from Presbyterian Scotland', *Social & Cultural Geography*, 3:1 61–80.

Manzi, M., dos Santos Figueiredo, G. C., Mourad, L.M. and de Miranda Rebouças, T. (2018) 'Neighbourhood planning and the right to the city: confronting neoliberal state urban practices in Salvador, Brazil', *International Journal of Urban Sustainable Development*, 10:1 1–15.

McClymont, K. (2018) 'They have different ways of doing things: cemeteries, diversity and local place attachment', *Journal of Intercultural Studies*, 39:3 267–285.
McCann, E.J. 1999, 'Race, protest, and public space: Contextualizing Lefebvre in the U.S. city', *Antipode*, 31:2 163–184.
Merrifield, A. (1993) 'Place and space: A Lefebvrian reconciliation', *Transactions of the Institute of British Geographers*, 18:4 516–531.
Merrifield, A. (2006) *Henri Lefebvre: A Critical Introduction*, London: Taylor & Francis.
Merrifield, A. (2011) 'The right to the city and beyond', *City*, 15: 3–4: 473–481.
Merrifield, A. (2013) *The Politics of the Encounter: Urban Theory and Protest under Planetary Urbanization*, Athens, GA: University of Georgia Press.
Merrifield, A. (2014) *The New Urban Question*, Pluto Press.
Mitchell, D. (2003) *The Right to the City: Social Justice and the Fight for Public Space*, New York: Guilford Press.
Mulíček, O., Osman, R. and Seidenglanz, D. (2016) 'Time–space rhythms of the city – the industrial and postindustrial Brno', *Environment and Planning A*, 48:115 131.
Neumann, T. (2018) 'Reforging the Steel City: symbolism and space in postindustrial Pittsburgh', *Journal of Urban History*, 44:4 582–602.
Oldenburg, S. (2018) 'Agency, social space and conflict-urbanism in Eastern Congo', *Journal of Eastern African Studies*, 12:2 254–273.
Parikh. A. (2019) 'Liminality of women's leisure in Mumbai, India', *Environment and Planning C: Politics and Space*, first published online at doi: 10.1177/2399654419859358.
Patel, S. (2018) 'Rethinking urban studies today', *Sociological Bulletin*, 67:1 1–19.
Pickvance, C.G. (ed.) (1976) *Urban Sociology: Critical Essays*, London: Tavistock Publications.
Purcell, M. (2002) 'Excavating Lefebvre: the right to the city and urban politics of the inhabitant', *Geojournal*, 58: 99–108.
Purcell, M. (2014) 'Possible worlds: Henri Lefebvre and the right to the city', *Journal of Urban Affairs*, 31:1 141–154.
Ross, K. (1983) Henri Lefebvre on the situationist international, interview conducted and translated in 1983 by Kristin Ross, access online at notbored.org/lefebvre-interview.html, date accessed, February 2018.
Saunders, P. (1981) *Social Theory and the Urban Question*, Abingdon: Routledge.
Schmid, C. (2011) 'Henri Lefebvre, the right to the city, and the new metropolitan mainstream', in: Brenner, N., Marcuse, P. and Mayer, M. (eds) *Cities for People, Not for Profit: Critical Urban Theory and the Right to the City*, London: Routledge.
Schmid, C. (2008) 'Henri Lefebvre's theory of the production of space: towards a three-dimensional dialectic', in: Goonewardena, K., Kipfer, S., Milgrom, R. and Schmid, C. (eds) *Space Difference and Everyday Life: Reading Henri Lefebvre*, London: Taylor Francis.
Schmid, C. (2014) 'The trouble with Henri: urban research and theory of the production of space', in: Stanek, L., Schmid, C. and Moravánszky, A. (eds) *Urban Revolution Now: Henri Lefebvre in Social Research and Architecture*, London: Routledge.
Shields, R. (1999) *Lefebvre, Love, and Struggle: Spatial Dialectics*, London: Taylor Francis.
Simone, A. (2019) 'Maximum exposure: making sense in the background of extensive urbanization', *Environment and Planning D: Society and Space*, first published online at doi: 10.1177/0263775819856351.
Smith, N. (1998) 'Antinomies of space and nature in Henri Lefebvre's *The Production of Space*', in: Light, J. and Smith, J.M. (eds) *The Production of Public Space*, Oxford: Rowman and Littlefield.
Smith, N. (1984) *Uneven Development: Nature, Capital, and the Production of Space*, Oxford: Blackwell.
Soja, E. (1989) *Postmodern Geographies: The Reassertion of Space in Critical Social Theory*, London: Verso.
Soja, E. (2010) *Seeking Spatial Justice*. Minneapolis, MN: University of Minnesota Press.
Soja, E. (1996) *Thirdspace: Journeys to Los Angeles and Other Real-and-Imagined Places*, Oxford: Blackwell.
Stanek, L. (2011) *Henri Lefebvre on Space: Architecture, Urban Research, and the Production of Theory*, Minneapolis, MN: University of Minnesota Press.
Stanek, L., Schmid, C. and Moravánszky, A. (eds) (2014) *Urban Revolution Now: Henri Lefebvre in Social Research and Architecture*, London: Routledge.
Stewart, L. (1995) 'Bodies, visions, and spatial politics: a review essay on Henri Lefebvre's *The Production of Space*', *Environment and Planning D: Society and Space*, 13:5 609–618.
Sun, X. and Huang, R. (2018) 'Spatial meaning-making and urban activism: two tales of anti-PX protests in urban China', *Journal of Urban Affairs*, doi: 10.1080/07352166.2018.1443010.

Unwin, T. (2000) 'A waste of space? Towards a critique of the social production of space…', *Transactions of the Institute of British Geographers*, 25:1 11–29.
Willging, J. (2019) 'A critique of everyday life in Michel Houellebecq's extension du domaine de la lutte', *Modern & Contemporary France*, first published online at doi: 10.1080/09639489.2019.1604500.
Wilson, J. (2011) 'Notes on the rural city: Henri Lefebvre and the transformation of everyday life in Chiapas, Mexico', Environment and Planning D: Society and Space, 29: 993–1009.
Zingale, N.C., Liggett, H. and Heinen, D. (2014) 'Trial by space: articulating value in the context of a shrinking city', *Administration & Society*, 46:9 1045–1070.

Part 1

Globalised neoliberal urbanism: Hegemony and opposition

John P. McCarthy and Michael E. Leary-Owhin

This part of the book is intended to assist the reader by placing the key ideas of Lefebvre in context. Essentially, Lefebvre addressed the problematic of the big state/private corporation nexus, or neoliberalism, in a range of his writings. This part serves two major roles. First, it provides an overarching context for the rest of the book by placing various ideas and policy and practice implications regarding the production of urban space in the context of the global phenomenon of neoliberalism. Second, the chapters as a whole illustrate how Lefebvrian ideas and scholarship have been taken up in response to perceived neoliberal threats to urban space. The chapters in this part also highlight the hegemonic but contested nature of neoliberalism, and also explain how Lefebvre's spatial ideas point to the potential for significant opposition which may make use of inherent cracks in the neoliberal façade.

Leary-Owhin provides the opening chapter which seeks to: provide insights into Lefebvre's intellectual development towards his ideas regarding cities, urbanisation and urban society. It does this mainly by charting the importance of Lefebvre's theoretical *and* empirical contributions to our understanding of cities under conditions of capitalism transformed into what has become known as neoliberalism. Over the last 40 years, neoliberalism often appears omnipotent partly because of the tendency to obscure alternatives or render them seemingly impossible. The chapter explains how Lefebvre's notion of transduction, underpinned by his regression-progression research model, provides intellectual and practical capacities to think beyond the straightjacket of powerful neoliberal forces, allowing a journey to a future possible-impossible urban society.

The spatial ramifications of the production of space under conditions of contemporary neoliberal urbanisation are examined in the next chapter by Haas. Lefebvre's ideas regarding the spatialisation of Marx's understanding of capitalism (and the right to the city) are deployed by Haas and brought up to date in a new research context. His empirical focus is the Palestinian reaction to the Israeli government's intention to establish a new 'Arab City' – the city of Tantour – in the Galilee region. In particular, Haas works with Lefebvre's arguments regarding anti-capitalist struggles for de-colonisation. Important historical details are provided by Haas and the government's modernist plans for the new settlement of Tantour are shown to contrast with existing Palestinian settlements. These are portrayed as backward; in the mould of shanty towns (see Huchzermeyer in Part 6). Research interviews were carried out to understand the Palestinian resistance to the ethnocratic state-led production of space. In their position as a spatially and politically peripheralised group, the nature of the Palestinians' right to the city is as a key issue in the residential urban development of the area. Alongside this, Haas contends that the state reformulates ostensible demands by neoliberal elite groups for homeownership into a rationale for certain kinds of urban development. Neoliberal urbanisation may be politically powerful but the right to the city, infused local resistance, allows inhabitants to undermine ethno-class segregation.

Jorge in the next chapter moves the focus to the Global South and considers pericentral self-produced neighbourhoods within Maputo in Southern Africa, reflecting the wider commodification and financialisation of urban space, albeit with resistance to the dominance of use-value and the corresponding implications for processes of expulsion and gentrification. She shows how intervention strategies have often involved the demolition of extensive areas to allow the development of megaprojects for 'urban renewal'. Such actions, she suggests, are particularly harmful when they occur within areas where tenure is indefinite or perceived as illegal. These spaces, often self-produced, are easily appropriated by local government or private investors. As with other chapters (such as Hesketh's), this illustrates the application of broader neoliberalist approaches within wider governance, which, within Maputo, has its ultimate expression in the simultaneous processes of centralisation and fragmentation, linked to the contradictions between use-value (via the behaviour of users over time) and exchange-value. Thus the public interest is used as the rationale for interventions via expropriation which does not benefit local users. As within many other chapters, Jorge also highlights the development of micro-resistances to such processes, illustrating the potential of an alternative future which privileges use-value over exchange-value.

Moving focus to contemporary Latin America, Hesketh's chapter explores the relevance of Lefebvre's writings in relation to class struggle as evidenced by the production of space. Hesketh's chapter shows that, in spite of the introduction of neoliberalism as a governing regime, resistance has emerged to create counter-projects and counter-spaces, as part of a broader struggle for utopian space situated within demands for democratic participation and collective rights in Latin America. In so doing, however, he identifies the underplaying in Lefebvre's writings of rural space, with agrarian struggles often being a primary element of wider social movements.

Mukhopadhyay's chapter then introduces the case of an urban renewal plan for an artists' colony in Kolkata in India to illustrate the rationale for an alternative form of production of space. In this case, it is clear that many of the artists feel that, even though their space was messy, dirty and slum-like, it was also distinctive and contributed to their unique identity. In addition, it was also a space that was well-adapted to the primary function of creating clay idols. This led to resistance to the state-sponsored plan for renewal and gentrification, illustrating the potential for an alternative mode of production which might offer more humane and relevant spaces.

Filion's chapter complements others by taking an overview of Lefebvre's writings in relation to their influence on policy and practice. He ascribes this influence to a series of factors, including the translation of his books into English, his thinking by association which leads to a multiplicity of ideas (linked perhaps to the influence of surrealist free-association), the potential for these ideas to be interpreted and adapted in different ways and his foresight over future developments in urban life. He foregrounds Lefebvre's vision for an urban context in which use values (based on human needs and fulfilment) predominate over exchange values, and in which centrality can allow everyone to enjoy rich multi-functionality, diversity, social interaction, creativity and festivity. He also highlights Lefebvre's focus on the linkage between human alienation (resulting from the exploitation inherent within capitalism) and the right to the city. Mechanisms such as land use planning, he suggests, are unable to address the root causes of problems, and what is needed is universal access to centres which encourage social interaction – the right to the city being essentially the right to centrality.

Nevertheless, he reflects the views of many other authors within the book by pointing out the lack of clear and focused implications in this context, in part due to the associative nature of his writing with many disparate ideas linked together, and the extensive use of analogy. Consequently he foregrounds the difficulty inherent in proposing clear implications from his writings (for instance in terms of concrete forms of development) as well as the lack of evident empirical evidence. But again he suggests that it is perhaps this very nature of Lefebvre's writing which has significant advantages in terms of flexibility of use and capacity to encompass many interpretations – which are elements that have perhaps assisted with the longevity of his ideas. Filion goes on to show how, in the contemporary context, Lefebvre's vision for a city which displays characteristics which assist the development of human fulfilment align with some elements of contemporary urbanism such as the provision of urban amenities and festivals. More broadly, however, he suggests that Lefebvre's vision can be contrasted with the common contemporary model of urban revitalisation based on emphasising elements such as recreation, diversity, culture and spectacle, applies these essentially to the narrow aim of stimulation of economic development, with the 'creative class' being the main beneficiary.

Spacek's chapter considers processes of urbanism in India and highlights again themes of dispossession and displacement resulting from the actions of the state government. He shows how this has led to increasing homogeneity of urban space, related to a dominant focus on functional approaches to land use, and he illustrates the symbolic use of space such as that in relation to airports and related infrastructure developed by the state. As with other cases in the book, however, he also shows the emergence of challenge and resistance to such processes, with the ultimate outcome remaining uncertain.

Marinic's chapter shows how a process of suburban abandonment, via disinvestment in obsolescent 'dead malls' at the Texas/Mexico border, has allowed an emergent heterotopia which offers opportunities for the appropriation of space by historically under-represented groups. This means that areas of previously high-end shopping were subsequently used for carnivals, clandestine pop-up shops, storefront churches and social clubs, showing how alternative occupiers can germinate and thrive in such areas.

Finally, Jorge's chapter shows how processes of gentrification, linked to use and exchange value, have played out in Maputo. She uses the specific example of pericentral self-produced neighbourhoods, which form 'prohibited places' as defined by Lefebvre, reflecting wider processes of commodification and financialisation of urban space, and simultaneous effects in relation to centralisation and fragmentation.

Preliminary conclusions and thoughts

Lefebvre's ideas and suburbs

To what extent do we need to focus on city centres in isolation in the context of this book? The conclusions begin with a rural example, and many authors in the book's substantive chapters problematise the rural/urban nexus. This is illustrated by the compelling case of 'dead malls' in the chapter by Marinic, showing how these can become spaces for a wide variety of informal use catering directly to the outcast and underprivileged. Can we apply such ideas in other parts of the suburbs, or even in the context of agricultural uses outside the city?

Lefebvre's ideas and spatial inequalities

Following from the above, how can we ensure that the means and mechanisms which Lefebvre's ideas encourage can speak directly to those most in need? This links of course to the operationalisation of his ideas for real political change as referred to elsewhere, but also to the centrality of notions of the iniquitous city, fundamental to persistent urban problems faced globally. Ultimately, some might argue that this might lead to destruction/insurrection on a large scale to the detriment of all. Who is addressing this in any meaningful way and how can Lefebvre be leveraged here?

Lefebvre's ideas and a 'concrete utopia'

As set out elsewhere there is a compelling need to communicate, translate and make reality the notions explored in this book. How can we engender a 'concrete utopia' via these ideas in the absence of empiricism in the ideas of Lefebvre as currently conceived? Should we simply point to positive examples? Should we compile checklists? Should we propose techniques? Often the desired outcome (multifunctional, diversified, festive and inclusive cities) is clearer than the means (none really exist to apply for instance rhythmanalysis to mainstream urban planning).

1

Lefebvre's transduction in a neoliberal epoch

Michael E. Leary-Owhin

> Not so many years ago, the word 'space' had a strictly geometrical meaning: the idea it evoked was simply of an empty area... and the general feeling was that the concept of space was ultimately a mathematical one. To speak of 'social space', therefore, would have sounded strange.
>
> *(Lefebvre 1991a: 1)*

> The urban (an abbreviated form of urban society) can therefore be defined not as an accomplished reality, situated behind the actual in time, but, on the contrary, as a horizon, an illuminating virtuality. It is the possible, defined by a direction, that moves toward the urban as the culmination of its journey. To reach it – in other words, to realize it – we must first overcome or break through the obstacles that currently make it *impossible.*
>
> *(Lefebvre 2003: 16–17)*

Introduction

Breaking through the obstacles, referred to by Lefebvre in the second epigraph above, means in part, I argue in this chapter, seeking to overcome the more pernicious aspects of the production of urban space inherent in presently existing neoliberalism. Henri Lefebvre is associated with a range of fascinating but challenging ideas related to cities and the urban. His ideas have endured severe criticism and fierce changes in academic and political fashion over the last five decades. They endure partly because they rest on a solid Humanist Marxist foundation and partly because Lefebvre tended to eschew the detailed prescription of ready-made solutions. That said, I argue he does provide guidance for empirical research – the spatial triad being an obvious point of entry. His ideas about space and a possible urban society were informed by his upbringing and later by his sociological empirical research. The chapter charts briefly how his Marxism influenced his approach to urban research. These issues are examined in new ways that challenge the critiques asserting Lefebvre's paucity of utility for empirical research. The chapter closes with a focus on Lefebvre's grand urban project, facilitated by transduction and his concrete utopian thoughts on cities and urban society.

Enigmatic, constant, mutable

Throughout his life there was not one Henri Lefebvre but many. At various times he was: a precocious Sorbonne student, surrealist, poet, Parisian taxi driver, schoolteacher, municipal councillor (in Montargis), Resistance fighter, champion of peasants and proletarians, contract researcher, university professor, art critic, government advisor, utopian thinker, media pundit, world traveller and bonne vivant retiree. We can add, in his own words: tormented youth, rebellious anarchist, militant thinker and balanced 30-something Marxist philosopher-sociologist (Lefebvre 1950 in Elden 2004: 1). Despite the superabundance of words written about him, Henri Lefebvre remains an enigma. This only adds to his charm and attraction. He is enigmatic not just because of his challenging, some would say inscrutable ideas, but also because of his unique Humanist Neo-Marxism, baffling for some scholars. Even his birthdate is shrouded in mystery. Anglophone readers of his work can be forgiven for thinking he was born in 1905, since this date appears in the copyright page of his most famous book, *The Production of Space* (TPOS), and several others. This date also appears in many library catalogues. Why the mystery? Opinions differ, as discussed below. During World War Two (WW2) he may have carried out daring acts of sabotage against the occupying Nazis, or he may have been more of an armchair Resistance strategist. I return to these points below. He was certainly hostile to Fascism, and his 1938 book, *Hitler au pouvoir*, criticising Adolf Hitler and National Socialism, marked him as a subversive and copies were burned by the Nazis during the occupation of France.

Lefebvre's protracted life and his extended metaphorical and actual journey from the rural to the urban provides too much fascinating detail to document in full here. Insights can be gained from the biographical-style authors of the third wave mentioned in the Introduction to this volume. Indeed, the journey from the rural to the urban is a key theme in much of the literature. Lefebvre was not just an accomplished philosopher and writer, he was a skilful archival researcher, who augmented archival data with his own observations of the impact of capitalism and industrialisation on rural and urban everyday life. Notwithstanding this, he became Professor of Sociology in his 60s, and he is endowed with a mutable range of identities and claimed by a variety of academic factions, especially since the 1990s. To some he was an adventurous dialectician or: a Marxist political scientist, Marxist geographer, professional sociologist, political geographer, postmodern geographer or oppositional heretic. Gottdiener (2018) is perhaps the most strident in asserting the unfairness of some disciplinary claiming of Lefebvre. He is especially scathing about the claims of geographers. It is apparent that Lefebvre was not a uni-disciplinary dogmatist, and he advocates the necessity of an interdisciplinary approach to the understanding of cities and the urban. And he was aware his spatial ideas would make some disciplinary specialists feel threatened because they 'view social space through the optic of their methodology and their reductionistic schemata' (Lefebvre 1991a: 108). For me, he is the academic *provocateur* par excellence.

Lefebvre liked to refer to himself as a 'political thinker', and it is not for me to try to adjudicate matters of his identity. On the contrary, I prefer to complicate them. It seems to me that from the age of 60, Lefebvre's sustained critique of urban planning qualifies him as 'honorary planning critic'. In recent research I suggest it is rather unfortunate that planning theorists and practitioners have, with a few noteworthy exceptions, ignored the potential contributions that Lefebvre's ideas can make to planning theory and practice (Leary-Owhin 2018). Given his constant engagement with planning since the 1960s, it is surprising that a leading Anglophone planning theory book, *Readings in Planning Theory* (2015), did not mention him until its fourth edition and then only in passing. This is despite one of the first Anglophone articles on Lefebvre and planning being published over two decades ago (Allen and Pryke 1994) and a steady trickle of publications since then (e.g. Carp 2008; Leary 2009; Buser 2012; Leary 2013; Holgersen 2015; Honeck 2017). Similarly, Lefebvre is ignored in recent leading urban regeneration textbooks (e.g. Tallon 2013 but see Leary and McCarthy 2013). Perhaps this is partly because Lefebvre is regarded by many, including urban policymakers and practitioners, as a tough and enigmatic read leading to 'bewilderment' for many first-time readers (Pierce 2017: 1).

It was not until Lefebvre was in his 70s that he started to become famous in the Anglophone academic world for his startling ideas about the production of space. He was aware that the proposition at first sight may have appeared strange, as the first epigraph to this chapter illustrates. Representations of space are at the heart of the production of space and notably also such disciplines as architecture, geography and planning. His writing encourages us to think differently about cities and the urban. Its power lies partially in the ways cities and the urban are represented in his various texts and his spoken word. That *his words* are, in his own terms, representations, is so obvious it is usually overlooked. Why? Perhaps because he rarely if ever presents them in this way. It is worth reminding ourselves of what Stuart Hall says about representations:

> In part, we give things meaning by how we *represent* them – the words we use about them, the stories we tell about them, the images of them we produce, the emotions we associate with them, the ways we classify and conceptualize them, the values we place on them.
>
> *(Hall 1997: 3)*

Hall draws our attention usefully to visual representations about which, in the context of representations of space, Lefebvre unfortunately has little to say, although he is complementary about some fine art painters but disparaging about guidebooks, photographs and some architectural representations of space (Lefebvre 1991a). Hall also indicates how emotions and values are important in how we interpret representations, and this is something of which Lefebvre was keenly aware. With these ideas in mind, I now turn to Lefebvre's early years, picking out moments in his life journey to elucidate his approach to cities and the urban.

Glory, money and women

Unlike some other French philosophers, Lefebvre was conscious that he made the study of and attempts to change cities and the urban his life's work, saying in an oblique jab at his rival Jean-Paul Sartre that he gave up 'the ideas of the *Manifesto*' and 'glory, money and women, for a hard and mediocre life, for militant thought working on real problems' (Lefebvre 1945 in Elden 2004: 20). Lefebvre was born in the town of Hagetmau just outside the French Pyrénées but grew up nearby in the Pyrenean town of Navarrenx. This town acquired great importance for the development of his ideas about space (see Leary-Owhin in Part 6). The town is located on the eastern bank of the River Oloron, where a large stone bridge was built in the 13th century, making Navarrenx more accessible but also of more strategic military importance (see Figure 1.1). Enlargement and fortification in the 1300s rendered Navarrenx an orthogonal medieval 'new town' (see Figure 1.2). Due to wars between the pre-French city states, it was remodelled and the fortifications reinforced in the 1500s to form encircling defensive Italianate ramparts. Thus historic Navarrenx became one of the first *Bastide* towns in what later became France. The ramparts are a treasured heritage feature of the town, now classified as historic monuments by the French planning system.

Navarrenx lies in the Department of Pyrénées-Atlantiques (the former traditional province of Béarn) in the French Basque area. It is close to the Lacq commune where the 'new town' of Mourenx is located. Historically, Béarn enjoyed a bucolic, organic landscape dominated by

Figure 1.1 The 13th-century bridge over the River Oloron at Navarrenx. *Source:* © Michael Leary-Owhin (2017).

Figure 1.2 Medieval fortifications for Navarrenx 'New Town'. *Source:* © Michael Leary-Owhin (2017).

agriculture, dotted with traditional farmhouses and buildings (see Figure 1.3). This socio-geography of Lefebvre's early years proved instrumental for his intellectual development and attitudes to the rural and urban. Lefebvre's family was relatively affluent and middle class, his mother being a staunch Catholic, his father, secular. This religious divide may well have encouraged him to embrace dialectical contradiction. Lefebvre grew up in a large, comfortable 18th-century family house in the centre of Navarrenx. When I visited in spring 2017 to meet Catherine Régulier-Lefebvre, it looked resplendent, swathed in flowering wisteria (see Figure 1.4). Lefebvre left

Figure 1.3 The Béarn countryside features many historic buildings constructed of stone and red clay rooves in the traditional vernacular design. *Source:* © Michael Leary-Owhin (2017).

Figure 1.4 Lefebvre's family home in Navarrenx, enveloped in wisteria. *Source:* © Michael Leary-Owhin (2017).

Navarrenx in his early teens, eventually travelling extensively in France and worldwide, but he spent a large part of his working life in Paris. Importantly, for his research, he returned regularly for extended stays at the family home. It became a kind of summer retreat where he entertained academic friends. In his 70s he retired there, living with his last wife, Catherine Régulier-Lefebvre.

At the age of 18 he went to study philosophy at the Sorbonne. There he encountered the writings of Karl Marx, but he may well have been a surrealist before a Marxist and he certainly read and met the leading French surrealists of the day. Evidently, any surrealist nihilism did not survive his multifaceted Neo-Marxism nor his opposition to Fascism. In his 20s while studying at the Sorbonne, he also discovered the writings of Hegel, Nietzsche, Engels and Lenin, developing over the decades his unique heterodox Neo-Marxism with its strong Humanist constituent. For Lefebvre, Marxism was less a blueprint for state communism and more an approach to knowledge and a set of ideas for political action by ordinary citizens rather than political parties. He rejected orthodox Marxist doctrine (and Stalinism), saying, 'Marxism is, above all, a method of analysing social practices; it is not a series of assumptions, postulates, or dogmatic propositions' (Lefebvre in Burgel et al. 1987). Gravitating to the political left, he joined the French Communist Party in 1928, becoming a leading theorist over the next 30 years. He was expelled from the Party in 1958 for his opposition to the Party's 'unrepentant Stalinism' (Elden in Lefebvre 2016 [1972]: xiii).

During WW2 we can be sure he joined the French Resistance, to oppose the Nazi puppet Vichy government, rising to the rank of captain. His fluent German was no doubt extremely useful then, as it was for reading original versions of Marx, Engels, Nietzsche, Hegel and Kant. During the war he spent some time in Marseille but mostly he was cloaked by the Pyrenean mountains. What he did there is somewhat shrouded in mystery. For Merrifield (2006: 3) he was a daring saboteur, derailing enemy trains, tracking down collaborators. While Elden (in Lefebvre 2016 [1972]: x) is more circumspect, venturing only as far as to suggest his exact involvement is unclear; roles as active agent or theoretician appearing feasible. Does Lefebvre himself help? Only a little. He says that during his time with the Resistance he had conspiratorial meetings including with railwaymen in the area of Aix-en-Provence. He says *there were* train derailments and collaborators *were* executed. But he also says, *we* worked to provide the Resistance with an ideology to counter the powerful Vichy ideological offensive (Lefebvre 1991 in Latour and Combes 1991). We know that wartime combatants, even decades later, can find it difficult to talk about their experiences. So, for now, his precise role continues to be enigmatic. Even so, this wartime episode is important because, although war is punctuated with short periods of frenetic danger laced with violence, there are also protracted phases of inactivity. It was during these times that Lefebvre was able to conduct empirical research that facilitated the development of his dialectical research methodology.

Opinions differ about Lefebvre's age and his attitude to it (Gottdiener 1993). For some 'He did not like to be reminded of his date of birth' (Shields 1999: 8). Apparently, he 'was always playful about his age' and enjoyed 'teasing his chroniclers' with different lived timelines. While visiting the University of California in the 1980s, he would 'correct' the date of his birth in the library catalogue to 1905 (Soja 1996: 28). Elden affirms that Lefebvre's birth date is disputed and that 'by 1965 he could have been 60, 64 or even 67' (Elden in Lefebvre 2016 [1972]: xiii). Without the benefit of contrary documentary evidence, I accept the consensus that Lefebvre was born in June 1901 and passed away in June 1991. Whatever his age, his extended life allows for much thinking, reading, writing and controversy. Chief among controversies are, how can we do and what does it mean to do Lefebvrian urban research?

Being Lefebvrian: research theory and practice

From his Marxist philosophical position, Lefebvre was concerned with persuading his students and readers of the primacy of the Marxist approach to investigating the past in order to understand the present and create a fairer future. Alongside this, he occasionally provided detailed research methods guidance before this was *de rigueur*. Even though he does so in his own inimitable style, few students of cities and the urban find the study of methodology stimulating per se. His philosophy was a springboard for empirical research and political action. This is evident from 1947, for example, in the first edition of the *Critique of Everyday Life*, in which:

> 'philosophy no longer scorns the concrete and the everyday'. By making alienation 'the key concept in the analysis of human situations since Marx', Lefebvre was opening philosophy to action: taken in its Kantian sense, critique was not simply knowledge of everyday life, but knowledge of the means to transform it.
>
> *(Trebitsch in Lefebvre 1991b [1947]: x)*

Transformation required knowledge and understanding of the theoretical *and* the empirical. Lefebvre was therefore, along with his students and colleagues, concerned intensely with the issue of how to conduct rural and urban sociological research based on sound theory, methodology and methods. He appreciated that the 'facts' of empirical reality can of course be interpreted differently. Long before Castells' 1970s criticisms of a lack of an empirical base, Lefebvre was aware of this danger, asserting that the second volume of *Everyday Life* 'ran the risk of becoming a work of philosophy rather than a piece of concrete sociological research' (Lefebvre 1991b [1947]: 98–99). He realised there was a vast archive of popular literature that could inform his analysis of everyday life, 'Moreover, this press represents an extraordinary sociological fact, which cries out to be analysed'. (Ibid.). So Lefebvre was a philosopher *and* a data-conscious researcher of grounded everyday life.

The Introduction to this Handbook outlines how Lefebvre's ideas are often regarded as confused, confusing, unfinished and therefore not useful as frameworks for empirical research. Such comments and criticisms relate in particular to theories of the spatial triad in TPOS. Some are more generalised (Schmid et al. 2014). Paradoxically, it is the triad that has stimulated the largest volume of empirical research worldwide. Schmid's (2014: 35) observations typify comments about the challenges of doing Lefebvrian-inspired empirical research:

> This brings us to the next source of possible trouble: while his theory has gradually found better understanding in recent years, and many of his concepts have been debated and clarified, the question of empirical application has long remained opaque. Lefebvre did not really offer clarification here, as his books remain elusive when it comes to this question, and examples he gives are often more illustrative in character than exact in presenting detailed results of concrete field research.

Schmid, who acknowledges Stanek's 2011 book for the voluminous and meticulous revelations regarding Lefebvre's empirical research (see also Stanek 2017), nevertheless asserts:

> However, Lefebvre did not develop any sophisticated methodology. He and his colleagues and assistants used the existing methods that were available at the time, based mainly on qualitative methodologies. It follows that there are no simple recipes or models that would allow us to apply his concepts.
>
> *(Schmid's 2014: 35)*

There is some credence to Schmid's position on this, but it is not the full story. Schmid is correct as far as Lefebvre may not have developed new research methods, but he did appreciate the importance of qualitative and quantitative research and contribute greatly to methodology and research theory, as explained below. The point is that these aspects of his work are largely ignored in the Anglophone academy. One consequence is that most production of space researchers seek empirical contrasts between representations of space and spaces of representation (see e.g. Fyfe 1996; Lapina 2017; Harwood et al. 2018).

If Lefebvre's urban ideas are so confused and confusing, how could they have been adopted for rigorous and high-quality empirical research by so many academics? Perhaps his intricate philosophy and written style overshadowed his guidance for researchers. Paradoxically, Unwin (2000) offers severe criticism of Lefebvre but accepts that the production of space idea engendered a trove of empirical research. Researchers from the early 1990s had to divine in Lefebvre an appropriate methodological approach. One would hope it involves more than citing Lefebvre early on, then leaving methodological issues implicit in the published research. That said, there is little consensus about how to do Lefebvrian research, which creates headaches but conveniently allows scope for a range of intriguing approaches and interpretations.

A range of Lefebvrian scholars offer insights into how to do research inspired by his ideas. Merrifield (1993: 522) thinks Lefebvre's framework provides a flexible device which can illuminate the nature of space and its relations with a broader social whole. Soja sees in Lefebvre the potential for a method based on 'trialectics' that stresses the interweaving of the three spatial elements (1996: 10) and the history of representations. Kofman and Lebas (1996: 8–10) maintain that being Lefebvrian 'is more a sensibility, rather than a closed system', but also many find his theoretical insights difficult to apply due to the fluidity and openness of his thought. Kofman and Lebas are still able though to deduce a Lefebvrian approach to production of space research based on observation, investigation of concrete reality and historical analyses. Borden (2001: 11–12) was one of the first Anglophone academics to articulate Lefebvrian guidance for empirical research and postulates eight tenets which, although useful, are more conceptual than concrete. They do not constitute a patented system but are an approximation of a method that provides a theoretical and methodological compass and helps keep researchers on the right track. Borden seems inspired by Lefebvre's claim that:

> The theoretical conception we are trying to work out in no way aspires to the status of a completed 'totality', and even less to that of a 'system' or 'synthesis'. It implies discrimination between 'factors', elements or moments. To reiterate a fundamental theoretical and methodological principle, this approach aims both to reconnect elements that have been separated and to replace confusion by clear distinctions; to rejoin the severed and reanalyse the commingled.
>
> *(Lefebvre 1991a [1974]: 413)*

Lefebvre left sufficient signs, scattered liberally throughout his major works, for how to research the urban, especially but not exclusively in TPOS. It is this sprinkling, rather than a complete lack, that creates problems for the researcher. I contend he provides guidance for how to do research at four differing levels ranging from the particular to the general: 1) specific research methods; 2) his heterodox Marxism; 3) a regressive-progressive theoretical model; 4) the theory of transduction. The last two are crucial for research regarding the RTC, production of space, planetary urbanisation and rhythmanalysis.

Specific research methods

Contrary to what several commentators say, Lefebvre does provide specific guidance for doing empirical research related to urban issues. In considering the issue of housing or habitat in the 1960s, he criticised the functionalist approach of Le Corbusier and others on epistemological *and* methodological grounds. In an introduction to a piece of research called *Preface to the Study of the Habitat of the 'Pavillon'* (1953) he critiques sociological positivism that rejects in-depth analysis and takes a superficial quantitative approach to 'the study of man, the city or society in general' (Lefebvre 1966 in Elden et al. 2003: 121). This empirical research project was carried out under his supervision while he was director of the Institut de Sociologie Urbaine. In this *Preface* Lefebvre presents philosophical and methodological justification for what we would now call semi-structured in-depth research interviews. He argues that although the sociologist's questionnaire is frequently used for researching housing issues, it is rarely sufficient. It produces numbers, percentages and correlations, it is 'precise but narrow, and may also be suspect' (Ibid.: 127). Questionnaires are sometimes used to match pseudo-concepts to pseudo-facts. Alternatively, the non-directive or semi-structured interview is indispensable because it gives a voice to research subjects through free expression. Unfortunately, it seeks to collect that which lies psychologically deep and puzzlingly 'cannot be collected' (Ibid.: 128).

Lefebvre then faces up to the methodological research bind between quantitative and qualitative approaches or 'the problems of steering a course between metaphysics and positivist triviality' (Ibid.). He concludes there is merit in both approaches. His way out of the bind is to advocate a twofold system, or what is now called mixed method research. In doing so Lefebvre was ahead of the time. Mixed method research approaches are now well established, although they remain controversial for some and the debate continues (Creswell and Creswell 2018). In the 1953 *Preface*, Lefebvre reveals an academic intellect and methodological astuteness and is adept at guiding the empirical researcher through the epistemological and methodology challenges of designing and implementing urban field research. We should not be surprised since such qualities were and are an absolute requirement for the director of a sociological research institute. But it was precisely because he was a Marxist philosopher that he was able to switch gears smoothly to provide powerful ontological and philosophical underpinning for researching the urban.

Marxism, dialectical research and transduction

By the mid-1980s Lefebvre was aware the 19th century, laissez-faire capitalism analysed by Marx had transmuted, due partly to the growth of the state, into neoliberalism, i.e. a tightening of bonds between state and big capital, especially regarding urban development:

> The official [French] planning body responsible for regional development, a powerful centralized organisation, lacked neither resources nor ambitions: to *produce* a harmonious national space – to bring a little order to 'wild' urban development, which answers only to the pursuit of profit. Today nobody is unaware that this innovative planning initiative... was wreaked, reduced to practically nothing by neo-liberalism and since put clumsily back together again.
>
> *(Lefebvre 1985 in Elden et al. 2003: 207)*

Neoliberalism in the Global North and South tilts power away from local government, local inhabitants and small business towards big business and the state. For Lefebvre, the locus of political power was a key factor in his Marxism. For example, regarding aspects of political power he provides this high-level and particular guidance for empirical research:

> Inasmuch as the quest for the relevant productive capacity or creative process leads us in many cases to political power, there arises the question of how such power is exercised. Does it merely command, or does it 'demand' also? What is the nature of its relationship to the groups subordinate to it, which are themselves 'demanders', sometimes also 'commanders', and invariably 'participants'? This is a historical problem – that of all cities, all monuments, all landscapes. The analysis of any space brings us up against the dialectical relationship between demand and command, along with its attendant questions: 'Who?', 'For whom?', 'By whose agency?', 'Why and how?'
>
> *(Lefebvre 1991a [1974]: 116)*

At the philosophical level, Lefebvre's research was guided by his heterodox Marxism, foregrounding the dialectical approach which shaped his ontology and epistemology, but it did not exist in isolation and was always in dialectical tension with the mass of phenomenological detail regarding neo-capitalism, institutions, everyday life and understanding of cities and the urban. Therefore, Lefebvre explored the 'urban problematic within the theoretical framework of historical materialism' (Elden in Lefebvre 2016 [1972]: viii). From the 1960s Lefebvre turned his research attention from the rural to the industrial and the urban. In the post-WW2 period, sociology in France and elsewhere was vying with established subjects, such as history, to become recognised as a university discipline. Lefebvre appreciated the tensions between these two ways of knowing the world but accepted the importance of history for sociology. In TPOS he is adamant a historical approach is essential for revealing the production of space. The history of space 'but also the history of representations along with that of their relationships' would need to be studied along with 'their links with the spatial practice of the particular society' (Lefebvre 1991a [1974]: 42).

This is one reason why time must be at the core of our appreciation of the production of space and why history is important for appreciation of the present in order to instigate change. Accordingly, Lefebvre builds time into what he calls a manual of sociology and 'a very simple method' (Lefebvre 1953 in Elden et al. 2003: 116). This simple empirical research manual appears in a paper called 'Perspectives on rural sociology' published in an academic journal called *Cahiers Internationaux de Sociologie* (International Sociological Notebooks). Publication in this prestigious journal is important because it was founded in 1946 to promote the study of post-WW2 French society and to create *new research instruments* and to do so through an international dimension.

'Perspectives on rural sociology' is a kind of research manual with precise instructions, but it is also more general and complex than that, so the term 'theoretical model' is more appropriate (Elden 2004: 38). Lefebvre's Marxism informs the model especially where it stresses the importance of historical analysis and social 'reality', and in the to-and-fro between the past and the present. It throws light on possible-impossible futures, or in today's parlance, socially just futures:

a) *Descriptive.* Observation, but with an eye informed by experience and a general theory. In the foreground: participant observation of the field. Careful use of survey techniques (interviews, questionnaires, statistics).
b) *Analytico-regressive.* Analysis of reality as described. Attempt to give it a precise *date* (so as not to be limited to an account turning on undated 'archaisms' that are not compared with one another).

c) *Historical-genetic.* Studies of changes in this or that previously *dated* structure, by further (internal or external) development and by its subordination to overall structures. Attempt to reach a genetic classification of formations and structures, in the framework of the overall structure. Thus an attempt to return to the contemporary as previously described, in order to rediscover the present, but elucidated, understood: *explained.*

(Lefebvre in Elden et al. 2003: 116–117)

The model encompasses some well-established research approaches such as: induction, the observation stage, quantitative and qualitative research. It is another example of an analytical triad: description, regression-progression, explanation. Merrifield notes shrewdly that the model 'became a methodology and political credo', privileging the quotidian (2006: 5–6). This regressive-progressive method or theory has the power to show history in a new light and in so doing present alternative understandings of the present and possible futures (Stanek 2011: 159). Lefebvre is clear that 'We can also say that transduction goes from the (given) real to the possible'. (Lefebvre 2002: 117–118). In studying the past, especially from the time of the modern era, archives and archival methods become essential. It is the space of Navarrenx and the history of its production which 'embodies the forms and actions of a thousand-year-old community' (Lefebvre 1995: 116). Interestingly, archival research has been pursued by a number of researchers inspired by Lefebvre (e.g. Fyfe 1996; Hubbard et al. 2003; Leary 2009). It is salient to note how Lefebvre accepts, in non-dogmatic fashion, a range of research methods and data as being useful for rural and urban research, and his mixed method approach is now part of research orthodoxy.

Far from being simply the thought experiments of a detached philosopher, Lefebvre was driven to ensure that his research was founded on 'empirical reality' – a range of data sources constructed through what we now call a mixed methods approach. Moreover, this concrete research and associated methods were underpinned by his profound transduction model, so much so that one of Lefebvre's strongest critics, Jean-Paul Sartre, explained graciously regarding the regressive-progressive method:

> we have nothing to add to this passage, so clear and so rich, except that we believe that this method, with its phase of phenomenological description and its double movement of regression followed by progress is valid – with the modifications which its objects may impose upon it – *in all the domains of anthropology.*
>
> *(Sartre 1960 in Elden 2004: 38)*

Lefebvre elaborates how this approach can work in the 1968 essay 'Right to the city' (in Kofman and Lebas 1996: 113). He considers the methodological tools and research focus required to undertake urban analysis are: 'form, function, structure, levels, dimensions, text, context, field and whole, writing and reading, system, signified and signifier, language and metalanguage, institutions'. And in the context of the RTC he argues two approaches are needed. The first goes from the general to the most specific, 'and then uncovers the city as specific and (relatively) privileged mediation'. The second 'constructs the general by identifying the elements and significations of what is observable in the urban'. It is possible thereby to reach 'the concealed daily life: its rhythms, its occupations, its spatiotemporal organization, its clandestine "culture", its underground life' (Ibid.).

When the research regression-progression model is applied to exploring the potential for the production of utopia or urban society, Lefebvre understands this as an imaginative process that challenges the neo-capitalist status quo and the ways in which it elides desirable futures by what he called a blind field (Lefebvre 2003: 41). This mask leads people away from a search for radical alternatives, an urban society, because at present they seem unattainable, impossible. Urban society is developed conceptually mainly in his book *The Urban Revolution* (2003), and its potential arises from the contradictions inherent in neoliberal capitalism and the abstract space of emerging planetary urbanisation. His Marxism had two consequences for Lefebvre's conceptualisation of urban society. First, it rendered him suspicious of collusion between big business and government. He favoured the idea of the withering away of the state in positive processes that would allow the emergence of community self-determination, *autogestion* or self-management (Harvey 2012; Purcell 2014).

Lefebvre calls his regression-progression approach to knowing, reasoning and researching 'transduction'. In fact, he had developed this concept by 1961 and refers to it in the second volume of *Everyday Life*. The need for transduction lies partly in Lefebvre's dissatisfaction with overly quantitative empiricist, trivialised approaches to urban research:

> Unlike a fact-filled empiricism with its risky extrapolations and fragments of indigestible knowledge, we can build a *theory* from a *theoretical hypothesis*. The development of such a theory is associated with a *methodology*. For example, research involving a virtual object, which attempts to define and realize that object as part of an ongoing project, already has a name: *transduction*. The term reflects an intellectual approach toward a possible object, which we can employ alongside the more conventional activities of deduction and induction. The concept of an urban society, which I introduced above, thus implies a hypothesis and a definition.
>
> *(Lefebvre 2003: 5)*

His focus on theory and methodology, rather than the detail of research methods, is a major strength because it is too easy to become wedded to particular methods that either go out of fashion or provoke acrimonious exchanges between rival research method camps. His term 'virtual object' predates the computing era and refers to that which does not yet exist fully. It is inchoate but is discernible, on the horizon. Transduction as research practice is difficult to execute and concrete utopias difficult to identify. Lefebvre added that transduction is a serious challenge partly because neo-capitalism veils using homogeneity and technology, 'the utopian part of urbanist projects' (Lefebvre 2003: 161). An urban society is one where difference is tolerated and celebrated, but it is also one where people and communities are able to realise they have more in common than they may think (Merrifield 2018). Furthermore, it is a society where the right to be different is a fundamental RTC (Leary-Owhin 2016). However, Lefebvre warned against expecting an immediate epistemology to develop that could bring forth research, practice and political action for an urban society:

> the argument I have developed would claim the contrary. For the moment, for a long time into the future, the problematic will outweigh our understanding. What is most needed is that we categorize, that we prepare concepts (categories) we can verify, that we explore the possible-impossible, and that we do so through transduction.
>
> *(Lefebvre 2003: 162)*

Purcell is one of several researchers to grasp the centrality of transduction for Lefebvrian research, which begins:

> by closely examining actual-but-inchoate practices that are currently taking place in the city, and then we extrapolate them using theoretical reflection to produce, in thought, a

> more fully developed version of them, a virtual idea (which he [Lefebvre] calls 'urban society') that shows us what kind of world they would produce if they were allowed to flourish and pervade the city. Once we have extrapolated this concept in thought, we then use it as a lens to help us better see those actual practices as they exist today, struggling to emerge and flourish. We need this lens, he says, because the fledgling urban society is difficult to see in the blinding light of the industrial city.
>
> *(2013: 319)*

Lefebvre admits his approach to research theory and practice does not aspire to the status of a completed totality but seeks to 'reconnect elements that have been separated and to replace confusion by clear distinctions' (Lefebvre 1991a [1974]: 413). Nevertheless, at the philosophical level Lefebvre signals a project, and it is this lifelong project to which he directs research effort and encourages others to do likewise. It is a grand project, the purpose of which is no less than conceptualising and striving to bring about a different kind of society, an urban society predicated on difference, on differential space, from that inherent in neo-capitalism:

> By seeking to point the way towards a different space, towards the space of a different (social) life and of a different mode of production, this project straddles the breach between science and utopia, reality and ideality, conceived and lived. It aspires to surmount these oppositions by exploring the dialectical relationship between 'possible' and 'impossible', and this both objectively and subjectively.
>
> *(Lefebvre 1991a [1974]: 60)*

An urban society is therefore a paradoxical characteristic of neo-capitalist abstract space, inherent in its contradictions. So we arrive at the ultimate contradiction – the possible-impossible. This is the root of and the route to Lefebvre's utopianism. Like all political activists, Marxist or otherwise, Lefebvre understood its attainment requires: patience; principles; pragmatism; wisdom; courage; humour; and, above all, protracted struggle through political action.

Conclusions

In this chapter I have argued that part of Lefebvre's enduring attraction and charm are the enigmas that shroud his persona. Even something as mundane, to many of us, as his date of birth is enveloped in mystery and a sense of Lefebvrian fun. He appears to have resented the criticism that he was only a detached philosopher, however accomplished. This chapter elucidates research ideas regarding the importance for Lefebvre of research theory and practice. It links the development of Lefebvre's understanding of theoretical and empirical research issues to his Marxism and to his wartime and post-WW2 experiences. So following that route, this chapter outlines some of his important interventions regarding not just ontology and epistemology but also methodology and research methods. The chapter demonstrates that Lefebvre provides a range of signs and clues for research theory and practice, and of paramount importance is the regressive-progressive transduction model. The chapter closes with an explanation of the salience of transduction, RTC, urban society and the necessary challenges to the capitalist status quo. Lefebvre's intellectual and political struggle, agonism, is important per se but the chapter asserts they are more imperative for their contribution to his grand project, a project that is more relevant in today's hyper-neoliberal context. His struggles and those of many other academics, politicians and activists offer hope in venturing from the real towards the possible-impossible.

Acknowledgements

I am most grateful for the hospitality shown me by Catherine and Léa Régulier-Lefebvre when I visited Navarrenx during Easter 2017 and for their assistance in researching this chapter.

References

Allen, J. and Pryke, M. (1994) 'The production of service space', *Environment and Planning D: Society and Space*, 12:4 453–475.
Borden, I.M. (2001) *Skateboarding, Space and the City: Architecture and the Body*, Oxford: Berg.
Burgel, G., Burgel, G. and Dezes, M.G. (1987) 'An interview with Henri Lefebvre', *Environment and Planning D: Society and Space*, 5:1 27–38.
Buser, M. (2012) 'The production of space in metropolitan regions: a Lefebvrian analysis of governance and spatial change', *Planning Theory*, 11:3 279–298.
Carp, J. (2008) '"Ground-Truthing" representations of social space: using Lefebvre's conceptual triad', *Journal of Planning Education and Research*, 28:2 129–142.
Creswell, J.W. and Creswell, J.D. (2018) *Research Design: Qualitative, Quantitative, and Mixed Methods Approaches*, London: Sage.
Elden, S. (ed.) (2004) *Understanding Henri Lefebvre: Theory and the Possible*, London: Continuum.
Elden, S., Lebas, E. and Kofman, E. (eds) (2003) *Henri Lefebvre: Key Writings*, London: Continuum.
Fyfe, N. (1996) 'Contested visions of a modern city: planning and poetry in post war Glasgow', *Environment and Planning A*, 28:3 387–403.
Gottdiener, M. (1993) 'A Marx for our time: Henri Lefebvre and the production of space', *Sociological Theory*, 11:1 129–134.
Gottdiener, M. (2018) 'Who owns Lefebvre? The forgotten sociological contribution to the new urban sociology', *Contemporary Sociology*, 47:3 266–271.
Hall, S. (ed.) (1997) *Representation: Cultural Representations and Signifying Practices*, London: Sage.
Harvey, D. (2012) *Rebel Cities: From the Right to the City to the Urban Revolution*, London: Verso.
Harwood, S.A., Mendenhall, R., Lee, S.S., Riopelle, C. and Browne Huntt, M. (2018) 'Everyday racism in integrated spaces: mapping the experiences of students of color at a diversifying predominantly white institution', *Annals of the American Association of Geographers*, 108:5 1245–1259.
Holgersen, S. (2015) 'Spatial planning as condensation of social relations: a dialectical approach', *Planning Theory*, 14:1 5–22.
Honeck, T. (2017) 'From squatters to creatives. An innovation perspective on temporary use in planning', *Planning Theory & Practice*, 18:2 268–287.
Hubbard, P., Faire, L. and Lilley, K.D. (2003) 'Contesting the modern city: reconstruction and everyday life in post-war Coventry', *Planning Perspectives*, 18:4 377–397.
Kofman, E. and Lebas, E. (eds) (1996) *Writings on Cities: Henri Lefebvre*, Oxford: Blackwell.
Lapina, L. (2017) '"Cultivating Integration"? Migrant space-making in urban gardens', *Journal of Intercultural Studies*, 38:6 621–636.
Latour, P. and Combes, F. (eds) (1991) *Conversation avec Henri Lefebvre*, Paris: Messidor.
Leary, M.E. (2009) 'The production of space through a shrine and vendetta in Manchester: Lefebvre's spatial triad and the regeneration of a place renamed Castlefield', *Planning Theory & Practice*, 10:2 189–212.
Leary, M.E. (2013) 'A Lefebvrian analysis of the production of glorious, gruesome public space in Manchester', *Progress in Planning*, 85:1 1–52.
Leary, M.E. and McCarthy, J. (eds) (2013) *The Routledge Companion to Urban Regeneration*, London: Routledge.
Leary-Owhin, M.E. (2016) *Exploring the Production of Urban Space: Differential Space in Three Post-Industrial Cities*, Bristol: Policy Press.
Leary-Owhin, M.E. (2018) 'Henri Lefebvre: planning's friend or implacable critic', *Urban Planning*, 3:3 1–4.
Lefebvre, H. (1991a [1974]) *The Production of Space*, Oxford: Blackwell.
Lefebvre, H. (1991b [1947]) *Critique of Everyday Life Volume One*, London: Verso.
Lefebvre, H. (1995 [1962]) 'Seventh Prelude: notes on the new town', in H. Lefebvre (ed.) *Introduction to Modernity: Twelve Preludes*, London: Verso.
Lefebvre, H. (2002 [1961]) *Critique of Everyday Life Volume Two*, London: Verso.
Lefebvre, H. (2003 [1970]) *The Urban Revolution*, Minneapolis, MN: University of Minnesota Press.

Lefebvre, H. (2016 [1972]) *Marxist Thought and the City*, Minneapolis, MN: University of Minnesota Press.
Merrifield, A. (1993) 'Place and space: a Lefebvrian reconciliation', *Transactions of the Institute of British Geographers*, 18:4 516–531.
Merrifield, A. (2006) *Henri Lefebvre: A Critical Introduction*, London: Taylor & Francis.
Merrifield, A. (2018) 'Planetary urbanisation: *une affaire de perception*', *Urban Geography*, 39:10 1603–1607.
Pierce, J. (2017) Review of *Marxist Thought and the City*, *Antipode*, accessed online via AntipodeFoundation.org at radicalantipode.files.wordpress.com, accessed March 2017.
Purcell, M. (2013) 'The right to the city: the struggle for democracy in the urban public realm', *Policy & Politics*, 43:3 311–327.
Purcell, M. (2014) 'Possible worlds: Henri Lefebvre and the right to the city', *Journal of Urban Affairs*, 31:1 141–154.
Schmid, C. (2014) 'The trouble with Henri: urban research and theory of the production of space', in L. Stanek, C. Schmid, and A. Moravánszky (eds) *Urban Revolution Now: Henri Lefebvre in Social Research and Architecture*, Abingdon: Routledge.
Schmid, C., Stanek, L. and Moravánszky, A. (2014) 'Theory, not method – thinking with Lefebvre', in L. Stanek, C. Schmid, and A. Moravánszky (eds) *Urban Revolution Now: Henri Lefebvre in Social Research and Architecture*, Abingdon: Routledge.
Shields, R. (1999) *Lefebvre, Love, and Struggle: Spatial Dialectics*, London: Taylor Francis.
Soja, E. (1996) *Thirdspace: Journeys to Los Angeles and Other Real-and-Imagined Places*, Oxford: Blackwell.
Stanek, L. (2011) *Henri Lefebvre on Space: Architecture, Urban Research, and the Production of Theory*, Minneapolis, MN: University of Minnesota Press.
Stanek, L. (2017) 'Collective luxury', *The Journal of Architecture*, 22:3 478–487.
Tallon, A. (2013) *Urban Regeneration in the UK*, 2nd edn, London: Routledge.
Unwin, T. (2000) 'A waste of space? Towards a critique of the social production of space…', *Transactions of the Institute of British Geographers*, 25:1 11–29.

2

Lefebvre in Palestine

Anti-colonial de-colonisation and the right to the city

Oded Haas

Introduction

In 2014, the Israeli Government announced the establishment of a new 'Arab City' – the city of Tantour. Conceived as a solution for housing shortage in the country's 'Arab sector', the new urban development is planned to increase housing supply in the Galilee region, where the majority of Palestinian Citizens of Israel (PCI) live. By applying a settler-colonial approach to urbanisation processes in Israel/Palestine, a complex reality of ethnicised spatialisation emerges as the context of this supposed shift in the state's approach towards its non-Jewish citizens, whereby territorialised construction of Jewish national identity simultaneously includes and excludes 'others' (Tzfadia and Yacobi 2011: 90). Specifically, the capitalist production of the nation-state in Israel/Palestine has been analysed as an 'ethnocratic' regime (Yiftachel 2006): ethno-national ideology articulated through an ethnic logic of capital, which produces a spatialised ethno-class system. In other words, the Zionist colonisation of Israel/Palestine as a whole follows an ethnic logic of space. This has been referred to as 'Judaisation of space': the pledge to settle a Jewish majority in Palestine while negating local Palestinian identity and its presence on the landscape (Jabareen 2014a; Peled and Shafir 2002; Tzfadia and Yacobi 2011; Yiftachel 2009).

While researching new cities in Israel and Palestine, in 2016 I interviewed local activists, academics, planners and politicians involved in the project of Tantour. It outlines resistance to the 'Arab City' as a determinedly anti-colonial struggle for de-colonisation of space. The local opposition to urban development elucidates how urban ethnocracy creates a neo-liberal battleground where the right to the city – as a political class struggle to de-colonise everyday life – may be threatened by the state's de-politicisation attempts and thus integrated as strategy for overcoming inherent contradictions in the production of space. In light of recent assertions about total global urbanisation (such as 'planetary urbanisation'), the meaning of the right to the city may become idle. However, urbanisation always produces new possibilities for resistance. Yet, as Lefebvre argued, anti-capitalist struggle for de-colonisation must be part of a broader social revolution (1991: 228). Considering 'colonisation' in a neo-colonial context, the struggle against Tantour certainly proves the right to the city is very much alive but only insofar as it suggests an alternative ideology to state-led urbanism; in this case, as an anti-colonial struggle against the ethnocratic regime.

The 'Arab City' in Israel: a new fact of colonial urbanism

In *De l'Etat* (1976–1978), Lefebvre explains that political regimes enable capitalism to overcome its inner contradictions through the production of space. Using the concept 'state mode of production', he describes specific strategies through which the state produces itself and its space simultaneously. These include the production of 'national territory': a concrete abstraction of the hierarchal order of state institutions and laws, state-imposed division of labour and uneven development (on all scales) that is intrinsic to capitalism (Lefebvre 2009 [1978]: 224). Producing the spaces of capitalism, such as housing estates and suburban enclaves, the state is thus at the core of Lefebvre's spatialisation of Marx's *Capital* (Lefebvre 2009 [1979]: 186). Indeed, Jessop (2008) reminds us that economic elements (e.g. national currency, taxation, mortgages) are always political and therefore the capitalist mode of production cannot be analysed without the (extra-economic) politics of the capitalist state (164). Moreover, the state creates a spatial power matrix, spatially selective interventions for maintaining control and countering resistance (Brenner and Elden 2009: 359). It is then impossible to interpret Lefebvre's right to the city (RTC) without considering the specific ideology with and against which space is produced. In the settler-colonial state, the spatial power matrix seeks to homogenise nation with territory (Poulantzas 2003: 74). And in Israel/Palestine, neo-liberal urbanisation is reproducing colonial relations of domination (Hanieh 2013). Hence 'Judaisation of space' is the spatial power matrix of a continuous colonial project.

It is worth highlighting two principles in Zionist ideology from the pre-state period that are still influential in Judaisation of space today; both were instrumental for the vision of creating a new, modern Jewish society in Palestine, by rejecting European bourgeois culture as well as Oriental past and present. First, the 'conquest of labour' was originally about annexing all local jobs to Jews, thus creating a nationalist Jewish working class (Nitzan and Bichler 2002). Second, 'conquest of land' meant working the land and settling the land in order to redeem it from Palestinian-Arab peasants who were perceived as primitive and reconnect the Jewish people with their historic landscape (Peled and Shafir 2002: 113). In the British Mandate period (1920–1947), the control and management of land in Palestine was necessary for incorporating it in the Empire's economy, as in other colonies (Goswami 2004: 56). When British officials facilitated the sale of peasants' lands to Jewish settlers, Palestinian-Arabs were made further vulnerable to dispossession (Kimmerling 1983: 38). The conquest of labour and conquest of land thus prevented an Arab-Jewish working-class alliance in Palestine. After the Palestinian Nakba – and the founding of the state of Israel – in 1948, this political economy generated an ethnically demarcated national market with a spatialised 'Arab sector'. Palestinian-Arab labour has been since partially excluded from the national market by being marginalised into confined spaces and specific occupations in Israeli economy (Hever 2012; Nitzan and Bichler 2002; Peled and Shafir 2002). At the same time, PCI have been marginalised in the national territory: in Arab-Jewish 'mixed cities', where Palestinian neighbourhoods are systematically neglected; in Palestinian towns that pre-date the state, whose development is persistently limited by the national planning institutions; and by a legal land system that enables expropriation of lands and prevents the selling of state land to non-Jews, to name a few prominent strategies for Judaisation of space. Israel's ethnicised class system that is maintained through spatial segregation is the foundation for its 'ethnocratic' regime: an ostensible democracy based upon ethno-national ideology (Yiftachel 2016). Colonial practices include subaltern classes in the national economy, but at the same time they objectify and bound particular groups into demarcated territorial and social spaces (Goswami 2004: 38). Lefebvre (2003: 125) noted that centrality orders differences in a hierarchy, through simultaneous segregation and integration. Spatial segregation is thus intrinsic to the settler-colonial capitalist homogenisation project, delineating zones for capital accumulation according to an ethnic logic of space. PCI are informally restricted to 'villages', that are systemically marginalised by the state's planning institutions with poor infrastructure, lack of available land for development and extreme over-crowdedness (Khamaisi 2013: 198). They are practically confined to a separate housing market, and most of those who are employed must commute into Jewish urban spaces (Hever 2012).

Therefore, when the government approved in 2014 the plan for the new 'Arab City' in the Galilee region, where the majority of PCI live, it seemingly signified a shift in policy. Since 1948, the Galilee region has been produced through multiple strategies of Judaisation of space; most explicitly, by a 1979 plan commonly known as 'Judaisation of the Galilee'. That plan determined the establishment of exclusively Jewish gated communities in strategic locations, to ensure a Jewish majority in the region and block the sprawl of Palestinian communities. Built mostly on state lands expropriated from Palestinian towns, these gated communities are permitted to exclude 'others' by using market-developed neo-liberal practice of admitting certain residents that fit a particular lifestyle, thereby reproducing ethnicised spatial separation (Rosen and Grant 2011: 785). Consequently, PCI have recently started moving into predominantly Jewish cities in search of housing and employment, where they face both popular and institutionalised racism and discrimination. It is within this threat to Jewish domination in the region that a right-wing government approved the construction of an 'urban zone for the non-Jewish population' (the instructions of the National Committee for Planning and Construction, first published in 3 February 2009). The very first 'Arab City' to be built by the state, Tantour was designed as a modern, orderly space, a spatial antipode to the poor Palestinian localities of the Galilee region (Mebel, 2016, personal communication, 7 September). However, the site that was chosen is located on state lands that were expropriated in the 1970s from those very localities. This decision situates the new 'Arab City' in conflict with existing Palestinian localities, and it has instigated local resistance.

The planning process of Tantour illuminates how neo-liberal urbanisation articulates the ethnic logic of space. Originally, the government sanctioned a plan for 10,000 apartments for approximately 40,000 residents to be sold in the private market as 'affordable housing', with some limited areas for commercial and industrial uses. Tantour was devised as a new suburban neighbourhood to be built as an extension to the adjacent Palestinian town of Judeida-Makr, one of the poorest towns in Israel. The architects' vision, in accordance with existing long-term plans of the national planning institutions, was to improve the town's economy, public services and physical infrastructure by adding a large suburban neighbourhood that would appeal to a socioeconomically stronger Palestinian population in the region (Mebel 2016, personal communication, 7 September). Their plan ('National Masterplan #44') includes detailed urban design schemes for mid-size apartment buildings with semi-private public spaces that are expected to facilitate incremental transition of close-knit communities from Palestinian towns and villages into the new suburb. The plan aspires to eliminate the stigma of Judeida-Makr as an unplanned, backwards Palestinian 'village' and reproduce it as a new, modern 'Arab City'. It is to do so by emulating the modernist space of Jewish cities in the Galilee. While the plan has been slowly moving up the bureaucratic channels of the centralised planning system, in 2016 Tantour was declared by the government's newly established Housing Cabinet as a 'special housing area'. This newly formed legal apparatus responds to the 2011 mass demonstrations in Israel that swept the nation with demands for social justice, and specifically affordable homeownership. The state resorted to increasing the supply of available land for development and to fast-tracking plans with a high volume of housing units, though arguably with deficient regulation on housing affordability. Plans promoted as 'special housing areas' may override existing plans and are essentially exempted from public scrutiny, in favour of swift approval. A new plan for Tantour was devised by a new architect's firm and exploded to 25,000 apartments for approximately 120,000 residents, with its territory circumscribing the existing town of Judeida-Makr. The revised plan for Tantour relies on an orthogonal grid pattern, completely foreign to the region, which is meant to encourage openness and continuity of urban space while allowing self-division of inhabitants to separate areas according to religious or family affiliation (Kolker 2016, personal communication, 27 September). With housing density and prices as high as in the predominantly Jewish metropolitan areas of the region, Tantour was conceived to be an antithesis to the Israeli myth of the 'Arab village' as unplanned and backwards, that is therefore fertile ground for resistance (Khamaisi 2013; Yacobi 2008). And so, could state-conceived urban development engender de-colonisation?

The Palestinian question is an urban question

Harvey (2012) defines the right to the city as a move towards greater democratic control over surplus that is concentrated in the city and an un-alienated right of urban inhabitants to produce for their own wants and needs (xvi, 23). It is the right to become collective users rather than capitalist consumers of space, and therefore requires anti-capitalist struggle. However, in the state mode of production, RTC must be understood as a political class struggle. Alternative organisation of production cannot be defined in capitalist terms alone, but must address the various specific forms of alienation imposed by the state. RTC thus becomes an ideological struggle over the logic that produces state-space. And since the spatial selectivity of nation-states is intertwined with class preference that excludes the masses from political power, concrete abstraction of ethno-national ideology is negotiated through class struggle and as such remains permanently fragile (Brenner and Elden 2009: 370; Poulantzas 2000: 127, 82, 120). Accordingly, national territory is always a relational product of struggle over the appropriation of social space (Kipfer and Goonewardena 2013: 110). Ethnocracy thus necessarily generates conflict which carries a potential to weaken the ethnocratic regime itself, if mobilising around different issues (such as housing) generates discontent with state politics and undermines ethno-national differentiation (Anderson 2016: 12). However, as Kipfer (2008) emphasises, claims of peripheralised groups against segregation can only be transformed to claims to the right to the city if they bring centrality to the margins (204); meaning, not a struggle for inclusion in state-space, but a transformation in control over state power. Mitchell (2003) contends that rights are always proven in practice and never guaranteed in the abstract (4). In other words, RTC – the appropriation of space and its production process by inhabitants – cannot be delivered by the state, but rather must be actively taken by inhabitants. The 2011 mass protests in Israel, despite some radical exceptions, focused on neo-liberal demands for affordable housing and did not succeed, or rather attempted to undermine ethno-national ideology (Marom 2013). It was not a right to the city moment. Unsurprisingly, the state responded by increasing housing supply in the market. In the case of Tantour, the state is providing a housing 'solution' to the 'Arab sector' by increasing the rate of homeownership and rendering PCI consumers of housing. Therefore, if it is to resist co-optation by the state, RTC must be the right of inhabitants for self-determination (Schmid 2012: 59). Resistance cannot come from state-initiated urban development, since urbanisation – even democratic, participatory – is a counter-revolutionary strategy (Harvey 2009: 279). RTC is therefore a struggle against integration into state-space as well as against the appropriation of such demands as a governance principle by state institutions. In order to revolutionise urbanisation, RTC must be a claim for de-colonising everyday life and to produce a radically different world (Goonewardena 2012: 98). Following the events of 1968 in Paris and the failure of sociospatial peripheries to converge at a world-wide scale, Lefebvre develops the concept of 'internal colonisation' as a particular form of alienation (Kipfer and Goonewardena 2013: 173–174). It illustrates territorialised domination of everyday life in post-colonial capitalism, sustained by spatial segregation of underdeveloped regions and urban neighbourhoods (Lefebvre 2009 [1970]: 181). A neo-colonial economic system of uneven development is manifested in dispersal, migration and settlement strategies that render various sociospatial peripheries – gendered and racialised – the margins of metropolitan centres. De-colonisation therefore takes on clear anti-colonial meaning (Kipfer and Goonewardena 2013: 97). At the same time, since segregation is intrinsic to the state mode of production, anti-colonial struggle takes the form of an urban struggle for RTC.

In that sense, the Palestinian question is an urban question (Jabareen 2014b): political class struggle must view relations of production as intertwined with contemporary manifestations of colonial ideology. In *The Critique of Everyday Life* (1991: 37–38), Lefebvre warns that an abstract notion of class struggle that neglects historic modifications in capitalism would be blind to the particular contents of capitalist relations, and therefore futile. Explaining those particularities in colonial context, Kipfer (2011) shows how direct territorial rule has been shaping the subjectivity of both the colonised and the coloniser, by referring to counter-colonial texts from after the Second World War by French writers such as Césaire, Memmi and especially Fanon (95). A spatial hierarchy underpins the colonial imbalance of power: while the coloniser is interested in labour, the colonised is looking for self-determination (Austin 2010: 22; Thobani 2007: 13). Israel thus has a clear interest in sustaining a spatialised 'Arab sector' in the nation's sociospatial margins. In fact, some writers have identified 'creeping apartheid' as a specific ethnocratic strategy for Judaisation of space, whereby strategies of domination are transferred from the 1967 occupied territories into Israel (Angotti 2013: 81; Yiftachel 2009: 93). For example, delegitimising PCI presence in the national territory; selective enforcement of 'illegal' construction in Palestinian localities within Israel; and militarisation of police force, especially in response to annual Day of Land demonstrations that commemorate the 1976 protest against expropriation of Palestinian lands in the Galilee region. Significantly, apartheid is also creeping through market-oriented policies that are part of global urbanisation processes (Yiftachel 2016: 35). As Gilbert (2009) shows in the Canadian settler-colonial context, a combination of neo-liberal values of free market, individual liberty and protection of national culture results in treating certain citizens as undeserving (35). The most banal acts of city planning correspondingly serve the ethno-national ideology of the ruling ethno-class, who sees itself as deserving of the national territory that is its homeland. Other groups in the same space are dominated and marginalised by racialised urban politics (Yacobi 2016: 112–113).

Still, urban regimes shaped by ethnocracy are potential sites for subverting the ethnic logic of space. Urban ethnocracy is where capitalism requires segregated, hierarchal spaces on the one hand, and free flow and flexibility of labour on the other, making it the space where ethno-national hegemony is most vulnerable to everyday life insurgency (Anderson 2016: 15; Yiftachel 2016: 34–35; Yiftachel 2006: 189). Privatisation of democratic planning procedures creates spaces for marginalised groups to exercise liberal claims within local urban development and contest the concentration of power within the ruling ethno-class. As planning becomes a site for political struggle, RTC may potentially be expressed as minority right to self-determination (Jabareen 2017: 18). Hence the recent move of Palestinian families into predominantly Jewish towns, especially in the Galilee region, is an active taking over of space and a threat to the Judaisation project. Meaning, Jewish cities becoming 'mixed' is as much a risk for as it is a result of the ethnocratic regime. Consequently, the 'Arab City' can be seen as a subtle attempt to re-establish ethnocratic hegemony through seemingly voluntary, market-driven suburbanisation of the Palestinian middle-class. A solution to acute housing distress, advanced by well-meaning professional planners through manipulation of municipal boundaries within an existing ethnically divided economic development, the new 'Arab City' is the reincarnation of demographic engineering and a new version of colonial urbanism. The state cannot be expected to solve social problems, since it was mutually produced with them: the state is not taken over by a pre-given class formation but rather produced along with it, according to a specific mode of production (Jessop 2007: 7; Poulantzas 2000: 17). Ergo, urban development conceived by the state as an 'Arab City' cannot transform the conditions that spatialise the 'Arab sector'. Instead, Tantour may help solve a potential political crisis that the government seeks to sidestep with its recent 'special housing areas' laws. A suburban alternative to Palestinian spaces of resistance that is still disconnected from Jewish spaces of employment, the project demarcates space for Palestinian nationality and responds to a set of obstacles in the conquest of labour and conquest of land: demands for intervention in housing supply; claims for investment in the 'Arab sector'; and ethnic 'mixing' of urban spaces. Tantour will presumably draw the relatively stronger Palestinian population from surrounding localities, thus further marginalising Palestinian towns and somewhat reversing the 'mixing' of predominantly Jewish cities. The 'Arab City' thus becomes essential in delineating a 'Jewish state'; and challenging urban development – essential in the Palestinian struggle for liberation.

Colonial fear and urban resistance

Out of tensions in urban ethnocracy in Israel, potential alternative appropriations of space may emerge, such as: public participation in the planning process; NGOs advocating planning rights; and legal recourse against the discriminatory land system. However, these cannot substantially undermine the ethnic logic of capitalist urban development. In fact, declaring the 'Arab City' a 'special housing area' shows how the state uses ostensible demands for RTC, reformulated by neo-liberal elites as demands for increased homeownership, as grounds for urban development. Development thus replaces demolition, neglect and disinvestment as an ethnocratic practice. Since the industrial city leads to decline in everyday life, Lefebvre was concerned with a social-cultural revolution wherein class struggle is crucial but not disconnected from broader de-commodification of society (1996: 149; 1991: 228). RTC is therefore required to overcome a risk of appropriation by the state. In response to the government's announcement on the approval of a plan for a new 'Arab City', a group of activists in the adjacent Palestinian town of Judeida-Makr organised in opposition to the plan. The group is called 'Khirak', meaning 'movement' in Arabic, and it targets the central planning institutions as well as the local municipality, which they see as complicit with state strategies. In interviews, activists in the groups say Tantour is not meant for the local inhabitants of poor Judeida-Makr, who are in dire need of housing and infrastructure. 'This is our land, but we cannot use it to build housing for ourselves' (Khirak activists 2016, personal communication, 15 September). The Khirak is fighting not just against the project of Tantour, but also for basic everyday municipal services. Despite being a small group of 10–15 activists, they have been gaining local residents' trust by becoming a de-facto municipality in their town and taking care of various local issues, thus actively replacing what they describe as a corrupt, incompetent and strikingly violent local government that most residents are afraid to resist. Such community-level institutions and interventions may form a resistance that represents the Palestinian identity and interest before state authorities (Khamaisi 2010: 68). The Khirak activists indeed fight for recognition from the state; for the right to use the lands that were expropriated from their town and to shape urban development according to local inhabitants' housing needs. They contextualise their activism – demonstrations, meetings with planners, lobbying with politicians – within explicit fear from expulsion and displacement emanating from the Nakba:

> People are still afraid. They can't tell the difference between the Shin-Bet [the Israeli security service] and the central planning institutions. Everything that comes from the state is perceived as potential threat. Resisting the state is considered risky.... Our struggle is successful, regardless of its outcome, because we fight against fear. By showing residents [of Judeida-Makr] there is someone who stands against state planning, we show them they should not be afraid.
>
> *(Khirak activist 2016, personal communication, 15 September)*

Moreover, the Khirak accuse the Jewish planners of Tantour for not paying attention to local needs during the first stages of planning that included public participation meetings. They report that only certain groups were included, such as Palestinian planners and academics who do not necessarily represent local communities but are considered 'Arab sector'. The explanation is simple: those who were invited to be heard were the potential future residents of the project; that is to say, middle-class families who may hinder exclusively Jewish cities in the Galilee. The struggle of the Khirak is therefore a political class struggle that is threefold: against their own municipal government that depends on the state for resources and cannot oppose the imposed development; against the expropriation of lands by the Zionist ethnocratic regime; and against neo-liberal gentrification that would turn their town to 'the back yard of a new suburb' (Khirak activists 2016, personal communication, 9 September). Thus it is an anti-colonial struggle against the ethnic logic of space. Fitly, the Khirak actively appropriates space by holding social and political gatherings on the lands designated for Tantour (Khirak activists 2016, personal communication, 29 September). Urban ethnocracy hence provides the Khirak an opportunity to articulate their fight for self-determination, for sustaining Palestinian presence on the landscape and against the homogenising logic of the ethnocratic regime as a legitimate fight against urban development. Although urban ethnocracy implies new, inconspicuous ways to segregate and new strategies for co-opting RTC, at the same time it creates new sites for struggle to undermine ethnocratic hegemony. Indeed, local inhabitants use their citizenship to demand alternative planning from the state as well as appropriate state-space by using it for their own political mobilisation. The Khirak appeal that the lands will be given back to the local authority, they reject high-volume residential development in Tantour and favour a low-density neighbourhood that would provide housing for the growing families of Judeida-Makr (Khirak activists 2016, personal communication, 29 September). Their struggle thus re-politicises urban development that was intended to provide housing 'solutions' in a neo-liberal context.

Recognising that colonisation operates through territorial organisation, Fanon conceived of de-colonisation as a practice of reappropriating space (Kipfer and Goonewardena 2013: 99). Anti-colonial struggle should therefore provide an alternative logic to that which segregates colonised space. The rejection of massive urban development and the demand for localised intervention instead may seem a refusal by the local community to open up to modern, progressive development. Some Jewish and Palestinian planners involved in the Tantour project disregard Khirak resistance for representing traditionalism and for not trying to stop the inevitable (Khamaisi 2016; Kolker 2016; personal communications). By insisting on reproducing their space, the activists are seemingly risking reproducing their own oppression by the ethnic logic of space. Yiftachel (2006) argues that any territorial partitioning of space according to ethno-national principles would simply generate new forms of segregation (263). Clearly, homogenising ethno-national collectivities is the project of the nation-state and cannot be a strategy for de-colonisation (Stasiulis and Yuval-Davis 1995: 32). However, the Khirak insistence on development that, in their view, puts the lands back in the hands of local Palestinians, must be examined in the context of the ethnic logic of space. In the asymmetric landscape that is Israel's national territory, Palestinians are effectively excluded from most urban spaces. In the privatised housing market founded upon the ethno-class system, urban regeneration as in the larger plan for Tantour would trample on the local community and re-marginalise those in the lower ethno-class who are most struggling. In light of Judaisation of space, the Khirak demand that the state invest in an existing Palestinian town instead of building a new suburb is certainly revolutionary, as it undermines the colonial underpinnings of the ethnocratic regime.

Conclusions

Acknowledging that anti-capitalist struggle for de-colonisation must be defined in political terms, this chapter alleged that multiple, alternative ideologies that produce space both with and against the capitalist state must be considered inherent in what is otherwise a qualified epistemology of urbanisation. First, the chapter has shown that in the case of the new 'Arab City' in Israel, state-initiated urban development ostensibly for the benefit of the marginalised Palestinian ethno-class is in fact promoting Zionist 'conquest of land'. Moreover, recent transformations in the national planning system provide a countering strategy to potential opposition, justified by a neo-liberal approach to relieving housing distress. Second, as evident in the planners' vision of Tantour as a middle-class suburb to replace a Palestinian 'village' as well as in the state's strategy of increasing Palestinian homeownership, a neo-liberal understanding of RTC threatens the 'Arab sector' in Israel and the Palestinian liberation struggle itself with integration into public demand for housing supply. Third, the Khirak's anti-colonial struggle against state-led urban development in Tantour proves that while a neo-liberal understanding of RTC may easily be appropriated by Zionist ideology of Judaisation of space, the inhabitants' imaginary of their right to the city may produce a political class struggle that seeks to undermine ethno-class spatialisation.

RTC as a political movement therefore must not be understood as a fixed struggle. It is rather best defined in relation to particular urban regimes, where everyday life and attempts at its depoliticisation by state interests are shaped. In 1968, Lefebvre laid the foundation for considering RTC as an ongoing social struggle against integration into the spaces of global capitalism (see: Lefebvre 1996), but it becomes politically useful when forged in the context wherein it is infringed. As the concept becomes contested and the struggle itself vulnerable to usurpation by the urbanising ideology, RTC needs to be brought back from liberal demands for inclusion to the everyday life of inhabitants, challenging strategies such as 'affordable housing'. Similarly, academics who focus on global urbanisation or a globalised understanding of RTC risk treating it as always already defeated by neo-liberal urbanisation. Instead, understanding urbanisation and resistance to it requires focusing on particular contexts of colonisation, where RTC is endangered and potentially redeemed. Indeed, de-colonisation may be possible if RTC becomes a demand that subverts state-space. Therefore, political resistance to urbanisation that undermines its ideological foundation is more urgent than ever.

References

Anderson, J. (2016) 'Ethnocracy: exploring and extending the concept', *Cosmopolitan Civil Societies Journal*, 8:3 1–29.

Angotti, T. (2013) *The New Century of the Metropolis: Urban Enclaves and Orientalism*, New York, NY: Routledge.
Austin, D. (2010) 'Narratives of power: historical mythologies in contemporary Québec and Canada', *Race and Class*, 52:1 19–32.
Brenner, N. and Elden, S. (2009) 'Henri Lefebvre on state, space, territory', *International Political Sociology*, 3:4 353–377.
Gilbert, L. (2009) 'Immigration as local politics: re-bordering immigration and multiculturalism through deterrence and incapacitation', *International Journal of Urban and Regional Research*, 33:1 26–42.
Goonewardena, K. (2012) 'Space and revolution in theory and practice: eight theses', in N. Brenner, P. Marcuse, and M. Mayer (eds) *Cities for People, Not for Profit: Critical Urban Theory and the Right to the City*, London, UK; New York: Routledge.
Goswami, M. (2004) *Producing India*, London, UK; Chicago, IL: University of Chicago Press.
Hanieh, A. (2013) *Lineages of Revolt: Issues of Contemporary Capitalism in the Middle East*, Chicago, IL: Haymarket Books.
Harvey, D. (2009; 1st edn 1973) *Social Justice and the City*, Athens, GA: University of Georgia Press.
Harvey, D. (2012) *Rebel Cities: From the Right to the City to the Urban Revolution*, London, UK: Verso Books.
Hever, S. (2012) 'Exploitation of Palestinian labour in contemporary Zionist colonialism', *Settler Colonial Studies*, 2:1 124–132.
Jabareen, Y. (2014a) 'Jaffa-Haifa-Beirut', paper presented at The Arab City in Israel Conference, Jaffa, June.
Jabareen, Y. (2014b) '"The right to the city" revisited: assessing urban rights – the case of Arab Cities in Israel', *Habitat International*, 41: 135–141.
Jabareen, Y. (2017) 'The right to space production and the right to necessity: insurgent versus legal rights of Palestinians in Jerusalem', *Planning Theory*, 16:1 6–31.
Jessop, B. (2007) *State Power*, Cambridge, UK; Malden, MA: Polity.
Jessop, B. (2008) 'Spatial fixes, temporal fixes and spatio-temporal fixes', in N. Castree and D. Gregory (eds) *David Harvey: A Critical Reader*, Malden, MA; Oxford: Blackwell.
Khamaisi, R. (2010) 'Resisting creeping urbanization and gentrification in the Old City of Jerusalem and its surroundings', *Contemporary Arab Affairs*, 3:1 53–70.
Khamaisi, R. (2013) 'Housing transformation within urbanized communities: the Arab Palestinians in Israel', *Geography Research Forum*, 33: 184–209.
Khamaisi, R. (2016) Interviewed (one hour) on 11 September in Tamra, Israel.
Khirak activist group (2016) Interviewed (for several hours) on 9, 15 and 29 September, Acre, Israel.
Kimmerling, B. (1983) *Zionism and Territory*, Berkeley, CA: Institute of International Studies, UCLA.
Kipfer, S. (2008) 'How Lefebvre urbanized Gramsci', in K. Goonewardena, S. Kipfer, R. Milgrom and C. Schmid (eds) *Space, Difference, Everyday Life: Reading Henri Lefebvre*, New York: Routledge.
Kipfer, S. (2011) 'The times and spaces of (de-)colonization: Fanon's counter-colonialism, then and now', in N. Gibson (ed.) *Living Fanon: Global Perspectives*, Basingstoke: Palgrave.
Kipfer, S. and Goonewardena, K. (2013) 'Urban Marxism and the post-colonial question: Henri Lefebvre and "colonization"', *Historical Materialism*, 21:2 76–116.
Kolker, A. (2016) Head Architect on the Tantour Project at the time of writing this chapter. Interviewed (one hour) on 27 September in his office, Tel Aviv, Israel.
Lefebvre, H. (1969) *The Sociology of Marx*, trans. Norbert Guterman, New York: Random House.
Lefebvre, H. (1991; 1st edn 1947) *The Critique of Everyday Life*, trans. John Moore, London, UK; New York, NY: Verso Books.
Lefebvre, H. (1996) *Writings on Cities*, trans. Eleonore Kofman and Elizabeth Lebas, Cambridge, MA: Blackwell Publishers.
Lefebvre, H. (2003; 1st edn 1970) *The Urban Revolution*, trans. Robert Bononno, Minneapolis, MN: University of Minnesota Press.
Lefebvre, H. (2009) *State, Space, World: Selected Essays*, trans. Neil Brenner and Stuart Elden, Minneapolis, MN: University of Minnesota Press.
Marom, N. (2013) 'Activising space: the spatial politics of the 2011 protest movement in Israel', *Urban Studies*, 50:13 2826–2841.
Mebel, E. (2016) Chief Planner of 'National Masterplan #44'. Interviewed (about one hour) on 7 September in his office, Kiryat Tiv'on, Israel.
Mitchell, D. (2003) *The Right to the City: Social Justice and the Fight for Public Space*, New York: Guilford Press.
Nitzan, J. and Bichler, S. (2002) *The Global Political Economy of Israel*, London, UK; Sterling, VA: Pluto Press.
Peled, Y. and Shafir, G. (2002) *Being Israeli – The Dynamics of Multiple Citizenship*, Cambridge, UK: Cambridge University Press.

Poulantzas, N. (2000; 1st edn 1978) *State, Power, Socialism*, London, UK: NLB.
Poulantzas, N. (2003) 'The nation', in N. Brenner, B. Jessop, M. Jones and G. MacLeod (eds) *State/Space: A Reader*, Oxford: Blackwell.
Rosen, G. and Grant, J. (2011) 'Reproducing difference: gated communities in Canada and Israel', *International Journal of Urban and Regional Research*, 35:4 778–793.
Schmid, C. (2012) 'Henri Lefebvre, the right to the city, and the new metropolitan mainstream', in N. Brenner, P. Marcuse and M. Mayer (eds) *Cities for People, Not for Profit: Critical Urban Theory and the Right to the City*, London, UK; New York: Routledge.
Stasiulis, D. and Yuval-Davis, N. (1995) 'Beyond dichotomies – gender, race, ethnicity and class in settler societies', in D. Stasiulis and N. Yuval-Davis (eds) *Unsettling Settler Societies: Articulations of Gender, Race, Ethnicity and Class*, London, UK: Sage.
Thobani, S. (2007) *Exalted Subjects*, Toronto, ON: University of Toronto Press.
Tzfadia, E. and Yacobi, H. (2011) *Rethinking Israeli Space: Periphery and Identity*, Abingdon, Oxon; New York, NY: Routledge.
Yacobi, H. (2008) 'Architecture, orientalism and identity: the politics of the Israeli-built environment', *Israel Studies*, 13:1 94–118.
Yacobi, H. (2016) 'From "ethnocracity" to urban apartheid: the changing urban geopolitics of Jerusalem/al-Quds', *Cosmopolitan Civil Societies Journal*, 8:3 100–114.
Yiftachel, O. (2006) *Ethnocracy: Land and Identity Politics in Israel/Palestine*, Philadelphia, PA: University of Pennsylvania Press.
Yiftachel, O. (2009) 'Theoretical notes on 'Gray Cities': the coming of urban apartheid?' *Planning Theory*, 8:1 88–100.
Yiftachel, O. (2016) 'Extending ethnocracy: reflections and suggestions', *Cosmopolitan Civil Societies Journal*, 8:3 30–37.

3

The urban revolution(s) in Latin America

Reinventing utopia

Chris Hesketh

Introduction

Since the region-wide debt crisis of the 1980s, Latin America has provided a key window with which to observe Lefebvre's (1991: 55) famous contention that 'Today, more than ever, the class struggle is inscribed in space'. The debt crisis marked a watershed in the continent's developmental history. The previous spatial order of the post-war years – based on nationally-scaled development – was remade in favour of a new neoliberal, global orientation. This was accompanied by a rollback in state-provisioning and a growth of poverty. However, whilst neoliberalism was a process engineered from above, it has, concomitantly, been rigorously contested from below. With access to land and vital resources made increasingly precarious, social movements in the region have frequently sought to assert their right to 'differential space'. This chapter explains the origins, practices and contradictions involved in these revolts from lived spaces, demonstrating how Lefebvre's writings can be extended to frame these movements to remake utopia but also reflecting on the need to rethink certain elements of his work in light of contemporary struggles.

It is important to note that the exercise of examining a specific locale to explore the potential of a theoretical body of work (and to highlight its possible limitations) is itself a highly Lefebvrian exercise. A leitmotiv of Lefebvre's work was the role of contradiction. He was interested in the dialectic between theory and practice so that the latter consistently informed the former, avoiding ossification into dogma (Lefebvre 1976). Expanding on Marx and Engel's (1848/2000) analysis of capitalism, Lefebvre (1976; 1991) was clear that capitalism survived through the production of space. However, an effervescent element of Lefebvre's Marxism was the accent that he placed on struggle and contestation, and the need for what he referred to as 'differential space' (defined as a collective oeuvre). This was formed in opposition to the 'isotopy', or sameness of capitalism, that he would later refer to as 'abstract space' (Lefebvre 1970: 1991). While the production of space was the element that explained capitalist survival, the new spatial forms it engendered simultaneously created the conditions for revolt. To explore how class struggle has been inscribed in space in the Latin American context it is necessary to examine the major shift in developmental practice that emerged in the 1980s, involving a shift from Import Substitution Industrialisation (ISI) to neoliberalism. This altered both the dominant form of spatial production and the terrain of resistance.

Lefebvre (1991) famously argued that three dialectically related elements comprised the production of space. These are spatial practices, representations of space and representational spaces. Spatial practices refer to the spatial norms of any given social formation that ensure a degree of continuity and cohesion. Thus, property relations, the physical layout of areas including factors such as roads and infrastructure as well as housing would be included under spatial practices, as would generalised work-patterns (see also Harvey 1990; Lefebvre 2003). This is closely associated with perceived space. In other words, it relates to our reflexive awareness of our surrounding environment. Representations of space on the other hand are tied to ideology, signs and codes. This is the realm of conceived space. Representations of space are thus related to the dominant ideology of society and this element is therefore synonymous with class rule. Lastly, there are representational spaces. This is the realm of directly lived experience. Spatial practices and representations of space can combine to 'facilitate the manipulation of representational space' (Lefebvre 1991: 59). However, this component of space is associated with subjective feelings or thought and can be linked to the more clandestine side of life where resistance can begin to emerge from. It is a cultural sphere concerned with our imagination and therefore has the ability to change and appropriate space through our everyday practices of 'habiting' (Lefebvre 1970/2003).

Spaces of ISI

ISI can be thought of as a Latin American variant of the 'state mode of production' that dominated Western capitalism after the Second World War. In other words, the space of growth was one clearly managed and controlled by the state (Lefebvre 1975/2009). This era of development mapped onto the above-mentioned triad of spatial production as follows. With regards to spatial practices, the focus on the growth of an internal market clearly represented a break from the previous model of export-led development that dominated up to the 1930s. ISI ushered in a wave of urbanisation in Latin America and subsequently new rhythms of work and daily life. In relation to 'spaces of representation', nationalism and the representation of 'national space' were utilised as an elite class strategy for capital accumulation. The bourgeoisie thus came to 'articulate the imagined community of the nation' (Radcliffe and Westwood 1996: 15). There was, however, a limited degree of incorporation of the demands of the popular classes such as spending on social services, subsidised consumption, increasing employment opportunities and rising real wages (Robinson 2004). In terms of the 'representational spaces' of ISI, the previous two elements were able to exert a powerful influence in creating a model for incorporation. Contestation in this era largely took place with the state defined as the horizon of political action (Zibechi 2012). However, incorporation was far from a complete process and differential spaces remained. On the one hand, the uneven nature of state-formation had left many groups, most notably indigenous communities, with de facto autonomy in a variety of countries (Yashar 2005). On the other hand, ISI suffered from a problem of structural unemployment owing to the use of imported technology that was labour-saving. The failure to meet expectations for social mobility would lead to tension and conflict as subaltern classes battled to maintain their precarious inclusion and urban slums proliferated (Davis 2006). As a spatial project, ISI had numerous contradictions. The redistribution of wealth was too limited to provide viable consumer markets. Inflation often resulted when governments resorted to printing money to cover their deficits. Finally, development was lopsided as rural areas were neglected in favour of urbanisation (Perrault and Martin 2005).

As a response to the economic contradictions of ISI (most notably the failure to consolidate an internal market and the continued dependence upon capital goods), Latin American states turned abroad for foreign finance. This coincided with the oil crisis of the 1970s in which large amounts of 'petrodollars' had been deposited in Western banks following the rapid raising of oil prices. These 'petrodollars' not only presented an opportunity for Latin American elites to offset the contradictions of ISI's by borrowing abroad (whilst also generating surpluses for a degree of social redistribution to offset rising labour militancy), but at the same time provided a 'spatial fix' for over-accumulated capital in Western banks, as opportunities for investment were limited there due to the onset of stagflation. Recycling these 'petrodollars' into Latin America thus became a way to productively put this capital to work and stave off domestic inflation and devalorisation (Lipietz 1984). The accumulation of debt within Latin America thus needs to be firmly situated within the very different socio-spatial relations contained within diverse geographical regions of the world. The assumption was that loans would be repaid through increased export earnings, the creation of profitable new markets and the further recycling of loans back to the centre to purchase capital goods, helping to stimulate Western economies (Lipietz 1984). Latin America in other words became a vital site for the reproduction and stabilisation of global capitalism. For Latin America, international debt was to become the very foundation of domestic economic growth.

Crisis and the export of devaluation

During the 1970s the composition of foreign capital flows to Latin America radically altered (along with the levers of power). Hitherto dominated by bilateral and multilateral lending, syndicated bank loans now emerged to provide the majority of liquidity (UNCTAD 2003). By 1980, 80 per cent of Latin America's debt was held by private banks, and the region held the largest accumulated debt stock in the world (Ffrench-Davis 1994). The viability of debt-led growth in Latin America was conditional, however, upon the persistence of three factors: (1) the continuing availability of foreign capital, (2) the maintenance of low interest rates and (3) rising commodity prices to help service accrued debt. However, the election of Paul Volcker as chairman of the Federal Reserve in 1979 precipitated a new monetarist policy in the United States in response to domestic fears of inflation. This cancelled out all the above premises. First, the unilateral raising of interest rates markedly increased the value of Latin American debt. Second, the raising of interest rates caused a contraction in international liquidity, leading, third, to declining demand for primary products as recession became a feature of the central economies. Financial markets became aware that Latin America could not repay its vast loans, and thus foreign capital began to dry up. These elements helped precipitate the debt crisis in Latin America that erupted in 1982. This would be used to profoundly reshape space.

The debt crisis marked a watershed in Latin American state formation and developmentalism. It would be used to redefine the trajectory of development, with an outward-looking neoliberal economic model emerging to replace the inward-looking one of ISI. International financial institutions (IFIs) such as the International Monetary Fund (IMF), the World Bank, and the Inter-American Development Bank were key levers of power in this regard, reshaping spatial practices and dominant representations of space. Although external forces had never ceased to influence Latin America's state formation and development, this tendency became ever more pronounced after the debt crisis. As countries in the region could no longer service their debts, and sources of private lending had ceased, they had to look to IFIs as a means of obtaining much-needed foreign exchange. These loans came with key conditionalities attached to them, including the reduction of public spending, exchange rate stability, import liberalisation, privatisation, deregulation and the opening of their economies to FDI (UNCTAD 2006). This was in line with the emerging Washington Consensus, which sought to reduce (in reality, alter) the role of the state in the economic affairs of developing countries and move them toward export-oriented models of growth. In terms of spatial production this was informed by what Lefebvre (1991) refers to as 'savoir': abstract, non-place-based forms of knowledge concerned only with facts and figures.

The huge debt burden meant that countries were forced to create the conditions necessary to service this debt. Practically, this entailed increasing exports while trying to reduce domestic demand. This quickly led to a disastrous recession, while the rapid opening of these countries' economies to foreign competition helped destroy local research and development (UNCTAD 2003). As a method for dealing with the debt crisis, economies were restructured to become more investor friendly. Tariffs on foreign trade, for example, dropped from 42 per cent in 1985 to just 14 per cent in 1995 (Robinson 2008). This decrease led to FDI replacing portfolio investment and commercial bank loans as the greatest source of capital (UNCTAD 2004). A significant proportion of this expansion was in mergers and acquisitions and the takeover of privatised state enterprises. Whereas under ISI state banks were the key providers of credit (in keeping with the national spatial strategy of development), FDI is 'to an increasing extent intended to serve global and regional markets often in the context of international production networks' (UNCTAD 2006: 10). The reality of this has been to confirm Latin America's spatial location as a subordinate region in the global political economy.

It is also important to view the debt crisis not simply as a crisis of Latin American capitalism but rather as a potential crisis of capitalism seen as a totality. Following the Mexican default in 1982, 13 American banks were owed $16.5 billion. Had other countries followed suit in defaulting, the financial system of world capitalism could well have collapsed, as it did in 1930, precipitating a global depression (Green 1995). As Duménil and Lévy (2004) note, by 1983 23 other countries had to reschedule debt repayments, and the four most indebted nations in the world (Mexico, Brazil, Venezuela and Argentina) owed 74 per cent of the debt held by developing countries. However, rather than becoming a *crisis of capitalism* and threatening the social relations upon which the system is based, the debt crisis simply became a *crisis within capitalism*, thus acting as a necessary precondition to drive the system forward and begin a new round of accumulation. This new round of accumulation, however, involved a process of highly spatialised class struggle. Regarding this process, Harvey (2003: 151) states: 'Regional crisis and highly localised place-based devaluations emerge as a primary means by which capitalism perpetually creates its other to feed on'. This was achieved through a massive privatisation of Latin American public resources and SOEs, as well as large-scale reductions in social welfare provisions. The countryside was also opened to large scale commercialisation (Hesketh 2013). This illustrates how resources went from being state-owned and geared toward national development to exclusive private property rights devoted solely to surplus value extraction. Latin America's transition to neoliberalism thus seems to support the view expressed by Duménil and Lévy (2004) that it is a class project designed to reconstitute the wealth of the upper fractions of capital at the expense of the subaltern classes (see also Harvey 2005). Evidence for this can be highlighted by the fact that average urban incomes in all Latin American countries (except for Chile) stagnated or declined since the onset of neoliberal reforms. This decline was especially pronounced in Uruguay and Venezuela, where income declined by 30 per cent and 50 per cent, respectively. The dominant classes, meanwhile, increased their income faster than average (Portes and Hoffman 2003). Business also came to be increasingly privileged over labour (Grugel 1998). During the 1980s, the number of people living in poverty increased by sixty million. Concomitantly, there was a massive growth in unemployment and underemployment, with new jobs largely being created in the informal sector (Veltmeyer 1997). With this transition, the very term development also came to be redefined. Rather than being concerned with the transformation of the productive structure as it was in the past, development came to be focused on issues such as poverty reduction, the provision of minimal needs, and individual advancement, eviscerating its most salient content (Chang 2010). Duménil and Lévy (2004: 82) are therefore surely correct when they conclude: 'That it was necessary to manage the crisis was an undeniable fact. That the neoliberal strategy was particularly harmful is another one'.

Resistance

The above factors meant that the viability of neoliberalism in Latin America as a new incorporation strategy was always inherently fragile, as there was a growing tension with the social polarisation that the accumulation strategy has caused, as well as a distrust of traditional political parties and elites that engineered this (Luna and Filgueira 2009). The viability of neoliberalism was therefore dependent upon generating a base of political support beyond the privileged few who have benefited from privatisation, deregulation, and the move to export-oriented growth (Cameron 2009). However, this sits at odds with neoliberalism's inherent nature as a class project. In fact, Latin America has been at the epicentre of resistance to neoliberalism worldwide since the twilight of the 20th century (Goodale and Postero 2013). This contestation necessitates thinking about how this resistance is best theorised and articulated. Beyond his work on the production of space, it is here that Lefebvre can offer an important set of intellectual resources, notably through his ideas about autogestion and urban revolt.

A key question is how an alternative political project that seeks to overcome alienation and that aims at the genuine inclusion of the subaltern classes can be formed that challenges established hegemonic practices. Here the issue of state power looms large. Castañeda (1994) acknowledges that the very things that give rise to the Left, such as poverty, discrimination, inequality and so on, have not disappeared, and thus Left-oriented governments are likely to remain a feature of Latin America. However, he also argues that historically the Left has failed to change any of these issues in a meaningful way, especially through armed revolution. He therefore advocates an approach to political transformation that seeks to combine free-market economic principles with social redistribution as the best means for taking the continent forward. Surveying the Pink Tide movement that returned left or left-of-centre governments to power, Castañeda (2006) sought to identify both a 'right' Left and a 'wrong' Left. The 'good', or right, Left is defined by a market-oriented 'third way' approach and is associated with countries such as Brazil, Uruguay and Chile. The 'bad', or wrong, Left, by contrast, is said to represent a threat to the region's future. This version of the Left is associated with the model of change in Venezuela and Bolivia and the whole legacy of the Cuban Revolution. The problem with such an analysis, however, is precisely the fact that it ignores the different conditions in which these movements have emerged and grown. First, classifying as a good Left those countries that accept market-oriented policies is to ignore the lessons of why neoliberalism failed as a project of incorporation (Cameron 2009; Luna and Filgueira 2009). Second, this analysis (shared but inverted by others) fails to imagine that other institutional arrangements and political practices could exist beyond the nation-state. However, as numerous interpretations have highlighted, so-called progressive governments of the region have largely been reformist rather than revolutionary and have often served to demobilise social movement activism (Hesketh and Morton 2014; Webber 2011). A tension thus exists between social movements seeking greater autonomy and the absorptive capacity of state power (Dinerstein 2015; Gutiérrez Aguilar 2008). This relationship between social movements and the state has been further strained by the model of neo-extractivism that has been pursued in large parts of

Latin America. This model has functioned as a new development paradigm, focusing on natural resource extraction and primary commodity exports as the major means of growth (Burchardt and Dietz 2014; Veltmeyer 2012). In the analysis of the 'two Lefts' the state becomes reified and defines the limit of political action (Luna and Filgueira 2009; Motta 2006). This analysis also ignores a hugely important feature of contemporary Latin American resistance. Rather than formulating just 'two Lefts' in Latin America, we must in fact postulate a 'third Left' in the form of social movements that seek to effect change through autonomous action. As opposed to a centralisation of forces concentrated on the state, such movements focus on the dispersal of political power (Zibechi 2010). Lefebvre (1976: 125) took a clear position in this debate, arguing that the choice we face is to 'either reconstitute society as society or reconstitute the state: either action from below or acts from the top down'. Such acts from below were theorised as a process of autogestion, which he explicitly defined as an anti-statist strategy of self-management (Lefebvre (1966/2009; 1979/2009). Rather than an end condition, autogestion should be conceived as a process that at the same time serves a reflexive, auto-pedagogical function. Thus, 'Each time a social group... refuses to accept passively its conditions of existence, of life, or of survival, each time such a group forces itself not only to understand but to master its conditions of existence, autogestion is occurring' (Lefebvre 1979/2009: 135).

This new modality of resistance is intimately connected to the way in which neoliberalism restructured space and social relations in Latin America. Everyday concerns and needs had to be attended to within the conditions of repression, poverty and state withdrawal from public provisioning (Dinerstein 2015). As a result, struggles were often moved beyond workplace issues around the means of production and were instead linked to 'minimal access to the means of collective reproduction, such as transport, water and basic services' (Portes 1985: 31). The territorialisation of social movements' struggles in various forms has thus been a major contemporary feature of the Latin American political landscape (Zibechi 2012). These social movements have roots in spaces that have been recuperated or maintained through political action as a means for providing a secure environment. Lefebvre referred to this as the 'right to space', which transcended work and non-work-based struggles, but rather concerned itself primarily with everyday life (Lefebvre 2003). For Lefebvre (1970/2003), urban reality was always more than simply the reflection of capitalism. Rather it was the realm of possibility, of encounters and lived experiences that had the power to subvert the dominant order. In terms of alternative spatial production, this was to be governed by 'connaissance' (contrasting with neoliberal 'savoir'). Opposed to a purely abstract knowledge, 'connaissance' is a place-based form of knowledge, informed by action against power (Lefebvre 1991).

Numerous recent examples attest to this struggle for the right to urban space in Latin America. In Buenos Aires, the piqueteros – who would later coalesce as the *Movimientos de Trabajadores Desempleados* or Unemployed Workers Movement, MTD – arose in response to the economic collapse of the Argentine economy in 2001 (which had been previously decimated by the neoliberal transition and a limited recovery that failed to provide meaningful job growth). Following the economic collapse, more street demonstrations were seen in the following year than had been witnessed in the previous 15 years. Issues of urban space then became vital to the unemployed movement. Confined to poor neighbourhoods, agency was exercised through the setting up of roadblocks to stop traffic and disrupt daily life. This action explicitly advanced the claim that the wealthy parts of the city could not continue whilst the poor were ignored. During this time, key neighbourhood associations were set up to attend to everyday needs. These emerged in conjunction with the recuperated factory movement which managed to reclaim 200 factories, including Zanon ceramics, which produced 20 per cent of the country's ceramic exports (Dinerstein 2015; Zibechi 2012).

In Bolivia, cities such as Cochabamba and El Alto have been the major 'rebel cities', rejecting the privatisation of key resources such as water and gas (respectively) that had been mandated by neoliberal IFIs (Perrault 2006). Looking down on the major commercial city of La Paz, El Alto provides not only a stark spatial reminder of the excluded but also their power of collective organisation (Lazar 2008). The city owes its current size and identity to a wave of migration that took place when former state-owned tin mines were closed and part of the countryside privatised during the neoliberal transition in Bolivia. The Aymara and Quechua migrants have reconstituted in an urban setting the communitarian organisation of the Ayllu to administer justice and provide for democratic deliberation (Dinerstein 2015). This was often as a necessity to provide for key elements of daily life that the state or municipal authorities were not delivering (Zibechi 2010). Such forms of organisation were integral to the insurrectionary activity that followed natural gas privatisation.

Finally, there is the example of Oaxaca City, Mexico, where in the summer of 2006 no official government functioned for five months, as an array of trade unions, social movements and civil society groups attempted to declare the city ungovernable. Instead, informal neighbourhood organisations sprang up with popular 'people's councils' replacing official political parties as the local centres of power. This was a response to the perceived authoritarian neoliberalism that was claimed to have reached its apogee under governor Ulises Ruiz Ortiz (Martínez Vásquez 2007). His regime sought simultaneously to extend the commodification of space in Oaxaca whilst cracking down on social protest (Hesketh 2013). These examples give credence to Lefebvre's (1947/2008) suggestion that whilst the city can lead to atomisation of social life, it can also create the conditions for the reinvention of community. However, while such examples demonstrate the possibilities of urban revolt and transformation, they also highlight what Lefebvre (1976) rightly viewed as the limitations of pre-figurative action that did not have a more wide-ranging counter-project to change space permanently. In each case, autonomous political practices have been restricted and in some cases entirely rolled back and absorbed by the state. It was for this reason that autogestion was never considered a ready-made programme by Lefebvre, but instead was viewed as 'itself the site and the stake of struggle' (Lefebvre 1979/2009: 134).

What potentially is lacking from Lefebvre's writings to understand contemporary Latin America? Although extending his idea of the 'right to the city' to include the right to space more broadly, Lefebvre (1976) was undoubtedly focused on the urban as the primary locale for resistance. Whilst Lefebvre did not neglect the rural in his writings, it was often framed as something that had been lost (Elden and Morton 2016). As we have seen, there are multiple examples of urban revolutions beginning in Latin America, giving credence to Lefebvre's ideas. However, we should also note that agrarian struggles, often with demands to retain access to land and territory, have been a major feature of the social movement landscape, including notable groups such as the Zapatistas in Mexico, the *Movimento dos Trabalhadores Rurais Sem Terra* (MST) in Brazil, and the *Confederación de Nacionalidades Indígenas del Ecuador* (CONAIE) in Ecuador. Another new trend that can be observed in Latin America is the rise to prominence of indigenous activism and resistance. This mobilisation must be understood with reference to processes of changing state formation as the transition to neoliberalism slowed or ended policies of land redistribution. It also undercut state support for agriculture, as well as opening land to global capital. All of this threatened the communal basis of indigenous life (Hesketh 2013; Yashar 2005). These are arguably unique elements to contemporary Latin American struggles, that whilst according with Lefebvre's broader notions of autogestion and the right to space, sit uneasily with his more resigned claims about the corrosion of agrarian life (1970/2003). As Elden and Morton (2016: 59) document, following the loss of a key manuscript, *Manuel (or Traité) de sociologie rurale*, Lefebvre's focus shifted from the rural to the urban 'at the expense of approaching urban and rural sociology together'. This, however, is a vital task at the current conjuncture, especially in light of the fluid relationship between town and countryside resulting from recent migration. Despite its practical difficulties, the search for utopian space retains a vital pedagogical function in practical experimentation (Lefebvre 1976). As the very term 'utopia' suggests, such spaces are still not a fully-fledged reality in many cases, but rather should be thought of as 'the non-place that has no place and seeks a place of its own' (Lefebvre 1970/2003: 38). However, as Dinerstein (2015: 60) asserts, 'The "not yet" occupies a significant place' for the politics of Latin America.

Conclusions

The chapter has detailed the relevance of Lefebvre's ideas about class struggle being waged through the production of space. It has done so by considering the transition from ISI to neoliberalism in Latin America. However, in opposition to the isotopic or abstract space that capitalism has sought to construct, numerous revolts have sprung up from everyday life that seek to create counter-projects and counter-spaces, defined as a collective oeuvre. The struggle for utopian space is thus a clashing of spatial projects to define the very meaning of utopia. For capital, this means creating new markets and new opportunities for realising profit. For the multiple movements from below, this is a broader struggle to define democratic participation and collective rights. This is not a battle that has a definitive end point (which is all the more important given the contemporary return of right-wing forces in parts of Latin America). Rather the struggle for utopian space is likely to remain a vanishing point on the Latin American political horizon for some time to come.

References

Burchardt, H.J. and Dietz, K. (2014) '(Neo-)extractivism – a new challenge for development theory from Latin America', *Third World Quarterly*, 35:3 468–486.

Cameron, M. (2009) 'Latin America's left turn: beyond good and bad', *Third World Quarterly*, 30:2 331–348.

Castañeda, J. (1994) *Utopia Unarmed: The Latin American Left After the Cold War*. New York: Vintage.

Castañeda, J. (2006) 'Latin America's left turn', *Foreign Affairs*, 85:3 28–43.

Chang, H. (2010) '*Hamlet* without the Prince of Denmark: how development has disappeared from today's "development" discourse', in S. Khan and J. Christiansen (eds) *Towards New Developmentalism: Market as Means rather than Master*, Abingdon: Routledge.

Davis, M. (2006) *Planet of Slums*, London: Verso.

Dinerstein, A.C. (2015) *The Politics of Autonomy in Latin America: The Art of Organising Hope*, Basingstoke: Palgrave Macmillan.

Duménil, G. and Lévy, D. (2004) *Capital Resurgent: The Roots of the Neoliberal Revolution*, trans. D. Jeffers, London: Harvard University Press.

Elden, S. and Morton, A.D. (2016) 'Thinking past Henri Lefebvre: introducing "The Theory of Ground Rent and Rural Sociology"', *Antipode*, 48:1 57–66.

Ffrench-Davis, R. (1994) 'The Latin American economies 1950–1990', in L. Bethall (ed.) *The Cambridge History of Latin America*, vol. 6, part 1, Cambridge: Cambridge University Press.

Goodale, M. and Postero, N. (2013) *Neoliberalism Interrupted: Social Change and Contested Governance in Contemporary Latin America*, Stanford: Stanford University Press.

Green, D. (1995) *Silent Revolution: The Rise of Market Economics in Latin America*, London: Cassell.

Grugel, J. (1998) 'State and business in neo-liberal democracies in Latin America', *Global Society*, 12:2 221–235.

Gutierrez Aguilar, R. (2008) *Los Ritmos del Pachakuti: Levantamiento y Movilización en Bolivia (2000–2005)*, Buenos Aires: Tinta Limón.

Harvey, D. (1990) *The Condition of Postmodernity: An Enquiry into the Origins of Cultural Change*, Oxford: Blackwell.

Harvey, D. (2003) *The New Imperialism*, Oxford: Oxford University Press.

Harvey, D. (2005) *A Brief History of Neoliberalism*, Oxford: Oxford University Press.

Hesketh, C. (2013) 'The clash of spatializations: geopolitics and class struggles in southern Mexico', *Latin American Perspectives*, 40:4 70–87.

Hesketh, C. and Morton, A.D. (2014) 'Spaces of uneven development and class struggle in Bolivia: transformation or *trasformismo*?', *Antipode*, 46:1 149–169.

Lazar, S. (2008) *El Alto, Rebel City: Self and Citizenship in Andean Bolivia*, London: Duke University Press.

Lefebvre, H. (1947/2008) *Critique of Everyday Life Volume 1*, trans. J. Moore, London: Verso.

Lefebvre, H. (1966/2009) 'Theoretical problems of autogestion', in N. Brenner and S. Elden (eds) *State, Space, World: Selected Essays*, London: University of Minnesota Press.

Lefebvre, H. (1970/2003) *The Urban Revolution*, trans. R. Bononno, London: University of Minnesota Press.

Lefebvre, H. (1975/2009) 'The state in the modern world', in N. Brenner and S. Elden (eds) *State, Space, World: Selected Essays*, London: University of Minnesota Press.

Lefebvre, H. (1976) *The Survival of Capitalism: Reproduction of the Relations of Production*, trans. F. Bryant, London: Allison and Busby.

Lefebvre, H. (1979/2009) 'Comments on a new state form', in N. Brenner and S. Elden (eds) *State, Space, World: Selected Essays*, London: University of Minnesota Press.

Lefebvre, H. (1991) *The Production of Space*, trans. D. Nicholson-Smith, Oxford: Blackwell Publishing.

Lefebvre, H. (2003) 'Space and state', in N. Brenner, B. Jessop, M. Jones and G. Macleod (eds) *State/Space: A Reader*, Oxford: Blackwell.

Lipietz, A. (1984). 'How monetarism has choked third world industrialization', *New Left Review*, 1:145 71–87.

Luna, J.P. and Filgueira, F. (2009) 'The left turns are multiple paradigmatic crises', *Third World Quarterly*, 30:2 371–395.

Martínez Vásquez, V.R. (2007) *Autoritarismo, Movimiento Popular y Crisis Política: Oaxaca 2006*, Oaxaca: Universidad Autónoma Benito Juárez de Oaxaca.

Marx, K. and Engels, F. (1848/2000) 'The communist manifesto', in D. McLellan (ed.) *Karl Marx, Selected Writings*, Oxford: Oxford University Press.

Motta, S. (2006) 'Utopias re-imagined: a reply to Panizza', *Political Studies*, 54:4 898–905.

Perreault, T. (2006) 'From the Guerra Del Agua to the Guerra Del Gas: resource governance, neoliberalism and popular protest in Bolivia', *Antipode*, 38:1 150–172.

Perreault, T. and Martin, P. (2005) 'Geographies of neoliberalism in Latin America', *Environment and Planning A*, 37:2 191–201.

Portes, A. (1985) 'Latin American class structures: their composition and change during the last decades', *Latin American Research Review*, 20:3 7–39.

Portes, A. and Hoffman, K. (2003) 'Latin American class structures: their composition and change during the Neoliberal era', *Latin American Research Review*, 38:1 41–82.

Radcliffe, S. and Westwood, S. (1996) *Remaking the Nation: Place, Identity and Politics in Latin America*, London: Routledge

Robinson, W. (2004) 'Global crisis and Latin America', *Bulletin of Latin American Research*, 23:2 135–153.

Robinson, W.I. (2008) *Latin American and Global Capitalism: A Critical Globalization Perspective*, Baltimore: John Hopkins University Press.

UNCTAD (2003) *Trade and Development Report*, New York: United Nations Publications.

UNCTAD (2004) *World Investment Directory: Volume IX Latin American and the Caribbean Part 1 and 2*, New York: United Nations Publications.

UNCTAD (2006) *Trade and Development Report*, New York: United Nations Publications.

Veltmeyer, H. (1997) 'Latin America in the new world order', *Canadian Journal of Sociology*, 22:2 207–242.

Veltmeyer, H. (2012) 'The natural resource dynamics of postneoliberalism in Latin America: new developmentalism or extractivist imperialism?', *Studies in Political Economy*, 90:1 57–85.

Webber, J. (2011) *From Rebellion to Reform in Bolivia: Class Struggle, Indigenous Liberation, and the Politics of Evo Morales*, Chicago: Haymarket Books.

Yashar, D. (2005) *Contesting Citizenship in Latin America: The Rise of Indigenous Movements and the Postliberal Challenge*, Cambridge: Cambridge University Press.

Zibechi, R. (2010) *Dispersing Power: Social Movements as Anti-State Forces*, trans. R. Ryan, Oakland, CA: AK Press.

Zibechi, R. (2012) *Territories in Resistance: A Cartography of Latin American Social Movements*, trans. R. Ryan, Oakland, CA: AK Press.

4

Contesting spaces of an urban renewal project

A study of Kumartuli's artist colony

Rishika Mukhopadhyay

Introduction

As the clay-smeared streets of Kumartuli get washed out by the torrential rains, the leaky roof of the artist's workshop covered with tarpaulin and plastic sheets gives up protecting the interior. The artists work night and day to finish the unfinished clay idols as the city gears up to celebrate its biggest festival – Durga Puja. Kumartuli, the abode of god-makers, tucked in the winding lanes of North Kolkata by the serene river Hooghly, is closely connected with the initiation of Durga Puja in the city as well. But each year as the colourful processions with trumpeting sounds of dhak and dhol (drums) take the idols to their *puja mandap* (place of worship), there is a melancholy behind this welcoming note. The city knows that after five days there will be another procession to take the deity back to the banks of the river for immersion, where it will meet its end. During the initial days of festivity, the lifeless clay idol is invoked with life through chants and hymns. The process transforms an idol into a deity, despite knowing the ephemeral nature of its existence. Invocation and immersion, two seemingly divergent yet rhythmic processes, are not only rituals of Hindu worship but signify the transient nature of life itself. Life, where permanence is an anomaly and where each creation comes with the precondition of destruction. Living close to the idol-makers' colony of Kolkata, where the abstract god takes a material form, and the river Ganges, where it meets its end, this philosophy is well known to people. But no one thought that the very existence of this colony could come under threat and the artists could face the same destiny as the deity that they create.

The chapter unfolds in the backdrop of a State-initiated urban renewal project which intended to give Kumartuli a quintessential 'modern' look with multi-storied apartments and artist's studios. Though initially people supported the cause of the project, after the first batch of artists was displaced and their homes were demolished, resistance started to grow in places. Since then the project has stalled. This evokes the question as to how not only the project but space itself has been envisaged both by the State and the inhabitants of the area. More importantly, when in the age of capitalist production space has been conceptualised by State as a commodity which needs 'renewal', how are other agencies of society looking at the same? Through the lens of Lefebvre's production of space in the neo-liberal age, the chapter delves into the reasons behind the project's unforeseen termination. Here, I would locate urban renewal projects taken up by the State government and other civil actors as a neo-liberal articulation of capital through the register of land. I would further contextualise the Kumartuli project's impasse within the overall political situation of the State. Is the response from the people a reflection of the circumstances in the State regarding the land right issue, or is their insecurity not only immanent to the loss of entitlement of land but emanating from a deep attachment with their ancestral land? Therefore, the attempt of this chapter is to see whether there is a scope to look beyond the capitalist production of space and acknowledge other forms of production of space following Lefebvre's spatial triad.

Urban renewal projects and capitalist production of space

The concept of urban renewal dates back to the 1950s during the era after the Second World War. It gradually evolved in concept and terminology. After the war, many countries embarked on rebuilding efforts in the war-devastated cities. While the initial process had a different motive, it eventually turned out to be based on demolition and reconstruction. Irrespective of the location of the city, destruction of old, dilapidated areas, clearance of city slums and construction of modern high-rises became a norm worldwide. In the 1950s it was urban reconstruction, rebuilding the destroyed portion of the city. Gradually the concept started evolving and government policy started changing. So in the 1960s it was urban revitalisation, in the 1970s urban renewal, the 1980s urban redevelopment and finally in the 1990s it was urban regeneration (Roberts and Sykes 2000). This is an inherently Western experience, which characterises un-slumming, beautification and large-scale investment. But it was adapted quickly by rapidly growing East Asian cities as well. In 1960s the Government of Seoul undertook a clearance program specifically with the motive of squatter eviction. Urban space was categorically segmented on the basis of its income generation. The project saw low-income spaces of the city as a threat to development and beautification. Therefore, with the help of bulldozers and policemen, large-scale eviction and relocation was undertaken (Ha 2001). Vast metropolitan cities in China have also gone through the same process where redevelopment came in the guise of un-slumming, eviction, relocation and beautification of the city. Shanghai's exemplifying story of transformation is another story of urban renewal which brought in foreign capital to rejuvenate the urban land market, boost industrial restructuring, initiate housing reform as well as infrastructural rebuilding and in the process provided a massive alteration to the physical appearance of its historic city (Wu 1999).

Hence it can be seen that the regeneration process is an inherently economic neo-liberal process which often targets dilapidated city cores and replaces the valuable old buildings with new structures. This process involves the displacement of underprivileged classes to make the city more sanitised, commercialised and global (Wu 2004). As Le Corbusier stated, 'The old city dies and the new city rises on its ruins – not gradually, but in a burst, suddenly – as the butterfly emerges from the cocoon of the caterpillar' (cited in Murray 2008: 41). If we try to explain the phenomenon from a Lefebvrian perspective, we can say that the statement indicates that redevelopment produces a kind of space which can be articulated only through the register of capital. These phenomena are explained as accumulation and regulation of capital in space (Harvey 1989; Smith 1996). Lefebvre's scholarship on production of space specifically says a surplus value is generated through commodification of space. In this space there is no tolerance for spaces inhabited, dwelled and experienced by the urban poor. In this capitalist setup, the position of the State therefore stops being a regulator or mediator of the market (Harvey 1990). Instead it becomes an active agent in promoting inequality and uneven distribution. During this period, the State expresses its autonomy and power through spatial planning, financial regulation, industrial policy and infrastructural investment, which scholars have called the 'state mode of production' (Brenner 2000). This modern territorial State under capitalism performs as the most instrumental agent in creating uneven development which then the state itself tries to level by carrying out spatial restructuring. An artificial unevenness is often created and perpetuated to start new rounds of redevelopment, which leads to gentrification. Regenerated neighbourhoods tend to attract more investment and capital, gradually leading to gentrified neighbourhoods, which displace the previous residents and socio-economic activities. By investing in new sites of deprivation and devalorisation, over-accumulation is avoided, and an elevated, nuanced and filtered level of capital accumulation is articulated as spatial fix (Brenner and Theodore 2002).

Following a somewhat similar pattern, after the 1990s when India opened its door to the global economy, large investment in real estate in Kolkata started gaining momentum. Private players invested more in high-rise residential buildings, shopping malls, multiplexes and new service and commercial centres, rather than renewal of its old districts. But a different procedure took place in old districts where informal redevelopment resulted in categorical demolition and erosion of old historic as well as old ordinary buildings, settings and open spaces. From April 2005 to March 2010, over five successive financial years, the total number of building proposals sanctioned for construction in the already congested municipal corporation area alone was 17,819. Obviously, most new structures were built by demolishing the old ones (Bose 2012).

Although Lefebvre has also talked about three ways of production of space, often his work has been appropriated only in terms of capital, giving less attention to the other ways of production. His spatial triad has talked about spatial practice (societal space), a representation of space which is often envisaged as an absolute space perceived by the planners, urbanists, technocrats and engineers. This again conceives space as an extension of materiality which can be consumed, produced and reproduced in a loop. The chapter will seek to articulate the ongoing struggle of the potters of Kumartuli under the ambit of representational space which is directly lived and experienced. Lefebvre mentioned this as the space of inhabitants, users and sometimes of artists, writers and philosophers. They actively or passively try to put a meaning on the space, imagine the spatial form beyond its physicality and bring in images and symbols to make it more experiential (Lefebvre 1991: 38). Increasingly, from the 1980s the reassertion of space in academic writing came to embrace various aspects of human subjectivity and everyday life to understand the spatial component in social life. Lefebvre proposes a trialectics of spatiality essentially because absolute space and space of capitalism cannot solely explain spatial phenomena. Only an enmeshing of cultural practices, imaginations and representations, or in other words, perceived, conceived and lived space, can explain the nature of spatiality.

The chapter first sets out a brief history of the neighbourhood of Kumartuli to contextualise the artists' standpoint regarding the project which will be explained later. Then it lays out the aspects of the State-initiated renewal project and its consequences in Kumartuli. The next analytical section discusses different reasons for the impediment of the project. This section elucidates the nature of production of space by the State and the community. How is space being produced under the State regime and what are the alternative ways of envisaging space? This will make sense of the renewal syndrome of the city authority and see how different actors of society perceive space. The chapter is based on both primary and secondary data. Ethnographic field work through detailed interviews in 2014–15 in Kumartuli and the nearby rehabilitation site has largely shaped the argument of the chapter. Newspaper reports were used as a secondary material to gather information about the project.

Genealogy of an artist colony

Kumartuli, an artist colony of idol-makers in the northern part of Kolkata, is tucked between the river Ganges on one side and the oldest road of Kolkata, Chitpur Road (renamed as Rabindra Sarani after independence), on the other. The main livelihood of the artists living in the neighbourhood is from manufacturing and sale of clay deities which are worshipped by Hindus annually for various religious festivals. Over the years, some auxiliary industries have sprung up in the area which revolve around the idols' ornamentation. They are mainly centres of related crafts, of Zari-work, Shola-work, Daker-Shaj and imitation hair of jute fibre, for instance.

Multiple stories consider the initiation of the colony. Cotton has mentioned that the potters were initially displaced families from the flourishing Gobindapur village, which was the chosen site for Fort Williams. The Company's Calcutta Zamindar John Zephania Holwell was given the instruction to distribute separate districts to these workmen (Cotton 1907; IANS 2007). It can be said that there is no doubt that, in the beginning of the 18th century, artisans of the surrounding villages of Bengal were the first to give the colony of artisans a proper semblance of *para* (neighbourhood). It all started when the landed and moneyed gentry, zamindars and landowners settled down in the newly built city in search of their fortunes (Banerjee 1989: 31). One way of showing their prosperity, influence and power was through the grand celebration of religious functions. Therefore, Durga Puja started in this city more as a social occasion for the neo-rich among the rising business (baniya) class in the colonial era than a religious one. It is a widely circulated story that Raja Nabakrishna Deb of Shovabazar was the first one to bring the artisans.

> Kumbhakar from krishnanagar of Nadia district of West Bengal, was famous for its clay products. The Bengal Consultations, a journal published in 1707 AD, gives an account of the presence of Kumartuli's artisans who occupied 75 acres of land at Sutanuti, which is a constituent of present day north Calcutta.
>
> *(IANS 2007)*

From the first half of the 19th century, the artisans of Kumartuli were in high demand. Eventually, potters from nearby districts of Banshbede, Nabadwip, Shantipur and Krishnanagar, which were famous for this art, started settling in the potters' quarter of the city (Banerjee 1989; Dutta 2003). The nature of settlement was migratory. Throughout the 18th and early 19th centuries, potters used to stay in their quarter for three to four months. Idol making was not a year-long profession, and during the off season they used to go back to their villages and indulge in their ancestral occupation of making potteries (Goldblatt 1981). By the end of the 19th century, there were 50 image-making workshops in Kumartuli and permanent settlements were seen in the locality (Goldblatt 1981). Leading the first settlers were Madhusudan Pal, Kanalicharan Pal, Kashinath Pal, Haripada Pal and Annadacharan Pal (Banerjee 1989).

After independence, partition brought new sets of artists with different artistic skills in the locality. They mainly came from Dhaka and Bikrampur. Kumartuli by this time witnessed two separate organisations for these two groups, one for the original settlers and the other from East Bengal. This made the already crowded Kumartuli more congested. This congestion, unavailability of basic infrastructure and amenities, unhygienic environment and inhospitable place for living gave the much-loved potters' quarter a reputation of *basti* or *notified slum* in the post-independence era.

Logistics of an urban renewal project

Kumartuli geared up for a structural change proposed by the government under the leadership of Buddhadeb Bhattacharya during late 2000. The Kolkata Metropolitan Development Authority (KMDA) proposed a plan to rejuvenate this unique neighbourhood with modern amenities. Despite being a place where clay goddesses come to life, Kumartuli is a notified slum and it has been repeatedly associated with filth, dirt and unhygienic living condition (IANS 2007). The potters' miserable conditions, with no adequate space for accommodating their large families, leaky roofs in their dark workshops, makeshift stalls with tarpaulin and bamboo, made the locality incompatible to produce artwork.

Indeed, Kumartuli has been depicted by scholars, writers, journalists and travellers as a place where excellence is produced in the most inhospitable environment (IANS 2007; P. Banerjee 2010). The residents who have been staying here for generations were therefore convinced that Kumartuli needed renovation sooner or later. An area was identified of 1.56 hectares in Kumartuli to build four blocks of high-rise apartment blocks. Two types of buildings were to be constructed, first with a ground floor plus three storeys and second with a ground floor plus four storeys. In both cases, the ground floor would have catered as the artists' workshops, while the upper floors were proposed as residential units. The dwelling units were to be 27 m^2 with a multipurpose room, a bedroom, a kitchen, a bathroom and a balcony. Some added facilities were also introduced in the plan. There was an exhibition-cum-sale hall spread over 325 square metres, a dormitory for workers, a health centre and a training hall, two parks, a stage and a community hall (Mazumdar 2009).

Also, there was to be a separate workplace for clay, shola and other ancillary works. The 'Kumartuli Urban Renewal project' sanctioned under the Jawaharlal Nehru urban Renewal Mission was a joint venture of both the state and the central government. The total cost of the project was 607.67 crore (Parmar 2013). Initially, 600 residents signed up in favour of the renovation project. But from the beginning, the project got delayed due to several reasons. A revised blueprint of the Kumartuli facelift was drafted in September 2008, and chief minister Buddhadeb Bhattacharjee laid the foundation stone in January 2009 (Mazumdar 2009). But the project did not take off again for several months. First, there was the issue of choosing an alternative suitable rehabilitation site to temporarily relocate the workshops as well as the family of the artists. Three sites – Strand Bank Road, Pran Krishna Mukherjee Street and Nivedita Park in Bagbazar – were considered by the KMDA. But problems seem to have been a continuous companion with this project. In the first phase of the project, two blocks were supposed to be relocated to a new spot, block B (Ganga Prasad Lane) and C (part of Banamali Sarkar Street).

The second phase will see the relocation of block A, comprising residents of Durga Charan Banerjee Street, and block D, containing the residents of Rabindra Sarani and other parts of Banamali Sarkar Street. The proposed stretch between Strand Bank Road and Sovabazar was found unsuitable for the purpose because of its congested nature (Mazumdar 2009). KMDA officials delayed the process further from survey to plan submission and deed processing, and irritation and impatience in the population grew manifold.

Rusted fate: segmentation and relocation

Nimai Pal, a veteran artist at the age of 67 and the president of the Kumartuli Mritshilpa Sanskriti Samity, never thought he would have to leave his ancestral home and take refuge in a godown (warehouse), which was an old flour mill locally known as 'Maida Kal' in the colonial era. But he hoped for a better future for the whole community and volunteered to be the first one whose home was demolished. On 23 April 2007 the first phase of demolition started and Nimai Pal, the president of the idol-makers organisation, witnessed his home getting smashed under the rolling bulldozer in front of his eyes: 'Our family has been staying in that house for four generations. Seeing that getting bulldozed in front of your eyes is painful. But we were hopeful something better will happen' (Pal 2015).

He talked about his 'home', Kumartuli, which is a few steps away from his current 'shelter' at Bagbazar while sitting on a long piece of unfinished wooden timber. Opposite him was another wooden appliance, a raised wooden spread known as a 'Khat' (bed), which is meant for sleeping but here used for all other purposes apart from sleeping. The room is filled with unfinished or somewhat finished clay and fabric idols of various sizes. Another room just opposite to this also belongs to him. In the hallway between these rooms his wife was busy in household chores. The other room is his bedroom as well as dining room and an extension of the kitchen as well.

This is the temporary shelter for 170 families which have been displaced from their ancestral home in Kumartuli in the hope of a better tomorrow (Banerjee 2010). After much inspection a spot was selected to relocate the residents of C block, five minutes from Kumartuli, at 541 B Rabindra Sarani. Asit Mukherjee, an elderly idol-maker who has been shifted from Kumartuli and is currently a resident of this shelter, said:

> This godown (warehouse) was full of heaps of cement and iron rods which were not suitable for living, but KMDA cleaned it and temporary rooms were made for us. We thought it is a temporary shelter and we will get a better place. But now our entire future is in crisis.
> *(Mukherjee 2015)*

Everybody thought these rooms were temporary when they first moved here in May 2010. But the project has hit a setback for the past six years. From 2010 to 2016, a lot has happened in Bengal politics. Trinamool Congress took an overwhelming majority of seats on the Bengal assembly and took control of the left front government. Subsequently, they own the seats of Kolkata Municipal Corporation as well, and the much-anticipated project of Kumartuli's revival has been difficult since then. The grand colonial building with a wavy roof, green windows and red brick walls by the riverside, a building presently owned by Public Health and Engineering Department and the Department of Health and Family Welfare, has become the shelter for nearly 200 families with no future hope. For the past five years the project is in stagnation. Only the rusted iron gate in front of the mill bears the mark of 'Rehabilitation for Kumartuli Urban Renewal Project', and the project is in irrevocable uncertainty. The fortune of the potters who left their homes anticipating a better future is in utter jeopardy.

Land as capital

The banner at the entrance of Kumartuli, the potters' quarter, seems to proclaim Kumartuli's position loud and clear. It is the banner, like many such banners in the locality, which announces Kumartuli's resistance to the renovation process. It declares loudly that Kumartuli finds its essence in the labyrinth of lanes and tottering huts. Between the relocation site and the original site there are two different worldviews. While the dislocated wanted fast completion of the project so that they could go back to Kumartuli again, those staying in Kumartuli refuse to be part of the project. Multiple narratives and multiple realities came up in the fieldwork as the reason behind the stalling of the project.

There was also inter-party conflict, which accelerated the whole dispute. The main two organisations comprising artists from East and West Bengal started meeting with the officials of KMDA, who were backed by the government. This instigated those who were supporting the opposing party, the Trinamool Congress.

The larger political scenario of the State was also marred with the land right and land accusation issue starting from Singur to Nandigram at this point in time. In May 2006, the government of West Bengal, led by CPI (M), selected 403.472 hectares of prime agricultural land in Singur. They wanted to acquire the land for the purpose of leasing it out to one of the biggest automobile companies of India, Tata Motors. Tata Motors intended to build a car manufacturing unit in Singur, with the motive of producing its most affordable car of the world, Tata Nano. The government acquired the land in spite of the fierce resistance of the unwilling farmers. The protest, which eventually turned into a movement of the farmers, continued against the land acquisition and, in late 2008, Tata Motors decided that it could not operate a factory in this hostile local environment. Therefore, they decided to shut its operations in Singur and move the entire factory to Gujarat. The opposition leader of TMC, Mamata Banerjee, took a lead role in advocating the cause of the farmers and eventually Tata Motors had to withdraw. But on the ground the situation remained in deadlock, as the acquired land remained vacant and it was not being returned to its owners. A similar incident with more political velocity happened in Nandigram in 2007 where the government wanted to acquire 4046.85 hectares of land for a Special Economic Zone (SEZ). The land was to be given to the Indonesia-based Salim Group to develop a chemical factory. As resistance began to brew in the area, the main organisation which was resisting took charge of the area and the police could not enter the villages for three months. In the meantime, several families who are supporters of the ruling left party were attacked and evicted from their land. At last, the State government cancelled the project, but they decided to send police in the region to break the standoff. In a clash between villagers and police, 14 people were killed and several were injured as per the official count. The impact of both the incidents and the government policy of industrialisation in the exchange of prime farm land acquisition resulted in a massive electoral defeat of the government after 34 years of its rule in West Bengal.

As stated above, in each scenario the left front government faced considerable opposition from the people who were backed by the opposing party, TMC. Things took the same course in Kumartuli as well. The landowner artists with the support of TMC formed two parallel associations, 'Kumartuli Adhikar Rakkha Samiti' and 'Kumartuli Pragatishil Mritshilpo o Sajshilpo Samiti'. They simply said that Kumartuli's renovation was stalled because of the land ownership issue.

Most of the land in Kumartuli is 'thikajami', which is leased land. Out of 600 applicants, 50 first resisted the handover of their own land to the government. Most of them were the owners of their land allocated for the workshop. They were unsure of getting back a place which would be equally valuable to their owned land, though, as a solution, KMDA was ready to allocate one extra flat to the land owners, but people still refused to give their land. Kartik Pal, a veteran idol-maker in his 60s, says,

> Historically there have been instances where the government did not keep its words. We cannot trust them. No matter which government it is. They can take up our lands and use it for something else. We are not sure if we will be relocated back here. The project can take very long to complete. What will happen to our business during that time if they displace us and put us somewhere far away from Kumartuli. If we move from here we will face huge loss because we won't get any order. People associate idol-making only with Kumartuli.
>
> *(Pal, 2015)*

Kartik Pal's last line, 'people associate idol-making only with Kumartuli', leads us to the other side of the story regarding the renewal project, namely its stagnation and how people see their ancestral land. Ownership of land, their insecurity of not getting back the lost land and distrust towards the government were reasons, but not the only ones. The land issue was the legal, logical and most easily understandable reason, which was communicated from the artists' side to put an impediment on the whole process of displacement, rehabilitation, relocation and construction of a modernised Kumartuli. This official narrative was put forward to stop the displacement process. But the one narrative which was completely unreported in the media was people's attachment with their land and how they see land as more than a vector of exchange. I discuss this in the next section.

Overall, in the country, the general trend of India's developmental paradigm involves large-scale transfer of land from rural communities and disposed groups to private corporations. It is noticeable in the State initiative at Kumartuli that for this quarter of the city, 'renewal' projects are the only answer for the State to show it is serious about the development of the old city. The proposed restructuring project by the State is intended to legitimise and recognise aesthetically pleasing and sanitised renderings of space as developed space. This proposed restructured landscape was, therefore, essentially uprooted from its temporality, which propagates the State's modernist agenda. As many scholars have already said, what we study in the 21st-century metropolis is largely a study of modernity (Robinson 2006). This modernity is largely an emulation of Western expressions and experiences, which does not remain confined in the scale of society but also gets reflected on space.

Space as lived

Narratives were presented by the people about their deepest attachment with the land. More than the exchange value of the land, some people talked about the 'character' of the place. Following Molotch, they talked in favour of a place where aspects like the economic, social, physical and intangible elements are all existing together (Molotch et al. 2000). The maze of lanes and bylanes, the small homes for large families, the deep dark workshop, the shaky bamboo scaffolding, the smell of riverine clay, everything associated with old Kumartuli is so dear to these people, so deep-rooted in their minds, that they could not consider a place with shining state-of-the-art buildings, broad boulevards and glittering art galleries.

In Kumartuli, fear of displacement as a result of a proposed restructuring project was overpowered by fear of losing one's home, fear of unfamiliarity, fear of losing one's identity and most importantly, fear of losing the mundane ways of life. Sunil Pal, an artist and a long-time resident of Kumartuli, proclaims,

> After old Kumartuli will be modernised I am sure nobody from outside world will come to see our way of work. The hardship, struggle and darkness of our lives have made us extraordinary. I will miss the shutterbugs and foreigners who find our way of work so intriguing that they spend days with us drinking tea from this roadside stall, from this clay pot. I don't think these shops will have any place after renovation.
>
> *(Pal, 2015)*

Sunil Pal's concern is not only his. The people of Kumartuli know that their way of living made them unique in front of the whole world. Once they become 'modern' and 'developed' they will lose that particular essence – that which makes them different, makes them one of a kind and unlike any other artists' colony in the world. Certainly, they feel they belong in this messiness, in this soil, and if something disrupts their familiar environment they won't be able to recover the loss. The loss here is not merely in economic terms but involves a loss of self and the loss of an artist's core existence.

Those people who were already displaced have got a covered workshop, which protects them from torrential rains in the monsoon and scorching heat in the summer. They don't have to run or hide, and they don't have to cover their unfinished clay idols – issues they constantly used to face in Kumartuli. They are secured from potential loss of any kind. Their homes and workshops are safe from nature's wrath. They should feel content. They should feel a significant improvement in their standard of life. Asit Mukherjee, a resident of the shelter says:

> It's true we don't have to cover our idols like we used to do in Kumartuli during rainy season. But our idols used to get dried up in the nature by sun, in the wide open. Here we are dependent on fans. Our electricity bills are huge. We constantly miss the vibrant activity of Kumartuli streets.
>
> *(Mukherjee, 2015)*

The artists work quietly inside their impenetrable workshops, but they feel a sense of detachment from the soil. They miss the buzzing activity in the narrow lanes of Kumartuli, they miss the sudden need to cover their idols from rain and they miss the natural sunlight where the idols got soaked up.

These narratives were completely missing from the official version that we heard before. How the piece of land acquired a certain meaning over time, which is incommensurable by an economic standard value, is worth noting. Here, people expressed their emotion about Kumartuli, which is much more intimate, pervading and in some ways ordinary too. It emerged from their daily life experiences, interactions with neighbours, friends, family and even strangers, which to a large extent defines who they are. I am not suggesting that the way the State viewed the land in Kumartuli and the way they wanted people to measure its disposability is not economic, but the idea that I am grappling with here highlights the need to acknowledge other forms of production of space. Similarly, Kumartuli's natural and built landscape is where they have grown up; they have ancestral roots attached to the place. Over the years, the space of Kumartuli itself has acquired a meaning because of its long tradition and its past connection. The very process of a facelift may uproot Kumartuli from its living traditions. There will be a complete rupture in the existing 'sense of place' once the low-lying tally huts are demolished and the narrow lanes are broadened.

Conclusions

In the case of Kumartuli, it can be seen how the State conceptualises space. First, it assumes that space is absolute and physical where a price can be put on it. Second, this spatial understanding has been positivist in nature; hence it has remained impersonal. In Kolkata, the mushrooming of redevelopment efforts in ties with the private real estate sector has shown utter disdain towards the existing design, architecture, sense of place, feeling of locality and historic value of a neighbourhood. Planned reconstruction has torn down old houses and in its place built high-rise structures where more people can be confined. The plan and vision has always been to make the city quintessentially modern, disregarding the experiential landscape. Therefore, more than the concern for crumbling infrastructural decay, the intention of the Kumartuli Urban Renewal Project was to make it worthy of a major tourist attraction centre in the city. To make it consumer worthy, a demolition was planned, which came at a certain cost: the cost of people's belonging, pride, identity, local distinctiveness and experience that made the spatial understanding more complex as well as fluid and dynamic. Here, Lefebvre's insistence that rather than two opposing tenets, only three moments of the spatial triad can produce urban social space makes this scholarship more insightful. To mark a crack in the neo-liberal façade of capitalist modes of production, there is a need to adhere to Lefebvre's representational space which gives precedence to everything that is passionate, humane, imaginary, yet lived.

References

Banerjee, P. (2010) Idol capital shifts base, *Telegraph India*, accessed online at http://www.telegraphindia.com/1100427/jsp/frontpage/story_12381667.jsp, accessed May 2015.

Banerjee, S. (1989) *The Parlour and the Streets: Elite and Popular Culture in Nineteenth Century Kolkata*, Calcutta: Seagull Books.

Bose, S. (2012) 'Historic buildings at Chitour in Kolkata: problems and prospects through urban conservation and planning', *disP-The Planning Review*, 48:1 68–82.

Brenner, N. (2000) 'The urban question as a scale question: reflections on Henri Lefebvre, urban theory and the politics of scale', *International Journal of Urban and Regional Research*, 24:2 361–378.

Brenner, N. and Theodore, N. (2002) 'Cities and the geographies of "actually existing neoliberalism"', *Antipode*, 34: 349–379.

Cotton, H.A.E. (1907) *Calcutta Old and New*, Calcutta: W Newman & Co.

Dutta, K. (2003) *Calcutta: A Cultural and Literary History*, New Delhi: Supernova Publishers and Distributers Pvt Ltd.

Goldblatt, B. (1981) 'The erosion of caste monopolies: examples from West Bengal', *Social Analysis: The International Journal of Social and Cultural Practice*, 7: 99–113.

Ha, S.K. (2001) 'Developing a community-based approach to urban redevelopment', *GeoJournal*, 53:1 39–45.

Harvey, D. (1989) *The Urban Experience*, Baltimore: Johns Hopkins University Press.

Harvey, D. (1990) *The Condition of Postmodernity*, Blackwell: Cambridge.

IANS (2007) Potters town Kumartuli may finally get Makeover, *DNA India.com*, accessed online at http://www.dnaindia.com/india/report-potters-town-kumartuli-may-finally-get-makeover-1127477, accessed May 2015.

Lefebvre, H. (1991) *The Production of Space*, trans. D. Nicholson-Smith, Oxford; Cambridge: Blackwell Publishing.

Mazumdar, J. (2009) Home and studio for idol makers, *Telegraph India*, accessed online at http://www.telegraphindia.com/1091215/jsp/calcutta/story_11865099.jsp, accessed 4 May 2015.

Molotch, H., Freudenburg, W. and Paulsen, K.E. (2000) 'History repeats itself, but how? city character, urban tradition, and the accomplishment of place', *American Sociological Review*, 65:6 791–823.

Mukherjee, Asit (2015) Interviewed (about half an hour) on 17 February at his shelter home.

Murray, M.J. (2008) *Taming the Disorderly City: The Spatial Landscape of Johannesburg after Apartheid*, Ithaca: Cornell University Press.

Pal, Kartik (2015) Interviewed (about half an hour) on 20 February at his workshop in Kumartuli.

Pal, Nimai (2015) Interviewed (about one hour) on 15 February at his shelter home.

Pal, Sunil (2015) Interviewed (about half an hour) on 5 February at his workshop in Kumartuli.

Parmar, A.S. (2013) On Account Payment of Additional Central Assistance (ACA) for the Sub-Mission on Basic Services to Urban Poor (SM -BSUP)- capital Assets under JNNURM for the States Annual Plan 2012–13, New Delhi: Government of India, Ministry of Finance, Department of Expenditure, Plan Finance 1 Division, accessed online at http://jnnurmmis.nic.in/writereaddata/Sanctions/BSUP/2012-2013/BSUP_137.pdf, accessed May 2015.

Roberts, P.W. and Sykes, H. (2000) *Urban Regeneration – A Handbook*, London: Sage Publications.

Robinson, J. (2006) *Ordinary Cities: Between Modernity and Development*, London: Routledge.

Smith, N. (1996) *The New Urban Frontier: Gentrification and the Revanchist City*, New York: Routledge.

Wu, W. (1999) 'City profile: Shanghai', *Cities*, 16:3 207–216.

Wu, W. (2004) 'Cultural strategies in Shanghai: regenerating cosmopolitanism in an era of globalisation', *Process in Planning*, 61:3 159–180.

5

Lefebvre and contemporary urbanism

The enduring influence and critical power of his writing on cities

Pierre Filion

Introduction

It is unusual for the influence of writers in fields related to the social sciences and humanities to outlast their own generation. Not so with Henri Lefebvre. More than 25 years after his death, his influence shows no sign of decline (Elden 2001; 2004a). Not only has he been a predominant intellectual figure during much of his long and productive life, but the impact of his work is still widely felt. The chapter examines reasons for this lasting influence. While the chapter considers the overall intellectual contribution of Lefebvre, its primary focus is, consistent with the objective of the book, on his main urban writing (Lefebvre 1968; 1992 [1974]; 1996; 2003 [1970]; 2004 [1992]).

The chapter maintains that Lefebvre owes his enduring influence to five factors. There is first the relatively recent translation of his urban books into English. For many English-speaking urban academics, exposure to the thinking of Lefebvre, therefore, dates mostly from the 2000s. And given the prolific output of Lefebvre over his lifetime, we can expect many more English translations of his work to follow. Second, because of the tendency of Lefebvre to think by association, his writing proliferates with ideas. The books of Lefebvre can thus be mined by urban researchers for ideas at different stages of their development, which they can further pursue in their own research. Third, given the embryonic nature of many ideas present in the writing of Lefebvre, they can be interpreted in different ways by researchers and thus be adapted to their own purpose. The large number of ideas and the multiple ways in which they can be understood make it possible for researchers to project their own perspectives on the work of Lefebvre. The fourth reason for the persistent influence of Lefebvre is his foresight regarding societal and urban circumstances to come.

The fifth and final reason for his lasting impact concerns the capacity of the perspectives he advanced to interpret contemporary urban issues and critique responses to these issues. Among the numerous themes that can be identified by interconnecting related ideas in the writing of Lefebvre, the chapter relates enduring interest in his work to the framing of arguments weaving together a concern for the alienation of the individual, a critique of both capitalism and technocratic state arrangements, adherence to a post-scarcity hedonistic vision, urban centrality and the right to the city. Although in a characteristic abstract fashion, his writing proposes a form of urbanism that promotes human fulfilment and thereby counters the alienating features of contemporary society. It would be an urban form where everyone can enjoy the rich multi-functionality, diversity, social interaction, creativity and festive atmosphere afforded by urban centrality. In such a place, use values, which cater to the necessities of life and to human fulfilment, would predominate over purely economically driven exchange values.

The chapter closes by opposing this liberating vision of the city to the present urban reality, in order to verify the relevance of the ideas of Lefebvre for late 2010s urbanism. Cities have become more oriented towards culture, art, leisure and overall human self-realisation, as evidenced by the transformation of downtown areas from purely functional central business districts to tourist attractions, cultural and recreational centres and high-density residential areas. But this transition remains driven by market processes and hence use values in an increasingly neoliberal economic and social climate. Moreover, the appeal of transformed city centres fuels gentrification, a factor of social segregation contradicting Lefebvre's calls for a right to the city. In a contemporary context, the conceptual instrumentation advanced by Lefebvre serves to evaluate the evolution of present forms of urbanism from a human fulfilment and social justice perspective, and to confront current urban trajectories to his humanist vision of the city.

The intellectual foundations of Lefebvre

Critical evaluations of the work of Lefebvre as well as a biography have charted the course of his intellectual pathway through most of the 20th century (Butler 2012; Elden 2004b; Hess 1988; Kipfer et al. 2008; Merrifield 2006; Shields 1999). They show how a long life and intellectual eclecticism have exposed him to the main intellectual currents of the century. Few intellectual figures, if any, synthesised as well as he did the thinking of the past century. His intellectual journey begins with the surrealists, the French successors to the Dada movement. He was close to core surrealist poets and like some of them converged on the French Communist Party. His adherence to Marxism had a lifelong influence on his thinking and his work. His was not, however, the official version of Marxism broadcast by the French Communist Party and its intellectual apologists.

The Lefebvre variant of Marxism emphasised the philosophically oriented young Marx and his concern for human alienation and the achievement of the fulfilled 'total person' over the more economically minded older Marx, whose vision constituted the official ideology of communist regimes. Lefebvre was also averse to the state technocratic apparatus, both in its Western capitalist and Eastern communist variants. Little wonder that, in these circumstances, his affiliation with the French Communist Party came to an end in 1958, albeit after 30 years of membership. He found himself more at home with the free-thinking Situationists, who inspired the French May 1968 movement (Lefebvre 1998). In France, May 1968 values represented a victory on the left for less dogmatic thinkers like Lefebvre over the communist doctrine of the time. There is a strong utopian flavour to the urban visions of Lefebvre, which while sympathetic to Marxist thinking, break with the then prevailing economist interpretations of Marxism (Ajzenberg 2001). His attention was instead on how use values could substitute themselves to exchange values and technocratic state intervention. Lefebvre proposed self-management (*autogestion*) as a way of achieving his urban vision (Bitter, Derksen and Weber 2009).

As his work demonstrates, Lefebvre was also influenced by the main schools of thought of his time. His writing on the city is filled with references to urban theorists from mid-20th-century decades (1920s to 1970s), such as Le Corbusier, Jane Jacobs and those belonging to the Bauhaus movement. He also supplemented his urban reflections with contributions from geographers, political scientists and sociologists. What is more, in a fashion that underscores his syncretism, he drew from disciplines that are not normally associated with the study of cities. Structuralism, linguistics and semiology, anthropology, psychoanalysis, the work of Michel Foucault, all bore on his reflection on the city.

Henri Lefebvre can be defined as both a sociologist and a philosopher (Shields 1999). The numerous allusions to philosophers in his urban work are therefore not surprising. He referred extensively to the ancient Greeks, but he devoted most of his attention to Marx. His work was also inspired by Nietzsche, who was used by Lefebvre to validate the prevalence of the lived urban reality over representations of the city – a perspective that is central to the urban work of Lefebvre. Heidegger, along with Sartre, also figure in this work.

As we will see in more detail in the next section, there is an accretive dimension to the influence the schools of thought to which Lefebvre had been exposed exerted on his thinking. These different intellectual outlooks have left an imprint on his writing. His urban work is largely about relying on numerous schools of thought, many not normally tied to the city, to extend the understanding of the urban phenomenon. It is of course legitimate to question the enduring relevance of an intellectual endeavour that relies so heavily on forms of thinking from the middle decades of the past century.

The thinking of Lefebvre was not only shaped by different schools of thought. The societal reality prevailing when he was writing his urban books also had an influence on him. He was writing this work during *les trente glorieuses*, the French Fordist-driven prosperity that spans broadly the three post-Second World War decades. This was a time of robust economic expansion and growing state intervention in the economy and civil society (Fourastié 1971). The full-employment prosperity of the time may explain the shift in the thinking of Lefebvre from conflicts stemming from the production sphere, the key concern of Marxists, to tensions within cities deriving in large part from technocratic interventions of the state.

The thinking and writing style of Lefebvre

Hegel and Marx have impacted the thinking and writing of Henri Lefebvre. Many of his positions are advanced in a dialectical fashion, involving a proposition and a counter-proposition and the exposition of how the tension between the two results in an outcome that transcends the two original propositions (Soja 1980). There is also a Socratic flavour to Lefebvre's written expression, which accounts for the interspersion of the text with a profusion of questions.

Lefebvre thought by association. Instead of concentrating on the elaboration of a few core ideas, he tended to jump from one idea to another, leaving many of them in an embryonic stage. Such a tendency was consistent with his general reliance on short sentences, which contrasts with philosophers' predilection for long compounded sentences. The writing style of Lefebvre involves skipping from one idea to another, steered by associations between these ideas. In the writing of Lefebvre, one idea leads to another on the basis of similarities in themes and historical or geographical contexts. This associative approach has the advantage of delivering wide perspectives on the urban phenomenon, exposing it to the interpretations of a multitude of ideas. Meanwhile, the downside is a scarcity of ideas sufficiently described by Lefebvre to lend themselves to an empirical investigative agenda, without further development. The wide range of ideas appearing in four pages taken at random from *The Urban Revolution* ([1971] 2003) illustrate the ease with which Lefebvre moved from one idea to another and the resulting abundance of ideas in his urban books: writing and the city, early urban plans, depictions of cities in paintings, commercial capital, passage to the industrial city and the history of this type of cities, city and anti-city, dialectical thinking, signs in the city, implosion-explosion of cities, different types of exchanges, the world-scale nature of urban systems and cities becoming a transformative force (Lefebvre 2003: 12–15).

I have already alluded to the cumulative and syncretic nature of the thinking of Lefebvre. His writing reflects the numerous schools of thought he was exposed to over his long career. Even the influence of his surrealist phase could be seen, without exaggerating too much, in his associative writing style. While obviously not as manifest as in the case of stream-of-consciousness surrealist poets, one can nonetheless detect a hint of resemblance with surrealist free-association in the frequency with which Lefebvre moved between ideas.

Consistent with his associative style of thinking and writing, Lefebvre does not generally integrate the different perspectives on the city (stemming from the schools of thought he adduces) into his urban books. In his writing, the urban phenomenon is exposed to these schools of thought one at a time. Sometimes, this leads to new ways of conceiving the city, while in other instances it serves to make a case against the theoretical perspectives that are raised. Targeted in this fashion as objects of his criticism are mainstream economics, economistic interpretations of Marxism and ideologies linked to technocratic statism, which include, in the view of Lefebvre, structuralism (Lefebvre 1971). While one can justifiably deplore insufficient conceptual integration in the work of Lefebvre, he can, at the same time, be praised for his ability to address the urban phenomenon from different perspectives. In doing so, he exposed this phenomenon to a uniquely broad range of interpretations and thus expanded the scope of the understanding of the city. It is indeed unusual to consider the city through a linguistic or psychoanalytic lens as he does, and of course, the use of such uncommon interpretive tools has the potential of yielding fresh perspectives on the city. Herein lies a major explanation for the breadth and originality of the perspectives advanced by Lefebvre. Reliance on these schools of thought to address the city in unexpected ways also underscores the unorthodoxy of Lefebvre in his use of conceptual approaches, in other words, his capacity to stretch the applicability range of these schools of thought.

Just as concepts are often summarily presented in the urban books of Lefebvre, so are, to an even greater degree, empirical references. He offered the minimum of information required for empirical references to serve as examples illustrating the concepts he introduces. These examples can be historical, such as those from the Greek and Roman antiquity, the European middle ages, the Renaissance or late 19th-century Paris, as in the case of the Haussmann reconstruction of Paris. Or they can originate from the 1960s and 1970s when Lefebvre wrote most of his urban books. Among examples he repeatedly made use of are: *Les Halles* redevelopment in Paris, the construction of public housing projects (*les grands ensembles*) in Paris suburbs, the gentrification of inner Paris, the dispersed urbanisation of the US. The critic Manuel Castells' focus on Lefebvre's scant concern for empiricism was justified (Castells 1997; Elden 2004b: 142). References to empirical reality were insufficiently developed to provide a systematic validation of the ideas that were advanced by Lefebvre. There was no case study or quantitative data that could grant empirical validity to the ideas advanced in his books.

The urban books of Lefebvre have much in common. They put forth similar ideas about cities, albeit with differences in emphasis. His latest urban book, *Rhythmanalysis*, a posthumous work, is an exception (Lefebvre 2004 [1992]). Although drawing heavily on the previous urban books, it does have a strong focus of its own – the importance of rhythms in the organisation of cities and daily life.

The thinking style of Lefebvre plays an important role in the enduring popularity of his urban writing. The profusion of ideas present in this work makes it possible for numerous researchers to latch onto these ideas and thereby claim a lineage between their own investigations and the reflections of Lefebvre. It is possible for them to build a research agenda around an idea that came from Lefebvre, even if the description of this idea occupied only a few paragraphs in his books. There is also the fact that the often summarily developed ideas in the writing of Lefebvre lend themselves to multiple interpretations. Researchers can assign different meanings to these ideas and operationalise them accordingly in their empirical investigations. Hence the 'Rorschach' test (whereby people reveal inner thoughts when interpreting ink spots) quality of the writing of Lefebvre. The urban books of Lefebvre could be likened to a supermarket of ideas for left-leaning urban researchers, were the concept of supermarket not so alien to his vision of the city.

Alienation and the Right to the City

With the worldwide proliferation of cities and their concentration of economic activity, social interaction and creativity, it is easy to understand the importance Lefebvre gave to the urban phenomenon. Not only do cities reflect present societal tendencies but they also point to the future trajectory of modern societies. And they can play a leading role in the fulfilment of the human potential of individuals, as this is where daily life for a majority (a large majority in economically advanced societies in the global North) of the population unfolds.

For Lefebvre, space is the dominant force driving the evolution of modern society. He applies the Marxist dialectic, resulting from the dynamic between the mode of production and the contradictions it generates, to an understanding of transformative forces centred on space and urbanism. For Lefebvre, such a view reflects the colonisation by capitalism and state technocracy of virtually all aspects of urban life, causing the city to become the sphere where daily life meets with most intensity the capitalist and state modes of production.

One can find connections between multiplicities of ideas appearing in the urban writing of Lefebvre, making it possible to identify overarching themes. Lefebvre rarely pulled these ideas together himself, so it is largely left to the reader to interweave compatible ideas into broad themes. I identify one such theme, chosen for its relevance to the present evolution of the city discussed in the next section.

A major theme that runs through the urban books of Lefebvre ties human alienation with the right to the city. It blends philosophical reflections on alienation with ways of achieving the total person, in a fashion that optimises human potential, well-being and happiness. From the perspective of Lefebvre, human alienation is the result of exploitation inherent in the capitalist mode of production and, increasingly, of an organisation of urban life that contributes to profit-making and reflects the technocratic control of the state on society (Brenner 2008). Planning is described in medical terms as a means of dealing with urban ailments (Lefebvre 1961). But, within the prevailing societal context, it is depicted as only capable of providing partial and temporary relief and unable to address the source of problems. Ultimately, as an instrument of the state, planning belongs to its ideological and societal control apparatus and thereby contributes to the reproduction of the capitalist/technocratic city and of the human alienation it causes. At the same time, however, Lefebvre saw conditions for human fulfilment in the essence of the urban phenomenon. But it is only by shifting from the predominance of exchange values to that of use values that the city, corrupted by the capitalist and state modes of production, can reconnect with its essence. According to Lefebvre, to promote human fulfilment, cities must provide universal accessibility to centres, which are diversified, festive, and multi-functional, and which encourage social interaction and creativity, all conditions for the realisation of human potential. In the Lefebvrian view, the right to the city corresponds to the right to such centrality.

Consistent with the tendency for Lefebvre to downplay empirical evidence and the practical implications of his ideas, his urban books offer scant evidence of the concrete form a city promoting human fulfilment would take. Proposals of such a transformed city remain at a philosophical utopia stage. Nevertheless, Lefebvre highlighted the need for such a city to stress the lived experience, in contrast to market-driven capitalist space and to abstract models put forth by the technocratic state. The emphasis on the lived environment reflects the influence of Nietzsche's Dionysian streak on Lefebvre. Both Nietzsche and Lefebvre indeed focused on the rich complexity of the lived experience rather than on simplifications inherent in rational explanations.

There is wide appeal to such a broadly defined alternative vision of the city. Its lack of precision makes it applicable to a multitude of contemporary urban situations, hence the popularity of the right to the city concept. The many interpretations of this concept are mirrored in its adoption in varying circumstances. Critiques of cities based on the right to the city can pertain, for example, to the social consequences of gentrification, the lack of services in self-built settlements, the high cost of housing, and splintered access to urban infrastructure and services (see for instance Graham and Marvin 2001; Newman and Wyly 2006; Weinstein and Ren 2009).

The Right to the City in the contemporary urban context

The passage of time makes it possible to verify the anticipatory capacity of the urban writing of Lefebvre, which has for the most part taken place four to five decades ago. Have the world and cities evolved in the directions foreseen in these books?

The predictive record of Lefebvre is far from perfect. For example, although he mentioned neoliberalism, which was pointing its head in the 1970s, he did not foresee the extent to which this ideology would alter economic and social policymaking across the world. As a result, his critiques of state power, especially of the technocratic interventions of governments, do not resonate as loudly as they did several decades ago. Indeed, with the retrenchment of the state and resulting direct exposure of individuals to economic volatility, many would welcome a return to Fordist redistributive programs, regardless of their technocratic proclivities. Likewise, his vision of a post-scarcity economy, which promotes human fulfilment, is confronted with uneven and overall sluggish growth in global North industrial nations. The post-scarcity vision was more credible in an era of rapid Fordist expansion than it is in the present slow-growth economy. Sputtering economic performance breeds social polarisation and marginalisation. The possibility of advancing social disparity was not an important object of discussion in the urban work of Lefebvre, which may have been due to the favourable economic and redistributive conditions prevailing when writing his urban books. Moreover, Lefebvre did not grasp the economic and social consequences of globalisation, especially in industrial nations in the global North, the main focus of his reflections. In a similar vein, while he acknowledged the environmental damage inflicted by an all-out exploitation of nature, he did not project these consequences at a planetary scale, as is currently done. The absence of planetary environmental vision is not, however, inconsistent with the fact that most of his urban writing preceded awareness of global warming. He also failed to predict the importance of global population flows, mirrored by the place minorities now occupy in cities and their predominant role in urban conflicts. The class-based urban struggles depicted in the writing of Lefebvre now intersect closely with ethnic and racial categories. Furthermore, as Shields (1999) noted, Lefebvre was curiously silent on women's movements.

Nevertheless, many expectations aired by Lefebvre did materialise. Lefebvre can be credited for anticipating the present age of universal urbanisation. Not only does most of the planet's population live in urban areas, the vast majority of the global economy is based in cities. What is more, urban settlements across the world are interconnected by intensely used transportation and communication networks, giving rise to a worldwide urban system. While in the past, the main population movements were from the countryside to the city, increasingly, migration is from city to city. It is true, however, that in the 1960s and 1970s the worldwide urbanisation process was already in motion and its present advanced stage could therefore be extrapolated. Lefebvre was also right in predicting, contrary to the Marxist orthodoxy of the time, a decline in the importance of production-based conflicts relative to those affecting other spheres of society, especially the urban sphere. The present reality in global North societies indeed points to an apparent demobilisation of production-based forces, while agitation around issues of social equity and identity, along with space-based conflicts, are gaining momentum. It is, however, important not to associate a lesser visibility of conflicts with reduced tensions in the production sphere. It may well be that diminished worker reaction to increasingly precarious employment conditions in the wake of globalisation, de-regulation and automation are a consequence of growing obstacles to mobilisation within the production sphere.

Seemingly aligned with the thinking of Lefebvre is the present celebration of urban life in general and more specifically of urban amenities and hedonistic lifestyles. On the surface, these transformations can indeed be interpreted as compatible with the Lefebvrian utopian vision of the city as an agent of human fulfilment, the vision at the heart of the right to the city concept. The proliferation of urban festivals can be interpreted as the expression of a hedonistic urban turn. The same can be said of the change in the vocation of downtowns of large North American metropolitan regions. Traditionally, these sectors were confined to a strict central business district role, consisting for the most part of office employment and retailing. These were nine-to-five downtowns. Over the last decades, however, downtowns have undergone a major transformation by attracting recreational and cultural activities. Downtowns are now populated around the clock by a much-expanded residential population along with masses of visitors including tourists. Such a transformation gives rise to a downtown lifestyle emphasising social interaction, recreation, diversity, culture, the arts and festive events (Filion and Gad 2006). Thus depicted, downtowns could be seen as corresponding to the centrality imagined by Lefebvre and therefore to the human fulfilment features of the vision of the city he put forth.

This form of downtown has become the symbol of the adaptation of the city to contemporary self-fulfilment values. If smokestacks represented the industrial city and high-rise office towers, the technocratic city, cafés, art galleries, running and cycling trails and condo towers symbolise a consumerist city promoting self-realisation. This is the new people-oriented depiction of the city.

Although these images of the contemporary downtown seem to mirror the utopian interconnection Lefebvre proposed between the total person and human fulfilment, on the one hand, and a form of centrality promoting social interaction, physical well-being, arts, culture and social interaction, on the other hand, the contemporary downtown reality (and indeed the urban reality) is different. For Lefebvre, an actual transformation of the city that would make it more respectful of human needs and wants (while sustaining the fulfilment of individuals) cannot take place without a deep societal transition. Otherwise, in his view, the city remains an instrument reproducing the inequality and alienation embedded in the capitalist and state modes of production. Not only has such a societal transformation not happened, but in global North industrial countries exploitation and polarisation have intensified with the advances of neoliberalism.

Even if it plays for some a human fulfilment role of sorts, the contemporary transformation of urban reality is deeply ingrained in the economic tendencies of the current neoliberal phase and contributes to the entrenchment of its defining features. The connection between the reorientation of large-city downtowns in North America and neoliberal economic features reveals the thinness of the utopian veneer of this reorientation. First, the transformation of these sectors has been economically motivated, not driven by the vision of places that allow persons to become more fulfilled. In the Lefebvre terminology, it reflects the predominance of exchange over use value. Facilities and activities targeted at downtown residents and visitors are either meant to generate profits or improve the image of an urban area and, thereby, its economic competitiveness. The economic development purpose of an apparent reorientation of urban sectors towards human fulfilment has been most blatantly conveyed by the creative class perspective. For Richard Florida (2002; 2007), cities must create an urban environment that corresponds to the tastes of young professionals belonging to the creative class in order to attract them. In this view, the interest of cities in this social category relates to its capacity to generate economic growth. We encounter here an approach that shares features of the urban vision advanced by Lefebvre, insofar as the means to attract the creative class involve art, culture, physical activity and an overall festive atmosphere. But, contrary to the vision of Lefebvre, these urban amenities are motivated by an economic development objective and targeted narrowly at the social category with the potential to achieve this development.

Second, even if the centrality that has emerged in North American downtowns did fully conform to the Lefebvrian understanding of the term, it still could be easily accessible only to a minority because of the existence of only one downtown per metropolitan region. Centrality has indeed proven to be notoriously difficult to reproduce, as illustrated by the absence of multifunctional centres in the North American suburban realm. As we well know, in a market system scarce goods go to the highest bidders. Hence the tendency for the rich to monopolise the advantages of centrality as evidenced by ever-advancing gentrification. Thus, in a capitalist society, and one could equally argue that in a technocratic state regime where the distribution of rare goods and services mirrors political influence, centrality is a factor of social division in blatant opposition with the right to the city upheld by Lefebvre.

Conclusion

The chapter has identified reasons for the enduring interest in the urban writing of Lefebvre. There is first the relatively recent nature of the English translation of his books. Enduring interest is also a function of the profusion of ideas present in his writing and of the possibility of interpreting them in different fashions due to their frequent embryonic state. One can also raise the fact that Lefebvre was able to foresee some contemporary societal and urban outcomes. And, finally, we have noted the relevance to the present urban reality of the right to the city theme, focusing on the connection between human fulfilment and centrality.

In its last section, the chapter has demonstrated the critical potential relative to the current urban reality of themes emerging from the writing of Lefebvre. Themes raised in his work indeed make it possible to see beyond the ideology serving to justify prevailing urban trends and policies. The enduring impact of the urban books of Lefebvre has benefitted from their concentration on big fundamental urban ideas rather than on the specific circumstances prevailing at the time of their writing.

References

Ajzenberg, A. (2001) Henri Lefebvre et l'idée communiste, *L'Humanité* (4 December) www.humanite.fr/node/256435, accessed February 2017.

Bitter, S., Derksen, J. and Weber, H. (eds) (2009) *Autogestion or Henri Lefebvre in New Belgrade*, Vancouver, BC: Fillip.
Brenner, N. (2008) 'Henri Lefebvre's critique of state productivism', in K. Goonewardena, S. Kipfer, R. Milgrom and C. Schmid (eds) *Space, Difference, Everyday Life: Reading Henri Lefebvre*, New York: Routledge, 231–249.
Butler, C. (2012) *Henri Lefebvre: Spatial Politics, Everyday Life, and the Right to the City*, New York: Routledge.
Castell, M. (1997) 'Citizen movements, information and analysis: an interview with Manuel Castells', *City*, 7: 140–155.
Elden, S. (2001) 'Politics, philosophy, geography: Henri Lefebvre in Anglo-American scholarship', *Antipode*, 33: 809–825.
Elden, S. (2004a) 'Quelques-uns naissent d'une façon posthume: La survie de Henri Lefebvre', *Actuel Marx*, 35: 181–198.
Elden, S. (2004b) *Understanding Henri Lefebvre: Theory and the Possible*, London: Continuum.
Filion, P. and Gad, G. (2006) 'Urban and suburban downtowns: trajectories of growth and decline', in T. Bunting and P. Filion (eds) *Canadian Cities in Transition: Local through Global Perspectives*, 3rd edn, Toronto: Oxford University Press, 171–191.
Florida, R. (2002) *The Rise of the Creative Class and How It's Transforming Work, Leisure and Everyday Life*, New York: Basic Books.
Florida, R. (2007) *The Flight of the Creative Class: The New Global Competition for Talent*, New York: HarperCollins.
Fourastié, J. (1971) *Les trente glorieuses, ou la révolution invisible de 1946 à 1975*, Paris: Fayard.
Graham, S. and Marvin, S. (2001) *Splintering Urbanism: Networked Infrastructures, Technological Mobilities and the Urban Condition*, London: Routledge.
Hess, R. (1988) *Henri Lefebvre et l'aventure du siècle*, Paris: A. M. Métaillé.
Kipfer, S., Goonewardena, K., Schmid, C. and Milgrom, R. (2008) 'On the production of Henri Lefebvre', in K. Goonewardena, S. Kipfer, R. Milgrom and C. Schmid (eds) *Space, Difference, Everyday Life: Reading Henri Lefebvre*, New York: Routledge, 1–24.
Lefebvre, H. (1961) 'Utopie expérimentale: Pour un nouvel urbanisme', *Revue française de sociologie*, 2–3: 191–198.
Lefebvre, H. (1968) *Le droit à la ville, suivi de Espace et politique*, Paris: Anthropos.
Lefebvre, H. (1971) *L'idéologie structuraliste*, Paris: Anthropos.
Lefebvre, H. (1992) *The Production of Space*, London: Wiley [English version of Lefebvre, H. (1974) *La production de l'espace*, Paris: Anthropos].
Lefebvre, H. (1998) *Mai 1968, l'irruption?* Paris: Éditions Syllepse.
Lefebvre, H. (2003) *The Urban Revolution*, Minneapolis, MN: University of Minnesota Press [English version of Lefebvre, H. (1971) *La révolution urbaine*, Paris: Gallimard].
Lefebvre, H. (2004) *Rhythmanalysis: Space, Time and Everyday Life*, London: Continuum [English version of Lefebvre, H. (1992) *Éléments de rhythmanalyse*, Paris: Éditions Syllepse].
Lefebvre, H., Kofman, E. and Lebas, E. (1996) *Writings on Cities*, London: Wiley.
Merrifield, A. (2006) *Henri Lefebvre: A Critical Introduction*, London: Routledge.
Newman, L. and Wyly, E.K. (2006) 'The right to stay put, revisited: gentrification and resistance to displacement in New York City', *Urban Studies*, 43: 23–57.
Shields, R. (1999) *Love and Struggle – Spatial Dialectics*, London: Routledge.
Soja, E.W. (1980) 'The socio-spatial dialectic', *Annals of the Association of American Geographers*, 70: 207–225.
Weinstein, L. and Ren, X. (2009) 'The changing right to the city: urban renewal and housing rights in globalizing Shanghai and Mumbai', *City and Community*, 8: 407–432.

6

Neo-liberalism, extraction and displacement

Abstract space and urbanism in India's 'tribal' belt

Michael Spacek

Introduction

This chapter draws on the work of Henri Lefebvre for an analysis of the changes currently unfolding across eastern and central India. The region's hills and forests, largely populated by a variety of indigenous groups collectively known in South Asia as Adivasi, are rich in natural resources such as coal, iron ore and bauxite. Following a balance-of-payments crisis in 1992, the Indian state began a period of liberalisation which led to a growth in demand for these resources. Subsequently, there has been a massive influx of capital into the region, accelerating processes of population displacement. Two sites make visible emergent processes of neo-liberal accumulation by dispossession. Naya Raipur is the new capital of the eastern Indian state of Chhattisgarh. Conceived as a 'smart city', it is a space designed to house state institutions, corporate offices and the emergent middle classes of the region. Belgaria, on the other hand, is a purpose-built housing colony in the neighbouring state of Jharkhand. Built as a settlement for the relocation of displaced villagers from the nearby Jharia coal mines, it functions as a container for the surplus population created by neo-liberalism.

Naya Raipur, Chhattisgarh

Chhattisgarh was formed in 2000 out of 16 eastern districts of the state of Madhya Pradesh. At approximately 31 per cent, the state has one of the highest percentages of indigenous people in India. The Adivasi are concentrated in the south and north of the state, areas which also contain the majority of the state's large reserves of natural resources. The economically and politically dominant central plains are populated largely by Hindi/Chhattisghari speakers who began settling in the area following the onset of British colonialism. While Chhattisgarh has a rapidly growing economy due to the boom in natural resource extraction and processing, most of its population continues to be impoverished, reflected in an HDI ranking placing it at the very bottom of all Indian states (United Nations Development Program 2011). While overall socio-economic indicators are low, extreme poverty and illiteracy are particularly acute among the Adivasi.

The newly created state government moved to quickly liberalise its mining license policies in order to attract private capital. Cities such as Raipur, the old state capital and largest city, have become booming frontier towns replete with new airports, roads and exclusive communities for the emergent elites of the state. These elite spaces exist alongside, and in relation with, the impoverishment of the actual extractive zones populated by Adivasi.

It was during the colonial period that the contemporary contours of Chhattisgarh's political economy were first established. While the region remained isolated from the colonial heartlands of Madras and Bengal, Chhattisgarh was being created as a zone of settlement and resource extraction, primarily timber. Large areas of land were reserved as state forests, effectively preventing the local population from practising traditional activities such as swiddening and the harvesting of forest resources (Véron and Fehr 2011).

The enclosure of forest land and the transformation of space which began in the mid-19th century had significant ancillary effects on the region. New opportunities arose with the commercialisation of forests and forest products. The result was a sudden influx of 'outsiders' from the plains who were, in part, able to capture the newly emergent positions in the political economy as a result of their knowledge of colonial law and their ability to navigate through the new political and administrative dispensation. The emergence of an elite group of outsiders occurred simultaneously with an assault on traditional economic and political power, proving disastrous for the Adivasi. It was the beginning of a long period of sedentarisation and spatial transformation in the region that continues to unfold to this day.

With independence, India pursued national 'modernisation' along the lines of the developmental statism that was prevalent in the post-colonial period. The Nehruvian state's drive for industrialisation and massive public infrastructure projects, seeking to increase India's economic self-sufficiency, often had detrimental consequences for the Adivasi. While the British had occasionally resorted to population relocation and corvée labour, the Nehruvian state undertook spatial transformations rooted in a regime of mass displacement. The state began large-scale iron ore extraction and processing, such as the Soviet-funded Bailadila project in southern Chhattisgarh, whose purpose was to industrialise and, more broadly, 'develop' the region, reflecting what Henri Lefebvre theorised as the State Mode of Production, in which the state sought to control both markets and space (Lefebvre 2009a: 106–112). Ultimately, the Bailadila project and other large-scale industrialisation projects created displacement and environmental destruction, benefiting only a few.

Following India's liberalisation, the most significant contemporary transformations of the political economy and spatiality in Chhattisgarh are connected to resource access and the dispensation of land by the state. Under the Nehruvian development state, the central government exercised tight control over extraction. This began to change after 1993. By the 21st century the central government had delegated the power to grant mining license powers to individual state governments, lifted foreign equity limits and removed controls on 13 minerals whose ownership and right to extraction had previously been exclusively reserved for public sector companies (Government of Chhattisgarh 2001: 4–5). Consequently, in the previous decade the number of licenses issued has drastically increased as private capital has flowed into the region (Bijoy 2008: 1772). The transition to neo-liberalism and the emergence of a new state logic is reflected in changes to space in Chhattisgarh.

The initial impetus for the building of Naya Raipur was the formation of Chhattisgarh in 2000. Unlike the chaotic and rapidly growing city of Raipur, the new capital, located approximately 20 km distant, was conceived as a highly planned, liveable city integrated with global networks of trade and circulation. Notably, Naya Raipur is integrated with the newly upgraded Swami Vivekananda Airport, whose passenger terminal was modernised in 2012 and which inaugurated Chhattisgarh's first cargo terminal in 2016 (Drolia 2016). The airport and the planning of the city reflect both the increased economic significance of the region to foreign and domestic capital as well as local elite and state aspirations to become a significant force in global networks of exchange. State marketing emphasises its business friendliness, which is also a tourist destination for those interested in exploring the region's 'natural' beauty and aboriginal culture (Chhattisgarh Tourism Board 2017). State/elite aspirations and neo-liberal political economy have merged to create an abstract space that aspires to become a local node of accumulation within a globalised network of capital (Lefebvre 1991: 33). The state has built symbolic spaces that have destroyed the space of what existed before. These are the airports and the state infrastructures, the spaces which Lefebvre refers to as, 'Dominant spaces ruling over dominated spaces (for example, through the planning of airspace, airlines, runways, etc.)' (Lefebvre 2009b: 246). Naya Raipur is conceived of a city which exists spatially separate from, but proximate to, India's extractive periphery.

The materialisation of Naya Raipur represents neither a complete rupture nor complete continuity from earlier urban forms. Independent India has a history of creating state-planned cities embodying modernist aspirations. Foremost among these is Chandigarh, the capital of the states of Haryana and Punjab. Built in 1966, the city was conceived and planned by a team of architects led by Le Corbusier, whose responsibility was to realise Nehru's vision of an architecture that symbolised India's emergence as a modern, egalitarian, democratic socialist state and society (Fitting 2002). Like the other planned cities of the country, the conception, development and management of the city are led by the Naya Raipur Development Authority (NRDA), a government body comprised of senior bureaucrats and government officials (Naya Raipur Development Authority 2017). Consequently, Naya Raipur bears some resemblance to planned cities built under the state-led developmentalism of the Nehruvian period. Thus, as a space which is conceived and materialised by the state, as both the earlier planned cities and Naya Raipur are, 'broken down into separate spaces, occupied by functions that are exercised within these distinct spaces: labor, housing, leisure, transit and transportation, production, consumption' (Lefebvre 2009c: 214). Both have little resemblance to the chaos of the largely unplanned, organic cities of India. Both Naya Raipur and Chandigarh are rigidly divided into 'sectors' which separate and demarcate the spaces designed for differing moments and functions of everyday life. Furthermore, like Chandigarh before it, Naya Raipur reflects Lefebvre's observation that: 'Wherever the state abolishes chaos, it establishes itself within spaces made fascinating by their social emptiness: a highway interchange or an airport runway, for example, both of which are places of transit and only of transit' (Lefebvre 2009b: 238). Images of hypermodern road networks and vast spaces of regimented and planned orderly emptiness permeate the promotional material of the NRDA. It is the imaginary of an urban space held together by a state space stripped of the everyday and life itself.

There are, however, significant differences between Naya Raipur and the earlier state-planned cities. Naya Raipur is not imagined as the spatialisation of a nationalist state-making project that seeks to materialise a liberatory, post-colonial state. Rather, Naya Raipur is intended to function as a regional industrial, financial and political hub. According to the NRDA's Mission Statement, Naya Raipur will 'Not only serve as Administrative Capital but also cater the infrastructural needs of industries in the region. Naya Raipur would not only encapsulate modern infrastructure but also push the envelope of eco-friendly city' (Naya Raipur Development Authority 2017). The new city is imagined as an emergent node in a global network of exchange, a space that mediates between extractive sites and domestic and international markets. Its function is that of a key regional node of capital and an integral part of abstract space in a peripheral region (Lefebvre 2009d: 187). Naya Raipur is a place which is geographically proximate to the industrial sites of the region but exists apart from them in a virtual nether space. This particular form of urban spatial planning and its existence as both an imaginary of state ambition and materialisation as a fragmented space of global capitalist exchange is reflected in its spatial production. It is a fragmented and functionally separated space, a post-modern new town where, like Lefebvre's modernist Mourenx (see Leary-Owhin in Part 6):

> Every object has its use and declares it. Every object has a distinct and specific function. In the best diagnosis, when the new town has been successfully completed, everything in it will be functional, and every object in it will have a specific function: its own.... In a sense, the place is already nothing but traffic lights: do this, don't do that.
>
> *(Lefebvre 2011: 119)*

Earlier planned cities in India reflected a nationalist, post-colonial ethos and assertion of statehood. These cities were wholly conceived, planned and built using the resources of the state and functioned almost exclusively as political centres. Naya Raipur reflects a different logic. The NRDA plan envisions its infrastructure as built and managed by private capital. Public-private partnerships are central not only to its financing of the city, but are fundamental to its logic as a city nested in global networks of exchange and as the primary nodal point in the mining regions of eastern and central India. Large areas have been zoned as industrial and commercial spaces, designated as Special Economic Zones where normal rules of taxation and regulation do not apply (Naya Raipur Development Authority 2017). Naya Raipur is imagined as more than a nationalist state space. It is a space of global capitalism and a state which functions as its handmaiden.

Furthermore, the conceived space of Naya Raipur reflects the postmodern pretensions of the 'smart city': techno-utopia that has selectively culled elements of new urbanism and fused them into a vision of a globalised, bourgeois capitalist city freed from the tyranny of demography and politics. In 2015, the Government of India announced the Smart Cities Mission (SCM), which provides substantial infrastructure, primarily for transport and surveillance/security, to those cities. The SCM reflects changes in the political economy of India and an attempt to spatially transform urban regions of the country, developing them as nodal points in global networks of exchange. As Söderström et al. argue, the 'smart city' has at:

> Its core... utopian storytelling. First, the smarter cities story is a univocal one: nowhere in the campaign are other approaches or solutions to urban problems mentioned. Second, the smarter cities campaign hinges on a before–after demonstration closely related to the above mentioned 'if (data-mining and systems thinking) then (cities will become smarter)' argument. The third aspect of a utopian rhetoric, the smarter cities story, depicts a model of a perfectly functioning urban society but, in contrast with classical utopianism, it is governed by code rather than spatial form.
>
> *(Söderström et al. 2014: 315)*

This contemporary capitalist utopianism claims to have designed a universal urban form able to solve the problems of the city. It is a conceived space whose individual components form a universally applicable system independent of context: 'The focus [of SCM]... is to create a replicable model which will act like a lighthouse to other aspiring cities' (Ministry of Urban Development 2017). Drawing on a mélange of new urbanist ideas and a corporate techno-utopian vision, the NRDA has planned an extensive network of bike lanes, a bus rapid transit system and has committed to ensuring that all public buildings will follow green building concepts which will 'use less water, optimize energy efficiency, conserve natural resources, generate less waste and have minimum impact on [the] environment' (Pande and Dua: March 2017). Naya Raipur also mimics earlier urban planning tenets which emerged in the late 19th and early 20th centuries. Spaces of leisure, culture and recreation have been produced to ensure that the residents of the city are able to access healthy living in the built environment. These spaces of leisure, recreation and nature include athletic facilities, parks and museums dedicated to Adivasi culture. It is, as Ola Söderström et al. write, a discourse that 'Promotes an informational and technocratic conception of urban management where data and software seem to suffice and where, as a consequence, knowledge, interpretation and specific thematic expertise appear as superfluous. This is a rather dangerous fiction' (Söderström et al. 2014: 317).

A key element of the 'smart city' *Weltanschauung* is the belief that information technology will eliminate urban infrastructural and social problems. According to the NRDA, Naya Raipur's design:

> Comprises IT enabled land management system, city surveillance… intelligent lighting systems, pay and use parking systems… intelligent transport system, city level wi-fi touch screens across the city, display boards across the city for providing real-time information, emergency alert and crisis response systems, traffic re-routing applications based on real time traffic data.
>
> (The Pioneer *27 May 2016)*

It is an urban space where the myriad infrastructural and 'social' failures of urban India can be resolved through the use of a technological 'fix' built and managed by an alliance of the state and technological firms. It is the spatialisation of a managerial and neo-liberalisation which envisions an urban outpost of global capital located in a peripheral frontier zone of extraction.

Naya Raipur is both an aspirational and concrete space which reflects and materialises changes in the political economy of the region. This political economy is rooted in a regime of accumulation by dispossession and the production of the region as an extractive periphery and site for investment by highly mobile domestic and global capital. Naya Raipur is a nodal space which blends both state-led development and private capital, creating an abstract space which materialises a regime of neoliberal accumulation by dispossession in an extractive periphery of global capital and the state.

Belgaria, Jharkhand

Carved out of 18 southern districts of Bihar in 2000, Jharkhand shares geological, geographic and demographic characteristics with neighbouring Chhattisgarh. It is a hilly state with significant forest cover also rich in natural resources. The state has extensive deposits of iron ore, copper, coal and uranium and produces more mineral wealth than any other state in India. Like neighbouring Chhattisgarh, the state has a large Adivasi population which stands at around 30 per cent of its total population.

Jharkhand, however, has a longer history of engagement with and integration into regional political and economic systems than does Chhattisgarh. Large-scale extractive industries were first established by the British in the late 19th century (Corbridge 1987: 254). Expansion into the region was conditioned by Britain's desire to control access to its coal and iron ore deposits. In Chhattisgarh the British ruled through a Princely state that mediated between the state and the local populations. In Jharkhand, the British established direct rule, placing the region under the administrative purview of the Bengal Presidency (Ekka 2011: 44–45). In the face of stiff resistance to land enclosure and taxation, coupled with the region's strategic importance as a source of resources for Empire, the colonial state was compelled to make compromises with the local populace. These compromises created special protected zones in Adivasi areas through legislation, such as the Chotanagpur Tenancy Act, which prevented the sale of Adivasi-held land to non-Adivasi and entrenched customary rule (Raza 2015: 16–23).

In the long term, the British pursued policies of accommodation that sought to create conditions securing their access to Jharkhand's natural resources without the prohibitive costs of permanently stationing large numbers of military and police forces. Subsequently, these political accommodations led to the creation of spaces which preserved customary rule and 'traditional' land ownership patterns.

The other significant spatial transformation that emerged in Jharkhand during the latter half of the 19th century was the emergence of parallel spaces of modernity existing alongside the zones of exception. These were spaces of industrial and state concentration, connected via transport corridors and railways to the great colonial cities and ports of the Bengal and Madras Presidencies.

Perhaps the best example of these new spaces of modernity is Jamshedpur, a centre of iron and steel production. With a population of over a million, it is also the world's largest company town:

> Governed by the Town Division of Tisco [Tata Iron and Steel Company].... At its head is a manager of town services, under him is a uniformed security force of 1,000 men, an Intelligence Wing to report on labor disorders and political activities, and a Vigilance Wing that investigates corruption, primarily in the steel mill itself. In 1924, the state government set up what is called a 'Notified Area Committee,' a form of government designed for small towns not yet capable of self-support. In Jamshedpur this committee is made up of representatives of the various large manufacturing companies. The committee is dominated by the Tisco representatives and has the authorization to collect taxes and fees of various kinds from the companies and individuals of the city.
>
> *(Kling 1998: 81–82)*

Jamshedpur was built using 'modern town planning principles, ushering in modernity through new modes of spatiality and lifestyles associated with industrialization' (Sinha and Singh 2011: 1). In 1919 construction began on the city and the steel plant. What emerged was a space wholly unlike anything that had been seen in Jharkhand: an angular, rigidly demarcated and wholly planned space in which functional separation existed between the various parts of the whole. Jamshedpur reflected and materialised a segmented and routinised mode of being composed of the house, the park, the road, the shops and the workplace. It was an ideal space of planned capitalist modernity, a place in which clock time and everyday life was spatially segmented and made concrete in the urban form, all of which was planned, built and controlled by Tisco (Kling 1998: 69–70).

Belgaria is an immanent spatial form of contemporary neo-liberal capitalism in Jharkhand. Much of the country's coal lies on land which is 'protected' by legislation such as the CNT. During the period of Nehruvian developmentalism, the large state firms were able to plausibly argue that dispossession under the principles of eminent domain was necessary for nation-building. This is less true in the period of neo-liberalism where much of the extraction is being done by private domestic and foreign firms. Eminent domain has become a less useful device than it was in the past.

In the decades following independence, one of the core pillars of state practice was Nehruvian developmentalism. Inspired in part by the Soviet state's industrialisation in the 1930s and the import-substitution industrialisation (ISI) widely adopted by numerous countries from the 1950s to the 1970s as a means of rapidly overcoming the legacies of colonialism, poverty and economic backwardness, ISI was a global political project that sought to create a self-generated modernity in the post-colonial world. Industrialisation as both practice and ideology required the extraction and processing of the natural resources found in abundance in Jharkhand (Corbridge 2000: 67). During the period of Nehruvian developmentalism, the nascent industrialisation and extraction which began under colonialism in the region accelerated. Large public sector firms established mines and factories in Jharkhand. In particular, the Second Five Year plan (1956–1961) sought to rapidly industrialise India through the establishment and capitalisation of public sector firms:

> In the late 1950s the Government of Bihar, in competition with the Government of West Bengal, persuaded New Delhi and the sponsoring Governments of Czechoslovakia and the Soviet Union to site a cluster of publicly-owned heavy engineering industries in Chota Nagpur.
>
> *(Corbridge 1987: 254)*

The production of industrial space in Jharkhand, which began in the late 19th century, significantly expanded in scope and scale under the Nehruvian developmental state (Corbridge 1987: 264). This had two primary effects on space in the region. The first was that it deepened and accelerated longer-term historical patterns of in-migration to the state as most firms were reluctant to hire local labour which they saw as less skilled and malleable. Thus, large numbers of workers were imported from outside of the state with only the most menial jobs reserved for local Adivasi populations.

Second, increased production and the expansion of industrial infrastructure, coupled with growing demand for resources, led to the growth of industrial spaces and urban centres which were established in the late 19th century. The result was a rapid increase in population and land pressures on both protected and non-protected Adivasi communities. The dualistic space established by the British began to break down as the contradictions inherent in the practices of statecraft seen in Jharkhand further deepened. Specifically, while the CNT and the fifth Schedule of the Constitution sought to prevent the alienation of Adivasi from their land, provisions for seizure existed under the principle of eminent domain. This principle was widely used for the intensification of industrial activity by state mining and processing companies and the construction of dams and other large infrastructural projects for the 'modernisation' of the country (Meher 2009: 469).

The Jharia fields, established in the late 19th century, are the largest coal fields in India. For over a hundred years, a number of deep level coal fires have been burning under the earth, creating a space of catastrophe in which a logic of violence and destruction is embedded into the production of space (Lefebvre 2009b: 246). If Jamshedpur is an example of a particular form of urban space emergent from modernist industrialisation, the contemporary spatial changes occurring in the region represent a form of neo-liberal accumulation by dispossession, creating a large surplus population. It is a vivid example of a place where 'The substitution of one space, by another including the destruction of antecedent spaces by subsequent spaces' (Lefebvre 2009b: 248). Extraction has literally destroyed and made permanently uninhabitable at least 11 villages and affected over four million people (Nirala: 16 May 2017). There are allegations that British Coking and Coal Limited (BCCL), the state firm which owns the Jharia fields, and other coal companies in the region have done little to extinguish the fires, as destruction and the subsequent depopulation is in their interest (Al Jazzera 2014). Much of the land in the region falls under the rubric of 'protected'. Specifically, protected land in the region cannot be transferred to anyone who is non-Adivasi and whose family has not historically inhabited the village. Allowing the fires to spread (thereby making more of the land uninhabitable) clears local villages, allowing for the expansion of mines. This would help realise the current government's plan of doubling coal production in the country by 2020 and enable BCCL to move from underground mining to opencast mining, a practice which is far more profitable (Daball: 1 February 2017).

Thus, the processes of displacement in Jharia through the production of spaces of catastrophe (making villages uninhabitable) have created a large surplus population. Furthermore, unlike during the period of Nehruvian developmentalism or the early 20th century, when state and large private firms felt a degree of paternalistic responsibility to hire local populations for at least menial positions, there is little or no employment for the displaced. The solution to the problem of what to do with surplus populations has been the creation of resettlement colonies. Under the direction of the central government, BCCL has developed the Jharia Action Plan with the intention of relocating a total of 70,000 families to the new town of Belgaria, eight miles from the coal fields in the middle of the jungle.

BCCL claims that resettlement will provide a new township with modern amenities and housing for the displaced, improving all facets of life However, Belgaria has provided little more than tiny flats for large families. Electricity is intermittent and BCCL has built little social infrastructure beyond a primary school – there are no hospitals, markets or even functioning sewer systems. Consequently, health and sanitation in Belgaria is poor, resulting in widespread endemic tuberculosis (Munshi: 16 November 2012).

In addition to the lack of infrastructure and social services for the residents, no tangible attempt has been made to create employment or facilitate productive activities. Unemployment is rife and the few who are employed need to make a long (and costly) daily trek to Jharia, about an hour distant by rickshaw (Munshi: 16 November 2012). Unsurprisingly, alcoholism and suicide are rife. It is a population with nothing to do and that has been provided only with the bare minimum necessary to sustain life.

Naya Raipur exists in symmetry with Belgaria, created by and hiding the ugliness of neo-liberalism and displacement. Both are materialisations of a form of neo-liberal, globalised capitalism structured by accumulation by dispossession. On the one hand, Naya Raipur exists as an urban form that aspires to be a central node in a globalised space of exchange. It mimics other spaces of capital across the globe through its pretensions as a 'smart city'. The materialisation of this is a regimented and fragmented space, one which provides its residents with security and moments of spatially segmented leisure and work. It is a utopian new city structured by the logic of neo-liberal capitalism and post-modernity. Belgaria is Naya Raipur's twin. It is a space stripped of everything but the bare essentials of habitation and life. In this way it is a non-space, a space of capitalism that exists outside of capitalism. Its function is to contain and provide minimal life for surplus populations displaced by spatial transformations currently unfolding across the region. It is a space of mimicry, containing the form of a modern urbanity, without any of its concomitant functions – a space of emptiness that exists as a mirror image of the 'smart' aspirations of Naya Raipur. Both spaces are the new spaces of neo-liberal capitalism in eastern and central India.

Conclusion

Naya Raipur and Belgaria are used as analytical devices that are evocative of the production of space across eastern and central India. Specifically, the region has been historically produced as a peripheral space in the process of being integrated with the broader Indian state and global political economy. They are, as Lefebvre states, the edges of space which are in the process of being absorbed (Lefebvre 1991). The process of this absorption is violent as the state is seeking to create homogeneity out of difference (Lefebvre 2009e). The creation of spaces such as Naya Raipur is rooted on dispossession and displacement. Naya Raipur is imagined as a space of power that not only contains the institutional space of state power but also the spaces of global capital. It is rooted in an angularity which, in spite of itself, mimics the nature of the new towns of modernity with its rigid demarcation of functional space (Lefebvre 2003). Furthermore, just as it creates a desacralised time, the irony is that Belgaria follows this form; however, it functions as little more than a mimicry device. It is a settlement for a surplus population that has nothing 'to do' which needs 'doing'. In Belgaria, the form of doing is followed, but there is nothing that needs to be done (Lefebvre 2004). This regime holds within itself the possibility for transformation and a potential for creating what Lefebvre refers to as 'differential space' (Lefebvre 1991: 52). Just as:

> In this same space there are, however, other forces on the boil, because the rationality of the state, of its techniques, plans and programmes provoke opposition. The violence of power is answered by the violence of subversion. With its wars and revolutions, defeats and victories, confrontation and turbulence.
>
> *(Lefebvre 1991: 23)*

The state, in eastern India, is confronted by numerous challenges: from the Maoist insurgency in the isolated forests of the region to popular mobilisations against displacement. The contradictions of an emerging neoliberal regime of accumulation by dispossession have created their own responses. What will emerge in the end is likely something which has yet to be imagined.

References

Al Jazeera (2014) Living with fire in India's Jharia, accessed online at http://www.aljazeera.com, accessed May 2017.

Bijoy, C.R. (2008) 'Forest rights struggle: the Adivasis now await a settlement', *American Behavioral Scientist*, 51:12 1755–1773.

Chhattisgarh Tourism Board (2017) accessed online at http://cgtourism.choice.gov.in/view/about.php, accessed May 2017.

Corbridge, S. (1987) 'Industrialisation, internal colonialism and ethnoregionlism; the Jharkhand, India, 1880–1980', *Journal of Historical Geography*, 13:3 249–266.

Corbridge, S. (2000) 'Competing inequalities: the scheduled tribes and the reservation system in India's Jharkhand', *The Journal of Asian Studies*, 59:1 62–85.

Daball, M.C. (2017) A city aflame: India's coal rush, accessed online at https://www.opendemocracy.net, accessed May 2017.

Drolia, R. (4 June 2016) One umbrella cargo terminal for Raipur airport, accessed online at http://timesofindia.indiatimes.com, accessed May 2017.

Ekka, A. (2011) *Status of Adivasis/Indigenous Peoples Land Series 4*, Jharkhand, Delhi: Aakar.

Fitting, P. (2002) 'Urban planning/utopian dreaming: Le Corbusier's Chandigarh today', *Utopian Studies*, 31:1 69–93.

Government of Chhattisgarh (2001) Mineral policy, accessed online at http://www.chips.gov.in, accessed May 2017.

Kling, B.B. (1998) 'Paternalism in Indian labor: the Tata Iron and Steel Company of Jamshedpur', *International Labor and Working-Class History*, 53 69–87.

Lefebvre, H. (1991) *The Production of Space*, Oxford: Blackwell.

Lefebvre, H. (2003) *The Urban Revolution*, Minneapolis: Minnesota University Press.

Lefebvre, H. (2004) *Rhythmanalysis: Space, Time and Everyday Life*, London: Bloomsbury.

Lefebvre, H. (2009a) 'The state in the modern world', in N. Brenner and S. Elden (eds) *State, Space, World: Selected Essays*, Minneapolis: Minnesota University Press.
Lefebvre, H. (2009b) 'Space and state', in N. Brenner and S. Elden (eds) *State, Space, World: Selected Essays*, Minneapolis: Minnesota University Press.
Lefebvre, H. (2009c) 'Space and mode of production', in N. Brenner and S. Elden (eds) *State, Space, World: Selected Essays*, Minneapolis: Minnesota University Press.
Lefebvre, H. (2009d) 'Space', in N. Brenner and S. Elden (eds) *State, Space, World: Selected Essays*, Minneapolis: Minnesota University Press.
Lefebvre, H. (2009e) 'The worldwide and the planetary', in N. Brenner and S. Elden (eds) *State, Space, World: Selected Essays*, Minneapolis: Minnesota University Press.
Lefebvre, H. (2011) *Introduction to Modernity*, London;Verso.
Meher, R. (2009) 'Globalization, displacement and the livelihood issues of tribal and agriculture dependent poor people: the case of mineral-based industries in India', *Journal of Developing Societies*, 25:4 457–480.
Ministry of Urban Development, Government of India (2017) What is smart city?, accessed online at http://smartcities.gov.in, accessed May 2017.
Munshi, N. (2012) Burning issues, accessed online at https://www.ft.com, accessed May 2017.
Naya Raipur Development Authority (2017) accessed online at http://www.nayaraipur.gov.in, accessed May 2017.
Nirala (2017) Jharia's infinite inferno, accessed online at http://www.tehelka.com, accessed May 2017.
Pande, S. and Dua, M. (2017) How Naya Raipur is emerging as one of the most well-planned smart cities in India, accessed online at http://www.thebetterindia.com, accessed May 2017.
Raza, A. (2015) *Chotanagpur Tenancy Act: A Handbook on Tenancy Law in Jharkhand*, New Delhi: Socio-Legal Information Centre.
Sinha, A. and Singh, J. (2011) 'Jamshedpur: planning an ideal steel city in India', *Journal of Planning History*, 10:4 1–19.
Söderstöm, O., Paasche, T. and Klauser, F. (2014) 'Smart cities as corporate storytelling', *City*, 18:3 307–320.
The Pioneer (2016) Naya Raipur can expect funding bonanza in 'Smart City' project, accessed online at http://www.dailypioneer.com, accessed May 2017.
United Nations Development Program (2011) Chhattisgarh economic and human development indicators, accessed online at http://www.in.undp.org, accessed May 2017.
Véron, R. and Fehr, G. (2011) 'State power and protected areas: dynamics and contradictions of forest protection in Madhya Pradesh, India', *Political Geography*, 30 282–293.

7

Constructed otherness

Remaking space in American suburbia

Gregory Marinic

Introduction

In *The Production of Space*, Henri Lefebvre interrogates the impact of production, consumption, and multiplicity of authorship in the built environment (Lefebvre 1991). He asserts that cities, buildings, and interiors are hybrid productions achieved not only by architects and designers, but also through the cultural traditions, social practices, and autonomous interventions of users. By reframing the design of the built environment with the inverse – the quotidian impact of people making space – Lefebvre celebrates the commonplace and unschooled actions that cities, buildings, and interiors receive apart from the top-down hand of formally trained professionals. He fixes his gaze on the lives of buildings well beyond the moment of their completion.

For immigrants and refugees, establishing a rooted narrative in a new country often begins within appropriated buildings and urban forms. Interventions are primarily spatial and temporal gestures that collectively reveal enduring perceptions, practices, and methods of production. In recent years, substantial demographic shifts in the United States have transformed conventional shopping malls into more complex environments supporting emergent forms of urbanity. As such, adapted commercial environments demonstrate Lefebvre's notion of heterotopia, or liminal social spaces of *potentiality* serving the quotidian needs of communities. Viewed through post-structuralist theory, these reprogrammed structures have not 'failed' by losing their conventional national chain retailers and department stores. Rather, they have embraced more nuanced and modest ways to adapt architecture and shape cities.

Latent borderlands

The United States was established by migratory flows that channelled immigrants from the densely populated cities on the East Coast to the hinterlands of the West. Early overland connections to the American West were forged by primitive routes including the Santa Fe Trail (1821), Oregon Trail (1840), Pony Express (1860), and later, with technologically advanced transcontinental railroads after 1869. Throughout history, the nation has been defined by intersectional cultural influences – Native American, African, Spanish, French, British, Mexican,

Creole, Cajun, and Anglo-American – as well as shifting political boundaries that ignored people while uniquely privileging the frontier for Anglo-American and European settlement. Since the 1990s, however, the heartland of America has been increasingly influenced by trans-border migration from the Global South facilitated on interstates rather than railroads – and most significantly – Interstate 35 (I-35). Linking Laredo, Texas, at the border with Mexico to Duluth, Minnesota, on the edge of the Great Lakes, I-35 illustrates that these shifts are not merely demographic. As the United States absorbs immigrants from more diverse origins, its ageing buildings and infrastructures are being re-made to address emerging socio-cultural needs.

In the mid-20th century, the *Immigration and Nationality Act of 1965* initiated transformative socio-cultural influences on the architecture and urbanity of American suburbia. It abolished a racially, ethnically, and religiously exclusionary quota system dating from the mid-1920s and shifted the country toward greater cultural diversity. Pivoting away from Europe, the resulting multiculturalisation of American society has incrementally changed the ways that buildings are used and occupied, as well as how interior spaces are being adapted.

The enclosed shopping mall exists as one of the most widespread and large-scale of America's increasingly obsolete building typologies. American malls generally operate between two extremes – the upscale super-regional and the increasingly obsolete – while being challenged by the growth of newer forms of virtual and physical retail. On the one hand, this study reveals how ageing shopping malls support the socio-economic well-being of immigrants in American society. On the other, it opens a window on to the long-term resilience and economic viability of increasingly 'dead' retail environments that have historically catered to a shrinking white middle class. Today, upscale malls serve a racially and ethnically diverse, upwardly mobile and assimilated upper-middle class, while so-called 'dead' or dying malls and big box stores have come to serve a largely disadvantaged underclass.

In the changing suburban periphery of Texas cities, the impact of Latin American immigration on shopping mall adaptation and hybridisation has become increasingly evident. Appropriated by an assimilated lower class of senior citizens, ethnic minorities, and recent immigrants, these places have been largely rejected by the white middle class, yet offer an important socio-economic support system to their regions. Positioned at a critical pivot point between Anglo and Latin America, as well as a mixing zone for the Global North and Global South, the southern sections of the I-35 and I-45 interstate highway corridors illustrate how abandonment, migration, globalisation, and suburban obsolescence coalesce to enable new opportunities for city-making.

Reflecting on Henri Lefebvre's seminal work in *The Production of Space*, non-hegemonic spaces and placelessness frame a view on 'rights to the city' and spatial adaptations seeded within obsolescence (Lefebvre 1995: 179). This chapter applies a Lefebvrian perspective on spatial production to assert that cast-off dead and dying malls have forged an alternative path toward community vitality. Using field studies of the Texas segments of I-35 and I-45 as a geographical focus, this research examines obsolete retail environments in Laredo and Houston to mobilise Lefebvre's notion of spatial production and heterotopia. Evidence collected along these corridors forecasts 21st-century retail futures across North America and a new American Dream. Emerging conditions provoke the reconsideration of suburbia, while questioning what gentrification means in the transnational peripheries of mid-continental America. Characterised primarily by modifications to existing buildings, interior interventions facilitate low-cost and participatory actions supporting socio-economic resilience. In an unlikely turn, commercial obsolescence in American suburbia has allowed immigrants to organise, participate, and prosper. These transformational shifts have fostered the growth of a *new* periphery, a dynamic ersatz-urban place of remarkable demographic diversity.

From utopian origins

Much has been said about the dramatic collapse of American cities and the social cohesiveness that they fostered alongside the simultaneous rise of a homogeneous, trans-continental suburban culture. After the Second World War, urban life in the United States began to fracture and disperse along demographic lines to facilitate the rise of limitless consumer opportunities in suburbia (Baumgartner 1988). Suburbs witnessed the emergence of entirely new and autonomous forms of public-commercial space shaped by urban retail fragments floating within a boundless landscape. Conventionally *urban* commercial and social experiences were displaced to the suburbs and became internalised; communal space was rapidly translated into controllable, transactional, and primarily *interior* worlds. Through this process, the developmental patterns of shopping malls became synonymous with automobile dependency and suburban sprawl (Rees 2003: 93). For a generation of Americans raised in the post-war suburbs, expectations of civic space were incubated in the encapsulated worlds of shopping malls, and thus, orphaned from any tangible connection to downtown.

Suburbia has historically epitomised the American Dream, a vision of opportunity and prosperity expressed in greater autonomy, single-family home ownership, and car dependence (Beauregard 2006: 107–108). As Americans left urban neighbourhoods for the periphery, mid-20th century suburbia imparted profound racial fragmentation and influence on popular culture, mass media, consumer tastes, and housing. Since that time, the United States has continuously exported its suburban ideals to the rest of the world, while American suburbia has simultaneously become increasingly globalised. Within this peripheral 'utopia', the shopping mall and strip centre became uncontested icons of American culture – exemplars of social and commercial values which represent the United States to the world and Americans to themselves. Although suburban consumer expectations were initially forged in a throwaway culture of post-war largesse, rising suburban retail abandonment in the early 21st century rivals the demise of downtown commercial districts in post-WWII American cities.

Concurrent with these shifts, the Texas cities of Houston and Laredo have attempted to compete with suburban malls by co-opting suburban typologies for downtown. Although these appropriated forms have generally failed, central cities have re-emerged through widespread reassessment of their cultural venues, building typologies, authentic urban configurations, and civic life. This generational turn has placed suburbs and their utilitarian buildings into a state of increasing disinvestment and abandonment (Alba and Nee 2003). At the same time, changing demographics in the expanding metropolitan areas of Texas demonstrate that suburbia has become considerably more diverse and complex, as well as undeniably influenced by urban influences from the Global South. Undercurrents of racial, economic, and social segregation persist, yet suburbia is being appropriated and hybridised into a place of opportunity for immigrants. Rejected by national retailers, suburban disinvestment has fuelled the growth of emerging places of *otherness*, radical shifts in usage, less formal occupancies, and alternative consumption patterns which fall far beyond conventional expectations. Although buildings are built to last for generations and expected to hold the habits and values which they embody, homogeneity has rapidly diminished to allow for greater diversity and a dramatically altered built environment (Abramson 2016: 16). Fading monocultural utopias have given way to a multicultural landscape of fragmented *heterotopias* that colonise, exploit obsolescence, and establish new normative conditions in suburbia.

Heterotopian peripheries in transition

Across time and cultures, the built environment has represented a dynamic social construct based on values, practices, perceptions, and production (Teaford 2008). Contemporary metropolitan development in Texas assumes car dependence and is characterised by informal methods of urbanisation and obsolescence. Within such seemingly adverse conditions, *heterotopias* – or spaces of a shared ethnic, social, or cultural experience – have emerged to fuel bottom-up approaches to adapting an overbuilt suburban landscape. Expanding disinvestment in suburbia has created undeniable challenges for urban connectivity, long-term sustainability, assumptions for uniformity, as well as mainstream perceptions of upward mobility. Small-scale interventions act as autonomous agents of resilience within a globalised city reflecting more diverse human needs, daily routines, and social desires. Contrasting obsolescence with the conventional notion of suburban stability, impermanence and internationalisation have encouraged more informal occupancies and forms of consumption. Furthermore, an implied devaluation of social uniformity and commercial homogeneity increasingly celebrates diversity.

Today, the increasingly globalised peripheral regions of Texas cities percolate up rather than trickle down, arguably offering higher self-sufficiency, more flexible organisation, and greater long-term viability. Intermingled within abandoned 1970s retail strips, shopping malls, and housing subdivisions, adaptation challenges the notion that such places are dead. Physically and psychologically distant from gentrified city centres, emergent heterotopias confront conventional assumptions about suburbia. Growing into spaces of otherness and modest vitality built upon compromise, these places enable economic freedom and social assimilation.

Lefebvre proposes a critical shift in spatial analysis by identifying multi-faceted processes of production that demonstrate multiplicity of authorship in the built environment (Lefebvre 1991). As a critic of economic structuralism, Lefebvre argues that the *everyday* manipulation of space is fundamental to the growth of society, and thus, to the shaping of cities. An embrace of such theory implies distrust of the heroic, formal, and fashionable, as well as suspicion of architecture that acts as an agent of commodification. By reframing architecture and urbanism with the inverse – the everyday impact of people reshaping buildings – Lefebvre celebrates the everyday adaptations that buildings and cities acquire from their users (Lefebvre 2005). Disorganised and fragmented, Lefebvrian actions are difficult to quantify due their irregularities; they exist as contradictions, contributions, and collaborations (Lefebvre and Régulier 1985). Accordingly, Lefebvre employs the term *heterotopia* to describe social spaces where something different is not only probable, but fundamental to their existence.

Blending immigration and the suburban neighbourhood, Lefebvrian heterotopias satisfy the basic human desire to mark and redefine space. Houston, the most amorphous of Texas cities, exemplifies globalised suburban informality; its unstructured and visually disorganised fragmentation reflects subtleties of past usage patterns and the marked transformation of recent demographic shifts. For marginalised populations including refugees and immigrants, survival is based upon the ability to socially, economically, and spatially organise. Immigrant communities in Houston have forged new American identities amidst the finely grained retail strips, enclosed shopping malls, abandoned warehouses, and residential streets of suburban sprawl. Emblematic of emergent heterotopias, anti-heroic and autonomous transformations reflect limited means, as well as the impact of time and collective memory. In Houston, a disinvested periphery subverts conventional characteristics to support occupancies that share common ground with working-class urban neighbourhoods in developing countries. Undervalued existing buildings act as *the* primary building block of places of otherness leveraged through an available and accessible infrastructure supporting quick-start and low-risk mercantile activities.

Otherness and resilience

The idea of fostering resilience in urban infrastructure is a strategic theme and operational goal for many cities around the world (Jha, Stanton-Geddes, and Miner 2013: 9–20). Researchers, scholars, and practitioners in various disciplines struggle to define and pursue *resilience* in their respective fields. What does resilience mean in terms of existing building infrastructures, and more specifically, how might it be accomplished in *suburban* peripheries? In suburbia, resilience can acquire a social dimension related to existing building stock, re-use, and re-investment, as well as associated redundancies that mitigate the potential for economic collapse. As architects, designers, and planners develop prescriptive models that guide resilient practices at the metropolitan scale, the ecological and socio-economic dimensions of resilience have become increasingly relevant within unideal suburban forms (Jha, Stanton-Gedde, and Miner 2013: 22–23). In recent years, the focus on resilience has shifted away from the anticipation of risk and mitigation, and toward a more integrated and incremental model that promotes protective and preventative strategies. Sharing common ground with Lefebvrian thought, conventional or low-tech approaches to resistance respond to regenerative aspects of resilience. Among the most vital aspects of resilience is the ability for people to support their own residential, economic, social, spiritual, and cultural needs.

In Houston and Laredo, the adaptive potential of existing and undervalued buildings systematises a bottom-up framework of social support for the economically disadvantaged. The following case studies illustrate retail change and emerging heterotopias tied to migrational influences in the US-Mexico Borderlands. Beginning in Laredo at the international border on I-35, so-called dead malls demonstrate ongoing changes in suburban production and consumption. Here, a landscape originally defined by 1970s homogeneity has become a place of multicultural hybridisation, urbanity, and resilience.

Laredo: *the portal*

Since its founding in 1755, Laredo has been a majority-Hispanic city blending two identities, well known for its ethnically hybridised schools, churches, and civic institutions that serve a populace in flux (Morales 2002: 221–222). The historical mixing of this region traces its origins to the founding of Mexico, and later, to the development of commercial connectivity with the opening of a trans-continental railroad (1881) and subsequent influx of European immigrants and American migrants from the east. Since the passage of the North American Free Trade Agreement in 1994, Laredo has become a centre of global commerce and the most important US-Mexico border crossing. In recent years, the intensification of border controls and a secondary security barrier north of the city have created physical and psychological boundaries between Laredo and the rest of the United States. In response to these constraints, undocumented immigrants now find it much harder to travel beyond Laredo. Accordingly, the city's appeal as a destination for migrants instead of a temporary landing point has become far more commonplace. *Colonias* – or unregulated settlements characterised by informal housing – have grown substantially in size since the mid-1970s as an outgrowth of the inadequate housing supply in the Rio Grande Valley (Arreola 2002: 90–91). More recently, Laredo has become one of the border cities receiving the highest flows of unaccompanied child migrants from Mexico and Central America. The current humanitarian crisis underscores a dramatic shift that has positioned Laredo as an international city and pivot point between the Global North and Global South.

Interstate 35 (I-35) is one of the most important highways in the United States. Beginning in Laredo and forming its path along the historic Chisholm Trail, it conveys a massive flow of consumer products, car parts, processed foods, and other goods along the NAFTA corridor from Mexico to the United States and Canada. Laredo serves as a hinge between two autonomous nations – a blurred zone, a third country, and a hybridised place – among the most Mexican of American cities housing a bilingual population that often prefers Spanish to English. Both cultures hold historic political and cultural claims to the region, while their parallel stewardship supports the socio-economic integration of new residents with identities that straddle the border.

Immigration is the primary reason for the rise of several private bus lines that use Laredo as a hub for operations connecting cities in Mexico and the United States. In Laredo, bus fleet route maps illustrate the extent to which Mexico has blurred into America. Daily coach services to Illinois, Indiana, Iowa, Kansas, Oklahoma, Alabama, Georgia, Tennessee, Kentucky, and the Carolinas reveal the places where an increasingly cross-cultural American experience is taking shape. Tornado, El Conejo, Turimex, Senda, and other bus lines weave a web connecting Mexican cities in the north and east of the country with small towns and big cities across the United States. Threaded through Laredo, these private bus companies provide a lifeline for families divided by the border. Migration has slowed with closer scrutiny of undocumented immigrants; however, buses depart daily in steady intervals supporting a connective network that gains little notice from mainstream American society.

Many cities in southern Texas trace their origins to the colonial administration of Spain and later to Mexico before the establishment of the independent Republic of Texas. In 1845, the Texas lands north of the Rio Grande River became part of the United States, whereby Anglo-American norms were imposed on Spanish-speaking society to cast Hispanics as 'others' by means of systematic political and socio-economic disenfranchisement (Collier, Galatas, and Harrelson-Stephens 2017). Notwithstanding territorial changes, Laredo has welcomed migrants and immigrants for many years, while cities in the Midwest have, until recently, reflected 'difference' primarily in the conventional American polarity of black and white. For many recent immigrants from Latin America, Laredo serves as a portal to the Midwest – a new Ellis Island – while I-35 acts much like the 'ferry to Manhattan'.

Immigrants arrive in Laredo in various ways – some hold green cards while others cross the border through clandestine means. Temporary visitors arrive with border-crossing cards allowing them to stay in the United States within 25 miles of the border for up to 72 hours. Of these, some travel north as tourists and overstay their visas. The historical bilateral openness of the Rio Grande Valley international border zone became more restricted in the 1990s, when undocumented migrants began to use freight railways and private bus companies to move further into the United States (Hernandez 2010: 221–225). At the same time, the population of Laredo doubled between 1990 and 2012 to 244,000, largely attributed to the flow of migrants from Mexico and Central America (Cave 2014). After 9/11, border security tightened with additional American officers, security cameras, powerboats, and a secondary checkpoint barrier farther north on I-35 (Anderson and Gerber 2008: 212–215). Regardless of shifting entry regulations and immigration policies of the American federal government, Laredo serves as a primary gateway for overland migration into Texas and the mid-continental United States via I-35 – a quintessential American highway that hosts several dying malls stretching from Laredo to Duluth.

An edge condition: River Drive Mall in Laredo

Located within a day's drive of major cities in northeastern Mexico, Laredo is a primary shopping destination for the middle- and upper-middle classes based in the sprawling Monterrey metropolitan area. Four bridges connect Laredo, Texas, and Nuevo Laredo, Mexico; the oldest of these, the historic Gateway to the Americas International Bridge, remains an important and critical crossing point. Originally built in 1889 as the Convent Street Bridge, this busy, diminutive bridge connects the downtown commercial cores of Laredo, Texas, and Nuevo Laredo, Tamaulipas, with a four-lane roadway, pedestrian paths, and international customs and immigration stations. Unlike most trans-border bridges, the pedestrian paths of the Gateway to the Americas International Bridge are typically busier than the automobile roadway.

Downtown Laredo is a bustling commercial district that serves local, regional, and international consumers and merchants. Many downtown retailers serve Mexico-based merchants as wholesale supply agents. Situated directly adjacent to the Gateway to the Americas International Bridge, River Drive Mall opened in 1970 as the first enclosed shopping mall in Laredo with anchors JC Penney, Frost Brothers, and Weiner's. The mall is only a five-minute leisurely walk from the downtown central plaza of Nuevo Laredo, Mexico, and was originally popular due to its pedestrian accessibility. River Drive Mall represents a common urban renewal strategy of the era that sought to attach suburban shopping centres and enclosed malls to established downtown commercial districts in small-to-medium-size cities. The popularity of the mall would be relatively brief, however, due to the construction of Mall del Norte that opened in 1977 outside of downtown on the I-35 corridor. Continued retail development along the freeway contributed to the decline of River Drive Mall through the mid-1980s. By the early 2000s, this downtown mall housed independent mom-and-pop retailers serving a pedestrian-based clientele from Nuevo Laredo. During the interstitial years, River Drive Mall represented a heterotopia supporting very modest needs and micro-entrepreneurship.

Shifting security concerns exacerbated the permanent closure of River Drive Mall in 2003, at which time Morgan Stern Realty purchased the property with the intention of converting it into an outlet centre serving a wealthy trans-border market (Herrera 2013). Although the mall offered proximity to a large consumer catchment zone of wealthy Mexican nationals, its location in the wholesale district of downtown Laredo has been largely avoided by the class-conscious elites of Mexico. Morgan Stern renamed River Drive Mall as the El Portal Center and embarked on major renovations in 2003, but abandoned the project by 2005 citing security concerns and marketing issues. In 2013, the City of Laredo and Horizon Group Properties announced redevelopment plans for River Drive Mall. The existing structure was demolished in mid-2015 and a new three-story open-air building with 90 outlet stores was built in its place. Forces of gentrification strategically decoupled River Drive Mall from the modest resilience that emerged within its faded obsolescence – independent merchants were evicted, and the lower-middle-class needs that they provided for have been ignored. The new Outlet Shoppes at Laredo opened in 2017 to serve the trans-border elite of northeastern Mexico. The municipal government believes that the new outlet mall will help to revive downtown Laredo by attracting wealthy Mexican nationals who typically travel further afield to outlet shops in San Marcos, Texas. Although the new outlet mall employs over 2,000 people and includes 80 national retailers, it no longer supports the microbusinesses that kept it viable for nearly 20 years. Here, an emergent heterotopia has been permanently erased.

A 'ruin': Indeterminate Façade Building in Houston

The Best Products Company of Richmond, Virginia, was a catalogue retailer founded by Sydney and Frances Lewis in 1958. As merchants and art and design aficionados with a desire to draw interest to their stores, they commissioned James Wines and SiTE Architects to design a series of unorthodox and irreverent retail showrooms. The Houston showroom opened in 1974 on a site near Almeda Mall – a regional shopping centre built by the James Rouse Company in 1968 that was the premier mall in Houston at the time. Named the Indeterminate Façade Building, the showroom was one of nine radical, prototype big box stores designed by SiTE (Sculpture in The Environment) for Best Products that rethought utilitarian suburban retail architecture. Wines described the project as de-architecturisation of the façade achieved by extending the brick veneer beyond the leading edge of the roofline; its 'finished' appearance alluded to something between construction and demolition (Drexler and Balch 1979). The resulting subversive and displaced form evoked ruins or deconstructed buildings, a vision of neglect that would ironically predict a future of 1990s disinvestment and decline in the Kingspoint neighbourhood. As a commentary on both the strip shopping centre and the suburbs, Wines mobilised everyday buildings to work against conventional expectations, as well as the reigning social, psychological, and aesthetic notions of 1970s suburbia. The ambiguity of their form, a subversive decay within a context of perceived 'normativity', distorted its relationship with site, formality, proportion, scale, history, and nature to evoke tentativeness and instability. In short, its architecture subverted suburban utopia to foretell a story of dystopia on the horizon – a provocative narrative that challenged the increasing irrelevance and complacency of post-1968 architecture.

As the first of a series of buildings that James Wines would design for Best Products, the Houston showroom was perceived as a conceptual statement in the art world, however, it was viewed with deep disdain by mainstream architects. Wines believed that the Houston showroom was an 'architecture of information', retelling a site narrative which straddled the line between art and utility, environmental and consumption, permanence and deconstruction. Similarly, the Indeterminate Façade Building's design emerges from a liminal place between art and architecture – its purposely ruined and fragmented profile has ironically succumbed to the cyclical rise and fall of suburban retail. Furthermore, the building has been radically altered to appear more conventional over time – its subversive façade has been diminished, intentionally levelled, and rebuilt by the current owners to convey a more ordinary and finished appearance.

As the neighbourhood surrounding this iconic retail store began to decline in the 1990s, the Indeterminate Façade Building assumed a more conventional appearance and banal purpose. Today, this historically important building serves as a warehouse and lapses in a tentative state of partial abandonment. Surrounded by a security fence and parking lot filled with shipping containers, the Indeterminate Façade Building has been unapologetically stripped of both its architecture and memory as a design icon. A building that forecasted the future of suburban obsolescence has been profoundly violated by the throwaway culture that it criticised. National retailers have left the nearby Almeda Mall and its vicinity for upscale malls – the middle ground has been gutted and mom-and-pop retailers have filled the void. In place of national brands and suburban homogeneity, immigrant merchants have carved out a Lefebvrian heterotopia by providing relevant services such as car repair, pawn shops, nail salons, clothing stores, and restaurants serving multi-ethnic consumers. Although neighbourhood retail has not been entirely abandoned, it has transitioned from a bastion of 1970s white middle-class conformity to a contemporary working-class Latin American immigrant heterotopia of aspiration. Today, the Almeda Mall, Indeterminate Façade Building, and their adjacencies have been significantly altered to host more informal uses and new patterns of consumption.

An interstitial space: Greenspoint Mall in Houston

In *The Architectural Uncanny*, Anthony Vidler considers architecture as the *uncanny* – a metaphor for the modern condition (Vidler 1992). He analyses the state of Architecture in an era of increasing obsolescence in everyday building typologies of the recent past. He interprets the

Figure 7.1 A nearly empty concourse of the Greenspoint Mall in Houston houses a closed department store, decommissioned fountain, community room, and hair salon, 2017. *Source*: author.

ungrounded qualities of abandoned or under-used shopping malls as the wasted territories of consumerism, corporate disinvestment, and post-industrial culture. Vidler analyses the problems inherent to such architectures as they lose their novelty and fade into the past, proposing a future vision of retail environments built upon the ruins of alienation, suburban exile, and obsolescence. Forgotten by their intended users and appropriated by others, these so-called *dead malls* are the cast-off orphans of a recent past civilisation. They exist along the undefined margins of our suburban landscapes as scars and dead zones, yet there are indeed qualitative nuances in their so-called deadness. Existing in various states of informality and disinvestment, 'deadness' ranges from total erasure to emerging, vibrant, and alternative communities of otherness.

Conventional malls built during the first and second waves of mall development are increasingly a lost generation. One prominent Houston example of this phenomenon is the Greenspoint Mall located at the junction of Interstate 45 (I-45) and Beltway 8 (Figure 7.1). It has been diminished by the re-emergence of downtown, new retail forms, web-based means of consumption, and shifting demographics. Opening in 1976, Greenspoint Mall was once the largest mall in the Houston metropolitan area, but by the late 1980s, it began to wane due to the oil recession, rising crime, and newer malls at Willowbrook and Deerbrook. Over the last ten years, most of the original six anchor stores have closed, while mall shops have transitioned from national retailers to mom-and-pop retailers serving a more diverse, predominantly African-American and Latino clientele. A permanent flea market was built across the street and the substantial parking lot hosts new activities including carnivals, motorcycle meet-ups, and clandestine pop-up retail shops housed in vans, moving trucks, and tents. These informal occupancies offer new layers of life that contrast significantly with the past.

Conclusions

Ageing strip malls, big box stores, and enclosed shopping malls afford immigrant communities an existing infrastructure that houses ethnic shops, storefront churches, and social clubs within former bastions of suburban conformity. Transitioning from solidly middle class to greater socio-economic diversity, suburban neighbourhoods and their long-time residents must negotiate various polarities – chaos/order, ambivalence/adaptation, resistance/resilience – while continually learning from and adapting to a new context. The obsolete and considerably altered retail environments adjacent to these neighbourhoods offer a window on the future of suburbia in Texas and across the United States. On the one hand, suburbs are increasingly less affluent – but more urban, heterogeneous, demographically diverse, and socio-economically blended. On the other, they demonstrate core attributes of resilient systems – resource diversity, resource availability, and institutional memory – and thus, offer the potential for even greater socio-economic layering and enhanced urbanity.

For Lefebvre, dead zones along the suburban edges of cites are charged with latent support systems offering a form of organic resilience to new users. Processes of disinvestment and fragmentation create voids that carve out physical space for alternative occupancies to germinate and thrive. Although enclosed shopping malls and their adjacencies have been conventionally perceived as privatised – and to a certain degree both secure and homogeneous – the increasing subversion of their formality and conformity by the 1980s gave rise to a significant shift in usage. Malls and their environs are vast spaces, so their abandonment and cannibalisation create unique issues for urban planning and architecture. Private, but perceived as public space by their users, the interior worlds of dead and dying malls give rise to unplanned and unofficial communities. Left empty, these de-programmed infrastructures evolve into transgressive places; they can no longer support hegemony and are being appropriated by socio-economically disadvantaged immigrants and minorities.

American society has become desensitised to the post-industrial abandonment in former sites of production and consumption; suburban typologies that have lost their original functions are generally perceived as both subversive and dystopian. These undefined zones – or *terrain vagues* – have always been acquired by marginalised communities and occupancies. Viewed through the lens of impending gentrification, Sola-Morales proposed the notion of *terrain vague* to describe the organic magic of obsolescence (Solà-Morales 1995). Dead mall heterotopias do the opposite by juxtaposing various seemingly incompatible functions – private micro-investment, national retailers, independent merchants, social services, and immigrant community activities – whereby their *collection* creates occupancy, stability, and emergent resilience. In a remarkable turn, these so-called dead zones enable the emergence of heterotopias by offering a foundation upon which delicate ethno-socio-economic configurations are constructed (Gennochio 1995). Blending the peripheral fringe of Sola-Morales's *terrain vague* with Lefebvre's perspective on radical new occupancies, these retail heterotopias reveal how conventional spaces are reborn. Heterotopias built upon so-called dead malls do not conceal difference, but rather, reveal the inherent flaws of social fragmentation, political polarity, and economic upheaval germinating within the late capitalist system.

Today, mid-continental suburbia in the southern tier of the United States is resoundingly less shaped by developers, architects, planners, and the exclusionary motivations of enclaves defined by uniformity. Adapted through the needs of laypersons, entrepreneurs, and makers, this temporal and globalised landscape is increasingly contingent upon compromise and connectivity far beyond the physical boundaries of the country. An emerging periphery sabotages conventional wisdoms and shifts expectations for what is possible in suburbia. The changing socio-cultural geography of Texas suburbia offers exceptional affordability and a near limitless opportunity for the adaptive re-use of overbuilt existing retail infrastructure and housing stock. A formerly homogeneous suburban utopia has been replaced by a far more complex, globalised, and resilient *urban* heterotopia with strong connections to Latin America and beyond.

References

Abramson, D.M. (2016) *Obsolescence: An Architectural History*, Chicago: University of Chicago Press.
Alba, R.D. and Nee, V. (2003) *Remaking the American Mainstream: Assimilation and Contemporary Immigration*, Cambridge: Harvard University Press.
Anderson, J.B. and Gerber, J. (2008) *Fifty Years of Change on the U.S.-Mexico Border: Growth, Development, and Quality of Life*, Austin: University of Texas Press.
Arreola, D.D. (2002) *Tejano South Texas: A Mexican American Cultural Province*, Austin: University of Texas Press.
Baumgartner, M.P. (1988) *The Moral Order of a Suburb*, New York: Oxford University Press.
Beauregard, R.A. (2006) *When America Became Suburban*, Minneapolis, MN: University of Minnesota Press.
Cave, D. (2014,) 'Deep ties, tested on Mexico's border', *The New York Times*, 17 May.
Collier, K.E., Galatas, S. and Harrelson-Stephens, J. (2017) *Lone Star Politics: Tradition and Transformation in Texas*, London: SAGE Publications, Ltd.
Drexler, A. and Balch, B. (1979) *Buildings for Best Products*, New York: Museum of Modern Art.
Genocchio, B. (1995) 'Discourse, discontinuity, difference, the question of other spaces', in S. Watson and K. Gibson (eds) *Postmodern Cities and Spaces*, Oxford: Blackwell.
Hernandez, K.L. (2010) Migra!: *A History of the U.S. Border Patrol*, Berkeley: University of California Press.
Herrera, C. (2013) 'A hot topic once again', *Laredo Morning Times*, 15 June.
Jha, A.K., Stanton-Geddes, Z. and Miner, T.W. (2013) *Building Urban Resilience: Principles, Tools and Practice*, Washington: World Bank.
Lefebvre, H. (1991) *The Production of Space*, Oxford: Blackwell.
Lefebvre, H. (1995) 'Right to the city', in E. Kofman and E. Lebas (eds) *Writing on Cities*, Oxford: Blackwell, 179.
Lefebvre, H. (2005) 'Critique of everyday life', in *Critique of Everyday Life*, London: Verso.
Lefebvre, H. and Régulier, C. (1985) 'Le projet rythmanalytique', *Communications*, 41:1 191–199.
Morales, E. (2002) *Living in Spanglish: The Search for Latino Identity in America*, New York: St. Martin's Press.
Rees, A. (2003) 'New urbanism: visionary landscapes in the twenty-first century', in M. Lindstrom and H. Bartling (eds) *Suburban Sprawl: Culture, Ecology, & Politics*, Essay, Lanham: Rowman & Littlefield, 93–114.
Solà-Morales, I. (1995) 'Terrain vague', in C. Davidson (ed.) *Anyplace*, Cambridge: MIT Press.
Teaford, J.C. (2008) *The American Suburb: The Basics*, New York: Routledge.
Vidler, A. (1992) *The Architecture of the Uncanny*, Cambridge: MIT Press.

8

Prohibited places

The pericentral self-produced neighbourhoods of Maputo in the neoliberal context

Silvia Jorge

Introduction

Maputo, the capital of a country considered one of the emerging economies in Africa (see, for instance, United Nations 2013), as a result of the discovery and exploitation of strategic natural resources, is crossing what Rolnik (2015: 15) describes as a 'colonisation of the urban land and housing by the global finance'. Following the language of finance and economics, the territory is reduced to its exchange-value and to the prospective of future capital gains, serving the relations of ownership or property more and more to define – or block – rights, such as the right to the place or, in a general way, the right to the city, as understood by Lefebvre ([1968] 2012). The Mozambican government, as a condition for being granted loans by the World Bank and the International Monetary Fund, as happened with many of the neighbouring countries (Lesotho, Madagascar, Malawi, Mauritius, Zambia and Zimbabwe), was forced to adopt a program of structural adjustment in 1987, under the umbrella of 'international aid', setting the stage for the neoliberal paradigm.

Regardless of the particularities of each historical-geographic context, the great role played by governments in the conduct of this process, the generalised increase of the insecurity of occupation or possession by lower income groups, the advance of forced evictions and, consequently, the violation of human rights tend to be recurrent. The demolition of extensive areas for megaprojects of infrastructure and urban renewal (see Figure 8.1) represents, in this sense, one of the main intervention strategies driven by the ongoing commodification and financialisation of urban spaces in Southern Africa. Any of these actions in inhabited spaces proves to be particularly violent when it occurs in the territories whose tenure is indefinite or read, by the new legal-urbanistic instruments, as illegal or irregular. More easily therefore appropriated by the local governments or private investors, these spaces, much of them self-produced, represent a vast territory waiting for the 'right moment' (Rolnik 2015: 174).

Immersed in the neoliberal paradigm, particularly in its expression in Maputo, this chapter is born of a set of questions and concerns about the present and the future of what Lefebvre designated as 'prohibited places' ([1974] 1991: 332, 366): spaces that, by their strategic location

Figure 8.1 Chicala/Kilombo (Luanda) in early 2012, before the demolitions. *Source*: Photographed by the author (2012).

from a market point of view, are reduced to the one-dimensionality of their exchange-value and fall into a situation of increasing vulnerability. With a focus on the pericentral self-produced neighbourhoods of Maputo, which have attracted a growing number of private investors and promoters in recent years, we analysed the new city model, the urban interventions and the everyday practices, as well as the interactions established between the different agents involved in their transformation, starting from a recent larger investigation about these places (Jorge 2017).

Within the tradition of critical theory about the socio-spatial inequalities generated by the capitalist mode of production, we follow Lefebvre's perspective on the production of space ([1974] 1991) and on the right to the city ([1968] 2012), crossing his vast work with other authors with a similar perspective. Starting from his dialectic approach, we seek to perceive and to analyse how the double movement of centralisation and fragmentation develops in the context of Maputo, understanding the contradictions formed between use-value and exchange-value, as well as the possibilities of resistance and social change. Taking into account global trends, we put forward the hypothesis that the pericentral self-produced neighbourhoods constitute 'prohibited places' due to the one-dimensionality of their exchange-value, but also 'interdicted places' to this same commodification, when voices and practices of resistance increase their use-value.

> 'A deurbanising and deurbanised urbanisation'
>
> *(Lefebvre 2012[1968]: 17)*

The Mozambican capital, today with more than one million inhabitants, about 70% of them living in self-produced spaces, is a creation of capitalism, a result of the unequal relations of exploitation that marked the colonial period, not a 'beautiful oeuvre' in the sense attributed by Lefebvre to some cities prior to industrialisation ([1968] 2012: 17). As in other African cities, but also as in many other regions of the world (see for instance Davis 2006), their periphery expresses what Lefebvre described as 'a deurbanising and deurbanised urbanisation' ([1968] 2012: 30).

In the Maputo case, the city expands around the old planned centre, of colonial origin, drawing a periphery progressively more distant, that ends up finding other neighbouring agglomerates and forming what some authors define as the Metropolitan Area of Maputo (Jenkins 2012) or Greater Maputo (Macucule 2016). Although we may consider the existence of new small centralities, such as Zimpeto, Xiquele or Xipamanine, the KaMpfumo district – the old planned centre – persists as the city centre par excellence. It concentrates the economic activities of greater volume and capital, the head offices of big companies and (multi)national agencies, the main services and administrative buildings, as well as a greater number of social equipment and infrastructures, as a result of public and private investment in their maintenance, recovery or reconstruction over time (Raposo 2007). In line with Lefebvre ([1968] 2012), the centrality assumes here its dual character: the place of consumption and the consumption of place. That means that, on the one hand, it is established as a centre of consumption and decision, wealth, information and knowledge, and, on the other hand, parallel to the concentration of relations of consumption and power, it fragments the space, generating 'recommended places' and 'prohibited places' (Lefebvre [1974] 1991: 332, 366).

Following this double movement of centralisation and fragmentation, the spaces around the KaMpfumo district assume a clear and defined socio-economic character: to ensure the continuity of 'abstract space' described by Lefebvre. In other words, the instrumentalisation of space, expressing the strategies of capital reproduction and the perpetuation of the social relations that constitute it, expels those who threaten its purely economic function ([1968] 2012; [1974] 1991). In this sense, the pericentral self-produced neighbourhoods reflect simultaneously the incapacity of the complete subordination to hegemonic rationality, which Santos ([1996] 2008: 309) calls 'anti-rationality' – and which generates the place of the poor and excluded – and the incessant demand for production of 'abstract space'.

Although these places have always occupied a strategic location (see Jorge 2017), it is in the neoliberal context that the production and transformation of space by capital tends to become more expressive in the pericentral self-produced neighbourhoods. At the end of the 1990s, Araújo (1999: 178–179) warned of their 'rapid process of transformation', particularly intense in the vicinity of Sommerschield, one of the city's noblest neighbourhoods. More recently, other authors, such as Vivet (2012) and Jorge (2015), emphasise the intensification of renewal and gentrification actions, both near Sommerschield, as well as along the coastline and the main access roads to the city centre. The pressure on the pericentral spaces therefore increases, especially in this new millennium, as a result of the consolidation of the neoliberal policies adopted since the late 1980s.

A hegemonic model of abstraction

The adoption of the structural adjustment program marks the beginning of a set of deep economic and political changes with a strong impact on the territory, enabling the establishment of networks of multinational companies and non-governmental organisations (NGOs) which today operate in the country and with whom the State has shared the responsibility of urban planning and management. In a first phase, its implementation focused mainly on monetary and financial instruments (devaluations, fiscal restraints) and the basic economic framework (privatisation, liberalisation of markets), contributing to rising levels of unemployment and poverty (Oppenheimer and Raposo 2007). Nevertheless, the program also triggered the passage from a

'Populist State' to a 'Democratic State' (see Serra 2013), reflected in the constitutional amendments introduced in 1990 and 2004. While the 1975 Constitution enshrined a single-party system and a planned and centralised economy, the 1990 Constitution introduced a democratic regime and recognised the market economy, encouraging private initiative and the freedom of enterprise and investment, even though the land remained as State property. In turn, the 2004 Constitution reaffirmed the guidelines enunciated in 1990, reinforcing the dimension of land value and the decriminalisation of its speculation.

According to Francisco (2013), the new function attributed to the State, apparently regulatory, allowed for the conversion of public policies into a new form of interventionism in economic activity, created and fomented, in large part, by the political party Frelimo, which has been in power since independence. Controlling the State, the party controls the natural resources, which are globally strategic and have a high market value, as well as the main productive asset of the population – the land – also the target of a valorisation process, particularly intense in the pericentral self-produced neighbourhoods. Some authors, such as Francisco (2013) and Waty (2011), consider that the current Constitution subordinates political power to economic power through the control of access to land, a fact perhaps related to the increase of corruption and conflicts of interest in the neoliberal context (see, e.g., Hanlon 2004). Following Lefebvre's reflections ([1968] 2012; [1972] 2001), the State and capital are allied in the creation of the unitary system and total urbanism, inducing a technological and technocratic rationality.

Indeed, under the influence of international agencies, Mozambique, like other African countries, has incorporated in the last decades, through deep legal reforms, new models of access, planning and management of urban land, distinct from pre-existing spaces. Although these models are presented by some academics and international agencies as essential for the resolution of urban problems, they have had a perverse effect. According to authors such as Souza (2011) and Maricato ([2000] 2013), the positivist and technocratic view of urban planning underlying these models ignores the incapacity that they have revealed over time in solving the socio-spatial inequalities and masks the conflicts and real causes of urban problems. In the 1970s, Lefebvre went further in considering that in 'the transparent air of functional and structural legitimacy' lay the technological and technocratic rationality of 'abstract space' ([1974] 1991: 317–318). The notion of planning as an instrument of political and economic power is reinforced by Sposito (2011), who considers that it is mainly the result of the adoption of these recent models that in the last decades have spread capitalist relations of production, domination and property on a global scale, as well as the idea of a unique society – the occidental one – devaluing other specific societies with different values, practices, ways of living and of producing space. This occidental ethnocentrism represents, in line with Lefebvre, 'a hegemonic model of abstraction' in the sense of ignoring concrete 'social space' and, consequently, everyday practices.

The new constellation of powers, responsibilities and interests reignited the market logic (see Jorge 2017). The city centre, already urbanised, gathered most of the interventions assumed by the government and private investors and promotors. Meanwhile, the role of the NGOs was strengthened in self-produced neighbourhoods, especially at the level of local urban management, but, as Raposo and Ribeiro (2007) reveal, their action tends to be occasional and of a charitable nature. The continuous population growth and the inoperability or inability of response by the entities responsible for urban planning and management led to: the extensive development of the territory in the 1990s, characterised by an increase of the average growth rates in the more peripheral districts; the densification of the pericentral neighbourhoods through the progressive subdivision of plots and the occupation of environmentally sensitive areas; and the deterioration of living and housing conditions, mainly as a result of overcrowding and the absence and/or deterioration of basic services and infrastructures (Oppenheimer and Raposo 2002).

Zones of natural risk

The large floods of 2000 represented a turning point in the way that technicians, politicians and civil society in general perceived the pericentral self-produced neighbourhoods, particularly affected by the heavy rains that were felt in February of that year and, already before, in 1998. Many families were resettled in different peripheral neighbourhoods and, from then, the pericentral ones were considered 'zones of natural risk'. This general classification has served as support for some actions of urban renewal, but it has not prevented the reoccupation of the same spaces by groups with greater resources, following on the idea that the land and nature are inexhaustible and that the risks are always acceptable as long as economically profitable.

Since then, urban renewal has assumed increasing preponderance, being associated with most of the partial urbanisation plans proposed for the pericentral self-produced neighbourhoods, with the (re)construction of structural roadways and megaprojects that imply the total or partial demolition of houses, as well as various projects launched in the city centre, where some buildings of historical interest gave way to high towers. The preference for this kind of intervention is due to the real estate interests involved and the primacy of the urban model in which they are inscribed, as well as, in the particular case of the pericentral neighbourhoods, to the negative image that accompanies them (Jorge 2017). This urban renewal has therefore implied the removal and resettlement of the families affected by these actions to peripheral districts through expropriations for 'public interest', compensated at best with the attribution of a land plot and some money, usually far from the legally defined 'fair compensation' (Jorge 2017).

However, in most of the peripheral neighbourhoods, the local government's engagement in the division of land, urban upgrading and land regularisation reveals another perception of the territory and other interests and strategies (see Melo 2015). Although the basic infrastructures and equipment are slow to arrive and environmentally sensitive areas are being occupied (for instance Chihango), the perception of these territories tends to be more positive, mainly due to their morphological characteristics and the rural nature that, in some cases, they maintain. The factors that determine or influence these different intervention proposals range from a more or less positive perception of a particular space to its location and commercial valuation. These factors denounce the ambiguity and arbitrariness that surround the current legal framework and the State's action, which legitimises some occupations and not others, depending on market logic (Jorge 2015).

Nevertheless, despite the strong investment in planning, the implementation of the so-called partial urbanisation plans is rarely concluded, unlike the megaprojects and the improvements in the road network, such as the (re)construction of the Great Circular of Maputo and the Julius Nyerere avenue, which have been achieved through partnerships with private investors and promotors, other countries and/or international alliances. Generators of expropriation, urban renewal, valorisation and gentrification processes, these large-scale interventions foster the parallel real estate market along the intervened spaces, taking into account the same unspoken objectives of the renewal plans: the expansion of 'abstract space' described by Lefebvre with the resulting expulsion of those who threaten its economic function. Considered as a company, the city tends to be conceived and thought of as an economic agent, which operates for the (global) market, considering planning as a mere accessory, resorted to or dispensed with according to the dominant interests and the market dynamics. This does not imply a renunciation of planning, but rather an elimination or reduction of the long-term rules, norms and limitations that could prevent or delay capitalist accumulation, favouring 'informal' agreements and procedures that facilitate and benefit this accumulation, within the spirit of negotiations and partnerships of circumstance (Souza 2006). It is what Harvey (1989) designates as 'entrepreneurship', characterised by an uncritical submission to market interests and an obsession with attracting new investments.

> 'The ideas out of the place and the place out of the ideas'
>
> *(Maricato [2000] 2013: 121)*

As in so many other contexts dominated by the neoliberal paradigm, the State assumes an economic role, producing, as Rolnik (2015: 14) underlined, its urban margins, to thereby 'unlock its territorial assets' and 'to increase the market frontiers'. In the case of the strategically located pericentral self-produced neighbourhoods of Maputo, the local government: (i) does not guarantee the safeguarding of acquired rights by occupation in good faith, as foreseen by the law; (ii) encourages the elaboration of partial urbanisation plans inscribed in the market logic, preventing or hindering the formulation of alternative interventions; (iii) uses expropriation for 'public interest' and subsidises the construction of infrastructures and megaprojects in order to attract companies/agents interested in investing and locating there; and (iv) ignores the parallel market, denied by law, and subsequently assigns titles of land use to the new occupants, denied to the inhabitants who lived there, in many cases, for several decades (Jorge 2017). On the one hand, we are witness to the (in)voluntary displacement of a growing number of inhabitants to more and more peripheral neighbourhoods, and on the other hand, to the conversion of self-produced spaces into commercial and residential spaces destined to groups with greater resources, strengthening 'the ideas out of place and the place outside the ideas' that Maricato ([2000] 2013: 121) alludes to.

This play on words refers to the gap between the matrixes that substantiate the urban planning and legislation and the local reality and specificity of self-produced neighbourhoods: 'the ideas out of place' enunciated in the greater part of the current legal-urbanistic instruments correspond to nothing and, simultaneously, 'the places outside the ideas', which are the self-produced neighbourhoods, have no presence in the city model advocated by the dominant class. Underlying this reading are the two dynamics of space enunciated by Lefebvre ([1974] 1991) – the 'abstract space' and the 'social space'. The first, as we referred to earlier, is associated with strategies of capital reproduction and with the perpetuation of the social relations that constitute it, while the second is forged from everyday practices, translating the plurality of life modes and appropriated forms of the space.

In a context where the land, from a legal point of view, is the property of the State and cannot be sold, mortgaged or pledged, the everyday practices are far removed from the principles set forth in the Constitution and in the reigning legislation (see Jorge 2016). As Jorge demonstrates (2016, 2017), the analysis of the transformation of the pericentral self-produced neighbourhoods reveals that not only is the access to land and to housing hostage of market logic, but also that the local, district and municipal structures are involved in the parallel real estate market, facilitating contact with private investors and promotors. A strong alliance between the political power and economic power is also present in the (re)construction of structural roadways and megaprojects and, finally, in the elaboration of the partial urbanisation plans. The use of expropriation for 'public interest', a concept coated with contradictions and ambiguities (Habermas [1973] 2007), aims at the improvement of the accessibility and, with it, the promotion of the parallel real estate market, opening new ways and possibilities for business. Furthermore, the existence of a plan, beyond legitimating market logic and the ongoing transformation, reiterates the intention of the local government to extend the urban renewal and gentrification actions to other spaces, consuming, at the limit, the whole extension of these neighbourhoods in the name of the 'public interest'. In this sense, there are many inhabitants, with different aspirations and perceptions of their rights, that choose to sell their houses, moving away from the city centre, that is, from the main work opportunities and means of survival.

Another face of the 'prohibited places'

Although it is difficult for there to be any type of individual or collective mobilisation against dominant interests and practices, in the face of intimidation, lack of information and the individualisation of the processes of acquisition, it is possible to identify some practices and focuses of resistance (Jorge 2017). At the level of inhabitants, we highlight: (i) the refusal of any kind of negotiation by inhabitants that, instead of the exchange-value, privilege the use-value of the neighbourly relations built over time, and the pleasure of living in the neighbourhood; (ii) the setting of very high sales prices by inhabitants to demotivate the purchase and pushing away potential buyers or intermediaries, thereby impeding or delaying their removal from the neighbourhood; and (iii) the denunciation, to the media, of forced evictions and of situations of discrimination and socio-spatial exclusion, in an attempt to prevent the continuation of some particularly violent actions, some of them realised with the consent of the municipal, district and local structures.

In addition, among technicians and members of NGOs involved in these processes, despite the many contradictions, we highlight those who: (i) confronted with partial urbanisation plans based on the tabula rasa of pre-existing urban fabric, criticise and deny the underlying market logic imposed by the local government; (ii) continue to create and implement awareness-raising projects on the right to housing and the right to the place after the approval of the renewal plans; and (iii) seek to counteract the risk of gentrification inherent in a future intervention proposal and to defend that which they consider to be the interests and needs of the current inhabitants, reflecting critical and reflexive thinking.

These are micro-resistances – forged in everyday life, in a situation of confrontation or professional practice committed to the inclusion and democratisation of the urban space – capable of reversing or contradicting, even if occasionally and often in a silent way, the relations of power and domination (Certeau 1998). In some cases, especially in those promoted and conducted by the inhabitants, it has been possible to prevent the expansion of the borders of commodification and financialisation of the urban space. In others, it has not been possible to go beyond good intentions and sweet words. However, the risk of a more extensive action of resistance or revolt by inhabitants, before the possibility of a total renewal of these places, can be one of the factors that justifies the delay or refusal by the local government to implement the renewal plans in full. Simultaneously, this same risk promotes the use of other strategies by the local government, namely ignoring the urban plan and closing their eyes to the parallel real estate market, while investing in infrastructure, which enhances market dynamics and triggers gradual renewal actions. Implicitly or explicitly, the practices and foci of resistance of the inhabitants show that, despite their limited room for manoeuvre, the iniquities of the dominant system continue to generate opposition and its contradictions continue to be exploited, revealing that another future is possible.

Conclusions

The pericentral self-produced neighbourhoods of Maputo reveal a market dynamic which seeks to generate 'prohibited places' and fomenting gentrification processes, simultaneously destructive and indifferent to the local realities, by supporting the *tabula rasa* of the pre-existing fabric and by imposing a city model far removed from local necessities. As we saw, the processes of renewal and gentrification are both inscribed in several of the urban plans, but also in the parallel real estate market and in the ambit of megaproject construction, implying in turn massive resettlements. Transversal to all of these processes is the recourse to violence, dissimulated by the exchange-value – especially when the associated transactions involve values never before imagined by the inhabitants – or evident in the use of force and coercion or by the more or less silent defence of the use-value.

However, in a context where the political party Frelimo and the State fuses and, more recently, in which emerges a strong alliance between the political power and the economic power, the self-produced neighbourhoods of Maputo also constitute 'laboratories' of resistance. Here, new ways of solidarity and critical discourses are engendered, and the consciousness of rights, ideas, strategies and new organisational structures germinates. Although everything tends to become a 'product' and is conducted by the language of commodities, the use-value resists irreducibly against the centralisation-fragmentation movement that marks the current transformation of these places. Practices and foci of resistance emerge among inhabitants, but also among technicians and members of NGOs involved in these processes. Far from triggering a collapse of the dominant system, the moment is now one of experimentation and learning from the *praxis*, even though the resistance is territorially circumscribed and, at a first sight, modest. Following the ideal of the right to the city, we are present before a struggle, in and through space, centred on the appropriation and collective achievement. Rescuing the hypothesis proposed, the 'prohibited places' carry a double meaning: the one-dimensionality of their exchange-value, but also the voices and practices of resistance that raise their use-value.

References

Araújo, M. (1999) 'Cidade de Maputo, espaços contrastantes: do rural ao urbano', *Finisterra*, 34:67/68 175–190.

Certeau, M. (1990/1998) *A Invenção do Quotidiano: Artes de Fazer*, trans. E. Alves, Petrópolis: Vozes.

Davis, M. (2006) *Planet of Slums*, London/New York: Verso.

Franscisco, A. (2013) 'Por uma Nova Constituição Económica em Moçambique', in C. Serra and J. Carrilho (eds) *Dinâmica da Ocupação e do Uso da Terra em Moçambique*, Maputo: Escolar Editora.

Habermas, J. (1973/2007) *Legitimation Crisis*, trans. T. McCarthy, Cambridge: Polity Press.

Hanlon, J. (2004) 'Do donors promote corruption? The case of Mozambique', *Third World Quarterly*, 25:4 747–763.

Harvey, D. (1989) 'From managerialism to entrepreneurialism: the transformation in urban governance in late capitalism', *Geografiska Annaler. Series B, Human Geography*, 71:1 3–17.

Jenkins, P. (2012) 'Sumário executivo do relatório de contexto', in J. Eskemose (ed.) *Home Space in African Cities*, Copenhagen: The Royal Danish of Fine Arts, School of Architecture.

Jorge, S. (2015) 'The re-emergence of urban renewal in Maputo: importance and scale of the phenomenon in the neoliberal context', in C. Silva (ed.) *Urban Planning in Lusophone African Countries*, Farnham: Ashgate.

Jorge, S. (2016) 'A lei e sua exceção: o caso dos bairros pericentrais autoproduzidos de Maputo', *Direito da Cidade*, 1:3 37–50.

Jorge, S. (2017) 'Lugares interditos: os bairros pericentrais autoproduzidos de Maputo', unpublished thesis, University of Lisbon.

Jorge, S. (1972/2001) *A Cidade do Capital*, trans. M. Ramos and M. Jamur, Rio de Janeiro: DP&A.

Jorge, S. (1974/1991) *The Production of Space*, trans. D. Nicholson-Smith, New York: Wiley-Blackwell.

Macucule, D. (2016) 'Processo-forma urbana. Reestruturação urbana e governança no Grande Maputo', unpublished thesis, New University of Lisbon.

Lefebvre, H. (1968/2012) *O Direito à Cidade*, trans. R. Polo, Lisboa: Letra Livre.

Maricato, E. (2000/2013) 'As ideias fora do lugar e o lugar fora das ideias', in O. Arantes, C. Vainer and E. Maricato (eds) *A Cidade do Pensamento Único, Desmanchando Consensos*, Petrópolis: Vozes.

Melo, V. (2015) 'A produção recente de periferias urbanas africanas. Discursos, práticas e configuração espacial: Maputo versus Luanda e Joanesburgo', unpublished thesis, University of Lisbon.
Oppenheimer, J. and Raposo, I. (2002) *A Cooperação Direccionada para os Grupos Vulneráveis no Contexto da Concentração Urbana Acelerada, 1. Pobreza em Maputo*, Lisboa: Ministério do Trabalho e da Solidariedade.
Oppenheimer, J. and Raposo, I. (2007) 'Entre os constrangimentos estruturais e a margem de manobra dos citadinos africanos: reflexões conceptuais e metodológicas', in J. Oppenheimer and I. Raposo (eds) *Subúrbios de Luanda e Maputo*, Lisboa: Edições Colibri.
Raposo, I. (2007) 'Instrumentos e práticas de planeamento e gestão dos bairros peri-urbanos de Luanda e Maputo', in J. Oppenheimer and I. Raposo (eds) *Subúrbios de Luanda e Maputo*, Lisboa: Edições Colibri.
Rolnik, R. (2015) *Guerra dos Lugares. A Colonização da Terra e da Moradia na era das Finanças*, São Paulo: Boitempo.
Santos, M. (1996/2008) *A Natureza do Espaço*, São Paulo: Edusp.
Serra, C. (2013) 'Transmissibilidade dos direitos de uso e aproveitamento da terra em Moçambique', in C. Serra and J. Carrilho (eds) *Dinâmica da Ocupação e do Uso da Terra em Moçambique*, Maputo: Escolar Editora.
Souza, M. (2006) *Mudar a Cidade: Uma Introdução Crítica ao Planejamento e à Gestão Urbanos*, Rio de Janeiro: Bertrand Brasil.
Souza, M. (2011) 'A cidade, a palavra e o poder: práticas, imaginários e discursos heterónomos e autónomos na produção do espaço urbano', in A. Carlos, M. Souza and M. Sposito (eds) *A Produção do Espaço Urbano, Agentes, Processos, Escalas e Desafios*, São Paulo: Editora Contexto.
Sposito, M. (2011) 'A produção do espaço urbano: escalas, diferenças e desigualdades socioespaciais', in A. Carlos, M. Souza and M. Sposito (eds) *A Produção do Espaço Urbano, Agentes, Processos, Escalas e Desafios*, São Paulo: Editora Contexto.
United Nations (2013) *Making the Most of Africa's Commodities: Industrializing for Growth, Jobs and Economic Transformation*, Addis Ababa: United Nations.
Vivet, J. (2012) *Déplacés de Guerre dans la Ville: La Citadinisation des Deslocados à Maputo (Mozambique)*, Paris: Karthala.
Waty, T. (2011) *Direito Económico*, Maputo: W&W.

Part 2

Rethinking the spatial triad and rhythmanalysis

John P. McCarthy and Michael E. Leary-Owhin

The spatial triad is perhaps Lefebvre's most popular conception, forming an innovative and powerful tool for understanding and interrogating apparently inert container-like urban space. Nonetheless, as with most Lefebvrian concepts, it is susceptible to a variety of interpretations, and – some would argue – misinterpretations. While each element of the triad is simultaneously intuitively simple and easily grasped, it also assumes increasing complexity and ambiguity on closer inspection. Spatial practice for example includes routine ways of life that are qualitatively different from the everyday life of spaces of representation. In turn, everyday life is woven by Lefebvre into ideas regarding rhythmanalysis. This part provides some radical rethinking and reinterpretation, while not seeking to denigrate previous research that counter-poses representations of space and spaces of representation. It highlights the range of approaches to Lefebvre's core ideas including the spatial triad and rhythmanalysis, and demonstrates ways in which the spatial triad and rhythmanalysis can provide intellectual inspiration for research on the production of space.

Olivier-Didier's opening chapter considers the role of language (via the common use of 'the Bronx is burning' in the context of the history of fire incidence in the South Bronx, and the more recent occurrence of gentrification. He includes the results of interviews with key figures, showing that the rebuilding of many parts of the area after fire damage has resulted in much lower-density development, with new houses replacing apartment blocks. He adds that this shift has been encouraged by the behaviour of federal mortgage insurance providers who stipulated that funding was to be applied to suburban single-family houses. Nevertheless, he shows how residents of the Bronx have overcome the prejudiced use of language and turned this into a narrative of resistance.

Viegas's chapter then focuses on the example of Luanda to illustrate the paradoxical effects of grassroots resistance to neoliberal development approaches. Specifically, she argues that self-produced settlements, while sanctioned by government, have no access to proper infrastructure and are denied many of the benefits of urbanisation such as employment and health. At the same time, she shows how the government has applied policies of forced eviction in other contexts, linked to resettlement and re-housing. Consequently, she suggests that the application of 'micro-resistances' to neoliberalist development strategies may actually divert needed energy and attention from larger and more important claims, though she also highlights a potential progressive way forward in this context is via the co-production of space. Interestingly in the light of other chapters in this book, she also highlights the need to take account of context dependency, arguing that the Angolan case in particular is clearly context-specific and so conclusions cannot readily be transferred to other situations.

Lee's chapter explores how Lefebvre's work might contribute to the understanding of urban anthropology. She outlines her view that Lefebvre has been overlooked by anthropologists in part because of the historic division between sociology and anthropology. She highlights several areas for which Lefebvre's work might be of interest to anthropologists, including for instance in relation to the theory of ground rent and rural sociology, and the critique of everyday life. Such conceptual frameworks, she suggests, have implications for anthropological research via for instance migration, or the impact of welfare provision on individuals.

Hanson and Rainey's chapter focuses on the current UK context, including the implications of the decision to leave the European Union. They make use of Lefebvre's notion of the 'concrete abstract' as the bridge between reified concepts and everyday social life, linking the infrastructural with the ideological. As with other authors, however, they highlight the potential problems in interpreting Lefebvre's ideas, but they focus on the need to promote political change rather than policy and practice implications. This perhaps aligns with the corresponding views of many observers who point to the way in which Lefebvre's ideas in some cases (particularly in relation to the right to the city) have been rather diluted in terms of radical edge.

Torres Garcia's chapter makes reference to Lefebvre's spatial triad in the context of rural settlements (colonisation towns) developed by the Spanish Francois regime from the 1940s to the 1970s. He makes specific reference to the notions of bricolage, ritualisation and fetishisation to explain how space evolves over time. The colonisation towns were set within an era and ethos of modernist planning as well as authoritarian power. In fact, he points out that the aesthetics of modernism were regarded as associated with the left and so were not emphasised. Around 300 new settlements were built across Spain, applying a compact urban model and design guidance that favoured curved alignments rather than long perspectives. The traditional single-family home dominated, reflecting, he suggests, the dominance of the State and Catholic Church. He suggests that future research could usefully focus on how and why the towns' residents use, appropriate and adapt the used spaces within them.

Cassegard's chapter considers the issue of homeless activism in encampments in Tokyo and Osaka. He shows how the number of homeless people in Japan increased significantly in the early 1990s, with many settling in new encampments or 'tent villages' which could afford a degree of protection. However, this led to the authorities attempting to evict them, often associated with planned regeneration, with consequent anti-eviction struggles often involving theatrical performance as a component. He uses Lefebvre's concept of representational space in this context, arguing that further classifications are needed where the homeless appropriate space in a way that goes against the dominant classification. Specifically, he suggests the use of 'counter-space', where activists refuse to conform to mainstream norms, and 'no man's land', where there is public neglect which allows for behaviour that would not normally be approved in mainstream public life. This analysis, he suggests, is important in assisting understanding of the dynamic of anti-eviction struggles throughout the world as well as the wider relation between space and social movements, which might, he suggests, lead to more inclusive notions of publicness.

Revol's chapter focuses on Lefebvre's notion of rhythmanalysis as a form of what she calls urban poetics. She foregrounds in this context the linkage between Lefebvre and the Situationist International, particularly Guy Debord, leading to ideas of an 'experimental utopia' which might include for instance the appropriation of space by alternative cultures.

Finally, Batuman's chapter considers political conflict and resistance in Turkey emerging over use of public space in Gezi Park in Istanbul as widely reported in media throughout the world. He highlights the importance of identity in relation to shared space – a theme echoed by many other authors in this book – and also shows the potential for resistance via occupation and appropriation. Nevertheless, in the context of the failure of such resistance, he proposes the potential for 'non-occupy' strategies of protest which might offer more potential for fundamental political change to address issues of inclusion and democratic urban land use.

Preliminary conclusions and thoughts

Lefebvre's ideas and self-build

The notion of 'autogestion' is central to many of the chapters in the book and reflected here for instance in the chapters by Viegas and Cassegard. How can we apply these more widely, for instance in ensuring not only that built space is more participative and reflective of individual needs, but is also safe, serviced adequately and fit for purpose? The example of the UK's Grenfell tragedy is perhaps instructive here.

Lefebvre's ideas and their concrete application – the case of rhythmanalysis and 'right to the city'

Many authors in this book highlight problems in operationalising Lefebvre's ideas, with each practitioner needing to 'find their own way'. How can they be assisted? Notions such as psychogeography (see for instance Revol's chapter) may be instructive here, showing the power of casual encounter and chance meeting/experience as methods not only of experiencing the city but evaluating its performance with the potential to cater for and encourage the potential of all its citizens. But how can this be encouraged/prioritised in an environment of formalised and functionalised land use regulation? Equally, how can the 'right to the city' be applied more clearly to policy and practice? And how can rhythmanalysis more directly inform urban planning, for instance?

Lefebvre's ideas and broader political change: how to politically operationalise?

Again returning to Hanson's chapter, how can Lefebvre's radical edge be revisited so as to fully encapsulate the transformative potential of notions such as the right to the city? How can political change (necessary at a deep level if Lefebvre's utopian aspirations are in any way to be accommodated) be assisted? Is there indeed the potential for damage to be done by treating his ideas in a tokenistic manner – with for instance the possibility of broader tactical urbanist strategies leading inevitably to neoliberal realities? More prosaically, how can we encourage effective resistance to inappropriate development of space and more differential spaces?

Lefebvre's ideas and urban culture/entertainment

The key element of Lefebvre's privileging of the city as festival runs throughout many chapters of this book. Certainly the idea of amenity and entertainment leisure is key to many urban strategies, even (as set out above) in the potential for planning to encourage informal uses. But how can we go further to encourage truly popular culture and even that which is transgressive? Can we find alternative uses for urban public space which fall outside traditional (or even non-traditional) plans and mechanisms for the organisation and production of space in our cities? Are there methods/means/techniques to be developed/suggested here which planners and other urban professionals should be aware of? What is the role of researchers here? How can the burgeoning informality of space (pop-ups/temporal space) be harnessed more widely?

These questions will be revisited in the conclusions to the book.

9

Still burning

The politics of language in the South Bronx

Oscar Oliver-Didier

Introduction

In game two of the 1977 World Series a helicopter broadcasted into the living rooms of millions of American homes a live image of a building ablaze just a few blocks from Yankee Stadium. At that same moment, the sports broadcaster Howard Cosell announced: 'There it is, ladies and Gentlemen, the Bronx is burning'.

On 16 May 2010, more than 30 years after, *The New York Post* (Flood 2010) published an article that stated that Cosell had never pronounced such words. The reporter, in going through all the tapes of the game, never actually heard Cosell use that famous phrase. Instead, it was likely that journalists covering the game, and for that matter the blaze, had placed afterwards the catchy phrase in Cosell's mouth.

What is even more surprising is that *The New York Post* is writing about this more than three decades after the fact. The truth is, to this day, the phrase is still recognised and greatly utilised, having been used as the title for a book published in 2006, *Ladies and Gentlemen, The Bronx is Burning* (Mahler 2005), and a miniseries based on the book and aired on ESPN in 2007, *The Bronx is Burning* (Chechik 2007).

Another theory is that the phrase was first employed in an episode aired in 1972 of a BBC series entitled *Man Alive*. The episode was called *The Bronx is Burning* (Morton 1972). The short documentary feeds its content from interviewing the mostly white firefighting members of Engine Company 82. The end result is an episode filled with racial assumptions that mainly places the blame of the fires on young ethnic minorities – who supposedly behave outside the accepted cultural norms and ideals of the US white suburban population. In the only brief moment where some of the young people of colour are interviewed, they are asked for the reasons why they started the fires in the first place – assuming they are the sole culprits and without any other empiric evidence or mention of the other reasons why the Bronx was really burning.

Midway through the documentary, the only black firefighter of Engine Company 82 is interviewed. He is only asked if being black made a difference when interacting with the residents, who again are assumed to be to blame for the fires. This underlying sense of racism and Otherness that is present throughout the documentary is also symptomatic of how government officials and other white Bronx residents were referring to and treating the mostly black and Puerto Rican populations living in the borough. The Bronx-born philosopher Marshall Berman states, 'The defamation of the Bronx helped to create a language for the much more extensive and profound defamation of New York' (Berman 2016: 126–127). This speaks volumes to the way in which language is used as an ideological tool for racism and prejudice. Berman goes on to say:

> Observers of the Bronx's troubles, including many New Yorkers, developed an elaborate vocabulary of deflection and denial, which very soon would be used against New York itself. 'What's wrong with these people? Why are they doing this to themselves?' Magic words like these transform victims of misery and misfortune into perverse perpetrators of malice. Social scientists got millions of dollars in grants, from foundations and federal agencies, to explore the character defects of poor people from the Bronx that led them – here was another dehumanizing cliché of those days – 'to foul their own nest'.
>
> *(Berman 2016: 127)*

The charged language utilised to describe these people of colour ignores the fact that between the end of the 1950s, and to a great extent until the 1980s, these communities were confronted with public policies that, some because of inaction and others stemming from direct measures, provoked the abandonment of great swaths of the Bronx. This led to a political and economic climate that provoked the burning of buildings in the South Bronx.

Dissecting some of the reasons why the Bronx burned

By simply partaking in a brief historical inquiry, one quickly realises that these fires were caused by a long list of reasons, among them that fires were started by landlords who, amongst other things, were looking to collect property insurance and come out with at least some profit for what was already considered a lost cause. In that sense, not only landlords, but financial and insurance institutions and the local and federal governments were also to blame for these fires via a complex web of actions: nationwide cuts on federal spending and national economic recession; an almost bankrupt city implementing austerity measures to deal with deindustrialisation; banks redlining areas to exclude them from property improvement loans; lack of resources and interest from fire marshals to investigate cause of fires and identify arson; quick and easy payoffs of insurance claims for burned down buildings; looters burning abandoned or inhabited buildings to facilitate the extraction of valuable metals; a huge influx of welfare recipients that, because of lack of heat or state of disrepair of building, got bumped up on the list to relocate if the building had burned down; and political manoeuvres by city elected officials that gave tenants the right to pay only one dollar in monthly rent if any, however minor, building violation was reported (Jonnes 2002: 202). In that sense, a perfect storm for destruction was set in motion, and its path crossed straight through the heart of the South Bronx and most of its geographies. The tenants that stayed, or that arrived here because they did not have any other alternative, suffered and went through unimaginable circumstances.

Arguably, during this period, no other municipality in the United States was more associated with urban detriment. The borough suffered a huge drop in population in the late 1960s and the 1970s that culminated in a tide of arson. When the flames finally subsided in the early 1980s the outcome was stark. The South Bronx had lost 60 per cent of its population and 40 per cent of its total housing units (Gonzalez 2004: 122). It is no surprise that so many people would start saying that the Bronx had burned down.

Statement and methodology

Language is a powerful tool. It can be used as a means to shape and manipulate knowledge, and it can be co-opted and employed as a political mechanism for empowerment and resistance. The verbal economies that nurture the perception of the South Bronx are highly volatile and can be used in different ways by its diverse actors. To this day, that infamous phrase, 'the Bronx is burning', is still overheard at Bronx Community Board gatherings, tenant association meetings, public hearings and government outreach sessions to describe what is still considered one of the darkest periods in the history of the city. Even though development and improvements have occurred since the 1980s, the physical and social impacts of this period are still very present in the minds and everyday lives of its residents.

The current scepticism from the community towards any project or government endeavour is very real and is part of a political climate that one has to keep in mind before initiating any kind of conversation or exchange with residents. For too much time backs were being turned towards Bronx residents and to their clamours for social and urban improvements. Instead, for decades, the only investments in infrastructure the Bronx would receive were ones that benefitted its close neighbours in Manhattan (water treatment and trash plants, food distribution centres with fresh products for populations outside the Bronx, amongst many others) or institutions that the privileged sectors did not want in their vicinities (for instance juvenile detention centres and homeless shelters, etc.). However, a lot has changed since this period, and today one can see affordable multifamily residential development and relatively small, yet significant increases in market rate housing – all standing in stark contrast to the handful of empty lots that are still left from this period and the low-scale redevelopment that occurred immediately after the fires.

In utilising a Lefebvrian approach, the juncture of what physically remains of this period of blight – now visually represented by new contrasting developments that fill formerly empty sites, public housing towers placed on vast green surroundings or one- to two-story residential homes that at one moment in time were thought to be the only alternative for development – and the verbal economy that is reproduced by its urban actors helps to continuously remind these communities of a not-so-distant period of neglect and racial injustice that for far too many continues to this day.

Through recent literature, interviews and field work, it is the aim of this chapter to illustrate how the constantly narrated space of the 'Bronx is burning' days, when combined with the collective memories of the everyday social rhythms of the period just before the fires, nurtured the will of these communities to rebuild. Today, the physical/visual disjunction of multiple and contrasting building types and spaces not only contribute to a recurrent visual reproduction of the rebuilding process after the fires, but also provide the physical traces to the ideological, economic and political underpinnings that shaped the rebuilding process itself. In unravelling this intersection of language and space – studying what is/was narrated, and dissecting what is/was seen – we can start to understand how charged memories of the near and distant past became perennial tools for the political action of minorities and communities of colour in the South Bronx.

Introduction to key Lefebvrian terms

The production of space is simultaneously physical – things in space – and at the same time, charged with assumptions, meanings and prejudices – constructed thoughts about space. One could argue that the South Bronx is a great example of how the machinations of the triad of space operate. Language plays a dual role in this understanding; on the one hand it is employed as a powerful tool to justly or unjustly describe and relate to a place, and on the other, the prowess of language is also reflected and solidified in the idiom of the built environment. Both interpretations work together to possibly subject a place to disinvestment, abandonment and neglect, but can also be repurposed into a language of empowerment, reconstruction and community prosperity.

This interplay between language and the built environment can also be further understood via another less well-known Lefebvrian concept: rhythmanalysis. Lefebvre saw rhythmanalysis as a pulse of the everyday, which stood in stark contrast to systems of power – such as that of a capitalist system. The everyday is shaped by the rhythm of biological bodies, but these bodies in turn produce social rhythms. However, Lefebvre believed that the rhythmanalist does not simply analyse the body as a subject but uses the body as the first point of analysis and a tool for subsequent investigations – the body serves as a metronome.

In a similar way that Walter Benjamin described the flâneur (see Benjamin 1999), Lefebvre talks about the rhythmanalist. However, in contrast to the flâneur, who wanders and passively observes the new social order of the city, the rhythmanalist tries to unravel, through the *present* image, the *presence* of objects – its historical, political and ideological realms.

This stress on the *mode* of analysis rather than analysis of rhythms is what is meant by rhythmanalysis (Lefebvre 2004: 6). In his book, by utilising a Mediterranean town as a case study, Lefebvre discusses how through routines and everyday rhythms people are able to resist political domination of space. In times of turmoil, these rhythms are determined by alliances, and space is appropriated from oppressive political powers.

A third concept – architectonics – is key to this particular chapter. Lefebvre explains that 'The task of *architectonics* is to describe, analyse and explain this persistence, which is often evoked in the metaphorical shorthand of strata, periods, sedimentary layers, and so on' (Lefebvre 1991: 229). Through an architectonic exercise, the current physical outcome of the Bronx is burning days will be briefly studied, and in the process unravel the political and social frameworks that led to this specific physical outcome.

Finally, there is the narrated. Time and repetition, when applied to language, bring to the forefront the importance of what is verbally expressed, how it reverberates in memory and in turn is structured in space. In a recent article published in *City and State New York* titled 'Re-brand-ing the Bronx', it is discussed, from mostly a public relations perspective, how the Bronx's past is verbally handled. The article has multiple quotes from elected officials, demonstrating different rhetorical approaches to the Bronx is burning days. For example, Bronx Borough President Ruben Diaz Jr. is quoted as saying, regarding 1970s-era stereotypes (Trangle 2016: 16):

> you don't want people to forget either. You don't want people to forget the struggles… you have to strike a balance. You want people to recognize the gains, rather, and you don't want them to forget the struggles. How do you recognize the gains? It's only by reminding folks of what it used to be like.

Diaz is of course recognising how powerful both the memories of the fires and of actually overcoming and rebuilding are. He wants to ensure generations of newcomers know about these events. However, in that same article, Derek Mueller, a board member of the Digital Rhetoric Collaborative, speaks of the political motives behind this utilisation of language: 'It's an exercise in contrast, and contrast underscored with political credit to current leaders and their deeds' (Trangle 2016: 16). Language ends up being a mechanism for people in power to frame their policies against a backdrop of past detriment – in order to highlight and differentiate their decisions from any form of negative criticism. Through contrast, the stories behind the fires and subsequent rebuilding are verbally co-opted to position current processes and decisions.

The article proves how the past can be interpreted in multiple manners – transformed into rhetorical constructs – in which alluding to the days that the Bronx was in blight and decay helps you gain control of a present narrative. In today's case, politicians allude to the fires precisely because so much has changed since they occurred. Redevelopment today is situated in a very different political and spatial context than when people were rebuilding decades ago.

Another recent example of how the Bronx is burning's meaning can be readily interchangeable was evidenced with a private party hosted by developers in a site where three market rate towers are slated for development. The party took place in the Port Morris area of the South Bronx in October of 2015. Officially titled *Macabre Suite*, star-studded guests such as Adrien Brody, Kendall Jenner and Naomi Campbell were encouraged to use the hashtag #bronxisburning in their social media updates. The party included bullet-riddled cars and dumpster fires that guests could take pictures with. Although Diaz, Jr. was in attendance, Speaker Melissa Mark-Viverito quickly pointed in her Twitter account to the insensitivity of theming and naming the party in this manner (Ellman 2015). This just proves how problematic any reference to the fires can still be to residents, especially when it brings any allusion of being pushed out again – in this case not through fires, but through the spectre of gentrification.

How it got rebuilt: interviews with key figures in the Bronx

As we stated before, it can be argued that when the community groups that decided to rebuild the Bronx were confronted with such a behemoth of a challenge, they were both thinking of the social arrhythmia and physical detriment of the Bronx during the fires, and the period of strong and vibrant everyday rhythms before they started to occur. In this section we will discuss, through a handful of conducted interviews, the simultaneities at play with the narrated processes in and of the Bronx.

Angel Garcia is a researcher on the Bronx and resident of the public housing project right next to the Jackson Avenue station of the two and five lines. When asked about the physical remnants of the Bronx is burning days he narrates the scene that unravels when riding the elevated train along Westchester Avenue from East 149th Street east to the Bronx River. Angel speaks about what he termed a 'dissonance' of the built environment from the pocket of 'nice homes' – as he describes them – on Intervale Avenue to the empty lots along Simpson Street and Freeman Avenue. Angel asks himself, 'Why are these [empty] lots here? It's a legitimate question. It looks from another area'. He goes on to talk about the contrasts between the old Bronx churches and the ranch houses at Charlotte Gardens: 'It's an unusual thing! Well kept, but odd… Dissonant… Not a consistency among them' (Garcia 2017)

Angel is of course well-versed in the history of the Bronx. While talking to him he went into great detail about the political players and policies that led to the arson, but he can also speak at length about the urban dissonance previously mentioned. This occurred for various reasons including: lack of federal funding for dense urban typologies, a belief that home ownership – not rentals – would create a broader sense of neighbourhood ownership, bureaucratic procedures that complicated and extended the process and timeframe for building housing, a genuine interest from these community builders to invest what little money they had to also provide tenant organisation and social services to residents – thus limiting the monetary reach of what could actually be built (brick and mortar) – and lastly, but more importantly, the end

Figure 9.1 View of contrasting building types in Charlotte Gardens in the Bronx. *Source*: Kingsnyc di Wikipedia in inglese – Trasferito da en.wikipedia su Commons. Publicco Dominio.

result of an area that saw huge drops in population after the fires. In the end, the new housing being built was mostly catering to the small population that remained. As Jonnes says, 'Equally important, rebuilding with houses meant population density plummeted. There may be as many people living in one block-long section of six-story apartment houses near Charlotte Street – perhaps several thousand – as there are in block after block of the new houses' (Jonnes 2002: 430).

Another interviewee, Dana Driskell – who was born and raised in the South Bronx, attended the locally renowned PS 31 or *Castle in the Concourse* and was District Manager of Community Board 3 during the late 1970s – talks in even more detail about the reasons behind this dissonance. He goes on to say, 'Federal mortgage insurance programs were tapped to fund the ranch style houses in Charlotte Gardens – the caveat being that these funds were to be used for suburban single-family houses' (Driskell 2017). The sheer lack of government programs to fund urban development in the Reagan years – outside his urban policy of tax incentives for urban developers – did not provide any other alternative. As Dana goes on to say during the interview, 'it was the only way to redevelop' (Driskell 2017).

Community organisers such as Genevieve Brook of the Mid-Bronx Desperadoes and Ed Logue – who was director of the city's South Bronx Development Office – had the vision to rebuild in one of the most impacted areas by the fires, Charlotte Street (Christie 2009):

> They wanted single-family homes; critics wanted density and multi-family dwellings, saying it would promote a lively, safe neighborhood and attract merchants. Brooks, though, knew most of the families in the area were African Americans from the South, Caribbean blacks and Puerto Ricans, and she was convinced that the long home-owning traditions of these groups would help make a community of single-family homes work.

This reconstruction and development process was in most cases supported by the triad of: Community Development Corporations (CDCs); the financial support of the Community Preservation Corporation (CPC), 'a nonprofit arm of New York City's major commercial banks and thrifts, lending money to small landlords in working-class neighborhoods for rehabs' (Jonnes 2002: 395); and the expertise provided from the Ford Foundation-backed Local Initiatives Support Corporation (LISC), which helped these groups get loans, grants or technical assistance to navigate through the bureaucratic and financial complexities required to get money for rebuilding. When these three entities were paired with federal programs such as the Community Reinvestment Act of 1977, 'enforcing by law… that banks had a responsibility to lend money where they garnered deposits' (Jonnes 2002: 369), and the Low Income Housing Tax Credit passed in 1986, 'a program that provides tax write-offs to corporations and individuals who invest in creating housing for the poor' (Jonnes 2002: 397), one is able to gain a reasonably complete grasp of how redevelopment occurred in the Bronx.

As was mentioned, the idea that home ownership was also a way to guarantee that a population stayed and felt a sense of ownership with their neighbourhood was also very informative to the way things got rebuilt. In order to ensure that people would be able to afford their new homes, a two-story housing model was dispersed all throughout different parts of the Bronx – the lower story having the potential to generate much needed income to the new buyers (Jonnes 2002: 396, 438):

> The basic format was a small two- or three-family row house whose one or two rental apartments could help pay the mortgage.… The Partnership houses were a brilliant way to offer home ownership to people who could *not* afford a house without a rental unit to subsidize their mortgage payments. It was also a brilliant way (in once-chaotic neighborhoods) to create rental-housing units that would be tightly supervised by vigilant resident owners.… The houses have three bedrooms with two baths on the second floor. The first floor is a separate unit with two bedrooms, two baths that they can rent for a $1,000 a month to help pay the mortgage.

Yet another interviewee, Xavier Rodriguez – who recently retired from his role as District Manager at Community Board 5 and was raised in the Bronx in the 1950s after he and his mother, a Puerto Rican, left East Harlem after their building was condemned to make way for an Urban Renewal project – can speak at great lengths not only about the days of the fires and the subsequent rebuilding, but of the everyday life and social rhythms of the Bronx before the period of neglect and abandonment. Again, this is key to understanding how language was a vital force for the community rebuilding that took place. With almost the exact same level of detail, Xavier's anecdotes can jump through all of these three important historical periods. Not only was Xavier there throughout, but he was also the Chair of Community Board 3 during the 1970s, the worst period of the fires, and in one of the most impacted areas in the South Bronx. Xavier was also present when President Carter visited, and he can intermix his anecdotes with specific knowledge of whom the political players were after this area gained all the notoriety, and what programs subsequently worked and which did not.

Additionally, what was unravelled during the conversation with Xavier was that not only is the repetition of language – both spatial and narrated – of the Bronx is burning days a key mechanism for communities to decide to fight back and rebuild but, additionally, the verbal economy that describes what the Bronx was like before the fires – full of healthy and everyday social rhythms – is a missing component to understand why people decided to stay and improve in the first place. Yet, as Xavier describes, it is assumed that the people who drove these everyday rhythms of social life before the fires were the Jewish and white European descendants that decided to leave once the fires started, yet people of colour such as Xavier were also here before the blight. He goes on to describe this period (Rodriguez 2017):

> This was the best part of my life. We moved into a two-bedroom apartment on Fulton Avenue in 1957 or 1958 in a lower class Russian Jewish community. My mother was able get credit at the local grocery store that Mrs. Rosenberg owned. We were part of the community.

However, by the time Xavier was 12 years old and had reached middle school, Fulton Avenue had changed drastically. It was the 1960s and the arson had started, and in fact, in 1969, Xavier's own building on 173rd and Fulton was set ablaze while he was living in an apartment there. A lot had transpired since his mother was able to get groceries on credit a little over a decade before.

Afterwards, when the Bronx started to rebuild, mostly the Catholic institutions such as Nehemiah Homes and Father Gigante's group, the Southeast Bronx Community Organization (SBCO), were building small houses and duplexes. As Xavier says, although the quality was shoddy and the aesthetics secondary, these houses created a sense of pride and perimeters of five to ten blocks without any crime.

Additionally, the community's will to rebuild – nurtured by how life in the Bronx was before the fires – not only energised those who stayed, but even drove some former Bronx residents to return. 'Another good sign is that many people who fled the area in the 70s are moving back' (Worth 1999).

Even the reasons for rebuilding Charlotte Gardens stem from an odd interplay of factors, including a powerful symbolic one. President Carter had stood there, so rebuilding here would only mean things stood to improve elsewhere. The physical outcome of how Charlotte Gardens looks stems from a desire to visually represent the ideal of the suburban home with white picket fences (Jonnes 2002: 376). It could be argued that even though federal housing programs were contracting, and this was the only way housing could get built here, the contrast with the surrounding context was done almost purposefully. Charlotte Gardens had to represent wholesome ideals: home ownership and pride, and a safe and protected environs. The fact that Charlotte Gardens was showcased as a great success all over newspapers, magazines and different TV shows, and as a symbol of things to come, speaks volumes to the capacity of architecture to represent a new vibrant social order that was similar in nature to the one present before the fires. Just gutting and rehabilitating an old tenement building was not as potent a signifier in this case. The fact that later developments incorporated typologies with higher densities not only points to an improving New York City housing market, but also to just how exceptional and unique the symbolic nature of Charlotte Gardens was (Ryan 2012: 29).

Much like previous endeavours before and after this, what was coveted was reestablishing the social vibrancy of neighbourhoods – not a literal recreation of how things looked in the past. In any case, this would have been impossible given the economic limitations of the time. A few examples: Irma Fleck's empty lots as urban gardens; Ramon Rueda's sweat equity, rehabbing of abandoned buildings and an idea for an urban village with his People's Development Corporation; and Ramon Velez spearheading the Model Cities program (with limited success) in Mott Haven, with its mix of open spaces, building rehabs and the offering of social services, are not only examples of urban models that are a product of the economic and political climate of the time, but also of an urge to create new vehicles that promote and are conducive to the social vibrancy that existed before the fires broke out.

Professor Mark Naison of Fordham University also had a chance to interview Dana Driskell in the fall of 2015 (Driskell 2015). Similar to Xavier Rodriguez, Dana describes his time growing up in the Bronx before the fires in the 1950s and early 60s. His descriptions pretty much follow the script of how life was around this period. His father was a union worker for a sheet metal manufacturer, and his mother a homemaker until she and his father separated and she began working at different garment centres across the city. Before that, his grandfather had come from Florida and ended up owning three buildings in the Bronx – one of which Dana grew up in on 165th and Brook Avenue. When encountering Dana's description of his neighbourhood, one can quickly tell it was a vibrant sector full of public life (Driskell 2015):

> But in general I think it was a pretty stable environment at least maybe through mid-sixties, late-sixties....and there was always a neighbor looking out the window and if you didn't do the right thing they would report to mom or they would feel its ok to discipline you or warn you away. So it was a neighborhood, a family type of environment.

While growing up in the neighbourhood, Dana loved all kinds of music, but he had a special interest in salsa music – his current office at the Department of City Planning is covered with old salsa concert posters from the 1970s. His interest in musicians such as Eddie Palmieri and Tito Rodriguez comes from his godmother, who was Puerto Rican and who also lived nearby. Jazz and doo-wop also fascinated Dana, and even though he was too young to go to clubs such as the Boston Road Ballroom and the ones along McKinley Square, he can describe the social and public scene along this stretch (Driskell 2015):

> I can remember the clubs that were up in McKinley Square: the Blue Morocco and the 845. Now of course I was too young to attend,... But I can remember just the impression – you going through McKinley Square and its Saturday night and it was, not like a Times Square but everyone had their neon lights and some of their music would be coming –...it was a hot spot... it was just a kaleidoscope of the different clubs... and the music coming out and the well dressed guys and the cars parked in the front... you know that this was something happening. And there would always be the, the posters up and down the blocks during the week advertising this person is coming in.... a lot of the big acts would come through.

Dana goes on to describe how this scene started to fade out in the early 1970s, and how it also coincided with some of the blue-collar jobs disappearing due to deindustrialisation in the northeast region and the strong sentiment of dispossession this produced in the community. The Cross-Bronx Expressway had also cut through the heart of the Bronx, disconnecting neighbourhoods and negatively impacting the social life of these areas. Yet there were also wider systemic racial issues that were additionally affecting these communities of colour – that also led to the subsequent nationwide Civil Rights Movement. These all coincided with a time when Dana was becoming much more politicised; assisting in a teacher's strike and starting a political action committee at his high school, and attending anti-war protests. Dana is but one of many examples of Bronxites who became more politically active around this time. This, in conjunction with the charged memories of daily social rhythms, public life and music on the streets helped feed the will to resist and rebuild in the South Bronx.

Said Bronx music could be considered the perfect metaphor for the social vibrancy and diversity of the Bronx. The birthplace of hip-hop and 1970s salsa, the Bronx's rhythms continued through the fires and were also significantly strengthened by them. An urban language of sorts, the music of the Bronx speaks volumes of the will to persevere, to rebuild and to reestablish the former social rhythms of the Bronx. Marshall Berman states when describing his impression of 'The Message' by Grandmaster Flash and the Furious Five (Berman 2016: 131):

> The prospects look bleak. And yet, and yet, the rapper can't help but see: sometimes, somehow – we don't know how, but somehow – it's possible to emerge from the vortex of horror and violence. You can come from ruins, yet not yourself be ruined. Social disintegration and existential desperation can be sources of life and creative energy. Our first hit rappers know something that Hegel said modern men and women had to learn: they know how to 'look the negative in the face and live with it'. They have looked the ruins in the face, and they have lived with them, and they have come through. Now they can see and feel their way to new life.

In that sense, the Bronx rapper is the rhythmanalist par excellence. These young black, Puerto Rican and mostly minority groups of kids were able to take the language of the streets – or for that matter, even the lack of an actual street – and turn it, in the spirit of Lefebvre, into a vigorous example of the intersection of language and space. In the same manner that hip-hop culture has spread and become popular all around the world, the stories of the Bronx is burning have also traversed the globe – simultaneously being renowned as a place of destruction and re-flourishing.

Conclusions

Today, the Bronx's comeback is a case study that could serve communities being afflicted by extreme fiscal austerity measures, exploitative speculation of land values, deindustrialisation or displacement though gentrification or conflict. In its process of rebuilding, Bronx residents created a diverse and novel array of visual, rhetorical and sonorous patterns of language that are still narrated and physically evidenced throughout different parts of the borough.

As discussed before, Bronxites overcame the charged language and prejudiced set of actions being utilised against them and turned the 'Bronx is burning' phrase into a repurposed narrative for resistance and activism. These communities now live surrounded by the powerful remnants of the co-opted rhythms of the time before the fires, and the new idioms that sprung from the aftermath of destruction and the subsequent rebuilding of the South Bronx.

References

Benjamin, W. (1999) *The Arcades Project*, trans. R. Tiedemann, H. Eiland and K. McLaughlin, New York: Belknap Press.

Berman, M. (2016) 'New York city: seeing through the ruins', in R. Solnit and J. Jelly-Schapiro (eds) *Nonstop Metropolis: A New York City Atlas*, Oakland: University of California Press.

Chechik, J.S. (2007) *The Bronx is Burning* (TV Mini-Series), ESPN Home Entertainment.

Christie, L. (2009) 'The greatest real estate turnaround ever', *CNN Money*, accessed online at http://money.cnn.com/2009/11/09/real_estate/greatest_neighborhood_turnaround/index.htm?section=money_realestate, accessed February 2016.

Driskell, D. (2015) Interview with the Bronx African American history project, BAAHP Digital Archive at Fordham University, 29 October.

Driskell, D. (2017) Research interview, 17 January, it lasted about an hour.

Ellman, B. (2015) 'Developer throws tin-eared "Bronx Is Burning" Halloween theme party in the Bronx', *New York Magazine*, accessed online at http://nymag.com/daily/intelligencer/2015/11/developer-throws-bronx-is-burning-theme-party.html, accessed February 2017.

Flood, J. (2010) 'Why the Bronx burned', *New York Post*, accessed online at http://nypost.com/2010/05/16/why-the-bronx-burned/, accessed December 2016.
Garcia, A. (2017) Research telephone interview, 28 January, it lasted about an hour and a half.
Gonzalez, E. (2004) *The Bronx*, New York: Columbia University Press.
Jonnes, J. (2002) *South Bronx Rising: The Rise, Fall, and Resurrection of an American City*, New York: Fordham University Press.
Lefebvre, H. (1991) *The Production of Space*, trans. D. Nicholson-Smith, Oxford: Blackwell Publishing.
Lefebvre, H. (2004) *Rhythmanalysis: Space, Time and Everyday Life*, trans. S. Elden and G. Moore, London: Bloomsbury.
Mahler, J. (2005) *Ladies and Gentlemen, the Bronx is Burning: 1977, Baseball, Politics, and the Battle for the Soul of the City*, New York: Picador.
Morton, B. (1972) 'The Bronx Is Burning', *Man Alive*, Season 7, Episode 1, BBC Television, 27 September.
Rodriguez, X. (2017) Research interview, 15 February, it lasted about three hours.
Ryan, B.D. (2012) *Design After Decline: How America Rebuilds Shrinking Cities*, Philadelphia: University of Pennsylvania Press.
Trangle, S. (2016) 'Rebranding the Bronx', *City and State New York*, 19 September, 10–16.
Worth, R. (1999) 'Guess who saved the South Bronx? Big government: the silent partner in community development', *The Washington Monthly*, 31:4 26–32.

10

Spaces of resistance in Luanda

'How do [small] gains become prisons?' an analysis from a Lefebvrian perspective

Sílvia Leiria Viegas

Introduction

The writing of this article was triggered by an email received during the recording of the findings of a larger investigation entitled: *Luanda, Cidade (im)Previsível? Governação e Transformação Urbana e Habitacional: Paradigmas de Intervenção e Resistências no Novo Milénio* (Luanda, (un)Predictable City? Government and Urban and Housing Transformation: Paradigms of Intervention and Resistances in the New Millennium). As I was cross-checking the theoretical and empirical data and preparing the ground for original conclusions, sharing ideas with José Patrocínio (26 September 2013), leader of the Angolan non-governmental organisation (NGO) OMUNGA, forced me to deepen an unexplored aspect of my research, namely an unpredicted and paradoxical consequence of the emerging and growing grassroots resistance to the current neo-liberal dominant system. This questioning will be further developed in this chapter; as Patrocínio (ibid.) states:

> which victories and losses [truly exist] in the apparent victories (I need you to explore this subject). I mean, there were gains but not victories, is that it? What do these gains cause in terms of the loss of energy to fight? When will the energy to new claims be back? How do [small] gains become prisons? (Freely translated by author.)

The uncertainties of Patrocínio suggest that, by strategically recognising the cracks in the current neo-liberal facade, the Angolan government seeks to weaken the resistance of the activists and inhabitants that contest official urban and housing policies and practices, while defending everyday self-production of space so as to solve some socio-spatial problems caused both by market-driven interventions and the absence of governmental positioning aimed at improving living conditions for deprived communities. As such, this chapter addresses specifically the Angolan capital in the 21st century and seeks to encourage the discussion about (1) the top-down and bottom-up production and transformation of space; (2) the 'in-between spaces co-induced' amongst both forms of space production, with reference to current co-produced practices; (3) the dynamic forces created by the interaction of the government, in line with its urban and housing political agenda, and the militant resistance of some segments of organised civil society encouraged by impoverished social groups that contest official practices; and (4) the socio-spatial consequences of these ever-changing and very unstable aspects regarding the quality of urban life, especially in Luanda, and the access of urban society to a complete form of citizenship.

Taking into account the scope of Part 1, *Globalised neo-liberal urbanism: hegemony and opposition*, my argument is basically built around two of Henri Lefebvre's most distinguished books regarding cities and urban society, specifically *La Révolution Urbaine* (The Urban Revolution) ([1970] 2003) and *La Production de l'Espace* (The Production of Space) ([1974] 1991), although also being guided by other important texts of the author, namely the well-known *Le Droit à la Ville* (Right to the City) ([1968] 2009). Nonetheless, I also considered the work of scholars who have critically commented on Lefebvre's thoughts, such as David Harvey, Neil Brenner, Christian Schmid and Andy Merrifield, so as to facilitate and reinforce a precise understanding of the reasoning. On the topic of the empirical framework of Angola, my comprehensive research about Luanda (Viegas 2015; 2016) sets the foundation for this theoretical-based discussion, whereas my previous theoretical-empirical discussions (ibid.) are used as a basis for a deeper exploration. All together, these reflections will carry the article to its conclusions and, hopefully, will pave the path for further questioning.

Theoretical-methodological and empirical framework

From its foundation, the Angolan capital always developed spatially in both 'official and un-official' ways. All of the management models (colonial, 1576–1975; socialist, 1975–90; and neo-liberal, 1990 to the present) were unable to promote urban and housing instruments capable of predicting, regulating and controlling the large occupation of land and its use. Together, local groups sustained the self-production of their space in *musseques* (i.e. un-official housing settlements) through what I call, in line with Lefebvre's thought ([1968] 2009; [1974] 1991), the 'unique appropriation of everydayness'. Given this, regarding the theoretical-methodological framework of this chapter, its structure follows central topic-analysis developed by Lefebvre (ibid.) in his writings *The Urban Revolution* and *The Production of Space* as referred to, among other relevant works. These selected key themes support the perspective that the ongoing urbanisation of the society is an uncompleted and uncertain process. This view takes into account conflicts between different modes of production of space and, regarding Luanda, will be analysed with reference to the guiding notion of the *Right to the City* (*idem* [1968] 2009).

Considering that Lefebvre addresses the problematic of neo-liberalism, including the theme of the 'big state' in line with the powerful and very engaged logics of the capitalist global system, the next section of empirical data refers to the urban and housing strategies, policies and practices of the Angolan regime in the present. Let us not forget that, according to the recent official national survey (INE 2016: 15), this African country has approximately 25.8 million inhabitants, and 6.9 million of its urban population lives in Luanda. In addition, the Angolan capital faces a growth rate of 1.9 per cent (United Nations 2014: 20), while 80 per cent of its population lives in self-produced settlements (Governo da República de Angola 2011: 1561). In such a challenging context, this section also introduces the perceived neo-liberal threats to urban space. In addition, the following section explains both (1) the procedures of the self-production of space, including those that are a consequence of current state interventions, and (2) the most recent processes of co-production of space within the framework of the capitalist mode of production where, according to Lefebvre ([1972] 2000), the participants of everyday life are reduced to their functionalist role as inhabitants.

The last two sections before the conclusions focus on the developing revolutionary contestation to the hegemonic nature of neo-liberalism and its consequences, although clarifying how

Lefebvre's spatial ideas point to the opportunity (and urgent need) to take advantage of the flaws in the system in order to overcome it. But then again, my considerations go further. The section referring to what I designate 'co-induced in-between spaces' – to be precise, spaces resulting from actions of resistance – focuses on the materialisation of the attempt to surpass the ideological and institutional structure that, according to Lefebvre ([1970] 2003), needs to be overcome so as to promote the construction of another greater rationality, namely true urban democracy, as this author (ibid.) defends. Nevertheless, empirical research will also reveal the overwhelming power of global capitalism by presenting the indicators of its efforts to imprison current militant resistance within the framework of its own line of action, namely of how the system tolerates militant resistance only to succeed in absorbing it, whilst seeking to predict the tendencies of future gains and losses for the socio-spatial and economic development of urban society towards the process of forging complete urbanisation, as stressed by Lefebvre ([1968] 2009].

Top-down production of space

After the end of the civil war (1975–2002), during the 'national reconstruction', the Angolan administration decided to produce a new middle class by boosting economic growth. To accomplish this ambitious endeavour, the government encouraged the enlarging of the construction sector, with preference given to the real-estate market. As such, space is now frequently targeted for a more affluent society, mostly in Luanda. Some policymakers inclusively argued that, along with the export of hydrocarbon-based resources (e.g. oil), it was fundamental to adopt this strategy, as the Angolan state was profoundly engaged in promoting social welfare. Let us recall that, in general terms, in the early 70s, Lefebvre ([1970] 2003) had already identified up-and-coming forms of spatial conquest (including: land purchase and sale and real estate speculation) that led to the contemporary on-going reorganisation of urban and/or housing space. These restructurings are now, as they were before, largely subordinated to the centres of power and primarily aim to create surplus value. As recently underlined by Brenner (2008), the analytical lens provided by Lefebvre's theory is a powerful and critical decoder of prevailing state productivism.

Presently, Angola is a presidential republic. Dos Santos, leader of the party in charge since 1975, concentrates power while being head of state and of the government, and commander in chief of the military as well. Furthermore, the long announced political-economic decentralisation has not been implemented. As indicated by Orre (2010), central administration has extended its authority to the entire country instead. In such a context, I would argue that alternative political and/or local solutions aimed at contesting the government's neo-liberal strategy and effects (such as deprivation, social exclusion and spatial fragmentation, as identified by Viegas (2015)) are often discouraged. Regarding urban matters, official and/or 'public' (namely of alleged public interest) production of space in Luanda is often narrowed to the decisions of an elite harshly motivated by the capital's broader urban renewal, expansion and internationalisation, as expressions of the economic rise of Angola in the contemporary global context. These private and/or governmental leading players act on behalf of the state, reinforcing its hegemony by means of implementing urban strategies and policies that follow the logics of class society, while seeking to control social practices, as Lefebvre ([1970] 2003) would surely argue. Harvey (1998: 55) refers to analogous dynamics in *The Geography of Class Power*, sustaining that 'the geographical meaning of the bourgeoisie is the reproduction of class'.

The large number of political instruments (including: legislation, comprehensive and detailed plans and programmes) recently created by the Angolan government is in deep consonance with its determination to forge a bureaucratic and standardised urban society aimed at supporting its neocapitalist agenda, being what Lefebvre ([1974] 2000) would classify as official representations of space. As such, (1) the *Lei de Terras* (Land Law), *Lei do Ordenamento do Território e Urbanismo* (Law of Spatial Planning and Urbanism), *Lei de Fomento Habitacional* (Law of Housing Development) and their specific Regulations, approved between 2004 and 2009, alongside with the (2) *Planos Integrados de Expansão Urbana e Infraestruturas de Luanda-Bengo* (Integrated Plans for the Urban and Infrastructural Expansion of Luanda-Bengo) from 2011 and the (another) *Plano Director Geral Metropolitano de Luanda* (Metropolitan Masterplan for Luanda), in preparation, follow the major effort of the (3) *Programa Nacional de Urbanismo e Habitação* (Million Houses Programme), in progress since 2009, specifically the extinction of self-produced housing settlements and the weakening of deprived communities inhabiting these areas, including their urban and housing rights. Harvey (2001) criticises the monopoly of legality sustained by the neo-liberal state, in general terms, by means of denouncing its instrumentalisation for the promotion of violent processes.

Through the use of these 'public' instruments, among several others of lesser prominence, the Angolan government is leading a myriad of socio-spatial practices, while using (the discipline of) urbanism both as a mask and a tool of political power. As argued by Lefebvre ([1970] 2003), the spaces that are produced according to their (and the ruling regime's) logics of representation, instead of responding to the particular logics (and needs) of the society which they claim to be working for, are eminently political spaces. As such, it is perceptible that, by claiming proprietorship over most of the country's land, also by demarking state land reserves for forthcoming official urban and housing practices, the Angolan government is imposing maladjusted urban and housing models, especially on the deprived populations. These models are based on grid street plans and on acritical imported post-modern architectural typologies, even while being in harmony with the Angolan elite's idealised image of an international city. Then again, the analysis of space and policies that support the Angolan state mode of production may well reveal tendencies and dangers related to the spacialisation of unrestrained capitalism and its escalation (Lefebvre [1972] 2000). Examples include neglecting access to appropriate urban infrastructures, basic services and employment.

Bottom-up production of space

At present, as stated for the political production of space in Angola, with greater expression in Luanda, the inhabitants of self-produced settlements also have no access to proper urban infrastructures. Additionally, these populations are deprived of most of the 'benefits of urbanisation', such as employment, health, education, formal trading and leisure. Together, these parameters somehow correspond to what Lefebvre ([1968] 2009) calls a 'product', specifically referring to the production of space as value, in line with the *Right to the City*. Yet, despite these limitations and with minor official assistance, these inhabitants transform urban reality through everyday practices of silent resistance, a performance that Lefebvre would recognise as production of (social) space by social agents (ibid.; *idem* [1974] 2000). As recently stressed by Merrifield (2006), for Lefebvre, the solutions for the contradictions of everyday life must inevitably be found in everyday life. As such, these communities renovate their way of living through the appropriation of space, while reinforcing the relations established with space. This occurs near the heavily saturated city centre, also due to the unceasing influx of residents flowing from peripheral areas for informal trading, and in the outskirts as well, where daily, and at times in conflict, the self-production of space takes place rapidly and massively, and in multiple and diverse ways.

Since the take-off of 'national reconstruction' in the early years of the 21st century, the impoverished social groups began to be reprimanded by the government, because of the materialisation of their daily practices. This tendency follows 'the [current economic-based] development drive that seeks to coloni[s]e space for the affluent' (Harvey 2008: 39) and is aggravating land tenure insecurity while promoting several different urban conflicts, particularly in Luanda. In fact, the government proclaimed co-production of space through self-construction, either assisted or not, of housing, outbuildings and so on, within the unclear framework of the Million Houses Programme. But then again, the provision of state land for such flexible and permeable practices has been largely ignored, to say the least, by most government agents who prefer to encourage the participation of private entrepreneurs in the forging of urban space. Even so, the self-production of urban and housing space persists in central and peripheral areas. Given this, the real drawing of the city follows the occupation and experience of users-producers of space, embracing both 'urbanity', a notion associated with production of urban culture, and rurality, as connected to rural life, and both rooted in the specificities of everyday life, as emphasised by Lefebvre ([1970] 2003).

Recently, when the Angolan administration decided to demolish self-produced areas within a context of urban renewal, *tabula rasa* and gentrification of the centre and its coastal districts (for instance Chicala), those who were being evicted sought to stay, whenever possible, in neighbouring areas. Similar routines of 'accumulation by dispossession', as labelled by Harvey (2001), arose in the 1990s and are at the moment being reinforced in peripheral areas, along with practices of mercantilist urban expansion (for instance Talatona and Kilamba, Cacuaco), some related to processes of clientelism. Concerning these actions, the perception of social space, in line with Lefebvre ([1974] 2000), as a theoretical-methodological research tool, also contributes to underline the costs of replacing the existing self-produced spaces by others resulting from the current capitalist mode of production; namely it helps to emphasise the space of losses. Given this, deprived social groups tend to occupy other lands close to the official housing projects, trying to obtain benefits from this proximity. Sometimes the urge becomes so great that it forces these needy populations to invade and occupy recently demarcated state land reserves. As emphasised by Lefebvre ([1970] 2003), everyday life also incorporates the exclusion of certain moments of life and of numerous activities.

Resettlement and/or rehousing interventions (for instance in Zango, Panguila and Sapu) are paradoxically accepting and even encouraging gradual occupation of delimited spaces through everyday practices. However, this executive change in action only occurs in peripheral areas, far from the 'metropolis-to-be' and its international brand image. Regarding the rehousing projects, the homogeneous space produced by the government (i.e. the housing) officially connects with many social spaces produced within the limits of the plots dispensed by the state (e.g. walls, carports, outbuildings). This overlapped co-existence sets up a mediation between quantitative and qualitative produced spaces, as Lefebvre ([1970] 1978) would describe, while permitting the strengthening of newly 'co-produced' areas, a concept explored by Melo (2015), specifically through official incentives. As such, in spite of all of the numerous constraints, such as proper access to urban infrastructures and resources, these housing projects (i.e. 'neighbourhoods', as Lefebvre [1970] 1978 would define them) persist beyond urban growth, along with the problems arising from their multiple social practices. Nevertheless, these interventions also mirror urban renewal and violent social exclusion practices in the centre, looking, to a certain extent, like a growing wound in a peripheral detached site.

Militant resistance and awareness

In Angola, the actions of a specifically organised civil society (e.g. non-governmental organisations, NGOs) are foremost related to the fight against the urban policies and practices of the government, especially the aggressive methods through which the administration takes over self-produced spaces. These NGOs, such as SOS HABITAT (started by Luiz Araújo in 2002), based in the capital, and OMUNGA (started by José Patrocínio in 2005), based in Lobito, are generally driven by the socio-spatial penalties of these 'public' activities, that occur above all (but not only) in Luanda. As pointed out by Lefebvre ([1968] 2009) in other contexts, the *Right to the City* is imprisoned by private or semi-private interests. Therefore, its formulation implies a democratic management of urban processes, hence it involves greater control over the production and distribution of capital surpluses, fuelled by the discontent and by the collective demand for power by (and for) the people. Lefebvre (ibid.) refers to this process as 'revolutionary reform', encouraging a permanent cultural and economic revolution with a policy of self-management consistent with social needs, and consequently with greater democratic control of the state apparatus. Accordingly, these NGOs insist on a major shift of power in Angola, in an attempt to formulate inclusive bottom-up paradigms of intervention.

The opposition between the Angolan state and these critical NGOs mirrors what Lefebvre ([1968] 2009) refers to as the (hidden or expressive) socio-spatial conflicts existing between the implosion of the old (or new) morphologies of the urbanised areas and some persistent islands of rurality. In such a context, some NGOs decide not to participate in the material production and transformation of housing spaces, although others chose, instead, to mediate housing conditions in self-produced areas and/or 'co-produced neighbourhoods' (such as Panguila), seeking to provide access to infrastructures and to the non-existing 'benefits of urbanisation' (ibid.). Unlike the former, these NGOs are trying to overcome flaws within the dominant system with regard to urban issues. Nonetheless, they similarly aim to contribute to the transformation of common-sense thought-frames opposed to the everyday self-production of space, a common sense shared by both the elite and leading political agents alike. Equally, their line of action, in some measure, points to 'an alternative socialist project grounded upon anti-[state] productivism and radical grassroots democracy' (Brenner 2008: 241). In this position of resistance, these NGOs encourage the active participation of the impoverished inhabitants in the production and/or transformation of their urban life, also as a viable path to build their own citizenship.

The distinction between the production of (social) space by social agents and the production of (political) space by the neo-liberal administration, in line with Lefebvre's work ([1974] 2000), helps to present social movements as acting on behalf of a revolution aiming at the renovation of the dominant urban system, with reference to the consequences of one and another mode of production, within the demanding agenda of the *Right to the City* (Lefebvre [1968] 2009). On the one hand, everyday practices of self-production of space, without appropriate 'official' upgrading, cannot overcome much of the socio-spatial constraints. On the other hand, forced eviction, even those linked to resettlement or rehousing processes, strengthen the rising criticism and claims of these NGOs. In effect, the gigantic scale of interventions (for instance approximately 30,000 people dispossessed in the Iraque-Bagdad 'neighbourhood' and surroundings, in 2009), aroused the empathy of activists towards the victims of 'public' interventions, thereby exposing and leading to the ethical condemnation of these events, both nationally and internationally, so as to strengthen the elaboration of a worldwide anti-exclusion strategy. As pertinently asked by Merrifield (2006: 128): 'what would be the point of *any* global politics if it... isn't rooted in some particular context?'

The activities of these local organisations (aiming at the support of those evicted and mediation with the government as well) contributed to increase the growing voices of disapproval that rise up against the neo-liberal system in the Global North, with its hegemonic routines and consequences. Militant players seek to promote resistant and emancipatory movements, all the while enabling larger grassroots organisation and assistance from the deprived groups in a new wave of bottom-up urban policies and practices of (real) public interest. These include issues such as the appropriation of power and space (as power) – not proprietorship – and, likewise, participation – not just consultation – in everyday decision-making, namely the 'work' as the motion that equals the 'product', concerning Lefebvre's ([1968] 2009) *Right to the City*. Some forms of urban struggles, such as those encouraged by these Angolan NGOs, perhaps have the ability to effect transformations outside the capitalist mode of production (Lefebvre [1974] 2000). Others also occur, frequently motivated by the strongly critical manifestations against the regime by the famous rappers MCK and Ikonoklasta. In this context, urban space combines and mirrors the processes and conflicts that shape that space. Nonetheless, the reading, *per se*, of these heterogeneous spaces does not permit a complete understanding of the processes that are at the origin of its form.

Spaces of resistance: 'co-induced in-between spaces'?

As identified, top-down political spaces are now being produced by the Angolan neo-liberal regime, particularly in Luanda. On the other hand, bottom-up social spaces are being self-produced by social agents, progressively. Moreover, the militant resistance of some NGOs supports everyday strategies and practices of survival as a prospective path to overcome the inequalities intrinsic to the capitalist mode of production as it leads to new 'uneven geographies' of class power (Harvey 2001). According to Lefebvre ([1968] 2009; [1974] 2000), the processes inherent to the global structure are permeable and influenced by the activities of local groups, including their existences and rhythms. As such, 'in-between spaces' are being 'co-induced' within these dynamics. These are naturally different from spaces co-produced by means of official incentive (such as Panguila) where changes in the forging of space are accepted or even introduced by the dominant system as a technique of co-opting opposing processes and practices. These tend to be spaces of micro-resistance instead, namely spaces produced within the struggles that emerge at the side-lines of the official context and that may be later enlarged, and thus are now being fought against by the regime.

The arena where the urban struggles occur is complex and challenging, therefore the process of eliminating any form of resistance is difficult to analyse since it has several permeable and overlapping layers. On the one hand, urban spaces are being co-produced through state initiatives, considering everyday processes and practices (such as Panguila). This corresponds to a first approach regarding the preservation of power by the dominant system. The space produced in this context is a variation of the capitalist mode of production. On the other hand, some public agents are responding positively, even though only occasionally, to a few of the demands of the militant resistance, while seeking to limit its strength. This routine corresponds to another attempt at preserving the existing structure. Nevertheless, the spaces 'co-induced in-between' these dynamics, i.e., the spatialisation of the regime's positive response to the demands of NGOs that care for the self-production of space are, in fact, (micro-)spaces of resistance. These have the potential to induce a revolution to overcome the capitalist mode of production and the forging of a new urban model with specificities yet to be identified and explored. However, Schmid (2008) recalls that an essential aspect of Lefebvre's approach consists in passing from theory to action.

Expressive examples of this two-stage process in defence of dominant structures are the supplying and/or the improvement of subsidised housing (such as in co-produced areas) without proper living conditions, as the government basically only invested in housing at the expense of urban and social infrastructures and resources. Similarly, when the regime temporarily suspends demolitions and evictions during the period of elections, just to restart them later on (for instance Chicala/Kilombo in Luanda – see Figure 10.1 – and the 27 de Março neighbourhood in Lobito). Given this, the 'gains' that Patrocínio (26 September 2013) mentioned are, on the one hand, an asset for the people who had almost no rights and occasionally enlarge their 'right to place'. Then again, these 'gains' weaken the resistance of activists and inhabitants. In fact, in these

Figure 10.1 Chicala (Luanda) in early 2012, before the demolitions. *Source*: Photographed by the author (2012).

circumstances, resistance tends to restrain further claims and, in so doing, reduces the room for manoeuvre of the most impoverished social groups regarding what was not properly taken into account by the government, such as full access to infrastructures, to the 'benefits of urbanisation' and, also, to a collective participation in the construction of a transformed urban life, namely full access to the *Right to the City* (Lefebvre [1968] 2009) both as process and goal. This is a fierce strategy of the government to sustain its hegemonic capitalist mode of production.

As a matter of fact, these 'co-induced in-between spaces' of resistance are being annulled for the very same reasons that self-produced spaces are being demolished. They are an effective threat to the Angolan regime in particular and, considering the growing international network of whistle-blowers, a global threat to the capitalist system itself. Brenner (2000: 376) mentioned that 'politics of scale might have powerful implications… oriented towards… socially just forms of urban society'. So, in order to promote better socio-spatial conditions in Angola, one must continue to demand the *Right to the City* as a whole (Lefebvre [1968] 2009). This claim includes widespread access to: (1) housing with infrastructure, (2) urban resources, (3) appropriation of space, because each individual or family inhabits space in specific ways, and (4) participation in the political agenda. Then again, the analysis of the socio-spatial consequences arising from the latest 'public' interventions in such a ferocious neo-liberal context discloses the downgrading of social groups in need, often inhabiting precarious self-produced areas and reveals extensive spatial fragmentation, as lately identified by Viegas (2015). Altogether, these limitations tend to ground new urban struggles, with new participants and motivations.

Conclusions: 'gains and prisons'

The opposing modes of production of urban and housing space existent in Angola, namely top-town and bottom-up, fall within the scope of the theoretical-methodological framework selected for this chapter, as they correspond to what Lefebvre ([1972] 2000) identified as official production of political spaces and social production of social spaces in a neoliberal context. Regarding the current market-driven approach, state and/or private interventions, supported by a myriad of political instruments, are restructuring urban and housing space according to the old powerful logic of the accumulation and over-accumulation of capital, whilst aggravating pre-existent socio-spatial imbalances. Nevertheless, despite the aggressive nature of this dominant process or, paradoxically, because of it, the potentials of daily urban life – under conditions of neoliberalism – have persevered while complexifying the urban processes and/or experiences in Angola by means of grounding anti-capitalist struggles, particularly in Luanda. This occurs through two different, although complementary, ways: (1) the actions of a silent resistance inherent to everyday self-production of urban and housing space, as embraced by most inhabitants in need; and (2) the counter-actions of a particular militant resistance encouraged by daily experiences of production and transformation of space critical of, and thus opposed to, the government's urban policies and practices with their socio-spatial consequences.

On the one hand, the administration inconsequently proclaimed political and administrative decentralisation, therefore local communities remained without local official representation, having a reduced capacity to negotiate with the government's leading players, hence with limited opportunities to participate in the official urban scenario. On the other hand, the actions of some resistant NGOs are being blocked, because they are stimulating public criticism while encouraging urban society's access to a complete form of citizenship. Until now, the Angolan regime has been, occasionally, acquiescent to some of the demands of the emerging resistance while appearing to be more flexible on the subject of accepting the self-production of urban and housing space. But then again, for the time being, the accomplishments are very modest (for instance, mostly related to the 'right to place') and they tend to become 'prisons' since they are likely to drain the energy of activists fighting for larger claims. As suggested by Patrocínio (26 September 2013), in Angola 'there were gains but not [true] victories'. For that reason, regarding the post-war transition period (2002 to the present), I conclude that the forces in charge of the production of space are clashing, unbalanced and thus point to an extremely uncertain outcome.

Nevertheless, in such a context, further questioning is necessary: What must these NGOs do to stabilise and enhance the positive effects of their militant resistance? And how can they annul the negative consequences of the counter-resistance of the state? These and other questions must be asked so as to identify alternative paths, new innovative horizons, for the improvement of the quality of urban life in Angola. Concerning this subject, this chapter stresses an important theoretical contribution, namely regarding the 'co-induced in-between spaces' of resistance. This was possible due to recent empirical field work (Viegas 2015) in a context of very difficult access and poorly documented. Though the analysis of the Angolan data is context-specific and therefore cannot be transferred to other situations, the presentation of a precise socio-spatial context, together with a shift to the theoretical field, might be of interest for the reading of other sites, and *vice versa*, as together they enlarge the existent and growing voices against the capitalist mode of production and its uneven geographies. As such, I may also conclude that, by identifying and sharing examples of production of spaces of micro-resistance in Luanda, as opposed to examples of co-production of spaces (co-opted) through official incentive, this chapter may contribute to the drawing-out of a suggestion for a global alternative system resulting from the multiplication of many 'co-induced in-between spaces'.

Acknowledgements

A first version of this article was published in the minutes of the 12th International Conference on Urban History: Cities in Europe, Cities in the World Lisbon, Portugal. The author thanks the organisers of the session *A Multitude of 'In-betweens' in African Urban Spaces*, Luce Beeckmans and Liora Bigon, for their helpful comments during the discussion.

References

Brenner, N. (2000) 'The urban question as a scale question: reflections on Henri Lefebvre, urban theory and the politics of scale', *International Journal of Urban and Regional Research*, 24:2 361–378, Oxford/Malden: Joint Editors and Blackwell Publishers. Online. Available HTTP: <http://as.nyu.edu/docs/IO/222/2000.Brenner.IJURR.pdf> (accessed 16 January 2017).

Brenner, N. (2008) 'Henri Lefebvre's critique of state productivism', in Goonewardena, K., Kipfer S., Milgrom R. and Schmid, C. (eds) *Space, Difference, Everyday Life: Reading Henri Lefebvre*, New York/London: Routledge/Taylor and Francis Group.

Governo da República de Angola (2011) *Planos integrados de expansão urbana e infraestruturas de Luanda-Bengo: Decreto presidencial nº 59/11, 1 de Abril*, Luanda: Diário da República.

Harvey, D. (1998) 'The geography of class power', *The Socialist Register*, 34 49–74, London: Merlin Press. Online. Available HTTP: <http://socialistregister.com/index.php/srv/article/view/5700/2596#.WLRHM_mLSUk> (accessed 22 February 2017).

Harvey, D. (2001) *Spaces of Capital: Towards a Critical Geography*, New York: Routledge.

Harvey, D. (2008) 'The right to the city', *New Left Review*, 53 23–40, London: NLR. Online. Available HTTP: <https://newleftreview.org/II/53/david-harvey-the-right-to-the-city> (accessed 3 February 2017).

INE (2016) *Resultados definitivos do recenseamento geral da população e da habitação de Angola 2014*, Luanda: INE. Online. Available HTTP: <http://www.info-angola.ao/attachments/article/4654/Publica%C3%A7%C3%A3o-Resultados-Definitivos-Censo-Geral-2014.pdf> (accessed 12 August 2016).

Lefebvre, H. (1968; 3rd edn 2009) *Le Droit à la Ville*, Paris: Económica–Anthropos.

Lefebvre, H. (1970) *Du Rural a l'Urbain*, trans. Javier González Pueyo (1978) *De lo Rural a lo Urbano*, Barcelona: Ediciones Península.

Lefebvre, H. (1970) *La Révolution Urbaine*, trans. Robert Bononno (2003) *The Urban Revolution*, Minneapolis/London: University of Minnesota Press.

Lefebvre, H. (1972; 2nd edn 2000) *Espace et Politique: Le Droit à la Ville II*, Paris: Económica–Anthropos.

Lefebvre, H. (1974; 4th edn 2000) *La Production de l'Espace*, Paris: Anthropos.

Melo, V. (2015) 'A produção recente de periferias urbanas africanas. Discursos, práticas e configuração espacial: Maputo versus Luanda e Joanesburgo', unpublished thesis, Universidade de Lisboa.

Merrifield, A. (2006) *Henri Lefebvre: A Critical Introduction*, New York/Oxon: Routledge/Taylor and Francis Group.

Orre, A. (2010) 'Entrenching the party-state in a multiparty era. Opposition parties, traditional authorities and new councils of local representatives in Angola and Mozambique', unpublished thesis, University of Bergen.

Patrocínio, J. (2013) 'Ida ao Lobito'. E-mail (26 September 2013).

Schmid, C. (2008) 'Henri Lefebvre's theory of the production of space: towards a three-dimensional dialectic', in Goonewardena, K., Kipfer S., Milgrom R. and Schmid, C. (eds) *Space, Difference, Everyday Life: Reading Henri Lefebvre*, New York/London: Routledge/Taylor and Francis Group.

United Nations, Department of Economic and Social Affairs, Population Division (2014) *World Urbanization Prospects: The 2014 Revision*, New York: United Nations. Online. Available HTTP: <https://esa.un.org/unpd/wup/publications/files/wup2014-highlights.Pdf> (accessed 14 January 2017).

Viegas, S. (2015) 'Cidade (im)previsível? Governação e transformação urbana e habitacional: Paradigmas de intervenção e resistências no novo milénio', unpublished thesis, Universidade de Lisboa.

Viegas, S. (2016) 'Urbanisation and peri-urbanisation in Luanda: a geopolitical and socio-spatial perspective from the late colonial period to the present', *Journal of Southern African Studies*, 42:4 595–618, London: Routledge/Taylor and Francis Group. Online. Available HTTP: <http://www.tandfonline.com/doi/full/10.1080/03057070.2016.1192893> (accessed 12 August 2016).

11

Reading and applying Lefebvre as an urban social anthropologist

Siew-Peng Lee and Ho Hon Leung

Introduction

Urban anthropology has been around for a long time, although there is the ongoing debate as to whether much of anthropological research in cities is actually anthropology *of* the city, or if it is merely anthropology *in* the city (Fox 1977; Low 1996; Pardo and Prato 2012). If Aronowitz (2007) felt that Lefebvre had been ignored as a philosopher and social theorist, his absence from anthropology is even more pronounced. Though Lefebvre is beginning to feature in recent anthropological work, particularly in research with a reference to social space (Heer 2015; Zeng 2009; Maguire and Saris 2007) and rhythms (Jalas et al. 2016), there is still a lot of his work that has yet to be fully translated, examined and appropriated by anthropologists.

This chapter will explore why and how anthropologists might benefit from the use of conceptual systems and methods proposed by Lefebvre. The first section that follows examines the current state of urban anthropology to identify some of its shortcomings and the possible reasons for the noticeable absence of Lefebvre in anthropology. I then examine some of his writings on the city to highlight how relevant they are in designing and theorising anthropological research, relating them back to the issues raised in the first section. Finally I consider how Lefebvre may have some answers to how we might develop an anthropology *of* the city.

To many people, sociology is the study of urban industrial society, often using quantitative methods, while anthropology concentrates on rural and non-industrial society with the emphasis on qualitative research. My own background was in philosophy and sociology, followed by a postgraduate research degree in sociology (in Singapore) before I embarked, after several years outside academia, on a doctoral degree in social anthropology (in London) where I researched elderly Chinese migrants living in sheltered housing in a northern English city. As such, my view is that there is a lot of overlap between sociology and anthropology, not least in their intellectual roots, although there are certain areas in which these disciplines are clearly distinct, and these distinctions will be noted as and when necessary.

The state of urban anthropology

A brief survey

Fox (1977) has been quoted often in highlighting how what purports to be 'urban anthropology' is nothing more than anthropological research conducted in city or urban settings. Low (1996) undertook a survey of post-1989 urban research, and one of the conclusions was such research tended to be particularistic (that is, not embedded in the larger context; not holistic in research and analysis) and was essentially research *in* the city. There has not been any major development in terms of theorising the city from an anthropological perspective. Nas (2011) edited a volume which concentrated on 'urban symbolic ecology'. While there was a huge emphasis on symbolism, which is the territory of anthropologists, it again falls short of theorising urban space *and* culture. A similar criticism was levelled at the next, and possibly most recent, compilation of urban research by Pardo and Prato (2012). While reviewers agree that this volume clearly shows that the ethnographic methods of anthropology translate very well into the urban context, the chapters were still lacking in any advancement of theory as proclaimed in the title.

It is worth scrutinising some of these criticisms. McDonogh (2013: 794) questioned why Pardo and Prato needed to allude to the rejection of urban anthropology as a valid sub-discipline in contrast to what 'real anthropologists' do. Sanchez (2014: 379) described this volume as a 'riposte to the perceived disciplinary hostility to urban anthropology'. Toulson (2015) suggested that this hostility is because the respondents are usually not 'Other' enough and therefore the research is not 'proper' anthropology. This may be so, but I will suggest that this perceived hostility is due in part to the massive 'pure-applied' divide within certain British anthropology departments. As an outsider I had observed the huge chasm between those who conduct traditional fieldwork and engage in theoretical debate versus those who also seek to apply their research data through whatever institutional means possible (for instance in advising government agencies, NGOs and corporations). As urban anthropology tends to focus on 'problems' (for instance ghettos and drunks) to understand urban civilisation and contemporary society 'as a whole' (Wirth 1940: 743), it appears to belong more naturally in the 'applied' corner. Note, though, that urban anthropologists had in fact been reluctant to participate in 'urban public policy debates' (Low 1996: 384). Jones and Rodgers (2016), who examined the origins of urban anthropology, argued that such particularism was not always the case. They credited the Chicago School (Department of Anthropology *and* Sociology till 1929) for the prolific output of ethnographic studies between 1917 and 1940 which provided a more 'embedded analysis' and were considerably more holistic compared to later (particularistic) studies which became 'intellectually disengaged' from their epistemological roots (Jones and Rodgers 2016: 21–2).

Perhaps this particularism is a 'design fault' in that anthropological research is usually conducted by lone individuals and therefore must be limited to small-scale studies or specific aspects (such as gender) as part of a larger-scale project, and thus leave us none the wiser as to how, anthropologically, a city is a city. Toulson, who reviewed Pardo and Prato (2012) and two other volumes together, noted how such research 'ignores both human agencies and the complexities of causality' (2015: 29) when it presumes that it is the city that shapes lives. She goes on to suggest that 'if the anthropology of the city is to be something distinctive, it should study urbanism as *process* rather than as fact' (Toulson 2015: 34, emphasis added). This chimes with what Lefebvre said about the creation or production of social space as not being 'the work of a moment', but is 'in fact, a *process*' ([1974] 1991: 34, emphasis in the original). We will return to this point later.

Missing in anthropological action?

I offer two reasons as to why Lefebvre has been overlooked by anthropologists. The first might be that Lefebvre is often described as 'a French Marxist sociologist'. For those students in university departments where sociology and anthropology are the fraternal twins separated at birth and never yearned to meet each other again, there is little impetus to pick up a book by sociologists, akin to what Tett calls the 'silo effect' (Stein 2016: 29). It does not help that Lefebvre seemed to have expressed some disdain for anthropologists (and ethnologists), describing their writing as 'long, circuitous meanderings' ([1968] 2016: 24), and in their focus on describing 'representational spaces' have ignored other properties of socially-produced space ([1974] 1991: 41). In contrast, he addresses sociologists directly, giving the sub-title 'Foundations for a Sociology of the Everyday' to the second volume of *A Critique of Everyday Life*.

A possible second reason for Lefebvre's absence in anthropological research could be his emphasis on triadic dialectics. I must confess that I did not entirely grasp the full meaning and significance of Lefebvre's exposition on the social space triad (spatial practices–representations of space–representational spaces) when I first read it as an anthropology PhD student. It was only through reading *Rhythmanalysis* ([1992] 2004) more than a decade after that PhD that I became enthralled by how the triadic dialectic of space–time–energy could resolve some of the fundamental issues I had in the analysis of my data. There was a 'light-bulb' moment when I realised that I could finally try to publish the paper that had been rejected so many times (Lee 2014); I had found an analytical framework that was theoretically acceptable to other anthropologists and social gerontologists.

As I revisited my research data, this time 'superimposing' *rhythmanalysis* over them, I came to realise how anthropologists, like many other social scientists, are prone to 'reductionistic schemata based on a binary opposition' (Lefebvre [1974] 1991: 123). I struggled with the issue of how time was defined: sacred versus profane, physical versus metaphysical (Hall 1983); public–private, work–free, women's–men's, individual–global and cyclical–linear (Novotny 1994). Durkheim, an influential figure in both sociology and anthropology, had bequeathed us with the sacred-profane dichotomy (and normal-pathological) upon which a lot of subsequent structuralist thinking was based. Lévi-Strauss also wrote of hot and cold societies, and of food being raw and cooked, fresh and rotten, moist and parched (1969). He believed that it is in understanding these binary oppositions that one might be able to find the 'mediator' to resolve these oppositions, or to 'unite' these oppositions in the 'thesis-antithesis-synthesis' dialectic. Somehow, many of us seemed to have been stuck at the binary state (for instance, culture-nature) and have failed to proceed to the triadic to resolve such tensions. In contrast, Lefebvre was convinced that a 'dialectic of opposites' always 'ends abortively' ([1947] 2014: 128).

Applying Lefebvre in anthropology

In this section I highlight those aspects of Lefebvre's writing that might be of interest to anthropologists, particularly in terms of methodology and analysis, and relate them back to the issues facing urban anthropology as stated above and its possible applications in conceptualising urban questions.

The theory of ground rent and rural sociology

The reason for Lefebvre's popularity in urban research is itself due to a process: the process of translating his works from its original French. In their commentary Elden and Morton rightly warn that this 'predominantly urban focus' risks 'marginalising another of Lefebvre's interests, which is the question of the rural' (2016: 57). Many anthropologists work in rural communities. The issue of ground rent 'as a socially determined category' (Elden and Morton 2016: 60) is not something I have encountered in anthropological literature. Perhaps this is because there are other (dyadic) relationships that preoccupy anthropologists (such as patron-client, gift exchange, gender, kinship) that have obscured the need to examine this 'social relation of production' and the process by which it has conferred landownership and the 'demand of a payment for their use' (Elden and Morton 2016: 60). Wherever a rural agrarian community comes into contact with capitalists, there is the need to elucidate the power relations between the various parties involved. It is possible that this power shift (imbalance?) is at the core of environmental 'hotspots' such as Indonesia where the burning of huge areas of forest causes misery not only to the humans and animals that draw life from the forest, but also to neighbouring countries.

In his exposition on ground rent Lefebvre observed how rural sociologists (and if I may add, anthropologists) must deal with very complex sets of structures originating from 'different historical Epochs' which means having to confront structures that are disintegrating as well as 'mixed with new forms and structures', and must therefore 'double as a historian' ([1956] 2016: 67, 68). Contact with capitalists does not only lead to a uni-directional depopulation of villages, there is also an opposite inflow of workers who 'replace the older population of peasants and artisans' (Lefebvre [1956] 2016: 71). As such a 'complex and contradictory process' (Lefebvre [1956] 2016: 72) that affects *both* the rural and urban contexts, it needs to be given focus. The theory of ground rent – or its urban contemporary equivalent in non-European contexts, together with this complex rural-urban dialectic – is possibly integral to finding the answer to the question as to what makes urban anthropological research *of* the city and not just *in* it.

Critique of *Everyday Life*

It will seem logical, if not at least superficially, that what Lefebvre has to say about the 'everyday life' must surely resonate with anthropologists who are ardent students of the mundane, the everyday. What exactly does Lefebvre mean by 'everyday life'? Why did he take three volumes over several decades (1947, 1961, 1981) to write a critique of something that seems so ordinary and taken-for-granted? I interpret these volumes as Lefebvre explicating the importance of separating the authentic everyday life from that which has been masked by the encumbrances of capitalism. Right from Volume I he works from the basis that alienation, as defined by Marx, has imprisoned us in a state of 'mystified consciousness' such that we are not able to live lives as our true selves. It is only in separating what is truly human from 'bourgeois decadence' that we can achieve the 'rehabilitation of everyday life' (Lefebvre [1947] 2014:147). He goes on to elucidate the futility in trying to find reality in art and philosophy, religion, festivals, communal meals and so forth. He asserts that we can only arrive at the 'truth' through the Marxist dialectical method which allows us to 're-establish order and reason in ideas' ([1947] 2014: 244); alienation cannot be resolved by 'inventing new rituals' ([1947] 2014: 245) but through a critique of life in its most mundane and everyday detail ([1947] 2014: 246–7). This requires a 'methodical confrontation' between '"modern" life' and '*the possible*' and an investigation of the 'exact relations' between everyday life and festival, triviality and splendour, reality, dreams and so forth. ([1947] 2014: 271). Only then can we achieve unity, the 'realization of the total man' ([1947] 2014: 272).

Reading Lefebvre has caused me to raise questions, but I do not yet have the answers. It will be safe to say that anthropologists work very much within the realm of everyday life. We participate in activities amongst the people we observe and then draw conclusions about their lives. How much of what we observe is the 'authentic' everyday life, free from the mystified consciousness that has been imposed by capitalism? We observe rituals and postulate reasons behind its genesis and evolution, but seldom relate this to the Marxist idea of alienation. Perhaps what anthropologists observe in their field does not (yet?) amount to a sense of the 'modern' (and capitalist), but is there any mileage in adopting Lefebvre's stance that we need to investigate the exact relations between everyday life and festivals? Were these rituals and festivals designed so that those who are usually oppressed have the chance for a day or two in the year to 'rebel' and act out their otherwise unspoken frustrations? Is the modern 'paid annual leave' of employees, or 'corporate fun days' (dinner and dance, Christmas parties) where hierarchy in the company is (temporarily) turned on its head, the corporate and urban equivalent to festival? To what extent do such rituals and festivals reproduce the society with its inherent hierarchies? What happens when these normal routines are disrupted?

Michel Trebitsch, who wrote its preface, said that Volume II can be read 'as a veritable "discourse on method" in sociology' ([1961] 2014: 278). As usual Lefebvre was meticulous in setting out his argument – and the 'implements' he uses – in a way reminiscent of a philosophical treatise. He is also consistent in giving a summary of his argument at the end of every chapter. In the first chapter he reiterates: There can be no knowledge of society (as a whole) without critical knowledge of everyday life and, conversely, there can be no knowledge of everyday life without critical knowledge of society. Knowledge also 'encompasses an agenda for transformation' ([1961] 2014: 392); *praxis* is at the heart of knowledge. Towards the end of this volume he presents a 'theory of moments' in which moments present themselves as 'duplicates of everyday life, magnified to tragic dimensions' ([1961] 2014: 650).

If there can be no knowledge of society (as a whole) without critical knowledge of everyday life, and vice-versa, then how valid is anthropological analysis on research that focuses on 'moments': political events and natural disasters? Where might the 'everyday' be situated, and more crucially, to what extent is what anthropologists observed mere 'duplicates' (imitations) of everyday life, overlaid by the spectre of mystified consciousness? With urban regeneration, who decides which aspects of 'everyday' are to be regenerated? Moreover, does Lefebvre's emphasis on *praxis* provide robust justification for British academic anthropologists to abandon the 'pure-applied' divide?

Volume III was published 20 years after Volume II. Noting that because there had been so much change in the world, the problem had also changed, and the question had become whether daily life is a shelter from, or 'a fortress of resistance' to, change, in whatever form it might take (Lefebvre [1981] 2014: 717). Noting how capital is the same everywhere and which both 'prescribes and imposes' homogeneity, fragmentation and hierarchisation in everyday life, he mulls over the continuities and discontinuities of Marxist thinking ([1981] 2014: 757–8). Intriguingly, though history had travelled much the way predicted by Marx, capitalism – far from disappearing – had persisted. Lefebvre attributes this to 'recuperation', which targets deliberately 'what might have changed, in order to prevent change' ([1981] 2014: 776). Lefebvre gives examples of how an idea that was 'regarded as irredeemably revolutionary… is normalized, reintegrated into the existing order, and even revives it' ([1981] 2014: 777). Take the ethical, fair trade and organic movement as a contemporary example. Whether it be in skincare, coffee or chocolate it started as being pro-worker, pro-environment and anti-capitalist to serve a niche but enlightened market. Soon its profitability led to huge international conglomerates buying out and taking over this sector. The subversive had become acceptable, is normalised again, and goes mainstream, with an even greater impact. Capitalism thrives once more. Technology also played a part in changing the face of capitalism, leaving us subject to the daily relentless grind of clock-time. Meanwhile the state intrudes into even more areas of our daily life (Lefebvre [1981] 2014: 794–800). With our lives thus regulated round-the-clock, he moves the discussion to space and time and introduces *rhythmanalysis*.

Such insight might apply in at least two areas of anthropological research. In a globalised world, migration is a frequent interest in urban research. To what extent are groups of migrants 'homogenised' by social forces such as immigration services, the border police and even people traffickers? How are these groups of migrants, at the same time, fragmented along lines of ethnicity, gender and labour leading to ghettos in their destination or transitional locations? On the other hand, how might the original and new hierarchies within these (simultaneously homogenised and fragmented) groups become hierarchised again at this new locus? How does one theorise, for example, the way foreign domestic workers 'gravitate towards the different parts within a public park according to the "map of the Philippines" which reinforces their regional differences' (Lee 2016: 15)?

Another aspect of modern and urban life that anthropologists do not usually address is the impact of the welfare state on individuals even though it can be construed as a form of gift exchange, not unlike the concept of *kula*. In my special interest area of ageing, not just anthropologists, but social scientists in general, seldom elucidate how welfare provisions affect how, when and why people choose to retire from paid work, where they live afterwards, and how they organise long-term care at the end of life, even though these issues are all tied in with state-sponsored welfare (and health) benefits in much of the developed, urban and capitalist world. My elderly respondents were keen to convince me that they were truly much happier living apart from their adult children. How different might their answers be if – as we confront the *possible* – there was not a welfare system that provided them with a sizeable weekly pension and generous housing benefit, courtesy of a capitalistic economy? How will their answers differ from individuals who also live in cities but do not receive comprehensive health and welfare benefits? Where the state – via the benefits and healthcare structure – has intruded into every area of our everyday life, can researchers afford to ignore its influence? Do we simply accept this as a 'given' or should we evaluate how welfare provision, as a component of neo-capitalism, alienates us from our authentic selves?

The *Production of Space*

The spatial triad in this volume derives from Lefebvre's conviction that space is not nothingness, 'free of traps or secret places' ([1974] 1991: 28). Capital and capitalism have an influence on 'practical matters relating to space, from the construction of buildings to the distribution of investments and the worldwide division of labour' ([1974] 1991: 9–10). This leads inevitably to a discussion on how hegemony is 'exercised over society as a whole, culture and knowledge included, and generally via human mediation: policies, political leaders, parties, as also a good many intellectuals and experts' ([1974] 1991: 10). Space is not only either real or mental, it is also imbued with a package of social relations that dictate how that space is used or abused, and has limitations on how those social relations might be reproduced.

In the unity of the spatial triad 'spatial practices–representations of space–and representational spaces', Lefebvre unpacked how these social relations relate to the way space is (respectively) perceived, conceived and lived. He lists these distinctions in different ways. My own distinctions are summarised as:

- *Spatial practice*: as the perceived normative practice for that space (for instance room, school, hospital).

- *Representation of space*: as conceived by the professionals including those who design the space (architects, town planners) and those who use them (teachers, social workers).
- *Representational space*: as lived by the users (inhabitants) in their everyday life according to how it has been conceived and perceived.

Lefebvre gives the example of the part of the Mediterranean as a locus of leisure and 'vast wastefulness' which could be perceived as an 'intense and gigantic potlatch of surplus objects, symbols and energies', a huge expanse of useless (purposeless) holiday space. In fact this 'seemingly non-productive expense is planned… to the *nth* degree' to serve 'the interests of the tour-operators, bankers and entrepreneurs' ([1974] 1991: 59). Translating this into the terminology of his spatial triad: in a neo-capitalist economy that requires leisure space and time (spatial practice, perceived), holiday centres have been planned and designed (conceived) to perform the functions of sport, relaxation and other rituals (representations of space) in order to allow the users to enjoy (live) in a seemingly purposeless manner (representational space) for the few days in the year that they are not working to an otherwise tight clock-time schedule.

Lefebvre seems convinced that anthropologists tend only to provide 'a purely descriptive understanding' ([1974] 1991: 122), are interested only in specific representational spaces ('childhood memories, dreams, or uterine images and symbols') and often neglect to see these alongside co-existing representations of space and social practices ([1974] 1991: 41). Looking back at my own research I see that I had, as Lefebvre accused, made detailed descriptions of the 'representational spaces': how the tenants lived and how they use their private and public spaces within and even outside the buildings, their routines or refusal to follow a set routine, their resistance to change in the way the night-warden system was run, and so forth. I might have ventured into pondering the 'representations of space' in trying to understand and explain the rationale for locating the buildings right smack in the middle of Chinatown but I had not paid too much attention to the 'social practices' of providing sheltered housing for older people as perceived by social workers and social housing providers. In fact, I was so embarrassed by my ignorance in the area of welfare benefits, and assuming – erroneously – that there had been some fraudulent benefits claims, I avoided confronting the issue of welfare entitlement.

Rhythmanalysis

Based on the triad of 'space–time–energy' Lefebvre explains that 'everywhere where there is interaction between a place, a time, and an expenditure of energy, there is rhythm' ([1992] 2004: 15). *Rhythmanalysis* begins with physiological rhythms. A body in working order is an example of a rhythm in perfect harmony or isorhythmia ([1992] 2004: 68). While every rhythm takes a cue from nature and society, the rhythms are at the same time individually 'owned' (Lee 2016: 9). We do not notice this steady state until there is a breakdown resulting in arrhythmia (Lefebvre [1992] 2004: 67). However, just as with recuperation (above), the system strives towards returning to isorhythmia. During this process, several rhythms – polyrhythmia – might be called into play (Lefebvre [1992] 2004: 67). Sometimes two or more different rhythms are meshed together to create eurhythmia (Lefebvre [1992] 2004: 67). There are few true isorhythmias but many more eurhythmias (Lefebvre [1992] 2004: 67).

In Lee (2014) I give examples of how my respondents seemed to always find a new 'equilibrium' in what they do, eat and sleep, and they professed to be happy in the freedom to make those decisions. What was crucial here is that *rhythmanalysis* takes its cue from the individuals or groups of individuals involved. There is a very clear sense of agency. People adjust their level of dis/engagement in various activities depending on the (physical, social, mental, other) resources they have. Hence the more intellectual respondents chose to remain socially and physically active because they could not cope with the idea of idleness, while those who had engaged in physically laborious work were pleased to be able to stop working altogether. With its emphasis on human agency, *rhythmanalysis* is a good candidate to counter Toulson's criticism (2015) that human agencies had been ignored in the theorising of urban anthropology.

In evaluating the triads in *Rhythmanalysis* we might also find the answer to the question as to why, following Lévi-Strauss, anthropologists have difficulty in establishing the mediating factors between structuralist oppositions. I will suggest that we have been using the 'wrong' dialectic. Instead of a Hegelian dialectic that pits a 'synthesis' against the oppositional forces of a thesis and anti-thesis, 'mediators' might be more easily found if the third component is found to relate *equally* to the other two components (Lefebvre [1992] 2004: 12). In a footnote (Lee 2016) I proposed that an intractable opposition between nature and nurture/culture could be made triadic by the addition of 'opportunities'. Imagine an innately talented footballer, poet or physicist who, without the opportunities of encountering the resources (ball, pen, education) needed to nurture these talents, might never get to fulfil their natural potential. Conversely less talented persons who are given these resources (opportunities) could mitigate their lack of natural talent to some extent. Nature/nurture is never 'either/or'; children who are given the opportunities, whatever their (lack in) natural talents, are more likely to reach their greatest potential.

I noted that *rhythmanalysis* can be used both as a conceptual framework (theory) in understanding social phenomena as well as a toolkit (method) (Lee 2016). It is easy to confuse or conflate the two. How does one 'do' *rhythmanalysis*? Lefebvre was not clear. I suspect that the meticulous recording of rhythms, or even the categorisation of these rhythms, *on their own*, does not constitute *rhythmanalysis*, which would otherwise be spelt as 'rhythm analysis'. Given Lefebvre's emphasis in his other expositions, for the paradigm to be useful in analysis (as distinct from method) there must be a *process* of disruption (arrhythmia) and return to isorhythmia (or a slightly different isorhythmia/new eurhythmia) via recuperation (Lee 2016: 11). Anthropologists who study events (moments), therefore, must first establish the point of isorhythmia or at least adopt a theoretical point of isorhythmia before their analysis can even begin. Similarly, in urban regeneration, there must be an agreement amongst those affected as to what the end-goal (the point of isorhythmia) might be. This gives rise to the question as to whose perspective are we to adopt to define 'success'? Enmeshed somewhere in this equation is the need to explicate the spatial triad of 'spatial practices–representations of space–and representational spaces' as 'conceived–perceived–lived'.

Lefebvre intended *rhythmanalysis* to be a transdisciplinary concept, capable of theorising everyday life 'from the most natural (physiological, biological) to the most sophisticated' ([1992] 2004: 18). Anthropologists might find it useful to question whether some of the traditional theories in analysing society, pre-capitalist and otherwise, could be interrogated using *rhythmanalysis* as I had begun to do with ageing theories (Lee 2014). Take Van Gennep's (1960) 'rites of passage', which is often misused by writers who neglect the constituent concepts of separation, liminality and re/incorporation in the original paradigm. Could we superimpose 'isorhythmia–arrhythmia–return to isorhythmia/eurhythmia' over these? Are there instances where a *rhythmanalysis* paradigm is a much better fit, particularly when there is no original group into which individuals are re/incorporated, such as old people moving into retirement housing (compare Barrett et al. 2012), or migrants adjusting to a new country being prime candidates in an urban context?

Conclusions

This chapter has only managed to consider a very small proportion of Lefebvre's prolific output, concentrating on insights that might be of special interest to urban anthropologists to address the question 'what makes anthropology truly *of* the city and not just *in* the city'?

In much of his writing, Lefebvre had stressed 'process': if social space is produced, then there is a process; if there is a process, there is also history. To avoid particularism, a criticism levelled at recent urban research, anthropologists must also, apart from researching the current static position of the research sample, elucidate the process/es by which individuals came to be in that particular urban context (such as analyse the multi-directional flow of migrants in the rural-urban dialectic). Whether it be the explication of ground rent (or its contemporary equivalent in a non-European context) to embedding the research in a wider rural/urban context, whether it be in isolating data according to the spatial practices–representations of space–representational spaces triad, whether it be overlaying *rhythmanalysis* over an event or a cycle of events, anthropologists must move away from the focus on the near and present – the moment – but view their respondents and their everyday life within the context of a much larger spatial and temporal canvas. In practice, this is an onerous task for lone researchers. Urban anthropologists might need to consider working more frequently as part of a team with expertise drawn from outside their discipline.

Lefebvre's starting point in his writings was capitalism and its faithful 'sidekick': alienation, as defined by Marx. This is not a concept that usually occupies the minds of traditional anthropologists. Perhaps anthropologists, particularly those working in areas where capitalism has taken hold, however tenuously, need to start looking at these (lived) 'representational spaces' differently, to separate the 'real' from the 'mystified'. Anthropologists will also do well to elucidate how people make decisions that might have been clouded by the all-encompassing influence of a welfare system within an urban capitalist context. With hindsight, I can see how some of my methodology, data and interpretation might be quite different had I also scrutinised those considerations that Lefebvre had termed (conceived) 'spatial practices' and (perceived) 'representations of space'. In view of Lefebvre's objective to set people free from the shackles of alienation, transformation is part of the agenda. Therefore, an anthropology *of* the city must also challenge its practitioners to use their research data to liberate their respondents from their bondage (of whatever kind), making engagement with policy discussion a necessity.

In summary, it is in learning to view space with the spatial triad (spatial practices–representations of space–representational spaces) and time in the space–time–energy triad that we can begin to understand, perhaps through *rhythmanalysis* and/or theory of ground rent, how, anthropologically, the urban differs from the rural. Perhaps it is only in adopting the analytical frameworks of a sociologist steeped in Marxist thinking that anthropologists can finally be able to attempt an anthropology *of* the city rather than one that is merely *in* it.

References

Aronowitz, S. (2007) 'The ignored philosopher and social theorist', *Situations*, 2:1 133–156.

Barrett, P., Hale, B. and Gauld, R. (2012) 'Social inclusion through ageing-in-place with care?', *Ageing and Society*, 32:3 361–378.

Elden, S. and Morton, A.D. (2016) 'Thinking past Henri Lefebvre: introducing "The theory of ground rent and rural sociology"', *Antipode*, 48:1 57–66.

Fox, R.G. (1977) *Urban Anthropology: Cities in Their Cultural Settings*, Englewood Cliffs, NJ: Prentic-Hall.

Heer, B. (2015) '"We are all children of God": a Charismatic church as space of encounter between township and suburb in post-apartheid Johannesburg', *Anthropology Southern Africa*, 38:3–4 344–359.

Jalas, M., Rinkinen, J. and Silvast, A. (2016) 'The rhythms of infrastructure', *Anthropology Today*, 32:4 17–20.
Jones, G.A. and Rodgers, D. (2016) 'Anthropology and the city: standing on the shoulders of giants?' *Etnofoor*, 28:2 13–32.
Lee, S.P. (2014) 'The rhythm of ageing amongst Chinese elders in sheltered housing', *Ageing and Society*, 34:9 1505–1524.
Lee, S.P. (2016) 'Ethnography in absentia: applying Lefebvre's *rhythmanalysis* in impossible-to-research spaces', *Ethnography* [currently only online, with restricted access].
Lefebvre, H. (1947, 1961, 1981/2014) *Critique of Everyday Life: One Volume Edition*, trans. J. Moore and G. Elliott (various years), London: Verso.
Lefebvre, H. (1956/2016) 'The theory of ground rent and rural sociology', *Antipode*, 48:1 67–73.
Lefebvre, H. (1968/2016) *Everyday Life in the Modern World, Bloomsbury Revelations Edition*, trans. S. Rabinovitch (1971), London: Bloomsbury.
Lefebvre, H. (1974) *The Production of Space*, trans. D Nicolson-Smith (1991), Oxford: Blackwell.
Lefebvre, H. (1992) *Rhythmanalysis: Space, Time, and Everyday Life*, trans. S. Elden and G. Moore (2004), London: Continuum.
Lévi-Strauss, C. (1969) *The Raw and the Cooked*, trans. J. Weightman and D. Weightman, London: Jonathan Cape.
Low, S.M. (1996) 'The anthropology of cities: imagining and theorizing the city', *Annual Review of Anthropology*, 25:1 383–409.
Maguire, M. and Saris, A.J. (2007) 'Enshrining Vietnamese-Irish lives', *Anthropology Today*, 23:2 9–12.
McDonogh, G.W. (2013) 'Review of Pardo, I. and Prato, G.B. (eds). *Anthropology in the City: Methodology and Theory*, Farnham: Ashgate', *American Ethnologist*, 40:4 794–795.
Nas, P.J. (2011) *Cities Full of Symbols: A Theory of Urban Space and Culture*, Leiden: Leiden University Press.
Novotny, H. (1994) *Time: The Modern and Postmodern Experience*, Cambridge: Polity.
Pardo, I. and Prato, G.B. (eds) (2012) *Anthropology in the City: Methodology and Theory*, Farnham: Ashgate.
Sanchez, A. (2014) 'Review of Pardo, I. and Prato, G.B. (eds). *Anthropology in the City: Methodology and Theory*, Farnham: Ashgate', *Journal of the Royal Anthropological Institute*, 20:2 379–380.
Stein, F. (2016) '"Anthropology needs to go mainstream": an interview with Gillian Tett', *Anthropology Today*, 32:6 27–29.
Toulson, R.E. (2015) 'Theorizing the city: recent research in urban anthropology', *Reviews in Anthropology*, 44:1 28–42.
Van Gennep, A. (1960) *The Rites of Passage*, trans. Monika B. Vizedom and Gabrielle L. Caffee, London: Routledge.
Wirth, L. (1940) 'The urban society and civilization', *American Journal of Sociology*, 45:5 743–755.
Zeng, G. (2009) 'The transformation of nightlife districts in Guangzhou, 1995–2009', *Chinese Sociology & Anthropology*, 42:2 56–75.

12

Towards a contemporary concrete abstract

Steve Hanson and Mark Rainey

Introduction

The 'concrete abstract' is a crucial concept within the work of Henri Lefebvre that has only recently been given the critical attention it merits, particularly through the writings of Łukasz Stanek (2008, 2011). Yet, while the 'concrete abstract' can be found throughout the work of Lefebvre, he was not particularly clear about what it was or is. The concept does, however, flash up in tantalising ways in texts such as *Dialectical Materialism* ([1940] 2009) and *The Production of Space* (1991).

It is our intention in this chapter to lay out our take on the concept as strongly and as clearly as we can. In its most basic form the 'concrete abstract' describes the ill-fitting relationship between reified concepts and everyday social life. It gives an account of the gap between the often-simplified categories we deploy to make sense of our everyday situations and the complex ways in which social life is actually produced. It also recognises how abstractions bear down on the quotidian textures of life in deracinating and malign ways, and how abstractions serve to conceal these very effects.

Lefebvre's notion of the 'concrete abstract' has its philosophical roots in the work of G.W.F. Hegel and Karl Marx, and while recognising its importance for Lefebvre's foundational work on space, a strong starting place to clearly view the 'concrete abstract' in relation to its philosophical roots is via Lefebvre's 1940 text, *Dialectical Materialism* ([1940] 2009).

In taking up the 'concrete abstract' as a critical concept, we also wish to re-insert a philosophical and highly political edge into scholarship on Lefebvre in response to what both Christian Schmid and Stanek have considered to be the increasing 'banalisation' of his work through the uncritical and de-politicised overuse of concepts like 'the production of space' (Schmid 2008: 28; Stanek 2008: 62).

A key point this chapter seeks to put across is that the 'concrete abstract' only gains meaning and relevance in relation to particular moments and particular situations. It is not something that is easily described in conceptual isolation. For this reason, towards the end of this chapter, we will draw out the concrete abstract by example. Our attention will turn most specifically to the 'British Isles' and the current political crises taking place there following the United Kingdom's decision to exit the European Union after a national referendum in June 2016.

The referendum, and the passionate debates leading up to it, exposed deep social and economic fractures within the UK. It exposed divisions between wealth and poverty, young and old, and metropolitan centres and peripheral towns and the countryside. It also exposed regional divides and deep-seated divisions over internal nationalisms within the UK as Scotland and Northern Ireland voted overwhelmingly to remain in the EU while England and Wales voted overwhelmingly to leave. These fractures often overlapped with each other and although they existed before the referendum, they were also exacerbated by it. Yet, at the same time as we recognise these divisions, we must also recognise that a re-emergent national chauvinism glosses them over – as it so often does.

'The idolatry of the state', Lefebvre writes in *Metaphilosophy*, is 'the greatest fetish on earth after God', and our argument here is that the 'United Kingdom', as does the very notion of the nation state itself, becomes reified and taken to be a neutral, natural, and fixed entity at the same time that it is rupturing on the ground (Lefebvre 2016: 143).

The UK, we will argue, is a Fata Morgana – a 'real mirage' – a physical form seen out at sea, which turns out to be an illusion. It is a declarative name, but one with a continuing and legally produced history that serves as a disperser for all sorts of social fractures, economic stratifications, and structures of feeling.

We wish to view this dialectically, as a concrete abstraction that often conceals the problems it produces.

Locating the concrete abstract

In his key text, *The Production of Space*, Lefebvre turns to the concrete abstract immediately after announcing his maxim, '(social) space is a (social) product' (Lefebvre 1991: 26). He poses the following questions:

> Is this space an abstract one? Yes, but it is also 'real' in the sense in which concrete abstractions such as commodities and money are real. Is it then concrete? Yes, though not in the sense that an object or product is concrete. Is it instrumental? Undoubtedly, but, like knowledge, it extends beyond instrumentality.
>
> *(Lefebvre 1991: 26–7)*

The structure of the concrete abstract emerges in relation to Marx and in particular Marx's writing on the commodity. As soon as Lefebvre declares that '(social) space is a (social) product' (ibid.), we are in the territory of the concrete abstract. This is further qualified by Łukasz Stanek:

> After asserting that '(social) space is a (social) product,' Lefebvre claims that the mode of existence of space is that of money and commodities, which are, together with labor, theorized by Marx as concrete abstractions.
>
> *(Stanek 2011: 137)*

It is crucial to point out that these ideas have their roots in Marx, as Andy Merrifield does:

> Marx makes an analytical distinction rather than a real-life separation and shows Lefebvre how to keep the link between the specific and the general, quality and quantity, use value and exchange value, and the concrete and the abstract in taut dialectical tension. You can't have one without its 'other'.
>
> *(Merrifield 2006: 133)*

It is also crucial to point out here that although the concrete abstract is not given a specific work in Lefebvre's oeuvre, as Stanek puts it, 'the concept of concrete abstraction brings together the most vital elements of Lefebvre's theory of production of space' (Stanek 2008: 62). Stanek explains that 'social abstraction' 'concretizes and realizes itself socially, in social practice'. It has 'a real existence, that is to say practical and not conventional, in the social relationships linked to practices' (ibid. 68):

> Following Marx's theorization of labor as a concrete abstraction, Lefebvre demonstrates that space is an 'abstraction which became true in practice' – produced by material, political, theoretical, cultural, and quotidian practices.
>
> *(ibid. 76)*

The concrete abstract is part of Lefebvre's project to at once defend Marxism and rescue it from those who seek to reduce it to a scheme, a party programme, a dogmatism. It blows open the base and superstructure metaphor to extend Marxism out into culture and space. This has ramifications for the contemporary diaspora of the Marxist tradition of the New Left in the UK and what became codified under the 'turn to culture' (Hall 2010).

But in order to arrive at the contemporary moment of the concrete abstract we need to first revisit its roots. Stanek argues that:

> a reading of Hegel and Marx, in particular the latter's discussions on labor, commodity, and money, allowed Lefebvre to examine space as the general form of social practice in capitalist modernities, characterized by distinctive features, such as its simultaneous homogenization and fragmentation and its blend of illusion and reality.
>
> *(Stanek 2011: xiii)*

What is very useful to cite before we proceed is Stanek's highly productive explanation of how the universal and particular are at play in Lefebvre's concept:

> Since space is itself socially produced and has historical conditions of existence, its universality can be conceived neither as a Platonic ideal nor as a Kantian transcendental form of sensibility. Rather, Lefebvre argues that space is one of the universal forms of social practice, as commodity and labor are in the analysis of Marx. Like commodity and labor, space has a paradoxical quality of being at the same time, and in many ways, both 'abstract' and 'concrete': space appears to be a general means, medium, and milieu of social practices, and yet it allows accounting for their specificity within the society as a whole.
>
> *(Stanek 2011: 133)*

The structure of the concrete abstract is taken from Marx's analysis of the commodity in *Capital*, particularly his explanations of use value and exchange value, that the everyday use of something and its abstract value on the market cannot be disentangled from each other:

> Space – just like commodity and labor – is a 'social,' 'real,' 'actual,' or 'concrete' abstraction; that is to say, its universality is produced by processes of abstraction attributed to a range of social practices and reflected in the specific 'abstract' experience of modern space.
>
> *(Stanek 2011: 134)*

The full breadth of Marx's *Capital* was never delivered. But the *Grundrisse* gives tantalising glimpses of its planned depth. Key to point out here is that the work appeared to be calculated to begin with the least abstract part of *Capital*, a particular object, the individual product, before working outwards to the highest level of abstraction. Some interpreters have suggested that the full *Capital* was planned to fill nine volumes, working from concrete to abstract (see Shortall 1994: 445).

Therefore the concrete abstract is not only a key part of Lefebvre's thought, but also of Marx. This concept reaches right down to the roots of both figures and in the philosophy of G.W.F. Hegel. The centrality of the concrete abstract to both Lefebvre and Marx is clear, as is its continued relevance in our time. It seems almost too obvious to point out that the crash of 2008 was a crash caused in part by excessive market abstraction.

Lefebvre's *Dialectical Materialism*

Stalin's narrow *Dialectical and Historical Materialism* (1938) is the target of Lefebvre's *Dialectical Materialism*, published in 1940. The book is attributed to Stalin but appears to have been written by committee under Stalin. Lefebvre was at odds with this work and published his own *Dialectical Materialism* in order to lay out a new version of the subject. Lefebvre's book is still fresh.

That we make this return through Lefebvre's 1940 book on *Dialectical Materialism*, written precisely to resist reductive Stalinist appropriations of the subject, is apt, as we are still faced with the potential for thin, state capitalist, instrumentalised attitudes to space in the West in 2017. Stefan Kipfer, adding a new preface to the book in 2009, points out that 'for Lefebvre, Marxism was above all a dynamic movement of theory and practice not a fixed doctrine' (Kipfer 2009: xiv).

Stalin's reading was based on a narrow interpretation of Engels' 'Dialectics of Nature', which is largely to be rejected in any case (ibid.: xvii). The dialectic produces movement, in content – which is inevitably held in language – through negation, but it does not operate outside language. There are some problems, then, with Lefebvre's reading of dialectics. In the forward to the fifth edition, Lefebvre claims it is possible to 'uphold' the idea of a dialectic in nature.

We only wish to raise these points before moving on, as they are the subject of another work. Here we need to focus on the moment when Lefebvre states that the concept of alienation cannot merely be confined to bourgeois societies, which was part of the implicitly Stalinist period: 'Institutional Marxists chose to reject the concept' of a more widespread alienation which had the effect on Marxism of blunting 'its cutting edge' (Lefebvre 2009: 39). It is clear in 2017 that alienation is much more widespread across social strata.

Lefebvre's *Dialectical Materialism* has two main sections, dealing with The Dialectical Contradiction and The Production of Man. First, Lefebvre deals with 'dialectical contradiction': He lays out how Hegel moves beyond Kant. He then critiques Hegel, moving on to Marx. He then outlines The Production of Man and it becomes very clear how *The Production of Space* emerges. Here, then, is a handy ladder down through the floor of Lefebvre's project to its origins in Hegel.

Lefebvre begins with formal logic and Kant. Hegel moved away from the split of object and subject, concrete and 'meta': Here are the root tips of the concrete abstract.

The universal of Kantian formal logic is a problem for Hegel as it floats free from the historical and social circumstances of the world. Hegel attempted to reunite a socially and historically specific 'concrete' with a knowing subject. But this knowing subject also has the limits of its knowing to face – as well as what it knows – when arriving at the social and historical moment it is inevitably framed by. Kipfer explains that Lefebvre's:

> Dialectical materialism refuses to enclose knowledge within a teleological search for the absolute idea, which for Hegel was eventually actualized in the Prussian state. In contrast to Hegel's dialectical logic, it [dialectical materialism] is no longer a dogma.
>
> *(ibid.: xxi)*

Having read this, it should become clear why Lefebvre aimed his own version of dialectical materialism at Stalin's instrumentalised version. Dialectical materialism is not some economic or philosophical algebra for Lefebvre. It is a flexible, ad hoc, and negative philosophy. It is interesting that the contemporary Soviet-style Marxist tradition still retains this 'algebra' version of the dialectic (see for instance Rees 1998).

Therefore this is an attempt to present a contemporary and particular version of the concrete abstract rooted in the political expediencies of the day: we must of course question Hegel's claim that the development of society and history entails a progressive search for 'freedom', and we must do so precisely through Marx, but we must resist in Marxism the dogmatic institutional or party rhetoric Lefebvre warned of in its Stalinist version in 1940.

The broader point to make here is that Hegel removes the universal of logic in order to re-install a proper universalism that is attached to historical specificity and the concrete social circumstances that arise from that.

But Hegel, too, over-reaches. For Lefebvre, human consciousness reveals his (sic) mastery of things but also its limitations. Perhaps only in this era do we see those limitations more fully. How would Hegel frame the 'absolute idea' now? With Europe fragmenting and the head of 'the geist' containing convincing evidence of man's extinction, along with his (sic) own mastery?

This is not a question to try to answer, but it does return us to the need to keep to dialectical materialism as a flexible cluster of very particular methods, techniques, and philosophies, to be applied to very specific contemporary circumstances, to be re-negotiated with each encounter.

But there is an opposite risk to reified dialectical materialism, and that has been the risk of spatial theory after the first two waves: that a concrete politics is sacrificed for abstraction. Lefebvre really begins again properly with the dialectic at Marx's *Critique of Political Economy (*1857–9) and *Capital* (1867). Exchange sets objects off on a 'second life' in which their value becomes abstracted precisely in order to move them through the market process. In Marx's account of capitalism, this process of abstraction itself becomes further abstracted.

Engels noted that in the Exchange in Manchester, in the late 19th century, traders became transfixed by the signifiers themselves, as the complex and horrible chains of signifieds vanished. The remains of those chains now sit under One Angel Square in Manchester (Engels 1890). Engels and some of the other early social explorers were really the only people from the class of the Exchange traders who were looking to reconnect the concrete and its abstract and to explain what that diremption of object and meaning was doing to the human social world in emerging industrial cities: we must hold to the specific concrete circumstances of the moment; the philosophy serves this concrete moment, the concrete moment should never serve the production of philosophy.

One big contemporary 'second life' for space is real estate, in which we can take the word 'real' as a relation between the concrete, literally, in terms of a concrete tower block, but also totally abstract, if one views the way that space appears in the name of its abstraction, in order to shift it into the realm of exchange and ultimately surplus value: 'In relation to individuals this new social whole functions as a superior organism' (Lefebvre 2009: 77).

The individuals begin to serve 'it' and not the other way around. The very obvious correlation to make for space is that individuals, tenants, begin to serve 'real estate', 'real estate' no longer primarily serves the tenant. Real estate has become, certainly by the mid-20th century, a 'superior organism' (ibid.). The tenant is primarily producing surplus value for landlords and estate agents. This process has reacted to urban population density increases by becoming more fragmented, short-term, and micro. The highly criticised Airbnb phenomenon is one example of this.

Of course it is perfectly possible to make the opposite claim and estate agents do, but it is clear that a functional alienation has risen, which is often sketched in using the media shorthand of 'the housing crisis' in Britain (see Wainright 2017).

Therefore a whole lifted-out alienated life that has become naturalised can be accounted for under the term 'concrete abstract'. It is real, literally objective, with car parking spaces, elevators, stairs, numbered flat doors and balconies. But it is abstracted, sold through fetishes and aesthetics, rhetoric and cultural myth.

A good example of the latter would be the Hacienda flats in Manchester, built on the site of the nightclub of the same name and sold back to that demographic. A particular generation of nightclubbers – those who went to the Hacienda in Manchester, named after the situationist Ivan Chtcheglov's invocation that the 'Hacienda must be built' – were sold mythologised, concrete spaces via the myths they produced themselves within a popular counterculture which was explicitly supposed to resist reification (Ward 2002). It is interesting, then, that the term 'real estate' emerges in the mid-17th century with all of its cultural explosions and cultural revolutions. Those explosions and revolutions, it seems, never halt either reification or abstraction.

Via his reading of Marx's *Critique of Political Economy* (1857–9) and *Capital* (1867), Lefebvre explains how 'there can be no pure abstraction': 'The abstract is also concrete, and the concrete, from a certain point of view, is also abstract. All that exists for us is the concrete abstract' (Lefebvre 2009: 36).

We all live, then, in a double reality: 'And it is with this double reality that the categories are linked together and return dialectically into the total movement of the world' (ibid. 77).

The first social reality is real yet abstract. The second social reality is abstract yet real. These should not be seen as two progressing historical stages, but phenomenological occurrences in the same whole reality. They are constellations, not linear trajectories. Christian Schmidt understands this (2008). Therefore a recourse to Hegelian dialectics is essential to understanding what is at stake in the 'concrete abstract'.

Use value is concrete, exchange value is abstract: 'The duplication of value into use-value and exchange-value therefore develops into a complex dialectic, in which we find once again the great laws discovered by Hegel: the unity of opposites and the transformation of quality into quantity and quantity into quality' (Lefebvre 2009: 79).

Because of these abstractions, inaugurated by wider circuits of exchange, society is distributed 'with a certain blind and brutal inevitability' (ibid. 78).

The concrete abstract

Gillian Rose (2009) comments on Hegel that if The Absolute cannot be thought then it is useless to social science. If The Absolute cannot be thought and so is useless to social science, then by extension it is also useless to geography and spatial theory, if, as in Lefebvre, space (*espace*) is a verb.

The concrete abstract must then always be grounded in the particular, if the particular is an idea or a wall. Even if that particular is shaped in a malign way by an abstraction.

In many ways language itself is a concrete abstraction. But at least in the spirit of Lefebvre – because he didn't make this clear – the concrete abstract should come together at the crossroads of the particular. In many ways a building is just communication. But it is also the tiny trace of a builder's mark in even the most flawless facade, and that facade itself and all the hyperbole surrounding it in the media and everyday life, taken in one whole.

But the concrete abstract is not simply the interplay of the subject and object. Kant and Neo-Kantianism cannot handle this interplay. What became referred to as Post-Structuralism cannot handle metaphysics, searching to expunge it, a pointless crusade.

It might be tempting to read the concrete abstract as spatial theory in the 1990s might: If the crossroads is a meeting of four paths then the concrete abstract is the sum total, the fifth place. If the crossroads is a meeting of seven paths then the concrete abstract is the sum of these, the eighth.

But this would return us to second- or first-wave Lefebvre scholarship. This would be to return us to the dialectic as thesis-antithesis-synthesis, a dialectic more attributable to Fichte than Hegel.

To explain this properly it is necessary to return to Marx's supposed 'inversion' of Hegel to reveal the hard rationality in the mystical drift. Yet *Capital*, particularly the opening sections of Volume 1, reveals its Hegelian theological dimensions as it attempts to conceal them. If, as Marx contends, the commodity is at once plastic and mystical, so then is architectural form in use, so then, is the street.

There has been a need to rediscover the idealism in Marx: so to give a fuller picture of Lefebvre is to return the 'theological niceties' of space to spatial theory. This has already been laid out:

> The 'third' wave of Lefebvre readings we propose links urban-spatial debates more persistently and substantively with an open-minded appropriation of his metaphilosophical epistemology shaped by continental philosophy and Western Marxism. In so doing, it also rejects the debilitating dualism between 'political economy' and 'cultural studies' that in effect marked the distinction between the 'first' and 'second' waves of Lefebvre studies, making it impossible for us to return to a simply updated or expanded earlier school of thought on Lefebvre.
>
> *(Gromark 2013: 18)*

However, we wish to go much further: 'Untruth' has been seen as waste material created in the pursuit of Truth and this can be traced back to Kant. Untruth is merely the dark matter of Truth and this revelation can be traced back to Hegel. The diremption of truth and untruth was a bad historico-philosophical divorce. This is work for another day, but it is crucial to state that for us the concrete abstract is always the relation of the Real and the Logos. The abstract cannot exist in any way other than as an abstraction from the Real and that is linguistic. The whole of human history is the gift relation of the symbolic and the Real.

The concrete abstract is not merely 'false consciousness' either. It is a real abstraction. You think space and space thinks you. We might bring in Latour's example here of driving through a series of breath-taking hills. The move through those curves and the perception of them are not to be separated (Latour 2005). Kant's completion of Enlightenment thought reduces it to human need, perception flattens so that the mobile human can navigate in order to conquer. You are the subject at the centre of an objectified world: Subject drives over object; GPS, now available on many mobile phones, is a good everyday example of this. For Latour, and for Lefebvre, this is not the case. (Nor is Latour Hegelian or Marxist.)

Lefebvre was no structuralist or existentialist. It is not the case that the concrete is the hard, irreducible real and the abstract is the language that inadequately describes it, as this would return us to neo-Kantianism.

The 'concrete abstract' is a combination of the two most opposite of words combined in a seething whole bursting with tensions and contradictions: But these internal pressure points are not 'difficulties' or 'problems' to be erased; they are where negation opens, where it fissures and moves the subject. They are the beginning of philosophy, not an arrival at its limits.

The concrete abstract can be a sort of spectacle in a situationist sense. What disturbed Baudrillard about simulacra was that negation was not possible (1981). This is what disturbed

Debord and the Situationists. It is surely due to the fact that the concrete abstract becomes surface, pure abstraction, and therefore the dialectic cannot properly move. We have not reached that situation despite Baudrillard's warnings. But the concrete abstract still saturates human affairs.

A Fata Morgana is a mirage, a physical form seen out at sea, which turns out to be an illusion. The Fata Morgana is a good way to illustrate the concrete abstract. Appropriately, 'Fata Morgana' refers to very real mirages, seen in the Strait of Messina, thought to be 'fata' or 'fairy' castles, luring sailors to their deaths. These mirages were named after the Arthurian enchantress, Morgan le Fay. We can look at Fata Morgana in dialectical terms; the 'real mirage', which is perhaps the ultimate Hegelian 'contradiction embodied', and again a good illustration of the applied concrete abstract. The island of Britain is a Fata Morgana. This is not just what Britain is in our globalised present, it is what Britain has always been (Ford and Hanson 2015).

Paul Mason's Channel 4 blog presented the most interesting map after the UK General Election of 7 May 2015 (Ford and Hanson 2015). It split the island into three: Scotland as 'Southern Scandinavia'; the southeast as the 'Asset Rich Southlands', swelling since the 1986 deregulation of financial markets, and, lastly; the 'Post-Industrial Archipelago', the Detroitified, abandoned middle, drawn as spiky red islands. What is fascinating about this map is that nation state and sovereignty are of little relevance to it. In some ways, this map is just as real as the standard maps of the island. But both maps are true, they are Fata Morgana, 'real mirages' (ibid.).

Paul Mason's Channel 4 map showed that the nation is not 'the island' (ibid.). It is those offshore rigs, pipelines, digital signals, and data cables. It is airstrips, ports, and satellites. 'Local' places are refilled everyday with power, petrol, food, and resources, from capital, arbitrarily creating ruin or regeneration. Capital's flows and circuits are utterly and necessarily hybrid and international. 'One nation' Toryism and fundamentalist Localism always conceal this, in order to hide capital's own interests and power. They are always already ideological (ibid.).

But the nation is also the simmering resentment of the Brexit vote, a vote largely made through emotion rather than hard facts. The concrete abstract, ultimately, is a figure that can help us to reconnect the infrastructural with the ideological. The concrete abstract should itself never remain abstract, it must be applied to the concrete and particular and ultimately the political out in the lifeworld.

Lefebvre's project itself is concrete abstraction, it resists total systematisation. We understand the need for a less disparate Lefebvrian scholarship – the political expediency is now desperately urgent – as the DSG put it: 'the post political = the most political' (DSG 2011).

But Lefebvre's project itself is concrete – he is dead – it is finite down to the last punctuation mark and yet it is also abstract, open to interpretation – of translators in the first instance – then readers and other academics. It is also lifted out, preserved, transformed, and cancelled in each historical moment, in this constantly collapsing present. It expands, and the arguments, alternative readings, and interpretations are part of that expansion. To slice these off as aberrant in search of a purer Lefebvre would be to return to Kant, not Hegel, and that would not be in the spirit of Lefebvre. It would in fact be to declare miners irrelevant to coal.

We need to take much of the critique seriously though. Gromark questions some of the relevance of Lefebvre's project, for example, through reading Stanek on Lefebvre, he also questions Lefebvre's focus on 'dwelling' and the domus when theorising space. This must be revised, and we must revise it for nomads. The idea of 'elective belonging' is only now owned by the elite, who in any case peripatetically attend to business across the globe.

To add to the critiques of Lefebvre and Lefebvrian thought, a reified version of spatiality is part of popular consciousness: We have had a referendum on space, but to suggest we have a referendum on time seems insane. Gillian Rose (1992) hits French thinkers for seeing the world as lack – Lefebvre is an exception – she doesn't say this, but for us the everyday life work is proof that he is an exception.

Conclusion

This chapter has explained the tricky and often obfuscated concept of the 'concrete abstract'. It has illuminated its main philosophical roots, in Hegel and Marx, but from there it has gone on to advocate, in the spirit of Lefebvre's work, an engaged dialectics of the 'concrete abstract' of the street, by turning to actual political examples in the world.

These are points to be developed further elsewhere. But what is absolutely clear in this moment is that for what is being described as 'post-truth', the concrete abstract is an essential philosophical figure. Ultimately, then, we give you notes towards a contemporary concrete abstract as an orienting litany that might be used to do real political work.

References

Baudrillard, J. (1981) *Simulacra and Simulation*, Ann Arbor, MI: The University of Michigan Press.
DSG (2011) *The Poster*. Online. Available HTTP: <https://deterritorialsupportgroup.files.wordpress.com/2011/08/postpolcolour.pdf> (accessed 28 February 2017).
Engels, F. (1890) *Letter to Conrad Schmidt*, trans. D. Torr. Online. Available HTTP: <https://www.marxists.org/archive/marx/works/1890/letters/90_10_27.htm> (accessed 28 February 2017).
Ford, S. and Hanson, S. (2015) 'New and accurate maps of the island', in *Memoria: Histories, Memories, Representations*, Leeds: Fold Press.
Gromark, S. (2013) 'A third wave of receptions: space as concrete abstraction – Łukasz Stanek on Henri Lefebvre', *Site*, 33 245–252.
Hall, S. (2010) 'Life and times of the first new left', *New Left Review*, 61, London: NLR.
Kipfer, S. (2009) *Space, Difference, Everyday Life: Reading Henri Lefebvre*, Oxford: Wiley-Blackwell.
Latour, B. (2005) *Reassembling the Social*, Oxford: OUP.
Lefebvre, H. (1940/2009) *Dialectical Materialism*, trans. J. Sturrock, Minneapolis: University of Minnesota Press.
Lefebvre, H. (1991) *The Production of Space*, trans. D. Nicholson-Smith, Oxford: Blackwell.
Lefebvre, H. (2016) *Metaphilosophy*, trans. D. Fernbach, London: Verso.
Marx, K. (1857–1859) *Critique of Political Economy*. Online. Available HTTP: <https://www.marxists.org/archive/marx/works/1859/critique-pol-economy/> (accessed 28 February 2017).
Marx, K. (1976/1867) *Capital*, vol. 1, Harmondsworth: Pelican.
Merrifield, A. (2006) *Henri Lefebvre: A Critical Introduction*, London: Routledge.
Rees, J. (1998) *The Algebra of Revolution*, London: Routledge.
Rose, G. (1992) *The Broken Middle*, London: Routledge.
Rose, G. (2009) *Hegel Contra Sociology*, London: Verso.
Schmid, C. (2008) 'Henri Lefebvre's theory of the production of space: towards a three-dimensional dialectic', trans. B. Goonewardena, *Space, Difference, Everyday Life*. 27–45.
Shortall, F.C. (1994) *The Incomplete Marx*, Aldershot: Ashgate. 445.
Stalin, J. (attributed to, 1938) *Dialectical Materialism*. Online. Available HTTP: <https://www.marxists.org/reference/archive/stalin/works/1938/09.htm> (accessed 28 February 2017).
Stanek, Ł. (2008) 'Space as concrete abstraction: Hegel, Marx, and Modern Urbanism in Henri Lefebvre', in Goonewardena, et al. (eds) *Space, Difference, Everyday Life: Reading Henri Lefebvre*, London: Routledge. 62–79.
Stanek, Ł. (2011) *Henri Lefebvre on Space: Architecture, Urban Research, and the Production of Theory*, Minneapolis: University of Minnesota Press.
Wainright, O. (2017) 'Britain has enough land to solve the housing crisis – it's just being hoarded', *Guardian*, 31 January.
Ward, D. (2002) 'Hacienda fans rave at plan for luxury flats', *Guardian*, 29 August.

13

Russian dolls

Trialectics in motion and spatial analysis

Miguel Torres García

Introduction: Spanish Francoist colonisation towns under the trialectic lens

Lefebvre's conceptualisation of space, similarly to de Certeau's (1990) or Arendt's (1958: 323) idea of *action*, relies on the socio-constructionist and post-Hegelian notion of trialectics. In his view, social space – place – is not a given entity, it is rather composed by objects and relations which in turn give shape to normative spaces, those of everydayness, and those ideal or imaginary (Lefebvre 1991: 77). In his terminology, these are representations of space, practised spaces and spaces of representation. The complex rapports between the three dimensions of place, far from being fixed, evolve in spatiotemporal configurations. It is for this reason that Lefebvre relies on historiography as a means of breaking down space into elemental parts. A combination of historical and hermeneutic enquiry of each component within and without its context – *découper et retourner* in de Certeau's terms – makes it possible to more accurately determinate their role in the production of space. A number of additional concepts, such as appropriation, domination, or *détournement* (Lefebvre 1991: 164–168, 2014: 98) refer to the time dimension of places and their study.

Lefebvre's often intricate, rich, and overwhelming style makes it difficult to elicit method, and in spite of the apparent intelligibility of the basic concepts of trialectics, they can unexpectedly turn elusive when examined in detail. In order to make it more operational, I propose in former writings a framework to break down the evolution of a social space into three basic dynamics by virtue of which components of the represented, representational, and practised spaces recombine. I draw on the social sciences concepts which illustrate such transitional states: *bricolage*, *ritualisation*, and *fetishisation*. Tracing how space evolves over time allows me to single out the materials, relationships, and agencies that are determinant of, for instance, a specific urban space (Torres García 2017), or a ritualised practice within the city (Torres García forthcoming). Moreover, a dynamic approach to the evolution of space allows grasping the different agencies involved in its production. An outstanding question remains regarding its boundaries. This chapter applies this analytical framework to the case of colonisation towns in mid-20th century Spain in order to bring forward how different spaces interact and are contained in each other, giving shape to an intricate geometry. Such complexity must nonetheless be overcome insofar as a proper delimitation of a spatial phenomenon is a necessary condition for its critical analysis.

The rural settlements developed by the Spanish Francoist regime roughly between 1945 and 1975 can provide for a particularly illustrative use of trialectics, because they are at the intersection of modernist planning and the exercise of power by an authoritarian regime. They are therefore an especially relevant instance of the concept of represented space, which Lefebvre most identifies with expressions of technocracy and ideology (Lefebvre 1991: 45), and with planning and architecture as disciplines.

During roughly the last two decades much attention has been dedicated to the spatial and architectural production of the Spanish Instituto Nacional de Colonización (henceforth INC), both from academia and, as a result, from institutions concerned with heritage study and protection. This chapter relies on this body of literature in order to reconstruct an alternative narrative around the three axes of the spatial triad. The following section characterises the colonies as represented space. Section 3 enhances the scope of the analysis so all three axes of space are represented. As a result, agents of production can be more accurately identified: the architects and their discipline. Sections 4 and 5 confirm this attribution, briefly outlining an additional instance of Francoist leisure architecture, and the more recent efforts for the heritagisation of the INC's legacy. Conflicts raised by the latter suggest that an alternative delimitation of these architectures is needed in order to better address the people's right to their space.

The emergence of a representation of space

The ideal of a productive settlement structure that smoothly articulates the Spanish territory was greatly developed during the second half of the 18th century. Following the experience gained during the expansion over the American territories, a rising entrepreneurial class conceptually coupled bringing to order the heart of the nation with wealth and moral health. Further to the effort to pace up productivity in underdeveloped agricultural land, populating vacant territories was to provide safer, smoother transit of goods and people. The idea pervaded spatial projects of all scales, from land division to the town square, from the strategic commerce land routes to the settler's house floor plan (Torres García 2016).

In the 19th century, the law *Ley de Colonias Agrícolas* (1868) insisted on the importance of a rational distribution of farmers and farming land, which on this occasion was considered to be one of dispersion. To this end, the law addressed the economic dimension of the territory through various incentives and tax exemptions. The 1868 law appears to have found its main limitations when applied to lands of limited agricultural yield. The later *Ley de Colonización Interior* (1907), which continued to support a dispersed territorial model, also faced shortcomings due to being devised rather as a means to combat social unrest than an effective territorial plan.

One of the most determining precedents of the particular focus of the INC's work was the irrigation policy fostered during Miguel Primo de Rivera's dictatorship (1923–1930) (Sambricio 2008). Former hydraulic policy had suffered from overconfidence in the private initiative and shortcomings in strategic planning on the part of the national administration. From 1926 onwards, the initiative on public works was delegated to federations operating on the different river basins, which yielded more significant advances but worsened regional unbalance. After Primo de Rivera's resignation and exile, the Second Spanish Republic passed the *Plan Nacional de Obras Hidráulicas* (1933) with a view to solving the disparities across the nation. The new plan envisaged transfers of water resources between the Mediterranean and Atlantic basins, which met strong criticism on both political and technical grounds (Ortega Cantero 1992). The overarching *Ley de Reforma Agraria* (1932) also faced hostility among large landowners due to its redistributive agenda. This law, though insufficiently articulated and underfunded, was the first to provide for the construction of colonies as a means to make land accessible to farmers. The Spanish Civil War (1936–1939) meant the demise of the Republic and water policy, at the intersection of technical progress, a utopia of transformation of underdeveloped land into wealthy territories, and a solution for social malaise, became a pillar of Francoist policy and a vehicle for the exercise and expression of power (cf. Swyngedouw 2015).

In this general context, the construction of settlements was a secondary component of a wider economic and socio-political initiative. Accordingly, the INC incorporated an Architecture Service in charge of designing and building towns following very few broad guidelines (Centellas Soler 2010). Granting settlers access to land generated in turn the need to 'lodge [them] in decorous dwellings' as well as to 'group them so as to cultural, religious, associational, etc. life can fully develop' (INC, Delegación Regional del Guadalquivir 1964: 6). The bourgeois yearning for addressing social unrest by means of providing the masses with a salubrious habitat met the fascist social project of indoctrination.

In terms of architectural design, during the period of autarchy characteristic of the first years of dictatorship, such objectives were sought following the officially sanctioned traditionalism. The advances of modernism were used inconspicuously, as the regime associated its aesthetics to the left wing. As a result, during this first stage, 18th- and 19th-century designs were a common reference, and the central square was considered a source of community and Spanish identity. In the early 1950s, however, architects openly welcomed the International Style as a suitable and much-needed answer to the pressing needs for urban development throughout the nation. The towns' designs swiftly became more sophisticated and innovative, while keeping the basic elemental spatial features: the single-family dwelling, the selective use of traditional materials and features, and an outspoken propagandist bent.

The colonisation experience also illustrates the characteristic divorce, in modernist spatial planning, of representations of space and spaces of representation. The approximately 300 new towns built across Spain reveal an effort, on the part of the authorities and the designers, to selectively translate everyday practice – farming and dwelling, production and reproduction – into a represented space of established spatial patterns. This is to say, in support of Lefebvre's argument, that the main elements in their spatial repertoire can be unpacked as an expression of the regime's power structure and ideological corpus. Moreover, in this case the exercise of power is virtually absolute because its context was one of an authoritarian regime, but just as much due to being a case of almost completely ex-novo design, extended to the re-arrangement of large territories by means of hydraulic engineering and irrigation.

The programme typically included a prominent religious centre, the town hall, and other official premises. Communal buildings and meeting places were usually under the control of sections of the fascist party Falange Española, such as the Sección Femenina, the Frente de Juventudes, the Sindicato Vertical, etc., thus framing social exchange within the parameters set by the Movimiento – the regime's ideological apparatus. Open spaces were an important part of this represented space; squares appeared associated to the overarching representations of authority: the church and the town hall (Centellas Soler 2010). There was a conscious choice of a compact urban model, which needed in turn that streets were devised. Their expression was nevertheless minimised by design guidelines which advised against long perspectives in favour of discontinuous or curved alignments.

Dwellings were designed following a principle of serialisation of the single-family home. Although this approach was unequivocally inspired in the advances of modernist architecture and planning (Álvaro Tordesillas 2010), agreements with the popular culture were sought to profit from readily-available, economic materials and construction techniques (cf. de la Sota Martínez 1955). Thus, the buildings incorporated much-sought elements of traditionalism in order to support both a nationalist narrative and the biopolitical (Foucault 2008) project of the farmers' attachment to the land. It is worth pointing out that rural migration was a pressing problem not only for the purposes of colonisation, but in cities at a national scale (Capel 1967: 37).

The designs, therefore, prescribed a population whose social identity was defined strictly along the lines of National-Catholicism. They were hemmed in the overarching boundaries of the Movimiento – family, municipality, and syndicate – an order which was monitored by the Church as a transversal agent. Thus, the radical processes of emergence which rise from within the spaces of representation were curtailed from logics of social control. Such rationales were not only inspired by fascist ideology; a wide range of references in play by the ensemble of colonisation experiences are drawn from the Enlightenment and bourgeois logics – social order, family, property, and a reflexive mistrust of the streets – and represent the meeting of right-wing traditionalism and the left-wing lens on the working class. Notably, the immediate precedent of the INC was the Second Republic's Instituto de Reforma Agraria, on a much more modest scale, and it is consistently pointed out in the literature that the staff of the INC Architecture Service was ideologically variegated.

Establishing a status quo involves social codes, social bodies, and the space over which they move; the practical, the performative (Schechner 2003: 44), and overarching discourses and knowledge structures (Foucault 1975). I use the term ritualisation in order to transcend a static portrayal. I see ritualisation as a derivative of Lefebvrian represented space; it is the process through which elements from performance and practice are constructed as memory (Ripoll and Veschambre 2005; Veschambre 2004) and closed down to contestation (Massey 1995 in Edensor 2005). Along with symbols such as the church or social spaces, the farmers' everyday practice was assimilated in the process of creating the new towns. Those leading to improved production and the support of the *Movimiento's* principles were adopted. It is the case of the use of carts for transport, which was stripped of any social significance and conceptualised as a basic distance module: a 45-minute trip, deemed optimal for the day labourers' purposes, ruled the distribution of settlements and plots. Traditional customs such as locally-sourced, cost-effective construction techniques were also utilised, which were in turn *détournés*, at a narrative level, into a praise of traditional know-how:

> [The town of] Esquivel is an attempt to turn those who always built the towns into masters, who did indeed a wonderful job: the rural bricklayers and master craftsmen.
>
> This is a theory for the architectural approach, not applicable to the town layout: to build a small town of one or two hundred houses is not to build one or two hundred houses together so a village results; it is a different problem. (de la Sota Martínez 1955: np)

The second part of this quote shows how, conversely, the practices that did not support the intended socio-technical schema were de-territorialised from the project. This was particularly true for spaces of casual encounter and of collective life. The primacy of represented space meant the eradication of spaces of representation.

Searching for the complete trialectic picture

Colonisation towns, as portrayed in the former section, seem incomplete spaces if the trialectic lens is to be applied. At first glance, they were not but represented space built out of officially-sanctioned representations of labour and dwelling. Deprived of elements of practice and of their spaces of representation, the farmers and settlers do not appear to carry out their daily life in a complete space. Granted, everyday use would soon trigger processes of appropriation and *détournement* (cf. Loren Méndez 2008), but let us examine this space in its pristine form, right off the hands of the designer, for a little longer: Can only represented space be produced? Must it be part of a space encompassing the three axes of the triad?

An aspect can be considered in order to frame these represented spaces into more comprehensive social space. It is the double role of the architect – both as a producer and as a cog within the mechanisms of production of space (cf. Stanek's discussion in Lefebvre 2014: xxxv and f.). If, instead of placing the focus on the dweller, we consider that architects are the agents of production, these towns appear again lived, their humanity restored, their existence to have a more natural texture. In turn, they reveal themselves a part of a wider spatial ensemble.

I use the term bricolage in reference to the creation of a practised space to gain a hold over otherwise undifferentiated space and to appropriate it. Levi-Strauss (1966) formulated bricolage as a mechanism for the creation of myths, which aim to give the (perceived) world a rationale (also Lévi-Strauss 1978). This concept has been revised in a post-structuralism context, mainly in order to weigh the effect of agendas behind myth studies (Patton and Doniger 1996). As a social structure extends over space, bits and pieces of its culture – materials, concepts, and symbols – are incepted in order to turn unfamiliar space-time into practised space-time. In the same way, elements that are found during the process can be incorporated into the dominant narrative, in a re-territorialisation process.

I refer to this process above, when I identify elements from the farmers' everyday practice which were either incorporated to the representations of labour and dwelling or discarded for not supporting the overarching socio-technical discourse. Much of the latest literature on the INC's new towns is dedicated to framing the phenomenon in history and within the evolution of contemporary architecture (for instance Álvaro Tordesillas 2010; ; Pérez Escolano and Calzada Pérez 2008). Applying the trialectic approach to this historiography allows identifying the constituent elements of this space, the rapports between them, and the agencies behind their interaction.

De la Sota's design for the town of Esquivel, in Seville, is widely regarded as one of the greatest exponents of this bricolage. Along with the symbols of the regime, he draws on international trends of formal abstraction, functional separation of pedestrian and cart traffic, and graceful expressionism. These are combined with shapes and spaces of popular and neoclassical inspiration. The important point to make here is that the producer of space, the author of this bricolage, and therefore the recipient of the resulting space is not the farmer but the architect. A wide range of elements, from local landscapes to international references in planning, were brought together to give shape to a space: that of the discipline.

If bricolage stands for the process of creation of practised space, and ritualisation of represented space, I use fetishisation to characterise the path to spaces of representation (see Figure 13.1). Originally coined to describe objects that were invested with magical powers (Ellen 1988; Pels 1998), the concept of fetish evolved within anthropology into having a definition grounded on cultural processes: 'Fetishism (…) is by definition a displacement of meaning through synecdoche' (Gamman and Makinen 1994: 45 in Dant 1996). Critical to Marxist formulations, the semiologist Baudrillard (1981: 92) sees the fetishised object as no longer a referent for itself, but instead standing for the system of values that originally produces it. Arendt's understanding of *action*, as opposed to *labour* and *work*, can clarify this point. Action is the sphere of the initiative, of the new. It starts irreversible processes because it exists in the transference between the individual and the collective (Arendt 1958).

Fernández del Amo's formal poetry (Cordero Ampuero 2014) and de la Sota's innovative designs are examples of how the practice of planning evolved one town at a time. Again, in

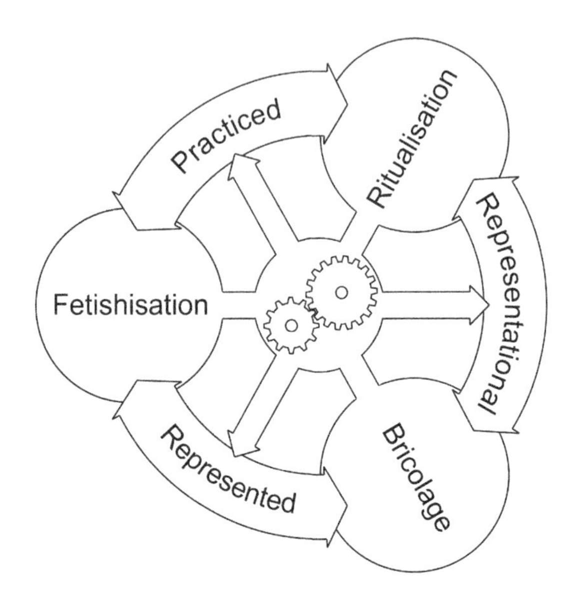

Figure 13.1 The elemental parts of space are in constant reconfiguration though processes of bricolage, by means of which elements from represented and representational spaces compose practiced space, ritualisation as a homologous pathway to represented space, and fetishisation, when they give shape to spaces of representation. *Source*: author.

Esquivel, de la Sota pushed the limits of the few prescriptions issued by the INC. The church and the town hall are thus arranged before the town, over which they dominate visually, detached from the square but still strongly asserting their symbolism. Each work, made from elements from practised and represented spaces, symbolised not only the regime's power over land, water, and bodies, but also architecture as a practice and a discipline. Thus a space of representation was engendered which created a performative sphere for a long list of 80 architects who worked for the INC Architecture Service, including other key emergent figures (Centellas Soler 2010: 110). Their individual takes on urban design re-elaborated the ensemble of the INC settlements and made a contribution to planning as a whole.

In summary, the settlements do not find a clear fit in the spatial triad when considered individually, but within a comprehensive, tessellated, nation-wide complex of architectural concepts, works, and professional figures. A complete social space appears when the focus is not on the farmers, but on the architects as producers. It is the architects' space, their playground.

The architect's playground

A second contemporary example helps make this point, which is also studied in similar terms – those of design excellence and, as of late, of heritage values. It is the *Ciudad Sindical de Vacaciones*, built under the initiative of the *Obra Sindical de Educación y Descanso* on the shore of Marbella, Málaga, between 1956 and 1963. This organisation, originally inspired by the Italian fascist Opera Nazionali Dopolavoro, was in charge of fostering educational, leisure, and sport activities for workers under the umbrella of the fascist Sindicato Vertical. It therefore responded to the same values that ruled the design of rural settlements.

Accordingly, the design programme was similar, in spite of their diametrically opposed purposes – work vs. leisure. As other vacation towns (for instance Perlora in Asturias, or Tarragona), it is a mostly self-sufficient complex where small villas and row houses are unequivocally designed for the single family. The church and the administration building – a foil of the town hall – assume much of the ensemble's expressive load. The architects resort to a Garden City model, which they combine with white-washed organic shapes meant to evoke an ideal Mediterranean village. Communal spaces are formalised and separated according to use, whereas the role of the streets is reduced to a minimum.

A few elements seem to puzzle the architecture critics. These are apparently whimsical and gratuitous features, such as a useless concrete cantilever resembling a springboard, or the monumentalised water tower which boldly emerges from within the management building and offers an incongruently rural skyline. Also, the church conveys the surreal and organic undertone of the whole complex, underlining its representative role in a gesture that García Vázquez (1999: 266–271) considers 'paradoxical'. In general terms, the design has been described as 'ironic' (Ramírez, Santos and Canal 1987: 57–62) or 'hardly classifiable' (VV.AA. 2002: 263).

The true nature of these features comes to light if, as I do above for the case of the rural new towns, the analysis considers not the vacationers but the designers at the centre of this space. To the former, if they hypothetically came from one of the INC settlements, the differences between their everyday environment and their leisure venue would be but superficial. The gentle curves of organic buildings and garden-city roads leading to the beach replace the straight lines of gabled roofs and field furrows. Still, church, administration, and family are ever present and, as their spaces interlock, leave few crevices for spontaneity.

But it is precisely in the in-betweens where Lefebvre locates enjoyment: the 'total body' emerges from within the dissolution of order that can be experienced in liminal spaces. As an example, he poses the beach, in the meeting of land, sea, and sky (Lefebvre 2014: 48–49). In the case of the Ciudad Sindical, it is the designers who revel in such a space. They take joy in the confusion between functionalism and fanciful ornament, in the eclectic choice of intellectual and formalist referents, and in the contamination of modernist layouts with textures drawn on tradition.

When writing about his own work on Esquivel, which Cordero Ampuero (2014: 46) somewhat dismissively describes as a 'surrealist pleasantry', de la Sota Martinez lets his exhilaration show:

> Sometimes, we must outdo our own knowledge and depths; on other occasions, it may be the case, our subtleties and frivolities: it is a matter of getting ready, of vibrating with the issue. When we want to show what we know, *we take pride in our work, we swell as peacocks, it is good that the architect arrives.* It is hard when, in order to be right, we actually need to forget everything, almost all we know, however little. Once achieved, as we then know nothing, we cannot but start copying…
>
> *(de la Sota Martínez 1955, emphasis added)*

Heritagisation of the architect's space

A 1950s aerial photograph shows the Ciudad Sindical on the then still-virgin coast of Marbella. The image reveals that this exercise of hedonism heralded the saturation of the Spanish Mediterranean shores with an overwhelming succession of self-centred, schizophrenic designs (Ramírez, Santos and Canal 1987). The success of the colonisation initiative is also questionable. Some 130 settlements were built in the region of Andalusia, yielding a success ratio of approximately one-third in terms of consolidation, integration within the regional urban system, and demographic and economic dynamics (Cruz Villalón 1996).

This evolution has been, in any case, contingent on the wider territorial policies, in which the quality of town planning played a limited role. It rather forms part of a complex geometry of spaces that reach the present day and, like Russian dolls, encase and are contained in each other (see Figure 13.2). The trialectic approach allows the interpreter to characterise the interactions between the components of such spatialisation and to unpack its boundaries.

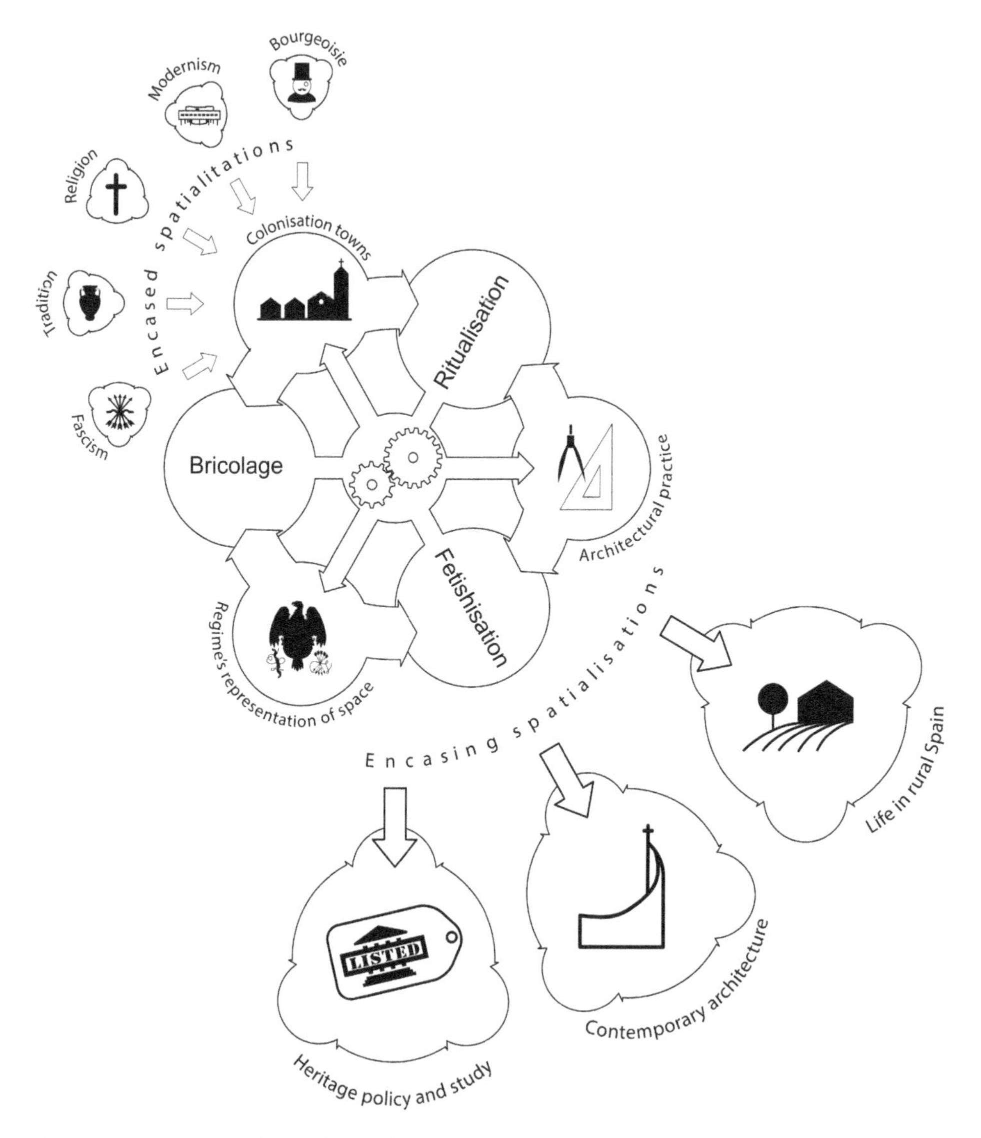

Figure 13.2 After applying the trialectic approach to colonisation towns, these appear to contain elements from different previous spatialisations, such as Christian Catholic, fascist, bourgeois, traditional rural life, and modernist architecture. In turn, these towns have become a part of overarching spaces, such as the ensemble of built heritage, contemporary architecture, and everyday country life. *Source*: author.

Architecture critics and heritage protection agencies consistently consider the new towns as built heritage deserving listing and protection. This is often argued in the same terms that originally ruled their design. The use of international references is praised as a sign of excellence and, not least, construed as the endeavour of architects who escaped the reviled Francoist aesthetics – an argument that tiptoes around the actual ideological affiliation of some of the most representative of them. The combination of modernism and tradition is also celebrated from a post-modern perspective, while their original propagandist purpose and the troubling rapport with the fascist social project are discussed as a separate issue. Finally, the experimental component is celebrated as a sign of the important role of the – then young – architects in the construction of modernity (for instance Junta de Andalucía 2006). As a token of recognition, the spiralling envelope of the Ciudad Sindical church forms part of the logo of the Andalusian Registry of Contemporary Architecture.

The Andalusian Institute of Historical Heritage (IAPH) produced a comprehensive monograph on the colonisation towns as part of the recent efforts towards their study and protection (Pérez Escolano and Calzada Pérez 2008). A dedicated issue in their journal advocates for their protection as heritage from a dedicated approach that encompasses 'disciplinary diversity', 'veteran as well as young authors', and 'comprehensive views next to specific cases', towards 'a more just project of society' (Pérez Escolano 2005: 38). The heterotopia of disciplinary rule, expert and innovative authorship, and of an understanding of society as a 'project', which was originally created by and for architects, technocrats, and ideologists, now extends to academia and critique and drives the formulation of the 'heritage regard' (Pérez Escolano 2005: 38).

My use of fetishisation is closely related to the concept of heritagisation (Veschambre 2004, 2005). Bernbeck (2013) exposes the relationship between the loss of cultural signs and their exaltation as symbols. Perceived at risk of deterioration, the colonisation towns are extolled as symbols of the development of modernist architecture in Spain, creating a space of representation. I favour the term fetishisation, in spite of its connoted and contested meaning, to avoid institutional formulations of 'heritage' (Edensor 2005). Its overlapping with administrative tools engenders, again, a representation of space meant to prescribe life and establish meaning:

> [The technicians] made me understand that they want to freeze the image... the photographs taken when the town was built... we cannot give our town to the architecture technicians because we need it to live in it.
>
> *(J. Caro, mayor of Vegaviana, in Coleto 2016: 51'40")*

Conclusions

The latest literature on the colonisation towns excels at identifying the INC experience as an example of the virtues and conflicts raised by modernism in Spain. This chapter's spatial analysis is buttressed on this wealth of knowledge, but aims to challenge the boundaries of this approach, which, in its current form, and in its translation into the tools currently in place for heritage management, perpetuates inequities of yesteryear. The space devised by the architects as a constellation of performances and authors, to a great extent for the representation of the National-Catholic regime, has outlasted the latter as a sphere of disciplinary practice and values that are recently put forward as drivers for heritage management. This overarching spatialisation is both concrete and abstract, and it contains in it the many instances of settlements which their inhabitants conceive, perceive, and live individually.

The introduction of the time dimension in the trialectic approach allows identifying objects, relations, and agencies operating along the three axes of social space. It also allows the analyst to approach the boundaries of an otherwise complex geometry of spaces that, like Russian dolls, encase and are contained in each other. After identifying the overarching agents in the production of this space, pressing issues rise as to the role played by the actual residents. Future research should focus on how and why the inhabitants of these settlements use, appropriate, *détournent*, and adapt them, in order to include additional views in framing the heritage regard. A more encompassing delimitation of the spaces that are the object of planning and management, one able to account for the intricacies of the production of space, can assist in determining who is at the centre, who is excluded, and who holds the right to it.

References

Álvaro Tordesillas, A. (2010) 'Referencias internacionales en los pueblos de colonización españoles', *Ciudades: Revista del Instituto Universitario de Urbanística de la Universidad de Valladolid*, 13 183–200.

Arendt, H. (1958) *The Human Condition*, Chicago, London: University of Chicago Press.

Baudrillard, J. (1981) *For a Critique of the Political Economy of the Sign*, New York: Telos Press.

Bernbeck, R. (2013) 'Heritage void and the void as heritage', *Archaeologies*, 9:3 526–45.

Capel, H. (1967) 'Los estudios acerca de las migraciones interiores en España', *Revista de Geografía*, 1:1 79–101.

Centellas Soler, M. (2010) 'Los pueblos de colonización de la administración franquista en la España rural', P+C: proyecto y ciudad, 1 109–26.

Coleto, M.Á. (2016) 'Los pueblos de colonización: la reforma agraria de Franco', *Documentos RNE*. Radio Broadcast. 25th August 2016. Available HTTP: <http://www.rtve.es/alacarta/audios/documentos-rne/documentos-rne-pueblos-colonizacion-reforma-agraria-franco-09-04-16/3560662/> (accessed 1 March 2017).

Cordero Ampuero, Á. (2014) 'Fernández del Amo. Aportaciones al arte y laarquitectura contemporáneas', PhD thesis, Universidad Politécnica de Madrid, Madrid.

Cruz Villalón, J. (1996) 'El mapa de la política de colonización en Andalucía', *Investigaciones Geográficas*, 16 21–34.

Dant, T. (1996) 'Fetishism and the social value of objects'. *Sociological Review*, 44:3 495–516.

De Certeau, M. (1990) *L'Invention du Quotidien, 1. Arts de Faire. Folio. Vol. Nouvelle Édition*, Paris: Gallimard.

De la Sota Martínez, A. (1955) *Pueblo de Esquivel (Sevilla) 1955* [draft statement], Madrid: Archive of Fundación Alejandro de la Sota, 52–X. Available HTTP: <http://archivo.alejandrodelasota.org/es/original/project/146> (accessed 1 March 2017).

Edensor, T. (2005) *Industrial Ruins: Space, Aesthetics and Materiality*, Oxford: Berg.

Ellen, R. (1988) 'Fetishism', *Man*, 23:2 213–35.

Foucault, M. (1975) *Surveiller et Punir: Naissance de la Prison*, Paris: Gallimard.

Foucault, M. (2008) *The Birth of Biopolitics: Lectures at the Collège de France, 1978–1979*, Basingstoke: Palgrave Macmillan.

García Vázquez, C. (1999) *Momo Andalucía: Arquitectura del Movimiento Moderno en Andalucía 1925–1965*, Sevilla: Junta de Andalucía, Dirección General de Arquitectura y Vivienda.

INC, Delegación Regional del Guadalquivir (1964) *Álbum Fotográfico XXV Aniversario*. Available HTTP: <http://www.mapama.gob.es/es/ministerio/archivos-bibliotecas-mediateca/mediateca/colonizacion.aspx> (accessed 1 March 2017).

Junta de Andalucía (2006) 'RESOLUCION de 21 de febrero de 2006, de la Dirección General de Bienes Culturales, por la que se resuelve inscribir colectivamente, con carácter genérico, en el Catálogo General del Patrimonio Histórico Andaluz, nueve bienes inmuebles del Movimiento Moderno de la provincia de Jaén', *Boletín Oficial de la Junta de Andalucía*, 51 53–64.

Lefebvre, H. (1991) *The Production of Space*, Oxford: Blackwell.

Lefebvre, H. (2014) *Toward an Architecture of Enjoyment*, Minneapolis: University of Minnesota Press.

Lévi-Strauss, C. (1966) *The Savage Mind: La Pensée Sauvage*, London: Weidenfeld and Nicolson.

Lévi-Strauss, C. (1978) *Myth and Meaning*, London: Routledge and K. Paul.

Loren Méndez, M.M. (2008) 'Asentamientos rurales metropolitanos. Colonización rural y segunda colonización del Valle del Guadalhorce en la aglomeración urbana de Málaga', in Víctor Pérez Escolano and Manuel Calzada Pérez (eds) *Pueblos de Colonización durante el Franquismo: la Arquitectura en la Modernización del Territorio Rural*, 450–66, Sevilla: Junta de Andalucía, Consejería de Cultura, IAPH.

Ortega Cantero, N. (1992) 'El plan nacional de obras hidráulicas', in Antonio Gil Olcina and Alfredo Morales Gil (eds) *Hitos Históricos de los Regadíos Españoles*, 335–64, Madrid: Ministerio de Agricultura, Alimentación y Medio Ambiente, Secretaria General Técnica.

Patton, L.L. and Doniger, W. (1996) *Myth and Method*, Charlottesville, London: University Press of Virginia.

Pels, P. (1998) 'The spirit of matter: on fetish, rarity, fact and fancy', in Patricia Spyer (ed.) *Border Fetishisms: Material Objects in Unstable Spaces*, New York, London: Routledge.

Pérez Escolano, V. (2005) 'Pueblos de la colonización franquista: objetivo patrimonial', *PH: Boletín del Instituto Andaluz del Patrimonio Histórico*, 13:52 38–42.

Pérez Escolano, V. and Calzada Pérez, M. (2008) *Pueblos de Colonización durante el Franquismo: la Arquitectura en la Modernización del Territorio Rural*, Sevilla: Junta de Andalucía, Consejería de Cultura, IAPH.

Ramírez, J.A., Santos, D. and Canal, C. (1987) *El Estilo del Relax: N-340. Málaga, h. 1953–1065*, Málaga: Colegio Oficial de Arquitectos.

Ripoll, F. and Veschambre, V. (2005) 'Introduction. L'appropriation de l'espace comme problématique', in Fabrice Ripoll and Vincent Veschambre (eds) *Norois. Environnement, Aménagement, Société*, vol. 195 7–15.

Sambricio, C. (2008) 'La 'revolución conservadora' y la política de la colonización en la España de Primo de Rivera', in *Pueblos de Colonización durante el Franquismo: la Arquitectura en la Modernización del Territorio Rural*, 61–72, Sevilla: Junta de Andalucía, Consejería de Cultura, IAPH.

Schechner, R. (2003) *Performance Theory*, London: Routledge.

Swyngedouw, E. (2015) *Liquid Power: Contested Hydro-Modernities in Twentieth-Century Spain*, Cambridge, MA: MIT Press.

Torres García, M. (2016) 'Cultural keys to the evolution of Spanish urbanism', *Journal of Urban History*. Online. Available HTTP: <http://journals.sagepub.com/doi/abs/10.1177/0096144216672657>.

Torres García, M. (2017) *Seville: Through the Urban Void*, Abingdon, New York: Routledge.

Torres García, M. (forthcoming) 'The evolution of a practice in trialectic space: an approach inclusive of norms and performance', *Outlines – Critical Practice Studies*.

Veschambre, V. (2004) 'Appropriation et marquage symbolique de l'espace: quelques éléments de réflexion', *ESO*, 21 73–7.

Veschambre, V. (2005) 'Le recyclage urbain, entre démolition et patrimonialisation: enjeux d'appropriation symbolique de l'espace', *Norois. Environnement, Aménagement, Société*, 195 79–92.

VV.AA. (2002) *La Arquitectura del Sol/Sunland architecture*, Barcelona: C.O.A. de Cataluña, Comunidad Valenciana, Islas Baleares, Murcia, Almería, Granada, Málaga y Canarias.

14

Counter-spaces, no-man's lands and mainstream public space

Representational spaces in homeless activism in Japan

Carl Cassegård

Introduction

Discussions of space and time have a long pedigree in Marxism. Lefebvre was, however, the first Marxist to make space and rhythm into centrepieces of his philosophy, using the sensory experience of space and rhythms as tools for systematically exploring capitalist society and its contradictions. His tripartite division of space in *The Production of Space* (Lefebvre 1991) into the spatial practices of the everyday use of space, the spaces of representation of planners and policymakers, and the representational spaces of clandestine experience and dreams has inspired researchers not least since it captures the antagonistic nature of space in capitalism. Sometimes this antagonism is openly visible – for instance, when 'dominated' spaces functionalised according to the needs of capital or government are challenged and transformed through acts of 'appropriation' when users redefine space according to their own heterogeneous needs and wishes (Lefebvre 1991: 165, 356, 363) – but the antagonism is present also in seemingly more tranquil situations where open conflict is lacking.

The 'rhythmanalysis' developed in Lefebvre's later writings helps us trace these sometimes half-hidden antagonisms through our sensory experience of different rhythms in modern life. Unfortunately, there is little in these writings to suggest how he thought of rhythm in relation to the tripartite division of space in *The Production of Space* (Lefebvre 1991). In this chapter, I seek to throw light on this relation by focusing on representational space among activists in the Japanese homeless movements, and on how rhythm connects up with the various ways in which space was experienced in the course of two anti-eviction struggles, in Osaka's Nagai Park in 2007 and Tokyo's Miyashita Park in 2010.

Lefebvre defines representational space as 'directly *lived* through its associated images and symbols, and hence the space of 'inhabitants' and 'users', but also of some artists and perhaps of those, such as a few writers and philosophers, who *describe* and aspire to do more than describe' (Lefebvre 1991: 39). Compared to spatial practices and spaces of representation, it is here, above all, that opposition to the officially sanctioned use of space is articulated and alternative spatial visions take form. But to understand this process, the category of representational space needs to be further elaborated, sensitising it to change and conflict. I will therefore focus on processes where activists attempt to appropriate space and redefine it for their own purposes. My argument will be that representational space as experienced during such processes can be seen as comprising three important modalities – mainstream public space, counter-space and no-man's land – which can be distinguished with the help of rhythmanalysis.

Below, I first provide a brief background on homeless activism in Japan. I then argue for the fruitfulness of distinguishing between the three modalities of representational space before summarising my findings in the conclusion. Throughout the chapter I will illustrate my arguments by drawing on my previous work on homeless activism in Japan, based on interviews with activists engaged in the Nagai Park struggle and interviews and participant observation during the Miyashita Park struggle conducted in 2009–2011 (Cassegård 2014: 117–179).

Why the homeless?

Activists in the homeless movement are seldom preoccupied solely with the publicly visible aspects of activism – the campaigns or protests through which social movements are usually thought to participate in the public sphere. While it is well-known that social movement activities to a considerable extent unfold not only in public but also in the 'invisible' networks of the everyday (Melucci 1989), the importance of having access to alternative spaces outside the mainstream public is especially crucial to homeless activists. This is because of their need to secure spaces for living and physical survival, because of the exclusion they usually suffer from the mainstream public sphere and because their mere existence in urban space is often felt by mainstream residents as a threat or rupture of the urban order. Both in daily life and in activism, their experience of space tends to be antagonistically structured around the interplay between dominated and appropriated space. That they are often forced to orient themselves simultaneously to several different forms of space in their struggles creates, I argue, a special dynamics that is one of the most important characteristics of the homeless movement.

In Japan, homeless people increased rapidly following the end of the so-called bubble economy in the early 1990s. Many settled down in encampments or 'tent villages' along rivers and in the parks of the big cities. There they could find water and relative safety, and make their living by collecting recyclables or other forms of simple work. The communities formed in the villages also provided a degree of protection since they could patrol the area, exchange information and look out for each other (Aoki 2006: 9; Cassegård 2014: 117). Starting in the mid-1990s, several anti-eviction struggles have taken place as authorities have attempted to evict homeless people and dismantle their villages – attempts that have often been made in conjunction with the planned urban 'upgrading' or gentrification of surrounding areas. Two of the most well-known of these were the struggles over the homeless encampments in Osaka's Nagai Park in 2007 and Tokyo's Miyashita Park in 2010. Both strikingly exemplify the increasing presence of cultural activism – including drama, dancing, paintings and music – in homeless activism.

The struggle in Nagai Park is best known for the theatrical play homeless villagers performed on the day of eviction as a last-ditch attempt to keep the guards and the city staff that had come to demolish the village at bay. The eviction took place on 5 February 2007 with the 2007 IAAF World Championships in Athletics as a pretext. On that day, more than 200 activists and other supporters had arrived to defend the village. Many were sitting in a protective ring around the stage, which had been erected in front of the remaining tents. Facing them were 200 city employees and 300 guardsmen. Outside, journalists and onlookers were flocking. The play was performed two times and had just resumed for a third time when the guards moved in and started demolishing the tents. It was over in a few minutes. The activists linked arms to protect the stage, but finally the stage too was torn down (for vivid descriptions of these events, see Kirokushū henshū iinkai 2007).

The protests around Miyashita Park were more drawn-out in time. They started in 2008 when it was revealed that Shibuya Ward had sold the park's naming rights to Nike Japan. According to the original plans, the corporation would provide the park – renamed 'Nike Park' – with new facilities requiring an entrance fee and the homeless were to be evicted. Rallying against this 'Nikefication' of the park, a coalition to 'preserve Miyashita Park' was established in June 2008. In 2010 artists and activists under the name AIR (Artists in Residence) occupied the park together with remaining homeless people to block the construction work. For six months, the park was used as a basis for public protest and as the stage for a variety of artistic activities, such as art workshops, exhibitions, concerts, filming and poetry readings. In September, the park was finally cleared and fenced in by the authorities. Reconstruction started shortly afterwards and the new park, complete with Nike-sponsored sports facilities, was opened to the public in late April 2011.

Rhythms and spaces in the homeless movement

A distinction often made in research on homelessness is between prime and marginal space (Duncan 1978; Snow and Anderson 1993). Prime spaces are spaces that are routinely used by people belonging to an area's host population or that have symbolic significance for such people, while marginal spaces are those that this population cares little about and consequently the least risky for the homeless to use – spaces such as alleys, dumps, space under bridges, abandoned buildings, vacant lots or left-over spaces around railroad yards. Duncan argues that this classification mirrors what the 'tramp' experiences to be the importance assigned to space by the host population (Duncan 1978: 28, 30). This means that the distinction works best when the homeless adapt so fully to mainstream classifications that conflict *as to the classifications themselves* tends to disappear. As Wright points out, however, space is an active relationship between city authorities and individuals, and struggles are constantly going on over the meaning of urban space. In particular, he directs attention to how homeless people resist the identities imposed on them by city and state officials by the creation of 'resistant heterotopias' (Wright 1997: 52).

This suggests that while the prime/marginal distinction works well enough for situations when established spatial divisions are in place and unchallenged (when space is securely *dominated*, to use Lefebvre's term), other classifications are needed when the homeless *appropriate* space, namely when they use space in ways that go against the dominant classification. While Lefebvre's spatial triad does not refer explicitly to resistance or contestation, it can easily be developed in that direction. Appropriating space does not simply mean asserting a given 'representational space' against the 'space of representation'. The struggle against the latter is itself part of the lived reality of homeless people and hence representational space is always indelibly marked by this struggle. The shifts and transformations undergone by representational space in the course of attempted appropriations can be traced with the help of rhythmanalysis.

Lefebvre defines rhythm abstractly – as repetition where the 'measure' is central and where differences can be introduced – and stresses that rhythms are apprehended by 'the most irrational', the body (Lefebvre 2004: 8). The rhythmanalyst, he points out, uses the body as an instrument, a 'metronome', listening to streets or towns as one listens to a symphony (Lefebvre 2004: 19, 22; 1996: 219, 229). In general, rhythms are multiple. Not only capital and state have the power to impose their rhythms on space. Rhythms are imposed also by nature; he therefore speaks of cosmic rhythms, lunar and solar rhythms, the climate and the sea (Lefebvre 1996: 232). Ordinary people also have the power to diffuse their autonomously formed rhythms, by challenging boundaries, carving out spaces for their own purposes and transforming them by their use. Lefebvre emphasises that each group, culture, religion, region and individual is a centre, and adds: 'Now, what is a centre, if not a producer of rhythms in social time?' (Lefebvre 1996: 239). The empowerment of marginal groups such as the homeless can, I suggest, be described as a process of getting rhythms of one's own going that are heterogeneous in relation to the more dominant rhythms of the surrounding mainstream society.

The ways activists adapt to dominant rhythms or try to impose rhythms of their own on space are reflected in how they relate to norms. I find it useful to think of spaces as structured by norms, which in turn are embodied in space. Space and norms together form a 'sens-escape', to speak with Degen (2010), or a distribution of the 'sensible', as Rancière (2004) calls it. In this context, a helpful distinction is between moral ideals and norms. Drawing on Durkheim, Jacobsson and Lindblom (2016) points out how activists in social movements often need to strike a precarious balance between their pursuit of various ideals and the danger of being perceived as norm-transgressors by mainstream society if they pursue their ideals too zealously. To avoid the latter danger, activists may try to minimise the norm-transgressive aspects of activism and adapt outwardly to mainstream norms. This does not necessarily mean compromising or abandoning their ideals, but often instead involves a struggle with how the ideals can be convincingly pursued within the framework of mainstream norms. This struggle, I would argue, produces a certain experience of space, which I will refer to as 'mainstream public space'.

Mainstream public space

When space is used by activists for staging their participation in the mainstream public sphere, as a platform for addressing authorities or the general public, it is part of *mainstream public space* – space for officially sanctioned dialogue. This is the way activists conceive of and experience space when they adapt to mainstream norms, for instance in regard to behaviour or outward appearance, in order to participate in public debate with mainstream audiences. Such spaces can be crucially important for political challengers since it helps them project messages to a wider public and to authorities. At the same time, there are limits to how radical the demands and the conduct permitted in mainstream public space can be, since they need to be considered legitimate or in tune with the normative expectations of authorities and mainstream society. To speak in Lefebvre's terms, such space is 'isotopic' in relation to mainstream society, attuned to the rhythms of mainstream life, in relation to which it poses few or no jarring elements.

In practice, this means that even in acts of protest, activists adapt to the rhythm of the official political system, synchronising with mainstream debates and political decision-making. It becomes important for activists to move in line with the political calendar of mainstream politics, e.g. elections, schedules for debates and voting in assemblies, scheduled negotiations or the debate in the major media. Protest becomes predictable. This does not necessarily mean that it is de-politicised or emptied of oppositional force: there are many examples of campaigns that have been successful precisely thanks to the ability of activists to utilise time and space strategically, playing to mainstream sensibilities and garnering support among established political actors. Moreover, the need to adapt to officially set schedules can often help speed up the rhythm of criticism and mobilisation. An example might be the frantic planning among activists in Nagai Park, and their hectic preparations to get the play ready in time, which was triggered by the official announcement of an eviction date. The rhythm of activism here should not be seen as a mere mirror of the more powerful rhythms of state and capital but is better seen as a rhythm of the public sphere, in which criticism against state and capital is allowed to play part.

Both in Nagai Park and Miyashita Park, great efforts were made by activists to participate in the officially sanctioned debate around the parks. Even when negotiations had proven futile, attempts to frame the encampments as mainstream public space remained. This was most evident in the case of Nagai Park, where the idea of dialogue was expressed by a huge banner over the stage with the words 'We seek dialogue, not fighting'. Activists in Nagai Park repeatedly expressed the idea that the idea of staging a play was to demonstrate the villagers' willingness to reach out and engage in peaceful dialogue with authorities and the general public. This was felt to be urgent in light of the experience of the anti-eviction struggles of Osaka Castle Park and Utsubo Park the previous year, where the media had highlighted conflict and confrontation (Kirokushū henshū iinkai 2007: 14). This goal of inviting dialogue, however, was hardly successful. Neither city officials nor the city staff on the ground responded to the play or the activists' invitation to dialogue. In the Miyashita Park struggle also, many activists and artists saw themselves as participating in mainstream public space, attempting to initiate a dialogue with politicians and the general public through street demonstrations, petitions and demands for negotiations. The appeal to the mainstream public could be most clearly seen in the often voiced arguments that entrance fees and evictions were inconsistent with the idea of 'public gardens' (the literal translation of *kōen*, park), which ought to be open to everybody, and that the renovation plans would transform the park into an 'ad for Nike', a place for consumers rather than citizens (see Cassegård 2014: 171).

Mainstream public space often invites dissatisfaction since attempts to participate often prove futile and since it is associated with norms that are felt to be discriminatory against homeless people and that limit the radicalness of demands and conduct. Such space is experienced as dominated, and hence as unfree and exclusive. Politics, however, is not limited to mainstream public space. There are also ways to appropriate space, by using space in ways that are contrary to mainstream norms. Examples of such use can be seen both in what I call counter-space and in no-man's land.

Counter-space

When activists refuse to conform to mainstream norms, space may shift into what I, following Lefebvre, call *counter-space*. Lefebvre describes these as sites for questioning the dominant organisation of space and as local resistance against the homogenisation of abstract space (Lefebvre 1991: 381ff). They arise by visibilising claims and behaviour normally suppressed in mainstream public space; in counter-spaces, mainstream norms are openly transgressed. One way of doing this is by trespassing on established division of space, occupying or publicly claiming space in provocative disregard of mainstream norms. As Don Mitchell points out, the struggles of homeless people have often involved the creation of new and alternative forms of publicness. Against the exclusion they experience in the mainstream public sphere – where participation is limited to an 'appropriate public' (Mitchell 2003: 51, 122) – they react by taking to the streets, plazas or parks to win the right to political participation, upsetting the spatial order by appearing where they are not meant to be. Using the language of rhythmanalysis, counter-spaces are places where autonomous rhythms are developed that aim at disrupting and, at least temporarily, overpowering those of the surrounding society, which for its part is usually quick to dismiss these rhythms as mere noise. Counter-spaces are thus prime locations for the study of 'becoming irregular', 'throwing out of order' and 'crisis', which Lefebvre points to as an important part of the analysis of rhythms (Lefebvre 2004: 44).

That counter-spaces tend towards a clash of rhythms could be seen both in Nagai Park and Miyashita Park, where activists used art and culture to challenge the restrictive norms of mainstream society. In both cases they did so by *taking* space where they were not meant to be, *making it public* by calling attention to themselves, thereby *upsetting* the official plans and schedules set by authorities. In both cases, activists did not rest content with delivering a political message for wider circulation in the mainstream public; they also sought to directly embody alternative norms that would prefigure a better society. Thus one aim expressed by activists concerning the play staged in Nagai Park was that of visibilising the 'culture of the excluded' through an act of 'cultural resistance' (Ōta 2009: 117, 120). In this case, the counter-space was short-lived and limited to a few hours on the day of eviction. Miyashita Park, however, exemplifies a more sustained challenge to the mainstream order. Here activists managed for a longer period to at least partially overwhelm and intrude on the rhythms of the surrounding society. The struggle entered an intense stage in March 2010, when the authorities announced the imminent closure of the park in order for construction work to begin. In response to this, AIR was formed and occupied the park for six months, turning the park into an exuberant playground for art and activism. During the occupation, the park was filled with dolls, banners, sculptures and a variety of other art objects assembled from the garbage that happened to be at hand in the park. Activists cooked food together and arranged workshops, rock concerts, rave parties, film screenings, outdoor *karaoke*, poetry readings and football. The park was also used for offensive forays into the surrounding public spaces. So-called 'home-made' street parties were arranged, in which hundreds of homeless, musicians and other participants paraded or danced through Shibuya while drumming, blowing trumpets and trombones or beating frying pans, metal cans or other sound-making items. The occupied park thus functioned as a generator of rhythms created by the activists themselves that spread into the surroundings, sometimes successfully disrupting the mainstream order and generating an empowering rhythm for the activists and their supporters. One example is when the activists retaliated against the planned 'Nikefication' of the park by marching to a nearby Nike store to do a sit-in, thereby 'parkifying' Nike. Activities like these often involved symbolical challenges to modernity, for instance via references to things and beings that had been discarded or sacrificed in the course of modernisation (such as the eel that had once lived in the now submerged river next to the park), and that now, through the protests, had come back to haunt modern Japan and seek revenge. These things were symbolised in the figure of 'Miyashita-san', a giant puppet which was carried around in the street demonstrations and in whose figure all discarded and downtrodden things achieved symbolic resurrection (Kindstrand 2012).

In both cases, activists tried to disrupt the calendar set up by the authorities and thereby change the course of the planned action, stealing time by occupations to prolong the life of the tent villages. In Nagai Park, the clash of rhythms was very literal as activists and city staff did their best to mutually disturb each other. Guards and city staff constantly interrupted the play with announcements and the noise of removal preparations, making it hard for onlookers to hear what the actors were saying. In return, the activists tried to get their own rhythm going with the help of bongo drums and shouts. The activists of AIR constantly tried to resort to the unpredictable, first in the timing of their sudden occupation of Miyashita Park and then again in the following series of actions that all seemed exhilarating since they came unexpectedly and caught Nike and the ward authorities off guard. The activists thus went by an old rule of war, avoiding getting caught up in the opponent's rhythm while looking for opportunities to impose their own on the opponent – 'knowing the rhythm of the opponents and using rhythms that they do not expect' (Miyamoto 2005: 21). City authorities too made use of unpredictability to weaken the activists, keeping them in suspense to wear them out. To the activists, the occupation was a time fraught with anxious anticipation: every day they kept watch for the city authorities who were bound to come, sooner or later, to close the park.

No-man's land

That activities in mainstream public space and counter-space seek public visibility makes them unlike activities in what I call *no-man's lands*. These are spaces that are experienced as enveloped in public neglect, but precisely for that reason are felt to allow for behaviour not normally approved in mainstream public life. Here the rhythms of mainstream society are too far to be clearly heard, and in their absence a relatively independent and heterogeneous world is allowed to develop. Even though the tent villages in Tokyo and Osaka were subjected to surveillance by park guards, they were experienced as sanctuaries where homeless people could withdraw into a community of their own, relatively free from the scrutinising view of mainstream society. Park guards attempted to prohibit new tents from being set up, but people living in these villages nevertheless described them as 'public' in the sense of being hospitable to marginal populations. Visitors to the villages as well as villagers themselves have described the time characterising them as tranquil and set off from mainstream society. To a large extent, rhythms were set by the seasons and by nature. Sharing space with birds, mosquitoes, cicadas and other animals, the homeless had to take part in their rhythms. The work necessary to secure subsistence also generated its own rhythms (for instance the need to get up early in the night to collect cans, or the weekly schedule of soup kitchens). In addition, some villagers took initiatives to joint activities like tea gatherings, painting circles and so on that had their own rhythms (for descriptions, see Ichimura 2006).

During the anti-eviction struggles, the encampments tended to lose their no-man's land character. Many villagers left the parks, leaving only a handful of activists to continue the protests. Instead, outside activists arrived to support the struggle. As the space of the parks shifted into counter-space, it attracted public light, destroying one of the premises of a no-man's land, namely public inattention. During these struggles, however, no-man's land continued to play a role as a memory and as a wistfully expressed ideal. In Miyashita Park, several activists were explicit in invoking no-man's land as an ideal. One of them stated that he liked Miyashita Park since it was 'more like a vacant lot [*akichi*] than a park'. It was because it had remained a vacant lot so long, he explained, that the homeless had been able to build their huts there. Furthermore, he added, vacant lots were birthplaces of culture and art (Ogawa 2009: 185). That the idea of vacant lot was held in high regard among many activists was also shown by cardboard signs in the park saying 'I love vacant lots!' Again, after the closure of the park, the activists on their homepage expressed the hope that in the future the park would again revert to a 'vacant lot' (MC Kasurippa 2010). In statements like these the idea of no-man's land is wedded to a Utopian imagination, linked symbolically to things discarded by modern society – like the eels of the submerged river or the homeless people themselves. The time of no-man's land no longer stands merely for the circular rhythms of nature or daily life, but becomes the seed of Utopian hopes and the resurrection of a past eradicated in the course of Japan's modernisation.

In Nagai Park, the ideal of no-man's land was just as crucial, although less explicitly expressed. As I have argued elsewhere, it was not the cultural performance on the day of eviction, but rather the memory of no-man's land that provided the image of Utopia for the core activists in Nagai Park (Cassegård 2014:166). Behind the choice of staging a theatrical play was not simply artistic desire or the instrumental idea of using the play to project a more appealing image to the general public. Many villagers realised from the start that resisting the eviction would be futile, and to them a prime reason for the play was the desire to bid farewell to the village in a satisfying way. A common argument for staging the play was that it would be a good way to end the village, since it would give them the opportunity to 'speak their mind' and 'say what's on their heart' without leaving regrets (Cassegård 2014: 161). The play was a way of paying respect to the lost community of the time when the park had been a no-man's land. What the core activists appear to have cherished most dearly was their image of that community, rather than the counter-space created in the attempt to resist the eviction.

In the struggles of both Miyashita Park and Nagai Park, then, no-man's lands were important reference points to activists, although less as a lived reality than as a memory, linked to new rhythms, namely those of mourning and nostalgia for the park as it used to be. At the same time, this memory became linked to a rhythm of Utopian anticipation.

Conclusion

What has the value been of using concepts such as mainstream public space, counter-space and no-man's land? First, the investigation into the role of different spaces in homeless activism has thrown light on the connection between space and rhythm. In mainstream public space, activists move in step with mainstream rhythms. In no-man's land and counter-space, by contrast, activists disengage from mainstream rhythms, but they do so in different ways. No-man's land is space that is felt to allow for behaviour considered contrary to mainstream norms since it is neglected or ignored by authorities. It thus offers space for alternative rhythms that unfold clandestinely in relation to those of the mainstream world, neither significantly disturbing nor being disturbed by the latter. Counter-spaces by contrast are spaces where such behaviour is intentionally made visible. Unlike no-man's lands, they bathe in public light, provocatively unfurling rhythms meant to clash with those of the surrounding world.

Second, they have helped us refine the notion of representational space and to see how it shifts in the course of contestation and appropriation. This, I suggest, gives us a better feel for the dynamics of the struggle and for how the spatial experiences of activists interconnect with larger contexts. Both of the two anti-eviction struggles on which I have focused, the ones in Nagai Park and Miyashita Park, are known as examples of how activists in the final stages of their struggles constructed spectacular counter-spaces. However, elements of orientations towards mainstream public space also existed and could be seen in the attempts by activists to initiate a dialogue with authorities or the general public. A central role in both struggles was also played by artists and activists oriented to the idea of no-man's land, the latter being idealised as a refuge and an arena of freedom and creativity. In both struggles, it was precisely the attraction many activists felt to the ideal of no-man's land that made them resort to counter-space as a last resort to defend their encampments. Only at a comparatively late stage did the idea of counter-space emerge as a major model for the parks. When it did, it was largely as a by-product of the activists' desire to preserve the parks as no-man's land. Ironically, although the counter-spaces had their origin largely in the desire to protect a no-man's land against outside forces, the latter became transformed in the process of being defended since what was typical of the no-man's land, namely its clandestine, neglected and unregulated character, was lost.

Finally, we have also gotten a better feel for possible tension within activism. The three forms of space are brought into play for different purposes and are attuned to different demands. Mainstream public space is a venue for peaceful protest and public dialogue, counter-space for joyful expression of alternative norms and rhythms and no-man's land is an object of devotion to which farewell must be bid properly and which holds a Utopian promise. Whereas no-man's land and counter-space can be seen as varieties of appropriated space, mainstream public space emerges through the (often futile) effort to make one's claims heard while remaining in tune with dominant norms. In contrast to the common portrayal in social movement literature of performances and other forms of cultural activism, I have tried to show that in homeless activism it is not necessarily the spectacular and eye-catching counter-spatial mobilisations that embody a utopian element, but rather the memory of the free and unregulated no-man's land which the encampment was once felt to have been. In both Nagai Park and Miyashita Park, a primary motivation for activists was the desire to protect a threatened no-man's land, to mourn it, to pay respect to it or to recreate it.

The distinction between mainstream public space, counter-space and no-man's land is, I argue, crucial to understanding the dynamics of anti-eviction struggles, not only in Japan but also elsewhere. It is also relevant to understanding the relation between space and social movements more generally. While social movement studies have tended to focus on the publicly visible aspects of activism, no-man's lands are arguably crucial for furthering empowerment, providing space for alternative lifestyles and discourses and serving as bases in times of mobilisation and places of retreat and hibernation in times of adversity. All three modalities of space are also, I suggest, important to keep in mind in order to understand how activists grope for more inclusive notions of publicness.

References

Aoki, H. (2006) *Japan's Underclass: Day Laborers and the Homeless*, Melbourne: Trans-Pacific Press.

Cassegård, C. (2014) *Youth Movements, Trauma, and Alternative Space in Contemporary Japan*, Leiden: Global Oriental.

Degen, M. (2010) 'Consuming urban rhythms: let's Ravalejar', in T. Edensor (ed.) *Geographies of Rhythm: Nature, Place, Mobilities and* Bodies, Farnham: Ashgate.

Duncan, J. (1978) 'Men without property: the Tramp's classification and use of urban space', *Antipode*, 10:1 24–34.

Ichimura, M. (2006) *Dear Kikuchisan – Burū tento mura to chokorēto*, Kyoto: Kyōtotto shuppan.

Jacobsson, K. and Lindblom, J. (2016) *Animal Rights Activism: A Moral-Sociological Perspective on Social Movements*, Amsterdam: Amsterdam University Press.

Kasurippa, M. C. (2010) 'Naiki pāku wa haikyo ni naru darō', Homepage of AIR. Online posting. published 2 November 2010. Available HTTP: <http://airmiyashitapark.info/wordpress/> (accessed 24 November 2012).

Kindstrand, L. (2012) 'Ghosts against Nike-ification: spaces of representation in Tokyo's Miyashita Park', unpublished master thesis, submitted 31 July 2012, Sophia University, Graduate Program of Global Studies.

Kirokushū henshū iinkai (ed.) (2007) *Sore de mo tsunagari wa tsuzuku – Nagai kōen tentomura gyōsei daishikkō no kiroku*, Tokyo: Birejji puresu.

Lefebvre, H. (1991) *The Production of Space*, Oxford: Blackwell.

Lefebvre, H. (1996) *Writings on Cities*, Oxford: Blackwell.

Lefebvre, H. (2004) *Rhythmanalysis: Space, Time and Everyday Life*, New York: Continuum.

Melucci, A. (1989) *Nomads of the Present: Social Movements and Individual Needs in Contemporary Society*, Philadelphia: Temple University Press.

Mitchell, D. (2003) *The Right to the City: Social Justice and the Fight for Public Space*, New York: The Guilford Press.

Miyamoto, M. (2005) *The Book of Five Rings*, Boston: Shambala.

Ogawa, T. (2009) 'Motto akichi o! Miyashita kōen ga naiki kōen ni', *Impaction*, 170:8 185.

Ōta, N. (2009) 'Haijo sareru mono no hyōgen – Nagai kōen tentomura kara', *Han* 2 114–121.

Rancière, J. (2004) *The Politics of Aesthetics*, London: Bloomsbury.

Snow, D.A. and Anderson, L. (1993) *Down on Their Luck: A Study of Homeless Street People*, Berkeley: University of California Press.

Wright, T. (1997) *Out of Place: Homeless Mobilizations, Subcities, and Contested Landscapes*, Albany: State University of New York Press.

15

Henri Lefebvre's rhythmanalysis as a form of urban poetics

Claire Revol

Introduction

At first sight, it is difficult to define the boundaries of the rhythmanalytical project and demonstrate precisely how it is related to Lefebvre's critical theory, notably in terms of the conceptual framework that he developed for urban planning and the production of space in general. Lefebvre did not detail the full implications of what remained a 'project', nor did he reveal the 'elements' of his 'secret garden', as his friend and colleague, Rene Lourau, who posthumously published Lefebvre's unfinished writings as *Éléments de Rythmanalyse* (Lefebvre 1992, 2004) would later describe it. Texts mentioning the rhythmanalytical project have been written over a long period of time and have been redefined according to evolutions in Lefebvre's thinking. In the second *Critique of Everyday Life* (Lefebvre 1961), inspired by Bachelard's book written in 1936, *The Dialectic of Duration* (Bachelard 1950), Lefebvre first mentions it as an approach to studying social time. He then alludes to it in *The Production of Space* (Lefebvre 1974), *Towards an Architecture of Enjoyment*, written in 1973 (Lefebvre 2014), and his writings on music, before finally describing it as a consistent project in the third *Critique of Everyday Life* (Lefebvre 1981), as well as the articles 'The rhythmanalytical project' in 1985 (Lefebvre 2004) and an 'Attempt at the rhythmanalysis of Mediterranean cities' in 1986 (Lefebvre 2004).

Elements of Rhythmanalysis: An Introduction to the Understanding of Rhythms (Lefebvre 1992; 2004) is a collection of chapters that provides no linear argumentation for the reader. It is rather a collection of fragments, a series of 'preludes', to appropriate Lefebvre's description of the structure of his book, *Introduction to Modernity* (Lefebvre 1962). It is hard to believe that these writings were unfinished because of the lack of time. Lefebvre is a prolific author. He has written more than 60 books, some of them in but a few months. As opposed to *Introduction to Modernity, Elements of Rhythmanalysis* is written in a lapidary way, with most of the terms referring to Lefebvrian concepts without any form of explanation. If we consider these texts to be a general view, or a synthetic picture of the rhythmanalytical quest – which could explain why some of the terms are written in bold or italic letters – we are still without an explanation of their status in his critical theory. Curiously, even though Lefebvre grounds his theory of rhythms in aesthetics, he rarely refers either to space or the urban question. If the *Elements of*

Rhythmanalysis is an 'introduction' to the knowledge of rhythms, how do we then complete it without over-interpreting it?

I shall seek to establish how rhythmanalysis fits into the general framework of Lefebvrian theory through an exploration of the particular texts in which he refers to the rhythmanalytical project. I assume that Lefebvrian critical theory unfolds along the lines of his development of a 'poetics' and that the rhythmanalytical project, which borrows its intuitions from Bachelard's poetics, can be read as a contribution to urban poetics. I have developed this point of view in my doctoral thesis in philosophy, *Henri Lefebvre's Rhythmanalysis as a Form of Urban Poetics* (Revol 2015).

Lefebvre develops rhythmanalysis by means of his contact with artistic practices, especially those of the Situationist International. Against his critique of abstraction as the inherent process of urban space and time, Lefebvre proposes rhythmanalysis as a way to construct an appropriate space-time, including the ability to transform urban society and reconfigure the 'total body' so that the urban can instead be considered as a work of art. Lefebvre thus provides more than a qualitative methodology for the observation of rhythms, as he outlines a rhythmanalytic experimentation that can be associated with the practices of an experimental utopia. Rhythmanalysis therefore offers the guidelines for an applied poetics; it goes as far as creating forms, textures and styles for urban living. This urban poetics, as a creative act steeped in knowledge, proceeds through experimentation; it restores the rhythmic game that enriches our aesthetic experience of urban space and time.

Thus Lefebvre's writings on space and time should not be considered as a purely social theory but rather as an integral part of urban poetics. Lefebvre developed the anthropological dimension of his poetics in his book *Métaphilosophie* (Lefebvre 1965b), which I will analyse as a combination of his writings on social theory and urban poetics.

Rhythmanalysis and the social production of space-time

The point of departure for the rhythmanalytical project can be found in the second volume of the *Critique of Everyday Life – Foundations for a Sociology of the Everyday* (Lefebvre 1961), in which Lefebvre defines his sociology as a radical and critical theory.

The critique of everyday life that Lefebvre explores is rooted in his project of structuring a critical knowledge of modern man and the modern world through his reading of Marx's first texts and his understanding of alienation. He developed this particular theorisation of dialectical materialism in the 1930s with Norbert Guterman, notably in the introduction to *Lenin's Hegel Notebooks*, (Lefebvre and Guterman 1938), *The Mystified Consciousness* (Lefebvre and Guterman 1936) and *Dialectical Materialism* (Lefebvre 1939b). The main thesis is that the banality of ideas in the functioning of daily society makes awareness of alienation difficult. Critical theory must study forms of consciousness in human practices (ideas, institutions, cultures) as concrete forms of the lifestyles and existence of human beings (Lefebvre 1939b: 67). Thus critical theory is a theory of *praxis* that dialectically exposes the production of ideas and things (Lefebvre 1935: 17–8) from the study of material human *praxis*, and aims to transform it. Revolution is not to be understood as the obtention of political power, or solely the change in the economic relations of production, but as the transformation of everyday life.

The publication of the first *Critique of Everyday Life* (Lefebvre 1947) in the context of the development of sociology as a leading discipline in post-war France provides the strategic background for Lefebvre's definition of the study of *praxis* on a sociological basis. But *praxis* can more broadly and originally be understood as the sum of the practical human relationships to the world and to others in society, or the practices by which human beings define their relationships to nature and their own nature. It is thus open to exploration by a plurality of disciplines. Critical theory aims to find ways by which to raise awareness, to orient and transform *praxis*. The Lefebvrian dialectic sees overcoming as an act that is based on a creative dynamic, a 'thinking-acting' ('*une pensée-acte*', Lefebvre 1939b: 58) inspired by his reading of Nietzsche's tragedy as the aspiration towards the 'total man' (Lefebvre 1939a). This creative dynamic takes life itself for its central goal and opens new possibilities for lifestyles considered collectively rather than individually. Lefebvre considers art as a way to accomplish the transformation of *praxis*.

This theoretical background determines the conceptual tools Lefebvre mobilises to study society and argue for the conscious transformation of life. To study *praxis* is to combine sentient and embodied experiences of daily reality, for example the act of gazing at an agrarian landscape, as described in the first version of the *Critique of Everyday Life* (Lefebvre 1947: 141). This experience is then replaced by the shock of the experience of the new town of Mourenx, as described in the *Introduction to Modernity* (Lefebvre 1962), and the transition from rural sociology to urban sociology becomes the focal point of the study of modern man. Yet Lefebvre does not abandon his intellectual project of a theory of *praxis*. Critical theory is fuelled by its intellectual context, and Lefebvre associates himself with *avant-garde* groups such as the Situationist International, especially Guy Debord, whose relations with Lefebvre were remarkably intense and difficult (Ross 1987; Merrifield 2006). Regardless of the conflict, their exchanges came to inspire a generation of urban theorists, critical of State Planning (Stanek 2011; Le Breton 2012).

Though Bachelard is far removed from critical theory, Lefebvre admires his attempt to reconcile science and poetry, particularly in his *Poetics of Space*, as he writes in his autobiography *La Somme et le Reste* (Lefebvre 1959: 142–3). Lefebvre does not just borrow the term 'rhythmanalysis' from Bachelard, he articulates his own scientific project with a theory of the poetics of the everyday experience as informed by Bachelard. Though critical theory produces empirical studies of urban society, Stanek notes that Lefebvre's bibliography for his courses in urban sociology includes descriptions of cities in philosophy and literature such as those of Heidegger, Sue, Steinbeck, Dos Passos and Lowry alongside the theoretical writings of planners such as Lynch and Alexander (Stanek 2011: 21). In the second tome of his *Critique of Everyday Life* (Lefebvre 1961), Lefebvre defines the principal categories of his sociology and considers a social rhythmology or rhythmanalysis to study the interactions of cyclic and linear periods of time or rhythms.

His enquiry on temporalities is directly linked with his critique of modernity. Cyclic rhythms compose the temporalities of agrarian and traditional societies, in which repetitions of periods in social life are ritualised to become part of symbolic exchanges between life and death, or the cosmic relation of the community to the earth (Lefebvre 1961: 52–4). Lefebvre sees modernity as a process of abstraction (Lefebvre 1962), as it introduces linear temporalities marked by the advent of machines, and above all clocks, representing a time that is fictitiously homogeneous and fragmentable. This measurable time becomes everyday time in accordance with the work time. Lefebvre shows that cyclic rhythms correspond to natural and vital rhythms, and to non-cumulative processes (Lefebvre 1961). They are predominant in archaic societies (Lefebvre 1961: 318–9), linked to ancient symbols that ritualise social practices, introducing rhythms in order to include individuals in a community.

In the modern world, where cumulative processes like economic growth are predominant, signs and signals replace older symbols, as Lefebvre claims in his theory of the semantic field (Lefebvre 1961). Because omnipresent signals are conditioning behaviours by means of their linear repetition and participation in the programming of life with all of its satisfactions, linear repetition can bring about monotony, boredom and tiredness. Lefebvre analyses alienation in everyday life in the modern world as a form of this integral programming of time that affects even leisure and rest, to which he opposes enjoyment. Rhythmanalysis is dedicated to the analysis of the persistence of the cyclic rhythms and their interaction with linear rhythms.

Bachelard's rhythmanalysis (Bachelard 1950) is alien to the study of social time. The contemplative philosopher prefers solitary time, free from the requirements of social life. Bachelard focuses on lived time and on how lived instants are dialectically related through rhythms, criticising Bergson's concept of duration. The temporal 'fortune' and 'efficiency' (Bachelard 1950) arise from the way instants are arranged in rhythmic patterns. Bachelard imagines rhythmanalysis as a personal practice aimed at orientating how energy is spent, especially in thinking activities (learning, memory, creative imagination…) and human relationships. It has a therapeutic or pedagogical goal, completing other practices like psychoanalysis, and responds to the fundamental human need for a temporal dialectic, alternating between full and void. But these temporal practices rely on ever-present social temporalities, isolation being possible in accord with rhythmic social modalities. Thus both can be included in the critique of everyday life as ways of transforming and creating moments.

Lefebvre did not extend his conception of the production of space to the production of time, but we can imagine that conceiving of social time as a social product would mean thinking the relationship between perceived time, conceived time and lived time. Rhythmanalysis is presented as a way to study 'concrete modalities of social time' (Lefebvre 2004: 73). Thus Bachelard helps Lefebvre think lived time, allying topo-philia with tempo-philia as *The Poetics of Space* (Bachelard, 1957) through the consideration of the space of representation, the space we live in and even 'happy' space. Each society produces its own space and its own time. The question 'How can appropriate time accompany appropriate space' becomes 'How can time be inhabited by a collective lifestyle?' and 'How can rhythmanalysis complete spatio-analysis in the quest for this appropriate space-time?'

Rhythmanalysis and the quest for the appropriation of urban space-time

By placing rhythmanalysis in the framework of the production of space, Lefebvre articulates it with the appropriation and the abstraction of urban space and time. How do rhythms found the dynamics of the appropriation of space and time? We can assume that these dynamics proceed from the living body, for Lefebvre names the body as the 'constant reference' for rhythmanalysis: 'the theory of rhythms is founded on the experience and the knowledge of the body' (Lefebvre 2004: 67).

Lefebvre evokes the rhythmanalytical project in the chapter 'spatial architectonics' (Lefebvre 1974), dedicated to the organic appropriation of space and the creation of forms by means of the material dynamic of the bodily occupation of space. The body plays a crucial role in appropriation dynamics; it secretes its appropriate and qualified space. The body's energy elaborates social space in successive layers and creates the cosmologies that anthropologists study. It is a creative *praxis* which proceeds from the sensitive and affective relationship of the body to space. Spatial architectonics is also temporal because space and time are inseparable in the living body. Concrete time is made of rhythms that emanate from the body: breathing, heart rate, hunger, walking, sexuality… Marcel Jousse's anthropology of gesture (Jousse 1974) shows that learning is nestled at the heart of a mimetic dynamic specific to human beings, and that gesture impregnates the human body with rhythm. This framework could allow for the micro-sociology of everyday practices, especially in public space, and contributes to the imagination of a spatial and temporal architectonics.

Yet another anthropologist, André Leroi-Gourhan, shows that rhythms contribute to our affective and emotional insertion in the world in *Gesture and Speech* (Leroi-Gourhan 1965). André Leroi-Gourhan does not think that social rhythms are imposed on individuals, but rather created by the activities of living bodies and their relationships with groups and living environments, thus constituting a fabric of aesthetic relationships. Lifestyles are both an emotional and affective insertion into the cosmos and society. He begins with the vital rhythms central to understanding gesture and the creation of human environments, and ends with the creation of symbols of domestic time and space. Days and nights, seasons and walking distances are folded into symbols: rhythms create inhabited space-times. Leroi-Gourhan's writings find their echo in Lefebvre's conception of lifestyle and the construction of a corresponding time and space through 'textures' (Lefebvre 1974). Social *praxis* appropriates rhythms in the creation of social time, and according to the different modalities of specific lifestyles. Situated and acting bodies have inventive capacities that are developed in time, space and through energy.

Lefebvre analyses the modern production of space and time as the concrete production of abstractions (Lefebvre 1974). The urban revolution is the disruption of a cosmic conception of space and time. Cities are no longer inserted into symbolic space and time, they are no longer an image of the cosmos, as was the model of the ancient Greek city. The modern city can no longer be appropriated through old symbols of space and time. This form of abstraction results in a production of repetitive spaces and time that is no longer rhythmic. These modalities transform the aesthetic experience, notably through the predominance of the visual. Bodily rhythms no longer foster the rhythmic appropriation of space and its extension into the creation of a cosmos.

To counter this form of abstraction, Lefebvre proposes rhythmanalysis as the quest to create an appropriate urban space and time. This proposition is correlated to his project for the restitution of the 'total body' (Lefebvre 1974: 465), which would use the knowledge acquired by rhythmanalysis. It can, for example, be found in the therapeutic search for eurhythmia to treat pathologies:

> the living body presents numerous associated rhythms (and we must insist on this crucial point); hence a eurhythmia, when in the state of good health. Pathology, in a word illness, is always accompanied by a disruption of rhythms: arrythmia that goes as far as morbid and then fatal de-synchronisation.
>
> *(Lefebvre, 2004: 68)*

This rhythmic diagnostic can also be used to conceive of urban environments. Because of Lefebvre's pedagogical view of rhythmanalysis as a tool for the appropriation of space (Lefebvre 1974: 237), rhythmanalysis can be thought of as a practical exercise based on listening to rhythms with each bodily sense. It is above all based on the aesthetic experience of rhythms. Furthermore, rhythmanalysis is not only a personal practice, it is a collective practice that is able to transform *praxis*. Lefebvre is interested in the fact that this creative *praxis* can create enjoyment. He develops the relation of the 'total body' to space and time through the perception of rhythms in *Toward an Architecture of Enjoyment* (Lefebvre 2014), demonstrating that the creative *praxis* of appropriation leads to artistic activity, notably in the construction of the urban as a work of art. As we will see, this is what Lefebvre means by *poiesis*.

Music and dance are 'privileged fields and experimental fields' (Lefebvre 1974: 237, 2004) of the creative experience of rhythms and of lived time. Lefebvre wrote on music and society (Lefebvre 1966, 1971) and meditated on Nietzsche's conception of style as the link between music and lived time (Lefebvre 1970, 1975). The ambivalent relationship between cyclical rhythms related to natural, cosmic and organic forces and linear rhythms, related to the measurement and quantification of time, prevails in Lefebvre's understanding of musical rhythms. Music appropriates the rhythmic body, as, in Lefebvre's terms:

> the relation between musical time and the rhythms of the body is required. Musical time resembles them but reassembles them. It makes a bouquet, a garland from a jumble. Through dance, first of all. Musical time does not cease to have a relation with the *physical*.
>
> *(Lefebvre 2004: 64)*

In musical rhythm, the appropriation of time and space occurs through the body, which is why semiology is insufficient in understanding music (Lefebvre 1971). Music gives its texture to the time that is lived in. Following this perspective, Lefebvre equates the conquest of everyday sound in concrete music and underground culture with urbanisation and globalisation ('*mondialisation*'). They provide an outlet for tensions in everyday life (Lefebvre 1981: 135). However, the aesthetic experience of the environment is not only a sonic experience. The rhythmic experience of architecture, urban space and landscape leads to an open inquiry (Maldiney 2012; Revol 2015).

The theory of moments marks another attempt on the part of Lefebvre to think appropriate space-time through the idea of repetition. 'Moment' is defined as a superior and richer form of repetition that is created from everyday forms of repetition such as gesture, behaviour, instants and disruptions in stable periods of time, objects and artworks, symbols and affective stereotypes (Lefebvre 1961: 344). Lefebvre describes different moments: love, knowledge, games… The 'moment' has temporal qualities of intensity, defined by paroxysmal fullness and a relationship to duration that Lefebvre qualifies as 'involution' and the accomplishment of a presence (Lefebvre 1961: 345). Lefebvre mystically describes presence as a creative act in *La présence et l'absence:*

> [Presence] supposes and implies an act: the *poietic act*. This also implies an adherence to *being*, to the fact of being and to the possibility of a fullness that is never fixed nor fully defined. Presence, as a moment and not a substance or pure form, rewards the risk-taking act. [author's translation]
>
> *(Lefebvre 1980: 226)*

The 'moment' is what integrates the everyday in order to transform it through the texture of space and time. And though 'moment' does not appear in Lefebvre's texts on rhythmanalysis, the figure of the poet rhythmanalyst is here described, along with the creative activity of this figure as a manner to include rhythms in daily life (Lefebvre 2004: 22–6). This poetic activity transforms everyday life and creates lifestyles comparable with the experimental utopias developed by the avant-garde group, Situationist International.

Rhythmanalysis as experimental utopia

How can rhythmanalysis be performed as poetic practice? Though Lefebvre did not develop specific descriptions of the poet rhythmanalyst, nor define the manner in which urban time and space were to be transformed through this activity, Lefebvre did develop the concept of an 'experimental utopia' in which an understanding of this poetic practice can be found. Utopia as a literary genre proposes to show a society in a spatiotemporal elsewhere in order to criticise the existing order. Lefebvre is inspired by utopian socialists like Charles Fourier and the conceptualisation of concrete utopias by Antonio Gramsci or Ernst Bloch. He considers utopia as a means of transforming reality, and therefore does not dissociate it from critical theory, but rather explores the possibilities of imagining a society that could be created by the Revolution and could fuel the imagination of counter-spaces. It is associated with the creation of an appropriate time-space.

Lefebvre uses the term 'experimental utopia' to describe the practices of the Situationist International that go beyond fictional writings to promote 'a utopia experienced on an experimental basis' (Lefebvre 1961: 336). At the end of the 1950s, the SI defines psychogeography as 'the study of the precise laws and specific effects of the geographic environment, consciously organized or not, on the emotions and behaviours of individuals' (Debord 1955). Its methodology can be found in the experiment of the '*dérive*', an urban exploration protocol that consists in a rapid crossing of urban space in the form of a game, thus rendering immaterial frontiers of neighbourhoods and highlighting the attractive and repulsive zones of the city. '*Dérive*' is inherited from the literary tradition of the '*flâneur*', recalling Balzac, Baudelaire and the surrealists, including Louis Aragon and André Breton, who are the direct ancestors of the SI groups. Walking without purpose is a way of contesting the reduction of the street to its utilitarian and rational purposes, as well as the practice of '*détournement*' (a form of hijacking, or of showing another use of an already-employed form or object). It allows for an analysis of the city by means of encounter and chance. The atmospheres of the city were seen as the possibility to construct new situations that would not define art as a specialised activity but include it in the creation of lifestyles. The city as an artwork was to announce the collapse of the society of the spectacle, as Debord would later call it (Debord 1967). The city was to become an artwork in and of itself, an '*oeuvre*'.

Lefebvre sets aside this experimental utopia to develop his critical theory in response to his dispute with Debord in the 1960s. He turns to the theorisation of the production of space. Evoking psychogeography in an interview (Ross 1987), Lefebvre considers it as evidence of the fragmentation of the city, meanwhile insisting on the necessity of a broader understanding of the urban phenomenon and urban practices. Moreover, he concedes that psychogeography has been unable to demonstrate any results, and few examples have contributed to the science of atmospheres. Lefebvre is sceptical of the revolutionary potential of the construction of situations, understood as punctual interventions, because the transformation of *praxis* takes time. The transition from experimental utopia and alternative practices to the right to the city brings forth a far broader program.

Lefebvre imagines the possibilities of an appropriate space-time and the city as an artwork, an '*oeuvre*', through the conceptualisation of differential space (Lefebvre 1974). Differential space overcomes the abstract space of critical theory through the definition of new centralities, counter-spaces that are appropriated by alternative cultures. They are therefore inseparable from utopian lifestyles. Lefebvre is inspired by Fourier's utopian conception of the phalanstery, a unified, collaborative architectural space that is cohesive with human sensibilities. But unlike Fourier's insistence on harmony and lifestyle, which he sees as ascetic, Lefebvre insists on the realisation of desire through enjoyment. Lefebvrian utopia requires the achievement of a total body, with its excesses and spent energy, as he describes it in *Toward an Architecture of Enjoyment*, written in 1973 (Lefebvre 2014). In this book, Lefebvre conceives of architectural work as the exploration of a concrete utopia that allows for the exploration of human desires. Utopian architecture here offers a key to differential space, since it imagines the possibility of overcoming the separation between product and *oeuvre* in a return to nature.

Evoking the rhythmanalyst as 'an enigmatic individual who strolls with his thoughts and his emotions, his impressions and his wonder, through the streets of large Mediterranean towns' (Lefebvre 2004: 87), Lefebvre recalls the practices of his old friends from the Situationist International. If Lefebvre imagines rhythmanalysis as an experimental utopia with creative ambitions, I consider it to be in line with the *dérive* and psychogeography in specific situations. As with psychogeography, rhythmanalysis requires a sensitive and aware experience involving body displacement or immobility. Rhythmanalysis 'separates as little as possible the scientific from the poetic' (Lefebvre 2004: 87): it has the ambition to constitute knowledge of concrete space and time through the form of a poetic activity. Its own practice should transform life and its effects.

Because rhythms lie at the heart of architectonics and the appropriation of space and time, rhythmanalysis is the experimental practice from which to renew their insertion into the cosmos and create an appropriate space and time in the urban world. Lefebvre seeks an experimental utopia in order to create the urban as an *oeuvre*.

Rhythmanalysis and Lefebvrian urban poetics

It is necessary to return to the conceptual framework that Lefebvre developed on modern time to understand how critical theory relates to a poetics. Lefebvre sets out his 'revolutionary romanticism' (Lefebvre ; 1962) during the same period in which he builds up his framework for critical theory and produces his first texts on rhythmanalysis. Revolutionary romanticism is anchored in the young Lefebvre's intellectual background, and is the basis from which Lefebvre constructs his personal critique of modern life as he gradually frees himself from the Communist Party (PCF). The idea of modernity as an abstract process is a romantic one (Lefebvre 1962). The romantic characteristics of this thought lie in its anti-systemic construction, its critique of quantification, mechanisation and his concern for the dissolution of communities (Löwy and Sayre 1992). To counter modernity's abstraction, Lefebvre defines revolutionary romanticism as a lifestyle and develops his theory of the city as an *oeuvre*, or artwork of a community. He exchanges with the members of the SI on this set of ideas and elaborates the elements for a poetics of the city.

The result is the definition of the city as an *oeuvre* in *The Proclamation of the Paris Commune* (Lefebvre 1965a) and *The Right to the City* (Lefebvre 1968). The implications of these theories are developed in other chapters in this book (see Parts 5 and 6). It enlightens us on the importance of a certain modality of *praxis* that Lefebvre call *poiesis*: the creation of lifestyles as artworks is a central concern among them. Thus Lefebvre's critical theory combines with poetics to open new possibilities for urban society. Theory cannot produce only a critical knowledge of the production of space and time; it must open new horizons for human *poiesis*.

This conceptualisation of *poiesis* is completed by the anthropological point of view Lefebvre develops in *Metaphilosophie* (Lefebvre 1965b), where he theorises more generally the relationship between humanity and its *oeuvres*. This opens the possibility for the poetic creation of the urban in terms of human artwork (Lefebvre 1965b: 117–18) and the creation of a lifestyle that transforms everyday life. How is rhythmanalysis related to this general program? I assume that rhythmanalysis, by placing lived experience at the basis of experimentation, is the outline for this poetic practice. But it has to be developed as an experimental device rather than an observation methodology.

Lefebvre did not expose rhythmanalysis as a fully elaborated program in his *Elements of Rhythmanalysis*: each practitioner would probably have to find his own way to practice it. But as we have seen, rhythms are essential to the appropriation of space and time, so Lefebvre's rhythmanalysis is engaged to globally rethink the human environment. His aims are related to the objective of building a differential space, in other words to establish an alternative *praxis* as *poiesis*. This *poiesis* is not only focused on the built environment as an architectural project but could encompass the creation of new institutions, new rituals, new ways of living and especially other temporalities... alternatives to neo-liberal urbanisation.

Conclusions

By placing the texts Lefebvre wrote on rhythmanalysis in the general framework of his theory, we have shown that the production of spatial theory is completed by the question of time in the urban era. The rhythmanalytical project thus adds a temporal dimension to the spatial one in its aim to consolidate our quest for the appropriation of time and space. It takes the form of an experimental utopia in which practice is not the application of a theory; it is elaborated exclusively from thinking and acting. This 'creative practice', arising from *praxis*, is what we have called *poiesis*. Thus the rhythmanalytical project invites us to consider the poetic aspects of Lefebvre's thought, and through creative practice, imagine devices to explore the urban as the human *oeuvre*.

References

Bachelard, G. (1950) *La Dialectique de la Durée*, Paris: Presses Universitaires de France.

Bachelard, G. (1957) *La Poétique de l'Espace*, Paris: Presses Universitaires de France; trans. Maria Jolas (1994) *The Poetics of Space*, Boston: Beacon Press.

Debord, G. (1955) 'Introduction à une critique de la géographie urbaine', *Les Lèvres Nues*, 6.

Debord, G. (1967) *La Société du Spectacle*, Paris: Buchet-Chastel.

Jousse, M. (1974) *L'Anthropologie du Geste*, Paris: Gallimard.

Le Breton, É. (2012) *Pour une Critique de la Ville: la Sociologie Urbaine Française, 1950–1980*, Rennes: Presses Universitaires de Rennes.

Lefebvre, H. (1939a) *Nietzsche*, Paris: Éditions sociales internationales.

Lefebvre, H. (1939b) *Le Matérialisme Dialectique*, Paris: F. Alcan; trans. John Sturrock (1968) *Dialectical Materialism*, London: Jonathan Cape.

Lefebvre, H. (1947) *Critique de la Vie Quotidienne: Introduction*, Paris: Grasset; trans. John Moore (1991) *The Critique of Everyday Life*, vol. 1, London: Verso.

Lefebvre, H. (1959) *La Somme et le Reste*, Paris: la Nef de Paris.

Lefebvre, H. (1961) *Critique de la Vie Quotidienne. 2. Fondements d'une Sociologie de la Quotidienneté*, Paris: l'Arche; trans. John Moore (2008) *Critique of Everyday Life 2. Foundations for a Sociology of the Everyday*, London: Verso.

Lefebvre, H. (1962) *Introduction à la Modernité: Préludes*, Paris : Éditions de Minuit; trans. John Moore (1995) *Introduction to Modernity: Twelve Preludes*, London: Verso.

Lefebvre, H. (1965a) *La Proclamation de la Commune: 26 mars 1871*, Paris: Gallimard.

Lefebvre, H. (1965b). *Métaphilosophie: Prolégomènes*, Paris: Les Editions De Minuit.

Lefebvre, H. (1966) *Le Langage et la Société*, Paris: Gallimard.

Lefebvre, H. (1968) *Le Droit à la Ville*, Paris: Anthropos.

Lefebvre, H. (1970) *La Fin de l'Histoire: Épilégomènes*, Paris: Éditions de Minuit.

Lefebvre, H. (1971) 'Musique et sémiologie', *Musique en Jeu*, 4 52–62.

Lefebvre, H. (1974) *La Production de l'Espace*, Paris: Anthropos; trans. Donald.

Lefebvre, H. (1975) *Hegel, Marx, Nietzsche ou le Royaume des Ombres*, Paris: Casterman.

Lefebvre, H. (1980) *La Présence et l'Absence: Contribution à la Théorie des Représentations*, Paris: Casterman.

Lefebvre, H. (1981) *Critique de la Vie Quotidienne 3. De la Modernité au Modernisme*, Paris: l'Arche; trans. John Moore (2005) *Critique of Everyday Life 3. From Modernity to Modernism*, London: Verso.

Lefebvre, H. (1992) *Éléments de Rythmanalyse. Introduction à la Connaissance des Rythmes*, Paris: Éditions Syllepses.

Lefebvre, H. and Guterman, N. (1936) *La Conscience Mystifiée*, Paris: Gallimard.

Lefebvre, H. and Guterman, N. (1938) 'Introduction', in V. I. Lénine (ed.) *Cahiers sur la Dialectique de Hegel*, Paris: Gallimard.

Leroi-Gourhan, A. (1965) *Le Geste et la Parole. La Mémoire et les Rythmes*, Paris: Albin Michel; trans. Anna Bostock Berger (1993), London: The MIT Press.
Löwy, M. and Sayre, R. (1992) *Révolte et Mélancolie: le Romantisme à Contre-courant de la Modernité*, Paris: Payot.
Maldiney, H. (2012) *Regard, Parole, Espace*, Paris: Édition du Cerf.
Merrifield, A. (2006) *Henri Lefebvre: A Critical Introduction*, London, Routledge.
Revol, C. (2015) 'La rythmanalyse chez Henri Lefebvre (1901–1991): contribution à une poétique urbaine', unpublished thesis, Université Lyon 3.
Ross, K. (1987) 'An interview with Henri Lefebvre', *Society and Space*, 5:1 27–38.
Stanek, L. (2011) *Henri Lefebvre on Space: Architecture, Urban Research, and the Production of Theory*, Minneapolis: University of Minnesota Press.

16

Space in representation

Dislocation of meaning from the Gezi Park protests to the new Turkish Presidential Compound

Bülent Batuman

Introduction

Lefebvre's concept of *representations of space* has generally been discussed in terms of meanings produced and attributed to space – how it is 'conceived'. The representations of space, as they stand within Lefebvre's triad of production of space, are associated with agents producing the conceptualisations of space and serving its organisation in tune with the dominant mode of production: 'scientists, planners, urbanists, technocratic subdividers and social engineers' (Lefebvre 1991: 38). That is, there is an implied negativity to the concept. Signs, tools and systems of meaning production have been defined within representations of space, and they are viewed as instrumental in the making of the *abstract space* of capitalism. What I would like to propose in this chapter is to expand the relation between *representations* and *space* by developing a 'trialectics' with the inclusion of *language*. This is an attempt to bring into consideration contingencies of historical circumstances and to show that representations of space also present us with a field of contestation. Thus, I would like to show that the social processes are not only related to space and language separately but also to the spatial representations produced through the interaction of these two components. I will develop my discussion through the analysis of recent political contentions over public space in Turkey. I will discuss the Gezi Park protests that took place in Istanbul in the summer of 2013 and their after-effects in terms of representations of space. My argument is that, while the Gezi protests represented a popular movement akin to the contemporaneous global protests based on the occupation of public space, they also witnessed the rise of representations of space as a field of contestation. With the suppression of the movement rendering 'occupy-style' protests virtually impossible in its aftermath, this field of contestation prevailed and proved highly efficient in a subsequent episode regarding the struggle over the new Presidential Compound in Turkey's capital, Ankara.

Occupation as production of space

The recent global tide of protests shared certain features, creating symbols (such as Tahrir Square), methods and names (such as *Occupy*) that created transnational links connecting the protests and the activists staging them (Castells 2012). Various scholars have pointed out the democratic desire at the core of these movements and the performances of citizens materialising this desire (Badiou 2012; Žižek 2012; Butler 2015). Perhaps one of the most cited common traits in these protests was their urban character and the key role of public space in their making (Harvey 2012; Weizman 2015). Lefebvre's concept of 'right to the city' was often a central theme in the ones raising urban issues of contestation (Mayer 2009; Long 2013).

The Gezi protests that took place in Turkey in the summer of 2013 have also often been analysed in relation to public space and the right to the city (Kuymulu 2013; Örs 2014; Batuman 2015a; Inceoğlu 2015). The events began as a small environmentalist demonstration against the destruction of Gezi Park adjacent to Taksim Square, the central public space of Istanbul. This was a response to the government's plans to regenerate the site as a sterilised tourist attraction. Taksim Square is the central hub of Istanbul, with historical significance: it was a major stage of Turkish modernisation since the late 19th century and an important political locale throughout the republican history of the 20th century (Batuman 2015a). The 70-year-old Gezi Park (literally 'esplanade') was a registered heritage site designed as part of Taksim's planned development, although this did not prevent the government from proposing to replace it with a reconstruction of the 18th-century Artillery Barracks that had existed on the site. When a small group of environmentalists camped in the park to impede construction work in the last days of May 2013, they were brutally evacuated by the police. This in return led to the growth of protests, first in and around Taksim Square, and then various quarters of Istanbul as well as the major public spaces of other cities. The square was occupied by the protestors during the first two weeks of June (together with the central squares of other major cities), until its evacuation by the police. Gezi Park was home to a communal encampment for two weeks, and during this time Taksim Square became the heart of the nationwide protests.

The Gezi encampment was similar in spirit to the protest camps which have been the trademark of the global tide of protest that has become effective since late 2010. While the organisation methods, the role of the unorganised youth and especially the role of social media were common traits, it is crucial to note that the major cause of the widespread riots was the government's Islamic interventions in everyday lives rather than economic hardships (Moudouros 2014). Such interventions had intensified under the Justice and Development Party (AKP) and its leader, Tayyip Erdoğan, after 2011. With the increasing number of legal cases against journalists, university students and even lawyers, the protests were defined by some scholars as a 'rebellion of dignity' (İnsel 2013).

The earliest protests literally aimed to defend the park, and the violence they faced rapidly marked the very public space as a locus and a focus to raise and represent diverse issues of contestation (Karasulu 2014; Lelandais 2016). Within the two weeks it prevailed, the Gezi encampment was a beacon of hope for the rest of the country, but it was also a real space that transformed those who participated in it (Karakayalı and Yaka 2014). With the lack of a central organisation, the young activists developed their own practices in accordance with what they were protesting for. While some distributed the leaflets of revolutionary organisations, some vocalised environmentalist concerns and even created a small garden. Solidarity tables were organised to provide food and basic needs for the campers, and a communal library was set up. A makeshift infirmary was also built where voluntary doctors tended to the injured. A container that had been used by the construction workers was occupied and transformed into a 'Revolution Museum' in which photographs of the moments of fighting with the police as well as trophies (such as helmets and pieces of armours that belonged to the police) were exhibited. There were workshops for different activities, including one for children. The future of the camp was discussed in forums organised inside the park. Soccer fan organisations protested police violence side by side with the clandestine revolutionary groups, and the LGBT associations protested homophobia together with anti-capitalist Muslims protesting the neo-liberal economic policies. Thus, Taksim was appropriated by diverse groups protesting distinct issues.

If we look at the demographics of the initial protests, the major actors were, similar to the recent mobilisations across the globe, young activists skilfully making use of social media to organise (Reimer 2012; Farro and Demirhisar 2014; Varnalı and Görgülü 2015). These were mostly young, white-collar professionals with college degrees and university students destined to occupy similar positions: a social stratum defined by some scholars as the 'new middle classes' (Keyder 2013). Although the demographics of the protests rapidly became heterogeneous, the influence of the young activists was felt throughout the events with their dynamism, use of technology and the sense of humour that produced a particular language of protest with slogans and graffiti (Sözalan 2013; Yalçıntaş 2015; Dağtaş 2016). The protests often embodied artistic creativity and involved spatial interventions to resist consumer culture (Velioğlu 2013; Çolak 2014).

While scholarly analyses of the Gezi protests mostly focused on the young protestors with little political engagement as the major actor, it is necessary to point out that without the complementary role of the experienced protestors politically affiliated with various revolutionary groups, it would have been impossible to hold Gezi Park and Taksim Square for such an extended period of time. The latter group was quick to construct barricades on the main thoroughfares as well as secondary streets and was continually on duty to maintain them. The architectural know-how of barricades was as important as the use of social media in defending the commons. This know-how was coupled by the geographical knowledge of rallying to Taksim, which contained practical information of meeting points, routes and topographic edges and gates that define the entire district.

The interesting point here is the role played by Gezi Park and Taksim Square as two distinct spaces. Although they are adjacent, they have separate meanings in collective memory. As it would be beyond the scope of this discussion to detail these differences, it should suffice to provide a brief history. Taksim as the emerging modern hub of the late Ottoman Istanbul was first organised with a monument at the centre of a roundabout in the early years of the republic. Later, a large processional ground was integrated to the square. Finally, Gezi Park was added to these two as a green extension (Batuman 2015a). These physical spaces, however, accumulated varying experiences and histories throughout the 20th century. While the park prevailed as a green area (albeit losing portions to several buildings along the way), the monument has served as the 'kernel' that defined the square as a *space of representation* (Lefebvre 1991: 42). Initially a symbol of nation-state dominating the square, the politicisation triggered by the monument soon led to the appropriation of Taksim Square by radical politics. A momentous event was a major student rally in 1969. The students protesting the US Sixth Fleet visiting Istanbul were attacked by a group of anti-communist counter-protestors; two of them were stabbed to death and more than 100 were wounded in the Square. This event, known as 'Bloody Sunday', marked a turning point in the history of the square and its association with left-wing movements. It soon became the major destination of May Day rallies, and the one in 1977 witnessed the death of 34 people in the chaos following the gunshots fired on the crowd. From then on, Taksim Square prevailed as the most important political space of the country associated with May Day, especially owing to the memory of 1977.

With different mental images in collective memory, the park and the square also gave way to different spatial practices of protest. While the camp inside the park prevailed as a secluded island of freedom, the square became the 'street': the site of slogans, barricades and violent clashes with the security forces. This distinction was not unnoticed by the government; the officials were quick to detect the coexistence of unorganised protestors and the militants from revolutionary groups. While the former mostly settled inside the encampment in the park, the latter were in the square fortifying barricades for the police assault that would eventually come. When the police raided the square on June 10, the governor of Istanbul announced that they would only disperse 'the marginal groups' resisting police forces in the square and the 'peaceful demonstrators' in Gezi Park were not a target (Anon. 2013). Nevertheless, five days later, the police stormed the park. The differentiation of the square and the park as spaces of two different groups appeared to be an effective strategy to destroy the diversity of the public space.

If we look at this in reverse, it is possible to see that the success of the protests was a result of the juxtaposition of these two spaces and the actors and practices they accommodated. Representations of space were key in closing the gap between these two spaces, physically adjacent but mentally distant from each other. An interesting illustration of this can be found in the different ways of naming the protests. If we look at the labels used to define the events during and immediately after, what we see is the constant use of two-word noun phrases: the first – descriptive – one was one of *Gezi*, *Taksim* and *Haziran* (June), while the second – the head noun – is either one of *direniş* (resistance), *olay* (event), *isyan* (uprising/ rebellion) or *protest*. A Google search has revealed that between the eruption of the protests and March 2014, the most frequently used Turkish phrases to label the protests (in terms of the number of Web pages in which they are found) were '*Gezi olayları*' (Gezi events) 1,030,000 times, '*Gezi Direnişi*' (Gezi resistance) 973,000 times, '*Gezi İsyanı*' (Gezi rebellion) 20,900 times, '*Gezi Protestoları*' (Gezi protests) 53,100 times; '*Haziran Direnişi*' (June resistance) 42,200 times, '*Haziran İsyanı*' (June rebellion) 9,200 times, '*Taksim Direnişi*' (Taksim resistance) 56,400 times, '*Taksim Olayları*' (Taksim events) 79,800 times and finally '*Taksim İsyanı*' (Taksim rebellion) and '*Taksim Protestoları*' (Taksim protests) only 3,500 times each.

These labels are important, for they constitute representations of space which define, limit and identify the space through inclusions and exclusions in its definition. In political terms, it may be argued that what is more important than who controls space is who defines it. In this respect, what is at stake is a political struggle over how to define the public space.

What we see in the frequency of these labels is that although the emphasis on time ('June') is used by socialist circles emphasising that the politics of the events had extended beyond the defence of the park, space is clearly the focus in most uses. If we look at the terms used to define the episode itself, *event* is probably the most ambiguous one. Reminiscent of the phrase 'May events of 1968', the term – although politically nebulous – implies an enduring historical significance, while the other three terms refer to direct political action. *Protest* is an action of shorter-term and limited violence in contrast to *uprising*, which suggests dissent out of control. Among the three, *resistance* requires a closer look. The term indicates a relatively long (not momentary) confrontation with a hegemonic power. In the literature it has been used to define armed struggles (such as anti-fascist French Resistance) as well as non-violent practices (for instance, cultural resistance). In recent years, it has often been used in the context of anti-globalisation movements to denote opposition to neo-liberalism. In Turkey, *resistance* has been used as a specific form of political action, especially in the 1970s. For instance, the momentous workers' strike of 15–16 June 1970 has been labelled as *resistance*. During the 1970s, marked by political turmoil, occupation of universities and factories parallel with strikes were also defined with this term.

Overall, it is possible to say that *event* refers to time (hence, History), implying a historical rupture in relation to the dominant order. *Resistance*, in contrast, refers to varying scales of a spatial niche; a persistent performance of occupation in a certain place. Yet, it is also important that *resistance* by definition suggests the existence of a greater (oppressive) power which it disputes. Meanwhile, *event* defines an imprecise, even semi-volitional political action which is believed to have led to a contingent break in historical continuity. In the case of the Gezi protests, the choice of *resistance* in naming the incident was very much related to the feeling of being treated as 'minority', the leftover from the majority which the government defines as 'the nation'.

Interestingly, while 'Gezi' was widely used in the naming of the protests, the most popular slogan was 'Everywhere is Taksim, resistance everywhere'. This spatial shift in linguistic representations of space is very much related to the politics of public space. In the first episode of the events, Gezi Park was the subject of protest; in semiotic terms, it was the signified to which political action pointed at. However, with the expansion of the protests and the multiplication of themes of opposition, it came to signify an increasing number of themes. Again, to use semiotic terms, the indexicality of political action disappeared here, precisely due to the shift of spatial signifiers: the park yielded to the square as the central public space, which was heterogeneous enough to become the signifier of a larger amalgam of (political) referents. This difference between Gezi Park and Taksim Square is crucial to understand the sudden jump in scale of protest: although they are adjacent physically, their mental co-ordinates within collective memory are not determined by their physical relation. Among these two mental spaces, Taksim Square, with its 20th-century history over-determining the multitude of meanings attributed to it, made it possible to become the void embodying possibilities, a feature Gezi Park did not have. Precisely because of the dislocation of the (physical) focus of contention, the protests could become about anything and everything. This dislocation led to the replacement of certainty with ambiguity in terms of wishes, demands and desires expressed via political action. This, in return, resulted in the rise of a practice of resistance as a performance for itself, surpassing concrete demands.

A typical linguistic illustration of this was the humorous slogan 'Down with some things!' (Dağtaş 2016). Here, it is not important what replaces 'some'; what matters is the counter-hegemonic potential of the slipperiness of signs within the struggle over meanings. In this regard, the slogan 'Everywhere is Taksim, resistance everywhere' can be interpreted as the spatial version of 'Down with some things'; the uncertainty avoiding a specific target in the latter slogan translates into space in the former. The slogan spatialises protest by linking it to Taksim, but simultaneously detaches it from Taksim by designating it to 'everywhere'. In this regard, it downplays the actual physical space as the subject of protest. Taksim is not the subject of a specific action or protest anymore; it becomes a spatial signifier recognised nationwide. Thus, the struggle over meaning (of Taksim) is also dislocated: the competition of (political) representations of Taksim gives way to the question of what Taksim as signifier represents. The power of the protests lies in its rejection of reducing the mental associations of political meanings embedded in Taksim Square and using language (in the form of slogans and graffiti) to embrace this irreducibility.

Representations of space in the post-Gezi moment

Among other things, the Gezi Park protests witnessed the rise of representations of space as a field of political contestation, where ironic displacement of meanings through humour assumed political character. With the suppression of the movement and the subsequent wave of oppression, a new phase – which I will define as the *post-Gezi* moment – began. This moment was defined by the fading of protests and the impossibility of 'occupy-style' political action based on physical occupation of public space (Batuman 2015b). Nevertheless, in the absence of conditions of street activism, representations of space would become effective in a conflict over another public space in the capital city of Ankara: Atatürk Forest Farm, where a new Presidential Compound was begun construction.

Being a project of then-Prime Minister Erdoğan, the new Presidential Compound in Ankara has been the most controversial construction in the country in decades. The most visible objection to the building was its excessive grandeur that found a place in international media (Arango 2014; Dombey 2014; Withnall 2014). However, its lavishness was far from being the only issue of controversy regarding the Presidential Compound, with a floor area of 300,000 square meters and sitting on a plot twice this size. Its location in Atatürk Forest Farm (AOÇ), a large green space created as part of the modernisation efforts of the early republican years, also created outrage in two respects. The secularists saw its location inside the AOÇ area as an attempt to suppress the legacy of Atatürk (the founding father of the republic) by conquering a site which bore his name and was his personal donation to the nation. Moreover, as the site was a registered conservation area, the courts ordered the construction to be terminated, to which Erdoğan responded: 'If they are powerful enough, let them come and demolish it' (Anon. 2014a). The building was inaugurated in late 2014, soon after his election as president.

In the eyes of his opponents, the compound has become the architectural symbol of an authoritarian presidential system associated with Erdoğan himself. The Turkish political system comprises a strong prime minister and a symbolic president as the head-of-state. In this context, the site was initially defined as a prime ministerial compound and reorganised with Erdoğan's ascent to presidency. Its architectural programme brought together most of the government bureaucracy and its transformation into the President's office went hand in hand with the gradual concentration of power in Erdoğan's hands after 2011. Erdoğan was not shy to announce that he was seeking a new system centred around a powerful president as head of the executive also controlling the legislature; this model would be implemented with constitutional amendments accepted with a controversial referendum in April 2017.

The site of the compound, AOÇ, had already become a hot topic of ideological struggle in 1994. In tune with the municipality's endeavours in eradicating traces of republican modernism, it was proposed to build Disneyland-style attractions with plan changes (necessary to bypass conservation regulations) to which civil organisations and particularly the Chamber of Architects responded with vocal opposition and took the matter to court (Chamber of Architects Ankara Branch 2014; Karakuş Candan et. al. 2015). With the rise of the AKP to power in 2002, the municipality increased its efforts to open the farm area to development. A plan for the area was approved by the municipality in 2010. Although this plan was also taken to court by the chambers, the court ruling was postponed for years, which allowed the municipality to implement its projects. Proposed roads cut through the farm area and disconnected its parts. The built core of the farm, which was designated as the historical recreation zone, was also destroyed with traffic organisation. Finally, a considerable amount of land was allocated to a large amusement park and the new Presidential Compound.

The interesting thing here is the political implications of the vagueness regarding the function of the unfinished building throughout this period. Between May 2012 (when excavation began in the site) and June 2014, when Erdoğan was elected president, the building under construction became a symbol of power although it was not yet occupied. The architectural programme of the compound was one of a not-yet-implemented presidential system from the beginning. While it brought together executive bureaucracy, it also contained a residential building which is typically not part of a prime ministerial compound. If this was envisaged as the new presidential seat, the immense office spaces were not a part of the presidency in the Turkish system. Hence, even before it emerged as an architectural edifice, the compound entered the field of political signifiers as one corresponding to none of the existing offices in Turkey. Yet, it did point to a particular person, Erdoğan himself, which made it all the more problematic in terms of political representation. Here, political influence emerged as an effect of the representational power of space. The compound under construction was a signifier of a position transcending the order of Turkish political system and also a referent of Erdoğan himself regardless of his official position.

This vagueness was reflected in the confusion on how to name the ongoing construction as well. Official documents referred to 'prime-ministry service buildings', while oppositional media labelled the compound as 'the palace'. Here, it is worth mentioning that the existing presidential seat built in 1930 was called 'Çankaya Mansion', referring to its location in Çankaya, the prestigious district overlooking the capital. Thus, the contrast between the old *mansion* and the new *palace* suggested the lavishness of the latter. After the Gezi protests, the compound gradually became the centre of political conflict. The opposition came up with a new name, '*Ak-Saray*' (White Palace), implying that Erdoğan was building himself a palace, playing on AKP's insistence that the party's acronym be pronounced 'AK Party' as well as a negative allusion to the White House, which is known in Turkey as 'Beyaz Saray' (White Palace). This implied that the compound was to become the base of a presidential system similar to the US presidency, that it was not a state enterprise but a personal one.

Initially the government attempted to appropriate this name. Erdoğan's successor Prime Minister Davutoğlu willingly used 'Ak-Saray' in a speech right after assuming position (Anon. 2014b). After this, the opposition began to use the term 'Kaç-ak Saray', with similar phonetics, where *kaçak* meant illegal, referring to the legal status of the ongoing construction defying court orders. To the dismay of the government, this name gained currency.

Although he welcomed the term in the beginning, the negative connotations of 'palace' later troubled Erdoğan. The more the government presented the complex as one that represented the nation and that its lavishness was a necessity, the more the oppositional campaign underlined its extravagance. The government tried to legitimise the cost by pointing at the restoration costs of Buckingham Palace (Anon. 2014c). Erdoğan even showed the Cuban presidential palace as an example, implying that such grandeur was found even in a socialist country, even though the building belonged to the pre-revolution era. Finally, Erdoğan began calling the complex a *külliye*. Külliye is an Ottoman mosque complex including educational and commercial facilities. Thus, Erdoğan's intention with this name might be argued to have been to point at the mosque within the compound, perhaps to refocus the debates on secularism. After that, this name has been in use especially in pro-AKP media, and *külliye* has begun to be used as an equivalent of 'campus' even in complexes without a mosque. Nevertheless, the new label did not succeed in suppressing prevailing use of '*kaçak saray*', which led Erdoğan to turn to a new title for the compound, *the house of the nation*: 'This building is not mine; today it is me, tomorrow another person will take office. This is the house of the nation' (Anon. 2015: np).

The struggle over the representations of space regarding the Presidential Compound had striking political outcomes. Opposition party representatives refused to attend events at the compound for a considerable time, denying its legitimacy. The Chamber of Architects initiated a campaign sending letters to foreign governments urging them not to visit the compound. It was also an important item on the political agenda during the June 2015 elections, which resulted in the AKP losing 20 per cent of its votes together with its majority in the parliament, although the renewed elections in November reinstated the party in power. Various surveys conducted on the election results pointed out the role of the lavishness of the 'palace' for the lost votes.

Conclusions

The recent tide of protests has raised interest in the work of Lefebvre, particularly in terms of themes such as the spatial making of protest and the right to the city. The case of Turkey presents a case to develop another one of Lefebvre's concepts: that of representations of space. As I have shown above, the consequent episodes of the Gezi protests in Istanbul and the struggle over the new Presidential Compound in Ankara illustrate the possibility of representations of space to become a site of struggle. Rather than a specialised field of professional activity functional to capital accumulation, it is possible to see representations of space as a field open to interventions and subversive appropriations.

In the case of Turkey, the Gezi protests followed a similar track of 'occupy-style' action based on the physical occupation of public space. Nevertheless, the protests also witnessed the use of language and humour as tactics of political subversion. These tactics also involved the representations of space, which prevailed as a remainder in the aftermath of the events. This line of political action soon resurfaced in a subsequent clash over another public space in Ankara. When construction of the new Presidential Compound was begun in AOÇ, the defence of public space turned to representations of space as the major means of struggle. In contrast to Taksim Square, AOÇ historically had never been a space occupied by heterogeneous social groups. The municipality's interventions in the area were never met with popular objection/protest, except for the legal battles pursued by professional chambers. While Taksim was a public space that everybody identified individually, AOÇ was a public space that no one felt personally related but considered to belong to 'everyone' due to its republican symbolism. It was only after the rise of the new Presidential Compound, that is, only after becoming a topic of 'high politics', that the area received popular attention. The debates around the prime ministerial compound evolved into a debate on political regime change after the Gezi events. Thus, the building was associated with an executive presidential system lacking public support and an authority figure transgressing legal limits. Here, the struggle over how to call the compound, that is, its representations, turned into a political issue.

Public spaces are defined not only through their physical uses but also their identity as intersection nodes allowing for contacts between physical and virtual domains. In a similar way, the politicisation of the representations of space made AOÇ a true public space for this first time. Thus, the farm became the forefront of political protest in the post-Gezi moment in two senses. First, the new compound became the focus of opposition to regime change. Second, it was the site for the emergence of a new – 'non-occupy' – strategy of protest in an era where the existing repertoire of protest performances have become impractical. The representations of space, here, became the essential component of this new strategy of contestation.

References

Anon. (2013) 'Marjinal gruplara firsat verilmedi', *Deutsche Welle Türkçe*, 11 June.

Anon. (2014a) 'Güçleri Yetiyorsa Yıksınlar', *Milliyet*, 5 March.

Anon. (2014b) 'Erdoğan'ın sarayının adını açıkladı', *Sözcü*, 5 September.

Anon. (2014c) 'Erdoğan ASKON'da Konuştu', *Hürriyet*, 7 December.

Anon. (2015) 'Cumhurbaşkanı Recep Tayyip Erdoğan'dan İftar Sofrası Açıklaması', *Hürriyet*, 25 June.

Arango, T. (2014) 'Turkish leader, using conflicts, cements power', *New York Times*, 31 October.

Badiou, A. (2012) *The Rebirth of History: Times of Riots and Uprisings*, London: Verso.

Batuman, B. (2015a) '"Everywhere is Taksim": the politics of public space from nation-building to neoliberal Islamism and beyond', *Journal of Urban History*, 41:5 881–907.

Batuman, B. (2015b) 'Gezi'nin Devamı olarak AOÇ Mücadelesi ya da Rejim Değişikliğinin (ve Buna Direnişin) Mecrası olarak 'Cumhurbaşkanlığı Sarayı'', *Dosya*, 34: 53–62.

Butler, J. (2015) *Notes Toward a Performative Theory of Assembly*, Cambridge, MA: Harvard University Press.

Castells, M. (2012) *Networks of Outrage and Hope: Social Movements in the Internet Age*, Cambridge: Polity Press.

Chamber of Architects Ankara Branch (2014) *Atatürk Orman Çiftliği'nde Hukuksuzluğu Meşrulaştırma!* Ankara: Chamber of Architects.

Çolak, E. (2014) 'Art in the street: the significant role of using the art, literature and humor in the Gezi resistance', *International Journal of Arts and Sciences*, 7:4 463–76.
Dağtaş, M.S. (2016) '"Down With Some Things!": the politics of humor and humor as politics in Turkey's Gezi protests', *Etnofoor*, 28:1 11–34.
Dombey, D. (2014) 'Price tag of Erdogan's new palace revealed: $600m', *Financial Times*, 4 November.
Farro, A.L. and Demirhisar, D.G. (2014) 'The Gezi Park movement: a Turkish experience of the twenty-first century collective movements', *International Review of Sociology*, 24:1 176–89.
Harvey, D. (2012) *Rebel Cities: From the Right to the City to the Urban Revolution*, New York: Verso.
Inceoglu, I. (2015) Encountering difference and radical democratic trajectory, *City*, 19:4 534–44.
İnsel, A. (2013) 'Haysiyet Ayaklanması', *Radikal*, 4 June.
Karakayalı, S. and Yaka, Ö. (2014) 'The spirit of Gezi: the recomposition of political subjectivities in Turkey', *New Formations: A Journal of Culture/Theory/Politics*, 83: 117–138.
Karakuş Candan, T., Hakkan, A. and Bolat, G. (2015) *Kaçak Saray: Kibir, İsraf, Hukuksuzluk*, Istanbul: Kırmızı Kedi.
Karasulu, A. (2014) '"If a Leaf Falls, they Blame the Tree": scattered notes on Gezi resistances, contention, and space', *International Review of Sociology*, 24:1 164–75.
Keyder, Ç. (2013) 'Yeni Orta Sınıf', The Science Academy, 34: 179/148. Online. Available HTTP: <http://bilimakademisi.org/yeni-orta-sinif-caglar-keyder/> (accessed 23 September 2016).
Kuymulu, M.B. (2013) 'Reclaiming the right to the city: reflections on the urban uprisings in Turkey', *City*, 17:3 274–8.
Lefebvre, H. (1991) *The Production of Space*, London: Blackwell.
Lelandais, G.E. (2016) 'Gezi protests and beyond: urban resistance under neoliberal urbanism in Turkey', in M. Mayer, C. Thörn and H. Thörn (eds) *Urban Uprisings: Challenging Neoliberal Urbanism in Europe*, London: Palgrave Macmillan, pp. 283–308.
Long, J. (2013) 'Sense of place and place-based activism in the neoliberal city', *City*, 17:1 52–67.
Mayer, M. (2009) 'The Right to the City' in the context of shifting mottos of urban social movements', *City*, 13:2–3 362–74.
Moudouros, N. (2014) 'Rethinking Islamic hegemony in Turkey through Gezi Park', *Journal of Balkan and Near Eastern Studies*, 16:2 181–95.
Örs, İ.R. (2014) 'Genie in the bottle: Gezi Park, Taksim Square, and the realignment of democracy and space in Turkey', *Philosophy and Social Criticism*, 40:4–5 489–98.
Reimer, M. (2012) '"It's the kids who made this happen": the occupy movement as youth movement', *Jeunesse: Young People, Texts, Cultures*, 4:1 1–14.
Sözalan, Ö. (2013) 'A few remarks on the lessons of Gezi uprising', *Badiou Studies*, 2:1 146–51.
Varnalı, K. and Görgülü, V. (2015) 'A social influence perspective on expressive political participation in Twitter: the case of #OccupyGezi', *Information, Communication and Society*, 18:1 1–16.
Velioğlu, H. (2013) '"Ghosts" of a situationist protest, deadly edges of Kemalism', *Hot Spots, Cultural Anthropology Online*, 31 October. Online. Available HTTP: <http://production.culanth.org/fieldsights/419-ghosts-of-a-situationist-protest-deadly-edges-of-kemalism> (accessed 12 March 2016).
Weizman, E. (2015) *The Roundabout Revolutions*, Berlin: Sternberg Press.
Withnall, A. (2014) 'Recep Tayyip Erdogan: the 'new sultan' now has a new palace – and it has cost Turkish taxpayers £400m', *The Independent*, 5 November.
Yalçıntaş, A. (ed.) (2015) *Creativity and Humour in Occupy Movements: Intellectual Disobedience in Turkey and Beyond*, Hampshire: Palgrave Macmillan.
Žižek, S. (2012) Occupy wall street: what is to be done next? *The Guardian*, 24 April.

Part 3

Representing and contesting urban space

John P. McCarthy and Michael E. Leary-Owhin

This part focuses on Lefebvre's deceptively simple idea of spatial representations. While this idea is apparently simple, Lefebvre suggested that official representations of space can be treacherous because they may serve to mystify social relations and the welding of power. Lefebvre presents increasingly sophisticated understandings of representations of space, including visual representations, under conditions of neo-capitalism or what may also be called neoliberal urbanism. Representations of space are problematised in three main ways: first, through examination of the idea that they can serve to exclude certain voices, communities and ways of seeing; second, by exploring how representations of space are contested within officialdom; and third, by investigating how urban space is moulded through processes of contestation where differing groups, including civil society groups, offer alternative representations of space. From the traditional paper map, plan, perspective and axonometric, this part of the book includes elaboration of how new technologies are impacting on the production of space in so-called smart cities and elsewhere.

Shaw's opening chapter focuses on the implications of Lefebvre's work for law and social justice. She asserts that the conceptualisation of public space is inextricably linked to what it means to be human, and that elements of public space such as surveillance clearly express regimes of control which, she suggests, are intensifying and in fact transforming the nature of public life. She also highlights how the law produces space and applies strategies (often subtle) of spatial regulation. She shows how authoritarian regimes (via the law and the state) often conceal the manipulation of power, and all buildings and land uses reflect the values of the dominant culture. While the legal framework might not determine spatial outcomes, it sets the context in which the privileged can dominate over others in the spatial realm, legitimised by the application and rule of law. Lefebvre, she suggests, attempts to render justice and injustice visible by critiquing grand narratives. She also highlights how there is a constant threat to the power of law, and the state, so that these institutions must constantly defend their power. Regulation of the built environment, she suggests, can be arbitrary and unjust, imposing penalties on some while withdrawing regulatory control in other contexts. While she acknowledges the importance of 'cyberspace' and the digitised environment in this context, she adds that sites of resistance against injustice continue (in spite of social media) to rely on an embodied component, which forms the essential basis for negotiations over power and control.

Heinickel and Meier Dallach's chapter then turns to consideration of how the dimension of time might be undervalued as part of the analysis of spatial practice, particularly in relation to the symbolic and emotional elements of space. They use aspects of landscape and mobility to support this contention, with linkage to concepts of the right to the city. In a similar way to many other authors in the book, they point to the need to operationalise the concepts of Lefebvre, for instance in relation to the right to the city, and show a way forward in applying ideas relating to his concept of rhythmanalysis, though they suggest that this has already led to many applications within space-time design as adapted by architects and planners.

Borden's chapter considers the case of skateboarding as a use of open space in cities, reflecting Lefebvre's notions of the city as ideally catering for uses without a (conventional) purpose and facilitating the city as festival. He shows how skateboarding as an activity is often transgressive and even illegal and that it fails to follow the frequent exhortation of planners and others for cities to be 'tidy' in terms of appearance and functionality. Nevertheless, it is gratifying to see how he also includes some positive examples where enlightened cities such as Montreal have grasped the potential of this form of use in terms of its contribution to broader aims for sustainable transport and leisure.

Jones's chapter focuses on cinema, illustrating its potential to explain how Lefebvre's ideas can be interpreted. He asserts that a range of genres can be helpful in this context, including for instance action cinema, which often illustrates how protagonists act in a subversive and transgressive way in terms of their use of public space. In such cases, not only does this show how weak points in the public realm can be used to exploit and appropriate space for the use of inhabitants, even if in an ephemeral or transgressive way, but he also shows how such use can point to possibilities for the alternative (popular?) use of public space as sites of creativity, initiative, leisure and entertainment.

Ellison's chapter explores the specific context of office workspace from a Lefebvrian perspective, comparing cases in London of an established organisation and co-working space for ethical start-up businesses. He shows how office workspace provision has become a dominant element of urban corporate real estate development in many cities. Thus this type of development affords a rich arena for considering the production of space in practice. He concludes that the examples illustrate the potential for a sophisticated approach to place-making, which can allow the development of contextually appropriate spaces which more accurately reflect the needs of the workforce. The latter, he points out, represents the dominant cost (human resource cost) for employers. Thus, he argues, the rationale for more humane workspaces can go hand in hand with straightforward business objectives.

Thompson's chapter applies Lefebvre's utopian-socialist revolutionary concepts in the context of housing in Liverpool, focusing on a comparison of the 1970s housing co-operative movement and the municipal socialist project of the Militant Tendency in the 1980s which essentially tried to 'municipalise' the co-ops. He points out that Engels's experience of Liverpool, as well as other industrial cities such as Manchester, inspired his writing, which doubted the capacity of the capitalist system to provide humane housing for the working classes. There followed a series of state-led intervention programmes which culminated in the municipal council housebuilding project of the Militant Tendency-controlled Labour council in the mid-1980s. Thompson suggests that all such state-led strategies apply a fallacy by treating 'dwelling' essentially as a noun (so as a static object) rather than a verb (and thereby reflecting a dynamic process). This, he suggests, misunderstands the key socio-economic causes of inadequate housing. He also shows how the object-based approach in this context was supported by experts such as geographer Alice Coleman, but also how, conversely, anarchist observers such as Colin Ward took a different view by emphasising the need for self-autonomy and collective control over housing. The latter approach is linked to Turner's ideas of 'user autonomy' and 'autoconstruction' based on his observations in South American urban settlements. However, these notions were later supplanted by the Militant Tendency's design determinism. Thompson concludes that what is needed is to focus on the process of housing production rather than the product itself – on the abstract *Eros* which privileges more open-ended approaches, rather than *Logos* which privileges absolute ends. In turn, this perhaps endorses Lefebvre's leaning towards the abstract rather than the concrete in his urban visions.

The chapter by Queiros, Ludovico and Malheiros explores the production of space in a migrant neighbourhood in Lisbon. Within this area, substandard housing was replaced via a new housing project. In their analysis of this project, they show how aspects of conflict and resistance have served to modify the space production process, illustrating aspects of creativity and resistance on the part of immigrants in this area who were able to adapt to the interstices of the city, both institutionally and physically.

Finally, the chapter by Yung and Leung considers the issue of contested cultural heritage in Hong Kong, using two case studies which illustrate a failure to preserve cultural identity and local history, and the loss of local businesses and social networks. The context of Hong Kong is interesting in this respect, since over the last two decades there has been an aggressive urban renewal regime which has involved the demolition of obsolete buildings in older districts, which have been replaced by what is seen as a better living environment. However, this process has also often destroyed social networks and cohesion. The authors show that such examples illustrate the concept of 'contradictory space' within which the pressures of global capital investment conflict with local notions of heritage space. The cases illustrate the fact that it is not only physical heritage assets that are worth preserving, but also traditional cultures and identities.

Clearly, this part of the book throws up many interesting cross-cutting themes and questions, including the following:

Lefebvre's ideas and urban planning

Again, what is the role for urban planning (regulation?) in terms of facilitating a vibrant, entertaining and liminal city space network for instance via skateboarding? How can planners bring this about without stifling the very innovation and creativity that this type of activity might seem to encourage? This also has implications for more general concerns of design since, for many, Lefebvre's ideas imply an attitude of anti-design and anti-planning. How can his ideas be popularised/communicated/applied in other professional realms where he is unknown/unrecognised? While this is not the case in architecture (or at least in some areas of this discipline and activity), it is the case in facilities management, increasingly important worldwide. Again, how can individual innovation and activity via self-build be effectively encouraged without wider loss of amenity?

Lefebvre's ideas and implications for identity, culture and place

How can planners and other urban professionals effectively take account of the need to privilege citizens' desire for the maintenance of identity and culture within local communities, for instance in relation to (but not in place of) notions of heritage and history? How can this be applied more widely in planning activity for urban space in the face of burgeoning city branding which often ignores or even displaces indigenous identity/ies, which further serves to further the interests of the powerful and privileged.

Lefebvre's ideas and implications for digital culture

How can we apply Lefebvre's ideas to the increasing pervasiveness of cyberspace and social media, with their implications for use and interpretation of space in the public and private realms? How can we apply elements of spatial justice in the context of spacelessness or cyber-spacelessness? How can the increasing need for more effective and democratic participation in land use planning be facilitated by the use of social media without furthering neoliberal interests and creating an underclass with a total lack of engagement or power?

These questions will be revisited in the conclusions to the book.

17

Lefebvre and the law

Social justice, the spatial imaginary and new technologies

Julia J.A. Shaw

Introduction

For Henri Lefebvre, space is both a material property and a process which enables social interrelationships and not simply a container or frame to be filled with content. Social space is explained as 'not a thing among other things, not a product among other products; rather, it subsumes things produced, and encompasses their interrelationships in their coexistence and simultaneously – their (relative) order and/or (relative) disorder' (Lefebvre 1991a: 73). How we conceptualise public space is inextricably connected to what it means to be human in a particular society. Interaction between the law and spatial imagery is a developing area of importance not least of all because the extent to which the prevailing culture either perpetuates or mitigates injustice can act as justification for intervening or intruding on space, for example to ensure its more equal distribution. Because metaphors, symbols and discourses articulate wider political and social transformations, an interrogation of the complex linkages and resistant boundaries between textual legal institutions and imagery is able to reveal the spatial framing of social and political organisation. In *Legal Emblems and the Art of Law: Obiter Depicta as the Vision of Governance* (2013), Peter Goodrich describes how the architecture such as courthouses, images and iconography of law are essential to the construction of legal authority and form a central component of legal discourse. Accordingly, the figuration of space and its elaborations is no less significant than the spectacle or event; 'the text circulates as an image and the power of its effect is largely resident in that aesthetic quality rather than its supposed rational content' (Goodrich 1991: 236).

The association of space with discourse, as 'rhetorical territories', depends on familiarity and understanding, because an individual knows about and can speak of what is familiar about the place they occupy; 'the character is at home when he is at ease in the rhetoric of the people with whom he shares life' (Augé 1995: 108). Understood as a form of *writing* – by cultivating a spatial imagination capable of interpreting the spatial structures of the past – the organisation of power is able to be *read* by those encountering it. This reconceptualisation of space, as both political and an aesthetic, has the capacity to disrupt dominant representations of space. Whilst avoiding a purely idealised or institutionalised conception, the fluidity of Lefebvre's spatial metaphor as structure and function has not only aesthetic but practical and political significance. It constitutes an imagistic way of thinking which transfers one idea or concept to another, and this cross-domain mapping can usefully represent the abstract in terms of the concrete. In this way, technologies of surveillance, policing strategies, areas of separation in the built environment are able to express regimes of order and control which, in turn, describe the nature of a society (Shaw and Shaw 2016: 45). In Lefebvrian terms, 'constructed space – a transparency of metal and glass – tells aloud of the will to power and all its trickery. It is hardly necessary to add that the "habitat" too shares in this spatial distribution of domination' (Lefebvre 1976: 88).

Spatial images are capable of articulating the hopes, dreams and fears of a society; in that 'wherever the hieroglyphics of any spatial image are deciphered, there the basis of social reality presents itself' (Kracauer 1997: 60). The government employment agency, for example, constitutes a barren space which expresses the desolation of those denied a place in the workforce, just as the prison cell acts as a tiny 'theatre of punishment' (Foucault 1977: 113). Since spatial imagery is particularly effective in encoding a variety of diverse types of information, its connection to the formulation of law and legal policy is an emerging area of importance and academic interest. This chapter will discuss how foregrounding Lefebvre's distinctive articulation of space can provide a better understanding of the modern socio-spatial phenomenon and a more robust theory of justice. It will also explore the language of spatial justice as an alternative methodological tool for analysing the influence of cyberspace and novel technologies as potential new sites of resistance and oppression. This is especially pertinent in the context of the emergence of a new aesthetics of security and surveillance which is in the process of transforming the nature of public life and all human interactions.

The influence of Lefebvre on law's spatial turn

Before the spatial turn in law began in the mid-1990s, legal scholarship only tackled spatial themes indirectly; focusing on the material characteristics of space while ignoring the social and theoretical aspects. Although spatiality as a critical idiom is still seldom used within jurisprudence, more recently Lefebvre's writing has influenced critical legal theorists working within the fields of law, state and the political; social justice and the city; and spatial justice. For example, Nicholas Blomley in *Law and Geography* (2003), Desmond Manderson in *Legal Spaces* (2005), Gerald Frug in *A Rule of Law for Cities* (2010), Chris Butler in *Spatial Politics, Everyday Life and the Right to the City* (2012) and Andreas Philippopoulos-Mihalopoulos in *Spatial Justice: Body, Lawscape, Atmosphere* (2015) discuss how central aspects of the law are socially produced and elucidate the ways in which law produces space and deploys subtle strategies of spatial regulation.

Following in the tradition of mathematicians and philosophers such as Euclid, Descartes, Spinoza and Kant, who all offered their own elaborations of space, Lefebvre's idiosyncratic contribution to the spatial discourse acts as a significant counterweight to representations of space by authoritative regimes such as law and the state. Such accounts can be deceptive, as too often they serve to obfuscate social relations and conceal the manipulation of power. The construction of legal space is implicit in ideas of, for example, the 'province' of law and law's 'empire', yet the legal profession resists acknowledging the spatiality of social life. Nonetheless, the production of space in everyday life depends upon a variety of social practices which order the material world where things have significance in relation to each other. These interrelationships are inevitably underwritten and transformed by law and regulatory structures. The production of public, civic and commercial places such as industrial developments, shopping malls, housing projects, entertainment zones, courthouses, post offices and places of worship, banks, libraries, parks and playing fields all reflect and reproduce the values of the dominant culture. For Lefebvre, '[a]ll we need to do is simply to open our eyes, to leave the dark world of metaphysics and the false depths of the "inner life" behind, and we will discover the immense human wealth that the humblest facts of everyday life contain' (1991a: 132). Socio-spatial relations are thus conceptualised in terms of their practical apparatus and their symbolic meaning; rendering social space as both a field of action and a basis for action.

As a set of material and ideological relations that are constituted by and constantly act on social relationships, space creates the conditions for playing out scenes of justice and injustice: 'Castles, palaces, cathedrals, fortresses, all speak in their various ways of the greatness and the strength of the people who built them and against whom they were built' (Lefebvre 1991a: 232). The agencies tasked with making a choice between the application of competing regulations and different legal principles, therefore, not only disturb and reshape material and social spaces, but also influence human responses to changing conditions. Even though the background legal framework does not determine the outcome of spatial negotiations, it affects those spatial relations which produce social relations and, in turn, engenders a set of just or unjust relationships as particular perceptions of reality are privileged over others. Lefebvre's distinct formulation provides a unique insight as to the significance of the spatial metaphor, spatial discourse and the discourses of law and the citizenry, as to what really matters for justice in space.

Lefebvre's spatial dialectic and social justice

Space is depicted as manifold; it has the character of an event in that it is created and constituted via a complexity of agencies and social relations, as well as through human and non-human relations. There can be no a-spatial social relations. The spatial ordering of the material world – where things exist in relation to each other – does more than simply reflect power and politics; it is itself a kind of power and politics. Since it is the ultimate place, medium and object of conflict and resistance, the acting out of power struggles within a spatial field means that, at heart, space is always a political issue: 'there is a politics of space because space is political' (Lefebvre 1974: 192). Instead of simply focusing on space and places which are marked by the longitude and latitude of power/knowledge, Lefebvre uses space as a broader category with which to disrupt and interrogate the theoretical constructs of disciplinary approaches. This usefully extends beyond the formal power of the state, by implicating the professions in their imposition of abstract conceptions of social relations. His critique also includes their authoritative disciplinary canons, which conveniently situate competing interests and give meaning to their hierarchies of beliefs, ideas and categories:

> No set of legal institutions or prescriptions exists apart from the narratives that locate it and give it meaning. For every constitution there is an epic, for each Decalogue a scripture. Once understood in the context of the narratives that give it meaning, law becomes not merely a system of rules to be observed, but a world in which we live.
>
> *(Cover 1983: 4, 5)*

By rendering both justice and injustice visible, grand narratives such as nation, society or judgement are subject to interrogation within the critical socio-spatial framework. Arbitrary categories of deviance and systems of exclusion which reference the underlying structural racism, sexism, ageism and classism that permeates contemporary neoliberal society are also held to account, along with lofty core values such as the rule of law – the purported purpose of which is to restrain the exercise of arbitrary power and, in so doing, protect the weak from the strong.

As an already powerful framing device, the legal institution has an almost limitless capacity to confer legitimacy on specific forms and categories of knowledge, operational principles and a range of entitlements and the conditions of conferring and owning such privileges. Abstract legal formulas (including the 'reasonable person', notions of equality and fairness, rights and duties, along with legal fictions such as the 'corporate personality' and 'consent of the governed') are the means by which the legal community exercises its control over the masses, and, notably, comprise law's 'narratives of truth'. Slavoj Žižek suggests in *Welcome to the Desert of the Real* (2002) that the terms designated to fundamental concepts such as democracy and freedom, human rights and, more recently, the war on terror have been co-opted by law and mask their origins. These 'false terms' only serve to mystify our 'perception of the situation instead of allowing us to think it. In this precise sense our "freedoms" serve to mask and sustain our deeper unfreedom' (Žižek 2002: 2). Accordingly, our freedoms themselves constitute a precarious autonomy, and the fabrication of such reality-framing untruths or partial truths (in support of law's transformative regimes of truth) means we often lack the language to articulate our 'unfreedom' (Shaw 2017: 95). Law's formative 'narratives of truth' are revealed, therefore, to be a potent stratagem for establishing the legitimacy of legal hierarchies and maintaining structures of power, as well as having the effect of stymieing public debate and frequently functioning as a vehicle of oppression.

The law is described as 'an object which can throw a shadow; a definite thing, which interacts with other things in legal space, and in whose shadow we dwell' (Manderson 1996: 1060). It acts as a domain which organises relationships, abstractly and exclusively, over a legal territory which has no authentic historical evolution or social context. Consequently, legal knowledge, dogma and law-making continue to be displaced from locality, and estranged from communication or encounter with local practice and experience. This results in the excision of difference, as people are reduced to 'identities' which attach to particular rights and duties set out in rarefied legal language. For Lefebvre this process of abstraction not only manufactures an unhappy consensus which legitimises state control and interference, it also affects the way in which people see themselves 'before the law', both as individuals and communities in their physical environment:

> The error – or illusion – generated here consists in the fact that, when social space is placed beyond our range of vision in this way, its practical character vanishes and it is transformed in philosophical fashion into a kind of absolute. In the face of this fetishized abstraction, 'users' spontaneously turn themselves, their presence, their 'lived experience' and their bodies into abstractions too. Fetishized abstract space thus gives rise to two practical abstractions: 'users' who cannot recognize themselves within it and a thought which cannot conceive of adopting a critical stance towards it.
>
> *(Lefebvre 1991a: 93)*

Nevertheless, social domination is constituted, constitutive and importantly resisted in space because abstract space is never absolute. As Lefebvre argues, this is because bodily and spatially, individuals always exceed the representations and images attributed to them by the overlapping and disciplining codes of contemporary social orders; not least of all because individuals hold different conceptions of spatial ordering and spatial justice. Where people act alone (by self-invention and self-determination) or collectively (by sharing words and actions), these spaces precede 'all formal constitution of the public realm and the various forms of government' (Arendt 1958: 199). The meanings and associations formed in spatial encounters are, therefore, always open to further processes of interaction and alterative interpretations. Just as spaces of marginality inhabited by the oppressed, exploited and colonised can be understood as more than simply sites of poverty, loss and deprivation. Rather, they can be construed as fragile but essential sites of openness, possibility and resistance, which provide sustenance for the dispossessed and alienated.

The constant threat of challenge to the power of law and the state means its social relations have to be continually reproduced, reinstated and defended. The perpetual process of verification and validation is likened by Lefebvre to a 'trial by space' in that 'nothing and no-one can avoid *trial by space* – an ordeal which is the modern world's answer to the judgment of God or the classical conception of fate' (1991a: 416). The trial acts as a legitimating procedure, wherein groups, individuals and ideologies must expose themselves and their beliefs to politic acknowledgement and affirmation. Whether it is philosophy, science, religion or law on the dock, the defendant is subjected to a rigorous cross-examination that places continual antagonistic pressure on established forms of spatial power. Since the encounter with other values, ideas and representations may result in a loss of distinctiveness or value, each subject in the courtroom seeks to promote and maintain their own position, make their mark on space and win that particular moment's trial by space. This is particularly important for transformative political strategies because their viability relies on their proficiency in producing, appropriating and organising social space.

Clearly, law and justice are intimately implicated in socio-spatial dynamics; for example, when imposing and refraining from imposing regulation in relation to the design, size, materials and location of the built environment. Legal frameworks impose a set of normative assumptions that underpin and inform policy and eligibility criteria, with discriminating social effects and fragmentary spatial outcomes; imposing penalties on some individuals and places whilst advancing the interests of other locales and actors. Spanning topics such as the control of public space, regulation of the city and the symbolic elements of spatial conflicts, the law is not simply a passive medium through which states can execute their spatial policies; rather, it is a productive force with a life of its own which shapes the physical, social as well as mental-subjective spaces. This is particularly pertinent in relation to regulating the impingement of information and communication technologies, social media, cybernetics and genetic augmentation onto the everyday lived space of the mundane, linear, repetitive and routine. On this matter Lefebvre asked in 1981, 'Will computer science with its repercussions and related disciplines go as far to transform everyday life… the social relations of production, reproduction and domination?' (2006: 236). The next section addresses this important question in the context of the cyber-spatial paradigm shift.

Lefebvre's spatial triad and the cyber-spatial metaphor

The term 'cyberspace' as a synonym for the Internet has become axiomatic. Attached to physical space via network nodes or intersection/connection points and interlinked by networks of power, it offers a metaphorical parallel to physical space. The characteristic notions of control versus unboundedness and materiality versus void express something of the polarity of this equivalent virtual reality. By hosting both symbolic and perceptible forms of information exchange, cyberspace presents a horizon of potentialities (Shaw 2015: 247). For that reason, the application of the Lefebvrian spatial metaphor usefully draws out the ideological possibilities, rather than only the material properties, of Information Communication Technologies (ICTs). For example, in common with the production of power in 'real space', the production of power in networked space also relies on techniques of both visibility and invisibility; being inhabited by real, embodied users and understood through experience.

Once the Internet was the paradigm of digital space; however, the interconnectivity of everything from urban transport to medical equipment to household appliances in the age of the

'Internet of Things' has transformed the ways in which physical bodies connect with other bodies, objects and spaces. As a synthesis of human and machine, the new technologies merge diffuse aspects of life into a single entity. Encapsulating and freezing human actions and making these available in contexts other than the ones in which they occurred allows for a complete projection and distancing of the self both within and beyond the lived urban space, as people inhabit multiple spaces at any given time (Latour 2002). The myriad invisible and instantaneous switches between diverse spaces have become a common site of everyday spatial habitation in most parts of the world. Even so, cyberspace is argued to be superimposed on or augment, rather than replace or add to, Lefebvre's conceptual triadic space. This is because, as Lefebvre explained, 'abstractions' are transformed into 'real' space by means of, for example, 'worldwide networks of communication, exchange and information… [and] that such newly developed networks do not eradicate from their social context those earlier ones, superimposed upon one another over the years' (1991b: 86).

The social models, relationships and practices facilitated by the Internet still contribute to the overall production of social space, as they are layered over large-scale policies and organising mechanisms such as calendars and maps that preceded the Internet (Shaw and Shaw 2016: 32). Despite being produced by a disembodied, unfettered, anarchic electronic network of connections mediating human life and abstract space, ICTs are still dependent on spatial fixity such as bandwidth and access points. Since the limitations of the physical body mean it cannot be fully absorbed into this new space – in the manner of Lana and Lilly Wachowskis' *The Matrix* (1999) – for the present, digital transactions between technology and humans still take place in the lived space. Moreover, the experience of Web browsing in Shanghai is very different from surfing the World Wide Web in Scunthorpe or San Francisco. It follows, therefore, that digital or cyber culture continues to be produced by dominant spatial forms comprising a web of multi-layered and contested discourses, and distinct social practices.

Cyberspace as a contested space

While ICTs impose their own teleologies, both networked space and embodied space are continually shaped and reshaped by the dynamics of power relationships. Although corporeal space determines the properties of virtual space, they are often presented as being in an antagonistic relationship, and this is problematic in terms of finding a holistic approach to the digital production of space. Material spaces are treated as subordinate to the deterritorialised landscapes of cyberspace, in that technological mediation allows sensations to be enhanced or repressed within digital spaces, such as through the immersive experience of the new virtual reality (VR) consoles and headsets. So, while cyberspace is often defined in relation to how its technologies directly or indirectly benefit physical space or according to what it lacks or modifies in relation to real space, our perception changes with each new media innovation. That these adaptations take place within social space is 'what permits fresh actions to occur, while suggesting some and prohibiting yet others' (Lefebvre 1991a: 73). As our technologies advance and reconstitute the relationship between architectural constructions, urban territories, media and social practices, our thinking and behaviour also adapts in response to the evolving technology and media-driven environment.

Maps, for instance, are more than just a mirror of the world or location device. They are important signifiers; representing particular cartographic discourses, political statements, conditions of authorship and rhetorical traditions. Associated with experience and memory, which remain notoriously difficult to integrate into digital interfaces, they can be interpreted as a type of language with which to mediate a specific view of the world. For Lefebvre, the cartographic articulation of 'castles, palaces, cathedrals and fortresses' offers 'a sort of instant infinity' of possibilities for mapping the distribution of power in space (Lefebvre 1991a: 85, 232). Yet the old metaphors of maps, movements and nations have been superseded. Online ICTs have, to a large extent, overwritten the traditional and authoritative depiction of space by means of a digitally produced version, which has transformed the relation between the global and the local. For example, electronic media are eliminating geographical space through the digitisation of maps, the digital projection of maps via Geographical Information Systems (GIS) and by the construction of virtual maps. Real-time information systems such as GPS with Google Maps are inarguably useful and have realised a wider process of historical change by connecting, in real time, 'conceived' and 'perceived' spaces of social interaction. Maps assist in framing reality by eradicating or altering details to help users see a chosen aspect of reality, and changes in mapping tools and technologies are in the process of transforming the socio-spatial. However, life forms mediated by technology tend to mask their cultural influences (due to the non-human nature of the design methodology); becoming non-linear and flattened out. Such technological innovations and their uses are dogged by an innate instrumentality which 'pushes us back towards a purely descriptive understanding, for it stands opposed to any analytic approach' (Lefebvre 1991b: 122).

Instrumentality is not the only problem with the digital mapping of space; users have been impacted by the medium in previously unimagined ways. Google Earth is one of the most extensively used GIS and has been downloaded more than one billion times, yet it operates what is essentially a corporate database masquerading as a photographic representation. As well as providing information to often non-technical users, it is used as a monitoring device to gather information on our home, car, health, personal habits and relationships. Ominously, new social practices are formed according to the panoptic and synoptic logic of innovative location-based technologies. Such novel inventions, and the interests of those who 'own' these vast networks of information, comprise new forms of exploitation and social control; and yet are too often unaccountable and exist beyond the reach of the law. It is even unnecessary to coerce the collective will into giving up their privacy as people passively accept particular ideological conceptions validated by the instruments of authority, and willingly collude in the internalisation of surveillance.

Foucault warned of the rise of the self-policing subject against the perpetual presence of 'an inspecting gaze' under which 'each individual under its weight will end by interiorising to the point that he is his own overseer... thus exercising this surveillance over, and against, himself' (Foucault 1980: 155). For example, people have become accustomed to allowing their mobile devices to track their coordinates. Also, the profound societal implications for privacy and freedom against the corporatisation of political power in the so-called snooper's charter (now enacted as the UK Investigatory Powers Act 2016) have been largely overlooked. The visibility of surveillance techniques has emerged as a forceful cultural theme and one that is inevitably linked with spatial control. Along with reappraising the complex interaction between legal regulation and technical design, confronting the nexus of radical spatial uncertainties requires a commitment to transdisciplinary research, including critical geographers and social anthropologists.

Spatiality, power and justice in a digitised world

The contemporary world of smart cities, big data and social media is highly complex with limitless connections between people and events, and manifold causes for effects that exert a profound influence over thought and activity. This has accelerated in late modernity, as the boundaries between human and machine have been transgressed to the extent that it is difficult to imagine human life without technological mediation. Consequently, humanity has a precarious existence, having become marginalised vis-à-vis technology (Shaw and Shaw 2015: 243). Time, space, knowledge and the body have all been subject to a reordering as a result of the intrusion of technological forms of life. In turn, this has impacted on our subjectivities, experiences and understanding of what it means to be together. Not least of all because by generating new capacities to act at-a-distance, technological mediation has reconstructed contemporary social life along the lines of what Scott Lash describes as 'sociality and culture *at-a-distance*' (2001: 107–108).

Lefebvre's classification of space provides a framework within which to understand modern society in its current state of being distanced from reality; characterised by inauthentic space, driven by capitalist consumer values and being reimagined by cyber technology. Every day, armies of tech zombies march along the street, headphones in place, eyes down facing a screen, busy tapping away, while gaming, listening to music or focused on the latest text, tweet or email; both physically present and at the same time existentially absent or at least remote. As Seegert explains, 'the orientations of cell phones and new media truncate our reciprocal relation with the Earth, stunting our senses and incarcerating ourselves in a technosoliloquy' (2011: 47). These spatial incongruities have significant implications for evolving theories of space and culture, but also in relation to our sense of distracted, distanced and distributed communities in the context of the new hypermobility of users and how this affects the construction of social spaces.

Cyberspace provides a context for the layering of fantasies upon earlier fantasies, the competition of one fantasy with another and the echoing of one fantasy, desire (and more recently fear, according to a new aesthetics of security in the wake of the so-called 'war' on terror), by another. Yet the social structures within which we express our identity – in terms of race, class, sex, gender and age – persist, in spite of the ethereal possibilities offered by social media and virtual reality. Access to essential resources such as food, shelter and a livelihood is still subject to dominant political values. So although cyber-resistance is useful – for example, social media enabled the 2011 'Occupy Wall Street' movement to go global – the sites of resistance against material injustices continue to rely on their embodied constituent. The power of technique or technology, defined as a 'calculated, quantified, and programmed' manner of operating, was one of the central problems that Lefebvre identified in modern society. Technological mediation within cyberspace, as an offshoot of capitalism, aligns with Lefebvre's characterisation and remains a powerful instrument of social discipline, which extends control from the economic to the cultural and social spheres of our existence (Elden 2004:144). It follows therefore that Lefebvre's philosophy of spatial justice can provide a framework for articulating the reasons for various incongruities and conflicts, and suggest the basis for ameliorative work towards producing a consensus of spatial-ethical values; even within the unruly and antithetical space of cyberspace.

Conclusions

A critical examination of urban space and spatial relations provides valuable insights into prejudices and discriminatory practices that too often go unnoticed and unchecked. In an increasingly complex world there is still much work to be done in relation to presenting a fully coherent and multifaceted articulation of spatialities, which explain how space is socially constructed and utilised within a multiplicity of diverse new contexts. The modern cityscape is not just a set of processes, infrastructures and networks. It is an essential mediating force (amplified by physical and virtual interconnectedness) and a discursive space which contains the possibility of enhancing sociability, where communities produce instinctive and tacit collective memories. Space has meaning because of the social relationships forged between things and people; and although a tension-filled and incomplete process, it is also the place in which they have learned to live and flourish. It is not something human activity fills up, rather space is something human activity produces; in other words, it is the very expression of society which has 'emerged in all its diversity' (Lefebvre 1991b: 83, 86).

Similarly, when understood as solely otherworldly and intangible, cyberspace is a dubious concept because, experientially, it is (like physical space) multiple. As it is not a unitary space, the impact of virtual space on individuals and communities is not identical or even the same in different parts of the world. To paraphrase Lefebvre, information technologies, whether viewed as purely techniques or tools, are 'not simply a means of reading or interpreting space; rather [they are] a means of living in that space, understanding it and producing it' (1991a: 47–48). Although society's public and private spaces of interaction are becoming more complex and human life has been irrevocably changed by technology; the cyber-conscious of the self remains physically situated in the lived space of pure subjectivity, of human experiences, imagination and feeling. It is, therefore, still within the borderlands of lived space that we encounter the past, the present and the future, the historical and the imagined, together with the endless negotiations of power and control.

References

Arendt, H. (1958) *The Human Condition*, Chicago, IL: University of Chicago Press.

Augé, M. (1995) *Non-Places: Introduction to an Anthropology of Supermodernity*, trans. J. Howe, New York: Verso.

Blomley, N. (2003) 'From 'what?' to 'so what?': law and geography in retrospect', in J. Holder and C. Harrison (eds) *Law and Geography*, Oxford: Oxford University Press.

Butler, C. (2012) *Henri Lefebvre: Spatial Politics, Everyday Life and the Right to the City*, Oxford: Routledge.

Cover, R.M. (1983) 'The Supreme Court 1982 term: foreword: nomos and narrative', *Harvard Law Review*, 97: 4–68.

Elden, S. (2004) *Understanding Henri Lefebvre: Theory and the Possible*, London: Continuum.

Foucault, M. (1977) *Discipline and Punish*, London: Penguin Books.

Foucault, M. (1980) 'The eye of power', in C. Gordon (ed.) *Power/Knowledge: Selected Interviews and Other Writings 1972–1977*, Brighton: Harvester Press.

Frug, G.E. (2010) 'A rule of law for cities', *HAGAR: Studies in Culture, Polity and Identities*, 10:1 61–70.

Goodrich, P. (1991) 'Specula laws: image, aesthetic and common law', *Law and Critique*, 2:2 233–54.

Goodrich, P. (2013) '"Conclusion" in virtual laws', in *Legal Emblems and the Art of Law: Obiter Depicta as the Vision of Governance*, Cambridge: Cambridge University Press.

Kracauer, S. (1997) 'On employment agencies: the construction of a space', in N. Leach (ed.) *Rethinking Architecture*, London: Routledge.

Lash, S. (2001) 'Technological forms of life', *Theory, Culture and Society*, 18:1 105–20.

Latour, B. (2002) 'Morality and technology: the end of the means', *Theory, Culture and Society*, 19:5/6 247–60.

Lefebvre, H. (1974) *Le droit à la ville suivi de Espace et politique*, Paris: Éditions Anthropos.

Lefebvre, H. (1976) *The Survival of Capitalism*, London: Allison and Busby.

Lefebvre, H. (1991a/1974) *The Production of Space*, trans. D. Nicholson-Smith, Oxford: Blackwell.

Lefebvre, H. (1991b/1947) *Critique of Everyday Life, vol. 1, Introduction*, trans. J. Moore, London: Verso.

Manderson, D. (1996) 'Beyond the provincial: space, aesthetics and modernist legal theory', *Melbourne Law Review*, 20: 1048–71.

Manderson, D. (2005) 'Legal spaces', *Law Text Culture*, 9:1 1–10.

Philippopoulos-Mihalopoulos, A. (2015) *Spatial Justice: Body, Lawscape, Atmosphere*, Oxford: Routledge.

Seegert, A. (2011) 'Ewe, Robot', in D.E. Wittkower (ed.) *Philip K. Dick and Philosophy*, Chicago, IL: Open Court Publishing Company.

Shaw, J.J.A. (2015) 'From *homo economicus* to *homo roboticus*: an exploration of the transformative impact of the technological imaginary', *International Journal of Law in Context*, 11:3 245–64.

Shaw, J.J.A. and Shaw, H.J. (2015) 'The politics and poetics of spaces and places: mapping the multiple geographies of identity in a cultural posthuman era', *Journal of Organisational Transformation and Social Change*, 12:3 234–56.

Shaw, J.J.A. and Shaw, H.J. (2016) 'Mapping the technologies of spatial (in)justice in the Anthropocene', *Information and Communications Technology Law*, 25:1 32–49.

Shaw, J.J.A. (2017) 'Aesthetics of law and literary license: an anatomy of the legal imagination', *Liverpool Law Review: A Journal of Contemporary Legal and Social Policy Issues* 38:2 83–105.

Wachowski, A. and Wachowski, L. (1999) *The Matrix*, Burbank, CA: Warner Home Video.

Zižek, S. (2002) *Welcome to the Desert of the Real*, London: Verso.

18

Interpreting the spatial triad

A new analytical model between form and flux, space and time

Gunter Heinickel and Hans-Peter Meier Dallach

Introduction

In reaction to the intensified urbanisation and suburbanisation processes after the Second World War, which proceeded very much under the mission statement of the city as a 'machine' in accordance with industrialised society, it was Henri Lefebvre's concern to save urbanity as a place of creativity, innovation, vibrant communication, close encounters and general exchange. His guiding principle always remained the recovery of the collectively used and formed public space (Lefebvre 2009: 108). He formulated this concern in the nowadays established claim of 'the right to the city' (Lamare 2015).

With this objective in mind Lefebvre was quite clear about the relevance of symbolic and attributed qualities for the contested urban settings. His famous model framework for a descriptive analysis of urban spaces, the 'spatial triad', already integrated conceptually the dimensions of the physical space, the spatial practices and the notion of concepts, perceptions and projections related to it. Among the three constitutive elements of his spatial triad: the 'representation of space' (concepts, plans, policy statements), the 'spatial practices' (daily routines of people) and the 'representational space' – it is foremost the latter which is the realm of symbolic and ascribed attributes, projected meanings and qualities (Goonewardena et al. 2008: 27–45).

In his later works, Henri Lefebvre also came to the notion that the qualities of physical-spatial situations and settings in the urban environment can be read as space-time-functions, as what he described as 'rhythms'. With his last book *Rhythmanalysis*, he substantially introduced the dimension of time as a mode or tool to analyse urban situations in an open, rather 'impressionistic' way (Lefebvre 2004; Mareggi 2013: 5). But despite Lefebvre's suggestions, the dimension of 'time' has found only limited analytical appliance in geographical and mobility research. Though the concept of 'space-time-design' is firmly established in geography and planning studies today, the topics evolve mainly around questions of organising time, or the social effects of organised time, in short: 'efficiencies versus inefficiencies' (Henckel et al. 2013: 99–117; 2007).

With reference to these opening observations concerning Lefebvre's work, we suggest a new model approach, by which Lefebvre's spatial triad can be linked in a functional way with his ideas about rhythmic time-functions in spatial, especially urban, situations. Though we developed our own model approach fully independent from Lefebvre's oeuvre, by revising the works of Lefebvre, his ideas indicate some resounding similarities to our own concerns.

However, we are convinced that our suggested descriptive and analytical model can be helpful to illuminate and to operationalise Lefebvre's spatial triad in certain respects for practical use, especially with regard to the proceedings in the sphere of the representational space and its relation to spatial practice. As an open and flexible, yet structured method, we hope that our model brings some of Lefebvre's ideas to life for planners and practitioners.

Nevertheless, the original aim of our own model approach was not so much the problem of urban space and its usage, or the right to the city (RTC), but improving our understanding of the contemporary reality of transport and mobility by incorporating individual and collective relations to spatial qualities as stimuli for mobile behaviour. Fundamental to our approach was the insight that 'mobility' is not simply the physical movement of people and goods, but a creation of social patterns in the fabric of society as in the individual life.

We recognise the parallels of our own approach to Lefebvre's reflections specifically in the understanding about the interrelationship of time perceptions, the projections of meaning and the experience of spatial settings. Quite similar to Lefebvre, we defined the motives for mobile behaviour between a triangle of objective (such as distances, physical conditions and obstacles), subjective (time qualities of spaces) and projective factors (needs, emotions, imaginations).

All in all, the time aspect became the central factor: in our opinion, and in our reading of Lefebvre, it is the dimension of time which is the critical variable in processing projections and symbolic appropriations to spatial settings. From our own research experience (to which we will refer on occasion) and the assessment of similar studies, it is very much the individual time experience and the reaction to it, which transports the passive evaluations and the active projections onto spatial conditions: cultural and collective memories, learned and consumed artistic interpretations of urban settings and realities are very much transmitted by time-related representations.

Transformation points between spatial practices and representational space

We can observe daily many such conflicting space-time functions between spatial practices and the representational space in the form of transformative situations in our urban routines and surroundings. We like to call such situations 'Transformation Points', because at those points our perceptions of spatial and mobility-related needs, wishes and realities may change, or become conscious to our mind. We just want to bring the five most prominent examples of such Transformation Points in our contemporary lives to attention, each representing one specific observational level. These observational levels we want to address as 'Landscapes', a terminology that will be explained in detail further down. It has to be stressed, however, that transformative processes affecting one of those observational 'Landscapes' inevitably influence the other layers as well.

Transformation Points regarding the Nature Landscape

Starting with the most archetypal geographical setting of any kind, the Nature Landscape, we can currently observe how urban mobility-related qualities and practices invade progressively into natural and rural areas. The resulting conflicts are very much about the question as to what extent the natural environment should keep its unadulterated qualities, or whether careful interventions for recreational and sportive activities should be allowed in line with social needs and practices: wilderness versus the parkland-ideal. The combat terms of 'ecology' or 'green' do not so much solve these conflicts, but rather obscure the fact of competing notions of nature. Notorious are the clashes in nature resorts between traditional hikers and new sports like mountain-biking – often resulting in violent outbursts and even assaults.

Transformation Points regarding the Mobility Landscape

The sphere of classical transport and traffic activities, the Mobility Landscape, represents itself today very much as the battlefield between established goals for reducing or abating transport activities, and the thrust to modernise public and private transport systems. Planners and politicians today are torn about whether existing trends and developments should just be adjusted by planning activities or redirected towards utopian ideals. Such unresolved targeting can be observed specifically in ambitious ventures to develop once again holistic drafts for the future mobilities of entire nations, like the *Vision Mobilität Schweiz 2050* (Stölzle et al. 2015). It goes without saying that such visions of mobility are intensely related to specific ideas about urbanity – notions of an assumed hyper-modernity to be realised in an agglomerated urbanised environment.

Transformation Points regarding the Settlement Landscape

Closely related to the speculations concerning the Mobility Landscape are therefore the topical debates on settlement structures: are new residential areas destined to become the homes of a commuting population, the backdrop of multi-local biographies, or new anchors of locally contained life-styles? To what extent should the workplace be integrated into the home? Are classical neighbourhoods hence to be recreated? Do new residents adapt to established communities or create new neighbourhoods? This discussion is reflected for example in new town planning schemes in the region of Zurich, Switzerland (Meier Dallach 2016).

Transformation Points regarding the Communication Landscape

The fourth level represents the most dynamic and expanding of mobility activities of our time: the Communication Landscape. The related potentials, challenges and even threats are still not yet resolved – popular reactions ponder between the still-powerful promise of an ever-greater expansion of personal communication skill and ideals of voluntary restraint, even ascetic usage. This communication layer of mobility activities is therefore today the centre of most foci scrutinising future life-styles and place-making. But, in addition, euphoric visions, scepticism and pessimistic outlooks are on the rise. The prospects of the new and still emerging Communication Landscapes are relevant here.

Transformation Points regarding the Societal Landscape

All these preceding topics culminate at last on the level of the Social or Societal Landscape: how valid are old neighbourhoods, how viable are new forms of familiarity? With what instruments could social relations be enhanced; which infrastructural conditions are needed? And what are the roles of institutions or private initiatives for these processes? Aspects of the spatial and architectural planning may promote or obstruct the formation of new and agreeable forms of cohabitation, but the actors involved are decisive in terms of their value orientations and plans for life (Meier Dallach 2016).

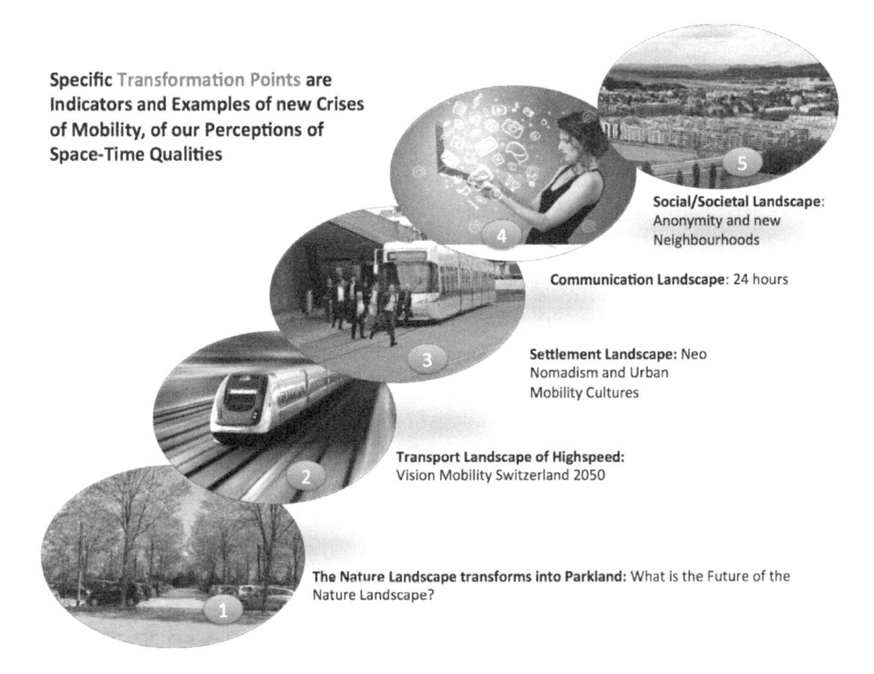

Figure 18.1 Mobility causes experiences of transformations and breaks by the effects of objective, subjective and projective impressions: the five most expressive Transformation Points of the landscape layers in our time. *Source*: cultur prospectiv Zurich Switzerland, (author) conference presentations and papers.

These five Transformation Points do indeed dominate consecutively or at the same time all our contemporary controversies about mobility and the urban space (see Figure 18.1). However, it is not to be forgotten that these phenomena are embedded in the currents of the great global transformation. The European migration crisis since 2015 highlights how mobility means so much more than transport. The latest images of an assumedly triumphant Euro-Atlantic modernisation paradigm – like opening the massive railway transversal of the Alps (NEAT) – are blatantly confronted with starkly differing images of another form of globalisation. The achievements of the technical mega-structure, defying all geographic-historical realities of the European collective memory, are suddenly taken into question. The still prevalent assumption in many fields of traffic and mobility research that accessibility and reachability are the most relevant concerns seems obsolete (Figure 18.2).

The model approach: representational space between form and flux

With reference to these presented five Transformation Points and their respective Landscape-layers, we now want to draft our suggested model approach in detail. For this we have chosen four main thematic areas or foci. In each of those focal areas we introduce some concepts and terminology. These focal points will build on each other, culminating in an overall theoretical and historical framework. We start with outlining our understanding of the relationship between individual spatial experiences and projective expectations, between the physical and the symbolic/emotional dimensions of mobility (Focus I). Then we will introduce our analytical modes of 'Form' and 'Flux' (Focus II). Our already introduced term of 'Landscapes' will be explained

Figure 18.2 Dramatic and conflicting new transformation points of contemporary mobility experiences. *Source*: wikicommons©AFP, wikicommons©copper.ch

in detail under Focus III; and eventually we will integrate all these terms and concepts in our summary as Focus IV.

Focus I: Mobility – a symbolic as much as a physical experience

By now it is well accepted and established in transport as in mobility research that mobility events are triggered not only by objective obligations and rational choices, but by emotional and symbolic meanings as well. However, there is still some confusion and unease among classical mobility researchers about how to describe and analyse this dimension in a systematic way. Though it is also accepted that these emotional motivations are somehow related to the physical qualities of spatial settings, this insight is still more based on pragmatic observations, less on systematic research. There is a lack of comprehensive understanding about *how* to link the experience of physical movement with emotional and symbolic needs and motivations. In our first focal area we therefore want to give some initial indications about our understanding of this interaction between individual physical mobility experiences and their interpretative sphere.

All of us are aware, though not always reflective, about how mobility creates sensations of movement, allowing us not simply to endure or to tolerate, but to enjoy the impressions of acceleration, braking or resting and relishing in situations of exertions and relaxation. Though in practice well known and understood by the motor industry, this passive emotional experience is hardly scientifically researched, with some exceptions for the public transport sector (Schiefelbusch 2008). Similarly, we are personally quite aware of the symbolic dimensions of travel and transport, emotions of familiarity, crossings of personal or public boundaries, visible or hidden, events of estrangement and arrivals. Personal perceptions of mobility situations usually oscillate between fluidity and arrestment, excitement and contemplation, even tensions and boredom. And such emotional and symbolic experiences are essentially caused, enhanced and transmitted by given spatial surroundings, like architecture, a busy shopping road, the quietness of a secluded residential area, by the impressive presence of a cathedral or the commotion of an illuminated tourism hotspot. Spaces are therefore firstly defined and rhythmised by physical sequences and material formations.

This array of physical factors we subsume under the notion of 'Form' (relating to 'Gestalt' in German), encompassing all observable images of shapes and patterns in our physical world: the size and design of houses and buildings, artistic details of architectural forms, the outlay of roads, the shape of squares and places, the composition and configuration of greens and wastelands. Geographers and planners are well aware of these factors, pondering about practical measurements and material devices to create the 'beautiful city' and to enhance the public acceptance of planning provisions or infrastructural implantations (Lynch 1960; Florida et al. 2009; Leslie et al. 2005: 227–236).

But we want to turn the attention to another, mostly overlooked category for the personal perception and evaluation of spatial and mobility situations: the time dimension, or better, the time experience. It is the time dimension by which the symbolic and emotional quality of places and spaces transmits itself most vividly. Time, as we all know, is nothing objectively fixed in our perceptions, not a universally and constant quality, but it is flexible, can expand and contract and can leave us in a mood of relaxation or distress, depending on exterior obligations, constraints and also by how comfortable we feel in our environment. Thus spaces, like the natural landscape, an urban environment or a specific transport infrastructure, are not just containers of physical objects – natural or artificial – but are laden with intrinsic time qualities as well. These time qualities vary between tranquillity and speed, sedateness and change, expressiveness and regression. Time is therefore an emotionally highly loaded quality and a main cause for spatially related sensations of attraction or repulsion, but also for the symbolic meaning for inhabitants, commuters or passers-by (Heinickel 2013: 201–203). These time-related factors for spatial appeal we subsume under the term of 'Flux'.

And there is a third dimension which influences the human valuation of spaces and accompanying mobility events: the personally and collectively motivated projections, which can transform any environment or setting either into a backdrop, into a screen or into an image, but also to an inner landscape. But these projections do not occur at random and are not detached from the respective settings they are directed at. These projections correspond closely with the endogenous material elements as with the intrinsic time qualities of those settings. Such projections are most easily to be observed in leisure and tourist travel activities (Heinickel 2013; 2005). They constitute the 'tourist gaze', which may charge surroundings and landscapes with varying meanings, depending on the individual, their own cultural background, but also on the cultural outlook provided by different historical epochs (Urry and Larssen 2011).

Now we can sketch these preliminary considerations about the interaction of space-time relations and the connected human projections and expectations into a first simple model. In this we can differentiate between the passive level of experienced spatial (Form/Gestalt: G) and time-related qualities (Flux: F), and a level of active projections, namely demands, wants, desires and expectations directed towards specific spatial destinations (Figure 18.3).

It is easy to imagine how these three kinds of differing factors – objective spatial, subjective space-time perceptions and active projections – shape our reality of mobility experiences and intertwine and may cause an array of clashes and conflicts, because the experience of competing space-time relations with social demands causes psychological stress. These arenas of conflict have been called 'Transformation Points' (see above). Transformation Points can be interpreted as functions of personal experiences within given spatial settings and projected wants and needs

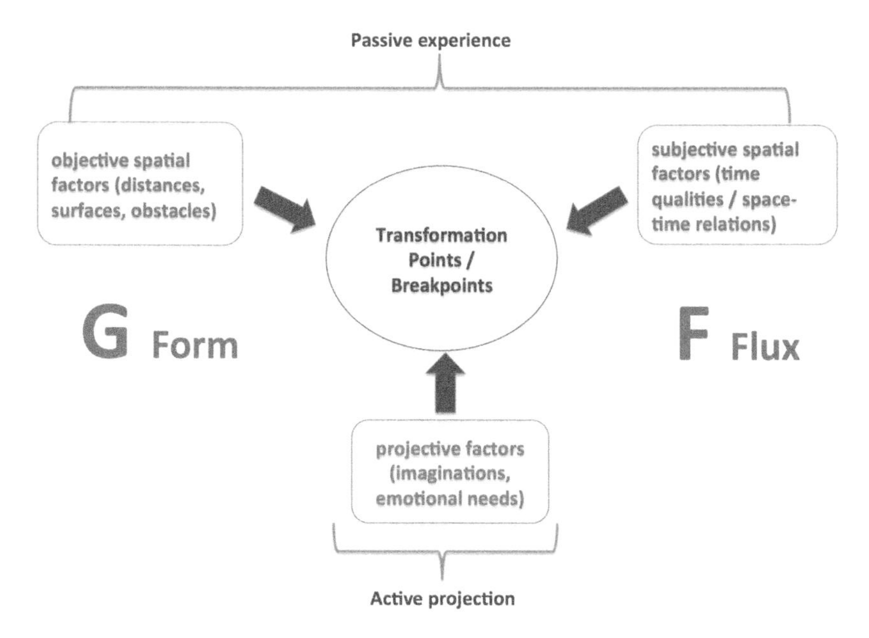

Figure 18.3 Mobility causes experiences of transformations and breaks by the effects of objective, subjective and projective impressions. *Source*: ScienceScapes Berlin Germany, (author) conference presentations and papers.

versus those settings. They provoke transformative processes, changes or alterations in the daily routine or demands of mobility and may influence subconsciously allegedly rational decision-making about mobility styles, strategies and adaptations.

Focus II: 'Form' and 'Flux': the central denominators for spatial qualities and mobility impressions

In our second focal point we want to develop the idea of 'Form' (G) and 'Flux' (F) even further and give illustrative examples to elucidate our terminology. As stated above, 'Form' and 'Flux' refer to objective as well as to subjective factors, which generate the impressions of a specific space-time relation in a given environment.

'Form', as we have seen, refers to the stable, the lasting qualities of environments, like the features of the natural landscape, but also of man-made structures of buildings and infrastructure. For the observer these various but constant features cause the impression of relative stagnation, a rather static, stable and motionless setting. In other words: an extreme *introverted* situation. 'Flux', in contrast, represents the factors of instability, of movement and ruptures. Most noticeable are motions caused by human activities: traffic, of course, but also other occupations like building and working situations, leisure activities like sports and even communication; these result in impressions of constant alterations and ruptures, transformations, pace, even general acceleration and restlessness, and are therefore extreme *extroverted* situations. 'Form' factors are part of the natural world as well as a result of human activities, whereas 'Flux' factors, according to our understanding, are created exclusively by humans.

As to the problem of space-time functions, uncompromised Form gives the environment, natural or man-made, the possibility to perform all its inherent potential of stability, which brings specific details and structures to the full awareness of the individual. A relaxed and concentrated focus on those elements and their quality is possible. Under the conditions of extreme Flux, however, the stable qualities of a landscape or urban situation may be totally lost to the subjective impression of time constrains, with impressions of constant flow and even speed. In general, though, the experience of mobility proceeds constantly between Flux and Form, and not only in a physical sense, but also symbolically and emotionally – as varying commotions will transmit sensations of time pressures or urgency, independent from our own personal state or social obligations: the contrast between sitting in a cathedral or at a busy train station.

According to our basic model, environments can therefore be assessed by differing gradients between those two factors, between introversion and extraversion. This flexible and open scale between Form and Flux allows us to equally evaluate urban environments, rural landscapes, building ensembles or street situations, using a coherent terminology and comprehension (Figure 18.4).

For the purpose of research and analytical operationalisation, it has to be emphasised, however, that the specific scale and criteria for Form and Flux factors have to be defined and adjusted to the specific environment under research. Flux and Form factors vary of course starkly when ascribed to an industrial city or to a medieval town of half-timbered houses; and introverted and extraverted situations in a mountainous region express themselves differently than at the seaside. And applying a defined scale between Form and Flux only makes sense if the settings under scrutiny *belong to the same spatial context* of a researched city or natural environment. A *direct* equation of single situations between *different* urbanities and even natural landscapes is therefore not advisable if not impossible. Nevertheless, this model approach does allow a comparison of one typified urbanity to another: such a comparison was ventured for instance for the urban contexts of Zurich and Berlin, and their respective consequences for stimulating regional and multilocal mobility patterns, or rhythms (Meier Dallach et al. 2003: 175–185).

Figure 18.4 Highly differing settings in the urban realities of Berlin: sometimes Flux factors are more prevalent, sometimes Form factors. *Source*: cultur prospectiv Zurich Switzerland, (author) conference presentations and papers.

Focus III: Landscapes – five observational and analytical layers

As already introduced and exemplified at the beginning of this chapter, we distinguish five significant observational and descriptive perspectives or layers, called 'Landscapes': the Nature Landscape (Naturlandschaft), the Mobility or Transport Landscape (Verkehrslandschaft), the Settlement Landscape (Siedlungslandschaft) and the Communication Landscape (Kommunikationslandschaft). At this point we want to explain our notion of 'Landscapes' further. Again we have to stress: this term of 'Landscape' does not refer to the classical term of a landscape as the totality of material elements in an environment. 'Landscape' in our sense is a descriptive and analytical tool to structure and group the various material elements of Form and the Flux factors alike in a given spatial setting. It is the term of 'spatial setting' with which we address the totality of a specific urban situation.

In our understanding, the Form (G) and Flux (F) factors express themselves on the various analytical Landscape levels within such settings. Therefore all five defined Landscapes can be identified and addressed in any given setting at the same time! Moreover, several of observed and identified elements of Form- and Flux-factors may even be attributed to different analytical Landscapes within one setting.

These Landscape-layers can be hierarchically ordered, bottom up, according to their assumed impact qualities on settings, starting at the ground level with the profoundest characteristics, the Nature Landscape elements, over which the other Landscape layers are positioned. What we receive is something like a tectonic model, and we can imagine how the lower layers or Landscapes influence the shape of the levels above, all together creating a characteristic and tangible 'surface' of space-time functions, namely urban rhythms. But it should not be forgotten that these layers with their respective elements are volatile and changeable over time. Landscapes in our sense are nothing fixed and defined for all times, but apt to change. That is why we like to represent them as broken lines in our sketch (see Figure 18.5).

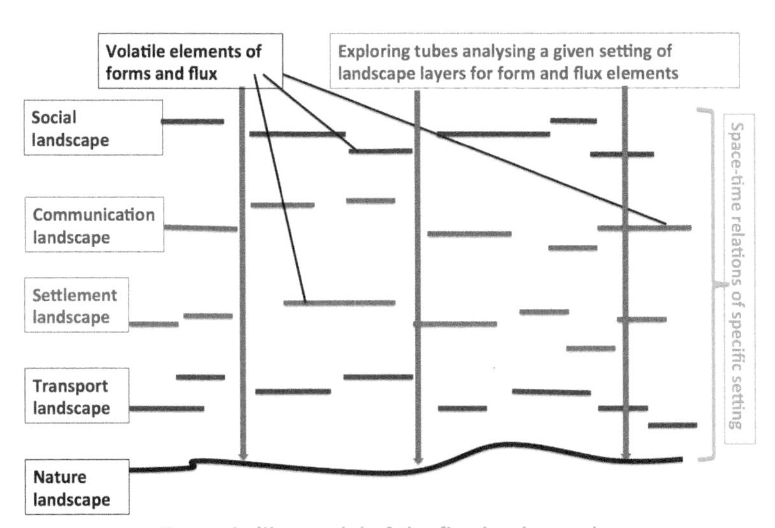

Figure 18.5 The five landscape layers can be imagined like a tectonic structure – and like a tectonic structure, these layers and their respective elements (symbolised by the broken lines) are volatile and changeable over time. *Source*: ScienceScapes Berlin Germany, (author) conference presentations and papers.

The Nature Landscape addresses the most profound spatial characteristics, because those qualities are the most lasting in any given setting. On top is the layer of the Mobility or Transport Landscape, as the transport infrastructure influences the quality of given settings like no other human intervention today. The shape of layers further up are influenced by the lower strata respectively, but still, as we send down some testing tubes into the tectonic ground, we would discover different relief formations of Form and Flux factors, depending on what layer we research. Therefore we would like to add some remarks about the respective effects of those Landscape layers for our contemporarily experienced space-time functions, or rhythms.

The Nature Landscape (Naturlandschaft)

The Nature Landscape contains the potentially highest grade of pure Form elements. Here we find the most constant physical conditions, promoting strong perceptions of introversion. Hence, the Nature Landscape offers to the observer the primal experience of 'being' – a quality in high demand in modern times! But even nature was and is transformed by human activities. Today, Natural Landscapes differ from each other not only due to their own inherent qualities. But also by the intensity by which human intervention introduced pathways and roads, farmed land, buildings and settlements, implementing considerable amounts of Flux factors: the transformation from untouched nature to the highly cultivated landscapes of agricultural and industrial society. Today, even many nature resorts rather resemble carefully tended and managed parkland, and the impassable characteristics made way for more targeted and defined movements.

The Mobility or Transport Landscape (Verkehrslandschaft)

The mobility and transport infrastructure appeared historically as the first human artefact upon nature, even before settlements – from the paths of hunter-gatherers to the great transgressional routes of the nomadic peoples. From then to now, with our modern motorways and railway lines, the Mobility Landscape carries the central notion of experiencing 'being here and there'. Because mobility means essentially movement between defined points, leaving and returning, and eventually staying in a destination of purpose, in other words: movement is a relational activity and creates a dialogue between fixed settings. Such an understanding is dissenting with the still dominating notion of mobility as an autonomous reality between locations. The infrastructural elements of the Mobility Landscape bring out the strongest Flux factors of all landscape layers. Hence, the Mobility Landscape affects all other landscape levels.

The Settlement Landscape (Siedlungslandschaft)

Similar to the Mobility Landscape, the Settlement Landscape has experienced dramatic changes over the ages, and especially during recent decades. Urbanisation today is in the full grip of globalising processes. Habitation as the classical setting for 'being at-home' – housing, sheltering and guarding the closest social relationships – increasingly gets under pressure. Historically the Settlement Landscape stresses sedentary and bonding qualities: 'being-at-home' or 'being-here' are its classical expressions. Today, this shell for securing the most personal existence becomes more and more fragile: various factors may disturb, compromise or even destroy this intimate bond. However, in the highly urbanised environments we still find the entire spectrum of potential living conditions, the most varied compositions of living quarters, business areas and production, entertainment and shopping, showing all the contradictions between the illuminated, silent, chaotic and wild city.

Historically the factors of Form signified most prominently in the Settlement Landscape, but since the industrial revolution with its modern transport means and infrastructures that has changed profoundly. Today, the extreme influence of the Mobility Landscape is almost at par with the stable characteristics in the urban environment. And what is more: currently even city planning and the architectural designs are often subjected to the logic and the requirements of Flux – for instance, housing estates are frequently structured accordingly to transport considerations. But still, buildings and quarters, despite their open and ill-defined interspaces, contradict and even break the forces of Flux. The resulting rhythms are often paradoxical.

The Communication Landscape (Kommunikationslandschaft)

Ever more important in our contemporary life is the dimension of the Communication Landscape. It is probably the most dynamic factor in our everyday life, introducing extreme levels of almost constant 'togetherness' in a virtual reality. The new communication technology makes us part of a network, of a grid, changes traditional forms of communion, superseding established forms of 'neighbourhood', and with it the familiar coordinates of time-space relations. Virtual communication surpasses frequently even the Mobility Landscape with its dynamic Flux factors. 'Being together' has become a highly obligatory form of existence, being player or subject of information channels, overbidding time and space. Conventional neighbourhoods and private encounters, with their necessity to coordinate time and space, are increasingly replaced by these new forms of socialising.

The Societal or Social Landscape (Soziallandschaft)

The Societal or Social Landscape unites all forms of (direct, not virtual) social activity, the social interaction of persons and institutions. The Social Landscape can be read as a function of all the other Landscape-perspectives. Because sociability is based on all forms of exchanges and activities, in contrast to the Communication Landscape, this form of 'togetherness' is neither separated from spatial conditions, nor segregated from the daily chores and obligations. 'Being with' is the central expression on this level of observation. In stark contrast to this, the dominating observable space-time dynamism is disintegration!

On the surface, the profile of the Social Landscape is most visible and most easily researched. Therefore, the Social Landscape reigns supreme for observational and analytical purposes, because all the effects and consequences of the various Landscapes converge on this highest level (see Figure 18.6). It is also the classical arena for all conflicts about the 'right to the city' (RTC). The Social Landscape therefore is ideally suited for studies comparing contrasting urbanities and their inherent rhythms.

Summary (Focus IV) – from description to conception

By combining our conceptual considerations concerning Form (G) and Flux (F) factors and the various observational Landscapes into one graphic depiction, a possible periodisation suggests itself. As the five Landscape layers present themselves in their relative dynamics between the two poles of Form (G) and Flux (F), the typified space-time functions, as described under Focus III, indicate a successive acceleration of Flux dynamics over time. And this acceleration corresponds with an evolutionary or historical logic, as demonstrated in Figure 18.6.

As a first historical period we may identify the chthonic age, the time of the most archaic forms of human life, when the conditions of the Nature Landscape reigned supreme even in

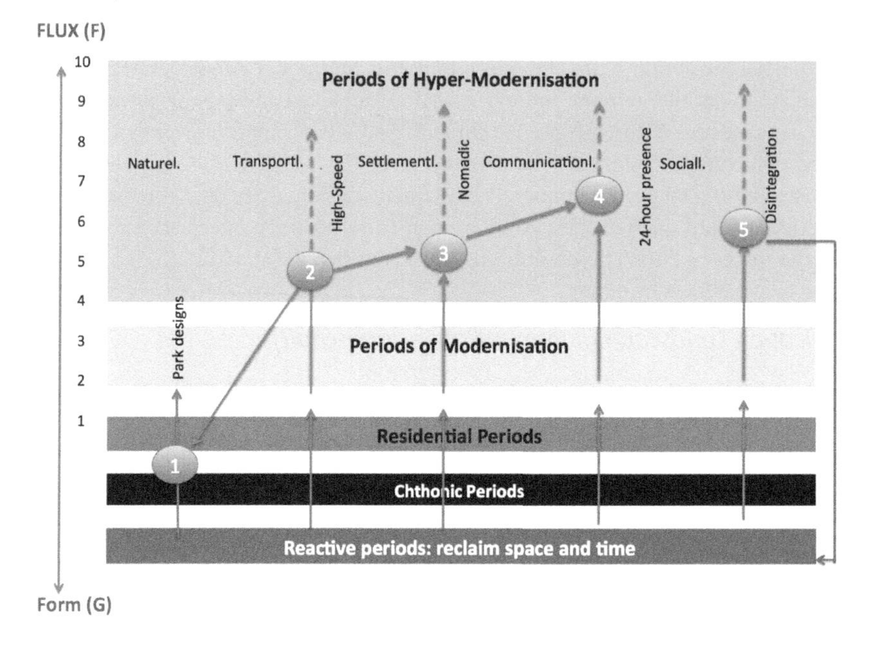

Figure 18.6 Landscape layers and historical periods. *Source*: cultur prospectiv Zurich Switzerland, (author) conference presentations and papers.

the artistic self-presentations of humans, as the impressive cave paintings of Ice Age Europe demonstrate. In that time period it was essential for humans to adapt to and to interact with the landscape, and to *communicate* something *with and through* the landscape, to *inscribe* themselves onto the landscape.

The second historical period could be called the time of residential existence. At this stage, constant settlements started to redefine the landscape but also agricultural activities. It is in this period that the Natural Landscape gradually gives way to garden or park-like structured nature, showing Flux elements increasingly as part of the natural setting. This acceleration of rhythm by Flux elements continues during the following third and fourth periods of modernisation. This leads to an increasing number of transformative situations, or Transformation Points in the daily human experience. The result is the experience of major social and cultural disintegration processes.

Nevertheless, it becomes apparent today that this thrust towards hyper-modernity and hyper-mobility is somehow broken, finds a new resistance, gets resented, objected and actively encountered: the intensified experience of disintegration provokes reactions, with an abrupt reorientation to reclaim space and time (Animento 2015) – a movement which in its ideals somehow indicates back to the chthonic period. The resulting new utopias, for instance the new urbanism, voice their concerns about losing the 'authenticity' of places, namely the disappearance of 'history'. Whereas imperfection, the irregularity of structures and places, and the existence of non-defined public spaces are interpreted as tokens of a genuine existence – and as a safeguard against too 'sleek' forms of routines and neighbourhoods (Heinickel 2013).

Conclusions

Despite the sentiments set out above for regaining more satisfying space-time relations, urban planners, architects, decision-makers, public transport and automotive providers are still setting their hopes predominantly onto the introduction of new digital communication and information devices – promoting even more acceleration! Being 'smart' is still marketed as 'doing everything at all places and times'. Data crunching seems therefore to be considered as the ideal solution for everything: not only for the introduction of alternative transport modes, the endorsement of new mobility routines, the establishment of alternative destinations, but also for guiding settlement choices. The urban dream is still sold as optimised space-time efficiency for everybody in their daily life, be it for transport or residential purposes.

Overlooked are the longings for retreat and intimate communication. While ignoring the deeper motives for movements and voices reclaiming the space, namely the urge to reclaim time, many new planning and mobility tools, organisational devices and infrastructural sanctions would seem to be in danger of becoming detached from the reality of individual mobility needs and daily routines. And indeed, unprepared planners and decision-makers are regularly running into resistance and protests against their well-meaning planning schemes.

On the other hand, debates, concepts and struggles for the right to the city (RTC) must address more than just accessibility and affordability of urban areas for the wider public. Hence, with the support of our model approach, we want to turn the attention to the importance of location-related space-time functions. Concepts of RTC have to find their own answers to the trends of *codification*, *virtualisation* and *technical deprivation of authentic and personal autonomies* in handling space and time.

References

Animento, S. (2015) 'Creativity in the "spaces of hope": interactions between mega-projects and social struggles in Hamburg', *Territorio*, 73: 30–38.

Florida, R., Mellander, C. and Stolarick, K. (2009) *Beautiful Places: The Role of Perceived Aesthetic Beauty in Community Satisfaction*. Martin Prosperity Research (Martin Prosperity Institute) (ed.) Working Paper Series REF. 2009-MPWP-008.

Goonewardena, K., Kipfer, S., Milgrom, R. and Schmid, C. (2008) *Space, Difference, Everyday Life: Reading Henri Lefebvre*, London: Routledge.

Heinickel, G. (2005) 'Berlin am Meer. Der historische Ostseetourismus zwischen Berlin und Usedom', in B. Gundler, M. Hascher and H. Trischler (eds) *Unterwegs und Mobil. Verkehrswelten im Museum*, Frankfurt/New York: Campus Verlag.

Heinickel, G. (2013) 'After the growth. Planning mobility culture in an environment of dynamic stagnation', in Institute for Mobility Research (IFMO) (eds) *Megacity Mobility Culture. How Cities Move on in a Diverse World* (Lecture Notes in Mobility), Berlin/Heidelberg: Springer.

Henckel, D., Pahl-Weber, E. and Herkommer, B. (eds) (2007) *Time, Space, Places*, Frankfurt am Main/Berlin/Bern: Peter Lang.

Henckel, D. and Thomaier, S. (2013) 'Efficiency, temporal justice, and the rhythm of cities', in D. Henckel and S. Thomaier, et al. (eds) *Space-Time Design of the Public City*, Stuttgart: Springer.

Lamare, M. (2015) *Le Droit à la Ville: Justifications, Apports, Manifestations et Portées*, Préface de Fabien Girard, Paris: L'Harmattan.

Lefebvre, H. (2004/1992) *Rhythmanalysis. Space, Time and Everyday Life*, London/New York: Continuum.

Lefebvre, H. (2009/1968) *Le Droit à la Ville*, Paris: Anthropos.

Leslie, E., Saelens, B., Frank, L., Owen, N., Bauman, A., Coffee, N. and Hugo, G. (2005) 'Residents' perceptions of walkability attributes in objectively different neighbourhoods: a pilot study', *Health and Place*, 11: 227–236.

Lynch, K. (1960) *The Image of the City*, Boston, MA/London: MIT Press.

Mareggi, M. (2013) 'Urban rhythms in the contemporary city', in D. Henckel and S. Thomaier, et al. (eds) *Space-Time Design of the Public City*, Stuttgart: Springer.

Meier Dallach, H.P., Hohermuth, S. and Walter, T. (2003) 'Heimat in der Großstadt und ländlicher Kontrastraum', in H.L. Dienel, C. Schröder and H.P. Meier Dallach (eds) *Die neue Nähe. Raumpartnerschaften verbinden Kontrasträume*, Stuttgart: Steiner Verlag.

Meier Dallach, H.P. (2016) 'Von der Bau- zur Lebensphase. Die ersten Zuzüger kamen in den noch unfertigen Glattpark. Die Bewohnerinnen und Bewohner fühlen sich aber nicht im "Irgendwo"', *Hochparterre*, Beilage March, 3: 22–25.

Schiefelbusch, M. (2008) *Reiseerleben. Die Gestaltung der Fahrt als neue Aufgabe für den öffentlichen Verkehr.* Nr. 45 der Schriftreihe A des Instituts Land- und Seeverkehr, Technical University Berlin.

Stölzle, W., Weidmann, U. and Klaas-Wissing, T., et al. (2015) *Vision Mobilität Schweiz*, IVT ETH Zurich and University St. Gallen. Available HTTP: <https://www.ethz.ch/content/dam/ethz/news/Vision%20Mobilit%C3%A4t%20Schweiz%202050.pdf> (assessed 12 March 2017).

Urry, J. and Larssen, J. (2011) *The Tourist Gaze 3.0*, London: Sage Publications.

19

Movement without words

An intersection of Lefebvre and the urban practice of skateboarding

Iain Borden

Introduction

Zero-degree architecture is a field of simple signs and complex instructions, a world of dogmatic regulation exemplified in films as diverse as Jean-Luc Godard's dystopian *Alphaville* (1965) and Marco Brambilla's sci-fi *Demolition Man* (1993), and recognisable in just about every high street worldwide. Yet our modern cities are not wholly constraining, for as Lefebvre explains there is a contradiction between, on the one hand, the economic homogenisation of space (where all space tends to be treated the same), and, on the other hand, the varied uses of urban space as a whole (Lefebvre 1991a: 18–9). And it is this contradiction that skateboarding works within; 'the act of street skating', states the skateboard video *Space for Rent* (2012: np), 'is in direct conflict with the ideals of society that places its principal emphasis on economic growth and profit'. 'There are no more white lines to stay within, sidewalks to conform to or bases to tag', asserted skateboard professional Stacy Peralta as street skating began to emerge in the mid-1980s: 'It's all an open highway with hydrants, curbs, bumpers, shopping carts, door handles and pedestrians' (Peralta 1985: 40).

Skateboarding creates new patterns and meaning, and in doing so counters the logic of signs and signals described by Lefebvre in 'Notes on the new town' (Lefebvre 1995). After all, these instructions are not there for their own sake, so that when skaters confront them these skaters are necessarily critiquing their underlying logic of control, efficiency, normalcy and predictability. This does not mean, however, that skateboarding is oppositional to all of society, and indeed, in stark contrast to the kind of aggressive street demeanour and public confrontations depicted in some skate videos like *Baker 3* (2005), many skaters act with care and respect towards their fellow urban citizens. For example, Joe Penny describes how the street skaters of Clermont-Ferrand carefully replace café furniture, cease skating in order to avoid creating annoyance and pass good-humoured banter with local police officers; 'I've never had any issues with people', states skater Joseph. 'But I pay attention, to how I skate, I'm considerate of other people' (Penny 2009: 34). Similarly, the *1st & Hope* (2006) video depicts the good neighbour demeanour of Brian Lotti and others in downtown Los Angeles, their chilled urban-drift incorporating friendly encounters with numerous fellow skaters, African-American pedestrians, white low-riders and Japanese restaurateurs. In *Albion* (dirs Kevin Parrott, Morph and Ryan Gray 2014), despite a few run-ins with disapproving members of the British public, skaters shake hands after colliding with a cyclist, converse with everyone from elderly pedestrians to religious weirdos and even repair a lighting bollard. And in videos like Volcom's *All the Days Roll Into One* (dir. Ryan Thomas 2008), actual skateboarding even takes a backstage role to skaters' various friendships and street encounters. All of this strongly invokes a sense of polite citizenship and a friendly co-existence between skaters, other dwellers and their shared environment, not of antagonism or confrontation but of affable respect and gentle belonging. Here, skateboarding stands not against the city, but interweaves another rhythm within the urban symphony.

It would be wrong, therefore, to portray skaters simply as a bunch of screw-you, self-centred rebels. Nonetheless, in its general logic and operations, in its speaking through performative movement rather than through words and texts, skateboarding challenges the notion that cities are to be resolutely obeyed, that we exist solely as passive dwellers, and that urban space is closed to social negotiation and diversity. Furthermore, if skateboarding suggests the move from things to energies, from design to experiential creativity, there should also be corresponding shifts in consumption, exchange and use. It is to these areas which we now turn.

Beyond the shiny product

One major purpose of architecture is to make things and services – either commodities in factories, knowledge in universities, workers in housing, decisions in offices and so on. Skateboarding as an urban act, however, offers little such contribution, consuming buildings without engaging with their productive activity. Consequently, it implicitly denies both that architecture should be directed toward that production and, more generally, that work should always be productive or 'useful' at all. We can see this kind of attitude whenever skaters like Chip Morton say that 'life's not a job, it's an adventure' (Anon. 1989: 124), or an image caption in *Thrasher* skateboard magazine declares that 'office politics mean nothing to Jamie Thomas as he rides the glass of a San Bernardino business complex' (*Thrasher* 2012: 56), suggesting that skateboarding produces neither things nor services, but is a pleasure-driven activity of its own.

Furthermore, this (seemingly) productive-of-nothing skateboarding is disruptive to highly ordered urban space. Skateboarding rejects the economic and efficiency logic of cities, undertaking an activity which has an entirely different rationale. 'In a culture that measures progress in terms of cost per square foot', noted skateboard advocate Craig Stecyk, 'the streetstylist takes matters into his own hands. He dictates his own terms and he makes his own fun' (Smythe 1981: 55). This is particularly evident in city centres, those concentrations of *decision-making* and power, where skateboarding appears as an irrational addition. 'In a society on hold and planet on self-destruct', added Stecyk, 'the only safe recourse is an insane approach' (Smythe 1980: 29). Why would one spend so much time balancing on a piece of wood with four wheels? Why would one confront the urban citizen's conventional mode of walking-and-looking by moving up as well as along, touching as well as seeing, striking as well as keeping distant?

In opposition to such actions, one critic railed that skateboarding 'appears to serve no known purpose in life and does nothing to raise national productivity' (Anon. 1980: 74). However, this is to miss the point, for although the act of skateboarding seemingly creates no tangible 'products', it nonetheless releases energies which create or modify space, thus espousing play, art and a sense of everyday festival – what Lefebvre calls Eros or the pleasure principle (Lefebvre 1991a: 177; Lefebvre 1996: 171). So when a skater summarily states that 'when they work, we'll skate' (Catterick 1997), or speaks of having 'moved beyond shiny products and consumerism' and possessing the ability to 'rise above the repressive, hassle filled, cess pit world' and so become 'higher types' (Powell 1996a), it is clear that skateboarders' labour is directed not at the production of saleable goods or services but at play and the ludic as positive and purposeful. And in doing so, skateboarding correlates with Kane's contention that alongside a work ethic we should also have an equivalent 'play ethic', an 'imaginative "re-form" of the basic timber of social humanity' in which play is considered not just personally pleasurable but also creative, politicised, collaborative and thoughtful (Kane 2004: 13).

One contradiction here, however, is that the extraordinary architecture of the city, from which skateboarding is born and upon which it relies, is itself a product of conventional labour. In this sense skateboarding is a revival of what Marx has called the 'dead labour' of the city (Lefebvre 1991a: 348). As *Space for Rent* notes, many buildings and urban spaces utilised by skateboarding are thought to be 'useless or abandoned, having no profits being derived from them', into which skaters 'breathe new life'. This might relate to the re-use of derelict sites, as with the industrial wastelands taken up by many DIY interventions, or the Bryggeriet skatepark in a Malmö brewery. Much more prevalent though is everyday street skating, which does not wait for a building to fall in disuse or dereliction, and which produces something which, to borrow Lefebvre's words, 'is no longer a thing, nor simply a set of tools, nor simply a commodity' but which creates 'spaces for play, spaces for enjoyment, architectures of wisdom or pleasure' such that 'use value may gain the upper hand over exchange value' (Lefebvre 1991a: 348).

There is also a different treatment of time at work here. Modern cities are commonly a mixture of production and speculation, alternatively sacrificing long-term social benefits for short-term profits or short-term social needs for programmed investment schedules (Lefebvre 1991a: 335–6). Skateboarding time, by contrast, is immediate, lasting no more than a second (single move), minute (run), weeks and months (repeated visits) or few years (a skater's individual activity). Skateboarding time is also discontinuous, composed of a few minutes here and there, spread over different parts of the city, and frequently runs contrary to conventional temporal arrangements. For example, the long time of property ownership, the medium time of lease arrangements or the short time of parking meters are all avoided by street skateboarders. While economic concerns in cities 'subordinate time' and political concerns expel time as 'threatening and dangerous' (Lefebvre 1991a: 95), skateboarding promotes an appropriative recovery of time as well as of space. Skateboarding reasserts the here-ness and now-ness of architecture. In short, skateboarding is what Lefebvre would call an alternating rhythm within the regular cyclical rhythms of the city (Lefebvre 1996: 221), or what skater Steve Shaw has called 'one rhythmical expression in a multitude of rhythmical expressions' (Shaw 1990: 38).

Gifts of freedom

If skateboarding critiques production and work in cities, then it also involves a critique of exchange and consumption in the modern city, and, above all, proposes a reassertion of *use* values as opposed to exchange values.

Modern urban space frequently exists for the purposes of exchange: either as a commodity which itself can be sold, bought, leased or rented, or to facilitate the exchange of other goods, as occurs for instance in shops, markets, malls, stock markets and trading floors (Lefebvre 1991a: 306–7). Thus by the simple act of reasserting use values – using space without paying for it – skateboarding is indifferent to space being used for the purposes of exchange. As skateboard magazine *Sidewalk* put it, skaters oppose 'the real criminals, who despoil the world in their never ending quest for capital' (Anon. 1997b). As Brad Erlandson argued in skateboard magazine *Slap*, skateboarding recognises that 'the streets are owned by everyone. Streets give the gift of freedom, so enjoy your possession' (Anon. 1997a: 53).

Over the last 20 years or so, many buildings and spaces have become treated as opportunities for retail and leisure expenditure, hence fulfilling Lefebvre's warning that 'exchange value is so dominant over use and use value that it more or less suppresses it' (Lefebvre 1996: 73). But it is precisely this intense focus on exchange which skateboarding rejects; by occupying those spaces immediately external to stores and offices, skateboarders refuse to engage in such processes and instead insert new, dynamic use values. For Marc Spiegler, skateboarders then are far more than mere 'secondary users' and instead 'essentially redefine business and governmental spaces' (Garchik 1994). This kind of attitude is particularly evident in street skaters' frequent refusal to pay skatepark charges, preferring to skate elsewhere in the city. As such, skateboarding is a small fragment of that utopian conception of the city as a place of rich and divergent uses, and not just as exchange of goods, services, products and commodities. It helps fulfil Lefebvre's contention that 'urban society has a logic different from that of merchandise. It is another world. The *urban* is based on use value' (Lefebvre 1996: 131).

The way this city of use opposes the city of exchange is further emphasised if we consider that not only cities but also society itself is being ever more organised for the purposes of consumerism, and that this consumption can be of tangible things, such as products and services, or it can be of less tangible things, such as ideologies, images and signs.

In architectural terms, the consumption of signs can be found in the heightened spectacularisation of iconic architecture, whereby the appearance of architecture – rather than its usage, spatial complexity, meanings or other less visible quality – is often emphasised, thus creating an urban realm more akin to a theme park than to a lived city.

Street skating, however, has an answer. Where modern architecture is often meant to be looked at, operating as a set of advertisements which we passively receive, skateboarding focuses on the physical, material nature of architecture and finds a way for the skater's whole body to engage with it. 'There was all sorts of craziness going on around me, all over the city', described an American street skater, 'but I skimmed above it on my skateboard. Just gliding along, protected by my board' (CSTR 1995: 60). Skateboarding in this sense is a reassertion of use values, of human needs, desires and actions. As Ewan Bowman explained, 'there are only a few routes to authentic happiness left that haven't been turned into theme parks for the brain dead' and that 'thankfully, skateboarding is one those alternative routes to fulfilment' (Bowman 1997).

The tactics here involve seizing specific spaces for small periods of time, so that skateboarding is rhythmically out-of-step with the dominant patterns of the city, and in Miki Vuckovich's words, is 'inconsistent with the adapted pace and uses of our molded environment' (Vuckovich 1995). Skateboarding here agrees with Lefebvre's contention that 'appropriated space must be understood in relation to rhythms of time' (Lefebvre 1991a, 166 and 356), and, specifically, is different to the time of ownership (longer term, pseudo-permanent) by seeking an active, more mobile time (short-term, transitory). For example, Bowman explained skating amid London traffic as a mixture of speeds and emotions, with 'the fear and the adrenalin mixing as you skate from spot to spot nearly being hit by cars' and with 'a mad rush going through your body, overtaking the cars, being overtaken, going through a red light in a junction, dicing with big metal f**kers that would probably kill you' (Bowman 1997: np).

City-based street skateboarding, then, is not so much a colonisation as a series of rolling encounters, an eventful journey. It is also a critique of economic ownership, realising that true social wealth comes not from exclusive possession as private ownership but from ability to, in Lefebvre's words, 'have the most complex, the "richest" relationships of joy or happiness with the "object"' (Lefebvre 1991b: 156). It is precisely this which street skating addresses, by asserting that, in Tod Swank's words, 'just because you own it doesn't mean you're in charge of it' (Anon. 1991: 90).

So if the relation between skateboarder and city is not one of production or exchange, what is it? As journalist Paul Mulshine noted in Philadelphia, a street skater's 'primary relationships are not with his fellow man, but with the earth beneath his feet, concrete and all' (Mulshine 1987: 120). In other words, skaters relate to the city, not through possession, production or consumerist consumption, but via bodily senses, and in the form of the 'mad rush' described by Bowman above. Similarly, skaters like the Vancouver-based 'Barrier Kult' crew (2003 onwards) have expressed their dissatisfaction with some of the more spectacular, commodified or star-skater tendencies of modern skateboarding by undertaking a particular form of skateboarding. Skating masked on 'Jersey' highway barriers, this act is in part a return to the appropriative tactics of 1970s backyard pool and the powerful architectural forms of 1980s ramp skating (Nieratko 2015).

Street skating on the physical architecture of buildings and urban spaces, then, helps to meditate between skater, other people and the city, and does so in a distinctive manner, such that, according to *Thrasher*, skaters are a 'breed that exists within a steel, asphalt and concrete framework' (Anon. 1983: 7). Cities frequently seek to control the social identity of their inhabitants through boundaries, public art and other pervasive gestures, and so operate an urban version of the 'marketing orientation' which, according to Erich Fromm, encourages people to adopt a specific role in society (Fromm 1967). By contrast, skaters use their mobile appropriation of the city to construct themselves and their relations with others. 'The skater is not a separate entity from his terrain', noted Peralta, for 'he is the terrain with all its intricate pieces' (Peralta 1985: 40). Rather than allowing architecture and the city to dictate who they are, the skateboarder responds with their own question of 'who am I?' and seeks an answer through their own actions.

The meaning of skateboarding, then, comes from its engagement with the city, together with a generalised critique of society. In terms of the kind of society this might indicate, evidently skateboarders do not create fundamental change; as Emily Chivers Yochim notes, following cultural historian Leerom Medovoi, skateboarding often correlates with the Cold War notion of America as being both anti-authoritarian and democratic, a place which positively welcomes rebels and nonconformists within its over-arching condition of middle-class suburbia (Yochim 2010, 33; Medovoi 2005). 'We're not out to fight the world', declared *Thrasher* (Anon. 1992: 4).

Nor do skateboarders undertake much self-critique). On the other hand, skaters undertake an 'ironic' assault on the rest of the world and so, as Joel Patterson realised, become 'aggressive whenever the opportunity arises' and hence defiantly 'irritate giants'. Thus through highlighting the conflict between, on the one hand, the law of private property and the logic of business efficiency, and, on the other hand, wider social uses of city spaces), skaters utilise their position of relative weakness to irritate officialdom and convention, and so to interrogate the city as a whole; as Patterson concluded, 'always question authority' (Patterson 1996: 104).

Above all else, skateboarding shows that pre-existing uses of buildings and city spaces are not the only possibilities, that architecture can instead be consumed by activities which are not explicitly commodified. Buildings, architecture and urban space, we might then propose, should be thought of as places of functions and experiences, logic and love, objects and ideas – all at once. Here, architecture and cities are not things, but part of our continual appropriation of the world, life and desires, space and time. And our freedom becomes not the bourgeois right to be separated from others, but Marx's much more complete sense of developing as human beings to our greatest potential (Lefebvre 1969: 22; Lefebvre 1991b: 170–1).

Skateboarding is not a crime

Skateboarding is antagonistic towards the urban environment, even if it causes little actual damage or disruption to the urban realm. In redefining space both conceptually and physically, skateboarders strike at the heart of what everyone else understands by the city, and so can 'hammer the panic buttons of those uninterested in this pursuit of thrill and achievement' (Vuckovich 1995: 46). As Arianna Gil comments of her New York street skating, 'we're here to add a little chaos' (Remnick 2016). This is the most overt political space produced by skateboarders, a pleasure ground carved out of the city as a kind of continuous reaffirmation of one of the central Situationist and Lefebvrian slogans of 1968, that 'sous les pavés, la plage', or beneath the pavement lies the beach.

Consequently, there are inevitable consequences of this kind of critical activity. In general, from the mid-1990s onward, skateboarding has been ever-increasingly controlled through myriad localised conventions, laws and reactions, and by 2011 was being included by the US Department of Justice as one of the 'problems' of 'disorderly youth' in public places, particularly when 'recklessly' practiced (Scott 2011). As *Sidewalk* commented, 'hardly a session goes by these days without someone hurling threats of bye-laws, cops and/or fines in our faces' (Anon. 1997c). Or as a UK skater commented after arriving in London, 'I hadn't counted on being moved on by the police every minute; had not expected to encounter so many skater-hating pedestrians and had not even begun to imagine that such ignorant gorillas could be employed as security guards' (Phraeza 1997). Today, skateboarding in public spaces is legislated against everywhere from Brisbane, Manchester and Quebec to the Bronx in New York, the general effect being to embed in everyday street skaters a fear of arrest, penalties and even imprisonment.

But treating skateboarding as a crime verges on the ludicrous, and such accusations are extremely tenuous. Consider *Sidewalk*'s comparison between a skateboard that 'runs on leg power, causes chips and scratches on bits of stone and metal' and a car that 'runs on poisonous shit, pollutes the air and water, causes the death of hundreds of thousands of people', while, despite all this society generally believes that 'cars are o.k. but skateboards are evil, objects of vandalism, a dangerous menace that *must be* stopped' (Powell 1996b: np).

Clearly, skateboarding is rendered criminal through what are essentially petty-minded laws. This is largely because skateboarding is aimed at the appropriation – and not domination – of city spaces. Nonetheless, because skaters care little of ownership, they do *implicitly* oppose this principle; 'All space is public space', asserted *Sidewalk* (Anon. 1997c). Thus although skateboarding seldom stops buildings from being built or used, it does run contrary to the implicit logic (business, retail, commuting, orderly behaviour) of these urban spaces. Anti-skateboarding legislature is perhaps then less concerned with a 'crime' as finding ever new ways to validate conventional society. According to Derby's City Centre Manager, one of the main reasons for banning skateboards was so councillors would not have to see untidy people skating, and in instances like these it is clear that skateboarding shares its supposed criminality with that of graffiti which, as geographer Tim Cresswell has noted, 'lies in its being seen, in its transgression of official appearances' (Cresswell 1996: 58). Rather than any real offence, 'disorder' as 'untidiness' – what Chris Long calls the 'cognitive dissonance' between skateboarding and the social norm (Long and Jensen 2006: viii) – is being targeted here, skateboarding being one of those 'false crimes' used to help legitimise the business- and commodity-oriented city (Lefebvre 1995: 23).

The conflict between skateboarding and conventional urban practices can also be representational. Although many street-oriented skate videos depict skateboarders squaring up to irate police, security guards, shop-owners or members of the public, everyday resistance by skaters to anti-skateboarding practices rarely involves direct contestation. More common are such campaigns as 'Skateboarding Is Not a Crime', first initiated by Powell-Peralta in 1987, in which stickers were plastered on urban surfaces. Similarly, the annual 'Go Skateboarding' day is not usually aimed at Occupy- or Reclaim the Streets-style mass seizures of city spaces, but rather at a general celebration that skateboarding can and should take place anywhere. Other actions include skaters simply removing 'no skateboarding' signs (and often displaying them at home), an act which Jeff Ferrell has called 'skate spot liberation'. As Ferrell concludes, all acts are 'skirmishes in an ongoing battle to liberate public space from legal regulation' and to 'reencode the meaning of public space within the experience of skating' (Ferrell 2001: 72–3).

Ultimately, being banned from the public domain becomes simply another obstacle to be overcome. As Steven Flusty concluded in his study of skateboarding in downtown Los Angeles, 'no matter how restrictively space is programmed, no matter how many "armed response" security patrols roam the streets, and no matter how many video cameras keep watch over the plazas, there remain blindspots that await, and even invite, inhabitation by unforeseen and potent alternative practices' (Flusty 2000: 156–7). Some skaters even remove 'skatestoppers' – small metal protrusions often added to ledges, benches and other low-lying horizontal surfaces – using angle-grinders, sanders or their own skateboard trucks; around 2005, the 'Skatespot Liberation Front' variously détourned anti-skateboarding signs, hacked away skatestoppers, smoothed cracks with automotive filler, and deployed QuickCrete to fashion ad hoc transitions (Vivoni 2009: 144). The underlying defiant psychology here is expressed by Ben Powell in *Sidewalk*, 'the point is f**ck 'em all, they can't touch us now' (Powell 1996a: np). Or in more legalistic terms, according to Carr skaters are here remaking property law, seeking 'to find seams within the law that enable them to circumvent exclusionary efforts' (Carr 2010: 991).

Such actions and attitudes are, of course, not without their problematics. As Simon Orpana and others have argued, the 'hyper-performing, predominantly masculine, individualised and active body' of the street skater aligns neatly with neoliberal patriarchal structures and its focus on risk, masculinist hierarchies and the denigration of women (Orpana, 2016: 160; Atencio, Beal and Wilson 2009; Beal and Wilson 2004). David Leonard has also remarked on how white street skaters 'violate societal laws without consequences', enjoying an impunity not extended to youth of colour (Leonard 2008: 99–101). Alternatively, altercations between street skaters and other road users can occasionally lead to violence and even death; to cite but one example here, in 2015 a cab driver was on trial for deliberately running down and killing Ralph Bissonnette, a 28-year-old chef who had been longboarding along a Toronto downtown street (Blatchford 2015).

Nevertheless, while our urban public realms have become increasingly privatised, unwelcoming and even hostile to citizens who are not directly engaged in shopping, tourism, work-focused or otherwise 'legitimate' urban activities, transgressive skateboarders have sometimes gained empathy from non-skaters, many of whom dislike these worrisome alterations to city spaces. In Australia, academic Elaine Stratford has called for 'geographies of generosity' which accommodate and even encourage street skating (Stratford 2016). Sometimes these pleas have been taken up by city authorities, as when Newcastle city officials, while seeking to prevent 'bad' street skaters in shopping areas, let 'good' skaters use their boards for local transport and leisure (Nolan 2003). Planning officers and academics like Stephen Lorimer and Stephen Marshall are increasingly considering how skateboarding might contribute to local transportation (Lorimer and Marshall 2016), and the city of Montreal has legalised skateboarding on bike paths and in Peace Park, as have Grand Rapids City in Michigan and Victoria in Canada for their downtown streets (D'Alimonte 2014). In a similar mood, 88 per cent of the public opposed a mooted ban on skaters in Coventry city centre, and in Green Bay, Wisconsin, police officer Joel Zwicky patrols on a longboard in order to enhance community interaction (Gilbert 2014). Black and ethnic Bronx skaters studied by White have even noted how they are *less* likely to attract police hassle when on their skateboards, presumably as skateboarding provides a legitimate reason to be in public space; as one skater remarked, 'You're not looking to cause trouble, you're just looking to skate' (White 2015: 74).

Conclusions

Whatever the solidarity with non-skaters, for skateboarders themselves, legislation and authority are there to be resisted, for reinterpretation, misuse and subversion of such regulations and legislation are key components of what many skaters see as their core values. In this respect, skateboarders are part of a long and important process in the history of cities, a fight by the disempowered and disenfranchised for a distinctive social space of their own. In this way, skaters engage city spaces, surfaces and buildings with their own bodies and skateboards, and do so in a highly creative and positive manner – they create not only a physical movement but a movement of ideas, and a critique of the urban spaces around them.

References

1st & Hope (2006) dirs Brian Lotti and The Mallous.

Albion (2014) dirs Kevin Parrott, Morph and Ryan Gray.

dir. Ryan Thomas (2008) *All the Days Roll Into One*.

Anon. (1980) 'Off the wall', *SkateBoarder*, 6:6 74–5.

Anon. (1983) 'Editorial', *Thrasher*, 3:2 7.

Anon. (1989) 'Trash', *Thrasher*, 9:1 124.

Anon. (1991) 'Trash', *Thrasher*, 11:11 90.

Anon. (1992) 'Editorial', *Thrasher*, 12:6 4.

Anon. (1997a) 'Sacramento', *Slap*, 6:1 53.

Anon. (1997b) 'Editor's response to letter', *Sidewalk Surfer*, 14:3 np.

Anon. (1997c) 'We are illegal', *Sidewalk Surfer*, 15:4 np.

Atencio, M., Beal, B. and Wilson, C. (2009) 'The distinction of risk: urban skateboarding, street habitus and the construction of hierarchical gender relations', *Qualitative Research in Sport and Exercise*, 1:1 3–20.

Beal, B. (1996) 'Alternative masculinity and its effects on gender relations in the subculture of skateboarding', *Journal of Sport Behaviour*, 19:3 204–20.

Beal, B. and Wilson, C. (2004) '"Chicks dig scars": commercialisation and the transformations of skateboarders' identities', in B. Wheaton (ed.) *Understanding Lifestyle Sports: Consumption, Identity and Difference*, London: Routledge, pp. 31–54.

Blatchford, C. (2015) Video snippet catches terrifying, fatal encounter between skateboarder and taxi, *National Post*, 25 September. Available HTTP: <http://news.nationalpost.com/full-comment/christie-blatchford-video-snippet-catches-brief-fatal-encounter-between-skateboarder-and-taxi> (accessed 1 February 2017).

BordenI. (2018) *Skateboarding and the City*, London: Bloomsbury.

Bowman, E. (1997) 'Comment', *Sidewalk Surfer*, 13: np.

Carr, J. (2010) 'Legal geographies: skating around the edges of the law. Urban skateboarding and the role of law in determining young peoples' place in the city', *Urban Geography*, 31:7 988–1003.

Catterick, G. (1997) 'Letter', *Sidewalk Surfer*, 15:4 np.

Cresswell, T. (1996) *In Place/Out of Place: Geography, Ideology and Transgression*, Minneapolis, MN: University of Minnesota.

CSTR (1995) 'Urban blight', *Slap*, 4:9 60.

D'Alimonte, M. (2014) 'Montreal's Peace Park has signs officially allowing skateboarding', 11 August. Available HTTP: <https://www.mtlblog.com/lifestyle/montreals-peace-park-now-has-signs-officially-allowing-skateboarding> (accessed 4 March 2015).

Ferrell, J. (2001) *Tearing Down the Streets: Adventures in Crime and Anarchy*, Basingstoke: Palgrave.

Flusty, S. (2000) 'Thrashing downtown: play as resistance to the spatial and representational regulation of Los Angeles', *Cities*, 17:2 149–58.

Fromm, E. (1967) *Man for Himself*, Greenwich: Fawcett.

Garchik, L. (1994) 'The urban landscape', *San Francisco Chronicle*, issue unknown: np.

Gilbert, S. (2014) 'Online poll: 88 per cent say skateboarding should NOT be banned in Coventry city centre after "nuisance" row,' *Coventry Telegraph*, 5 November. Available HTTP: <http://www.coventrytelegraph.net/news/coventry-news/online-poll-88-say-skateboarding-8053669> (accessed 9 November 2014).

Kane, P. (2004) *The Play Ethic: A Manifesto for a Different Way of Living*, London: Macmillan.
Lefebvre, H. (1969) *The Explosion: Marxism and the French Revolution*, New York: Monthly Review.
Lefebvre, H. (1991a) *The Production of Space*, Oxford: Blackwell.
Lefebvre, H. (1991b) *Critique of Everyday Life*, vol. 1, London: Verso.
Lefebvre, H. (1995) *Introduction to Modernity: Twelve Preludes September 1959–May 1961*, London: Verso.
Lefebvre, H. (1996) *Writings on Cities*, Oxford: Blackwell.
Leonard, D. (2008) 'To the white extreme in the mainstream: manhood and white youth culture in a virtual sports world', in M.D. Giardina and M.K. Donnelly (eds) *Youth Culture and Sport*, Abingdon: Routledge, pp. 91–112.
Long, C. and Jensen, T. (eds) (2006) *No Comply: Skateboarding Speaks on Authority*, Ventura County Star: FunNotFame.
Lorimer, S.W. and Marshall, S. (2016) 'Beyond walking and cycling: scoping small-wheel modes', *Proceedings of the Institution of Civil Engineers – Engineering Sustainability*, 169:2 58–66.
Medovoi, L. (2005) *Rebels: Youth and the Cold War origins of Identity*, Durham: Duke University Press.
Mulshine, P. (1987) 'Wild in the streets', *Philadelphia Magazine*, 78:4 119–26.
Nieratko, C. (2015) 'Barrier Kult is the anonymous elite black warrior metal skate crew here to jack your shit', *Vice*, 9 June. Available HTTP: <https://www.vice.com/en_uk/article/vancouvers-barrier-kult-are-the-anonymous-elite-black-war-metal-skate-crew-here-to-jack-your-shit> (accessed 4 July 2016).
Nolan, N. (2003) 'The ins and outs of skateboarding and transgression in public space in Newcastle, Australia', *Australian Geographer*, 34:3 311–27.
Orpana, S. (2016) 'Steep transitions: spatial-temporal incorporation, Beasley skate park, and subcultural politics in the gentrifying city', in K.J. Lombard (ed.) *Skateboarding: Subcultures, Sites and Shifts*, Abingdon: Routledge, pp. 152–68.
Patterson, J. (1996) 'Redeye', *Transworld Skateboarding*, 14:1 104.
Penny, J. (2009) '"Skate and destroy"?: subculture, space and skateboarding as performance', unpublished MSc thesis, University College London, Department of Geography.
Peralta, S. (1985) 'Skate of the art 85', *Thrasher*, 5:8 40.
Phraeza (1997) 'Fiction?', *Sidewalk Surfer*, 15 np.
Powell, B. (1996a) 'Not a toy', *Sidewalk Surfer*, 3 np.
Powell, B. (1996b) 'The number one four wheeler', *Sidewalk Surfer*, 9: np.
Remnick, N. (2016) 'Sisterhood of the skateboard', *New York Times*, 29 July. Available HTTP: <https://www.nytimes.com/2016/07/31/nyregion/brujas-a-crew-of-female-skateboarders-in-the-bronx.html?_r=0> (accessed 11 August 2016).
Scott, M.C. (2011) *Disorderly Youth in Public Places*, U.S. Department of Justice: Center for Problem-Oriented Policing.
Shaw, S. (1990) 'Club method air', *Skateboard!*, 39: 38.
Smythe, J. (1980) 'The history of the world and other short subjects', *SkateBoarder*, 6:10 28–31.
Smythe, J. (1981) 'No parking', *Action Now*, 8:2 52–7.
Space for Rent (2012) dir. Jeremy Knickerbocker.
Stratford, E. (2016) 'Mobilizing a spatial politics of street skating: thinking about the geographies of generosity', *Annals of the American Association of Geographers*, 106:2 350–57.
Thrasher (2012) *Maximum Rad: The Iconic Covers of Thrasher Magazine*, New York: Universe.
Vivoni, F. (2009) 'Spots of spatial desire: skateparks, skateplazas, and urban politics', *Journal of Sport and Social Issues*, 33:2 130–49.
Vukovich, M. (1995) 'Please use the handrail', *Warp*, 4:1 46.
White, K. (2015) '"We out here": skateboarding, segregation and resistance in the Bronx', unpublished MA dissertation, Fordham University, Department of Sociology.
Yochim, E.C. (2010) *Skate Life: Re-Imagining White Masculinity*, Ann Arbor, MI: University of Michigan.

20

Visual productions of urban space

Lefebvre, the city and cinema

Nick Jones

Introduction

From the 1990s, the humanities has undergone a 'spatial turn', with ideas from human geography finding purchase within humanities disciplines like literary studies, sociology, and history (Warf and Arias 2009: 1). As a discipline studying a medium predicated upon the depiction of space, film studies has equally alighted on the potential use of spatial and geographical methodologies. As geographer Doreen Massey suggested in 1999, 'the potential for creative dialogue between people in film studies and those in geography is enormous. It has already been productive, and I think could be more so' (Massey and Lury 1999: 233). Certain ideas sourced from spatial theorists have accordingly made numerous appearances in film studies, two notable examples being Marc Augé's (1995) description of the 'non-places' of postmodernity and globalisation, and Fredric Jameson's (1991) oft-cited concept of 'cognitive mapping'. However, even in this context of increased attention to space and spatiality, the work of Henri Lefebvre has only been fitfully employed. This is surprising, not only because of the tight interrelationship of the formation of film studies as a modern university discipline with Marxist methodologies, but also because of the attention long paid in the discipline to issues of how meanings can be spatially encoded and expressed through mise-en-scène.

In this chapter, I will ask to what extent Lefebvre's work might prove useful for understanding the spaces created by moving image media, in particular those representations of urban space with which cinema is often associated. Lefebvre's spatial triad delineates between spatial practice, representations of space, and representational spaces, with the second term naming abstract, often visual conceptualisations of space (Lefebvre 1991: 38–39). Lefebvre generally associates these with drawing boards, blueprints, and blank sheets of paper, and he repeatedly proposes that they are the tools of technocrats who would impose their instrumental will on space. Although quite different to these, cinema, with its visual privilege, would seem to fit such a rubric. Yet, as Lawrence Webb (2014: 24–25) notes, cinema's representation of space 'mediate[s] between the material profilmic environment of the city, the conceptual world of architectural theory and urbanism, and the socially experienced space of the city of which it is both a "representation" and a concrete instance'. Cinema thus opens up to the complexities and ambiguities of Lefebvre's spatial triad, a triad which Lefebvre himself (1991: 40) described not as an abstract typological model but a malleable critical tool. With this in mind, I will not delineate the space of the moving image purely as a representation. Instead, I will propose that Lefebvre's work on the devaluation of space in critical theory can reveal spatial biases in methods of film analysis; that his writings on urban form can enlighten readings of the city spaces represented in cinema; and, finally, that his emphasis on visuality and neocapital urban environments has only become more pertinent as screens and moving-image media increasingly permeate and define the contemporary city.

Foregrounding cinematic space

Film may consist of shifting patterns of light and colour upon a fixed screen, but it is also emphatically a spatial art. As Erwin Panofsky (2003: 71) described in 1959, narrative cinema of the kind that has been culturally dominant since the second decade of the 20th century is both a 'dynamisation of space' and a 'spatialisation of time'. Unlike theatre, space in film is quintessentially active:

> Not only bodies move in space, but space itself does, approaching, receding, turning, dissolving, and recrystallizing as it appears through the controlled locomotion and focusing of the camera and through the cutting and editing of the various shots.... This opens up a world of possibilities of which the stage can never dream.
>
> *(Panofsky 2003: 72)*

These possibilities produce a distinctive form of space, *cinematic space*. This space overwhelmingly seeks – whether consciously or automatically – to replicate something of our experience of embodied, lived space. But, as Panofsky's words imply, it remains ontologically and perceptually distinct. In film, flat compositions, edited views, and moving camera angles become, through the attention of the spectator, something like a space we can feel we are experiencing, even though at the moment of their reception these perceived spaces are solely imagistic.

In this, cinematic space is like real space, only perhaps more so: it is *produced*, created at the point of perception by a spectator/occupant. '*(Social) space is a (social) product*', Lefebvre (1991: 26–27) asserts, and is not reducible to 'mental space (as it is defined by philosophers and mathematicians)' nor 'physical space (as defined by practico-sensory activity and the perception of 'nature')'. Space is not fixed and inert, but a product that comes into being thanks to 'networks of exchange and flows of raw material and energy' (Lefebvre 1991: 85), its social character arising from its investment with meaning by living bodies, historical actions, and the imagination (Lefebvre 1991: 40–41). This is of course not the only way that space is produced for Lefebvre – it is also manufactured by the ruling powers of the corporation and the state. Lefebvre's Marxism leads him to call for at the very least a rethinking, and at most the overthrow of these powers and their spatial code, a code of visuality and abstraction that takes space out of the hands of its everyday user (see for example Lefebvre 1991: 57).

As a medium that puts space onscreen, and which provides movement through space rather than static or semiotic descriptions of space, film is not only connected with architecture (Vidler 2000: 99) but positioned as a peculiarly urban art form. David B. Clarke proposes that films and cities are 'imbricated to such an extent that it is unthinkable that the cinema could have developed without the city', and that the city has equally 'been unmistakeably shaped by the cinematic form' (Clarke 1997: 1; see also Koeck and Roberts 2010; McQuire 2008: 59). In 1936, Walter Benjamin (1969: 250) claimed that cinema was the only art form that was sufficient for the modernising city, inuring spectators to the 'profound changes in the apperceptive apparatus' that defined the early 20th century. Since the 1990s, scholarly literature in cinema and the city has grown apace, a phenomenon Charlotte Brunsdon (2012) links to the instability of the disciplinary objects at the heart of both film and urban studies. (That is, at the onset of the 21st century, urban studies has as much trouble defining 'the city' as film studies does in defining 'cinema'). If existing accounts stress the innate link between cinema and the city, they do not in doing so claim that all cinema exclusively depicts or is solely received within the urban environment. Instead, they highlight the way in which industrialised, mechanised processes associated with urban life have rearranged social organisation, how this converges with the emergence and functioning of the cinematic apparatus, and how cinematic images create and propagate ideas of specific cities and the city in general upon a global stage. (Rather than provide an overview of this literature here, interested readers are directed to Brunsdon's [2012: 214–215] extensive timeline of key publications.)

Yet, even as cinema's relationship to the city has received considerable attention since the late 1990s, this work has been for the most part slow to take up Lefebvre's own insights regarding contemporary urbanism and, particularly, spatial production. While Lefebvre might make semi-regular appearances in books and articles on film and urban space, he often does so only as a fleeting reference, functioning as a heavy-hitting scholarly bulwark testifying to the importance of analysing space and its impact upon those living within it. (There are exceptions, and these will be discussed in the next section.) Certainly, any formalist reading of cinematic urban space that privileges the manner in which the film text is producing a distinctive form of space and spatiality – and what the political consequences of this production are – might reasonably be considered Lefebvrian, and there is a wealth of material in this vein which looks at specific films and cities. But a concerted engagement with Lefebvre is for the most part lacking.

Nonetheless, one of the most influential accounts of cinematic space has intriguing, if subtle, links with Lefebvre's work. Stephen Heath's 1976 essay 'Narrative Space' (Heath 1976) sets out to show how space is part of onscreen actions, but describes this involvement as defined by support, economy, and legibility. Film is produced 'as the realisation of a coherent and positioned space', one in which views of reality have been coded in order to tame their potential 'excess'; that is, the way reality (or illusionistic representation) always escapes the limits and movements of the frame (Heath 1976: 74). If perspectival renderings of space in painting imposed a spatial code built around the eye – an idea Heath shares with Lefebvre (1991: 273, 361) – then much the same is true of film. The movement and visual variation introduced by the cinematic apparatus into this perspectival code threatens to potentially destabilise it as a system, and is for Heath tamed in narrative filmmaking through compositional staging and continuity editing techniques. Or, to put it a slightly different way, 'narrative ensures that film's mobile frames and figures remain consumable for a viewer weaned on Renaissance perspective' (Cooper 2002: 139). Being cinematically constructed through an intentional process ('space will be difficult'), Heath (1976: 79) asserts the *work* of cinematic spatial production, and how this work is ideologically loaded and how it binds the spectator within the (imaginary, but coherent and believable) space that is cinematically produced.

Heath's essay is firmly situated within 'the semiotic-psychoanalytic tradition' of 1970s film theory and its focus on Althusserian ideological interpellation (Clarke 1997: 8). But his consideration of how space's onscreen legibility is dictated by socio-cultural currents and a politically loaded concept of 'realism' all connects his essay to Lefebvre's contemporaneous concerns. Both Heath and Lefebvre reveal the effort required to subordinate space, what is at stake in this subordination, and how this process is culturally conditioned and even implicitly capitalistic. Lefebvre (2009: 170) describes how if space 'has an air of neutrality and indifference with regard to its contents and thus seems to be "purely" formal, the essence of rational abstraction, it is precisely because this space has already been occupied and planned, already been the focus of past strategies, of which we cannot always find traces'. This is equally applicable to filmed space, in which the 'strategies' of coherence, a kind of 'background-ing', and an orientation around character and narrative *support* are overwhelming but unquestioned. Film grammar, as it is employed in a mainstream context, is about privileging the movements and experiences of onscreen individuals, and space is used to serve this goal, usually through its marginalisation as an agential force (often being, quite literally, out of focus and in the background).

Indeed, in formal accounts of cinematic space – such as those by Richard Maltby (2003) and David Bordwell (1985; 2006) – its function as safe, legible background or support is accentuated to the extent that any alternative reading of cinematic space is effectively closed down. Maltby (2003: 354) may intriguingly point to *Psycho*'s (1960) shower scene as inaugurating the possibility of 'unsafe space' in mainstream cinema – space in which 'our comforting ability to predict what will happen in a space or a story can be arbitrarily violated' – but the nightmarishness, the psychic and visceral threat of such space comes from the very fact that it denies (but only at key, narratively motivated moments) the classical model of space built on coherence and subordination. Anthony Vidler (2000: 100–107) describes how film and built space had a problematic relationship, since cinema's potential for spatial experimentation and warped and distorted spaces (seen in German Expressionism of the 1920s) denied modernist narratives of clarity, poise, and functionalism. But those moments when space seems to occupy the foreground, is placed on the agenda as an unstable, formative agent in cinematic form, are always exceptions. Indeed, many scholars and critics call attention to the importance of space in any given film text by suggesting that it functions *like a character in itself*. Meant as a way of lauding space and of paying space its due, this compliment only elevates spatial concerns by endowing them with the mantle of the psychological or the narrational. Suggesting that space can only be viably studied by bestowing upon it the traits of an onscreen individual is problematic and dichotomises cinematic space into that which is character-like (and thus worthy of study) and that which is apparently just background (and which can therefore be safely considered a neutral support). Space is important even when it does fade into the background: when it is not encouraging us to see it as character in itself, it is still influencing characters, textual form, and spectatorship in crucial ways.

Lefebvre and cinematic urbanism

In listing problems that might be encountered when applying Lefebvre's theories to contemporary urban and political life, Neil Brenner and Stuart Elden ask how successfully his work might 'travel' into new geographical and temporal situations (Brenner and Elden 2009: 32). Similarly, questions may be asked around the usefulness of Lefebvre for analysing moving image media. As Webb (2014: 242) points out, Lefebvre had little to say about cinema directly, which as a result might be thought of as one of his own 'blind fields' (see Lefebvre 2003: 29–31). The application of Lefebvre's work to film is then made especially problematic thanks to his words on images and media culture, which are rarely celebratory.

Discussing the image as part of a broader semantic field in 1961's *Critique of Everyday Life* vol. 2, Lefebvre (2002: 289) credits it with expressivity and emotionality, and even proposes that 'the deepest communication of all is achieved through images'. This is a rare moment of openness towards the visual, an openness which is rarely in evidence. In *The Production of Space* in particular, Lefebvre offers a savage condemnation of images. Connecting images of any kind with the modernist 'trinity of readability-visibility-intelligibility', he attacks them as fundamental to a conceptual framework that may be normalised but which is highly deceptive (Lefebvre 1991: 96). Denying the body, Western culture's establishing of the faculty of sight as the primary sense

– to the extent that that which is true is synonymous with that which is or can be seen – is the root of abstract space and the political problems of urban space and social praxis in the late 20th century. For Lefebvre, images can never reveal, only reinforce:

> photographs, advertisements, films. Can images of this kind really be expected to expose errors concerning space? Hardly. Where there is error or illusion, the image is more likely to secrete it and reinforce it than to reveal it. No matter how 'beautiful' they may be, such images belong to an incriminated medium.
>
> *(Lefebvre 1991: 96–97)*

Images, in particular those of cinema and television, divert attention from real problems, and their emphasis on visibility encourages passivity in their audiences (Lefebvre 1987: 11), infantilising them and encouraging 'non-participation in a false presence' as a way of life (Lefebvre 2002: 223–224; see also Jones 2015: 66). Indeed, not only do images deceive, we may find the idea of imagistic representation (and abstraction) wherever we find illusion, all of which prompts Lefebvre (1991: 97) to proclaim that 'the image kills'.

However, for all this condemnation, there is hope: sometimes an artist's 'tenderness or cruelty' transgresses the limitations of the image, and, as a result, 'a truth and a reality' might thus emerge which avoid the image's normal emphasis upon 'exactitude, clarity, readability and plasticity' (Lefebvre 1991: 97). A brief footnote in this passage from *The Production of Space* points to photographer Henri Cartier-Bresson's work as an example of that which is able to enact such an overcoming. Just as Merrifield (2006: 26) suggests that Lefebvre 'could never comprehend modern capitalism as seamless', so too this minor admission indicates that the image may be capable of something other than deception and abstraction. Deploring the 'monopoly on intelligibility' granted written texts and semiotic coding, Lefebvre (1991: 62) identifies a series of non-verbal systems which are more ambiguous, including 'music, painting, sculpture, architecture', and even theatre; if, tellingly, film does not feature on this list, we might nonetheless suggest that it is not so different or so separate from those arts that do appear.

Accordingly, some film studies work has sought to explicitly use Lefebvre's ideas to discover how cinema might show the spatial strategies of neocapital through its visualisation of urban contestation. Perhaps the most widely known example of this is Edward Dimendberg's *Film Noir and the Spaces of Modernity*, in which Lefebvre's work on abstraction and the multidimensionality of space is used to reconsider the spatial strategies at work in films noir of the 1940s and 1950s. Dimendberg proposes that these films show 'far more of the city than is necessary to commence their narratives' (2004: 89), and that skylines and urban landmarks feature 'as simultaneously background and active agent' (2004: 92), words which bely the difficulty in getting away from the language of narrative surplus (and the dichotomising of narrative/activity on the one hand and background on the other). Approaching film noir as a 'process and product' of cultural negotiations and restructurings, Dimendberg (2004: 12) suggests that this genre or cycle of films responded to the rise in the US of 'abstract space', and acutely shows the consequences of the 'centripetal' changes that reshaped urban centres in the early 20th century into regimented, corporatised spaces of accumulation, and also reshaped beliefs around who the city belonged to and what it was for. Noir characters negotiate constant police surveillance and highly securitised urban streets, living in an environment of 'functionalized separation' in which 'urban forms connected to the life of a culture and charged with collective meaning' have been lost (Dimendberg 2004: 102). If the city in film noir was a place of threat, then this was not simply due to its lawlessness: the anxiety of noir urbanism arises not only from threats to personal safety, but also from what Dimendberg (2004: 91) calls 'the psychic hazards of dwelling in an urban space whose historical mutation yields real spatial gaps and temporal voids'. The spatial changes wrought by Lefebvre's bureaucratic society of controlled consumption are therefore felt cinematically in noir filmmaking.

Dimendberg's reading of this body of films, then, is Lefebvrian in as much as it highlights how space is influencing generic and narrative procedures as well as the actions of onscreen characters. It is not unusual for Lefebvre's (admittedly still expansive) commentary on specific spatial restructurings to be employed alongside his more theory-led critique of space in this way. Webb (2014: 235, 242–244), for instance, applies Lefebvre's work on the city in the 1970s to the way Paris was being represented cinematically by French New Wave directors like Jean-Luc Godard. To explore the intersection of cinema with processes of urban modernisation and transnational branding, Webb uses both Lefebvre's description of the historical re-shaping of urban space (itself inspired by Paris) and his more overarching and philosophical claims regarding the multidimensional and political nature of space. In my own work, I have likewise applied these two strands of Lefebvre's thinking, in my case to contemporary cinema. Action cinema, I have argued, depicts physical movements in urban space that both rely on and subvert spatial strategies of top-down control, as protagonists attend to their surroundings in concerted and inventive ways, exploring the nuances, weak points, and latent possibilities of contemporary urban space (Jones 2015: 55–66). Meanwhile, recent mainstream cinema that shows the manufacturing of false or duplicitous spaces (Jones 2016), or which employs extensive digital imagery to generate environments (Jones 2013), can likewise be fruitfully read alongside Lefebvre's ideas around strategies of spatial abstraction and their consequences (despite the historical and conceptual distance of these films from his writing).

In the same vein, Katherine Shonfield considers Lefebvre's work invaluable for indicating how cinema can reveal the normally disguised ideological impositions of contemporary urban space. In *Walls Have Feelings*, Shonfield (2000: 160) argues that fictions such as films should not be accepted as 'at best interesting, but subordinate, parallel commentaries to mainstream architectural history'. Seemingly inspired by Lefebvre's intense distrust of knowledge specialisation and the hallowed expertise of the technocrat, she puts films forward as 'non-expert' spatial interpretations that can 'make bold connections between apparently disparate circumstances' in a way denied more technical accounts of urban space (Shonfield 2000: 161–162). Indeed, her claims here echo Lefebvre's (2005: 28) historically-situated assertions that from the 1960s onwards the 'people' of French society, and Western society by extension, somehow sought a kind of 'counter-knowledge' that might amount to a 'counter-power'. For Shonfield, both films and their audiences evince a kind of spatial knowledge that is unappreciated and politically untapped. Inspiringly, she closes the book with a plea that more – and more serious – attention must be paid to cinematic fictions and their spatial representations, since moving this 'fictional insight to centre stage' will show the public their stake in 'discussions of the city's future' (Shonfield 2000: 173).

In all these cases, then, Lefebvre's descriptions of shifting spatial imaginations have substantiated claims around the onscreen urban space of a given set of films; but his conceptual framework for interpreting space and its constitution have also proven crucial in thinking about the production of (cinematic) space more generally. Films offer insight not only into historical urban formations, but also the way these formations shape action and different forms of spatial consciousness (see Jones 2015: 148). Moreover, such depictions might go beyond the representable and become habitable. This is the argument made by Giuliana Bruno, who contemplates film in relation to architecture, travel, memory, and mapmaking, and claims that films are part of a 'habitable, spatiovisual configuration' (Bruno 2002: 2). Lefebvre thus informs what she calls her 'cartographic reading of haptic space' (Bruno 2002: 255), a reading of cinema open to our subjective experience of living somehow within a film and experiencing the extended world of global tourism through this moving-image habitat and its psychogeography. She describes how Italian neorealist cinema – which shot on location in the war-torn urban spaces of Europe in the late 1940s and early 1950s – offers city walks in which we witness the interplay between city and body, urban space and social existence, and does so in ways that call to mind Lefebvre's thinking on the interplay of environment and subjectivity (Bruno 2002: 30). Like architecture, film is 'built as it is constantly negotiated by (e)motions, traversed by the histories both of its inhabitants and its transient dwellers' (Bruno 2002: 66). Referring to cinema as 'a vehicle for psycho-spatial journeys' (Bruno 1997: 23), her reading is explicitly Lefebvrian in its attention to the body and the involvement of the body and the embodied subject in perceiving space, and how this happens even in the seemingly immobile situation of viewing screened media.

Bruno (1997: 22–23) sees the 'mobile urban viewpoint' of film as a kind of in-between, transitional space, and argues that cinema and its psychogeographical journeys are 'a mobile map of differences, a production of socio-sexual fragments and cross-cultural travel'. This use of Lefebvre to highlight cinema's progressive potential – echoed in Shonfield's writing (2004) as well as my own – might be considered a sort of counter-reading of his work. For Lefebvre (1991: 33, 41), representations of space, being as they are 'tied to the relations of production and to the "order" which those relations impose', play their own social and political part in replicating and enforcing those relations. As Heath's writing on cinematic space indicates, narrative filmmaking's articulation of space certainly seems to function in this manner. However, Lefebvre's work, when applied to the images and media apparatus he often harshly condemned, can also helpfully *reveal* the same kinds of social shifts and spatial ideologies he sought to understand. Moreover, in many ways the spaces depicted by cinema cannot be described purely or even primarily as 'representations of space'. Many film scholars unsettle or outright reject such a typological distinction (see Dimendberg 2004; Jones 2015; Webb 2014), and films – even if they are in many cases commercial commodities – are not normally the products of the 'scientists, planners, urbanists, technocratic subdividers and social engineers' that Lefebvre (1991: 38) associates with such representations. Films are rather urban explorations that map the spaces they depict in ways that open them up to our inhabitations; these mappings may or may not be accurate in any strict sense, but they do reflect spatial ideologies and possibilities in insightful, useful, too-little-explored ways.

City of screens

In stressing the importance of the visual in the contemporary urban landscape, Lefebvre implicitly joins the images of television and the globalised media apparatus with this landscape. Both cinema and abstract space reduce 'three-dimensional realities to two dimensions' and seem to offer space 'no social existence independently of an intense, aggressive and responsive visualization' (Lefebvre 1991: 285–286). If the contemporary urban form functions according to visual dictates, then this only becomes more overt as the city itself becomes reliant upon mediation and screens to provide public information, social spectacle, and urban branding (Berry, Harbord, and Moore 2013; Pratt and San Juan 2012; Verhoeff 2016). Images form an increasingly intrinsic part of what Nigel Thrift (2004) calls the 'spatial politics of affect' in today's cities. They assist in neocapitalism's 24/7 functioning and its relentless soliciting and management of attention (Crary 2013). And the way these images are globally constructed, aesthetically engineered, and geographically situated all reveal the workings of the urban (Rose, Degen, and Melhuish 2016). As Scott McQuire summarises, in today's 'media city', digital moving image technologies take a constitutive part in 'the dynamic production of contemporary urban space, in Lefebvre's sense of binding affect and cognition to space' (McQuire 2008: vii).

So, the contemporary city is not only 'cinematic' insofar as it utilises visual logics to implicitly place space beyond the control of its inhabitants, but also for the (perhaps more obvious) reason that it is inundated with screens. And if the image functions in the powerful fashion that Lefebvre claims, then it is vital to undertake work deciphering the content of those images that circulate not only of but also *in* the city in an era of urban screens, camera-phones, and periscope. Lefebvre (1991: 188–189) suggests that representations of space confuse investigations into the foundations of social space because 'they offer an already clarified picture', and so he dictates that they must be 'dispelled'. But the images encountered by the city-dweller are a kind of spatial practice in and of themselves. If the individual in the city street was for Lefebvre (2003: 18) both 'spectacle and spectator', then these spectatorial relations are now often mediated and understood through screens, and these screens accordingly reshape urban space, becoming part of the urban semiology which Lefebvre (1996: 108) suggests must be read in any process that seeks to decipher the deeper, less visible, 'unconscious' aspects of the contemporary city.

If Lefebvre's work points most overtly to how these screens might maintain a kind of spatial status quo of visualisation and organised passivity, then we must note that they also have the capacity to intervene in our experience of the urban (Verhoeff 2016). We can thus propose that the right to the city as it might manifest today should take in not only the occupation and use of urban space, but also the right to disseminate images of this pedestrian activity, and should influence the kinds of images that circulate within urban spaces and what their content might be. In an era not only of digital surveillance but also of sousveillance – the recording and dissemination of filmed material on consumer platforms – the circulation of moving images of the urban environment can function not only to control and indoctrinate but also to stimulate debate and engender political activity. Images are not only observed, they are used, *enacted*, in ways that escape Lefebvre's negative considerations of them, and that include them within more radical and interventionist spatial politics.

Conclusion: the Lefebvrian cinematic city

Lefebvre may think of images as a lifeless medium and consider them to privilege visuality at the expense of the body in order to maintain an ideology of abstract space, but, despite this, films offer important interpretations of urban space that must be understood in order to account for the workings of the contemporary city. Film provides not only a vital record of how urban space is conceived at a given time – as in film noir's tracking of the centripetal changes to the city – it also manifests space itself in ways that reveal underlying spatial ideologies – as in mainstream narrative cinema's focus on legibility, coherence, and the individualised subject. Moreover, thinking about the Lefebvrian cinematic city as something that is not just confined to a screen (working in a unidirectional fashion, and solely representational) but rather as something embedded within the urban environment, bi-directional and processual, and as operating across all terms of Lefebvre's spatial triad, allows us to update his urban theories for present conditions. Images, then, have a more complicated relationship to the urban than merely providing a kind of abstract model for planners, or an example of the disembodied forms of experience and knowledge sought by neocapitalist states. Moving image media can highlight spatial strategies and even counter-strategies; offer representational spaces of meaning that are somehow psychologically inhabitable; and even re-shape the affective and experiential economy of the city itself. As such, it is clear that film studies can learn much from Lefebvre's writings, and vice versa.

References

Augé, M. (1995) *Non-Places: An Anthropology of Supermodernity*, trans. J. Howe, London: Verso.
Benjamin, W. (1969) 'The work of art in the age of mechanical reproduction', trans. H. Zorn, in H. Arendt (ed.) *Illuminations*, New York: Shocken.
Berry, C., Harbord, J. and Moore, R. (eds) (2013) *Public Space, Media Space*, Basingstoke: Palgrave Macmillan.
Bordwell, D. (1985) 'The classical Hollywood style 1917–60', in D. Bordwell, J. Staiger and K. Thompson (eds) *The Classical Hollywood Cinema: Film Style and Mode of Production to 1960*, London: Routledge.
Bordwell, D. (2006) *The Way Hollywood Tells It: Story and Style in Modern Movies*, Oakland, CA: University of California Press.
Brenner, N. and Elden, S. (2009) 'Introduction', in N. Brenner and S. Elden (eds) *State, Space, World: Selected Essays*, Minneapolis, MN/ London: University of Minnesota Press.
Bruno, G. (2002) *Atlas of Emotion: Journeys in Art, Architecture, and Film*, London/New York: Verso.
Bruno, G. (1997) 'Site-seeing: architecture and the moving image', *Wide Angle*, 19:4 8–24.
Brunsdon, C. (2012) 'The attractions of the cinematic city', *Screen*, 53:3 209–27.
Clarke, D.B. (1997) 'Introduction: previewing the cinematic city', in D.B. Clarke (ed.) *The Cinematic City*, London/New York: Routledge.
Cooper, M.G. (2002) 'Narrative spaces', *Screen*, 43:2 139–57.
Crary, J. (2013) *24/7: Late Capitalism and the Ends of Sleep*, London/New York: Verso.
Dimendberg, E. (2004) *Film Noir and the Spaces of Modernity*, Cambridge, MA/London: Harvard University Press.
Heath, S. (1976) 'Narrative space', *Screen*, 17:3 68–112.
Jameson, F. (1991) *Postmodernism, or, the Cultural Logic of Late Capitalism*, London: Verso.
Jones, N. (2013) 'Quantification and substitution: the abstract space of virtual cinematography', *Animation: An Interdisciplinary Journal*, 8:3 253–66.
Jones, N. (2015) *Hollywood Action Films and Spatial Theory*, London/New York: Routledge.
Jones, N. (2016) 'This is my world: spatial representation in the Resident Evil films', *Continuum: Journal of Media and Cultural Studies*, 45:1 477–88.
Koeck, R. and Roberts, L. (eds) (2010) *The City and the Moving Image: Urban Projections*, Basingstoke: Palgrave Macmillan.
Lefebvre, H. (1987) 'The everyday and everydayness', trans. C. Levich, *Yale French Studies*, 73: 7–11.
Lefebvre, H. (1991) *The Production of Space*, trans. D. Nicholson-Smith, Oxford: Blackwell.
Lefebvre, H. (1996) *Writings on Cities*, trans. E. Kofman and E. Lebas, Malden, MA: Blackwell.
Lefebvre, H. (2002) *Critique of Everyday Life, Volume II: Foundations for a Sociology of the Everyday*, trans. J. Moore, London/New York: Verso.
Lefebvre, H. (2003) *The Urban Revolution*, trans. R Bononno, Minneapolis, MN: University of Minnesota Press.
Lefebvre, H. (2005) *Critique of Everyday Life, Volume III: From Modernity to Modernism (Towards a Metaphilosophy of Daily Life)*, trans. G. Elliott, London/New York: Verso.
Lefebvre, H. (2009) 'Reflections on the politics of space', trans. N. Brenner and S. Elden, in N. Brenner and S. Elden (eds) *State, Space, World: Selected Essays*, London/Minneapolis, MN: University of Minnesota Press.
Maltby, R. (2003) *Hollywood Cinema*, 2nd edn, Malden, MA: Blackwell.
Massey, D. and Lury, K. (1999) 'Making connections', *Screen*, 40:3 239–38.
McQuire, S. (2008) *The Media City: Media, Architecture and Urban Space*, London: Sage.
Merrifield, A. (2006) *Henri Lefebvre: A Critical Introduction*, London/New York: Routledge.
Panofsky, E. (2003) 'Style and medium in the motion pictures', in A. Dalle Vacche (ed.) *The Visual Turn: Classical Film Theory and Art History*, Piscataway, NJ: Rutgers University Press.
Pratt, G. and San Juan, R.M. (2012) *Cinema and Urban Space: Critical Possibilities*, Edinburgh: Edinburgh University Press.
Rose, G., Degen, M. and Melhuish, C. (2016) 'Looking at digital visualizations of urban redevelopment projects: dimming the scintillating glow of unwork', in S. Jordan and C. Lindner (eds) *Cities Interrupted: Visual Culture and Urban Space*, London/New York: Bloomsbury.

Shonfield, K. (2000) *Walls Have Feelings: Architecture, Film and the City*, London: Routledge.
Thrift, N. (2004) 'Intensities of feeling: towards a spatial politics of affect', *Geografiska Annaler*, 86B:1 57–78.
Verhoeff, N. (2016) 'Screens in the city', in D. Chateau and J. Moure (eds) *Screens: From Materiality to Spectatorship – A Historical and Theoretical Reassessment*, Amsterdam: Amsterdam University Press.
Vidler, A. (2000) *Warped Space: Art, Architecture, and Anxiety in Modern Culture*, Cambridge, MA/London: MIT Press.
Warf, B. and Arias, S. (2009) 'Introduction', in B. Warf and S. Arias (eds) *The Spatial Turn: Interdisciplinary Perspectives*, London/New York: Routledge.
Webb, L. (2014) *The Cinema of Urban Crisis: Seventies Film and the Reinvention of the City*, Amsterdam: Amsterdam University Press.

21

Dominated and appropriated knowledge workspaces

A tale of two cases

Ian Ellison

Introduction

The ubiquity of the office as an organisational location for many workers in myriad 'knowledge-based industries' (Greene and Myerson 2011: 20) to undertake some or all of their work is without question. What seems also beyond question is that organisational space matters to both organisations and to their people. It matters in a wide range of different functional and symbolic ways, to the extent that its provision is now a specific management area. Since the early 1980s (Price 2003) facilities management (FM) has become the organisational function typically responsible for overseeing organisational workspace in use: 'ultimately, the practice of FM is concerned with the delivery of the enabling workplace environment – the optimum functional space that supports the business processes and human resources' (Then 1999: 469). From FM's unique organisational perspective, the extent to which organisational space matters is often experienced viscerally first-hand, particularly during workspace change initiatives.

Consequently, the provision of office workspace as a dominant type of urban corporate real estate has become a lucrative business. A complex built environment industry ecosystem including developers, architects, designers, engineers, consultants, surveyors, project managers, construction, product and service suppliers, FM specialists and many more has evolved to provide organisational workspace for clients. Parallel academic disciplines fuel knowledge production and transfer. Yet despite this elaborate and established framework of 'expertise', for many, the working environment remains objectively and/or subjectively mediocre. Worse still, it can have various negative effects on working lives and performance (Baldry 1999). Particular approaches to workspace provision like 'open-plan' and 'hot-desking' have become terms which often trigger perpetual dismay.

Systemic design solutions that promote spatial allocation by task rather than as a default user entitlement, with names including 'the non-territorial office', 'activity-based working', 'agile working', 'nomadic working' and the ever-vague 'new ways of working' are claimed as contemporary, despite a heritage now approaching 50 years old (van Meel 2011). But after literally decades of research, few organisations seem able to claim physical working environments that have a genuinely positive impact on both their staff and their organisational outcomes, despite the wealth of claimed expertise. Significantly, whilst objective performance measures of built environment utility, economy and efficiency might be achievable, accepted measures of effectiveness are far harder to achieve, often relying on subjective reporting of perceived impact and performance.

What lurks beneath many such debates is a lack of ontological and epistemological awareness, compounded by media-fuelled populist and often naively empirical accounts of science (Blaikie 2007). By reaching beyond the typical literature and research of the workspace industry to reconsider the socio-spatial, the bounded nature of such expertise becomes clear, alongside the political and economic ramifications of perpetuating such received wisdom. Essentially, the industry responsible for the provision of organisational space shows little evidence nor desire for a 'spatial turn', whereupon a discipline embraces the reciprocal and inherently political socio-spatial interplay. Lefebvre is widely associated as a seminal figure here (Dale and Burrell 2008), yet scant few in the workspace industry are aware of his contribution. Is this paradoxical, even alarming, given its core purpose of providing organisational space for users? This context became the departure point for the research that underpins this chapter, which sought to step clear of received industry wisdom and explore 'what matters about workspace?' to both providers and users (Dale and Burrell 2008) of knowledge workspace in two radically different contemporaneous case-settings. The study then considered the qualitative data analytically, through the holistic lens of Lefebvre's conceptual apparatus (Boano 2015), his doubly-designated spatial triad. The remainder of this chapter traces the literature, methodology, findings and implications of the study.

A case for Henri

Despite, as Crang and Thrift (2000: 1) contemplate, 'space being the everywhere of modern thought', different disciplines demonstrate differing spatial awareness. Built environment disciplines evidence decades of research concerning different workspaces, particularly the office, and its interrelationship with workers. To some degree clerical offices have existed for centuries, but for many the history of the organisational office commences substantively in the late 19th century. With the industrial revolution and the birth of the corporation, significant administrative functions became necessary to support organisational processes (Saval 2014). From here, different organisational ideologies and corresponding workspace design propositions evidence an ever-present socio-spatial interrelationship. Limited space prevents deeper exploration of this historical line of enquiry. One key observation though is that this wealth of literature, past and present, remains virtually divorced from far rarer socio-spatial consideration within mainstream organisational theory (Elsbach and Pratt 2007). A range of authors from different fields cite Mayo's 1920s/30s Hawthorne studies as a seminal reason for this. Here, revelations about 'human relations... eclipsed the physical environment' (Sundstrom and Sundstrom 1986: 47) and henceforth relegated it to a subordinate role, reinforcing 'an implicit assumption that the physical work environment can be disregarded in any analysis of work organisation' (Baldry 1999: 535). This position was likely reinforced by the likes of Maslow's (1943) and Herzberg's (1959) 'universalist' psychological theories.

While we ought to remain sceptical of such decisive and linear origin stories, the under-representation of spatially aware research in organisational theory resulted in calls from more sociologically inclined theorists like Halford (2008) and Kornberger and Clegg (2004) to bring space back into organisational studies. Consequently, Taylor and Spicer (2007) evidence a nascent field rich with the potential to, as Halford (2008: 393) puts it, 'stop seeing the spatialities of work and organisation as only supporting actors... [and] start taking space seriously as a starting point in its own right'.

Organisation theorists Dale and Burrell (2008) were amongst the first to respond substantively, drawing critical attention to FM and the managerial interests of workspace provision by the second paragraph of their book's preface! What at first seem like sociological outliers on the periphery of an established built environment intra-disciplinary knowledge base unfold to reveal a far broader, more philosophically informed and more interconnected socio-spatial constellation. Metaphorically, we might consider whether, until the built environment industry undergoes a spatial turn, its conventional/dominant wisdom is on the primitive side of a Copernican revolution. Gazing beyond this myopic field, one intellectual figure seems omnipresent: Henri Lefebvre. This chapter need not revisit ground covered elsewhere in this volume, but it is important to locate the study. As Merrifield (2006) notes, despite Lefebvre's spatial expositions during the 1960s leading to *The Production of Space* in 1974, the political implications of Castell's critique meant it didn't receive an English translation until 1991. Consequently, apart from early exceptions including Marxist geographer Harvey, it was its post-translation rediscovery by postmodern and cultural geographers including Soja and Massey that presented a seminal platform from which the spatial turn rippled through sociologically inclined disciplines.

In some ways, it seems so befitting. This is a theory which foregrounds the *production* of space, rather than space per se, for the industry responsible for the producing! Taking Lefebvre's oft-quoted line with its 'just right' brackets (Till 2009: 125), '(social) space is a (social) product' (Lefebvre 1991: 26), such a short, deceptively simple sentence represents hotly contested philosophical ground. To appreciate this requires an awareness of both Cartesian (space as a container for the social) and Kantian (space as socially constructed) perspectives. The existent locus of managerial and broadly positivistic workspace knowledge (Cairns 2008) underpinned by Cartesian duality (Kornberger and Clegg 2004) within those who provide organisational workspace promotes the notion that there are right (and wrong) workspace decisions in relation to intended outcomes.

Yet the broader interpretive, critical, trans-disciplinary awareness underpinned by the Kantian notion of socio-spatial dualism where 'the spatial characteristics of buildings... are both medium and outcome of actions they recursively organize' (Rosen et al. 1990: 71) implies that the socio-spatial relationship, or spatiality, is far from objective. Put simply, there are socio-spatial considerations beyond Cartesian notions of space as 'distance', which may include socially constructed elements relating space to 'materialised power relations' and 'experience' (Taylor and Spicer 2007: 327). For Till (2009: 126), this 'banishes any notion that space could be treated as an abstract matter... [and] once and for all scotches the myth that space is produced by a single person... space is "produced" by a complex set of overlapping societal agencies'. Lefebvre (1991) further unfolds a subtlety to such agency, acknowledging that we all produce space, whether 'experts' with authority displaying *savoir* knowledge, or 'inexpert others' with *connaissance* knowledge. By shifting attention from material workspace per se to our interaction with and within it and how we all produce space in different, mutually constitutive, interacting ways, we can begin to apprehend not just the rich diversity of experience, but the ongoing, unfolding *quotidienne* of spatiality and the power dynamics concealed within. The spatialities of provider intent and user enactment become entangled perspectives. Accordingly, Till (2009: 126) counsels architects as providers: 'remember that you too use buildings, occupy space... users, you included, are more than abstractions or ideals; they are imperfect, multiple, political, and all the better for it'.

What unites all perspectives though is the implicit acceptance that workspace provision and design decisions are intertwined with their impact. Elsbach and Pratt (2007: 181) eloquently invoke the notion of trade-offs and tensions, explaining from extensive review that no workspace solutions 'are exclusively associated with desired outcomes'. Further, they conclude that the physical environment serves 'aesthetic, instrumental and symbolic functions'. Consequently, workspace provision can be considered at a strategic level in terms of what it intends to achieve, as permutations of design decisions for change, control, commodity and community (after Cairns 2002 and Halford 2008). Embracing the totality of spatial production *and* appreciating that typical industry narratives foreground certain design elements, whereas more sociological commentators might foreground others, offers an opportunity to consider the interplay holistically through a philosophically ambivalent approach (Cairns 2008). In the following sections, Lefebvre's spatial triad is deployed analytically to this effect.

Locating Lefebvre

Following Cairns (2008) and answering Airo and Nenonen's (2014) call for more interpretive FM workspace research, the study was intentionally subjectivist, acknowledging knowledge, truth and validity of workspace issues as socially constructed. Some might associate Lefebvre more closely with Critical Theory, beyond interpretivism, given his ideological Marxist associations. However, there were three reasons for the interpretive positioning of the study. First, by intentionally attempting to cast aside received industry wisdom and reflexively stretch beyond, the research focus was a deliberately open question, 'what matters to providers and users about workspace?' to potentially challenge conventions, including my own. Second, the novel research technique utilised afforded an emergent and entirely subjective opportunity. Third, and arguably most importantly, in attempting to encourage a broadly positivistic yet naively empirical industry to consider alternative spatial perspectives, interpretivism had more tactical potential than Critical Theory. This need not be problematic. From a Giddensian perspective, sociology can be regarded 'as an inherently critical discipline in its capacity to undermine ideology and the capacity of dominant groups... in contrast to Critical Theory, Giddens's form of critique is incidental to, rather than an integral part of, his scheme' (Blaikie 2007: 162–3).

Accordingly, two organisational case-settings were accessed in London, UK: 'PropCo', a 'top twenty' built environment professional services firm that recently relocated into a contemporary office workspace, designed and built in consultation with 'workplace specialists' 'D&BCo' to industry 'agile working' principles; 'ShareCo', a brand-franchised co-working environment for ethical start-up businesses, designed and initially managed by architecture practice 'ArcCo', according to their own principles of 'placemaking for innovation'. Both case-settings exist as single-floor leasehold arrangements within larger multi-tenant landlord-managed premises, accessed through communal building entrances. Seventeen participants joined the study voluntarily. Provider roles included an architect, researcher, designers, consultants, workspace hosts and a chief operating officer. User roles ranged widely in occupation, although given Till's point above that all providers also use workspace, this distinction is somewhat arbitrary.

Participants were asked 'what matters to you about workspace?' and encouraged to use their own camera-phones to gather a small portfolio of images in response. This technique, called participant-led photography, is novel for built environment research and can be 'deployed with the aim of exposing the ways in which social positions and relations are both produced by, and produce, distinct [spatial] experiences' (Rose 2012: 299). The images were the catalyst for emergent qualitative interviews where empowered 'participants [explained] why they took the pictures and the meaning and significance they hold' (Vince and Warren 2012: 281). Interviews were regarded as co-constructions rich with value-laden *inter-views* (Kvale 1996) reflexively acknowledging 'the interview is a negotiated text... a conversation... not a neutral tool... [which] produces situated understandings' (Denzin and Lincoln 2000: 633). Following intelligent verbatim transcription, deep immersion in the audio recordings, images and transcripts afforded considered construction of participant vignettes (Tracy 2013). These were offered for participant consideration. Over half reengaged with the study at this stage offering further input from endorsement to additional information. The vignettes became the basis for a reflexive constant comparative manual coding approach (ibid.). As a visual sense-making framework, a matrix organised the resultant analytic codes according to case-setting and participant type similarities and differences. This emic, emergent process became the basis for an etic counterpoint, the reconsideration of the codes utilising Lefebvre's spatial triad as an analytic tool. Ultimately, sense-making was through iterative analysis (ibid.). Before exploring findings, however, this Lefebvrian approach warrants further methodological attention.

Moravánszky et al. (2014: 16) counsel three uses of Lefebvre, to guide research, as a source of inspiration and as an invitation to use and develop ideas via 'exchanges between theoretical experimentation and empirical research... not to search for catchphrases to decorate a text, but as an instrument of analysis and research'. There are now numerous Lefebvrian studies which mobilise his concepts and particularly the triad, some respecting Lefebvre's lifelong approach of sketching concepts and calling for their *active* use to promote enlightenment and change: 'The perceived-conceived-lived triad... loses all force if it is treated as an abstract "model". If it cannot grasp the concrete... then its import is severely limited, amounting to no more than that of one ideological mediation among others' (Lefebvre 1991: 40). That said, the applied degree of sophistication is often moot. The triad is typically articulated (through translation of *l'espace perçu, l'espace conçu* and *l'espace vécu*) as perceived, conceived and lived space. However, its social construction is actually doubly-designated, determined both *semiotically* (what space is being materially produced) and *phenomenologically* (how spatial production happens through experience) (Schmid 2014). This subtlety can get overlooked or ignored.

The relationship between these dialectic moments, or the *trialectic*, as Soja puts it (Dale and Burrell 2008), has a bearing on the 'present' spatial outcome (Merrifield 2006). Where the conceived dominates the lived, as is according to Lefebvre (1991) typically the case in capitalist modes of spatial production, the result is dominated, abstract present space. Such conceived spatial production, positioned politically as 'true space' by 'experts' (typically those in powerful positions of provision) is set up through the knowledge and authority of modernity and capitalism, and dialectically privileges the conceived whilst repressing the lived. The potential outcome is the alienation of 'inexpert others'. Alternatively, in situations where 'subaltern communities accomplish the physical manifestation of their socio-spatial particularities' (Carp 2008: 130), one can consider an alternative, perhaps more dialectically allied 'truth of space', where the appropriation of space for lived purposes affords myriad possibilities of differential present space. In this respect, Lefebvre's discourse tends to pit the provider and the user against each other. Spatial production therefore represents a perpetual interplay between the strategies of the dominant and the tactics of the dominated (De Certeau 1984).

Of particular relevance here is an aesthetic analysis of the new Israeli Ministry for Foreign Affairs, claimed to be the first study to use all three Lefebvrian spaces (sic) in a single organisation. The authors, Wasserman and Frenkel (2011), outline an extensive but rigid approach which considers both providers and users of the Ministry building and workspace. According to their analysis, the organisation is materialised within a clearly abstract, dominated, resisted and contested workspace, to the extent that one might wonder how the American Institute of Architects award-winning 'experts' of 'one of the ten most beautiful buildings in the world in 2004' (ibid.: 506) failed to recognise or consider the organisational implications of their architectural conceptions. Yet the authors mobilise the triad in an analytic fashion which seemingly serves to underline the conflict. By essentially locking the conceived to providers, the lived to users, then observing and critiquing the material practised outcomes, is the conclusion potentially a fait accompli?

Alternatively, Carp (2008: 130) conceives the triad's utility as 'an analytical tool for discerning imbalance between abstraction and difference in social space' which might reveal clues to facilitate greater understanding and interpretation. Her approach, and evidence-based testimonial to its success, catalysed a more holistic rendering of Lefebvre's triad in the present research. Crang (1999: 176), appreciating and critiquing Lefebvre's own positionality seems to agree: 'although there is an implicit drift in Lefebvre's work to privilege lived space... there seem other angles we can draw out. Looking at the categories together offers the chance to think about the reflexive interactions between the various components'. According to Moravánszky et al. (2014: 16), 'the three "moments" of space production, as theorised by Lefebvre, do not form a synthesis but rather exist in interaction, in conflict or in alliance with each other'. They caution against applying the triad as a schematic, 'but develop with a view to complexity and specificity of each case'. Thus, following the initial emic coding analysis, the themes in each case-setting were reconsidered according to their semiotic and phenomenological attributes within Lefebvre's doubly-designated etic tool (Tracy 2013: 184) by location, dialectic valency and interplay. The following section unfolds and explores the evident present spatialities of each case-setting.

Henri on the case

Fundamentally, by asking 'what matters?' and embracing emergent responses, a huge range of aesthetic, instrumental and symbolic functions (Elsbach and Pratt 2007) are offered, immediately challenging the Cartesian notion of space as a social container. Providers *tended* to foreground conceived elements and users lived, but by no means exclusively. Many participants reflexively explored different socio-spatial elements, thinking about the thinking behind workspace provision decisions. The majority also stretched beyond their workspace as a specific locale, considering work praxis, work/life balance, well-being and indeed humane activity.

Articulating each case-setting's socio-spatial dynamic in turn, the analytically-derived themes of PropCo evidenced an 'agile' workspace approach (van Meel 2011) conceived with D&BCo to materially reflect PropCo's brand, embodying a discerning quality experience for clearly valued staff and client users alike, managed by protocols within a hierarchical, traditional organisation. Resultant emotional responses embodied both organisational pride and individualism. Users displayed a range of perceived and lived behaviours, including resistance tactics such as tribalism, mischief, to some degree territoriality and notions of personal identity. An acute awareness was shown regarding trade-offs of the agile approach, including work continuity, interruption and distraction. A focal point for this was PropCo's clear-desk policy (as a hot-desking enabler), likely to become more stringently enforced due to increasing headcount. Beyond such organisational accounts, a minority more abstractly considered the inherent challenges of contemporary workspace wisdom including agile approaches.

Conversely, the ShareCo dynamic foregrounded a permissive, democratic ethos manifested socio-spatially. Reflecting the emphasis of co-working for mutual benefit through community engagement, ArcCo sought to enable agentive users through a range of conceived behavioural nudges. The well-being of ShareCo's members was integral to the community endeavour. An integrated socio-spatial conception sought to disarm preconceived workspace expectations and entrenched behaviours. Resultant emotional responses embodied tribal community pride through a range of perceived and lived behaviours, including ownership, trust, learning and togetherness. Rather than providing conceived protocols, the lived realities about working at ShareCo existed in the stories people shared. Tensions experienced by some users and providers were related to the inherent chaos some perceived in the socio-spatial dynamic. A minority of participants also discussed how this bespoke, purposefully experimental workspace was at risk of creeping managerialist practices eroding the carefully conceived socio-spatial system, jeopardising its inherent permissiveness.

Considering PropCo in terms of change, control, commodity and community (after Cairns 2002 and Halford 2008), change by means of 'agile' flexibility was sought thorough controlling protocols. The commodity value of space was clear from aesthetic brand conceptions and the importance of the user entrance experience to convey quality. A rhetoric of community was contested by the reality of incumbent hierarchy. Alternatively, ShareCo achieved change as flexibility with no espoused control or 'agile' nomenclature. Commodity value was also clear from aesthetic brand conceptions and the importance of the user entrance experience to disarm conventional expectations. The rhetoric of community was reflected in the rich socio-spatial interplay foregrounded by design.

Clearly, both case-settings are rich with socio-spatial complexity. Echoing Elsbach and Pratt (2007), variously perceived positives and negatives were inevitable, materialised and embodied. Space was always already produced, 'perceived individually and collectively, experienced, interpreted, contested and appropriated' (Moravánszky et al. 2014: 8), conversely conceived by users and lived by providers. It would be entirely inappropriate and arguably impossible to judge the case-settings in any absolute sense. But we might consider contextual appropriateness through holistic consideration of their qualitatively different 'present' Lefebvrian spatialities (Merrifield 2006). Accordingly, the dialectic interplay of PropCo's themes suggests an example of what Lefebvre would have considered a typically capitalist, dominated abstract organisational work*space*. Conversely, ShareCo provides an example of something less common, a differential work*place* conceived deliberately with appropriation by the user community in mind. The carefully introduced distinction between work*space* and work*place* from this point onwards is deliberate. It affords the introduction of two broad, qualitatively and ideologically different Lefebvrian ways to conceive of or 'do' organisational space (Awan et al. 2013) and does not, as is typical in space as distance parameters (Taylor and Spicer 2007) typical of the built environment, relate them by scale: workplace as essentially a collective container for various localised individual workspaces.

Dominated *space-making* and *place-making* for appropriation afford alternative organisational outcomes by approaching the inherently political production of organisational workplace from different socio-spatial perspectives. Through space-making, facilities are managed, typically from above. Space is regarded as a static product. Through place-making, communities enable themselves from within. Space is accepted as active process. Place-making is not a new term, and indeed has been mainstreamed in recent years for varying agendas. But the resonance here with 'placemaking' as originally conceived in the 1960s as a human-centric urban design movement is clear, and is indeed specifically acknowledged in ArcCo's narrative. Space-making as a counterpoint to highlight the qualitative differences in both design approach and outcomes appears to be a novel expression.

Lefebvrian implications

> We are not able to act within a space without having developed [a communicable] idea of what that space looks like... thus representations of space are defined in a twofold manner – as (conceived) ideas and (communicated) concepts. They signify something, they prescribe something, they guide our actions and give them a direction.
>
> *(Schmid 2014: 40)*

Lefebvre may well have disapproved of such dichotomous simplicity, given one aim of his triad was to challenge the binary nature of thesis/antithesis, instead embracing the possibility of one *and* the other. A more developed conception, and indeed a prime area for further empirical research, might be to explore whether different 'present' organisational spaces could be considered along a continuum from domination to appropriation according to their dialectic interplay, unearthing no doubt diverse reasons as to why. But the new nomenclature presented here is a canny, tactical play. As Merrifield (2006) attests, using Lefebvre is fascinating, unsystematic and perplexing. Attempting to invoke such esoteric notions to enlighten an industry yet to acknowledge the spatial turn is a move some might dismiss as foolhardy. But, echoing Lefebvre's Marxist sentiments, what is the point of research if not to challenge and potentially emancipate? 'Lefebvre's theoretical concepts therefore cannot simply be applied… explaining various empirical case studies. Rather, the purpose is to confront these concepts with reality… and thus make them productive' (Schmid 2014: 36). Analysis can therefore become an instrument of transformation, both project *and* critique, to open possibilities of practical change. Thus, relatively accessible conceptualisation, like place-making vis-à-vis space-making, is an attempt to afford a seed from which something far more significant can germinate.

We know, as did Giddens in 1979 with his principle of the under-determination of theory by facts, that promoting alternative knowledge is a tricky business (Blaikie 2007). Before the findings of the case-settings unfolded through the stages of analysis, there was no preconception of their dominated and appropriated spatialities, nor their different systemic interrelationships between the control, change, commodity and community facets of workspace design. For sure, both case-settings were radically visually different modern working environments. Turning to design briefs from D&BCo and ArcCo might have afforded siloed information regarding conceived intent, as it did for Wasserman and Frenkel (2011). But only by mobilising Lefebvre's triad analytically to ground truth, as Carp (2008) puts it, in each socio-spatial dynamic, did a holistic awareness of each spatiality unfold. In this sense, there is no absolute right *or* wrong for each case-setting, because both revealed socially-constructed unfolding, often transient rights *and* wrongs. Consequently, the value of this analytic approach seems clear, in that it affords an alternative perspective which helps bridge the dichotomy of providers and users to reframe what *are* often debates about rights or wrongs in organisational space design.

Ultimately, the study suggests three things of importance for the industry ostensibly 'responsible' for producing organisational space. First, embracing alternative research philosophies to positivism, along with unconventional research approaches, can yield unexpected, diverse and inherently valuable findings. Second, radically different socio-spatial dynamics are possible by moving beyond long-maintained workspace solutions and dominant industry narratives. As a consequence, we can make more contextually informed choices about how to provide workplace differently. Third, it is possible to conceive workplace with user appropriation in mind, but the stakes can be high. In a wider capitalist socio-economic system, the risk of erosion of such alternative ideology is ever present. Yet wherever workplaces like ShareCo exist, they can be celebrated as examples of 'doing' space differently, through and for users, not to them, by foregrounding place-making through Awan et al.'s (2013) fundamentally Lefebvrian principles of what they term 'spatial agency', summarised here:

1. The production of space is a shared enterprise. Professional expertise involves facilitating and enabling collective contribution, not exclusive expert authorship.
2. Social space is dynamic space, meaning its production continues over time. It is generative (Kornberger and Clegg 2004). There is no single moment of completion, project plan or otherwise.
3. As people live out their lives in social space, it is intractably political, charged with the dynamics of power/empowerment, interaction/isolation, control/freedom and so on.

Interestingly, architects DEGW, seminal advocates of FM (Price 2003), demonstrated early principles of spatial agency (Awan et al. 2013). Not unlike the documented transformational impact of the Hawthorne studies on workspace significance (Baldry 1999), Cairns's (2008) rhetorical questioning of whether the organisational contribution of FM might be different had the discipline followed a more sociological path comes to mind. The previously declared ideological assumption regarding the core purpose of the built environment industry as providing enabling organisational space for users may well mask a far messier and politically contested reality. Given that organisational property and workspace costs are typically dwarfed by the costs of human resource, yet efficiency decisions regarding the former regularly influence the effectiveness of the latter, alternative ways to articulate the potential organisational contribution of organisational space are invaluable.

Appropriated spatialities are often associated with transient external urban occurrences, such as carnivals, protests, festivals, skateboarding and graffiti. The study discussed here shows not only how they leave the pavement outside, move through corporate reception areas and unfold inside knowledge workplaces, but how contextually appropriate design decisions might encourage them. In summary, there are ideological possibilities in the Lefebvrian consideration of organisational space. Place-making vis-à-vis space-making offers a reflexive counterpoint to conceive of alternative workplace design, and place-making in particular represents a politics of hope for a reimagined and far more deeply appreciated work *place* industry in service of its users.

Conclusions

The study explored in this chapter shows how the considered application of Lefebvre's spatial triad as an analytic tool can reframe the perceived rights and wrongs of workspace provision. Conceived strategies will always coexist with lived, emergent tactics. Acknowledging this, the role for providers of workspace isn't to try and oppress one in favour of the other. Instead, it becomes an appreciation of how they are irreconcilably intertwined, and an endeavour to seek to conceive contextually appropriate spaces accordingly. Lefebvre's triad provides a conceptual tool to frame the sheer possibility of generative, empowered social space. Moreover, spatial agency embraces a Lefebvrian perspective, particularly a differential mode of 'doing' space, with appropriation as an aspirational goal. In seeking to embrace place-making and the deep-seated ideological differences of this alternative approach to spatial production, more humane workplaces can be the result. But being open to the potential of place-making requires awareness, tolerance, patience, inquisitiveness and experimentation.

The study presented here is the only known *application* of Lefebvre's triad as an *analytic lens* from the field of facilities management and workplace provision at this point. It seems fitting then to conclude with some reflexive questions to promote further 'spatial turn' work in this disciplinary area. In the spirit of Cairns (2008), what would a more ambivalent Lefebvrian approach to theorising the built environment look like? How might such findings challenge and enhance the practices of workspace production and provision? Finally, what possible spatialities could unfold if expert *savoir* and inexpert *connaissance* knowledges of spatial production were more complementary by design?

References

Airo, K. and Nenonen, S. (2014) 'Review of linguistic approach in the workplace management research', *Facilities*, 32:1/2 27–45.

Awan, N., Schneider, T. and Till, J. (2013) *Spatial Agency: Other Ways of Doing Architecture*, Abingdon: Routledge.

Baldry, C. (1999) 'Space – the final frontier', *Sociology*, 33:3 535–53.
Blaikie, N. (2007) *Approaches to Social Enquiry: Advancing Knowledge*, Cambridge: Polity.
Boano, C. (2015) 'Henri Lefebvre, toward an architecture of enjoyment', Journal of Architecture, 20:3 544–9.
Cairns, G. (2002) 'Aesthetics, morality and power: design as espoused freedom and implicit control', *Human Relations*, 55:7 799–820.
Cairns, G. (2008) 'Advocating an ambivalent approach to theorizing the built environment', *Building Research & Information*, 36:3 280–9.
Carp, J. (2008) '"Ground-Truthing" representations of social space: using Lefebvre's conceptual triad', *Journal of Planning Education and Research*, 28:2 129–42.
Crang, M. (1999) 'Globalization as conceived, perceived and lived spaces', *Theory, Culture & Society*, 16:1 167–77.
Crang, M. and Thrift, N.J. (2000) *Thinking Space*, vol. 9, London: Routledge.
Dale, K. and Burrell, G. (2008) *The Spaces of Organisation and the Organisation of Space: Power, Identity and Materiality at Work*, Basingstoke: Palgrave Macmillan.
De Certeau, M. (1984) *The Practice of Everyday Life*, trans. S. Rendall, Oakland, CA: University of California Press.
Denzin, N. and Lincoln, Y. (2000) *The Handbook of Qualitative Research*, Thousand Oaks, CA: SAGE.
Elsbach, K. and Pratt, M. (2007) 'Chapter 4: the physical environment in organizations', *The Academy of Management Annals*, 1:1 181–224.
Greene, C. and Myerson, J. (2011) 'Space for thought: designing for knowledge workers', *Facilities*, 29:1/2 19–30.
Halford, S. (2008) 'Sociologies of space, work and organisation: from fragments to spatial theory', *Sociology Compass*, 2:3 925–43.
Kornberger, M. and Clegg, S. (2004) 'Bringing space back in: organizing the generative building', *Organization Studies*, 25:7 1095–1114.
Kvale, S. (1996) *InterViews: An Introduction to Qualitative Research Inquiry*, Thousand Oaks, CA: SAGE.
Lefebvre, H. (1991) *The Production of Space*, trans. D. Nicholson-Smith, Oxford: Wiley-Blackwell.
Merrifield, A. (2006) *Henri Lefebvre: A Critical Introduction*, Abingdon: Routledge.
Moravánszky, Á., Schmid, C. and Stanek, L. (eds) (2014) *Urban Revolution Now: Henri Lefebvre in Social Research and Architecture*, Abingdon: Routledge.
Price, I. (2003) 'Facility management as an emerging discipline', in R. Best, C. Langston and G. De Valence (eds) *Workplace Strategies and Facilities Management*, Oxford: Butterworth-Heinemann, pp. 30–48.
Rose, G. (2012) *Visual Methodologies: An Introduction to Researching with Visual Materials*, 3rd edn, London: SAGE.
Rosen, M., Orlikowski, W.J. and Schmahmann, K.S. (1990) 'Building buildings and living lives: a critique of bureaucracy, ideology and concrete artifacts', in P. Gagliardi (ed.) *Symbols and Artifacts: Views of the Corporate Landscape*, New York: Aldine de Gruyter, pp. 69–84.
Saval, N. (2014) *Cubed: A Secret History of the Workplace*, New York: Random House LLC.
Schmid, C. (2014) 'The trouble with Henri: urban research and the theory of the production of space', in Á. Moravánszky, C. Schmid and L. Stanek (eds) *Urban Revolution Now: Henri Lefebvre in Social Research and Architecture*, Abingdon: Routledge.
Sundstrom, E. and Sundstrom, M.G. (1986) *Work Places: The Psychology of the Physical Environment in Offices and Factories*, Cambridge: Cambridge University Press.
Taylor, S. and Spicer, A. (2007) 'Time for space: a narrative review of research on organizational spaces', *International Journal of Management Reviews*, 9:4 325–46.
Then, D.S.-S. (1999) 'An integrated resource management view of facilities management', *Facilities*, 17:12/13 462–9.
Till, J. (2009) *Architecture Depends*, Cambridge, MA: MIT Press.
Tracy, S.J. (2013) *Qualitative Research Methods: Collecting Evidence, Crafting Analysis, Communicating Impact*, Chichester: John Wiley and Sons.
van Meel, J. (2011) 'The origins of new ways of working: office concepts in the 1970s', *Facilities*, 29:9/10 357–67.
Vince, R. and Warren, S. (2012) 'Participatory visual methods', in G. Symon and C. Cassell (eds) *Qualitative Organizational Research: Core Methods and Current Challenges*, London: SAGE, pp. 275–295.
Wasserman, V. and Frenkel, M. (2011) 'Organizational aesthetics: caught between identity regulation and culture jamming', *Organization Science*, 22:2 503–21.

22

Dwelling on design

The influence of Logos and Eros, nouns and verbs, on public housing renewal and cooperative alternatives

Matthew Thompson

Introduction

> An unequal struggle, sometimes furious, sometimes more low key, takes place between the Logos and the Anti-Logos, these terms being taken in their broadest possible sense – the sense in which Nietzsche used them. The Logos makes inventories, classifies and arranges: it cultivates knowledge and presses it into the service of power. Nietzsche's Grand Desire, by contrast, seeks to overcome divisions – divisions between work and product, between repetitive and differential, or between needs and desires.
>
> *(Lefebvre 1991: 391–2)*

This evocative passage conjures up some of Lefebvre's most fundamental insights in *The Production of Space* (1991), distilled into a single comprehensible idea – an eternal battle between two forces. *Logos* is that principle derived from the Ancient Greeks which Lewis Hyde (1979: xiv) usefully defines as 'reason and logic in general, the principle of differentiation in particular' – those forces which act to divide, isolate and abstract things from each other in order to signify, classify, compartmentalise and give order to the world around us. Though Lefebvre doesn't quite spell it out, that which opposes Logos, what he calls 'Anti-Logos', is perhaps more accurately described as *Eros* – which Hyde, writing about the 'erotic' life of the gift economy as opposed to the 'logical' life of market exchange, defines as 'the principle of attraction, union, involvement which binds together' (Hyde 1979: xiv). Eros describes the holistic unifying force that flows through everything and 'seeks to overcome divisions' – divisions which Logos imposes on an otherwise undifferentiated cosmic whole. Following this line of thought, we might say Lefebvre has a clear normative agenda for promoting the 'erotic' life, in this most expansive sense, as a tonic to the societal consequences of narrowly 'logical' thinking. Here, Nietzsche is a significant reference point for Lefebvre, especially his opposition in *The Birth of Tragedy* between Apollo and Dionysus – gods associated respectively with differentiation, order, clarity and calm rationality, and with oneness, excess, intoxication and overflowing creativity (Merrifield 1995). In citing Nietzsche's Grand Desire as the Anti-Logos, Lefebvre is implicitly positioning Apollonian forms of thought with Logos; Dionysian with Eros.

Ontologically speaking, Lefebvre's distinctly dialectical thinking in no way privileges Eros over Logos, nor Dionysus over Apollo, as these polarities are conceptualised dialectically as necessarily co-constitutive and interwoven with each other. However, in more political moments, such as in the opening passage and elsewhere in his writings, he suggests that the latter is winning what he sees as a grand historical battle, an 'unequal struggle'. We can interpret this as an allusion to his original insight into the accelerating historical dominance of abstract space over lived space; the 'devastating conquest of the lived by the conceived, by abstraction' (Lefebvre quoted in Wilson 2013: 3). Excessive Logos feeds abstract space; whilst Eros prefers to inhabit lived space.

In this chapter I want to elaborate on this central distinction in Lefebvre's dialectical thought, by introducing other related polarities: between ends and means, thing and flow, spatial form and temporal process; and between objects and activities, or nouns and verbs. I bring Lefebvre into a novel conversation with two broadly contemporaneous anarchist writers, John F.C. Turner and Colin Ward. Their ideas had a significant influence on the development of cooperative and self-build housing, which we could say channel the spirit of Eros, as alternatives to the kind of system-built mass housing delivered by technocratic bureaucracies, which on the contrary embody Logos. Although this may risk oversimplifying and overstretching these concepts and their interrelations, I hope to show in the following how this heuristic helps clarify our thinking on the way we treat our urban environments and our approach to dwelling as well as, in the conclusion, our understanding of Lefebvre's thought on utopian possibilities for experimental transformation of space. I do this by way of example, focusing on the history of public housing and regeneration in the British city of Liverpool and in particular on various approaches, from top-down comprehensive urban renewal to community-led cooperative projects, which can be seen to perpetuate Logos or Eros in varying combinations.

Liverpool: a brief history of treating housing as a noun or a verb

> Liverpool, with all its commerce, wealth, and grandeur treats its workers with the same barbarity. A full fifth of the population, more than 45,000 human beings, live in narrow, dark, damp, badly-ventilated cellar dwellings, of which there are 7,862 in the city. Besides these cellar dwellings there are 2,270 courts, small spaces built up on all four sides and having but one entrance, a narrow, covered passage-way, the whole ordinarily very dirty and inhabited exclusively by proletarians.
>
> *(Engels 1892)*

Engels' experience of Liverpool, and other northern industrial cities, inspired him to write *The Housing Question* (1872), raising grave doubts over the ability of capitalism to house the working classes in humane or minimally sanitary conditions. Conditions have however since improved since these darkest days of industrialisation, but Liverpool continued to suffer from inadequate housing for much of the 20th and even into the 21st century. Successive municipal modernist projects to improve such appalling conditions first targeted the speculatively built dockside slums and then the inner-ring of Victorian terraces, replacing each in turn with tenements and tower blocks. These state-led programmes were of a diverse ideological and architectural bent: the Tories' inter-war monumental art deco 'garden' tenement blocks; Labour's post-war modernist 'Slum Clearance Programme', redeveloping terraces as walk-up tenements and high-rise flats; the far left municipal socialist strategy of suburban house-building under the Trotskyist Militant Tendency controlling the Labour Council in the mid-1980s; and, in the 21st century, Labour's neoliberal public-private Housing Market Renewal Pathfinder programme, which has been characterised elsewhere as a tool of abstract space (Thompson 2017), in a similar vein to its American predecessor HOPE VI (Jones and Popke 2010), but which I lack the space to explore here. Yet despite such differences, all succumbed to the same fallacy – treating dwelling(s) as a noun, a static material object, rather than a verb, a dynamic lived process.

Such approaches misapprehend the structural and socioeconomic nature of problems associated with housing deprivation, dilapidation and neighbourhood decline. Housing becomes a fetishised object for planners and policymakers, who lay the problem at the door of the house itself rather than the complex background processes that produce space; treating the material building as the target of their intervention rather than the social relations that produce it. This is one way in which the dominance of abstract space is so problematic: it acts to fetishise objective, material space in a similar vein to how 'commodity fetishism' works to divorce products from their productive histories. Lefebvre (1991: 95–6) alludes to this in his insight that modernity is marked by the 'manifest expulsion of time' by fetishised space; that 'with the advent of modernity time has vanished from social space'. In being blinded by abstract space, technocratic planners reproduce, and often exacerbate, the very same problems they are trying to resolve. Indeed, in post-war attempts to address the housing question, Lefebvre (1991: 314) identifies the increasing incursion of abstract space in the discursive shift from 'residence' to 'housing':

> It was at this juncture that the idea of *housing* began to take on definition, along with its corollaries: minimal living-space, as quantified in terms of modular units and speed of access; likewise minimal facilities and a programmed environment. What was actually being defined here… was the lowest possible *threshold of tolerability*. Later, in the present century, slums began to disappear.

As the welfare state began to eliminate the worst conditions brought about by capitalist urbanisation, this was achieved through the imposition of standardised units measured according to the 'bare minimum' of acceptable standards, both in terms of material tolerability and the 'lowest possible *threshold of sociability* – the point beyond which survival would be impossible because all social life would have disappeared' (Lefebvre 1991: 314). Here, Lefebvre articulates the idea that housing delivered through impersonal state bureaucracies – which he refers to as 'habitat', exercised by the rationalising will of Logos, in contrast to 'habiting', infused with the 'erotic' flow of everyday life – *alienates* dwellers from their immediate living environments, disconnects them from others and thwarts the forces of Eros from bringing people together for even the most minimal of sociality that makes life at all meaningful, or even tolerable.

This is a thesis supported, if not expressed quite so strongly, right across the political spectrum. At one end we find the likes of geographer Alice Coleman, who argued that the design of concrete tower blocks in particular were responsible for social problems in deprived estates, by removing any real sense of ownership or pride and removing all obvious incentives to care and maintain property. Her work, notably *Utopia on Trial* (1985), was very influential in the development of prime minister Margaret Thatcher's policies, particularly the Right to Buy and the replacement of collectivist housing estates with more privatist family houses – a rare example of a geography academic enjoying real impact (Jacobs and Lees 2013). Her ideas also gained traction amongst the 'Militant' Labour Council in Liverpool, pitting socialist councillors against the growing housing cooperative movement, their Labour-voting constituencies – a strange twist of fate, which I explore below.

At the other end of the spectrum, we find anarchist arguments, particularly those of the so-called 'anarchist architect' John F.C. Turner and 'anarchist planner' Colin Ward, who emphasise the politically-empowering, spiritually-fulfilling and identity-forming qualities that flow from self-autonomy and collective control over living environments. The bureaucratic alienation of public landlordism, argues Ward (1985), is a kind of 'municipal serfdom' that treats tenants as 'inert objects' rather than active subjects and is responsible for the swift physical dilapidation of council housing estates.

This theory is borne out by Liverpool's experience with mass modernist housing. By far the most notorious of developments were known locally as 'the Piggeries'. Tony Lane (1978: 338–9) explains some of the motivations leading to the demolition of tenements and tower blocks like the Piggeries after only a few years of use:

> Who would have dreamt in the 1950s that a municipal housing department would have to invent the term 'hard-to-let? Who would have dreamt that some tenants would have been driven to a systematic destruction of their own housing as a means of forcing a change in policy? Who could have imagined a situation where tenants would have complained of the state of repair of their buildings – and then said that they did not want repairs carried out because they wanted the place to deteriorate to the point where they would have to be re-housed?

Ironically, the old 'slum' terraced housing was simply replaced with new slums. Modernist system-built housing produced simplistic, largely cosmetic *end solutions* to complex problems, proving too inflexible and unresponsive to residents' needs and their desires to change their dwellings in accordance with their everyday lives. In short, they placed too much emphasis on the building itself, seen as a static noun, inhibiting the vital flow of collective activities needed to sustain it. So what is the alternative?

Liverpool's cooperative revolution

In 1970s Liverpool, working-class communities began struggling against the 'Corpy' – the colloquial term for Liverpool City Council, the 'Corporation' – and their slum clearance programme which displaced tenants to new towns and outer estates. Residents campaigned for cooperative alternatives with direct involvement in the design, development, ownership and management of their housing. It was at the birth of this movement that Ward's, and by extension Turner's, ideas found their expression, as Ward retrospectively explains:

> The proudest moment of my housing advocacy was when the Weller Street Co-op chairman, Billy Floyd, introduced me at a meeting by waving a tattered copy of Tenants Take Over and saying: 'Here's the man who wrote the Old Testament.... But we built the New Jerusalem!'
>
> *(Ward and Goodway 2003: 74–5)*

Here, the Old Testament refers to Ward's (1974) book, *Tenants Take Over*, which articulates his radical manifesto for 'collective dweller control'; whilst the New Jerusalem is the Weller Street Co-op, the country's first co-op to be designed, developed, owned and managed by its working-class residents. Weller Street in turn ignited what has been dubbed Liverpool's 'new-build cooperative revolution' (CDS 1994), the country's largest and most innovative housing cooperative movement. It heralded a radical new model, the 'Weller Way' of doing things (McDonald 1986).

This new model of public housing, or 'Public Housing 2.0', as it was heralded, incorporated radical new ideas around dweller control, design democracy and participatory techniques – then being experimented with in what became known as the 'community architecture' movement

– and inspired groups of council tenants to develop successive waves of new-build co-ops across Merseyside. It represented an extraordinary shift from a situation in which most of Liverpool's working-class residents were housed by the Corpy without control over the design, location or management of their home, to one in which residents had for the first time genuine dweller control.

Whilst Ward had some discernible impact on the pioneering Weller Street Co-op, he in turn was strongly influenced by Turner's (1972) ideas of 'user autonomy', derived from the self-build 'autoconstruction' that Turner witnessed in South American informal settlements. In his prolific writings Ward often cites what he calls Turner's First Law of Housing: 'When dwellers control the major decisions and are free to make their own contributions in the design, construction, or management of their housing, both this process and the environment produced stimulate individual and social well-being' (Turner and Fichter 1972: 241).

Turner in particular draws our attention to the anarchist insight that *means* are just as important as *ends*; that dwelling is a verb as well as a noun – an active lived process of doing, as well as a static material resource, the building itself (Turner 1972). So how does treating housing like a verb rather than a noun, a lived process over a fetishised object, actually play out in the development of cooperative housing? And how did these cooperative experiments affect the lives of their inhabitants?

The leading secondary co-op agency, Cooperative Development Services (CDS), played a crucial role in the movement's development – working with the Weller Street community to co-produce the 'Weller Way', which was to set the trend for the rest. First, CDS suggested architects, developers and agents for co-ops to shortlist and then interview; the Weller Street committee insisted that 'the architects act as advisers and scribes' (McDonald 1986: 84). Architects worked with Weller Street Co-op to pioneer a radically democratic design process that would put flesh on the theoretical bones of Ward's 'dweller control' concept. Participatory techniques and 'planning for real' exercises, such as group modelling exercises, were innovated to traverse the wall separating technical architectural knowledge (namely Logos) from the lived experience of inhabitants (Eros). Spin-off benefits of such intensive involvement included individual empowerment through teaching new skills; tackling socioeconomic needs by producing more responsive designs, lowering long-term maintenance costs; and building better communities, in developing community confidence and sense of ownership, thereby instilling responsibility for housing, helping deter vandalism, crime and neglect and giving people a political voice in local decision-making.

Empowerment meant life-changing education for individuals – providing people with the knowledge and skills to find new employment, often in professional practices such as architectural firms – but also a deeper sense of mutuality, community togetherness and collective political power.

Communities were indeed politicised. Co-op campaigns were like a kind of 'political school' for many, who had cut their teeth on political campaigning and collective negotiation with key gatekeepers, and who were inspired and empowered to go into politics full-time, representing their communities and often becoming councillors and cabinet members, of which there are countless individual examples. In these various ways, therefore, the intensive campaign and design process was a vital move in making new co-op neighbourhoods more than just a collection of better-quality material dwellings: it also strengthened collective capacities for community self-government over the social activity of dwelling.

In providing the resources and skills for people to make significant steps towards housing themselves, the co-op movement in many ways bridged the growing chasm between the ends and means of dwelling, between lived process and end product, bringing the user and the producer of housing into closer synergy. The movement made real steps towards resolving the alienation at the root of the problems of housing deprivation and embedded the seeds of an alternative model of regeneration which would later inspire a new generation of grassroots action in immersive, participatory and democratic dwelling design: Liverpool's budding Community Land Trust (CLT) movement (Thompson 2015, 2017).

Importantly, the co-op campaigns protected communities from being broken up and displaced to the new towns and outer estates then being built on the metropolitan periphery. They not only brought communities closer together through the deeply political process of what many saw as a 'battle' or a 'war' with the Corpy over how they were housed – an exercise which helped highlight what exactly they were fighting for – but they also thereby preserved and enhanced existing ways of life, rooted in religious and clan identities, neighbourliness, trade unionism, local Labour Party organising, and various everyday practices of mutual aid, informal gift economies and networks of solidarity. These 'erotic' forms of life were effectively threatened by the 'logical' exercise of cutting through urban space and rationalising neighbourhoods in slum clearances and modernist reconstruction.

But no sooner had the movement begun to snowball – with 30 or so co-ops built or in the pipeline, having gained support from Council policy, led by a Liberal administration, which had turned away from comprehensive renewal – than a surprising new political threat emerged, dramatically curtailing the life of the movement.

Militant mono-mania for housing design

Within less than half a decade of the completion of Weller Streets, following the council election in 1983 of the Labour Party led by the Militant Tendency, the new-build cooperative revolution was dissipated by an agenda of centralised local state control over the means of social reproduction (Taafe and Mulhearn 1988). Militant saw co-ops as a 'bourgeois conspiracy' and a threat to municipal housing – much like how Engels saw Proudhon's anarchism as 'bourgeois socialism' or mere reformism – and forcefully quashed co-op development through 'municipalisation'. Gestating co-ops were either aborted or taken into municipal ownership, as part of a bold and ambitious £350 million housing renewal programme, the Urban Regeneration Strategy (URS), which met targets of 1,000 new homes built per year up to 1988 – a remarkable achievement for a time under prime minister Margaret Thatcher when, nationally, council housebuilding had come to a standstill (Grosskurth 1985; Mars 1987).

Militant believed that large-scale municipal house-building would revitalise Liverpool's economy and environment by providing jobs and decent homes for all, but became seduced by a form of design determinism – seeing dwelling as a noun rather than a verb. Their assessment of council housing designs revealed 'one bright spot' of 'problem-free' semi-detached housing built in the inter-war period; this 'insight was the germ of the URS housing programme' (Grosskurth 1985: 26). At around the same time, Alice Coleman (1985) was popularising her ideas on the 'design disadvantagement' of modernist council housing estates, which she had adapted from Oscar Newman's theory of 'defensible space' (Jacobs and Lees 2013). Ironically, despite the clear influence of Coleman's ideas on Militant's most despised ideological opponent – Thatcherism – they nonetheless came to the same conclusions. Coleman gave her seal of approval to Militant housing policy, publicly stating that "Liverpool has got it right", which leading Militant members were proud to report: 'she completely concurred with the main thrust of the URS and of the council's conviction that the majority of people preferred to live in traditional houses' (Taafe and Mulhearn 1988: 159).

The URS development principles that Coleman praised were published as new guidelines which prescribed that only houses and bungalows, semi-detached where possible, were to be built, laid out in rational, grid-like street patterns – inspired by Logos. No cul-de-sacs, clusters or inward-facing dwellings; no shared surfaces, common areas or play spaces; only conventional road layouts with private gardens (Grosskurth 1985). This was bad news for uncompleted co-ops – for co-op designs tended to favour denser, communal spaces that encouraged community interaction. Many of the more interesting courtyard designs with inward-facing neighbourhood spaces, community centres and focal points for collective gathering – essential to the social life of a co-op – were now in contravention of the URS guidelines. Not only did Militant take co-op developments that had yet to sign a council agreement under council control, but they also radically altered their design to reflect their belief in plain suburban housing, thereby threatening the social existence of these co-ops.

In many respects, however, Militant had accurately captured the mood of many tenants, alienated by several decades of living in dysfunctional and decaying council flats. However, the resulting products were perhaps less desirable, derided by many as 'Hatton houses' – a doll's house or 'story-book look' of a traditional family home, which in practice was often too small for tenants' furniture and which conspired against the neighbourliness and collective street-life that the dense terraces and tenements had at least facilitated and which the co-ops were explicitly designed to engender. Militant's monomania for 'logical' housing designs was found guilty of spatial determinism by critics – including CDS Chief Exec Catherine Meredith, who accused Militant of a 'megalomaniac belief in housing type' (quoted in Mars 1987: 27) – for failing to recognise the importance of dweller control in the management and maintenance of housing. Militant ultimately fell victim to a kind of product fetishism. The Coleman/Militant 'design modification' approach worked on the assumption that people wanted semi-detached houses, overseeing the fact that working-class co-op tenants had opted for terraces, enclosed courtyards, cul-de-sacs and communal features in the participatory design processes at the heart of the new build co-op movement.

Discussion: towards experimental utopias?

The similarities between Colin Ward's pragmatic, distinctly English anarchism and Lefebvre's romantic French utopianism are striking – though strangely overlooked in the literature. Both contend that Marx and Engels misapprehended the inextricable interrelationship between ends and means – that revolutionary or insurrectionary action cannot alone procure lasting change without first cultivating new forms of social life as the necessary socioeconomic and cultural basis for any substantively different future society. So, too, for the technocratic action of comprehensive renewal, which of course characterised Soviet state socialist planning.

Seen as technologies of abstract space, the modernist mentality of the slum clearance programme and URS privilege the thingness over the flow of space, focusing on the end-product, the final design, over the process of getting there – neglecting the lived space of inhabitants. The Corpy's successive municipal-socialist visions of good clean housing for all, rationally designed and executed, can in Lefebvrian terms be seen as 'utopist' (Pinder 2013) – abstract, transcendental visions of an ideal city, procured through spatial closure, and totalitarian in their prescriptions on social life – rather than truly 'utopian': concrete explorations of the possible in everyday life, or what Lefebvre calls 'experimental utopias' (Pinder 2013). This distinction captures a deep tension in utopian-socialist thought between openness and closure, change and fixity, or process and object, which David Harvey (2000: 183) diagnoses thus: 'To materialise a space is to engage in closure (however temporary) which is an authoritarian act.... The problem of closure (and the authority it presupposes) cannot be endlessly evaded'.

The paradox is such that utopias are created as endlessly open projects of reimagining and reinventing social relations through idealist visions, tending never to come to a point of closure, keeping possibilities open for constant evolution, whilst at the same time needing to realise and materialise this vision in a definite socio-spatial form, which, as Harvey astutely recognises, is inherently counter-utopian, foreclosing change, and therefore authoritarian. Interestingly, Harvey (2000) accuses Lefebvre of an 'agonistic romanticism' for refusing to make specific spatial recommendations or definitions of utopian futures for fear of falling into the totalitarian trap of reproducing technocratic abstract space. But Lefebvre's aversion to closure is not so much per se as with the authorial source of utopian design: insisting that utopian projects must flow from users and inhabitants themselves, from their quotidian experiences in experimenting with possibilities in practice, not from detached planners or visionaries – resonating with Ward's 'dweller control'.

Harvey's (2000) solution of a dialectical utopianism that acknowledges spatiotemporal interplay is succinctly expressed by Lefebvre (1991: 189–90):

> The idea of a new life is at once realistic and illusory.... The fact is that the space which contains the realized preconditions of another life is the same one as prohibits what those preconditions make possible.... To change life, however, we must first change space.

But for Lefebvre, just as for Ward and Turner, it matters precisely *how* space is changed. The Militant's URS made the very same mistake as the post-war modernist designs it critiqued. Supposedly alienating high-rise flats and council estates may have been replaced by more popular, human-scale traditional houses, but the distant paternalistic bureaucratic structure remained unmoved. These approaches foreclosed the possibility of ongoing change, lacking the necessary temporal openness and flexibility for genuine engagement of users with their environment. The promise of participatory design techniques and cooperative governance relations in co-ops and other mutual models like CLTs lies in greater degrees of interaction between users, and between dweller and dwelling, such that a dialectical spatiotemporal process of experimental-utopian change *may* occur.

In this way, change is not brought about from on high, through some bird's-eye blueprint plan, delivered in one fell swoop of comprehensive renewal. Rather, in the case of the co-ops, it was done incrementally and iteratively; the 'end users' themselves experimented with design possibilities in close collaboration with their architects to see what was desirable, workable and possible within physical and political limits. The co-ops are not perfectly realised 'experimental utopias' allowing for endless experimentation with spatial form for exploration of revolutionary ways of life – for the very act of choosing a design and constructing a building involves setting in relative permanency a spatial configuration which necessarily enhances or precludes certain ways of life over others. Moreover, the regulatory landscape determining the process of co-op development (they were after all publicly funded through the Housing Corporation) meant that designs had to accord with certain, often conservative and limiting, regulations which inevitably constrained the full exercise of residents' imaginations and collective agency. Indeed, architectural critics at the time derided the co-ops as 'ordinary' and 'uninspiring' suburban designs, out of place with their urban contexts (Mars 1987).

But these critics missed the point: the spatial structure of the co-ops reflected the democratic process of getting there and enabled the continued interaction of residents as a co-op community. However, this nonetheless highlights the constraints of conventional construction methods: that experimental change can only go so far when you are dealing with bricks and mortar, let alone rules and regulations. A higher degree of spatial closure marks the co-ops more than (say) the more truly self-build designs of the South American informal settlements studied by Turner. The emerging CLTs in Liverpool perhaps better demonstrate how material environments can be an organic evolving spatial expression of inhabitants' lived space and political imaginaries: old terraced houses radically reimagined and internally reworked in combination with the grassroots transformation of public space through guerrilla gardening and community art (Thompson 2015).

This relative flexibility stands in stark contrast to ready-made system-built modernist housing estates, which simply aren't flexible or malleable enough to respond to residents' desires for dweller control. In fact, they ossified around people's lives, much like a large-scale concrete dam holds back a reservoir of flowing water. The potential energy contained in by such a spatial (en) closure is huge – but if the dam cracks, then this frustrated kinetic energy is suddenly released as a deluge, much like the pent-up political energies unleashed amongst communities campaigning for co-op alternatives.

The scene I've just painted above focuses still too much on the housing product, the material object, suggesting how all too easy it is to get lured in by the fetishism of material objects. The true power of the co-op projects was their focus on the *process* of housing production: in the participatory design techniques which not only helped bring about these more sensitive designs, but also mobilised a process of empowerment which would infuse the lives of the participants and the collective life of the co-op long after the campaign, design and development 'phases' were completed.

The picture is of course not all pretty. Co-ops often turned inwards, adopting a defensive kind of urbanism which reflected the harsh treatment they received from the Corpy and the hostile socioeconomic environment to which they literally turned their backs, through cul-de-sac designs, but also the religious sectarianism that gripped Liverpool, then more than now. It is understandable, then, why the Militants' URS guidelines sought to rid their municipal housing of these kinds of design features which to them only propagated the elitism and nepotism that flew in the face of egalitarian principles of universal basic public services underpinning any form of socialism. Impersonal justice and rationality – Logos – does indeed have its place after all. Moreover, after the intense campaign process was over, many co-op residents developed a kind of 'post-development blues', where the mundane reality of collecting rent arrears and managing day-to-day maintenance whittled away the will, particularly of subsequent generations without the memory of collective action to sustain them, to cooperate and participate in the ongoing life of the co-op. As a result, many co-ops have outsourced their management requirements to professionals, the contemporary heirs of agencies like CDS, which could do it more efficiently and effectively – another example where Logos is essential. We might also see the more positive outcomes of the co-ops in these terms: lower maintenance costs, more efficient cost-benefits, individual empowerment to find jobs and political positions of power and so on – these are all examples of benefits flowing from more 'erotic' approaches to dwelling being translated into the language of Logos, and for good reason.

Conclusion

In this chapter, I outlined two opposing tendencies in the historical production of space – towards the abstract logic of Logos, privileging absolute ends and spatial closure, or the 'erotic' life of Eros, amenable to more open-ended and evolving interaction – and then sought to identify these forces in empirical developments over Liverpool's recent housing history. Whilst the technocratic mentality of top-down state-led regeneration programmes can be seen to personify Logos, the cooperative housing movement that flourished as a community-led do-it-yourself alternative to comprehensive urban renewal embodies to a much greater extent Eros. This distinction is not intended as a strict divide, but rather offered up as a dialectic, whereby each informs the other; a heuristic that might help us see how our treatment of dwelling – from its narrowly material reading as a noun to its most expansive sense as a verb – is shaped by our variously 'logical' or 'erotic' perspectives and the different approaches we thereby take to producing and reproducing our urban environments.

This chapter also warns that a militant belief in the end product and final design of housing – fetishising dwelling as a noun over a verb – can have so many damaging, even if unintended, consequences for inhabitants. However, although the balance between Eros and Logos may tilt too far one way or the other, they must always be, as Lefebvre enjoins, treated as part of a dialectical whole: without one, the other cannot be. On this final point I want to close this chapter as it opened, with the wise words of Lewis Hyde (1979: 38) who, in describing the 'essential polarity between the part and the whole, the one and the many' that animates the vital dialectic of Logos and Eros, has this to say: 'Every age must find its balance between the two, and in every age the domination of either one will bring with it the call for its opposite'.

Ultimately, I hope to have shown, by the light cast by Lefebvre's thought, how a world increasingly dominated by abstraction nonetheless contains the *preconditions of another life*.

References

CDS (1994) *Building Democracy: Housing Cooperatives on Merseyside. Update '94*, Liverpool: Cooperative Development Services (Liverpool) Ltd.

Coleman, A. (1985) *Utopia on Trial: Vision and Reality in Planned Housing*, London: Hilary Shipman Ltd.

Engels, F. (1872) *The Housing Question*, London: Progress Publishers.

Engels, F. (1892) *The Condition of the Working Class in England*, London: Penguin Classics.

Grosskurth, A. (1985) 'Bringing back the Braddocks', *ROOF*, 10:1 January/February.

Harvey, D. (2000) *Spaces of Hope*, Edinburgh: Edinburgh University Press.

Hyde, L. (1979) *The Gift: How the Creative Spirit Transforms the World*, London: Random House.

Jacobs, J. and Lees, L. (2013) 'Defensible space on the move: revisiting the urban geography of Alice Coleman', *International Journal of Urban and Regional Research*, 37:5 1559–83.

Jones, K. and Popke, J. (2010) 'Re-envisioning the city: Lefebvre, HOPE VI, and the neoliberalization of urban space', *Urban Geography*, 31:1 114–33.

Lane, T. (1978) 'Liverpool – city of harder times to come', *Marxism Today*, November: 336–43.

Lefebvre, H. (1991) *The Production of Space*, London: Blackwell.

Mars, T. (1987) 'Mersey tunnel vision?' *ROOF*, 12:6 November/December.

McDonald, A. (1986) *The Weller Way: The Story of the Weller Street Housing Cooperative*, London: Faber and Faber.

Merrifield, A. (1995) 'Lefebvre, anti-Logos and Nietzsche: an alternative reading of the production of space', *Antipode*, 27:3 294–303.

Pinder, D. (2013) 'Reconstituting the possible: Lefebvre, utopia and the urban question', *International Journal of Urban and Regional Research*, 39:1 28–45.

Taafe, P. and Mulhearn, T. (1988) *Liverpool. A City That Dared To Fight*, London: Fortress.

Thompson, M. (2015) 'Between boundaries: from commoning and guerrilla gardening to community land trust development in Liverpool', *Antipode*, 47:4 1021–42.

Thompson, M. (2017) 'LIFE in a ZOO: Henri Lefebvre and the (social) production of (abstract) space in Liverpool', *CITY: Analysis of Urban Trends, Culture, Theory, Policy, Action*, 21:4: (forthcoming).

Turner, J.F.C. (1972) 'Housing as a verb', in J.F.C. Turner and R. Fichter (eds) *Freedom to Build*, London: Collier Macmillan, pp. 148–175.

Turner, J.F.C. and Fichter, R. (eds) (1972) *Freedom to Build*, London: Collier Macmillan.

Ward, C. (1974) *Tenants Take Over*, London: Architectural Press.

Ward, C. (1985) *When We Build Again: Let's Have Housing that works!* London: Pluto Press.

Ward, C. and Goodway, D. (2003) *Talking Anarchy*, London: Five Leaves Publications.

Wilson, J. (2013) '"The Devastating Conquest of the Lived by the Conceived": the concept of abstract space in the work of Henri Lefebvre', *Space and Culture*, 16:3 364–80.

23

The consequential geographies of the immigrant neighbourhood of Quinta do Mocho in the Lisbon Metropolitan Area

Margarida Queirós, Anna Ludovici and Jorge Malheiros

Introduction

For Lefebvre (1991: 12), 'space considered in isolation is an empty abstraction'. Under a social-spatial dialectics, social practices produce spaces, and these influence social practices. So space is not neutral, neither a container of political life, nor a stage in which society acts (Soja 1996, 2010). There are significant differences regarding the privileged positioning of certain spaces in relation to the centres of power, that are contextually the richest in terms of social, cultural and economic opportunities. The production of space is therefore an outcome of a combination of exogenous geographies of power where political order and private property rights interact and impose from a top-down standpoint and also of endogenous geographies that are discriminatory and socially (re)constructed in a bottom-up perspective.

Lefebvre (1991) suggests that space plays a fundamental role to the lived experience of the world, comprising the triad composed by the conceived space (representations of space, abstract, symbols, codifications), the dominant space in society, the perceived space (spatial practices, characteristic spatial sets of each formation) and the lived space (space of representations of lived experience, of users and inhabitants). Each component of the triad represents a dimension of the social production of space (Pierce and Martin 2015).

A significant advance in the inseparability of power and space discourse has been proposed by Soja (1996). The author defined a spatial typology resulting in three spatial formations: (1) the firstspace, filled with material or physical objects that are subjects of measurement (for instance buildings, roads and bridges), can be mapped and has an exchange value; (2) the secondspace, reflecting the underlying plans, rules and codes, formulations defined by city planners, engineers, architects and others, as representations and paradigms of the places (ideological); and (3) the thirdspace, the space of users and inhabitants in which the material objectivity of the firstspace crosses the abstract creativity of the secondspace, filled with use values. So, the thirdspace is where tensions and conflicts arise, between the powerful and the less powerful, embodying human agency. The thirdspace is the one where 'both exchange value and use value comes to life as social actors with multiple interests fight over the control of social space' (Saatcioglu and Corus 2016: 236).

According to Buitrago Bermúdez and Aguirre (2016), space defines the initial characteristics of its interactions through which it is possible to understand relations of power that shape the modes of socio-spatial production. If space is assumed as a social product (Lefebvre 1991, 2012), power relations determine its production, leading to processes of openness and closure that will generate distinct social behaviours (for example, hegemonic and/or counter-hegemonic projects). Reflecting Soja, 'space is not a scientific object removed from ideology and politics; it has always been political and strategic. It is a product filled with ideologies' (Queirós 2016: 155). Accordingly, space has to be considered both as a dynamic product and a process, a terrain for resistance and human agency among its different actors.

The process of production of space is mainly defined by the reasoning of the dominant power that generates an abstract (public, urban) space that is insured by capitalist rules and characterised by restrictions or limits to access (Leary-Owhin 2016), in which Lefebvre's idea of 'representation of space' is dominant. However, the process of production of space is affected by other complex and contingent elements, which are forged and negotiated. The uses and practices reproduced in them also generate new 'spaces of representation' through the action of counter-hegemonic projects that are born under the initiative of civil society (such as grass roots movements), becoming spaces of resistance, between the exchange and use values.

Without adopting a lived space (or a thirdspace) perspective, expressing differential capacities, coherent or in conflict, it is extremely difficult to capture the levels of insight of human spatiality as does for instance the historical perspective. So space is 'consequential', not just a stage where social life materialises (Soja 2010). The production of space, therefore, remains a central theme in studies concerning the relation between social dynamics and urban planning. The producers of social space (urban planners, land developers and policymakers) assume its control and build its exchange value, while the users and consumers (civil society, residents, consumers, migrants and minorities) seek to increase the use value it provides. This process has an impact on urban landscapes, generates (un)just geographies and affects the ways people experience public space at different scales.

Contemporary neoliberal dynamics impact not only the formal and physical production of space, but also its symbolic production through the consolidation of new behaviours and socialisation practices. Public space is thus changed and conditioned in the possibilities of its fruition, being constrained and mediated by ideologies and norms mainly linked to consumption. This leads to the production of spaces of inclusion but also of exclusion and segregation. Conditioned by capitalism, space is transformed into a commodity – considering both the residential space *per se* and the involving public space – materialised in spaces traditionally linked to social living practices, such as traditional neighbourhoods. These transformations are likely to be more visible in city centres due to the higher levels of centrality and to the potential identified in the rent gap, and usually have a smaller impact on the peripheries and in rehousing neighbourhoods (Sobarzo 2006). This process generates an urban hierarchy of spaces and the production of socio-spatial inequalities, particularly in the dichotomy between centres and peripheries. The former is progressively at an advantage as it is transformed into a 'high-quality consumer product for foreigners, tourists, people from the peripheries and suburban' (Lefebvre 2012: 25), while the latter is at a disadvantage, as it is converted into 'undesirable areas', perpetuated through dominant discourses.

However, the progressive marginalisation, and a certain formal desertion of the undesirable areas, allows some peripheries to develop autonomous processes of bottom-up empowerment, gaining new meanings in terms of specificity and self-representation, namely the production of a sense of belonging or a spatial consciousness specific to neighbourhoods that differentiates them from the rest of the metropolis. In fact, if in the central areas the socio-spatial dynamics are strongly conditioned and promoted by the dominant order, logics of exchange value and consumption, in some peripheral neighbourhoods we can experience a way of life less conditioned by the representations of space. These spaces allow the production of informal, inclusive and spontaneous spatial practices and lived experiences, embodying resistance to the hegemonic power. In this sense, the space of appropriation, as lived space or thirdspace, puts into relation the spaces of representation – the appropriate spaces of subjectivity – and the representations of space – the abstract spaces of objectivity (of the dominant power).

In order to elaborate on Lefebvre's spatial triad (with a move towards a more contextual and agency-based thought of urban social space) and to illustrate its use as a powerful tool to understand the urban world, in this chapter we adopt the concepts of 'lived space' and 'consequential geographies' to emphasise that space is not just a container (Soja 2010). It generates an active outcome of coalitions between groups of interest revealing diverse stakeholders with different (sometimes conflicting) purposes producing human spatiality. In this way, by looking into 'marginal' neighbourhoods where informality and spontaneous social action apply, we explore the production of 'differential space' in a migrant neighbourhood of Lisbon's immediate periphery: *Quinta do Mocho* (municipality of Loures – North of Lisbon) where communities of migrants from the PALOP (Portuguese-Speaking African Country) are over-represented.

We use Lefebvre's production of space and Soja's approach on the consequential geographies to understand the urban production of space by migrant communities in metropolitan areas. In Quinta do Mocho, recent public intervention is assuring better living conditions and creating a new image and therefore new representations of space. In this community, different counter-hegemonic spatial representations seek visibility and advocate for spatial justice, especially in the phase of the first occupation. These are bottom-up initiatives not always concordant and compatible between them – an emergent characteristic in the production of the lived space.

We use a mixed methodology that allows one to reflect on the triad proposed by Lefebvre from an empirical perspective. On the one hand an historical analysis was carried out, considering the genesis and the development of the neighbourhood. This is important in order to reflect on the relationship between urban planning and social practices, and on the consequences that the first has on the latter. Simultaneously, it was possible to observe how the lack of minimum survival conditions, such as housing, requires performances and actions that impact directly in the production of space, both physically and symbolically. Thus, the chosen neighbourhood functioned as an example of these opposing and complementary forms of acting within and through the public space.

In addition, this research is also the result of a reflection resulting from several previous projects carried out by the authors at different times and over the last 20 years. The methodologies used in these projects were mostly qualitative, carried out either through semi-structured interviews or direct observation – participatory and non-participatory. In addition, the permanent monitoring of judicial and political vicissitudes, which characterised the growth of the neighbourhood, made it possible to draw a sufficiently complete profile.

Quinta do Mocho in the LMA (Lisbon Metropolitan Area) context

The history of Quinta do Mocho illustrates a part of the complex process of territorial and social reorganisation that characterised Portugal and especially the metropolitan area of Lisbon since the 1970s. In the previous decade, an important internal rural-urban migration affected Lisbon. A process of concentration of investment, linked to manufacturing and the subsequent development of services, contributed to the progressive widening of labour supply, making the city and its surroundings highly attractive to those who left rural areas of the country in the search for jobs.

The 1970s were marked by the April Revolution (1974) that put an end to the fascist-inspired dictatorship that lasted between 1933 and 1974, and also by the collapse of the colonial empire in Africa, a situation that resulted in a definitive break with the former political, economic, social and territorial organisation of Portugal. This rupture resulted in important migratory flows in the mid-late 1970s and 1980s, characterised by: (1) a significant increase in the urban population, mainly from the return of Portuguese from the African ex-colonies, to which should be added the return of emigrants and exiles from European destinations; and (2) a growing number of immigrants from the PALOP, followed by Brazilians, South Asians and Chinese (Malheiros 1996). This process generated a significant housing shortage that gradually worsened. The original neighbourhood of Quinta do Mocho is, precisely, a good example of an informal and bottom-up response to the strong demand for housing by immigrants.

In a context marked by low housing offered at affordable prices and great demographic pressure enhanced by the arrival of more than 300,000 returnees in the mid-1970s, Greater Lisbon saw the proliferation of informal housing responses that included shanty-neighbourhoods (*barracas*), self-constructed illegal houses (*clandestinos*) and a few examples of occupation of abandoned semi-finished buildings, as in the case of Quinta do Mocho.

Quinta do Mocho: representation of space, spatial practice and space of representation

Quinta do Mocho ('The Owl Farm'), renamed by a law approved in July/August 2008 as 'Terraços da Ponte' ('Bridge Terraces'), is located in the Municipality of Loures, which has a resident population of 205,054 inhabitants, bringing together a majority of white Portuguese with people from the PALOP, Brazilians and Indians, among others. It has about 20 social housing neighbourhoods (http://www.cm-loures.pt/).

According to the National Statistical Institute, in 1991, Loures was the third municipality in the Lisbon Northern Metropolitan Area with the largest total number of foreigners and with the largest number of people from African countries. During the second process of regularisation of foreigners (from June to December 1996), the municipality of Loures accepted about 13 per cent of all requests received by the Regional Directorate of Lisbon of the Foreigners and Borders Service (SEF), with a bigger demand coming from Angolans and Guineans, in addition to Cape Verdeans, Santomese, Indians and Pakistanis. This shows the ethnic diversity of the municipality, where already in 1995 the main ethnic groups (Africans, Gypsies and Indians) showed diverse geographical distribution patterns, a situation that led to segregation combining ethnicity and accessibility. Despite the identification of socio-spatial segregation patterns involving the various ethnic groups, several social housing neighbourhoods of Loures display a multi-ethnic character, and this is not considered as a factor of conflict, with social relations marked by bonds of solidarity and reciprocal support (Baptista and Cordeiro 2002).

The origin of the neighbourhood dates back to the second half of the 1960s when the *Jota Pimenta* construction company was granted permission by the Loures City Council to build four buildings of more than ten floors – a complex of 1500 private dwellings – to which were added 'another 400 social housing' (Malheiros and Letria 1999: 74). Despite the definitive approval of the project in 1969, the construction was delayed and then interrupted because of the lack of agreement between the builder and the political actors involved, and definitely blocked in the post-revolution period (after 1974). As Malheiros (1996) reported, in 1975 the land and buildings were declared of public utility and the project transferred to another construction company.

In the following four years, the situation reversed, but the passage from one construction company to another delayed the works, and in 1979 the land ceased to have public interest and the original company regained the property.

Bureaucratic problems related to property rights of property continued for another ten years, until in 1988 'the Banco Fonsecas & Burnay took possession of 2/3 of the property as a form of payment of bank loans contracted by the construction company' (Malheiros and Letria 1999: 74). However, this situation did not provide a positive outcome. The approval of the first phase of the allotment (planned for 480 dwellings) was not carried out, and there was no authorisation to carry out the second phase of the project (planned for 1477 dwellings) by the Commission of Coordination and Regional Development of Lisbon and Tagus Valley and the General Directorate of Spatial Planning (Esteves 2004).

Throughout all these years some of the planned constructions were partially erected. These unfinished building structures, abandoned by the construction company, were gradually squatted by families from the PALOP. Many others later joined the 63 families surveyed in 1989. The Census of 1991 counted 296 families (1093 individuals), but the survey conducted in 1993 under the National Rehousing Programme (PER) reported 447 families (1495 individuals).

With the growth of the population of Quinta do Mocho, the Loures Municipal Council was forced to broaden and accelerate the process of resettlement of populations under PER (Decree-Law 163/93), in order to include all residents in Quinta do Mocho. Thus, in 1997, the Municipal Housing Division of the City Council of Loures made a last assessment of the Quinta do Mocho's population that identified about 3500 people. The rehousing that took place in new social housing located in the immediate vicinity of the occupied unfinished buildings was completed in two phases (April 2000 and March 2002).

The original semi-finished buildings (see Figure 23.1) were demolished, and nowadays only the memories of the residents and a few photographs register the times they were occupied by migrants – that still call them 'Mocho Velho' (the Old Owl). In the following analysis that mentions the old demolished neighbourhood composed by the semi-finished buildings and the new

Figure 23.1 The Mocho Valley, Quinta do Mocho. *Source*: Pedro Letria.

social housing neighbourhood, the former will be always referred to as Mocho Velho and the later as Quinta do Mocho.

Malheiros and Letria (1999: 75) describe the ancient specific micro-pattern of ethnic-spatial organisation that existed in Mocho Velho:

> in the shacks Cape Verdeans, the older inhabitants of the neighbourhood, some of them settled in the neighbourhood for more than twenty years; in the building corresponding to lots 1, 2 and 3, that displays a more advanced construction stage, the Guineans are installed; the largest part of Santomese are in the blocks 5 to 8, and the Angolans in lots 9 to 10.

The rehousing process led, in fact, to the construction of a new municipal neighbourhood located a few yards from the original occupied buildings of Mocho Velho; in place of these, luxury condominiums were built. The set of these two housing complexes – the social housing neighbourhood of Quinta do Mocho (see Figure 23.2) and the private condominiums of Terraços da Ponte – is today globally renamed as Terraços da Ponte. However, the new designation is not entirely consensual, as the original residents still use the traditional name: Quinta do

Figure 23.2 Terraços da Ponte, Quinta do Mocho. *Source*: Anna Ludovici.

Mocho. The separation between Quinta do Mocho and Terraços da Ponte is evident from both a symbolic and a structural perspective.

Today the recent luxury condominiums, built in the former area of Mocho Velho, constitute an island of self-segregation. This division, physically materialised in the avenue that separates the two housing complexes and also in the different architectural typologies (municipal re-housing neighbourhood *vs.* closed condominium), is also expressed in the logics of self-representation that comes out from different ethnic and social compositions. Although the nomenclature Terraços da Ponte is extensive to Quinta do Mocho, the two areas are distinct in the geographical origin of the residents. While 92 per cent of the residents of Quinta do Mocho are native from the PALOP (according to the Sociological Study of the Resident Population in Municipal Districts carried out by the Local Intervention Office of Mocho, Loures Municipal Council in 2004–2005), Terraços da Ponte buildings are designed for a completely different model of dwellers, mainly white and wealthy, as it appears from some initiatives (see, for example, the 'Association of Residents and Merchants of Urbanization Terraços da Ponte'[AMCUTP], created in 2011, whose objectives are to 'defend, preserve and strive for the improvement of the urban conditions, environmental and quality of life of its members' [http://www.amcutp.pt/]).

However, in the rehousing process, the relations created during the coexistence in Mocho Velho were respected where possible, and this allowed the maintenance of a sense of solidarity bonds and a strong organisation capacity among the residents, which led to the maintenance and development of different organisations based on common culture and, sometimes, ethnic origin. The oldest association is the 'Quinta do Mocho United and Cultural Association', created in 1993, precisely to represent and advocate for the interests of residents in the access to housing and related services. Being one of the local organisations with a strong aim to support the daily needs of its members, it results from the fusion of the former four Resident Commissions constituted by the main ethnic groups existing in the neighbourhood to fight for better conditions of living (to demand for infrastructures such as basic sanitation, electricity and piped water). This association is related to other projects with specific goals and objectives (for instance to help foreigners in their demands for residence permits, certificates of residence and poverty) and is recognised and supported by the Office of the High Commissioner for Migration.

For the past seven years, Quinta do Mocho has been deeply affected by the massive intervention of the Municipality of Loures in partnership with the local associations. The new infrastructures and equipment existing in the neighbourhood allow the development of some commercial and associative activities for the residents, and the use of indoor spaces for leisure practices, education and religious functions (whether for Catholics, Christians or Muslims). In 2014, the decision to nominate the neighbourhood for the European Urban Art Festival 'Diversity Advantage Challenge', with the project *O Bairro i o Mundo* (*The Neighbourhood and the World*) – through the creation of more than 30 murals painted on the residential buildings – was a turning point in the history of Quinta do Mocho. Today, the representation of a dangerous and inaccessible neighbourhood is being replaced by the image of an open-air art gallery with important artwork of famous *grafitters*, such as Vhils, Bordalo II, Odeith, Slap, Adres ou Pantónio. The Calouste Gulbenkian Foundation awarded this place with the award named Autarchic Good Practices for Hosting and Integration of Immigrants.

The rationalities of perception and representation of space as a socially produced object are reflected not only in the ways in which territory is materially organised and constructed – its morphology (Lefebvre 2012: 24) – but also in the social relations it produces. Led by the City Council, the rehousing process that took place in Quinta do Mocho physically shaped the place based on the Western housing models, namely those applied in social housing spaces marked by orthogonal street structures that optimise space and rows of repetitive four-floor box-style buildings with rather poor construction materials. Nevertheless, this process of social integration attempted to respect the self-organisation and sense of belonging structured in Mocho Velho; as the original residents emphasise, the strategies developed at the time of occupation can be considered not only as daily survival practices, but also as instruments of community aggregation and solidarity, including party organisations and symbolic rituals. People from the first generations of immigrants reproduced personal forms of its main use as an extension of private domestic use. Simultaneously, younger people are more motivated to use the indoor areas dedicated to the associations (for sports and leisure), and they participate actively in the most recent intervention of the Municipality of Loures for the creation of the street art-gallery, also offering guided tours for visitors and tourists and other local cultural activities. And the transport company has overcome the stigma of the neighbourhood and has set up a bus route that crosses Quinta do Mocho, responding to the combined demands of the municipality and the residents.

According to Carp (2008), although physical places have social authority that affects socio-spatial practices in their environment, it is people who sustain and transform places through their interrelationship with space. Quinta do Mocho, in planning-related terms, underlines both the physical and the lived experience elements, since the contemporary main use of the place is quite conditioned by the public policies and the physical space these have produced. This is more evident among the younger population who grew up and adjusted to this spatial organisation, since informal uses of space are more evident among older people. Nevertheless, informal practices are by no means exclusive of the immigrants, as their offspring, the younger residents, are strongly linked to the local associations that have endowed specific cultural practices and products.

In Quinta do Mocho, spatial practices are not actually separable from representations of space or lived experience. We have observed that this mutual conditioning between space and practices is more evident by comparing the behaviours of the first wave of immigrants with the behaviours of their descendants and also of a few people that arrived more recently (Ludovici 2016). It can be seen that the first wave of immigrants, despite recognising improvements in the material conditions of the new social dwellings, is 'pushed' to reproduce informal practices of use of public space that reveal the subtler logics of the traditional territorialisation, in which the production and appropriation of space continues to be mainly a collective operation. Their ties with the daily practices were mainly developed in their country of origin, displaying the relevance that is given to the neighbourhood public space as an element that extends home and living practices. Consequently, informal practices such as hanging clothes to dry in the neighbourhood trees, cooking on the street pavement of the buildings' entrances or producing 'street living rooms', especially on Sundays (that join several neighbours together who sit in chairs brought from home and do not use the public urban furniture) are visible and common in the supposedly rationally produced space of Quinta do Mocho. The tradition of developing their domestic dynamics not only in the private space, but also in the public space (which prolongs the first), is currently reflected in the neighbourhood's soft informal practices. This was even more evident in Mocho Velho, where spatial/physical structures were more flexible and the transition from private domestic space to public common space was more blurred due to the lack of windows or the broken walls.

The relationship of the immigrants' offspring, who grew up in Lisbon, to the spaces of the neighbourhood, is quite different. Despite maintaining a strong socio-spatial identity construction linked to the image of the country of origin, the second generation is mostly included in the dynamics of the destination country, via consumer and leisure practices. This different attitude is also confirmed by a sense of belonging to their own neighbourhood (and to a lesser extent to their country of destination). These different ways of construing the connection of the various groups with the conceived space shows the importance of the relation between spatial planning and social practices within specific forms of socialisation and acculturation.

Therefore, if the space of the neighbourhood continues to be the main place for activities and representations for both generations, confirming for both a phenomenon of spatial segregation and ghettoisation, at the level of appropriation, the second generation experiences the public space through a more dynamic relation, that also involves a higher and more diversified practice of the urban space of Lisbon. In addition, the proactive presence of the local authority, whether through projects aimed at the 'opening of the neighbourhood to the world', or in projects exclusively developed within the community/local associations, contributes substantially to the dynamics of the neighbourhood's territory and partly determines its appropriation (for instance, it reduced the phenomena of spatial segregation and ghettoisation among the second generation).

We assert, therefore, that the elements of the process of producing space that are associated with daily practices and complex social relations taking place in this rehousing neighbourhood characterised by a great cultural and ethnic variety are conditioned by different factors. The cultural identity of the country of origin, when it does not degenerate into inter-ethnic conflicts, can form the basis of ties of solidarity, which nourish and improve the sense of belonging and well-being. In addition, the need to fight adversity, as happened more explicitly in the days of Mocho Velho, also forges ties and leads to forms of collective action. But the construction of a sociospatial identity is not a static process, being shaped and re-shaped through the presence of the immigrants' descendants, who grew up to identify more with the formal space of the neighbourhood and also bring to it new social dynamics from the host country. It is particularly for this group and in this dimension that the representation of space is more evident. Taking into consideration these processes, we may underline that there is a straightforward relationship between urban planning and social behaviour and that the daily practices – both the personal/individual and public/collective ones – need specific spaces to be realised.

Conclusions

In this chapter, we highlight the connection between conceived space/representation of space and lived space/space of representation/thirdspace, continuously interacting and redefining according to the norms, knowledge, identities and actions of the actors and their collective and individual use of urban space. Therefore, the spatial facets of the triad are not separated since via dialectical interrelations they produce a simultaneous space. Expressing differential capacities, in conflicted or coherent space, is where social life materialises.

This migrant neighbourhood provides insights into a latent conflict and presents the basis of a case for cooperation. In its origins, the place-making of Quinta do Mocho is a function of the formal development of the LMA within a suburbanisation process associated with the 1960s capitalist investment and the spatial concentration of activities that attracted labour and justified the expansion of new residential areas. If these dynamics have defined the original lines of space production, namely their destinations and uses, it is evident that the following decades showed how the informal action of its residents contributes to modify and redesign the urban landscapes.

The occupation of the old estate in the 1980s and the creation of Mocho Velho illustrates a narrative of conflict and resistance that is evident at different scales. At a global level, it is the result of the economic and political changes that included a long-lasting conflict in Angola and frequent military tensions in Guinea-Bissau, and the birth of new independent national states. At a local level, it expresses the creativity and resistance of immigrants in irregular situations, showing that these people were capable of producing a living space that was able to use the weaknesses of the formal conceived space and fit into its interstices, not only institutionally, but also in physical and territorial terms.

At the end of this process, the final production of Terraços da Ponte, involving the demolition of Mocho Velho and the construction of the present social housing estate, illustrates, as Pierce and Martin (2015) put it, a narrative of cooperation between the institutions that dominate/control space (the dimension of the representations of space or the conceived spaces that come from the technical knowledge and political power) and the practices/uses/appropriation of the space of its inhabitants in their daily routines.

In the case of the new Quinta do Mocho we may now identify different spatialities. On the one hand, older residents, mostly immigrants from the PALOP, show more contradictions in the uses of the neighbourhood spaces, blurring the distinctions between private and public spaces and reproducing some informal uses inherited from African territoriality. On the other hand, the second generation, mostly youth with an immigrant background, express cooperative cultural practices and an appropriation of the Quinta do Mocho space more in conformation with the contents of conceived space, showing how growing up and being socialised in a particular space contributes to shape people's territoriality, and enlightening the effects of the public policies in the territory.

All things considered, the evolution of Quinta do Mocho/Terraços da Ponte represents different phases of this complex process, where conceived space/representation of space and lived space/representational space intersect and influence each other in various ways with distinct levels of impact. The final rehousing stage determined a return to 'normality' via the restoration of the representation of space, arising from the regularisation of a process of clandestine occupation that culminated with the construction of the social housing neighbourhood. In this new space, where neighbourhood planning and architecture forms meet the other levels of formal reproduction (for example the school system and local policy regulations), lifestyles and daily practices have progressively become more homogenised, reflecting the consequences of daily involvement in these systems.

References

Baptista, L.V. and Cordeiro, G.I. (2002) 'Presentes e desconhecidos: reflexões socio-antropológicas acerca do recente fluxo imigratório no concelho de Loures', *Sociologia, Problemas e Práticas*, 40 23–43.

Buitrago Bermudez, O. and Aguirre, M.A. (2016) 'Analisis socioespacial de los humedales Guarinó y la Guinea (Municipio de Jamundí, Colombia)', *Finisterra. Revista Portuguesa de Geografia*, 51:103 3–24.

Carp, J. (2008) '"Ground-Truthing" representations of social space. Using Lefebvre's conceptual triad', *Journal of Planning Education and Research*, 28 129–42.

Esteves, A. (2004) *Imigração e Cidades: Geografias de Metrópoles Multi-étnicas - Lisboa e Washington D.C.*, Lisboa: Universidade de Lisboa (unpublished PhD thesis).

Leary-Owhin, M.E. (2016) *Exploring the Production of Urban Space: Differential Space in Three Post-industrial Cities*, Bristol: Policy Press.

Lefebvre, H. (1991) *The Production of Space*, Oxford: Blackwell (original publication 1974).

Lefebvre, H. (2012) *O Direito à Cidade*, Lisboa: Livraria Letra Livre.

Ludovici, A. (2016) *Perceção e Apropriação do Território. Os Guineenses em Lisboa: duas Gerações em Comparação*, Lisboa: Universidade de Lisboa. Available HTTP: <http://repositorio.ul.pt/handle/10451/26049>.

Malheiros, J. (1996) *Imigrantes na Região de Lisboa: os Anos da Mudança, Imigração e Processo de Integração das Comunidades de Origem Indiana*, Lisboa: Ed. Colibri.

Malheiros, J. and Letria, P. (1999) *À Descoberta dos Novos Descobridores*, Lisboa: Comissão Nacional para as Comemorações dos Descobrimentos Portugueses.

Pierce, J. and Martin, D.G. (2015) 'Placing Lefebvre', *Antipode*, 47:5 1279–99.

Queirós, M. (2016) 'Edward Soja: geographical imagination from the margins to the core', *Planning Theory & Practice*, 17:1 154–60.
Saatcioglu, B. and Corus, C. (2016) 'Exploring spatial vulnerability: inequality and agency formulations in social space', *Journal of Marketing Management*, 32:3–4 230–51.
Sobarzo, O. (2006) 'A produção do espaço público: da dominação à apropriação', *GEOUSP: Espaço e Tempo*, 19: 93–111.
Soja, E.W. (1996) *Thirdspace. Journeys to Los Angeles and Other Real-and-Imagined Places*, Cambridge, MA: Blackwell Publishers Inc.
Soja, E.W. (2010) *Seeking Spatial Justice*, Minneapolis, MN: University of Minnesota Press.

24

Contested cultural heritage space in urban renewal

The case of a dense urban city in Hong Kong

Esther H.K. Yung and Ho Hon Leung

Introduction

Cultural heritage is defined very broadly to include both tangible and intangible values. Tangibles include individual sites, buildings or structures, as well as urban or rural landscapes. Intangibles refer to 'the practices, representations, expressions, knowledge, skills – as well as the instruments, objects, artefacts and cultural spaces associated therewith – that communities, groups and, in some cases, individuals recognise as part of their cultural heritage' (UNESCO 2016 Article 2: 5). In this study, cultural heritage space refers to urban landscapes that contain intangible values that include rich local culture. According to Jacobs (1961), streets are regarded as the main public places of a city which can make a city look interesting.

Under the current trend of urbanisation, rapid and uncontrolled development is transforming urban areas and their settings, resulting in the fragmentation and deterioration of urban heritage and having a deep impact on community values. Urban renewal is an elementary measure intended to tackle the problem of urban ageing which occurs in all developed cities. It is increasingly recognised that conservation and redevelopment can be complementary (English Heritage 1999; Kearns and Philo 1993; While 2006) and that heritage conservation, as part of the urban renewal of old districts, can have a significant impact on enhancing sense of place and identity and community development (Steel and Slayton 1965). However, with the intense demand for housing and commercial spaces, there has been insufficient consideration of the impact of (re) development on historical settings and the original urban life, leading to the phenomenon of restructuring urban space (Logan 2002 in Wang and Lee 2008).

In Hong Kong, an aggressive urban renewal regime has been implemented in the last two decades. This has involved the demolition of decayed and obsolete buildings in old districts and the subsequent construction of new buildings with the intent to create a better living environment. Although it is claimed that preservation is one of the key principles of the urban renewal strategy, the exchange value of redevelopment has been more apparent than the use value of the sites that have cultural significance. In this chapter, we will study two public areas in Mongkok and Wanchai, in Hong Kong, which have a unique heritage and social and cultural values; namely, Bird Street market and Wedding Card Street. Unfortunately, both these public areas have been demolished to make way for commercial development and high-rise residential development. Inevitably, the social network, social cohesion and the unique local culture and characteristics of the two districts have as a result of redevelopment been destroyed. The relocation and recreation of heritage spaces in planned urban spaces demonstrate a contentious reproduction of space. In these two cases, Lefebvre's (1991) concept of the 'representation of space' offered from planners and professionals (the conceived) and inhabitants and users (the lived) is vividly portrayed. These two cases also illustrate the notion of 'contradictory space', in which contemporary global capital is contested with the localised meaning of heritage space.

Theoretical framework

Conservation of cultural heritage space with local significance

It is commonly recognised that cultural heritage connects people to their histories and collective memories (Halbwachs 1980). It relates to the everyday lives, communications and meanings attached to a district (Assmann 1995). Heritage conservation can also help develop cultural identity through the identification of common grounds, including location, history, aesthetics and religious beliefs (Ashworth et al. 2007). The evolving trend of heritage conservation shows that heritage does not only refer to preserving listed buildings or monuments of international or national significance, but also to those buildings associated with familiar and cherished local scenes (Delafons 1997; Lamei 2005) associated with the rich social values of a society as a whole. It is increasingly recognised that historic and local streets reflect traditions, local stories and beliefs and traditional ways of life (Yung and Chan 2015) which represent a sense of cultural identity (Orbasli 2000; Doratli et al. 2004) and attract tourism (Wang and Lee 2008). Many countries worldwide, such as Taiwan, Singapore, Japan, China and the United States, have designated conservation areas or historic districts to preserve historic streetscapes. In addition, many market districts in urban centres are vibrant places which have become important tourist destinations as they portray the everyday lives of people. The case of La Rambla Market in Barcelona and Pike Place Market in Seattle are cases in point.

Relationship between urban renewal and heritage conservation

Urban renewal usually involves large-scale demolition and redevelopment, which destroys sense of place, history and people's local memories. The conservation of historic buildings is increasingly recognised as contributing to the social well-being and sustainability of an urban city. Although urban renewal has gradually changed from the bulldozer approach to regeneration and revitalisation, towards a social consciousness which takes account of economic, physical and environmental conditions, numerous historic buildings located in old run-down areas are often demolished to make way for new development. As a consequence, social problems have arisen, such as the destruction of existing social networks and social cohesion, forced eviction of vulnerable groups and discontinuation of everyday community lives, loss of sense of identity and loss of collective memories (Hayden 1995). As such, there have been increasing demands from the public for retention of local characteristics and communities, particularly the preservation of sites and structures of historical, cultural and/or architectural interest, and preservation of the social networks of local communities.

Representation of cultural heritage space

Growth politics is often a common goal in cities and is a key determinant in the social production of urban space (Lefebvre 1991, 1996). Different interest groups, including preservationists, developers and politicians, increasingly working together to market historic sites for combined profits, have been considered to be 'heritage machines' (Barthel 1996). Although cultural heritage spaces and sites often bring a wide range of externalities, including cultural identity and sense of place, heritage is often seen as an economic resource with which to promote tourism, economic development and urban regeneration (Graham 2002). Heritage has also been recognised as an important element of symbolic economy (Zukin 1995) by which 'cultural strategies drive the production of commercialized urban space geared towards entertainment and tourism' (Reichl 1997: 515).

In addition, the propaganda of heritage conservation has increasingly reshaped and altered the true meaning of heritage assets in society. The notion of heritage conservation has been invariably used for different purposes by people with different interests. For example, cultural heritage is often associated with enhancing upper-class cultural life. At the same time, the commodification of heritage has become 'a nostalgic twist of an increasingly consumption-orientated society, turning history into a commodity to suit the tastes of the affluent classes (Lu 2002, quoted in Ng 2009). It is evident that the revitalisation of historic buildings is increasingly being used to promote consumption. These economic and cultural uses also help create the images with which places are marketed in economic and cultural terms (Graham 2002). In turn, the association of the upper classes and consumerism often results in the relocation of old traditional businesses due to dramatic increases in rents (Smith 1996; Lees et al. 2008).

Urban renewal in Hong Kong

Hong Kong is a very vibrant metropolitan city. The city is constantly under construction and development in order to meet the demands stemming from a population explosion and rapid economic changes. Due to a shortage of flatlands, in this tiny territory, packed with more than 7 million residents, there is an increasing need for housing, traffic, green areas and commercial space. Thus, urban renewal has become one of the most important issues in Hong Kong.

The British Administration in Hong Kong, in 1988, set up the Land Development Corporation (LDC), which was officially responsible for the renewal of urban areas in Hong Kong (Adam and Hastings 2001). However, due to severe public criticism of the slow progress of urban renewal and unfair compensation, the LDC was transformed into the Urban Renewal Authority in 2001 after the handover of Hong Kong to China under the Urban Renewal Authority Ordinance (URAO) (URA 2015). The URA is a statutory body responsible for undertaking, encouraging, promoting and facilitating the regeneration of older urban areas in Hong Kong. Compared to LDC in the past, the URA acts as an implementer that initiates redevelopment projects on its own instead of facilitating other parties' projects (URA 2011a). Apart from initiating projects by itself, the URA also responds to the joint approach, which means working with the buildings' owners or landlords to redevelop the owned lots or buildings. Along with the establishment of the URA, the Urban Renewal Strategy (URS) was published, which acts as a guideline for urban renewal measures. The approach the URA has adopted is 'people first, district-based, public participatory', and 4Rs (redevelopment, rehabilitation, heritage preservation and revitalisation) are the principles of the URS. Among the 4Rs, redevelopment and rehabilitation are at the core of the URA (URA 2011b).

However, criticisms have been levelled at the urban renewal movement in Hong Kong and communities have questioned whether the movement has achieved the URS's principles. Criticism has also been raised by the public over issues such as economic-based redevelopment and lack of cultural preservation. Furthermore, much of the revitalisation of historic buildings projects in the old districts has been censured for its bias for achieving economic objectives and overlooking the social impact on the local community. The URA has claimed that its emphasis on ensuring space for everyday businesses has helped to preserve local characteristics and maintain social networks and enhance district vitality (Development Bureau 2011). However, the claims that the regenerated neighbourhoods have had a positive impact on the original inhabitants and encouraged business remain contentious. In fact, many inhabitants have been forcibly evicted and have sold their properties to the URA for unilaterally agreed compensation. Moreover, many traditional trades and businesses have been disappearing from the Hong Kong streetscape. Commercial redevelopment, which claims to subsidise the revitalisation and reuse of historic buildings, often destroys the local characteristics of districts. In addition, in the process of redevelopment, the quantities and qualities of the original open spaces are lost. The demolition and relocation of traditional local open street markets to indoor buildings are cases in point. Given this dilemma, the question becomes: to what extent do urban renewal projects negatively affect the local culture and public spaces of local significance?

Case 1 – Bird Street (Hong Lok Street demolition and relocation)

The oldest and most lively area, Bird Street market, used to be located on Hong Lok Street in the centre of Mongkok. Bird Street market was crowded with shops, restaurants and residential areas. This open market street was a favourite meeting place for people who enjoyed watching birds and chatting with each other. Thus, the district was a local landmark and a tourist attraction. Hong Lok Street was formed around the 1960s and was popular due to an old restaurant named *Ki Heung tea-house*. It then became a gathering place for bird-keepers. It attracted bird hawkers who ran their stalls near the tea house. Hong Lok Street was chosen because of its very convenient location to major transport networks. The centre lanes bordered by stalls in Hong Lok Street were so narrow that no more than two people could pass through them at the same time. This car-free street housed more than 100 stalls of all types of businesses related to bird keeping, such as birds, cages, bird food and bird-feeding products. The diversity of products attracted many visitors and bird-keepers, even after the tea house was demolished. According to the Wong (1995) survey, those who frequented Bird Street market were more concerned with its unique characteristics rather than its negative environmental conditions, such as the lack of hygiene, the crowdedness and noisiness.

Bird Street market signified the soul and the long tradition of the bird-keeping culture in Hong Kong. The bird-keeping culture in Hong Kong, and in China in general, is very different from that in Western countries where birds are kept at home. In Hong Kong, a common hobby, especially among elderly males, is to walk the streets with their caged birds. Garisto (1992: 46), describing the history of bird keeping in the Chinese culture, remarks that birds and birdcages have been 'an essential element of Chinese domestic life'. Like people who keep dogs or cats, bird-keepers stroke and talk to their birds. As a 53-year-old retired civil servant commented, 'I like their singing. I can get a sense of success if I can keep them healthy and beautiful'. Both bird-keepers and their birds enjoy their gatherings in popular areas, and the former Bird Street market was one of them. When bird-keepers are with their birds, they interact with each other and also interact and talk with the bird-shop owners. The majority of elderly bird-keepers are retirees, so a key motivation for keeping birds is that they can make friends more easily when

Figure 24.1 Photo of Bird Garden, Yuen Po Street. *Source*: Photo taken by Esther H.K. Yung.

they have the birds. Other than personal reasons, bird-keeping behaviours actually reflect the highly dense environment and the housing situation in Hong Kong. Keeping birds rather than dogs or cats is preferable and more practical in a very small living unit, particularly in public housing which imposes stringent restrictions on keeping dogs and cats.

After a ten-year negotiation, Bird Street market in Hong Lok Street was demolished in 1998 and relocated to Yuen Po Street and became Bird Garden, next to the Flower Market in Mongkok (Figure 24.1). The former Bird Street market was replaced by the Langham Place Shopping Mall, an upscale shopping mall opposite Langham Place, and a five-star hotel. Although much cleaner, better organised and nicely decorated in the traditional Chinese garden style and with sufficient open space, the new Bird Garden can house only a few dozen stalls. The atmosphere and the ambience of the place are not as vibrant as before. The setting of Bird Street market was a lane in Hong Lok Street located in one of the busiest areas in the Mongkok district. After relocation, it was turned into a garden with tiled roofs, a pavilion, gates and trees.

Both bird-keepers and shop owners indicate that the former Bird Street market in Hong Lok Street had a stronger sense of belonging than the new garden. Now, bird-keepers tend to stay in the market for shorter periods of time (interviews conducted in 2016). To make matters worse, as it is at a distance from public transportation, the numbers of bird-keepers coming to the new Bird Garden is far less than before.

Case 2 - Wedding Card Street (Lee Tung Street/McGregor Street Redevelopment Project)

Lee Tung Street (Figure 24.2) is located in Wan Chai, which is in the middle part of Hong Kong Island and is one of the foremost developed areas of Hong Kong. Lee Tung Street was built around the 1910s to 1920s in the three-story tradition of Chinese tenement buildings. In the 1950s, the street added six- to seven-story buildings. Lee Tung Street was known for its large number of publishing companies; before the street was redeveloped, 28 out of 52 shops were publishers. In particular, from the 1970s, wedding invitation card manufacturing became the signature of Lee Tung Street, which earned it the nickname of 'Wedding Card Street'. The

Figure 24.2 The avenue in the former Lee Tung Street. *Source*: Photo taken by Esther H.K. Yung.

red invitation cards and red pockets created a special scene along the street. Apart from the publishing industry, other businesses were also available in the street such as tailors, shoemakers, property agencies, laundry service providers, groceries and Chinese restaurants.

The Hong Kong government began to consider redevelopment in 1999 and commenced the H15 project – Lee Tung Street/McGregor Street redevelopment project in early 2004. The project was the first government-initiated large-scale redevelopment project. The question of urban development and the unique cultural status of Lee Tung Street became the subject of great debate by the public. The project demolished 35 tenement buildings to make way for a number of residential towers. Within the site area of the project, only three old tenements buildings (186–190 Queen's Road East), built in the late 1930s, have been preserved and revitalised.

The H15 Concern Group was comprised of former business operators and residents formed to pursue the interests of the affected people and to convey to the government the importance of the community in urban renewal. During the planning stage of the H15 project, the Concern Group proposed keeping the old low-story buildings. It was believed that this would preserve local characteristics while boosting the economy. The Concern Group also initiated a number of hunger strikes in an attempt to persuade the URA to preserve the street and accept the alternative Dumbbell Proposal, which would conserve the local character of the neighbourhood while allowing original habitants and businesses to return after redevelopment (Wissink 2015), but it was rejected by the Town Planning Board. Despite strong protests against the demolition of Wedding Card Street organised in 2007, the 35 post-wars tenement buildings were demolished in the same year.

After the redevelopment project, due to increased rents, none of the 28 publishers remained in Lee Tung Street. Only shops which aimed to promote their brand names could afford the sky-high rents. Without the old residents and shop owners, the community was lost and, hence, Lee Tung Street no longer has a cultural identify. At present, there is only one publisher on the street located at the basement level; the local culture of Lee Tung Street is now completely different.

Redeveloped Lee Tung Street has become a 200-metre walkway named Lee Tung Avenue with a group of low-story buildings providing 70 shops in different floor areas that suit the majority of businesses. Instead of only ground floor shops being available, basement and upper-ground shops have been added. Therefore, the total number of shops has increased. The rental income of the avenue is estimated at HK$25 million per month. The businesses on the Avenue are now diverse, including Chinese and Western cuisine, fine dining and cosmetic and clothing retailers. High-end multinational companies are also located on the Avenue, which has attracted tourism and facilitated the overall economy of the Wan Chai district.

Lee Tung Avenue was designed based on the concept of integrating Eastern and Western cultures and combining both traditional and modern styles. For instance, buildings have been built in the European style, and the very few elements which recall the old Wan Chai are the window grilles, the mid-20th-century-style wall tiles and the street lamps in the style of 19th-century gas lamps. The name of Lee Tung Street has disappeared from the city's map, as the area now is traffic-free and not regarded as a street in the government land registry (Ng 2013). It has been renamed 'The Avenue' in English. The 'inside' of the street is totally different, and it is no longer a unique cultural spot in Wan Chai. The past characteristics of 'Wedding Card Street' have been demolished, together with the old tenement buildings. In reconstructing Lee Tung Street as an open shopping mall, no unique characteristics were created, nor the local history reflected.

In an interview with a member of the Concern Group, Ms. Wong, the representative commented:

> It is undeniable that the overall business environment is improved, shops with branding are set-up here. However, the lack of local business is a major problem. I don't see the reason for which a resident of Wan Chai would come here and purchase for the daily consumables. The business here is indeed improved comparing to the past but Lee Tung Street has no difference to a shopping mall now. The cultural value and community network has [been] lost (interview 2016).

According to Ms. Wong, other than the publishing industries, Chinese tailors and watch repairs also contributed to the spirit of Lee Tung Street. The H15 Concern Group representative claimed that the developers had no interest in maintaining the local wedding card production businesses (interview 2016). No rental discounts were given to past business owners on the street, nor was there any mention of any shop-for-shop scheme. Although some of the past business owners were eager to continue their businesses, they were forced to close because of high rents. Without the preservation of these unique businesses, Lee Tung Avenue has become another common shopping mall in Hong Kong where the value of the street is now similar to its counterparts after redevelopment. As a result of redevelopment, the attraction of Lee Tung Avenue for local residents has been reduced because of the lack of cultural identity and valuable traditional businesses. A resident mentioned that it was a pity for Wan Chai, if not Hong Kong, that a street with such a unique character was lost as a result of urban renewal.

Discussion

Bird Street market in Hong Lok Street provided a vibrant space for social integration and fostered a sense of community and identity in the district, as well as being a big attraction for visitors and tourists. It also provided space and opportunities for small and marginal businesses, which highly reflected the local character and local culture. Since the relocation of the bird market to the new Bird Garden, the majority of visitors and customers are those who frequented the old Bird Street market in Hong Lok Street many years ago. Unfortunately, the Bird Garden did not maintain the character and prosperity of the former Bird Street market. The case of the relocated Bird Street market to the Bird Garden is an illustration of a conceived space reproduced by planners and professionals. On the one hand, it relocated the hardware, since most of the stalls were relocated to an environment which was conceived to be better because of improved hygiene and reduced noise. On the other hand, it failed to maintain the software, namely the 'lived space', the bird-keepers and shop owners' experiences of using the space and the meaning of the space. The professionals thought that the Bird Garden would be a more pleasant gathering place for the bird-keepers and a new attraction for new visitors. However, the ambience of the space has been changed. What the visitors want to see is the traditional cultural lives of people holding their bird cages and walking along the streets and talking to each other. It was the prosperity of the street, the trading activities and the social networks which attracted people.

Lee Tung Street was the first and only street famous for its wedding card businesses. Lee Tung Street was involved in numerous marriages and should be one of the unforgettable memories of many couples. The cultural value of Lee Tung Street was invaluable, and it possessed unique meaning in the minds of the Hong Kong people. Unfortunately, its core value was demolished together with the old tenement buildings, and the redesigning of the street is totally different. The influx of foreign restaurants and shops has led to a mixture of Chinese and Western cultures. While the development in this case has helped the economy, the social benefits are questionable, and what was once the hallmark of Lee Tung Street – the wedding card shops – has been lost. Clearly, the redevelopment project failed to preserve local culture and characteristics. None of the characteristics of the past Lee Tung Street were kept. Wedding card businesses, publishers and, most importantly, community networks, have been lost. The bonding of the neighbourhood has been demolished together with the old buildings. It is undeniable that redevelopment is beneficial in terms of economy and environment. However, its negative impact on communities is worth addressing.

In the Lee Tung Street conservation and redevelopment controversy, the strong protests from the community questioned whether consideration of architectural and historical value is sufficient or if it is more important to respect the relationship between buildings, local culture and characteristics, people and social networks. As Jacobs (1961) states, what makes a street special, and therefore the city, is the people. In this case, the locals' lived space was replaced by the reproduction of space for consumption, dominated by commodification over the continuation of everyday lives, meanings and stories of the habitants (Lefebvre 1991, 2003).

In both cases, the streets contained traditional business activities, as well as being the mainstay for social interactions, stories and memories of the visitors and the businesses and stall owners.

In the reproduction of the Bird Garden, the planners' and professionals' intention was to provide a better physical environment for the bird-keepers and bird-traders. However, what they conceived was unexpectedly different from what was actually needed by the users who saw social linkage as more important than the comfort of the environment. The location of the Bird Garden in terms of proximity to other facilities, such as eating places, has been one of the major hurdles discouraging bird-keepers from visiting and staying in the place. The different spatial form of a lane and a garden has also made the ambience of the former Bird Street and Bird Garden very different from each other. Although there is little evidence of commodification, the local meaning and character of the space has been distorted.

In comparison, the reconstruction of Lee Tung Avenue reveals strong evidence of production of space for consumption. It can be shown that the use values of the former wedding card and publishing business street have been transformed to exchange values, as reflected in the high-rent shopping and catering precinct. Similar to the Bird Garden, it has changed the local meaning of the space derived by the users' experience. The meaning of the space has continued evolving, as other people, rather than the old tenants and business people of Lee Tung Street, are coming to use the space. The European-style shopping avenue is definitely a representation of consumption space using the heritage theme as a marketing strategy, and it can be seen as a contradictory space created under urban renewal.

Conclusions

This chapter illustrates that tangible built environment, the street itself, and intangible cultural roots, traditions and social networks are essential considerations for heritage preservation. It is not just the physical building fabric which is worth preserving, but also the traditional cultures and local characters. The two cases examined, Bird Street market and Wedding Card Street, are controversial examples of urban renewal where planners and professionals planned and designed the spaces to reproduce the conceived quality of the space. However, they both reveal that there is a mismatch between what is conceived and experiences of the everyday lives of the users. Intangible cultural heritage is an essential source of a community's cultural vitality and an important source of social cohesion. The complexities of community life, particularly in terms of cultural identity and ethnicity, often result in contestation of space regarding the designation and representation of local heritage. Inevitably, urban renewal threatens the legacy of lived space with the replacement of conceived space or, even worse, results in the recreation of 'contradictory space', which distorts the meanings and stories of the space. Has aggressive urban renewal been carried out at the expense of heritage space in past decades?

References

Adams, D. and Hastings, E.M. (2001) 'Urban renewal in Hong Kong: transition from development corporation to renewal authority', *Land Use Policy*, 18:3 245–58.

Ashworth, G.J., Graham, B. and Tunbridge, J.E. (2007) *Pluralising Pasts: Heritage, Identity and Place in Multicultural Societies*, London: Pluto Press.

Assmann, J. (1995) 'Collective memory and cultural identity', trans. J. Czaplicka, *New German Critique*, 65: 125–33.

Barthel, D. (1996) *Historic Preservation: Collective Memory and Historical Identity*, New Brunswick, NJ: Rutgers University Press.

Delafons, J. (1997) *Politics and Preservation: A Policy History of the Built Heritage 1882–1996*, London: Spon Press.

Development Bureau (2011) *People First – A District-Based and Public Participatory Approach to Urban Renewal Urban Renewal Strategy*, Hong Kong: Development Bureau. Online. Available HTTP <http://www.ura.org.hk/en/pdf/about/URS_eng_2011.pdf> (accessed 15 January 2017).
Doratli, N., Hoskara, S.O. and Fasli, M. (2004) 'An analytical methodology for revitalization strategies in historic urban quarters: a case study of the Walled City of Nicosia, North Cyprus', *Cities*, 21:4 329–48.
English Heritage (1999) *The Heritage Dividend: Measuring the Results of English Heritage Regeneration 1994–1999*, London: English Heritage.
Garisto, K. (1992) *The Birdcage Book: Antique Birdcages for the Contemporary Collector*, New York: Simon & Schuster.
Graham, B. (2002) 'Heritage as knowledge: capital or culture?' *Urban Studies*, 39:5–6 1003–17.
Halbwachs, M. (1980) *The Collective Memory*, trans. F.J. Ditter and V.Y. Ditter, New York: Harper Colophon.
Hayden, D. (1995) *The Power of Place: Urban Landscapes as Public History*, Cambridge, MA: The MIT Press.
Jacobs, J. (1961) *The Death and Life of Great American Cities*, New York: Random House.
Kearns, G. and Philo, C. (1993) *Selling Places: The City as Cultural Capital, Past and Present*, Oxford: Pergamon Press.
Lamei, S. (2005) 'Insights into current conservation practices', *Museum International*, 57:1–2 136–41.
Lefebvre, H. (1991) *The Production of Space*, trans. D. Nicholson-Smith, Oxford: Blackwell Publishing.
Lefebvre, H. (1996) *Writings on Cities*, Cambridge, MA: Blackwell.
Lefebvre, H. (2003) *The Urban Revolution*, trans. R. Bononno, Minneapolis, MN: University of Minnesota Press.
Lees, L., Slater, T. and Wyly, E. (2008) *Gentrification*, New York: Routledge, Taylor and Francis.
Logan, W. (2002) *The Disappearing 'Asian' City—Protecting Asia's Urban Heritage in a Globalizing World*, New York: Oxford University Press.
Lu, H. (2002) 'Nostalgia for the future: the resurgence of an alienated culture', *Pacific Affairs*, 75:2 169–86.
Ng, J. (2013) 'Wedding Card Street in Wan Chai vanishes off map for "high street" plan', *South China Morning Post*, 26 June 2013.
Ng, M.K. (2009) '"Kainos" renewal: promoting urban regeneration as a natural quality', in *Surveyors in Urban Regeneration, HKIS Annual Conference*, 26 September.
Orbasli, A. (2000) *Tourists in Historic Towns Urban Conservation and Heritage Management*. London: E & FN Spon.
Reichl, A.J. (1997) 'Historic preservation and progrowth politics in U.S. cities', *Urban Affairs Review*, 32:4 513–35.
Smith, N. (1996) *New Urban Frontier: Gentrification and the Revanchist City*, London: Routledge.
Steel, R. and Slayton, W.L. (1965) 'Urban renewal/a retrospect of addresses and discussions', at the Chartered Surveyors Annual Conference 1965 with Supplementary Information on Urban Renewal Procedures in The United States of America, Royal Institution of Chartered Surveyors, London.
UNESCO (2016) *Basic Texts of the 2003 Convention for the Safeguarding of the Intangible Cultural Heritage*, 6th edn, Paris: UNESCO. Available HTTP: <www.unesco.org/culture/ich/en/convention> (accessed 15 February 2017).
Urban Renewal Authority (2011a) *Urban Renewal: The Past 20 Years*. Available HTTP: <http://www.ura.org.hk/media/436552/pub_past20years.pdf> (accessed 15 March 2016).
Urban Renewal Authority (2011b) *Urban Renewal Strategy*. Available HTTP: <http://www.ura.org.hk/en/pdf/about/URS_eng_2011.pdf> (accessed 15 March 2016).
Urban Renewal Authority (2015) *About URA*. Available HTTP: <http://www.ura.org.hk/en/about-ura.aspx> (accessed 15 March 2016).
Wang, H.Y. and Lee, H.Y. (2008) 'How government-funded projects have revitalized historic streetscapes – two cases in Taiwan', *Cities*, 25:4 197–206.
While, A. (2006) 'Modernism vs. urban renaissance: negotiating post-war heritage in English city centres', *Urban Studies*, 43:13 2399–419.
Wong, M.L. (1995) 'Urban renewal and cultural heritage conservation in Hong Kong: a case study of Hong Lok Street renewal project', unpublished thesis, Master of Sciences (Urban Planning), Centre of Urban Planning and Environmental Management University of Hong Kong
Wissink, B. (2015) 'Fights to the city: class and difference in Mumbai and Hong Kong', *International Journal of Housing Policy*, 15:3 323–46.
Yung, E.H.K. and Chan, E.H.W. (2015) 'Re-examining the pro-growth ideology in cities: conservation of historic properties in Hong Kong', *Urban Review Affairs*, 52: 2 182–210.
Zukin, S. (1995) *The Cultures of Cities*, Cambridge, MA: Blackwell.

Part 4
Planetary urbanisation and 'nature'

Michael E. Leary-Owhin and John P. McCarthy

Between the publication of Rachel Carson's revelatory book *Silent Spring* in 1962 and the influential 1972 book *Only One Earth* by Ward and Dubos, Lefebvre was busy formulating and refining many of the ideas that resonate throughout this part of the book. Conceptualisations of nature or absolute space and its transformation are central to Lefebvre's theorising about the history of space, capitalism, industrialisation, the production of space and planetary urbanisation. This part of the book assembles some of the latest theoretical and empirical developments that seek to employ Lefebvre's ideas to further our understanding of urbanisation and nature in the early 21st century. In the late 1990s, Neil Smith commented that Lefebvre's rethinking of nature was rather poor. Smith returns to Lefebvre's engagement with nature in the Foreword to the English translation of *The Urban Revolution* (2003). Although Lefebvre's treatment of nature under capitalism is less nuanced than his consideration of space, from the 1960s it is clear that Lefebvre understood the importance of the related issue of the environment. While Lefebvre often laments romantically the ravage of nature by industrialisation, he notes the persistence or re-emergence of a version of nature, perhaps a simulation, in cities: open spaces, parks and gardens; such spaces enhance city life but may be a poor substitute for nature, simulacra. To those we could add urban wild-scapes, waterscapes and water margins. For Lefebvre a paradox is apparent. From the 19th century, nature appeared to be shrinking in the face of industrialisation, yet signs of nature and the natural were multiplying in cities and were bought and sold as commodities. Planetary urbanisation ideas imply that cities continue to infiltrate and dominate beyond the limits of their physical boundaries. However, the purpose of this part is not to attempt to settle the city/urbanisation controversy but to allow a re-consideration of the role of nature in the production of space regarding urbanisation and the city.

That said, Lefebvre argued controversially that a rather passive nature is destroyed relentlessly by the production of abstract space. But we are now acutely aware of how nature still impacts on cities both in negative and positive ways. In the last decade, ideas about nature and environmentalism have taken on global importance as the contested ideas of the Anthropocene are debated. Lefebvre's contributions regarding nature and environment are perhaps some of the least explored and developed avenues of his writing and theorising. They are, however, related to ideas about planetary urbanisation and the production of urban abstract space. Actual planetary urbanisation over recent decades in the Global North and South seems to be confirming as a concrete phenomenon what Lefebvre described in the early 1970s as something virtual and impending. Whether this means we can abandon the idea of the city, the rural and the wilderness remains, however, a moot point. An overriding contention here is that seeing the city through a Lefebvrian lens of urbanisation and nature allows us to move beyond the reignited debate as to whether cities are best understood primarily as bounded physical sites or complex economic and social processes and power relations. This part explores the implications of how processes of urbanisation interact with nature when they are unleashed globally via the tendency of neoliberalism urbanism to commodify and appropriate space globally.

The agricultural countryside blurs the boundaries between nature and human culture, the natural and the artificial. Food production became industrialised in the 20th century with agri-business perhaps creating as many problems as it solved. Yet food production has always occurred within cities and with the phenomenon of shrinking cities gained in importance in recent decades. Urban agriculture then is where this part of the book starts. The chapter by Granzow and Shields argues that urban agriculture should be viewed not so much as a descriptive category, but more a new ordering of urban natures, places and publics. They configure the emerging urban agricultural landscape as being produced by a 'trial by space'. The authors stress the importance of seeing urban agriculture as a global phenomenon that transcends the simple growing of food and urban gardening, containing within it transformative potential. This potential can be triggered by inhabitants asserting their right to the city by making claims on space. In the process, space is appropriated for its societal use value; the socio-spatial practice is re-spatialising food production. What may have assisted urban agriculture is the proliferation of 'terrain vague' in some shrinking cities. Granzow and Shields point to Detroit and New York as cities where the tactical urbanism of guerrilla gardening acts as a form of spatial resistance providing ways to contest the imposition of and maintenance of abstract space. Edmonton, Canada, is highlighted for the fact that urban food production issues are now incorporated into the city's municipal development plan, and it may be that an opportunity is emerging to re-think relationships between food, space and the city. The chapter challenges us to rethink urban agriculture as a process that provocatively mixes up the urban and the rural. A process that recasts how the urban, the built environment and the city might be reimagined.

We stay with Canada for the next chapter by Nicholas Scott. In this chapter Scott draws on Lefebvre's critique of the car to proffer some fresh thinking about the production of nature and mobilities. He urges us to destabilise the dominance of automobility through a case study of the ways in which nature is produced by cycling in Canada. And he paradoxically proposes that the 20th-century idea of the automobile parkway could ecologically inspire the expansion of 'cycling nature' in the city more generally. Scott challenges Lefebvre's contention that natural space is disappearing and looks ahead at a nature-filled horizon but builds on Lefebvre's ideas to offer insights into the manner in which cycling might produce alternatives to the nature created by automobility. Scott asks how might alternative urban mobilities cultivate more sophisticated representations of nature and what might be the role of cycling in an alternative production of nature. Through an analysis of empirical research, Scott proposes that 'cycling nature' produces nature as an embodied and ecologically valuable process. He does this while acknowledging the dangers created by the production of cycling-friendly environments principally for the benefit of global elites. He signals also that his approach could be applied to issues of cycling inequality and issues in the Global South.

Most of us may imagine that few things are more natural than air, but Ford in the following chapters sets out to disrupt our presumptions by concentrating on the production of the air. Although he sees his research as addressing a lacuna in Lefebvre's work regarding the production of space, he informs his analysis by drawing on analytical components of Lefebvre's spatial architectonics, developed in his book *The Production of Space*. Ford relies on a Lefebvrian dialectical understanding of the production and persistence of abstract space that seeks under capitalism to extend commodification into all realms of society and nature. Outdoor air quality in cities is an increasing concern around the world and while of great importance, Ford focuses on the production of air for its industrial and exchange value. He highlights struggles over who controls the circulation, abstraction and quality of air and for what purposes. The commodification of air creates contradictions and inevitable resistance from those suffering its harmful effects. Ford concludes that atmospheric resistance can be utilised as a weapon in the struggle against the domination of urban space.

Subsequent to Ford, Paiva invokes the Lefebvrian idea of second nature, for example, the city where nature and culture co-exist, but with culture dominating nature, and relates this to current debates about the Anthropocene. Appropriation and domination are key concepts put to work in this chapter as Paiva explores the production of nature by drawing on Lefebvre's concepts of rhythms. Paiva seeks to go beyond Lefebvre and critique second nature by applying the concepts of cyclical appropriation and linear dominance. However, Paiva deploys Lefebvre's understandings of nature as physical reality pre-dating human existence but including the human body, which can logically be dominated in the production of space. Cyclical rhythms of nature can interfere with linear rhythms prescribed by capitalist abstract space and with the linear rhythms of everyday life. There are also human practices that appropriate space in a cyclical manner in harmony with the rhythms of nature. Paiva reasons that some of the common binary distinctions such as natural/artificial could usefully be re-conceptualised through an understanding of nature as continuously produced and transformed by human and non-human agency. An analytical framework foregrounding linear dominance and cyclical appropriations offers an adjunct to dialectical thinking beyond the spatial triad.

Humankind's relationship with nature is a crucial factor in planetary urbanisation, and in the ensuing chapter Lehtovuori et al. present a revisiting of Lefebvrian urbanisation and urban society in the context of ideas about habiting. Along the way they challenge the notion of planetary urbanisation as developed by Brenner and Schmid. The chapter points to Lefebvre's reluctance to give concrete examples of a future urban society but suggests that the current critical phases see subordination of the global to the urban. It is claimed in the chapter that in addition to mobility, understood as migration, the local scale, bodily experience and everyday mobilities are vital for Lefebvre's thinking about the emerging urban society. So for example, mobility is not just about physical instrumental movement but also politics, power relationships and the uneven distribution of the possibilities for mobility. Lefebvre's urban society was a hypothesis – something to aim for – so the chapter asserts that we need a new theory to illuminate the self-organising order that urban society requires.

Chapter 6 by Liinamaa uses the expansive canvas of planetary urbanisation to re-evaluate the position of aesthetics in Lefebvre's thought, inspired it seems by Lefebvre's demand to let everyday life become a work of art. Aesthetics here concerns some aspects of the study of beauty but mainly refers to the form and senses of everyday life. Liinamaa seeks to bring out the complexity in Lefebvre's deployment of aesthetics, showing how its use serves differing purposes, especially how Lefebvre develops a critical spatial aesthetics. After this she provides a brief encounter with Lefebvre's relationship with art and artists and his consequent influence on recent urban art and culture scholarship. Empirically, the chapter explores the vulnerabilities of expansive urbanism through the discussion of a recent major work of art. The chapter argues that Lefebvre's theorising makes an essential contribution to the repositioning of artistic production and everyday life in Marxist theory. Liinamaa concludes by questioning whether a new triad can help recast our understanding of urban processes.

It is clear that Lefebvre has grave concerns about the harmful environmental impacts of capitalist industrialisation and urbanisation. This is the starting point for Wiedmann and Salama to explore Lefebvre's production of space theories alongside current ideas about sustainable urbanism. An analytical framework is introduced that juxtaposes the spatial triad with the sustainability triad of economic, environmental and social issues. A further triad is proposed to assist in developing the analysis, namely: diversity, efficiency and identity. The chapter authors then invoke Lefebvre's views regarding the crisis of cities and conflicted space rooted as it is in the contradictions of the abstract. Planetary urbanisation sees the development of globalised conflicted spaces. A damaging result is that consumption becomes disconnected from human needs. Conflicted space is then characterised in the chapter as a key threat to urban sustainability. Underlying this conflict is primacy of the search for profit and capital accumulation. In closing their chapter, Wiedmann and Salama emphasise the importance of what they call three divides related to decision-making, social groups and capitalist investment imperatives.

It is interesting to note that Wiedmann and Salama point to the importance of town (city and urban) planning in the history of space, especially abstract space. Marescotti concentrates on a critical analysis of town planning from a standpoint infused with Lefebvrian ideas. The chapter seems inspired partly by Lefebvre's planetary urbanisation thesis and partly by Marescotti's intellectual journey through planning academia and his dissatisfaction with elements of that. A number of themes are examined by Marescotti as the text moves from a discussion of the future of cities to the need to live peacefully on the Earth, the need for scientific independence and the need for political commitment. Environmental issues are central to Marescotti's chapter, and he laments that scientific reductionism prevents planning from addressing successfully global environmental and social problems such as poverty. In the chapter, useful comparisons are drawn between Doxiadis' concept of ecumenopolis and Lefebvre's planetary urbanisation. Marescotti updates Lefebvre by suggesting the era of planetary urbanisation might be seen as an Anthropocene ecumenopolis. In the chapter, Marescotti wonders if technology is killing art and if city centres are under siege by the standardised consumption wrought by abstract space. He likens his long-term attempt to rescue planning and see its policies and practices contribute to an urban society, to Lefebvre's project as outlined at the end of *The Production of Space*.

Preliminary conclusions and thoughts

Lefebvre's ideas and environmental issues

Despite the environmental havoc wrought by industrialisation and urbanisation, there are signs that communities and politicians and elements of the private sector are engaged in projects that offer hope of an urban society. How can we deploy Lefebvre's incisive thinking and subsequent Lefebvrian literatures to make sense of the implications for nature of planetary urbanisation?

Lefebvre's ideas and interdisciplinarity

This part (and other parts) of the book flag up the importance of being able to take an interdisciplinary approach to issues of nature and planetary urbanisation and a variety of concomitant societal issues, including: automobility and cycling, aesthetics, air quality, habiting and urban agriculture. To what extent is theoretical development possible by seeking to align, in useful tension, the Anthropocene and Lefebvrian urbanisation debates?

Lefebvre's ideas and total urbanisation

While we wait, in this critical phase, for the urban society to emerge from total urbanisation, how can we identify and document moments that may be opening up fissures to alternative futures? Given the scope that Lefebvre provides for interpretation of his ideas, and given the creative and penetrating scholarship on display here, how can we consolidate Lefebvrian-inspired research rather than see it dissipate?

25

Urban agriculture

Food as production of space

Michael Granzow and Rob Shields

Introduction

The landscape of urban agriculture (UA) extends beyond the fences of community gardens, backyard chicken coops, and beehives to entangle an increasing assortment of places and spaces – local and global; urban and rural; public and private; real and imagined. Defined simply as the growing of food within cities, UA has been a part of the urban environment since cities first emerged. However, the origins of a more narrowly defined community gardening movement are often traced back to the establishment of English allotment gardens towards the end of the 18th century. Originating as a response to land enclosures and the resultant widespread poverty, the idea of allotment gardens spread quickly across Europe and North America. Since then, the community gardening movement has expanded across the globe, while literature on the topic has tended to focus on ideas of community, social capital, and individual health and well-being.

In addition to literature on community gardening, there is a growing body of work tracing the rise of UA in the Global North as a grassroots political response to numerous food-related problems. While notions of UA have now come to include community gardening practice, the relatively recent discursive shift from 'gardening' to 'urban agriculture' signals a new emphasis on the cultivation of productive urban landscapes geared towards feeding cities through the cultivation of local, sustainably grown food (Viljoen 2005). Indeed, the focus of much of the UA research over the last decade has been around determining its productive potential. While proponents of UA often overemphasise its ability to feed cities, critics underplay its transformative potential (Davidson 2017).

While assessing the quantity of food that UA contributes to the food security of cities is important, there is, as Tomkins (2014) notes, 'a need to move beyond the notion that the objective or objects of harvest alone can represent practice' (13). Drawing on Lefebvre's dialectical model of spatial production, we contend that UA is not only a means of food production or community-building, but constitutes a new and contested socio-spatial practice that is re-spatialising food and reshaping cities across the Global North in new and meaningful ways. In this chapter, we consider the general relevance of Lefebvre's research to questions around the rise of UA as both spatial tactic and strategy within a broader production of sustainable urban space.

Lefebvre helps us consider UA beyond describing or cataloguing local food-growing practices to critically grasp it as a contingent moment in the co-production of urban space.

Introducing Lefebvre

In the nearly half-century since Lefebvre began his critical inquiries into questions around cities and urban space, an impressive and interdisciplinary literature has emerged on these topics. Whether such work is officially coded as sociology, geography, cultural studies, or urban planning, Lefebvrian studies of space invite interdisciplinarity. This is appropriate given Lefebvre's deep distaste for specialisation. However, at the heart of the heterogeneous literature coalesced around Lefebvre's work is a relational understanding of both time and space. As Harvey (1997: 22) puts it, 'this view is that space and time do not exist outside of process: process defines space/time'. Relational space takes a particular form with Lefebvre – one outlined most comprehensively in *The Production of Space*.

While a relational view of space is not unusual in academic scholarship, the term 'space' continues to denote a neutral void in colloquial English. Many continue to think of the spaces in which we live, from our homes, to our neighbourhoods and cities, as different settings of social life. Even when such spaces become meaningful or cherished 'places', they still tend to be perceived as external to who we are and how we live. In *The Production of Space*, Lefebvre (1991) outlines a conceptual framework that relates everyday life, space, and capitalism. Through his spatial triad, Lefebvre illustrates how even the most intimate spaces of our everyday lives are deeply shaped through a dialectical relationship between *representations of space*, *spatial practice*, and *representational space*. Before situating UA more specifically in relation to Lefebvre's concept of trial by space, we briefly outline his spatial triad.

Representations of space refers to 'conceptualized space, the space of scientists, urbanists, technocratic subdividers and social engineers… all of whom identify what is lived and what is perceived with what is conceived' (38). Such representations are contrasted with *spatial practices* – actually existing spatial arrangements and their intersection with everyday routines and perceptions. Finally, *representational space* is 'space as directly *lived* through its associated images and symbols, and hence the space of "inhabitants" and "users", but also of some artists and perhaps those, such as a few writers and philosophers, who describe and aspire to do more than describe' (Lefebvre 1991: 39). A dialectical thinker influenced by Nietzsche, Lefebvre vehemently rejected static categories, emphasising instead instability, conflict, and change (Merrifield 1995; Shields 1999). As tempting as it may be to use Lefebvre's triad to simply categorise or identify discrete spaces, these concepts do not map neatly onto actually existing spaces; rather, they are meant to be used dialectically to help grasp socio-spatial contradictions and potentially illuminate moments of difference.

Towards the end of *The Production of Space*, Lefebvre (1991) writes: 'nothing and no one can avoid *trial by space* – an ordeal which is the modern world's answer to the judgement of God or the classical conception of fate' (416). The idea of trial by space relates closely to the thesis that guided much of Lefebvre's work on space through the late 1960s and 1970s: A mode of production can survive and indeed thrive only insofar as it manages to produce a space that absorbs its contradictions. It is not only that capitalism expresses itself through particular kinds of spaces, but that capitalism *is* a production of space (Lefebvre 1976: 21). An early interpreter of Lefebvre for Anglophone audiences, Harvey argued that capitalism cannot be fully grasped without attending to its geographical dimensions. The myriad examples of geographical transformation, expansion, and destruction/ruination seen under capitalism are, for Harvey, not incidental or natural, but, on the contrary, fundamental to the survival and growth of capitalism itself. The specificities of this point are explored in Harvey's idea of the 'spatial fix'. Whether it takes the form of suburban development or downtown revitalisation, spatial fixes temporarily resolve the contradictions of capitalism by way of transforming space. Capitalism does not just happen *in* space, it happens *as* space. Furthermore, once capitalism stops producing its own space, it enters crisis.

The spatial fix presents one way of understanding how capitalism has been successful in its trial by space. Returning to Lefebvre's spatial triad, however, it is important to remember that the moments of the production of space are always in dialectical relation with each other, and that no moment can be considered in isolation. In the abstract, spatial fixes appear as tidy *representations of space*; however, actually existing spatial fixes might also generate unexpected moments of resistance or counter-space. Instead of focusing on the ways neoliberal capitalism dominates space, a Lefebvrian analysis must attend also to those smaller ideas and movements that attempt to produce their own space. In subjecting UA to a trial by space, we call it into question, asking how it is emerging as a contested social spatialisation. We argue that UA constitutes a re-spatialisation of the relationships between food and the city, and also between place, publics, and nature.

Coming to terms with urban agriculture

Discussion around UA first emerged in the context of the Global South as a way to address poverty, malnutrition, and food insecurity. For many in the US, Canada, and other countries in the Global North, the spectre of food insecurity and malnutrition largely vanished with the mass industrialisation of food that followed World War Two (WWII) (Morgan and Sonnino 2010). Thus, for the latter half of the 20th century, UA was viewed almost exclusively as a crisis measure for poorer nations. This is reflected in the academic literature, which continues to focus on the rapidly urbanising context of the Global South. Writing just over a decade ago, Mougeot (2005: 3) commented that UA 'has spread to become a critical source of food for urban populations in countries affected by natural disasters (Honduras), economic crisis (Togo), civil wars (Armenia) and disease epidemics (Malawi)'. In the last decade, with the industrial agriculture model that promised a bottomless supply of cheap, accessible food entering into crisis, UA has increasingly gained attention in the Global North.

UA is most commonly defined as the growing and processing of food in or near cities. Despite recent warnings of what Born and Purcell (2006) have called the 'local trap', it remains difficult to find detailed descriptions of UA that are not overdetermined by the question of distance between 'farm and fork'. While such considerations are certainly important, they cannot be meaningfully understood outside broader contexts.

If UA is considered as solely a question of form defined by the placing of food-growing practices in or near the city, we could identify it in different contexts across time and space, from ancient city gardens to Havana's organoponicos to the vertical farms springing up in cities across North America. Indeed, it is common for contemporary scholars to speak of the recent resurgence of UA. It should be emphasised, however, that even UA initiatives existing at the same period of time in a single city can be quite diverse, let alone urban food-growing initiatives across time and space. Foregrounding Lefebvre (1991), we theorise UA as an emergent, contested, and contradictory 'food spatialisation' implicated in the co-production of urban natures, places, and publics differently across times and space (see the chapter by Lopes de Souza in Part 5). Such a framework also means considering UA as a spatialisation that overlaps with, but also potentially challenges, dominant urban sustainability frameworks.

Whereas for Swyngedouw (2010), sustainability is post-political, Whitehead (2007) makes an important distinction between 'sustainability' and 'sustainable development'. With Whitehead, we acknowledge the need to critique sustainability in its dominant globalised form, while at the same time being open to 'actually existing sustainabilities' (Krueger and Agyeman 2005) and engaging in a meaningful way with UA as it exists in relation to the particularities of place. Whitehead (2007: 5) notes:

> My problem with these depictions of sustainability is that they tend to (often inadvertently) reduce sustainability to the historical emergence of a singular concept of social and ecological development – that of *sustainable development*. To consider the spatialities of the sustainable society, then, is to become aware of the stories, struggles and values which cut across the history of sustainable development.

We make a similar point regarding UA. While definitions of UA are usually broad, including everything from windowsill herbs to regional food systems, these diverse forms are held together by a discourse of sustainability that is producing similar spatialisations of UA in a growing number of cities. At a basic level, this sameness results from the fact that major cities often face similar practical challenges around land use, development pressures, and rents. Thus, while we see UA strategies making space for local food, very rarely are such strategies as radical as is often claimed by a growing popular literature. For example, UA strategies usually do not impinge on sacrosanct ideas of private property, at least not at the level of urban policy. Looking at official urban food strategies, representations of UA tend to seamlessly fit within broader urban and regional plans that repeatedly fall back on the familiar pillars of sustainability as played out within the dictates of capitalist socioeconomic relations. Moreover, UA is often positioned as a lower-level concern at the mercy of the priorities reflected in the productivist principle of 'highest and best use'.

Rather than subsume the question of UA under the umbrella of sustainable development, we see it as a contingent and contested spatialisation with the potential of disrupting not only the industrial model of agriculture but also our ideas about the nature of the urban. The explosion of UA in popular literature, academic scholarship, and on the agenda of city planners and policymakers is evidence of its recent claims on urban space. Lefebvre offers a way to critically consider the rise of UA dialectically, through its representations, practices, and imaginaries.

Lefebvre and urban agriculture: cultivating the right to the city

Lefebvre meant something quite specific by 'the urban'. For him, urban society is roughly akin to what is often called post-industrial society. In contrast to the setting of the city, the urban is a historical and spatial formation linked to a post-industrial capitalist mode of production. Lefebvre (2003) sees the urban as a productive force 'modifying the relations of production without being sufficient to transform them' (15). The urban reorganises internal city space, while also shaping the countryside, agriculture, and nature to its uses (4).

Lefebvre's conception of the urban allows us to look at cities not as located *in* the country, but as part of a larger production of space driven by accumulation and profit that has fundamentally shaped the relations between the city and country. Moreover, it produces a particular landscape where cities are given primacy and the countryside is positioned as backward and simple. The rise of UA helps to illuminate the dynamism of this spatialisation, presenting the relation between urban and agriculture as a *cultural topology* that is slowly being torqued and remoulded (Shields 2013).

Work establishing UA as a critical socio-spatial question is already underway (Blomley 2004; Shillington 2013; McClintock 2014; Tomkins 2014; Tornaghi 2014; Purcell and Tyman 2015), and a handful of authors have drawn on the work of Lefebvre, with particular emphasis placed on 'the right to the city'. Lefebvre's (1996/1968) formulation of the right to the city is closely tied up with the everyday practices of 'inhabitants', who he contrasts with the 'bourgeois aristocracy' who do not inhabit the city but 'go from grand hotel to grand hotel, or from castle to castle, commanding a fleet of a country from a yacht' (p. 158). Merrifield (2006) relates the alienation and drudgery often experienced by the inhabitant to 'a loss of the city as *oeuvre*, a loss of *integration* and *participation*' (69). 'Indeed', says Merrifield, 'it is to denigrate one of humanity's great works of art – not one hanging on a museum wall but a canvas smack in front of our noses, wherein we ourselves are would-be artists, would-be architects' (69). While some have argued that the concept of the 'right to the city' is underdeveloped (Attoh 2011), others value its openness (Mitchell and Heyen 2009; Harvey 2012). Mitchell and Heyen (2009) argue that the concept's 'capaciousness is valuable because it allows for solidarity across political struggles while at the same time focusing attention on the most basic conditions of survivability, the possibility to inhabit, to live' (616).

The few studies that have used Lefebvre to grapple with UA have tended to focus on issues of spatial appropriation, with an emphasis on 'guerrilla' practices. For example, Gillian Wales (2013) draws on Lefebvre's spatial triad to examine guerrilla gardening in Glasgow, finding that gardeners demonstrate 'alternative spatial practice' beyond the gaze of the state. The Spanish architect and philosopher Ignasi de Solà-Morales (2014) uses the term 'terrain vague' to refer to those spaces that exist 'outside the city's effective circuits and productive structures' (26). It is in these spaces that guerrilla gardening thrives, with plots often taking root in forgotten or unvalued land along railway lines and in abandoned or vacant city lots. It was the proliferation of 'terrain vague' in post-industrial cities such as Detroit that has allowed for UA practices to flourish.

Once the bustling heart of the global automotive industry, Detroit saw a dramatic socio-economic decline after the 2007–2008 financial crisis, becoming the largest municipality in the history of the US to declare bankruptcy. Since then, the former industrial centre has become the archetype of post-industrial ruination in the US, providing new spaces of urban experimentation and attracting artists and writers interested in the dystopian sublime. Detroit has also become a unique opportunity to rethink relationships between public space, food, and the city and explore alternative UA imaginaries. While many residents of Detroit have supported investment in UA as a potential way forward for the city, others have been less optimistic. Speaking informally about the rebuilding of Detroit, Florida (2011) queried, 'why would you want to turn a great city into a corn field?' This comment deems certain futures to be regressive and anti-urban. It seems that for Florida, Detroit's embracing of UA is as backward as trying to revive the region's manufacturing and auto industry.

UA is not necessarily about 'turning cities into cornfields' but presents a way of re-imagining the very idea of the city; it is not a return to an imagined past (though such imaginaries might certainly be a part of it), but an emergent site of struggle over relationships between cities, people, food, as well as the very idea of the urban. Considered as a spatialisation, UA is not merely a geographical expansion and entrenchment of older gardening practices in the city, but a reconstitution and politicisation of those practices. Even the recent shift from a discourse of community gardening to UA presents an important change in meaning and a reorienting of practice. With this discursive shift, we see the emergence of new places (e.g. vertical gardens, agri-hoods) and shifting subjectivities (urban farmers, food planners). Whereas post-WWII suburban gardening in North America was largely coded as a space of leisure and/or of family, UA discursively re-orients food-growing practice towards pressing social and ecological challenges that range from community alienation, to local food insecurity, to urban decline and ruination.

In another example linking 'terrain vague' and urban food growing, New York City saw an explosion of community gardens during the recession of the 1970s. Indeed, it was in this moment of inner-city decline and divestment that the term 'guerrilla gardening' was coined by the New York artist Liz Christy. Speaking of Christy's activist group the Green Guerrillas, Reaven and Zeitlin (2006) note, 'on the surface their mission was all about gardens. Just below the surface lurked the radical idea of bringing into the public domain land that once had been privately held but now had been callously abandoned' (274).

Radical moments of UA based on the spatial appropriation of city lands easily lend themselves to a 'right to the city' framing. As with other forms of tactical urbanism, guerrilla gardening can act as a form of spatial resistance – a way to contest space as conceived by planners and developers and present an alternate view of what the city might be. Through an analysis of contested community gardens in New York and the case of South Central Farm in Los Angeles, Purcell and Tyman (2015) consider UA as both 'spatial contestation' and 'spatial autogestion'. The authors emphasise Lefebvre's radical optimism, calling on other researchers to search out everyday moments of resistance already occurring, and illuminate the ways in which these moments work to break through the monotony and drudgery that pervades much of urban existence. In another recent and notable study of UA, Tornaghi (2014) draws on Lefebvre to sketch out what a critical geography of UA might look like. Working with Marcuse's (2009) formulation of 'the right to the city' and his related formulation of 'critical urban planning' (194), Tornaghi politicises UA through a social justice lens: 'given the political and strategic role which UA can play in the future, the development of its critical theorisation will set the parameters for evaluating what type of initiatives are fit for non-regressive and socially just urban food policies' (4). In yet another study, Shillington (2013) discusses local UA initiatives and patterns of consumption as socio-spatial practices that make a particular claim on the city. Shillington connects mundane practices of everyday life in San Augusto to a larger spatial politics. She argues that through the cultivation of fruit trees and everyday consumption of refrescos, local inhabitants intervene in urban metabolisms, producing a particular socio-natural space.

As illustrated by these studies, the existing literature linking Lefebvre to UA focuses on the spatial tactics of urban inhabitants, emphasising everyday practice, with less written about the changing nature of these tactics in relation to emergent UA strategies. In the remainder of the chapter we ask how Lefebvre might help us to better conceptualise the current UA moment, where it is increasingly cities themselves that are taking an interest in food and developing strategic pathways forward.

Urban agriculture on trial: reconciling tactics and strategies

We began the chapter by introducing Lefebvre's idea of trial by space as a way to orient the reader towards a critical questioning of UA. So far, we have theorised UA as a novel spatialisation that rests in part on challenging dominant socio-spatial orders. In other words, we have identified ways in which UA includes elements of counter-space that potentially resist not only the global regime of cheap food but also the assumption that cities are best organised by professionals (i.e. government officials, planners, and policymakers). At the same time, we must re-iterate that UA is not in itself an emancipatory spatialisation, a point well made by McClintock (2014). 'Contradictory processes of capitalism', writes McClintock, 'both create opportunities for UA and impose obstacles to its expansion. Identifying these contradictions requires analysis of UA's various forms and functions at multiple scales' (148). Where many have been quick to praise the expansion of UA as a move towards goals of urban sustainability, we emphasise the need to consider both the particular politics and limits of this expansion as it is taking place in different contexts. To subject UA to Lefebvre's trial by space is to critically evaluate the current 'urban agriculture moment' that is occurring in a growing number of cities across the Global North.

The recent UA moment is defined in large part by the acceptance of certain UA practices and imaginaries by cities. The fact that food was not really considered by urban planners prior to the first decade of the 21st century (APA 2007) is evidence of an ideological blind spot rooted in a constructed separation of the rural and the urban. Since 2007, however, food policy and planning have exploded onto the scene, with municipalities across North America and Europe enthusiastically taking up a food systems lens. The American Planning Association's 2007 report suggests a number of possible reasons why planners were so late to embrace food as a planning issue, the first of which is the view that food systems only indirectly affect the built environment of cities. Of course, it was not that food systems did not shape the built environment before the rise of UA, but rather that the way that these systems shaped urban environments was difficult to see and therefore easy to ignore.

Over the course of the last decade, many cities in the Global North have embraced UA as part of developing broader urban food strategies. Our own city of Edmonton, Alberta, established itself as a centre of UA in Canada in 2010 when the City officially incorporated a food and agricultural strategy into its municipal development plan. Momentum and support for the integration of UA into broader planning and governance frameworks has continued, resulting in the approval of an official food and UA strategy, the amending of local bylaws, and the creation of a food council. Emergent interrelations between food and the city, and the rise of urban food strategies, are part of what Morgan and Sonnino (2010) have termed 'the new food equation'. UA equally presents a new urban equation or, more accurately, an opportunity to re-think relationships between food, space, and the city in relation to our collective urban futures.

Emergent urban food equations re-cast UA as a legitimate and even desirable urban land use in a context of a faltering industrial, globalised food model that prioritises productivity above all else. Gaining prominence after WWII, the productivist model of agriculture subsumed any different or idiosyncratic practice to a singular emphasis on output and, ultimately, profit (Halfacree 2007: 130). The dominant turn-of-the-millennium industrial food spatialisation separated each stage of production and consumption. Refrigerated global shipping has linked together these sites into a complex globalised system. Harvest is often in a naturally-advantaged location or low-regulation region where food can be grown or harvested easily. Processing is in a low-wage factory location. Heavily packaged and often chemically or sugar-treated foods are sold via supermarkets located in specifically zoned and serviced city locations with a view to limit home preparation for consumption in the household. Industrial food production was largely invisible until the emergence of the food movement of the 1970s, when people started to be concerned with where their food came from and how it was produced. Increasingly this social movement has gained ground in relation to what is generally acknowledged to be a crisis of industrial, export-oriented agriculture, and food production under capitalism. This food crisis corresponds to a regime of 'cheap food' in the globally dominant societies of Europe and North America. Since WWII this regime has risen victorious in a global trial by space, transforming the rural into a productivist, profit-oriented landscape.

Industrial agriculture has overcome several crises through various agricultural revolutions, and it is attempting to overcome the latest through synthetic fertilisers, pesticides, and genetically modified crop 'fixes'. Food has had a dominant spatial footprint on late capitalism. Inherited from the feudal European divide between cities with their guild production and rural fiefdoms with peasant agriculture, industrialised agriculture has pushed toward oil-reliant, mechanised, and pesticide-dependent practices in increasingly depopulated rural areas. 'Cheap food' is a spatialisation – a global and overlooked geography that has reached the point of being unsustainable. That said, UA is by no means the natural or only answer to the multifaceted problems of cheap food. For example, while much writing on UA has emphasised the environmental benefits of UA, its actual impact is less clear. In their overview of the benefits and limitations of UA, Santo et al. (2016) found that the ability of UA to mitigate greenhouse gas emissions is overstated. There is, however, clear evidence supporting other benefits of UA, such as its 'ability to increase social capital, community well-being, and civic engagement with the food system' (22).

Speaking to the politics of UA in Cameroon, Page (2002) highlights the stark difference between dominant conceptions of UA and the reality of its implementation within a context of austerity and structural adjustment. As Page argues, Buea's urban farmers:

> set out to reduce their bills in food markets and ended up helping to diffuse the social discontent associated with the reconfiguration of the Cameroonian political contract that had been prompted by the forced shrinking of the government's wage bill.
>
> *(51)*

This 'darker side' of UA has also been highlighted by scholars studying cities in the Global North. Looking at the cases of Vancouver and Detroit, Walker (2015) raises questions about the radical potential of city-led UA strategies. Unlike most critical studies of UA, Walker focuses on the role of the state as opposed to the grassroots practices of urban farmers and organisers. While there exist important differences between contexts, Walker finds that both cities have pursued UA as a sustainability fix designed to foster economic development. This critical approach to the politics of UA, pursued also in the work of McClintock (2014) and Tornaghi (2014), highlights the ways in which UA is, in many cases, the forces that on the surface it seems to challenge.

UA has emerged as both *tactic* and *strategy*; it can equally be implicated in the production of counter-space (Lefebvre 1991: 349; 367; 381–3; Leary-Owhin 2016) as it can with that of neoliberal space. As McClintock (2014) writes, UA practice 'is not simply radical or neoliberal, but both, operating at multiple scales' (165). Recognising the contingency of UA practice is the first step, but a Lefebvrian analysis requires us to go further to critically consider such practices as they are actually taking place. By thinking through the conceived, lived, and perceived elements of UA space, we can begin to assess its production through the lens of trial by space.

Lefebvre's trial by space rests on the assumption that a mode of production can survive and thrive insofar as it manages to produce a space that absorbs its contradictions. On one hand, we can view certain instances of UA as consistent with abstract conceptions of space that ultimately prioritise creating economic value and the pursuit of capital. On the other hand, there are many ways in which grassroots UA challenges neoliberal urbanism, making claims on urban space that create counter-spaces and lend themselves to new urban imaginaries. Where much research on UA has privileged the more obviously radical, grassroots elements of the practice, more research is needed to understand how urban strategies are incorporating UA into their policy frameworks. We recommend caution in rushing to position the urban planners and the local state as the sole progenitors of abstract space. State-implemented UA policy does not necessarily mean the co-optation of more radical visions presented by grassroots UA movements.

Neil Brenner (2015) has detailed the dangers of equating tactical urbanism with radical urbanism, identifying specifically the ways in which tactical urbanism's 'anti-planning rhetoric' can unintentionally reinforce neoliberal logics:

> The anti-statist, anti-planning rhetoric of many tactical urbanist interventions may, in practice, significantly erode their capacity to confront the challenges of upscaling their impacts.

> To the degree that advocates of tactical urbanism frame their agenda as an alternative to an activist role for public institutions in the production of urban space, they are at risk of reinforcing the very neoliberal rule-regimes they ostensibly oppose.
>
> *(n.p.)*

Brenner's point is not to draw attention away from the complicity of the state in producing urban spaces according to neoliberal logics that alienate urban inhabitants. The question is, how can tactics be scaled up and institutionalised in ways that might better effect change? Brenner is clear that elements of neoliberalism can be found in tactics that might appear at first to be examples of counter-space. A version of this point was nicely articulated by Blomley (2004: 637), who argues, 'let us not take the politics of neo-liberalism at face value. Even neo-liberalism can contain, if you like, forms of neo-socialism' (637).

Critical geographers such as David Harvey and Peter Marcuse are right to emphasise the complicity of planning in shaping the city to serve the interest of capital; that said, the current moment of neoliberal urbanism presents a situation where even the most ostensibly radical moments of urban praxis do not necessarily exist outside neoliberal spatial logics. Where there has been a tendency in critical studies of UA to prioritise the everyday actions and experiences of inhabitants, we see value in examining more closely the recent UA moment as an emergent spatialisation that cannot be grasped without a dialectical approach. In other words, we must attend to not only *spaces of representation* but spaces of the dialectical relations between all three moments of spatial production identified by Lefebvre, as they emerge in sometimes unexpected places.

To view UA through Lefebvre's concept of trial by space is to shift emphasis to the ways in which urban food growing is recasting the city. Lefebvre (1991: 417) writes, 'ideas, representations or values which do not succeed in making their mark on space, and thus generating (or producing) an appropriate morphology, will lose all pith and become mere signs, resolve themselves into abstract descriptions, or mutate into fantasies'. It is not just historical modes of production that can be thought of through the lens of trial by space. Any idea or political project must undergo a trial by space. UA is not merely a mode of producing more local and sustainable food; it is also an opportunity to imagine and actualise alternative urban futures. In Lefebvre's emphasis on the dialectics of spatial production, we see a need to expand studies of UA to be sensitive to counter-space, wherever it may exist.

Conclusions

We have considered UA beyond questions of local food or urban sustainability. We frame UA instead as an emergent and contested food spatialisation. However small the plot, placing agriculture in cities is a spatial gesture that disturbs, recomposes, and questions the separation between the urban and rural, the built environment and natural environment. By provocatively mixing up the urban and rural, UA in its various forms challenges the established order dividing the civilised *polis*, as the heart of culture and rational economic public sphere, from rural hinterlands.

UA as a food spatialisation is practised and anchored in participatory production and everyday consumption. It recasts what can be imagined and located within the category of the urban and the built environment of the city, thus providing a living laboratory in which we can observe the shifting topology of food spatialisations. Applying Lefebvre's trial by space and emphasising the dialectical spirit of his thought, 'food spatialisations' emerge as a way to position food as an urban question. As an emergent spatialisation, UA represents a possible re-ordering of urban natures, places, and publics that challenge the now dominant spatialisation of 'cheap food'.

This spatialisation is in contest with other frameworks that source and process food globally and rigorously separate consumers' dinner tables from producers' fields. They separate the site of consumption, from retail supermarket, from processing in low-wage factories, and from harvest in naturally-advantaged, but pesticide-permissive and labour-exploiting regions.

Where most studies using Lefebvre to understand UA have focused on the tactics involved in claiming the 'right to the city' through gardening practice, we emphasise the need to open inquiry up to a broader, dialectical inquiry into the production of UA space. With many cities incorporating UA into sustainable development and food strategy frameworks, we must consider carefully the ways in which such moves may support or negate claims to urban space by urban inhabitants.

References

APA (2007) *Policy Guide on Community and Regional Food Planning*, APA.

Attoh, K.A. (2011) 'What kind of right is the right to the city?', *Progress in Human Geography*, 35:5 669–85.

Blomley, N. (2004) 'Un-real estate: proprietary space and public gardening', *Antipode*, 36:4 614–41.

Born, B. and Purcell, M. (2006) 'Avoiding the local trap: scale and food systems in planning research', *Journal of Planning Education and Research*, 26:2 195–207.

Brenner, N. (2015) 'Is "tactical urbanism" an alternative to neoliberal urbanism?', *Post: Notes on Modern & Contemporary Art Around the Globe*. (accessed 12 June 2017).

Davidson, D.J. (2017) 'Is urban agriculture a game changer or window dressing? A critical analysis of its potential to disrupt conventional agri-food systems', *International Journal of Sociology of Agriculture and Food*, 23:2 63–76.

Florida, R. (2011) *Detroit: The Next American Ghost Town?* Available HTTP: http://bigthink.com/videos/detroit-the-next-american-ghost-town (accessed 1 March 2017).

Halfacree, K. (2007) 'Trial by space for a "radical rural": introducing alternative localities, representations and lives', *Journal of Rural Studies*, 23:2 125–41.

Harvey, D. (1997) 'Contested cities: social process and spatial form', in N. Jewson and S. Macgregor (eds) *Transforming Cities: Contested Governance and New Spatial Divisions*, London: Routledge, pp. 19–27.

Harvey, D. (2012) *Rebel Cities: From the Right to the City to the Urban Revolution*, London: Verso.

Krueger, R. and Agyeman, J. (2005) 'Sustainability schizophrenia or "actually existing sustainabilities?" toward a broader understanding of the politics and promise of local sustainability in the US', *Geoforum*, 36:4 410–17.

Leary-Owhin, M.E. (2016) *Exploring the Production of Urban Space: Differential Space in Three Post-Industrial Cities*, Bristol: Policy Press.

Lefebvre, H. (1976) *The Survival of Capitalism: Reproduction of the Relations of Production*, London: Allison & Busby.

Lefebvre, H. (1991) *The Production of Space*, Oxford: Blackwell.

Lefebvre, H. (1996/1968) 'The right to the city', in E. Kofman and E. Lebas (eds) *Writings on Cities: Henri Lefebvre*, Oxford: Blackwell.

Lefebvre, H. (2003) *The Urban Revolution*, Minneapolis, MN: University of Minnesota Press.

Marcuse, P. (2009) 'From critical urban theory to the right to the city', *City*, 13:2–3 185–97.

McClintock, N. (2014) 'Radical, reformist, and garden-variety neoliberal: coming to terms with urban agriculture's contradictions', *The International Journal of Justice and Sustainability*, 19:2 147–71.

Merrifield, A. (1995) 'Lefebvre, anti-logos and Nietzsche: an alternative reading of the production of space', *Antipode*, 27:3 294–303.

Merrifield, A. (2006) *Henri Lefebvre: A Critical Introduction*, New York: Routledge.

Mitchell, D. and Heyen, N. (2009) 'The geography of survival and the right to the city: speculations on surveillance, legal innovation, and the criminalization of intervention', *Urban Geography*, 30:6 611–32.

Morgan, K. and Sonnino, R. (2010) 'The urban foodscape: world cities and the new food equation', *Cambridge Journal of Regions, Economy and Society*, 3:2 209–24.

Mougeot, L. (2005) *Agropolis: The Social, Political and Environmental Dimensions of Urban Agriculture*, London: Earthscan.

Page, B. (2002) 'Urban agriculture in Cameroon: an anti-politics machine in the making?', *Geoforum*, 33:1 41–54.

Purcell, M. and Tyman, S.K. (2015) 'Cultivating food as a right to the city', *Local Environment*, 20:10 1132–47.
Reaven, M. and Zeitlin, S.J. (2006) *Hidden New York: A Guide to Places That Matter*, New Brunswick, NJ: Rivergate Books.
Santo, R., Palmer, A. and Kim, B. (2016) *Vacant Lots to Vibrant Plots: A Review of the Benefits and Limitations of Urban Agriculture*, John Hopkins Center for a Livable Future.
Shields, R. (1999) *Lefebvre, Love and Struggle: Spatial Dialectics*, London: Routledge.
Shields, R. (2013) *Spatial Questions: Cultural Topologies and Social Spatialisation*, London: Sage.
Shillington, L.J. (2013) 'Right to food, right to the city: household urban agriculture, and socionatural metabolism in Managua, Nicaragua', *Geoforum*, 44 103–11.
de Solà-Morales, I.D. (2014) 'Terrain vague', in M. Mariani and P. Barron (eds) *Terrain Vague: Interstices at the Edge of the Pale*, New York: Routledge, pp. 24–30.
Swyngedouw, E. (2010) 'Impossible sustainability and the post-political condition', in M. Cerreta, V. Monno and G. Concilio (eds) *Making Strategies in Spatial Planning*, London: Springer.
Tomkins, M. (2014) *Making Space for Food: Everyday Community Food Gardening and Its Contribution to Urban Agriculture*, University of Brighton.
Tornaghi, C. (2014) 'Critical geography of urban agriculture', *Progress in Human Geography*, 38:4 551–67.
Viljoen, A. (2005) *Continuous Productive Urban Landscapes: Designing Urban Agriculture for Sustainable Cities*, Oxford: Architectural Press.
Wales, G. (2013) 'Gardens of transgression, spaces of representation: an analysis of guerrilla gardens using the works of Lefebvre as a theoretical framework', in *Faculty of Humanities and Social Sciences*, Strathclyde: University of Strathclyde.
Walker, S. (2015) 'Urban agriculture and the sustainability fix in Vancouver and Detroit', *Urban Geography*, 37:2 163–18.
Whitehead, M. (2007) *Spaces of Sustainability: Geographical Perspectives on the Sustainable Society*, London/ New York: Routledge.

26

Ecologising Lefebvre

Urban mobilities and the production of nature

Nicholas A. Scott

Introduction

Among global brands selling high-end urbanism across the planet, 'Vancouverism' stands out for the way it animates nature. According to city officials, 'Vancouverism is an internationally known term that describes a new kind of city living'. This new way of living:

> combines deep respect for nature with enthusiasm for busy, engaging, active streets and dynamic urban life. Vancouverism means tall slim towers for density, widely separated by low-rise buildings, for light, air, and views. It means many parks... combined with an emphasis on sustainable forms of transit.... No wonder city planners and urban designers come to Vancouver from around the world to find inspiration. (City of Vancouver 2017)

Taking advantage of the city's stunning coastal mountain backdrop, Vancouverism attracts people who love the city and love leaving it for the surrounding forests, mountains and beaches (Stoddart 2012). Vancouver's urban brand is part of a larger production of nature that travels on either side of city limits, and by different modes. While sustainable mobilities grow inside the city, Vancouverites mainly use the car to reach wilderness outside it and happen to use more luxury vehicles than anywhere else in North America (Azpiri 2016). On any weekend or Friday afternoon, many residents hop in their Mercedes-Benz, BMW, Audi, Lexus, Bentley, Rolls-Royce, Ferrari or Lamborghini and leave the city in style for wild nature. Many other residents drive to nature in more humble vehicles – including a small army of vintage Mitsubishi Delicas, Volkswagen Westfalia Campers and other new-age, diesel-powered hippie vans. Overall, this heavy reliance on the car highlights a contradictory production of nature, fragmented between work/recreation, urban/non-urban and banality/splendour. The luxurious automobility, watercraft and real estate (comparable to Hong Kong, Sydney and London) with which elite Vancouverites orchestrate nature, points to broader relations between car-capitalism and neoliberal planetary urbanisation. Such production of nature cries out for Lefebvrian analysis.

In this chapter I examine how Lefebvre's ideas elucidate the production of nature, using qualitative data on cycling from cities across Canada to explore nature's enactment in the city. Ecological Lefebvrian analysis is underdeveloped. After all, Lefebvre construed nature as succumbing to abstract spaces of capitalism: 'the fact is natural space will soon be lost to view. Anyone so inclined may look over their shoulder and see it shrinking below the horizon behind us' (1991: 31). Leaving behind this much-contested passive idea of nature, and looking ahead at a nature-filled horizon, I examine how Lefebvre's production of space ideas and historical approach to urban space enrich the analysis of the production of nature, wherein nature takes on transformative agency. This agency means nature cannot be reduced to a product or producer of car capitalism. In this chapter I ask, how can nature be produced otherwise?

To cultivate Lefebvrian ecological analysis, I put his production of space ideas in dialogue with the new mobilities paradigm (Sheller and Urry 2006). As Haraway (2016: 35) says, 'it matters what thoughts think thoughts,... what relations relate relations'. For example, it matters if humans continue to imagine and relate nature with the car and SUV rather than other ways of moving together. Thinking about the production of nature with mobilities can invigorate ecological analysis of 21st-century conundrums such as the way Canada orchestrates nature. Lefebvre's dim view of nature relates to his prescient critique of the car and its planetary impact (Scott 2013). In fact, Lefebvre's car critique helped inspire mobilities scholarship, which exploded ten years ago in part because scholars started scrutinising the self-organising, coercive, hegemonic and ecologically disastrous 'system of automobility' (Conley and McLaren 2009; Urry 2004; Sheller and Urry 2000). In this chapter I build on Lefebvre's car critique by arguing that cycling can produce nature differently.

The chapter has three sections. In the first section I review Lefebvre's car critique and situate new mobilities literature within the production of nature discussion. Drawing on Lefebvre's production of space ideas, I refine three questions for my subsequent analysis:

- How can *spatial practices* that enact nature not simply reproduce existing social relations, but transform them, along with what it means to be human?
- How can urban mobilities cultivate *representational space* alongside more sophisticated *representations of nature* that cultivate associations between humans and nonhumans?
- Whereas automobility sets nature apart from human activity in the abstract space of neoliberal capitalism, how can cycling, following Lefebvre's *historical approach to the production of urban space*, produce nature differently?

To explore these questions, in the second section of the chapter I analyse data from an ongoing ethnographic study of urban cycling in major Canadian cities, started in 2013, for which I am principal investigator (with funding from the Social Sciences and Humanities Research Council of Canada). These data include interviews with city planners and politicians as well as mobile video data collected using GoPro cameras during ride-alongs and follow-up video elicitation interviews. I present results from four ride-alongs and follow-up interviews with experienced urban cyclists that focus solely on biking to nature in the city: Theo and Dasha in Winnipeg, Colm in Toronto and Francesca in Ottawa (all pseudonyms). These interviews all lasted about an hour and a half and were conducted, recorded and transcribed during the summer of 2015. I then expand my analysis by applying Lefebvre's historical approach to the production of urban space to show how automobility, through early 20th-century parkways in North America, can ecologically inspire the expansion of cycling nature. My analysis finishes with a comparison of cycling nature in Canada and Finland. In the third, concluding section, I suggest that nature and wilderness stand a greater chance of flourishing if their production by cycling ameliorates rather than exacerbates planetary gentrification.

Automobility and the production of nature

Lefebvre's writing on the car, more than his observations on nature, cultivates the production of nature as a site for ecological analysis. For Lefebvre, the car is a critical tool for imposing the conceived and orderly space of engineers and planners upon historical, messy lived space. He decries how complex, knotted cities, like the old Paris obliterated by Haussmann, were 'sliced up, degraded, and eventually destroyed... by the proliferation of fast roads and places to park and garage cars'. Lefebvre laments 'tree-lined streets, green spaces, and parks and gardens' that were 'sacrificed to that abstract space where cars circulate like so many atomic particles' (1991: 312–359). Some take issue with Lefebvre's depiction of the car as a purveyor of abstract space and global capitalism. Edensor (2003), for example, complains that Lefebvre contributes to 'legion' depictions of car driving as 'inherently malign'. He protests Lefebvre's criticism 'that the driver moves through an "abstract", "flattened" space and is concerned only with reaching a destination'. Edensor (2003: 152) argues that car driving affords sensuous and creative possibilities for 'reverie', kinaesthetic skills and pleasure. But this is unfair; Lefebvre does not deny such possibilities, and it misses Lefebvre's larger point, that mass car travel engineers the dominant space in a growing number of societies for accelerating capital flows through new car-oriented development as infrastructure for neoliberal urbanism (Scott 2013). Lefebvre was onto something big.

Writing decades ago, Lefebvre shows how the car contributes to rigid nature/city dualisms, setting nature apart from city and society by enveloping it in the abstract space of neoliberal capitalism. This prescient car critique is borne out by research on the hegemonic system of automobility in the new mobilities paradigm (Conley and McLaren 2009; Sheller and Urry 2006; Sheller and Urry 2000). The car is not inherently malign but is caught up in significant ecological degradation, America's 'secessionist automobility' (Henderson 2009) and divergent forms of neoliberal populism, like the pro-car authoritarian populism of Rob Ford's government in Toronto, 2010–2014 (Walks 2015). What did become legion across the 20th century were inherently benign, freedom-loving depictions of the car by prolific commercial advertising as the only vehicle for nature. As Aronczyk (2005) puts it, 'without cars, wilderness as we know it could not exist'. Automobility assembles wilderness, from Iceland to Quebec, as a predefined destination for car travel across many sites of travel practice (Huijbens and Benediktsson 2007). New versions of the car, via digital sophistication and construction of ever more intelligent roads and vehicles, produce nature by focusing more attention on car consoles, media and software, with wilderness outside the car acting as a backdrop. In effect, neoliberal car capitalism re-romanticises and pacifies wilderness, producing nature as something pure, external and exclusive for people rolling in with the right wheels.

The production of nature discussion grew over the last few decades from a predominantly Marxist focus (Smith 2010) towards diverse theoretical perspectives, including science and technology studies (Haraway 2016). One way to characterise this shift entails a move away from strong social constructionism to approaches that recognise the material reality and agency of nature. For example, Fitzsimmons (1989: 106–10) argues that once capitalism and urbanisation abstracted nature as society's antithesis, nature gained a 'mystifying power' over urban intellectual life that shrouds the role of capitalist production and the 'material reality' of nature. Elaborating this line of thought, Castree (1995: 13) argues the materiality of nature must include 'both the ontological reality of those entities we term "natural", and the active role those entities play in making history and geography'. Fast forward to more recent discussion, and we see less emphasis on exceptionalising humans and more comfort with inviting diverse nonhumans into the ranks of nature's producers (Latour 2004; Haraway 2008). As Ingold (2011: 7–9) puts it, there are multiple ways of being alive in the world – 'there are human becomings, animal becomings, plant becomings, and so on'. Or, as Vannini and Vannini (2016: 215) put it, wilderness is a 'meshwork' of tangled lines of growth that 'force us to confront life as becoming, as movement, as something entangled in multiple currents of formation'.

Recent discussions on assembling nature and wilderness refocus attention on the significance of movement (Lund 2013), which Lefebvre emphasised in his car critique and the production of space. Lefebvre imagined social space as moving, among other ways, like the motion of water:

> Great movements, vast rhythms, immense waves – these all collide and 'interfere' with one another; lesser movements, on the other hand, interpenetrate.... any social locus could only be properly understood by taking two kinds of determinations into account: on the one hand, that locus would be mobilized, carried forward, and sometimes smashed apart by major tendencies;... on the other hand, it would be penetrated by, and shot through with, the weaker tendencies characteristic of networks and pathways.
>
> *(1991: 87)*

Hegemonic automobility and driving nature comprise a great interference, an immense wave with vast sociotechnical momentum. It begs the question, can other spatial practices of mobility that enact wilderness avoid reproducing existing social relations, like rigid nature/society dualisms and neoliberal urbanisms, caught up in the car? Will cycling get dragged along with automobility's production of nature, or can it escape the car's undertow sideways and reassemble nature differently? Can cycling cultivate representational spaces of nature alongside sophisticated representations of nature in which humans and nonhumans flourish? Can ecology actually challenge neoliberal capitalism? This last question, about nature's moral worth (Thévenot et al. 2000), speaks to the high stakes of politically challenging the car and its wilderness.

Producing nature differently

Spatial practices of cycling nature carry the potential to contest, if not transform, existing social relations and perceived spaces of nature and the city. Whereas automobility tends to produce nature as an unspoiled destination for the car bracketed apart from the ordinary urbanised space in which the majority of humans live, cycling insinuates nature and wilderness into daily spaces of practice. Participants in a multiyear ethnographic study of urban cycling in Canada demonstrate how cycling, on river trails and multi-use pathways set outside car traffic (yet in the orbit of work and home), brings nature, nonhumans and immersive weather into everyday life. It is already well documented (Spinney 2006; Furness 2010; Aldred 2013; Larsen 2014) that cycling exposes people to the nature of their surroundings. In what follows, to advance Lefebvrian ecological analysis I show how cycling transforms the production of nature.

Theo in Winnipeg captures a common sentiment about cycling nature in practice. Lamenting the high number of pickup truck drivers in the prairie city who seem dangerously detached from their surroundings, he avows:

> but that's the fun piece in biking in my mind, is that you have to be aware of your environment and how you interact with it, you don't have a choice.
>
> *(Interview with author 2015)*

Dasha says cycling, even in Winnipeg's brutal winter, gives her:

> 'the time to actually enjoy her surroundings'. She describes how her cycling to work also affords access to city green spaces, where 'there's a sense of removal from traffic, and then

Figure 26.1 Dasha cycling along the confluence of the Red River and Assiniboine River in Winnipeg, Manitoba, Canada (photo by author 2015).

> quiet, or like a different noise. Like by the river, there's spots where you can kind of get away [see Figure 26.1], even around other people'.
>
> *(Interview with author 2015)*

Theo elaborates how cycling on pathways in the woods by the water, separated from the car, does not separate 'nature' and 'daily life' into silos or what Lefebvre (1991: 329) calls different 'space envelopes'. For Theo, cycling:

> reminds you that the city has a natural or nature component to it. It's not all *jungle de béton*. Especially in a city like Winnipeg with so many spaces interacting with the river and natural riverbanks, there's something fun to allow yourself, especially if it [is] as simple as getting to work, to insert a bit of natural environment in your life, and combine it as part of something utilitarian.
>
> *(Interview with author 2015)*

The integration of work and play, concrete and nature, playing out in Winnipeg through cycling nature challenges tidy divisions between functional spaces of practice conceived by automobility, especially human society versus wild nature. Spatial contraventions by cycling show how prolific spaces of practice, like work and nature, become embroidered in the material world.

Cycling transforms nature by pouring its spatial practice into daily life, but also by prying the production of nature apart from neoliberal capitalism. Lefebvre (1991) criticises the car, not for creating routine practice per se, but for the way the car's practice coerces people into driving through commodified landscapes (Sheller and Urry 2000), increases compulsory consumption (Soron 2009) and:

> services the reproduction of global capital, in a manner that increasingly alienates us from the rhythms and desires of the human body… and the cycles of the natural world.
>
> *(Gardiner 2004)*

Cycling can, though, lubricate the flows of global capital by contributing to gentrification and racialised neoliberal urbanism (Hoffman 2016). But cycling, being less coercive and channelled into development than automobility, also opens up spaces of practice beyond market worlds of neoliberal capitalism, like wilderness pathways in the city. Ecological skills, such as noticing and following change in more-than-human environments, flourish in public spaces outside the marketplace. On wilderness pathways, people cycle with enough speed to satisfy daily mobility needs, yet with enough slowness and physical exposure to know, and feel like they are part of, their surroundings. On such paths, Theo and Dasha engage nonhumans, like rivers, birds and plants, as ends rather than commodified instruments. Such slower, closer enactments of nature may only comprise a small space of practice compared to driving nature with a car. But they render people more vulnerable to becoming human with nonhumans, and challenge anthropocentric notions of what it means to be human.

Representational spaces of cycling nature further contest abstract, human-centred notions of nature by folding moments of wonder and possibility into a narrow but growing space of cycling practice. Such cycling moments alter the directly lived space of nature. For Lefebvre, lived spaces, spaces of representation, are alive in the sense of folding time or duration into space. Lived space opens up the production of space to a 'radical outside' (Seigworth 2000: 248) past the familiar productions of the city that slice up space into society/nature, work/play and urban/wilderness. Representational space enters the production of cycling nature during fleeting moments, notably while riding and dwelling on wilderness pathways assembled through water, bridges, play and public art.

Colm in Toronto slips through a representational space that resonates with the experience of other people cycling through Canada's cosmopolitan ecology (Gandy 2013). While riding along a polluted if biodiverse and resilient river valley hidden in the ravines beneath Toronto's cement crust, Colm stumbles upon a man painting a mural on the concrete foundation of a gigantic viaduct arching over the top of the valley. Lurid colours and surreal animal figures, floating up the viaduct towards an invisible vibrating stream of cars, clash with the faded grasses and sumac trees around the river banks. The odd scene pulls Colm off his bicycle into conversation with the artist, who explains the mural aims to defamiliarise and celebrate Toronto's natural underworld as part of an international sporting event. What makes the familiar valley suddenly seem strange to Colm, however, is a feeling of awareness of the wider watershed as a whole piece of the city that overflows this space, if only for a moment. The ability to stop and imagine his surroundings with art allows Colm to notice nature differently, with more mindfulness (interview with author 2015). He contravenes and plays with his boundaries between city and wilderness, but also between the trivial and extraordinary, articulating an 'everyday utopianism' (Gardiner 2004).

While riding to work or just for pleasure, Theo experiences similar flashes of presence triggered by water crossings and public art. He calls them 'moments of Zen'. There is one bridge in particular that plays with his temporal, representational sense of space:

> There's one bridge in Winnipeg, the Arlington [1910], that was designed for the [River] Nile, and then something happened, so Winnipeg got it for like, half off kind-of-thing. There's something about that moment, where you cross a bridge that was designed to cross the Nile.
>
> *(Interview with author 2015)*

Moments of Zen or spatial duration emerge for Theo, especially where art escapes the spatial control of the gallery and moves into active transportation pathways along natural corridors. Such art ranges from human faces carved in trees that watch and startle passers-by, to furnishings in natural amphitheatres at ancient river crossings that point to constellations of stars. Cycling along nature pathways, says Theo:

> Many bridges now incorporate an art element, and it's fantastic to witness and be inspired by that and stop and actually look at the art. So I go, it's not necessarily nature versus non-nature. I don't know what the opposite of nature is... But cars go too fast, so they miss the art.
>
> *(Interview with author 2015)*

Dasha, too, finds Zen moments on bridges in the urban wild, dwelling less on the art than the special acoustics and big soundscapes along bridges and shores set apart from cars by water. She plays with this space, taking her feet off the pedals and standing on her bicycle frame while swerving around manhole covers, listening to how nature and the city enable each other. In fact, people cycling nature across urban Canada, from Vancouver to Halifax, report flashes of extensive presence with nonhumans during playful moments on wilderness pathways, wherein the regular risks and rules of the road relax. They may be ephemera, but representational spaces of cycling nature form an important part of the reason why cyclists search for nature outside in the city to begin with.

Spatial representations of cycling nature provide a powerful tool for expanding small spaces of practice and fleeting moments of Zen into a larger, concrete production of urban nature. Mobilising representations of cycling nature is politically complicated in Canada, where cycling activity across the board remains low. In 2011, the proportion of workers commuting by bicycle ranged from a low of 0.2 per cent St. John's, Newfoundland, to a high of 5.9 per cent in Victoria, British Columbia (Statistics Canada 2013). Whether cycling practice and banal utopianism grow into more-than-human nature outside the car depends on plans, technical models and conceived space. Simple conceived spaces of cycling nature helped cultivate the planning of Winnipeg's Churchill Parkway, on which Theo and Dasha enact nature in the heart of a continent, and Toronto's Don River Valley Trails, on which Colm lets his imagination wander off into an urban wasteland (Gandy 2013) teeming with nonhuman history and rhizomes. These and other wilderness pathways help assemble the Trans Canada Trail (renamed The Great Trail upon its completion in 2017), an ambitious 24,000-kilometre route aiming to provide a dedicated trail for persons-without-cars that links 15,000 communities across the country.

Models and representations of cycling nature, like Vancouverism and The Great Trail, wield a double-edged sword. On one hand, they picture planning and development for nature outside the car. On the other hand, because all representations of space 'are tied to the relations of production and to the "order" which those relations impose' (Lefebvre 1991: 33), representations of cycling nature may lead to the production of a nature suspiciously similar to the romantic and anthropocentric nature produced by the car-industrial complex. However, there are hopeful signs that some conceived spaces of cycling nature can help swim cycling sideways and escape the strong undertow of automobility into a more embodied, messier nature, one which cultivates closer relations with nonhumans.

Representations of cycling nature are not planned and plonked down on some virginal tabula rasa, but in the material reality of the city (usually around its knotty points and bends). This means in urban Canada that cycling nature involves tangoing with cars where, according to Francesca in Ottawa, 'cars think they own the place' (interview with author 2015). Just like her Winnipeg and Toronto counterparts, Francesca begins the process of cycling nature by leaving her street and negotiating fast-moving motorists on a linear arterial road conceived geometrically and instrumentally for car commuting. Before long, however, Francesca escapes into a fulsome, planned cycling habitat: Ottawa's Capital Pathway, a 600-kilometre network of multi-use, curvilinear paths linking parks with the rest of the city (Scott 2016). Both city pieces – the car artery and wilderness pathway – co-constitute cycling nature, and the latter is as much a conceived space as the former, meticulously crafted by federal city planners charged with maintaining Canada's capital as a beautiful Washington of the North.

Lefebvre's historical approach to the production of urban space illuminates how some representations of cycling nature, like Ottawa's pathways, over time bolster particular, embodied ways of practising and reimagining nature with cycling (Scott 2016). An historical approach also shows that while car driving and cycling diverge with respect to nature's production, driving nature can guide, if not ecologically inspire, the expansion of cycling nature. For example, Ottawa's wilderness pathways, Canada's most prominent, also constitute the nation's earliest network of such pathways. They were conceived and laid down starting in the 1970s along the capital's canals and rivers, in many cases alongside parkways that were designed for the same purpose – connecting people to nature into the city – but with motorists in mind. Ottawa's parkways enjoy an even longer, illustrative lineage. In 1888, Frederick Law Olmsted established, in the case of Buffalo, what:

> is to be understood by the term parkways. They plainly serve, not simply as branches or outworks of the park with which they connect but as a part of the general street system of the city'.
>
> *(Olmsted 1888, in Sutton 1997: 147)*

The parkway's value for Olmsted derives not from speed, but from the nature surrounding it. Ottawa took this idea to heart, elaborating slow and windy park roads across the city for motorists to experience remarkable vistas without the signage or industrial trucking that would distract them from contemplating nature (Scott 2012; Gordon 2015). The moral force of this production of nature is tempered by its romanticisation of nature and role in imposing colonial space on unceded Anishinaabe (Algonquin) territory. Notwithstanding these important limitations, the parkway offers ecological inspiration and material direction for the urban expansion of cycling nature.

As cycling nature in Canada's capital suggests, the production of nature is an international process, which can be further elucidated through international comparison. Canada affords diverse forms of cycling nature across a vast urbanising landscape, but levels of cycling activity, as in the United States, are generally low. One avenue for developing this analysis entails comparing Canada to another country with a more advanced cycling culture and more extensive experience with cycling nature. While the Netherlands, Germany and Scandinavia all offer fruitful possibilities, I suggest Finland, especially the northern Finnish city of Oulu, provides an analytically important comparison. As a capital of winter cycling in the Global North (Babin 2014), Oulu faces many of the same weather-related challenges to everyday cycling as Canadian cities, yet meets these challenges with innovative maintenance and effective educational campaigns. Winter, however, is not the most salient basis of comparison.

On the surface, Oulu's *pyörätie* or cycle pathways look very similar to multi-use pathways in Ottawa, Vancouver, Winnipeg, Toronto and Halifax. However, a critical difference emerges that helps explain why people cycle at much higher rates in Oulu and think of cycling nature, if they think of it at all, as simply part of everyday life. This difference, according to a city transport planner and a cycling campaign leader I interviewed in Oulu in summer 2015, lies in the way Oulu's pathways were planned and developed, beginning in the 1970s, as the most efficient way to travel in green spaces between city neighbourhoods and the urban core. Ottawa was lucky, also getting in on the ground floor of pathway planning, such that Capital Pathway network grew into efficient routes that combine nature with daily mobility needs. Most cities in Canada, however, like those in the United States, are now left scrambling to assemble wilderness pathways after decades of already implementing expensive car-based infrastructure, which, as Europe shows, is very difficult, politically and physically, to unbuild (Hommels 2008). As part of a larger, historical production of urban nature, the *pyörätie* offers a compelling opportunity for international policy transfer, with Ottawa already demonstrating a version in Canada.

Conclusions

In this chapter I demonstrate how Lefebvre's broader production of space ideas, more than his observations on nature, offer an effective conceptual toolkit for expanding the production of nature as a site of Lefebvrian ecological analysis. In particular, I show how spatial practices, representational space, representations of space and Lefebvre's historical approach to the production of urban space can be applied in combination to illuminate the production of nature, using cycling across urban Canada as a case study. To advance the production of nature research, I place Lefebvrian thought in dialogue with the new mobilities paradigm, noting how this paradigm draws inspiration from Lefebvre's prescient car critique. In the spirit of Lefebvre's biting critique of the car and its production of space, I examine how nature can be produced outside the parameters of hegemonic automobility and neoliberal car capitalism. I show how cycling nature, more than driving it with a car, produces nature as an embodied, exposed and ecologically valuable process of cultivating closer relations with nonhumans. I conclude that cycling can reassemble how humans value and engage with more-than-human nature.

Limitations of my analysis, including its focus on the Global North and my lack of attention to social cycling inequities, point to ways of advancing research on the production of nature. There lingers the fundamental, ethically complex question: who gets to cycle with nature? The preponderance of high-quality cycling facilities in Northern Europe and a growing number of wealthy cities in the Global North speaks to two pressing needs: transferring and adapting cycling nature knowledge to the Global South, while cultivating existing cycling cultures and expanding cycling in an equitable way within cities worldwide. Cycling equity is made all the more significant in light of megacities in the Global South expanding through automobility, but also because of the globaliation of gentrification, as Vancouverism illustrates, which threatens to transform cycling-friendly areas of the city into exclusive living space for global elites. An important avenue for future Lefebvrian ecological analysis entails examining how cities can produce nature as neither abstract nor absolute space, but as differential space with political possibilities for expanding rights to urban nature, in particular non-anthropocentric urban natures that contribute to the flourishing of nonhuman beings. Given the capacity of people cycling to notice, engage with and reimagine their surroundings, cycling, in combination with artful, off-car wilderness pathways, can mobilise ecologically good productions of urban nature.

References

Aldred, R. (2013) 'Incompetent or too competent? Negotiating everyday cycling identities in a motor dominated society', *Mobilities*, 8:2 252–71.

Aronczyk, M. (2005) '"Taking the SUV to a place it's never been before": SUV ads and the consumption of nature, invisible culture', *An Electronic Journal for Visual Culture*, 9 1–15.

Azpiri, J. (2016) *Metro Vancouver the Luxury Car Capital of North America*. Online. Available HTTP: <http://globalnews.ca> (accessed 21 February 2017).
Babin, T. (2014) *Frostbike: The Joy, Pain and Numbness of Winter Cycling*, Toronto: Rocky Mountain Books.
Castree, N. (1995) 'The nature of produced nature: materiality and knowledge construction in Marxism', *Antipode*, 27:1 12–48.
City of Vancouver (2017) *Urban Planning, Sustainable Zoning, and Development*. Online. Available HTTP: <http://vancouver.ca/home-property-development/planning-zoning-development.aspx> (accessed 15 January 2017).
Conley, J. and McLaren, A.T. (eds) (2009) *Car Troubles: Critical Studies of Automobility and Auto-Mobility*, Aldershot: Ashgate.
Edensor, T. (2003) 'Defamiliarizing the mundane roadscape', *Space and Culture*, 6 151–68.
Fitzsimmons, M. (1989). 'The matter of nature', *Antipode*, 21:2 106–320.
Furness, Z. (2010) *One Less Car: Bicycling and the Politics of Automobility*, Philadelphia, PA: Temple University Press.
Gandy, M. (2013) 'Marginalia: aesthetics, ecology, and urban wastelands', *Annals of the Association of American Geographers*, 103:6 1301–16.
Gardiner, M. (2004) 'Everyday utopianism: Lefebvre and his critics', *Cultural Studies*, 18:2 228–54.
Gordon, D.L.A. (2015) *Town and Crown: An Illustrated History of Canada's Capital*, Ottawa, ON: Invenire.
Haraway, D.J. (2008) *When Species Meet*, Minneapolis, MN: University of Minnesota Press.
Haraway, D.J. (2016) *Staying With the Trouble: Making Kin in the Chthulucene*, Durham: Duke University Press.
Henderson, J. (2009) 'The politics of mobility: de-essentializing automobility and contesting urban space', in J. Conley and A.T. McLaren (eds) *Car Troubles: Critical Studies of Automobility and Auto-Mobility*, Burlington, VT: Ashgate.
Hoffman, M. (2016) *Bike Lanes Are White Lanes: Bicycle Advocacy and Urban Planning*, Lincoln, NE: University of Nebraska Press.
Hommels, A. (2008) *Unbuilding Cities: Obduracy in Urban Sociotechnical Change*, Cambridge, MA: MIT Press.
Huijbens, E.H. and Benediktsson, K. (2007) 'Practising highland heterotopias: automobility in the interior of Iceland', *Mobilities*, 2:1 143–65.
Ingold, T. (2011) *Being Alive: Essays on Movement, Knowledge and Description*, New York: Routledge.
Larsen, J. (2014) '(Auto)ethnography and cycling', *International Journal of Social Research Methodology*, 17:1 59–71.
Latour, B. (2004) *Politics of Nature: How to Bring the Sciences into Democracy*, trans. C. Porter, Cambridge, MA: Harvard University Press.
Lefebvre, H. (1991/1974) *The Production of Space*, Oxford: Blackwell.
Lund, K. (2013) 'Experiencing nature in nature-based tourism', *Tourist Studies*, 13:2 156–71.
Scott, N. (2012) 'How car-drivers took the streets: critical planning moments of automobility', in P. Vannini, P. Jiron, O.B. Jensen, L. Budd and C. Fisker (eds) *Technologies of Mobility in the Americas*, New York: Peter Lang Publishing.
Scott, N. (2013) 'Like a fish needs a bicycle: Henri Lefebvre and the liberation of transportation', *Space & Culture*, 16:3 397–410.
Scott, N. (2016) 'Cycling, performance and the common good: Copenhagenizing Canada's Capital', *Canadian Journal of Urban Research*, 25:1 22–37.
Seigworth, G.J. (2000) 'Banality for cultural studies', *Cultural Studies*, 14:2 227–68.
Sheller, M. and Urry, J. (2000) 'The city and the car', *International Journal of Urban and Regional Research*, 24:4 737–57.
Sheller, M. and Urry, J. (2006) 'The new mobilities paradigm', *Environment and Planning A*, 38:2 207–26.
Smith, N. (2010) *Uneven Development: Nature, Capital, and the Production of Space*, 3rd edn, Athens, GA: University of Georgia Press.
Soron, D. (2009) 'Driven to drive: cars and the problem of 'compulsory consumption', in J. Conley and A.T. McLaren (eds) *Car Troubles: Critical Studies of Automobility and Auto-Mobility*, Burlington, VT: Ashgate.
Spinney, J. (2006) 'A place of sense: a kinaesthetic ethnography of cyclists on Mont Ventoux', *Environment and Planning D: Society and Space*, 24:5 709–32.
Statistics Canada (2013) *Commuting to Work: National Household Survey (NHS), 2011*. Online. Available HTTP: <https://www12.statcan.gc.ca> (accessed 14 December 2016).
Stoddart, M. (2012) *Making Meaning out of Mountains: The Political Ecology of Skiing*, Vancouver, BC: UBC Press.

Sutton, S.B. (ed.) (1997) *Civilizing American Cities: A Selection of Writings on City Landscapes*, New York: Da Capo Press.

Thévenot L., Moody, M. and Lafaye, C. (2000) 'Forms of valuing nature: arguments and modes of justification in French and American environmental disputes', in M. Lamont and L. Thévenot (eds) *Rethinking Comparative Cultural Sociology: Repertoires of Evaluation in France and the United States*, Cambridge: Cambridge University Press.

Urry, J. (2004) 'The "system" of automobility', *Theory, Culture & Society*, 21:4–5 25–39.

Vannini, P. and Vannini, A. (2016) *Wilderness*, New York: Routledge.

Walks, A. (2015) 'Stopping the "War on the Car": neoliberalism, Fordism, and the politics of automobility in Toronto', *Mobilities*, 10:3 402–22.

27

Lefebvre and atmospheric production

An architectronics of air

Derek R. Ford

Introduction

On 17 July 2014, Eric Garner's last gasps of air were captured on video by passer-by Ramsey Orta. Garner was attacked by police officers in Staten Island, New York City (for allegations of selling loose cigarettes), and was ultimately killed when he was choked by officer Daniel Pantaleo. On Orta's video we hear Garner's final words: 'I can't breathe. I can't breathe. I can't breathe. I can't breathe. I can't breathe. I can't breathe. I can't breathe. I can't breathe. I can't breathe. I can't breathe. I can't breathe'. His last breaths, used to issue urgent pleas about the availability of breath, resonated and were amplified across the United States. They became collectivised into chants of 'We can't breathe!' at protests and actions against racially-based police brutality in the country. They soon became internationalised as when, for example, Palestinians in the West Bank chanted them after a Palestinian Authority minister was killed from a tear gas canister fired by an Israeli soldier.

These collectivisations and appropriations of Garner's last breaths are so energetic because they so clearly link the ubiquity of racism and oppression to the most fundamental of bodily habits. To pose the fight against oppression as a fight for breath is to pose politics in a radical, incandescent manner. We cannot read or utter words without breathing, and as we read them, our relationship to our breath changes. This change is a process, as Peter Sloterdijk (2009: 84) would put it, of 'atmospheric explication'. That which was previously implicit is brought to articulation, entering the realm of signification. The course that this process of atmospheric explication takes is, to be sure, an open question. In this chapter, I want to propose a narrative of atmospheric explication by bringing Lefebvre's architectronics of space into the realm of the pneumatic. Doing so not only enhances Lefebvre's analysis by applying it to that which actually makes it possible to inhabit space, but it also provides us with insights to aid in struggles for spaces liberated from the constraints of private property and the demands of exchange value.

To perform this task, I begin with Lefebvre's architectronics of space. I delineate the four key concepts in this methodology: absolute space, abstract space, contradictory space, and differential space. Having laid this foundation, I turn to an explication of our contemporary atmosphere. I begin with the absolute space of the hut before turning to the capitalist abstraction of air. I locate this abstraction within the commodity form and the law of value with the help of Marx and Lukács, demonstrating that air conditioning both flows from and enhances capital's more general processes of abstraction and fragmentation. Locating contemporary air conditioning within the factory and city, I draw out the struggles over the production of air, with examples from factories and urban air-rights projects. This takes us through to contradictory air, which is where, in Lefebvrian fashion, I leave the analysis.

An architectronics of space

Lefebvre's theorisations of space have been profoundly influential across—and controversial in—a range of disciplines. What has not yet been considered, however, is that which must fill space for it to be habitable by life: air. Indeed, in his writings on space Lefebvre makes only a few specific mentions of air, and he never addresses air as a unique element of space. He generally only writes of it in a series of other natural elements and components like land and water. Indeed, this is how air is traditionally conceived of in both political economy and critiques of political economy (see the chapter by Lopes de Souza in Part 5). Thus, Marx (1967b: 745) considered air to be primarily a 'free gift of Nature'. Lefebvre (1991: 329), with a century on Marx, can recognise that air is no unproblematic gift:

> Those commodities which were formerly abundant because they occurred 'naturally', which had no value because they were not products, have now become rare, and so acquired value. They have now to be produced, and consequently that come to have not only a use value but also an exchange value. Such commodities are 'elemental' – not least in the sense that they are indeed 'elements'. In the most modern urban planning, using the most highly perfected technological applications, *everything* is produced: air, light, water – even the land itself.

Lefebvre's architectronics provide us with a methodology for uncovering just how it is that air is produced and the process by which it has come to be produced. Architectronics is a methodology intended 'to describe, analyse, and explain' the manner in and by which 'the preconditions of social space have their own particular way of enduring and remaining actual within that space' (229). Things have a way of remaining in space; 'In space', that is, 'nothing ever disappears – no point, no place' (212). The idea is that social relations remain etched in space throughout time, although to different extents. But this lasting presence means that we can uncover the history – and the present – of the production of space, with an eye toward the future. Lefebvre often uses the word 'deciphering' to describe the analysis of space that is architectronics. This is a transdisciplinary endeavour that 'embraces and seeks to reassemble elements dispersed by the specialized and partial disciplines of ethnology, ethnography, human geography, anthropology, prehistory and history, sociology, and so on' (299).

The four key analytical headings for Lefebvre's architectronics of space in *The Production of Space* are absolute space, abstract space, contradictory space, and differential space. Importantly, absolute space is not equal natural space. While absolute space can *appear* as natural space, it is precisely that space which has first been reconceptualised and transformed by human labour, in particular its origins are 'fragment[s] of agro-pastoral space, a set of places named and exploited by peasants, or by nomadic and semi-nomadic pastoralists' (Lefebvre 1991: 234). Absolute space is a lived space wherein the concept of representational space reigns. It is, as such a space of the inhabitant, the dweller who makes the space through their own use; spaces as they are directly lived in everyday life. As such, absolute space is not so much about epistemology or the realm of the Symbolic and is more ontological.

The move toward abstract space is concomitant with the rise of private property, which allows space to be striated and owned. Absolute space was abstracted through the process of entering into the realm of signification, as spaces were governed just as much by bodies and labour as by legal codes and tax regimens. The town's overthrow of the country as the motor of social and economic growth is crucial in this transition, what Lefebvre labels as the first spiral of abstraction, in which the logic of exchange value through mercantilism comes to dominate. A second spiral of abstractions comes later by way of the consolidation of towns into an overarching state formation. With the second spiral of abstraction the logic of capital, and not merely the merchant, is stabilised. Space becomes an exchange value, not a use value.

The capitalist logics of abstraction seek to make space a blank slate upon which the agents of capital – merchants, industrialists, financiers, real estate agents, corporate executives – can inscribe their plans and desires. It seeks to order space, thereby subjecting it to the 'rationality' of the market. It is, however, '*not* homogenous; it simply *has* homogeneity as its goal, its orientation, its "lens"' (287). Here we reach what I take to be a key principle of Lefebvre's pedagogy: use and use value endlessly persists and resists. Capital is never able to completely subject space to its demands because its internal antagonist, labour, cannot help but meddle in its plans, making space for itself and its social relations. Thus, the abstract space of capital sits beside and in fact depends on the contradictory space of labour in the same way that abstract labour sits beside and depends on concrete labour. They sit together in a contradictory unity.

A productive way to think about this contradictory unity concerns the use value and the exchange value of space. The use value of a commodity is its socially determined utility, the fact that it fulfils some need or want, some singularity, whereas the exchange value of a commodity is a quantitative relation of the commodity to other commodities. That which makes commodities exchangeable is the human labour that produces them, and this is also what makes commodities possess a use value and not merely a value or a use. In order for something to have a use value it must be exchangeable, must be the product of human labour, and in order for it to have an exchange value it has to have some utility. No one exchanges useless things. There is a unity, but each are opposed to the other, for neither can be realised at the same time; one cannot use and sell a commodity at the same time. Capital cares only for exchange value – and thus for abstraction – whereas labour cares only for use value – and thus for singularity, for difference. Once space is understood as a product of labour, then it too becomes the subject of this antagonism. Thus, capital's abstract space is not a mere *perspective* on space, it is actually a *mode of producing and realising* space, a mode inclined to commodify space. Lefebvre gives a clear example of this tendency and the ways in which space can be abstract and contradictory:

> When an urban square serving as a meeting-place isolated from traffic... is transformed into an intersection... or abandoned as a place to meet... city life is subtly but profoundly changed, sacrificed to that abstract space where cars circulate like so many atomic particles.
> *(Lefebvre 1991: 312)*

And yet even when cars circulate in an atomised and individualistic fashion, we find ways to engage with each other, to communicate and engage in a type of social intercourse.

Lefebvre is clear that whether a space is abstract or contradictory hinges upon its relationship with the mode of production and its position relative to the class struggle. As he puts it quite bluntly early on in the book, 'the class struggle is inscribed in space. Indeed, it is that struggle alone which prevents abstract space from taking over the whole planet and papering over all differences' (55). Here, we arrive at the final index of architectronics: differential space. This is the project toward which Lefebvre's book, and one might say, his life, was oriented: 'the project of a different society, a different mode of production, where social practice would be governed by different conceptual determinations' (419). Whereas abstract and contradictory space are locked in a relation of the unity of opposites, differential space is freed from this particular unity; it is not just use value against exchange value, but *use without any value whatsoever*. The logics of differential space are governed neither by production nor by consumption, but rather by enjoyment. Differential space is that which breaks us out of the triple dialectic of absolute, abstract, and contradictory space. As Lefebvre (2013) insists in *Rhythmanalysis*, in one of his typically counterintuitive moves, the dialectic is always triadic. The dialectic *contains* not only the thesis and antithesis, but also the synthesis, and the last term is not the passive product of the first two. Instead, it reacts back on both the thesis and antithesis. The struggle on this reading, then, is to harness contradictory space in order to break through into a new mode of production, a new set of social relations and conceptual determinations, that are proper to differential space: pure use, pure enjoyment. We might also put it like this in a more classic Marxist formulation: Contradictory space arises when the working class becomes as a class *in itself*, and the class struggle produces differential space when it, organised as a class *for itself*, accomplishes its self-abolition.

An architectronics of air

If we want to take up residence in contradictory space, mobilise the spatialised class struggle to inaugurate something radically new, an event that is congealed into a new set of relations, then we have to engage in an architectronics of air, as that which fills space and makes it inhabitable. This is not only strategically useful, as this chapter's introduction intimates, it is also an increasingly dire task. In December 2016, what is popularly referred to as an *airpocalypse* began in northern China, forcing those residents with the means to flee, and everyone else to remain indoors or, if they dare venture outside, to don a mask. Flights were grounded and classes were cancelled. Less dramatic (and sensationalised) realities of air pollution abound throughout the world, with many urban centres having smog indexes that are featured on regular weather reports and dictate movements across and through space.

Architectronics begins in the present, so I will start here, where I sit in my office in an academic building dedicated in 1938. At the highest level of abstraction, I am enclosed in an interior. This is the basic state of being. Thus, Heidegger, on whom Lefebvre relies quite heavily, missed the mark when he focused in solely on the question of *being*, for, as Sloterdijk (2011) shows, being is always a *being-in*. Sloterdijk's spheres project is an investigation into the fact that we are always contained in various types of enclosures:

> for humans, being-in-spheres constitutes the basic relationship—admittedly, one that is infringed upon from the start by the non-interior world, and must perpetually assert itself against the provocation of the outside, restore itself and increase. In this sense, spheres are by definition also morpho-immunological constructs. Only in immune structures that form interiors can humans continue their generational processes and advance their individuations. (46)

We are always contained within something, and we are constantly constructing interiors in which to contain ourselves and others. This is the primal situation, one that begins in the most intimate of spheres: the womb. In the uterine sphere one resides with one's mother and one's double, or placenta. When one is born and 'the cord is cut', one leaves this bubble and enters into a larger sphere, or series of spheres: hospitals, nurseries, houses, schools, and so on.

Interiors raise not only architectural questions but also questions of the atmosphere, for air is that which fills and determines the viability and enjoyability of enclosures. Being only takes place, after all, within a quite slim range of temperatures. Not only do we seek out the best microclimates, but we also construct these microclimates. The hut is a simple interior that transforms a landscape in order to produce more viable and enjoyable microclimates. As architect Lisa Heschong (1979: 8) notes: 'As soon as a simple square hut is built, at least six new microclimates are created'. There is (in the Northern Hemisphere) 'the south side warmed by a sunny wall, the north side in shade... an east side with its morning sun and perhaps protected from the prevailing breeze, and a west side warmed in the afternoon but buffeted by the wind' (8). Finally, we have 'the inside with its shelter from the rain and wind and sun, and the roof, raised above ground level, more exposed to wind and sun'. Heschong's insights are important for my purposes here because they demonstrate that the conditioning and production of the air is not a recent phenomenon, something that Sloterdijk's analysis gets wrong (he locates the beginning of air conditioning in 1915).

The hut is a part of the absolute space, a natural landscape transformed through human labour for the purposes of lived life. A manufactured and controlled fire provides another example, as humans transform wood into smoke, heating the air in the process (and polluting it, too). The lit fire helps us fulfil basic needs, allows us to cook food and escape the cold. But it also has social and affective functions. As a microclimate, it produces a totalising atmosphere that touches all of our senses: 'a flickering and glowing light, ever moving, ever changing. It crackles and hisses and fills the room with the smells of smoke and wood and perhaps even food. It penetrates us with its warmth' (Heschong 1979: 29). Our consciousness, too, comes alive, and 'memory and an awareness of time, are also bought into play, focused on the one experience of the fire'.

That which births the space of the abstract from the absolute is private property. As Lefebvre (1991: 252) notes, 'Once unshackled, the principle of private property did not remain sterile: rather, it gave birth to a space'. The abstraction of air likewise hinges on its production for the purposes of exchange rather than use. This is, in fact, the origins of our modern air conditioning technologies and apparatuses. Rather than in human comfort (use), they began for the purposes of production (exchange). In 1902 there were two air conditioning installations in New York City, one for the New York Stock Exchange and another for the Sackett-Wilhelms Lithographic Company in nearby Brooklyn. While the former was meant to increase economic productivity by cooling stockbrokers, the latter was designed with only materials in mind. The printing company contracted Buffalo Forge (which would later be the Carrier Corporation) to develop humidity controls to help level out fluctuations that occurred during the printing process. At the time, colour publications were produced by running the same paper through different printers successively (once through for black, another for red, etc.). As the humidity would fluctuate, the paper would expand or contract, misaligning the ink. This would then result in 'poor quality, scrap waste and lost production days' (Schultz 2012: 4). Similar systems were soon installed in factories producing all sorts of hygroscopic commodities such as cotton, candy, pharmaceuticals, film, macaroni, tobacco, bread, and munitions. Government contracts for engineering air conditioning systems in the production of weapons, ammunition, and gunpowder were critical for helping the still-nascent industry survive through World War I and the Great Depression. The abstraction of the air amounts to its reification, its domination by the law of value, and capital's drive to decrease socially-necessary labour time. Only through such an analysis can we truly see how air becomes not just an ancillary material, but an actual commodity, which will help us prepare for the investigation of the production of contradictory air.

As mentioned, use value and exchange value exist in a dialectical tension. What we have not yet covered, however, concerns that which makes different singular commodities exchangeable, which is the labour power embedded within them. The value of the commodity is thus determined by 'The labour-time socially necessary... to produce an article under the normal conditions of production, and with the average degree of skill and intensity prevalent at the time' (Marx 1967a: 47). In other words, even if two factories produce commodity *x* in different quantities of time, each of the total number of *x* commodity would contain the same value. The value of commodity *x*, however, will not be constant over time, for socially-necessary labour time:

> is determined by various circumstances, amongst others, by the average amount of skill of the workmen, the state of science, and the degree of its practical application, the social organisation of production, the extent and capabilities of the means of production, and by physical conditions.
>
> *(47)*

We could certainly add in air conditions and air conditioning technologies as determining factors, particularly for hygroscopic industries. Yet socially-necessary labour time is also determined through the operation of the coercive laws of capitalist competition:

> The rule, that the labour-time expended on a commodity should not exceed that which is socially necessary for its production, appears... to be established by the mere effect of competition; since, to express ourselves superficially, each single producer is obliged to sell his commodity at its market price.
>
> *(327)*

If factory *a* produces commodity *x* at a rate above the given socially-necessary labour time, then that factory will sell commodity *x* below its value, and the factory will either conform to the socially necessary labour time or will be shuttered. In this manner, if one film producer, say, Kodak, installed a process air conditioning system, it would, through the coercive laws of capitalist competition, compel other film producers to follow suit. This tendency was more than amenable to the burgeoning air conditioning industry for two reasons. First, it obviously increased their sales. Second, and relatedly, it would standardise the installations. As Gail Cooper (1997: 48, 44) observes, 'If, for instance, all rayon plants adopted the same process, they would present the same engineering problems', and this 'standardization within the industry [served] to make the heavy initial investment in engineering work pay off in subsequent installations'. The logic of standardisation was at work at the level of the air conditioning industries and their clients in amalgamated and individual factories. It was also, however, at work at the level of the individual worker's body. This is what brings us to the contradictory air of the modern industrial factory.

For Lukács, the commodity form and its accompanying fetishism structure the way that people under capitalism understand and experience themselves, others, and the world. For Lukács (1971, p. 83), this is why Marx begins volume one of *Capital* with the chapter on commodities, for the commodity is 'the central, structural problem of capitalist society in all its aspects'. In essence, Lukács' concept of reification is commodity fetishism—the notion that through the commodity social relations are present as relations between objects—extended throughout the totality of capitalism. This generalisation has two implications that negatively impact society and workers, and both have to do with abstraction. First, the object of production is fragmented in time and space; production processes are divided up and extended across the entirety of the globe. Second, the subject of the production process is abstracted by the developmental logic of capital, wherein 'human qualities and idiosyncrasies in the worker appear increasingly as *mere sources of error* when contrasted with these abstract special laws functioning according to rational predictions' (89). The working body has to conform to the individual machine and the totality of capitalist relations, including the coercive laws of competition and the law of value. The worker's entire being and sense is abstracted: the principle of rational mechanisation and calculability must embrace every aspect of life.

Air conditioning both flowed from and accelerated the abstraction inherent in reification. It standardised factories and industries and, most significantly, it introduced the atmosphere as an ontological category that could be subjected to the principles of mechanisation, standardisation, and calculability. Most notable in this regard was the formulation and, in 1911, publication of Willis Carrier's paper, 'Rational psychometric formulae'. The paper provided a theoretical discussion of the mechanical control of atmospheric temperature and moisture and, most importantly, proposed a chart that correlated temperature and humidity, a version of which is still in use today. Each industrial process had unique atmospheric requirements, as did the different stages of each process. Thus, Carrier's 'Mechanical Weather Man' – the company's animated persona – promised manufacturers that the corporation could make 'every day a good day' (Schultz, 2012: 34).

The abstract air conditions of modern industrial capitalism also worked on the worker, who in turn entered into the struggle over *who* would control the air and for what purposes. Cooper (1997: 45) presents us with compelling examples of this struggle as it unfolded in everyday practices of cotton workers and managers via the battlefield of the window:

> While closing windows could increase humidity within certain limits, that increase was almost always achieved at the expense of ventilation and personal comfort. Because of the conflict between the needs of production and the conditions of comfort, the manner in which workers chose to regulate the windows under their immediate control was not easy to predict. Those who were paid by piecework rates might decide to keep the windows closed in warm weather to preserve high humidity levels and increase their productivity despite the personal discomfort, or they might sacrifice wages for better working conditions.

While air conditioning engineers and advertisers promised their clients the ability to not only increase but better predict and calculate their production, workers pushed back, exerting their agency to undermine and disrupt the abstraction of their workplace air.

The battle between abstract and contradictory air plays out in the urban through air-rights projects. David Gissen, for example, analyses the Washington Bridge Extension Complex, five buildings hovering over the Trans-Manhattan Expressway in New York City. Air-rights refers to the ability to build over roadways, something that became more attractive as the density of urban development increased in the mid-20th century. Literature studying air-rights development 'suggests emerging links between risk analysis, cost-benefit analysis, and other forms of technocratic engagements' (Gissen 2014: 39). The Bridge Apartments capitalised on the polluted air and carbon dioxide emissions generated from the thousands of cars passing underneath it in its (purported) ability to seal and completely condition the interior using aluminium as a skin, which was thought of as both resistant to the invasion of pollutants and easily cleaned of them.

Many of the initial technologies proposed to help create an interior condition divorced from the exterior were scrapped due to budget costs, including 'a cap over the highway, an integrated ventilation system for the buildings' interiors and park spaces, and an air-conditioning system for the buildings' (49). Ultimately, the Bridge Apartments were unable to produce this kind of interior space, leading to rent strikes, a USA Environmental Protection Agency study, numerous lawsuits and litigations. Ultimately, the urban planners here transformed the air into 'a form that might be captured and isolated within a city' (65). This air was 'both an objectified thing that could be traded... and a substance that could be transfigured into something more inhabitable' (65). The former represents the abstraction of air and the latter the contradictions of commodified air.

Conclusions

Lefebvre never does tell us what a true differential space looks like. This, I proffer, is for the simple reason that he could not. It's not that he didn't have the intellectual proclivities to do so, but rather that he could not answer a question theoretically that could only be worked out through the practice of class struggle. What Lefebvre does give us, however, is a spatial methodology and a historical and social excavation of our present spatial configurations so that we can know that space is never just as it appears. Indeed, this is the trick of representations of space of which we must always be cautious: they 'confuse matters precisely because they offer an already clarified picture' (Lefebvre, 1991: 188–9). In a similar vein, I cannot offer up what differential air feels like, how it circulates through blood or carries affective dispositions between subjects and bodies. What I have done in this chapter, however, is to introduce air and the production of atmospheres into our understandings of planetary urbanisation and the production of 'nature'.

Air is not only produced through capitalist industrialisation. Indeed, the most basic and primal conditions of the human *being* require the creation of microclimates and aerial containers, and the innovation of thermal strategies. This is crucial to understand so that we do not construct a false history of a moment of natural, unadulterated air to which we can return or which we can reconstruct. At the same time, however, it is just as crucial that we grasp exactly how thoroughly capitalist industrial modernisation has conditioned our air in particular ways, and just how air has properly become a commodity, with all of the fragmentation and reification that commodification implies. This commodification has always been met with resistance. The perpetual expression of resistance is, I suggest, one of the main threads that runs throughout Lefebvre's entire body of thought. And resistance is not an abstract concept or a romanticised idea, but springs forth from our desire for use value as against exchange value. As Lefebvre (1996: 170) put it, 'use and use value resist irreducibly'. The task at hand is to utilise this atmospheric resistance as a weapon in the struggle against the domination of space and for a real appropriative movement.

References

Cooper, G. (1997) *Air-Conditioning America: Engineers and the Controlled Environment, 1900–1960*, Baltimore, MD: The Johns Hopkins University Press.
Gissen, D. (2014) *Manhattan Atmospheres: Architecture, the Interior Environment, and Urban Centers*, Minneapolis, MN: University of Minnesota Press.
Heschong, L. (1979) *Thermal Delight in Architecture*, Cambridge, MA: The MIT Press.
Lefebvre, H. (1991) *The Production of Space*, trans. D. Nicholson-Smith, Malden, MA: Blackwell Publishing.
Lefebvre, H. (1996) 'Right to the city', in E. Kofman and L. Lebas (eds) *Writings on Cities*, Malden, MA: Blackwell Publishing.
Lefebvre, H. (2013) *Rhythmanalysis: Space, Time, and Everyday Life*, trans. S. Elden and G. Moore, London: Bloomsbury Academic.

Lukács, G. (1971) *History and Class Consciousness: Studies in Marxist Dialectics*, trans. R. Livingstone, Cambridge, MA: The MIT Press.
Marx, K. (1967a) *Capital: A Critique of Political Economy*, vol. 1, New York: International Publishers.
Marx, K. (1967b) *Capital: A Critique of Political Economy*, vol. 3, New York: International Publishers.
Schultz, E. (2012) *Weathermakers to the World: The Story of a Company, the Standard of an Industry*, New York: Carrier Corporation.
Sloterdijk, P. (2009) *Terror from the Air*, trans. A. Patton and S. Corcoran, Los Angeles, CA: Semiotext(e).
Sloterdijk, P. (2011) *Spheres I: Bubbles: Microsphereology*, trans. W. Hoban, Los Angeles, CA: Semiotext(e).

28

Transforming nature through cyclical appropriation or linear dominance?

Lefebvre's contributions to thinking about the interaction between human activity and nature

Daniel Paiva

Introduction

The objective of this chapter is to discuss Lefebvre's perspectives on the transformation of nature. Although some authors argue that Lefebvre's conceptualisations about the production of nature fall short in comparison with his works on the production of space (Smith 1984; Kipfer et al. 2008), I would argue that we can find useful concepts to think about the transformation of nature in the philosopher's works. While some of the author's main concepts, such as the spatial triad – the most acknowledged Lefebvrian concept for the social sciences in general (Pierce and Martin 2015) – leave out the issue of the production and transformation of nature, Lefebvre's theorisations about spatial appropriation and domination, and the rhythms of these divergent modes of spatial production, are particularly useful for scholars that are addressing issues related to the transformation of nature in diverse ways. My main focus in this chapter will be the relation between the author's discussion about appropriation and dominance, and the discussion about cyclical and linear rhythms. By relating these discussions, we arrive at two modes of spatiotemporal production – cyclical appropriation and linear dominance – that can be useful for addressing the transformation of nature in the contemporary world, especially when it comes to thinking beyond traditional binaries such as artificial-natural, cultural-natural, or unsustainable\sustainable.

Academic debates about the transformation of nature have recently been framed under the notion of the Anthropocene. Crutzen (2002) initially defined the Anthropocene as a geological epoch characterised by the dominance of the Earth by human activity which has supposedly become the main geological factor of the planet. The term has since then become very popular, triggering further conceptualisations (Wilkinson 2005), as well as some sceptical critiques (Malm and Hornborg 2014). Despite the manifold ways in which this concept has been defined and the fragilities inherent to this type of classification, the core idea is close to some of Lefebvre's arguments. Lefebvre (1991) warned about the progressive destruction of nature by the dominance of human activity and what he called the possible complete urbanisation of the world (Lefebvre 2003), which would lead to what he called a 'second nature'. In this context, nature refers to the life of what Lefebvre (1991) calls the physical world, including geomorphology, vegetation, and the bodies of humans and other animals. While Lefebvre's (1991; 2003) conceptualisation of nature tends to focus mostly on geomorphology (rivers, mountains) and vegetation (trees, forests), his last works address in greater detail the bodies of animals, especially humans, as natural, biological, or part of nature (Lefebvre 2004). On the other hand, second nature is a characteristic of the city and can be defined as a union of nature and culture, where the artificial includes and dominates the natural (Lefebvre 1991; 2003).

My argument here is that in a context of 'second nature', or even the Anthropocene, it is useful to think about the spatiotemporal production of nature, and this can be done through the concepts of linear dominance and cyclical appropriation that I draw from Lefebvre's works.

This chapter is further divided in three sections. First, I will begin by outlining Lefebvre's definition of nature. After this, I will discuss the history of its transformation according to Lefebvre, and the concepts of appropriation, dominance, the linear, and the cyclical. Third, I will critique Lefebvre's notion of 'second nature' and explain how thinking in terms of cyclical appropriations and linear dominance can be useful for conceptualising the transformation of nature.

Nature

In *The Production of Space*, Lefebvre (1991) argues that 'nature', along with 'physical reality' and 'matter', is 'common parlance' for the 'substance' of this cosmos or the 'world'. In this work and others, Lefebvre uses the terms 'nature', 'physical reality', and 'physical space' to name the same thing. In general, Lefebvre's idea of nature refers to the matter that constitutes the universe and pre-dates human existence but includes human biology. Lefebvre draws upon the perspective of modern physics to define the substance of physical space as energy expended through space and time (1991: 13).

This physical space, or natural space, is distinguished from mental space, which refers to the human understanding of space, and social space, which refers to society's production of space and spatial representations. Physical or natural space is not fully grasped in everyday life. It is instead defined by the 'practico-sensory activity and the perception of "nature"' (1991: 27), although technologies have made it possible for humans to grasp aspects of the physical world not available to the senses, such as ultrasound (Lefebvre 1991). Nature includes the human. The biology of the human body is also a part of nature and, as we will see, the human body does not escape the domination of nature that the social production of space implies.

Lefebvre's conceptualisation of the transformation of nature is mostly an historical one, but at the same time his conceptualisation is ahistorical, in the sense that it does not outline a sequence of dated events but instead it defines two modes of spatial and rhythmic production that emerge from specific historical contexts: pre-industrial and modern (Stanek 2011). Mainly in the books *The Production of Space* and *The Urban Revolution*, but also in *Rhythmanalysis*, Lefebvre outlines the history of space as a progressive annihilation of nature by the social production of space. Nature was first appropriated by human societies, creating absolute space, and then dominated, leading to the production of abstract space. A decisive aspect in this process was the substitution of cyclical rhythms for linear rhythms, two conceptions that are further developed in *Rhythmanalysis*. However, this process does not imply a complete substitution of nature by human abstract space, neither does it mean the total destruction of nature's cyclical rhythms by the linear rhythms of contemporary capitalist societies. Instead, in our present world we can distinguish cyclical appropriations of space and the linear domination of space. Even if the latter tends to be the main form of spatial production at present, cyclical appropriations resist. I argue here that these two conceptions of spatial production are useful to aid understanding the mutual relationships between humans, technology, and nature, even if we must apply these concepts in ways in which Lefebvre did not. Before outlining the significance of cyclical appropriations and linear domination, it is necessary to present Lefebvre's history of the transformation of nature from which these concepts emerge.

An ahistorical history of the transformation of nature

Lefebvre's explanation of the social production of space is largely historically based. Yet, most likely due to the influence of dialectical thinking, Lefebvre's history of space is also based on dual oppositions. In *The Production of Space*, Lefebvre describes two types of spaces produced by human societies (absolute space and abstract space), which correspond to two types of spatial production (appropriation and domination) with two corresponding types of rhythmicity (cyclical and linear). These two types of rhythmicity are further developed in *Rhythmanalysis*. Therefore, on one hand, we encounter absolute space, appropriation, and the cyclical. On the other, there is abstract space, domination, and the linear. The first generally matches pre-capitalistic, agrarian modes of production, while the latter is related to capitalist, industrial modes of production.

The first production of space that occurs in the history of humankind is absolute space, which is:

> made up of fragments of nature located at sites which were chosen for their intrinsic qualities... but whose very consecration ended up by stripping them of their natural characteristics and uniqueness... (Lefebvre 1991: 48)

Absolute space, then, refers to spaces that are chosen by humans for a specific geological or morphological quality – most commonly relief or presence of water – that is useful to attain a certain spatial goal: defence, access to natural resources, food, and so forth. Lefebvre offers as examples the Greek temples and the Shintoist sanctuaries. Absolute space is then nature appropriated by humans and thereby made into human property. An appropriated space for Lefebvre is a space which is not possessed by definition, but by use, and through use it is marked and shaped, and eventually made into an *oeuvre* (Stanek 2011).

At this point, it is useful to recover Lefebvre's thoughts about everyday life. Lefebvre defined everyday life, or the quotidian, as:

> a certain appropriation of time and space, the body, vital spontaneity and 'nature', an appropriation prone to dis-appropriation or ex-appropriation (alienation)... which 'unfolds' and is constituted in a space and time distinct from natural space and time, as well as mental space and time...
>
> *(2005: 10)*

In absolute space, this appropriation of time and space occurs in a context of a close relation between the natural and the social. As a result of this close relationship, social life is mostly related to natural or biological phenomena. Events such as rites of passage between boyhood and adulthood, solstitial rites linked to seasonal agricultural necessities, or rites of fertility are examples of this relationship. In this context, the production of space by humans also reflects this close relationship. As Stanek (2011) argues, for Lefebvre appropriation must be symbolisable. Absolute space is marked with symbols of myths of natural creation, collective history, or personal identity over time. At this point, we must bring rhythms to the discussion. The issue of rhythms is present throughout Lefebvre's works (1991; 2003; 2005), but it is in *Rhythmanalysis* (2004) that this issue takes centre stage. The production of absolute space through appropriation implies cyclical rhythms. For Lefebvre (2004: 8), cyclical rhythms originate 'in nature', in the form of astronomical and geological rhythms: the day and night cycle, lunar cycles, waves and tides of the sea, the seasons, among others. These cycles shape the cycles of the practices of bodies in nature: 'cycles of trees, flowers, birds and insects' (2004: 17). Likewise, they shape the internal rhythms of these bodies: 'respiration, the heart, hunger and thirst, etc.' (2004: 9).

The perspective of rhythms leads us to consider how the various scales of existence are interrelated. Lefebvre draws upon Gaston Bachelard's (2016) readings of Lúcio Pinheiro dos Santos, a Portuguese philosopher who developed rhythmanalysis as a means to interpret the world from a perspective that is simultaneously physical and psychological (Cunha 2008; 2010). Therefore, in Lefebvre's perspective, social, biological, and physical rhythms are nested within each other (2004: 20; see also Vrobel 2013). They also follow similar scales of cycles: circadian, monthly, and seasonal (Meyer 2008). But this nesting occurs mainly in absolute space, where there is a certain harmony between the natural and the social. The calendarisation of social events that match yearly and monthly cycles in agrarian societies is an example of such harmony. In this sense, Lefebvre's thinking is close to that of Mircea Eliade (1987), who argued that some societies' sense of history is based in a cyclic temporality that is manifested in myths such as the eternal return or the golden age, while modern society's sense of history is based in a linear temporality that emerged from clock time (see also Adam 2005) and Darwinian evolutionary perspectives. Lefebvre's perspective on cyclical rhythms can also be linked to Fernand Braudel's (1995) concept of *longue duree* (Kipfer et al. 2008). Braudel's concept refers to the long-term changes that structure societies over time as opposed to short duration of societal events. The transformation of nature through cyclical rhythms occurs in the scale of a *longue duree*: space is slowly laboured, adapted, and manipulated over centuries. Lefebvre also argues that by looking at cyclical rhythms, we can see that there is '[n]othing inert in the *world*' (2004: 17), even if some changes, such as geological phenomena, are very slow and go unnoticed. This conception that everything is moving has become popular in recent times (Massey 2005; Bennett 2010).

Absolute space created through cyclical appropriation can also defined as concrete space. In *The Critique of Everyday Life*, Lefebvre (2005) defines concrete space as the space of habiting, where the work of bodies in everyday life slowly alters places through gestures, paths, and the creation of meaningful symbols.

In Lefebvre's thought, this harmony is a characteristic of pre-industrial societies. In *The Urban Revolution*, Lefebvre (2003) presents the political city, mostly based upon the ancient Greek model, and the mercantile city, mostly based in Mediterranean cities from the Renaissance such as Venice, as two archetypes of urban space in which economic production was linked to agriculture. These can also be seen as archetypes of absolute space. In the same work, Lefebvre presents what he argues to be the great historical moment in which the production of space shifts from appropriation to domination: industrialisation. The industrial city, for Lefebvre, carries along it 'a process of domination that absorbs agricultural production' (2003: 2). This view of Lefebvre parallels those of other authors, most notably E.P. Thompson, whose well-known argument was that the Industrial Revolution provoked a substitution of the flexible and non-standardised rhythms of agriculture, which were based in natural cycles, by industrialised rhythms based in human-made schedules, which were driven by capitalist theories of work value and profit rate.

For Lefebvre (2003), this transition meant the dominance of abstract space. Abstract space can be seen as a materialisation of human mental space. It functions in an objective way, 'as a set of things/signs and their formal relationships: glass and stone, concrete and steel, angles and curves, full and empty' (Lefebvre 1991: 49). One particular aspect of this rupture from nature is the evolution of measurement systems, 'which proceeded from measuring space with parts of the body to universal, quantitative, and homogeneous systems' (Stanek 2008: 71).

The materialisation of abstract space, which tends to be geometric, linear, and functional, erases what was there before, including nature and even traces of historical time. However, abstract space is not just a process of destruction of nature and past. Instead, the geometric spaces that derive from this spatial production are closely related to power (Prigge 2008). Abstract space is a set of forms constructed by and for technological and scientific possibilities of control. Recent works on airports (Adey 2009), shopping malls (Miller 2015), or city centres (Tulumello 2016) have provided descriptions of real spaces that illustrate Lefebvre's arguments. For Lefebvre (2003; 2005), the power generally meant the state. Today we should also account for the spatial production of private companies.

Abstract space is inscribed through practices of domination. Unlike appropriated spaces which are generated by long-term use of humans, dominated spaces are generated by technologies that allow fast and large-scale changes on the Earth's surface. Domination means the possibility of materialising mental projects in all aspects, and in dominated spaces everything is produced; not only buildings and surfaces, but also light, air (see the chapter by Ford in Part 4), sound, and water. Domination is a twofold process. On one hand, it means that space itself is dominated: sterilised, emptied, moulded, closed. But it also means that the space tends to dominate what dwells in it, to dominate nature.

We arrive then at linear rhythms, which are those that abstract space imposes. Linearity is a way of sensing time that is based in sequences and repetition. As mentioned earlier, clock time (Adam 2005) and the Darwinian evolutionary perspective (Eliade 1987) are two prime examples of this linearity. Lefebvre (2004) identifies socio-economic reasons as the main driver of this linearity and highlights the role of technologies in imposing its rhythms. The linear is described by Lefebvre in the most negative way, as 'times of brutal repetition' that are 'tiring, exhausting and tedious' (2004: 73). Linear rhythms are a cause for the receding of nature. If in the industrial phase the seasonal rhythms of agriculture were substituted by the linear rhythms of industrial production, a trend that affected the rhythm and pace of agricultural production itself (Moskowitz 2009), in the present day every living thing is subject to the dominance of linearity. In the words of Lefebvre, 'the rhythm that is proper to capital is the rhythm of producing… and destroying' (2004: 55). This linear dominance sets everything to a pace that is valuable for capital: rivers, vegetation, animal breeding, sleep, etc. Many studies have highlighted phenomena that match these concerns. Animal production is accelerated (Graham et al. 2007), public space is set to the rhythm of shopping (Kärrholm 2009), work and family time are being accelerated (Paiva et al. 2015). Linear dominance, therefore, is characterised by an imposition of rhythms by those in power without regard for the stability of natural rhythms. Nature ends up reduced to a set of resources that are held in stock for technological demands (Elden 2008).

In *Rhythmanalysis*, Lefebvre looks closer at the imposition of linearity upon the rhythms of bodies. While 'the domination-exploitation of human beings begins with animals, wild beasts and cattle' (Lefebvre 2004: 52), it soon turns to humans. In the chapter 'Dressage', Lefebvre

(2004) looks closer at how the human body mixes natural and imposed rhythms. Social rules and education force the human body to conform to certain schedules that in dominated space-times are not always in conformity with the body's circadian cycles, leading to a 'dispossession of the body' (Lefebvre 2004: 75). Much like geometry produces abstract spaces, rational and numerical rhythms 'superimpose themselves on the multiple natural rhythms of the body' (Lefebvre 2004: 9). The deregulation of work times and the 24-hour day that media technologies allow are prime examples of linear rhythms imposed upon the human body, a form of what Rosa (2013) calls the social acceleration of everyday life. Information technologies are one of the main drivers of this linearity. As Lefebvre argued, decades before the smartphone:

> It is convenient to have a technical device at home that seems to take the whole of everyday life in hand. One day it may well be that, sparing the unforeseen or some initiative, an array of bureaucrats, under the orders of a technico-political high command, will treat daily life not as an object or product, no longer as a semi-colony, but quite simply as a conquered country.
>
> *(Lefebvre 2005: 128)*

At the present time, what Lefebvre describes sounds familiar. Daily life seems to be mostly under the control of various technologies that belong to the entertainment-security complex (Thrift 2011), which imposes linear dominance. Nature, be it the climate, flora, fauna, or the human body, is subject to attempts of control by human activities and technologies that are able to alter natural processes, although not completely dominating their subjects.

In this way, what Lefebvre calls 'second nature' emerges. 'Second nature' refers in general to the life of the city, as a space which is at once a product and a work, combining 'art and science within itself' (Lefebvre 1991: 409). This second nature is a union of nature and culture, where the artificial includes and dominates the natural. A characteristic of spaces of second nature is the increase in the simulation of nature. While nature itself recedes, shrinks, or even seems to disappear from urban space, the representations of nature increase (Lefebvre 2003). As Neil Smith highlights, in Lefebvre's view 'the steady, violent death of nature is matched by an obsessive "ideological naturalization" of society and the parodic reproduction of nature as denatured open spaces, parks, gardens, images of femininity' (Smith 2003: xv). This takes form not only in the reproduction of images of nature (photography, film, geographical information systems, street views), but also in the green spaces of the city (gardens, parks, artificial lakes) in which space is created in a way that intends to look as if it was not conceived and constructed by humans. The garden is the paradigmatic example of the mixture of reality and illusion that defines this 'second nature' (Stanek 2011). At the same time, the human body is also subject to the same pressures. In *The Critique of Everyday Life*, Lefebvre (2005) pointed out that the natural needs of human bodies are moulded by social rules and sometimes artificialised. But it was in *Rhythmanalysis* that the philosopher looked closer at how the human body suffers from the imposition of linearity. In this work, Lefebvre (2004) argues that the imposition of social rhythms on the human body can ultimately lead to arrhythmia, when the linearity of the social forces itself on the natural cycles and causes harm to the body.

Looking at the dynamics between linear dominance and cyclical appropriation in a context of second nature

As we have seen, Lefebvre's history of the transformation of nature does not constitute a linear sequence of events (Stanek 2011). Instead, we can identify two forms of spatiotemporal production – cyclical appropriation and linear dominance – that stem from historically specific contexts but are mixed in everyday life. As Lefebvre wrote:

> in present daily life, the rhythmical is overwhelmed, suppressed by the linear. But the rhythmical cannot disappear, the repetitive cannot be reduced to the results of a combinatory, a prefabricated, imposed linearity.
>
> *(2005: 11–12)*

The cyclical, in this excerpt referred to as the 'rhythmical', cannot disappear because, despite the development of technologies that aim to alter almost every aspect of life, everyday life is still very much determined by elementary physical needs (Meyer 2008). Despite acknowledging that the cyclical and the linear in reality 'interfere with each other constantly (Lefebvre 2004: 8), what Lefebvre describes throughout his work is an ongoing substitution of nature by human technology, which imposes a social order of linear rhythms.

The views of Lefebvre often romanticise nature. The paradigmatic example is Lefebvre's recurrent description of the dawn as 'always new' (2004: 90) and having a 'miraculous charm' (2004: 73). On the other hand, for Lefebvre, linear rhythms are insistently described in a negative way and the receding of nature they cause is 'a cause for lament' (Smith 2003: xv). These descriptions of the cyclical and the linear are over-romanticised, and it has been argued that we must look beyond the interference of cyclical rhythms by linear dominance. Simpson has argued that it is important to address the 'always emergent interaction of these two modalities of the rhythmic' (2008: 824), as cyclical appropriations continue to subsist in spaces of 'second nature'.

Simpson (2008) provides an example of an encounter of a street performer with rain, in which the natural, cyclical element of rain comes to disturb the linear schedules of the street performer. It is not difficult to imagine other situations in which the cyclic rhythms of nature appropriate and interfere with linear space-times, such as dramatic climatic events (floods, tornados, or storms). But cyclical appropriations can be much more than that. If we listen closely, like Lefebvre listens to the rhythms of the courtyards and gardens below his window (2004: 31), to places that linear dominance has left behind such as ruins or urban vacant lands (Brito-Henriques 2017), we can see and hear that, although these spaces do not exactly 'return to nature' (Nassauer and Raskin 2014), animals and vegetation slowly appropriate these spaces and impose new rhythms. Likewise, there are manifold human practices that appropriate spaces in a cyclic manner that is harmonious to the rhythms of nature. Sometimes this may even take place in spaces dominated by abstraction, and slowly transform these places into something more similar to absolute space. In *Rhythmanalysis*, Lefebvre centres the analysis on the body, as 'the site and place of interaction between the biological, the physiological

Table 28.1 The main features of two modes of spatiotemporal production.

Linear dominance	*Cyclical appropriation*
Imposes new rhythms	Fits into existent rhythms
Generates space quickly	Adapts space slowly
Tends to be geometric	Multiple spatial forms
Segregates subjects and objects into functional spaces	Relates subjects and objects into ecological spaces
Promotes simultaneity and permanent availability	Creates specific recurrent spacetimes (schedules, calendars)
Promotes socio-economic change (commoditises)	Produces stable meanings (symbolises)
Exhausts resources	Renews resources

(nature) and the social (often called the cultural)' (2004: 81). However, an analysis of the production of space-times calls for a more-than-human approach that can understand the rhythms and agency of various bodies, as some authors have suggested (Simpson 2008; Edensor 2010).

Taking this common existence of linear dominance and cyclical appropriations into account, it may be useful to describe what characterises each type of spatial production. In Table 28.1, the main features of these two modes of spatial production according to Lefebvre (1991; 2003; 2004; 2005) are summarised.

Conclusions

When I suggest we should look at the dynamics between these two modes of spatial production, this does not mean to classify practices as one or the other. Instead, these concepts could help us take into account how diverse human and non-human practices produce and constantly alter specific space-times. Instead of defining what certain practices, spaces, or events already are (artificial-natural, cultural-natural, or unsustainable-sustainable), in a context of 'second nature' (or Anthropocene, if one prefers) it seems more useful to understand how nature is continuously produced and transformed, and attune ourselves to the consequences of the modes of spatiotemporal production. Most importantly, bearing in mind that linearity cannot absorb every practice or body in the world – it could only do so at the expense of life itself – this realisation must lead us to investigate the interactions between different modes of spatiotemporal production. These interactions are not one-sided. First, every space-time produced through linear dominance will be subject to cyclical appropriations over time, either by human or non-human agents. These cyclical appropriations may be defined by what linear dominance imposes, but they can also generate more harmonious rhythms and spatialities. On the other hand, in the present world, it is very difficult to encounter spaces that have not been subject to some kind of practice related to linear dominance, especially if we consider the fluidity of the atmosphere and water systems.

There is a danger in seeking closed categories. Instead, they should be seen more as opposite poles of the same scale. Surely no practice, space, or event can be undisputedly classified as uniquely linear or cyclical, as absolutely dominant or not, but they will likely be closer to a certain pole than the other. As two modes of production that coexist in everyday life, looking at linear dominance and cyclical appropriations offers another kind of dialectical thinking beyond the spatial triad. It offers a perspective on 'second nature' as a synthesis between two modes of relating human activity and nature.

Acknowledgement

This work was funded by the Fundação para a Ciência e a Tecnologia under grants SFRH/BD/108907/2015 and PTDC/ATP-EUR/1180/2014.

References

Adam, B. (2005) *Time*, London: Polity Press.

Adey, P. (2009) 'Facing airport security: affect, biopolitics, and the preemptive securitisation of the mobile body', *Environment and Planning D: Society and Space*, 27:2 274–95.

Bachelard, G. (2016/1950) *The Dialectic of Duration*, London: Rowman and Littlefield.

Bennett, J. (2010) *Vibrant Matter: A Political Ecology of Things*, London: Duke University Press.

Braudel, F. (1995/1966) *The Mediterranean and the Mediterranean World in the Age of Philip II*, vol. 1, Oakland, CA: University of California Press.

Brito-Henriques, E. (2017) 'Arruinamento e regeneração do espaço edificado na metrópole do século XXI: o caso de Lisboa', *EURE*, 43:128 251–72.

Crutzen, P.J. (2002) 'Geology of mankind', *Nature*, 415:3 23.
Cunha, R.S. (2008) *A Filosofia do Ritmo Portuguesa*, Vila Viçosa: Serra D'Ossa Edições.
Cunha, R.S. (2010) *O Essencial Sobre Ritmanálise*, Lisbon: INCM.
Edensor, T. (2010) *Geographies of Rhythms: Nature, Place, Mobilities and Bodies*, London: Ashgate.
Elden, S. (2008) 'Mondialisation before globalization: Lefebvre and Axelos', in K. Goonewardena, S. Kipfer, R. Milgrom and C. Schmid (eds) *Space, Difference, Everyday Life: Reading Henri Lefebvre*, London: Routledge, pp. 80–93.
Eliade, M. (1987/1957) *The Sacred and the Profane: The Nature of Religion*, New York: Harcourt, Brace.
Graham, J., Boland, J. and Silbergeld, E. (2007) 'Growth promoting antibiotics in food animal production: an economic analysis', *Public Health Reports*, 122:1 9–87.
Kärrholm, M. (2009) 'To the rhythm of shopping – on synchronisation in urban landscapes of consumption', *Social & Cultural Geography*, 10:4 421–40.
Kipfer, S., Goonewardena, K., Schmid, C. and Milgrom, R. (2008) 'On the production of Henri Lefebvre', in K. Goonewardena, S. Kipfer, R. Milgrom and C. Schmid (eds) *Space, Difference, Everyday Life: Reading Henri Lefebvre*, London: Routledge, pp. 1–24.
Lefebvre, H. (1991/1974) *The Production of Space*, London: Blackwell.
Lefebvre, H. (2003/1970) *The Urban Revolution*, Minneapolis, MN: University of Minnesota Press.
Lefebvre, H. (2004/1992) *Rhythmanalysis: Space, Time and Everyday Life*, trans. G. Moore and S. Elden, London: Continuum.
Lefebvre, H. (2005/1981) *The Critique of Everyday Life, Vol. 3: From Modernity to Modernism (Toward a Metaphilosophy of Daily Life)*, London: Verso.
Malm, A. and Hornborg, A. (2014) 'The geology of mankind? A critique of the Anthropocene', *The Anthropocene Review*, 1:1 62–9.
Massey, D. (2005) *For Space*, London: Sage.
Meyer, K. (2008) 'Rhythms, streets, cities', in K. Goonewardena, S. Kipfer, R. Milgrom and C. Schmid (eds) *Space, Difference, Everyday Life: Reading Henri Lefebvre*, London: Routledge, pp. 147–160.
Miller, J. (2015) 'The critical intimacies of walking in the Abasto Shopping mall, Buenos Aires, Argentina', *Social & Cultural Geography*, 16:8 869–87.
Moskowitz, M. (2009) 'Calendars and clocks: cycles of horticultural commerce in nineteenth-century America', in E. Shove, F. Trentmann and R. Wilk (eds) *Time, Consumption and Everyday Life: Practice, Materiality and Culture*, Oxford: Berg, pp. 115–128.
Nassauer, J.I. and Raskin, J. (2014) 'Urban vacancy and land use legacies: a frontier for urban ecological research, design, and planning', *Landscape and Urban Planning*, 125: 245–53.
Paiva, D., Cachinho, H. and Barata-Salgueiro, T. (2015) 'The pace of life and temporal resources in a neighborhood of an edge city', *Time & Society*, 26: 1 28–51.
Pierce, J. and Martin, D. (2015) 'Placing Lefebvre', *Antipode*, 47:5 1279–99.
Prigge, W. (2008) Reading the urban revolution: space and representation', in K. Goonewardena, S. Kipfer, R. Milgrom and C. Schmid (eds) *Space, Difference, Everyday Life: Reading Henri Lefebvre*, London: Routledge, pp. 46–61.
Rosa, H. (2013) *Social Acceleration: A New Theory of Modernity*, New York: Columbia University Press.
Simpson, P. (2008) 'Chronic everyday life: rhythmanalysing street performance', *Social & Cultural Geography*, 9:7 807–29.
Smith, N. (1984) *Uneven Development: Nature, Capital and the Production of Space*, London: Basil Blackwell.
Smith, N. (2003) 'Foreword', in H. Lefebvre (ed.) *The Urban Revolution*, Minneapolis, MN: University of Minnesota Press, pp. vi–xxiii.
Stanek, L. (2008) 'Space as concrete abstraction: Hegel, Marx, and modern urbanism in Henri Lefebvre', in K. Goonewardena, S. Kipfer, R. Milgrom and C. Schmid (eds) *Space, Difference, Everyday Life: Reading Henri Lefebvre*, London: Routledge, pp. 62–79.
Stanek, L. (2011) *Henri Lefebvre on Space: Architecture, Urban Research, and the Production of Theory*, Minneapolis, MN: University of Minnesota Press.
Thrift, N. (2011) 'Lifeworld Inc.', *Environment and Planning D: Society and Space*, 29:1 5–26.
Tulumello, S. (2016) 'Toward a critical understanding of urban security within the institutional practice of urban planning: the case of the Lisbon metropolitan area', *Journal of Planning Education and Research*, 37: 4 397–410
Vrobel, S. (2013) 'Measuring the temporal extension of the Now', *Progress in Biophysics and Molecular Biology*, 113: 92–96.
Wilkinson, B.H. (2005) 'Humans as geological agents: a deep-time perspective', *Geology*, 33: 161–64.

29

Drivers of global urbanisation

Exploring the emerging urban society

Panu Lehtovuori, Jani Tartia and Damiano Cerrone

Introduction

> But how can we know if urban society will enable the development of a new humanism, so-called industrial society, capitalist or not, having effectively rejected its earlier forms?
>
> *(Lefebvre 2003: 65)*

Pervasive urbanisation is a historically unique process, intertwined with industrialisation. For millennia, cities remained relatively small. Only industrialisation from the early 19th century made mass-urbanisation both possible and necessary, leading to industrial metropolises and later to regional, multi-centred urban forms (Gottmann 1961; Gottdiener 1985; Ascher 1995; Castells 1996; Byrne 2001). With the continuous urban and industrial growth, and the relative decrease of agriculture as a source of income and employment, the urban population, i.e. the percentage of the total population living in urban areas, has surpassed 90 per cent in many countries in the Global North and South. In Belgium, the share of urban population is 97.9 per cent and in Argentina 91.8 per cent (*The World Factbook* 2016, figures for the year 2015). Today, only sub-Saharan Africa and Southern Asia have significant rural populations. As people continue to move from countryside to urban areas, some countries in those regions currently have very high rates of urbanisation. In Laos, to take an example, the annual growth of the urban population is 4.9 per cent, in China 3.1, in Angola 5.0 and in Nigeria 4.7 (average 2010–15, ibid.). Because of immigration, some highly urbanised countries still show relatively high rates of urbanisation as well; an example is Australia's 1.5 per cent (ibid.). In Western Europe and the US, urbanisation rates are generally clearly below 1 per cent, while Germany and some Eastern European countries experience a decrease of urban populations, referred to as the 'shrinking cities' phenomenon.

In the complex picture of global urbanisation, industrial growth and job opportunities still do play a role, but industrialisation is not the only or even the major driver of urban growth. In what follows, we will discuss contemporary drivers. It is important to understand the qualitative aspect of *radical societal transition* the industrial urbanisation process has created and is creating. This transition changes everyday life, our ways of habiting, moving and socialising, calling for new humanism, or human-centred urbanism, as noted in the Lefebvre quotation above.

Lefebvre's work is valuable, because he foresaw that developed societies were moving from the industrial phase to something new. In *The Urban Revolution*, originally published in 1970, he discusses the relations between industrialisation and urbanisation: 'the industrial revolution and the urban revolution are two aspects of a radical transformation of the world. They are two elements (dialectically united) of the same process, a single idea, that of global revolution' (Lefebvre 2003: 144–145). Playing with the relation and wishing to propose a novel status for the urban, he continues that 'while it is true that the second aspect has increased in importance so that it is no longer subordinate to the first, this does not imply that the first suddenly ceases to have any importance or reality' (ibid.).

Lefebvre proposes a new perspective or hypothesis of an 'urban society' that results from the process of 'complete urbanization'. For him, urban society is not yet here. Rather, his conceptual proposal is about 'tendencies, orientations, and virtualities' (Lefebvre 2003: 1–2). Urban crisis and urban problematic are global issues, but Lefebvre does not focus on numbers and statistics. The 'planetary society' is for him a goal of action, a *virtual object* towards which his philosophical and political thinking is aiming. He writes that 'the urban (an abbreviated form of urban society) can therefore be defined not as an accomplished reality, situated behind the actual in time, but, on the contrary, as a horizon, an illuminating virtuality' (ibid.: 16–17).

This idea challenges the notion of 'planetary urbanisation', as developed by Brenner and Schmid (2012; 2015). The geographic explosion of mining, industry, energy, communications and war is phenomenal, for sure, but it does not nullify dense urban centrality as a key for economic innovation, local and regional social dynamism and meaningful political organisation (e.g. Jacobs 1969; Gottdiener et al. 2015; Davidson 2016). While it may be true that 'even spaces that lie well beyond the traditional city cores and suburban peripheries… have become integral parts of the worldwide urban fabric', as Brenner and Schmid (2012: 12) claim, both Lefebvre's ideas and more recent observations direct us back to urban places and urban dynamics (not necessarily to 'cities', though) when searching for the seeds of the urban society. While planetary technical infrastructures are important for the functioning of global capitalism, they have little meaning without the urban cores that produced the necessary innovations and organisations to build and manage those infrastructures. This was made clear by Jane Jacobs in *Economy of Cities* (1969). Today, we witness increasingly powerful examples where urban centralities drive new practices of resource-efficient circular economy, local hybrids in urban metabolism as well as ground-breaking digital services. To a degree, these new phenomena are characterised by social and organisational novelties that give insights into the emerging urban society. 'Peerism', sharing economy and the 'fourth sector' of commons and activisms (Jacobsson 2015) can be seen as examples of the new social and material relations of the urban.

Five eventful decades have passed since Lefebvre's original formulations. Now is the right time to critically revisit the notions of urban society and complete urbanisation, as well as the innovative methodological ideas of 'critical phase' and 'blind field'. The chapter, thus, will explore what the qualitative leap or societal transition after industrial urbanisation might really mean. We will explore Lefebvre's notions in the light of the latest knowledge on urban processes in selected places. The contemporary logic of urban processes differs markedly from the 19th and 20th century, warranting new approaches and analytic concepts. On one hand, the demographic decline of cities such as Detroit or Leipzig, that may seem an exception in the large picture of urban growth and is usually explained as a local effect of industrial disinvestment (Friedrichs 1993; Martinez-Fernandez et al. 2012), actually suggests important and systemic uncertainties regarding urbanisation in the future: we do not know what happens in our cities and urban regions after the rapid urbanisation phase is over. On the other hand, statistical and cartographic representations of urbanisation give a limited view of the new possibilities and unfolding realities of urban society, understood as a qualitatively new context and product of human activity. How to study that new context? Temporality and spatiality, concrete engagement and reliance on data produced by users, are answers we wish to explore and operationalise. It is essential to focus on body and place, on human experience and social life, on concrete materiality and the possibilities of the ephemeral. The notions of *rhythm* and *rhythmanalysis*, *temporary uses* that produce difference as well as *urban metamorphology*, an expression of socio-technical hybridities that challenge the link between space and its social use, are for us important concepts for helping to understand the urban (Lefebvre 2013; Cerrone et al. 2015; Lehtovuori and Ruoppila 2017).

Drivers of urbanisation and critical phase

Much of the recent literature gives a straightforward picture of the reasons why urban areas grow, with population growth, human mobility and local urban transformations as the main drivers. According to the United Nations (2016: 7), 'Africa's rapid urbanization is driven mainly by natural increase, rural–urban migration, spatial expansion of urban settlements through the annexation, the reclassification of rural areas, and, in some countries, negative events such as conflicts and disasters'. Annez and Buckley (2009: 21) further note that the high rates of urbanisation in many African countries are partly explained by their relatively small urban populations, to which the mobility of large rural populations can easily make big percentage additions. Thus, the period after World War II (WWII) has witnessed unprecedented relative growth. In countries of the Global South, urban populations increased by 188 per cent between 1950 and 1975 – a much larger increase than the 100 per cent for industrialising countries of Europe and North America between 1875 and 1900. Absolute growth numbers are of historic proportions right now: Between 1950 and 1975 the world's urban population increased by 790 million people, while between 2005 and 2030 the growth is expected to be 1.76 billion (United Nations 2005). The lion's share of the ongoing growth is in the Global South, reflecting its rapid transition to lower mortality rates in the post-WWII period (Annez and Buckley 2009: 20).

World-wide, urban demographic dynamics can be decomposed into three factors: home-grown population growth, net domestic migration and net international immigration (McKinsey Global Institute 2016: 10). In many European countries and cities, international immigration is the dominant or even the sole source of growth at the moment. Knowing the high uncertainties in the directions and scale of international immigration – caused by changing immigration policies, impacts of climate change, surprising crises and other factors – the future growth of urban areas seems uncertain, even vulnerable. This may sound surprising in the light of the above-mentioned growth figures, but the period of rapid urbanisation seems indeed to be coming to an end. Population growth in cities will decline as populations age and the transition from rural to urban areas runs its course (ibid.: 7). Thus, declining populations and shrinking cities are likely to be common phenomena in the future, paralleled with high growth in some cities and regions.

The demographic explanations remain descriptive, though. Surely, it would be interesting to know why, exactly, people move to cities? Or why they engage in intense efforts in transforming their inherited villages to makeshift towns and cities? To answer these questions, we need to study local and trans-local networks, personal motivations, will and pride. Doug Saunders in *Arrival City* (2010) gives a fantastic account of personal ambitions of newly arrived urban dwellers in 20 emerging urban localities across the planet. In highly diverse places, such as Berlin's city district Kreuzberg, Dhaka's squatter enclave Karail or the small village Liu Gong Li outside of Chongqing, Saunders finds determined individuals that use their skills and village-based social capital to shift from rural to urban society. One generation, or some 200 million people, is ready to sacrifice nearly everything to arrive to urbanity, working in substandard conditions, risking their fortune in shady real estate operations or enduring insults in poor jobs, for the benefit of their children. Saunders provides a journalistic view to emerging proto-cities, where people move well before the institutions are in place to accommodate an orderly urbanisation process, as Annez and Buckley (2009: 26) note. Typically, there are serious problems, but in surprisingly many new urban places, or 'arrival cities', they are solved. More often than not economies thrive, reducing poverty and providing urban amenities. Broad empirical evidence thus suggests that a significant societal transition is unfolding: 'to make modern cities work, a transformation, not incremental change, in fiscal and administrative institutions, is needed, and it often comes in response to a crisis of some kind.... Managing urbanization will affect politics, social norms, institutional change, and the broader financial system' (ibid.: 26, 32). Urbanisation has indeed helped millions escape poverty through higher productivity, employment, better quality of life via education and health and access to infrastructure and services (United Nations 2016: 34). Cities were historically the drivers of a developing economy, and they have that role today.

After these cautiously positive notes on urban growth and economy, we should stress that after the period of rapid urbanisation, we are entering a new phase, a 'critical zone' (Lefebvre 2003). The situation is characterised by the fact that the urban population of a given society is near 100 per cent and that a certain break-up of urban form towards regional, networked or multi-centred pattern – 'implosion-explosion' – has occurred. 'What happens in the critical phase?' Lefebvre (ibid.: 16) asks, and continues that the whole *Urban Revolution* is an attempt to answer that question. The key idea is that during critical phase, 'a second transition occurs, a second reversal of direction and situation. Industrialization... becomes a dominated reality during periods of profound crisis' (ibid.). We do not focus on the crises here (for account on the recent financial crisis see Gottdiener et al. 2015: 29–34, 50–56). For our purpose, it is enough to note that while locally and regionally there will be variations and fluctuations in urban futures, societies in general are entering a new epoch, an epoch of (nearly) complete urbanisation that puts industry and the industrial 'logic' to a subordinate position when thinking social practices and urban futures.

Studying habiting as embodied and situated processes

> What does our blindness look like?
>
> *(Lefebvre 2003: 29)*

In his writings, Lefebvre refrains from giving concrete examples or substantive projections regarding the urban society. He claims that we are in a 'blind field' between industrial and urban society, using concepts that do not help in seeing the possibilities of the emerging society and space. He nevertheless suggests that besides the subordination of industry to urbanisation, the current critical phase is about 'subordination of the global to the urban' as well as 'the urban to habiting' (Lefebvre 2003: 100).

Lefebvre discusses the notion of habitat and habiting across *The Urban Revolution*. At this junction, he clearly wants to stress the importance of the everyday life and the 'private level' (P), claiming that the ongoing fundamental societal reorganisation comes from grassroots, from changes in the everyday practices and spaces. In the current critical phase between industrial and urban society, 'a level that was always considered unimportant now becomes essential, namely habiting.... I assume that the urban is primary and priority is given to habiting' (Lefebvre 2003/1970: 89–90). Lefebvre stresses the freedom of invention and suggests that architecture (as against planning or urbanism) might have a role in making the new society possible and acting as 'social condenser' (ibid.; Stanek 2011).

Thus, we can say with some confidence that besides the planetary scale and mobility understood as migration, the scale of the local, everyday mobilities, the body and experience of the urban are central in Lefebvre's thinking regarding the emerging urban society. Discussing the three analytic 'levels' (global, intermediate and private), he for example says that:

> one point worth noting is that the social and professional mobility so desired by planners (primarily urban planners and moving companies) [companies that wish to relocate] is fundamentally superficial. It does not refer to the intense mobility that can only occur near a center.
>
> *(Lefebvre 2003: 97)*

What is this 'intense mobility'? How is urbanisation experienced on the local level, as part of the spatiotemporal routines and mobility patterns of everyday life? Here, a framework built around *rhythm* gives important insights.

Rhythms and mobilities of the urban

Cities contain different embodied mobilities, from walking and biking to driving and passengering. The urban is experienced and *lived* in motion. A mobilities paradigm 'is becoming increasingly central to contemporary identity formation and re-formation' (Elliott and Urry 2010: 7), as people connect everyday fragmented – both physical and digital – sites together in numerous ways. Mobility, in this context, is not only about physical, instrumental movement, but about the politics, meanings and power relations of being on the move, and how the possibilities for mobility are unevenly distributed and organised as various channels in and between cities (see Cresswell 2010). Our global world is one of speed but 'speed, after all, exists only in relation to slowness, and, if some things are speeding up, others must be slowing down (in both absolute and relative terms)' (Hubbard and Lilley 2004: 277).

In the popular imagination, a common way to visually represent city life is through time-lapse photography, which evokes imageries of the city in perpetual motion, presenting a certain kind of mobile and temporal structure and order (and its repetition); how the city breathes life on the street level. How are these mundane and ordinary patterns of movement perceived, practised, experienced and lived? In the city setting, mobilities are not only part of the subjective 'obliged time schedules' that encompass everyday lives thoroughly, but also result in collective spatiotemporal 'urban structures' through the different uses and temporalities of both public and private spaces (Mareggi 2013). Notions on mobility as embodied practices, performed in urban settings, are crucial in examining everyday mobilities as not only instrumental movements between different points, between different physical locations, but as practices that constitute, in addition to the body-environment relations, the *city* as well (see Simonsen 2004; Jensen 2013). The image can also be expanded from the local to the global network of cities, connect by flight paths, media connections and the like. How to grasp this kind of mobility is a key, although complicated, question that might reveal something of the nature of the *urban*.

Lefebvre's (2013) 'rhythmanalysis' is a methodological opening that sets focus on the interrelations between space, time and energy, or *rhythm*. Rhythm, here, refers to repetitions, in an analogue to music and sound, and the differences of these repetitions. Lefebvre, in formulating the rhythmanalytical approach in the study of urban space and its social production, was interested to examine the relations between the natural, *cyclical*, organic rhythms and the socially produced, *linear*, more mechanistic rhythms, and how the two intertwined (see Amin and Thrift 2002; Mels 2004). Lefebvre notes the general 'noise' of the city – forming, alongside other factors, of inter-crossing mobile trajectories – and how rhythms emerge, or are possible to differentiate, from this cacophony of complex urban life by listening to it attentively through the multisensory body (Lefebvre 2013; see also Hetherington 2013; Lehtovuori and Koskela 2013). The body is here the instrument and the main referent of urban rhythms. The rhythmanalytical framework highlights the ensemble-like character of the urban realm, the coming together of different rhythms, or in Lefebvre's (2013) terms, 'polyrhythm'.

What exactly these rhythms are, or how one could engage in practical rhythmanalysis, remains, however, quite elusive. Urban rhythms include mobile trajectories, ways of using of space, schedules and the temporal social changes of the day-to-day city, but is this all? Could rhythmanalysis reveal deeper questions in relation to how the *urban* is produced, or experienced? While Lefebvre's (unfinished) work does not say much about how rhythmanalysis could be conducted in practice or where the empirical analysis should focus (see e.g. Amin and Thrift 2002; Mels 2004; Edensor 2010), some cues can nevertheless be followed. For Lefebvre (2013/1992), the body has a central role in uncovering urban rhythms. Utilising the body as a point of origin for the analysis of rhythms, the everyday urban sites can be approached analytically, and examined as lived temporalities (Crang 2001). These notions on the centrality of the body also bring up another key point, that is the combination of both qualitative and quantitative research perspectives in the examination of the urban (Lefebvre 2013/1992; see also Mels 2004).

In a recent research project, the rhythmicities of everyday sites and embodied contexts of urban mobilities have been examined. These notions have brought new insight to the modes and processes of how spatiotemporal meanings are formed in motion, as part of everyday routes and mobility routines in the city. The various and immediate rhythms of moving, interacting and perceiving the environment come into the fore in the experiences on everyday walking and driving routes. Similarly, the modes of controlling-producing mobility practices and patterns in top-down-bottom-up approaches produce distinctive rhythms, pacing the environment through an ensemble of various power relations. (Tartia forthcoming). The multitude of spaces and places, revealed by examining urban rhythms, as prompted by Simonsen (2004) and others, might be a key notion of the urban. The question of forms of habiting in urban society, though, remains still partly open. Everyday practices reflect changes in the urban, and those changes influence practices, but the process is half-way. Lefebvre's blind field, thus, remains a relevant critique.

Metamorphology, the invisible form of the urban

Besides the rhythms of everyday life and experiential mobility, the networked digital realm warrants attention. Today, the ubiquitous personal mobile devices not only provide services and information but essentially change our agency as citizens and urban dwellers. There is a 'cyborgian' dimension to the current phase of urban society (Graham and Marvin 2001), a dimension that will be exacerbated in the near future with powerful networked AI, real-time monitoring of urban activities as well as user-created 'street data' and new place-based digital approaches. This development has radical influence in the social production of space, in our ways of habiting the urban realm and in the relations between state and market, space and society, as well as place and its use.

In architecture and urban planning, a rich tradition of analysis and practice foregrounds physical space and spatial configuration as the set of constraints that shape flows of people and direct the evolution of urban activity patterns. Space is seen as the 'machine' (Hillier 1996) that drives change and opens development opportunities. This urbanistic approach was rooted in critiques against the perceived problems of modernistic planning and architecture (e.g. Jacobs 1961; Alexander 1965; Hillier and Hanson 1984). In analysing urban society today, this historic approach has merits but also important limitations. Regarding the cyborgian agency and related socio-cultural shifts, we argue for a fresh view on urban dynamics, a view that puts the activities people engage in in buildings and public urban space on par with space. The concept of *metamorphology* communicates this novel understanding in the fields of spatial analysis, planning and urban studies.

We claim that both strategic and tactical actors, both government and citizens, can influence urban evolution through fostering new activities (temporary uses, events, mobile apps, social media memes) as well as through intervening in space (new developments, infrastructure, reconstruction). For the metamorphological reading, the distinction between 'place-shaping' and 'place-making' (Andres 2013) is useful. The agile place-shaping may be even more important for the emerging urban society than traditional processes of place-making that require long planning and big public investments. Activities – meeting someone, having lunch, taking care of your ageing mother – are not less important than space. Furthermore, activities are not dependent on space, but provide for equal domain of analysis and action. In the metamorphological frame, we are interested in new social, cultural and economic trends that create surprising and moving centralities, independent of the traditional centrality that is based on morphology of the street network, the 'space syntax' (Hillier and Hanson 1984).

Lefebvre's central idea of the *spatial dialectic* already points in metamorphological direction. In the spatial dialectic, centrality or 'urban rationality' aspires to be total, centres are understood as events of 'gathering-together' of various urban resources, and margins may become instant centres (Lefebvre 1991: 331–332; Lehtovuori 2010: 127–133). All this is very clear in contemporary cities and urban regions. Digitalisation exacerbates this dynamic, and we may be facing a transition phase. Social media, for example, has become a common feature of our everyday life. For researchers, social media provides large amounts of readily accessible, on-time and qualitatively rich data that can be used to study urban activities and interactions. Social networks such as Twitter and Instagram have become near-pervasive, in the Global North at least. The data made available by these social networks can be collected and analysed by third parties, providing a granular and detailed description of social processes through space and time.

These analyses tell that our ways of using the city are changing, influencing also the processes manifested as material urban patterns and configurations. Metamorphology, thus, is a way to understand Lefebvre's contention of the priority of habiting, telling about new spaces that may act as social condensers. On the other hand, social media is not innocent. As recent elections in the US and UK show, both state and market actors can influence its dynamics in multiple ways, directing attention and use. Power struggles and social divisions, thus, unfold in the hybrid urban-digital reality as they unfolded in urban space before. Rules are not the same, though. Serious workers' rights and customer protection questions around the taxi service Uber are a case in point.

Conclusions: coping with the blind field

> Would enjoyment correspond to urban society? It remains to be seen.
>
> *(Lefebvre 2003: 32)*

For Lefebvre, urban society was a hypothesis, a future state to which we should aim, a call for action. 'Habiting' was for him the main driver of the change, not so much 'information', 'networks', 'flows' or 'virtual spaces' (e.g. Castells 1989; 1996). While Lefebvre clearly acknowledges the importance of tourism, weekend homes, distant industrial estates, highways and airports, those are for him elements of the critical phase between the industrial and the urban society. They are results of the 'implosion-explosion' that starts to pave way for something new, but it is not that new, yet. The current interest in globally scattered urban and infrastructural fragments, thus, is not very helpful in finding the political power of space and transformative potential of urban society. Paraphrasing Lefebvre, we could suggest a horizon where *capitalism* becomes a dominated reality, dominated by the urban. Instead of oceans and deserts, we should look, again, in the countless close-knit urban places across the planet, many of which are not cities but do nevertheless show signs of social, political and spatial organisation for the better.

To do this, we need a new theory to illuminate the self-organising urban order as it emerges. Urban rhythms and metamorphology, outlined above, are some tools to address the temporal and spatial dimensions of the contemporary production of space. Metamorphological approaches can learn from theories of complex adaptive systems that originally stressed the impossibility of planning and control and foregrounded the networked human process. It can also enrich them by illuminating the heterotopian multiplicity of habiting the urban society, the multitude of people's activities and the richness of values people give to their environment. The excursions into urban rhythms of everyday mobilities provided insight as to how to conduct rhythmanalysis, helping to understand what kind of meanings are engraved into the urban fabric through daily (mobile) habiting. These notions bring up the importance of temporality and activity, alongside space, in the formation of the urban, as noted by Lefebvre as the triadic basis of rhythm: space, time and action.

In the emerging, arrival and sometimes shrinking places people shape their futures and our common urban future. We hope to have shown that through studying these multiple places we can gain understanding about habiting as a transformative activity. These insights should lead our focus from planning to peer-action and from top-down urban design to bottom-up co-creation. The countless local and place-based but essentially open, emergent and scalable processes are the drivers of urban society and global urbanisation, understood as a qualitative change from industrial to urban logic.

Acknowledgements

Sincere thanks to URMI project (Finnish Academy SA 303618), Turku Urban Research Programme as well as Aleksi Neuvonen, Kaisa Schmidt-Thomé and Otto-Ville Koste.

References

Alexander, C. (1965) 'A city is not a tree', *Architectural Forum*, 122:1 58–62.

Andres, L. (2013) 'Differential spaces, power hierarchy and collaborative planning: a critique of the role of temporary uses in shaping and making places', *Urban Studies*, 50:4 759–75.

Amin, A. and Thrift, N. (2002) *Cities: Reimagining the Urban*, Cambridge: Polity.

Annez, P.C. and Buckley, R.M. (2009) 'Urbanization and growth: setting the context', in M. Spence, P.C. Annez and R.M. Buckley (eds) *Urbanization and Growth: Commission on Growth and Development*, Washington, DC: The World Bank, pp. 1–45.

Ascher, F. (1995) *Métapolis, ou l'avenir des villes*, Paris: Odile Jacob.

Brenner, N. and Schmid, C. (2012) 'Planetary urbanization', in M. Gandy (ed.) *Urban Constellations*, Berlin: Jovis, pp. 10–13.

Brenner, N. and Schmid, C. (2015) 'Towards a new epistemology of the urban?' *City*, 19:2–3 151–82.

Byrne, D. (2001) *Understanding the Urban*, New York: Palgrave.

Castells, M. (1989) *The Informational City*, Oxford: Blackwell.

Castells, M. (1996) *The Rise of the Network Society*, Oxford: Blackwell.
Central Intelligence Agency (2016) *The World Factbook*, Washington, DC: Central Intelligence Agency. Online. Available HTTP: <https://www.cia.gov/>.
Cerrone, D., Lehtovuori, P. and Pau, H. (2015) *A Sense of Place. Exploring the Potentials and Possible Uses of Location Based Social Network Data*, Turku Urban Research Programme Research Report 1/2015.
Crang, M. (2001) 'Rhythms of the city: temporalised space and motion', in J. May and N. Thrift (eds) *Timespace: Geographies of Temporality*, London: Routledge.
Cresswell, T. (2010) 'Towards a politics of mobility', *Environment and Planning D: Society and Space*, 28:1 17–31.
Davidson, M. (2016) 'Planning for planet or city?' *Urban Planning*, 1:1 20–23.
Edensor, T. (2010) 'Thinking about Rhythm and Space', in T. Edensor (ed.) *Geographies of Rhythm: Nature, Place, Mobilities and Bodies*, Farnham: Ashgate.
Elliott, A. and Urry, J. (2010) *Mobile Lives*, London: Routledge.
Friedrichs, J. (1993) 'A theory of urban decline: economy, demography and political elites', *Urban Studies*, 30:6 907–17.
Gottdiener, M. (1985) *The Social Production of Urban Space*, Austin, TX: University of Texas Press.
Gottdiener, M., Budd, L. and Lehtovuori, P. (2015) *Key Concepts in Urban Studies*, 2nd edn, London: Sage.
Gottmann, J. (1961) *Megalopolis: The Urbanized Northeastern Seaboard of the United States*, New York: The Twentieth Century Fund.
Graham, S. and Marvin, S. (2001) *Splintering Urbanism: Networked Infrastructures, Technological Mobilities and the Urban Condition*, London: Routledge.
Hetherington, K. (2013) 'Rhythm and noise: the city, memory and the archive', *The Sociological Review*, 61:51 117–33.
Hillier, B. and Hanson, J. (1984) *The Social Logic of Space*, Cambridge: Cambridge University Press.
Hillier, B. (1996) *Space is the Machine*, Cambridge: Cambridge University Press.
Hubbard, P. and Lilley, K. (2004) 'Pacemaking the modern city: the urban politics of speed and slowness', *Environment and Planning D: Society and Space*, 22:2 273–94.
Jacobs, J. (1961) *The Death and Life of Great American Cities*, New York: Random House.
Jacobs, J. (1969) *The Economy of Cities*, Random House.
Jacobsson, K. (2015) *Urban Grassroots Movements in Central and Eastern Europe*, Aldershot: Ashgate.
Jensen, O.B. (2013) *Staging Mobilities*, Abingdon: Routledge.
Lefebvre, H. (1991/1974) *The Production of Space*, Oxford: Blackwell.
Lefebvre, H. (2003/1970) *The Urban Revolution*, Minneapolis, MN: University of Minnesota Press.
Lefebvre, H. (2013/1992) *Rhythmanalysis: Space, Time and Everyday Life*, trans. S. Elden, London: Bloomsbury.
Lehtovuori, P. (2010) *Experience and Conflict: The Production of Urban Space*, Farnham: Ashgate.
Lehtovuori, P. and Koskela, H. (2013) 'From momentary to historic: rhythms in the social production of urban space, the case of Calçada de Sant'Ana, Lisbon', *The Sociological Review*, 61:1 124–43.
Lehtovuori, P. and Ruoppila, S. (2017) 'Temporary uses producing difference in contemporary urbanism', in J. Henneberry (ed.) *Transience and Permanence in Urban Development*, Hoboken, NJ: Wiley, pp. 47–64.
Mareggi, M. (2013) 'Urban rhythms in contemporary city', in D. Henckel, S. Thomaier, B. Könecke, R. Zedda and S. Stabilini (eds) *Space-Time Design of the Public City*, New York/London: Springer.
Martinez-Fernandez, C., Audirac, I., Fol, S. and Cunningham-Sabot, E. (2012) 'Shrinking cities: urban challenges of globalization', *International Journal of Urban and Regional Research*, 36:2 213–25.
McKinsey Global Institute (2016) *Urban World: Meeting the Demographic Challenge*, New York: McKinsey & Company.
Mels, T. (2004) 'Lineages of a geography of rhythms', in T. Mels (ed.) *Reanimating Places: A Geography of Rhythms*, Aldershot: Ashgate.
Saunders, D. (2010) *Arrival City*, London: Windmill Books.
Simonsen, K. (2004) 'Spatiality, temporality and the construction of the city', in J.O. Baerenholdt and K. Simonsen (eds) *Space Odysseys: Spatiality and Social Relations in the 21st Century*, Aldershot: Ashgate.
Stanek, L. (2011) *Henri Lefebvre on Space: Architecture, Urban Research, and the Production of Theory*, Minnesota, MN: Minnesota University Press.
Tartia, J. (forthcoming) *The Temporality and Rhythmicity of Lived Street Space*, Forthcoming doctoral thesis, Tampere: Tampere University of Technology.
United Nations (2005) *World Urbanization Prospect*, United Nations. Online. Available HTTP: <http://www.un.org/>.
United Nations Human Settlements Programme (2016) *Urbanization and Development: Emerging Futures*. World Cities Report 2016.

30

The aesthetics of spatial justice under planetary urbanisation

Saara Liinamaa

Introduction

Castells, many years after his critique of Lefebvre in *The Urban Question* (1977) and subsequent reassessments, refers to Lefebvre as 'almost like an artist' (Castells 1997: 146), a characteristic that Castells deems 'invaluable'. However, at the same time, Castells takes issue with his 'intellectual style' and apparent distance from empirical research (see Leary-Owhin in Part 1). Accordingly, if one reads between the lines, Castells is in essence criticising Lefebvre for being *too* artistic in his approach. I am going to start with this premise of Lefebvre as 'almost like an artist' as a crucial point of departure for considering what Lefebvre offers to our contemporary understanding of the spatial aesthetics of planetary urbanisation. The deep aesthetic sensibility of his urban social theory recognises the dynamic interplay between research, imagination, and representation.

There is considerable weight behind Madden's deceptively plain assessment, 'these are interesting times for urban studies' (Madden 2015: 297). Urban studies in the contemporary moment is under heightened scrutiny and redefinition, and the vigorous debate around planetary urbanisation is an emblematic case. Certainly, in this regard, 'interesting' refers to two related circumstances. The first, as Madden explains: 'urbanization and urbanism are, in some sense, ruling the day'; that is, not only is urbanisation ever-progressing, but all things urban have captured the popular and academic imagination. Yet, we are reminded that this imagination has a particular shape. Despite the hard work of many committed and active urbanists, critical urbanism is *not* ruling the day. The second issue, one that shadows the first, is the urgency behind discussions of the adequacy of current urban theories and methods to engage with this voracious urbanisation. This certainly makes for turbulent and interesting times for urban scholars. In this chapter, I am responding to debates on expansive urbanisation by returning to Lefebvre's aesthetic legacy, using his work to reassert how the aesthetics of spatial justice are an important feature of the problematic of planetary urbanisation. The first part of this chapter will review key aspects of Lefebvre's aesthetic orientations. The second part will examine the deployment of visual metaphors in the discourse of planetary urbanisation, and how these unfold into issues of representation and imagination. The third section, through an analysis of a specific work of art, elaborates upon what art can contribute to our understanding of a critical spatial aesthetic under expansive urbanisation.

Lefebvre, art, and artists

This section briefly tours through Lefebvre's relationship to art and artists, and his subsequent influence on recent scholarship on urban art and culture. Artistic culture cannot be divorced from critical urban theory. As Léger (2006) demonstrates, Lefebvre's theorising of 'moments' makes a central contribution to the repositioning of artistic production and everyday life in Marxist theory. Similarly, as Shields makes clear, Lefebvre's 'experience with artistic avant-gardes seeking revolution through art, not politics, influenced him for the rest of his life' (Shields 1999: 1). Here, I am sketching out some of the key interests of the artistic Lefebvre and his important contributions to critical urban theory, mainly through his insights into spatial aesthetics and critical visuality. I am not presenting an exhaustive overview, but drawing attention to a few features of this work and its ensuing reception in English language academic discourse.

Lefebvre was frequently embedded in evolving artistic worlds. Shields' book *Lefebvre, Love and Struggle: Spatial Dialectics* examines the relationship between his early work and intellectual development by 'taking seriously his engagement with ideas and proposals generated by the artistic avant-gardes of the 1920s' (1999: 29). Lefebvre had a marked attraction to Dada, one compelled by Dada's: refusal to make sense, attention to productive disintegration, and embrace of moral disorder (53–8). These avant-garde values continued to interest Lefebvre, especially during his alliance with the Situationist International. As Merrifield asserts, 'The Situationists, and notably Guy Debord, exerted a strange grip on Lefebvre' (2006: 31). Lefebvre's energetic collaboration and eventual (inevitable) falling out with Debord, his dialogues with Constant Nieuwenhuys and ongoing interest in the automated utopia New Babylon – these are some of the many ways he connected aesthetic experimentation to urban possibility.

Generally, Lefebvrian-inspired analysis of urban artistic practices has assumed many shapes and guises, where his work is drawn on to account for everything from street art to new media projects. Lefebvre has been central for positioning art as an agent of change within urban spatial justice claims. There is an arts-based version of the well-known spatial justice work indebted to Lefebvre (e.g. Harvey and Soja), and with a consistent tenor to these art-focused discussions; Lefebvre is an essential building block for understanding the place of aesthetics within transformative action, even when the key works cited vary, from the *Critique of Everyday Life* to *Rhythmanalysis* to *The Production of Space* to lesser-known untranslated essays. For example, Papastergiadis draws on Lefebvre as a resource to explain how art draws on everyday life, where 'everyday life can illuminate the complex ways in which subjects exercise their potential to be emancipatory and critical' (2010: 24), and Pinder presents him as a model for a distinctly spatial critical pedagogy within the evolving tactics of hybrid urban art-activist intervention practices (Pinder 2008). One of the most influential English language art historical contributions to understanding these alignments, Deutsche's (1996) *Evictions: Art and Spatial Politics*, turns to Lefebvre to account for the multifaceted place of art and architecture's role within spatial dynamics. Importantly, this book provides numerous case studies around the relationship between art and spatial conflict, and begins a dialogue around the ways that artistic practices have appropriated the dominant space of capital. At the same time, Deutsche draws on Lefebvre to explain how public art, architecture, and design can impose aesthetic coherence and rationality onto spaces in ways aligned with the exercise of power, for example, in the case of the redevelopment of Battery Park City in New York City.

Lefebvre provides a framework adept at diagnosing existing conditions, and he is the theoretical hinge that informs the emancipatory possibilities of art. For example, Loftus' (2012) case study of the multi-city non-profit City Mine(d) is instructive here. Loftus draws on Lefebvre as one of the foremost theorists of everyday life, using his call to 'let everyday life become a work of art' (Lefebvre in Loftus 2012: 109) to examine City Mine(d)'s process-based aesthetic and participatory methods during an experimental community art project in London. He uses City Mine(d) to re-read Lefebvre in a manner that positions urban cultural praxis as a central Lefebvrian concept. However, Loftus develops this concept to redress one of the central limitations of Lefebvre's thought – his inability to 'see nature as an ally in the struggle for this better world' (110). Thus, within contemporary art discourses, Lefebvre provides both theoretical and methodological guidance. For example, rhythmanalysis is a key tool for recognising the aesthetic dimensions of urban life. As Highmore argues, rhythmanalysis is a methodology for understanding urban cultural representations as 'the *multiple* rhythms of modernity: the various speeds of circulation; the different spacings of movement; and the varied directions of flows' (2005: 11, emphasis in original). Fraser (2015) argues for a new model of urban cultural studies that better coordinates the key interests of urban studies and cultural studies. Again, Lefebvre is the most prolific exemplar and glue for such a project. Fraser draws attention to the often-undervalued place of culture within urban studies' social science orientations and makes clear that Lefebvre is an exception. As he explains, his 'engagement with the urban problematic is sustained, multidimensional, both intellectual and radical, interdisciplinary, historical, far-reaching, cultural, eclectic' (26). Further, Fraser argues for the essential role aesthetics plays in a Lefebvrian urban cultural studies methodology: 'Lefebvre's work suggests that the formulation of an urban cultural studies method requires an aesthetic theory as its base' (69).

But it is not just that academics use Lefebvre to explain existing urban art practices, contemporary artists and collectives also use Lefebvre as a resource. For example, Cohabitation Strategies (CoHaStra) is a non-profit multi-city 'socio-spatial research' collective that mixes collaborative art, design, and social science research to develop site-specific urban projects. They specially cite the 'right to the city' as a foundation for their aesthetic development of collaborative spatial justice practices. According, they place as one of their key principles, speculating with radically new urban imaginaries: stronger local solidarities, communal politico-economic subjectivities, social networks at larger scales, and parallel urban economies (CoHaStra 2015). For example, their recent examination of affordable housing in New York for the Metropolitan Museum of Modern Art (MoMA) project and exhibition *Uneven Growth: Tactical Urbanisms for Expanding Megacities*, 2014, started with interviews and a compilation of the spatial inequalities that animate the New York housing crisis. They used this material to propose an alternative affordable housing strategy, Cooperative Housing Trust, which they describe as a hybrid tenure model for New York City. Yet not to oversimplify this process, this exhibition has produced its own debate around the possibility of radical change offered by such practice-based curatorial initiatives. As Brenner (2015) queries in his review of the exhibition, 'tactical urbanism may be *narrated* as a self-evident alternative to neoliberal urbanism; but we must ask the question: is this really the case, and if so, how, where, under what conditions, via what methods, with what consequences, and for whom?'.

The above discussion demonstrates how Lefebvre's work informs contemporary discussions of critical spatial aesthetics, be it termed urban cultural praxis or rhythmanalysis, and that this sensibility developed alongside his reflections on art, space, and the everyday. Certainly, *The Production of Space* is full of references to movements, artists, and architects (Bauhaus; Constructivists; surrealism, Klee, Kandinsky, Le Corbusier, Picasso), to name but a few. For example, we stumble upon references to paintings (Picasso's *Guernica* and *Les Demoiselles d'Avignon*), ruminations on the development of perspective in the Italian Renaissance, and reflections on Greco-Roman architecture. And, of course, there is the notable place accorded to 'some' (Lefebvre 1991: 39) artists within the spatial triad. As he explains, 'Representational space is alive: it speaks. It has an affective kernel or centre… Consequently, it may be qualified in various ways: it may be directional, situational or relational, because it is essentially qualitative, fluid and dynamic' (42). Lefebvre recognises that artists can access this 'qualitative' character and play a pivotal role in dramatising the production of spatial understanding, something missed by the 'naivety' of art historians (305). Yet Lefebvre's comments on surrealism in *The Production of Space* are particularly relevant to my wider argument because of the way in which he describes the movement's shortcomings. He speaks of the 'scale of the failure' (109) of surrealism, noting a number of key limitations. As he explains, 'the intrinsic shortcomings of the poetry run deeper, however: it prefers the visual to the act of seeing, rarely adopts a "listening" posture, and curiously neglects the musical both in its mode of expression, and, even more, in its central "vision"'. Elsewhere, he accuses them of fleeing from reality, if not denigrating it altogether (110–11). While these are phrased as points of criticism, at the same time, we can read these comments as a proposition for an alternative version of critical artistic practices. This practice is reflective and embodied. It is responsive and can productively engage with 'the act of seeing'. With this description in mind, the next section asks: can we apply this distinction between the visual and the act of seeing to current debates on planetary urbanisation?

Planetary urbanisation and the search for a critical spatial aesthetic

In this section, I am looking specifically to how aesthetics, but especially visuality, animates the discourse on planetary urbanism in order to consider some of the conceptual and representational challenges that planetary urbanisation poses – be it termed 'complete urbanization' (Lefebvre 2003: 1), 'generalized urbanization' (17), or 'implosion-explosion' (15). I will place this entwining of questions of urbanisation, representation, and imagination that emerge in the recent literature alongside strategies in contemporary art that reconceptualise urbanisation and globalisation. Both contemporary art and critical urban theory share a commitment to the role representation plays in contemporary understanding.

The significance of developing new cartographies of planetary urbanisation is beyond dispute given the challenging task of representing the vastness of its processes. The section 'Visualizations – ideologies and experiments' in Brenner's (2014) edited collection starts with the visual problematic: 'How to develop appropriately differentiated spatial representations of historical and contemporary urbanization processes? What taxonomies are most effective for mapping a world of generalized urbanization, massive spatial development and continued territorial differentiation?' (396). Yet the response that I am developing here, in keeping with Lefebvrian thought, is not entirely straightforward. First, I return to the issue of blindness and other metaphors of sight as they emerge within discussions of total urbanisation's theoretical and methodological dilemmas. Second, I explore how this question of vision is twinned with a concern for our urban imagination. Third, I argue that understanding abstraction as an aesthetic genre can help us to better respond to the many violent and voracious abstractions of planetary urbanisation.

The visual culture of planetary urbanisation vividly demonstrates urban expansion and the blurring between city, region, and territory, from NASA photographs of satellite beehives, to global transit infrastructure maps, to transnational resource extraction circuits. At a glance, these images do not seem much different than the sort of illustration one expects from research on globalisation or international development. But it is important to stress their role: to render visible the diversity of urbanisation processes that remain so expansive as to be easily missed or misrecognised. At issue is how to practise a critical visuality that enables us to consider 'how we see, how we are able, allowed, or made to see, and how we see this seeing and the unseeing therein' (Foster 1988: ix). Or, as visual theorist W.J.T. Mitchell (2004) asks, how do we better reckon with the agency of images, their dynamic role in life and discourse that far exceeds their status as mere illustrations?

Benjamin's (2002) *Arcades Project* has long demonstrated that there is no one method for representing urban research that can adequately capture the complexity of urban constellations. While planetary urbanisation is a key contemporary example of this question of how to capture processes without constraining conceptualisation, concepts and techniques of representation have always occupied a place within definitional debates in urban studies. The literature on urban representations emphasises that understanding urban life must always negotiate the terrain of incompatibilities found at the intersection of the city and its representation. So, while we recognise that 'the city is inseparable from its representations' (Balshaw and Kennedy 2000: 3), at the same time, it is 'neither identical with nor reducible to them'. For Kittler (1996), it is not that we need to regard the city a representation *per se*, it is that we need to recognise the city itself as a medium. Or, as Shields (1996) argues, 'representations make the city available for analysis and replay' (228), but a perilous feature of this process is that 'representations are treacherous metaphors, *summarising* the complexity of the city in an elegant model' (229, emphasis in original). By this argument, we cannot trust urban representations, but we are at the same time reliant upon them. In more recent terms, Merrifield (2013) maintains that we fumble around without adequate descriptors, challenged by this new, boundary-less urbanisation. By virtue of such, it appears that our need to envision becomes even stronger. As he explains, 'to construe our field of vision as a container, is an exorable human need: the need to restrict reality so we can cope, so we can comprehend' (4). It is worth remembering that the second chapter of *The Urban Revolution* (2003) is entitled 'Blind Field'. The relationship between urbanisation and vision resonates on many levels as a mutually constituting problem of representation and research.

For Lefebvre, the term blind field is not just a description of an insufficient research approach. A blind field is 'not merely dark and uncertain, poorly explored, but blind in the sense that there is a blind spot in the retina, the center – and negation – of vision' (2003: 29). The blind field is the inability to see the emergent realities of urbanisation. As he continues, 'we focus attentively on the new field, the urban, but we see it with eyes, with concepts, that we shaped by the practices and theories of industrialization... and is therefore *reductive* of the emerging reality. We no longer see that reality' (29, emphasis in original). It is a matter of not being able to 'perceive or conceive multiple paths, complex spaces' (29). There is a conceptual faltering in the interplay between presentation and representation at the core of the blind field that is inseparable from but cannot be totally reduced to questions of ideology:

> Blindness, our not-seeing and not-knowing, implies an ideology. These blind fields embed themselves in re-presentations. Initially, we are faced with a presentation of the facts and groups of facts, a way of perceiving and grouping. This is followed by a re-presentation, an interpretation of the facts. Between these two moments and in each of them, there are misrepresentations, misunderstandings. The blinding (assumptions we accept dogmatically) and the blinded (misunderstood) are complementary aspects of our blindness.
>
> *(Lefebvre 2003: 30)*

Thus, to avoid both 'blinding' and being 'blinded' hinges on the unique qualities of representation itself. Given Lefebvre's description, it is not surprising why vision is so central to contemporary dialogues on expansive urbanisation, and to various ends.

Merrifield's discussion of the emerging politics of planetary urbanisation starts with a very different image, one drawn from pioneering science fiction author and biochemist Isaac Asimov's vision of the urban planet Trantor, a single-city-planet with 40-billion inhabitants. Lefebvre briefly references Asimov in 'Right to the City' (1996: 160), and for Merrifield, this is a telling spectre, a glance to the challenge of living with the immensity of urbanisation to come. Reminding us of Lefebvre's work as an ominous foretelling of the contemporary condition of planetary urbanism, Merrifield indicates the challenge of addressing an all-encompassing urbanism present and future: 'the complete urbanization of society, of something that's both here and about to come here soon' (2013: 2). Appropriately following Lefebvre, Merrifield remains thoroughly imagistic in his discussion. His book includes many visual analogies to account for the supersession of the urban over the city – cubist painting, abstract expressionism, wormholes – these are all employed to capture the challenge of organising around expansive urbanisation. Now, he explains, the urban is an:

> ontological reality inside us, a way of seeing ourselves and our world. Thus another 'way of seeing', another way of perceiving urbanization in our mind's eye, is to grasp it as a complex adaptive system, as a chaotic yet determined process. As a concept, even a 'virtual concept', the term 'planetary' already connotes a perspectival shift and conjures up more stirring imagery, maybe even more rhetorical imagery, something seemingly extraterrestrial and futuristic.
>
> *(Merrifield 2013: 3–4)*

Expansive, rhetorical, futuristic, chaotic: what Merrifield presents is the speculative image within the imagination of expansive urbanisation. And his formulation recognises both the ambivalence and critical potential of such speculation. Which brings me to my second point: the inseparability of representation and imagination in current discussions.

The dilemmas accompanying the representation of urban research under planetary urbanisation is a recurring theme in many texts. For example, Brenner's 'Theses on urbanization' proposes a 'reconstituted vision of the "site"' within urban studies (2013: 94). Or, in Brenner and Schmid's (2014) article 'The "Urban Age" in question', they explicitly tackle presumptions underlying the Urban Age approach to understanding contemporary urbanisation, referring to the 'extremely blurry vision of the global urban condition' (740) and the 'particularly obfuscatory vision' (744) of its representational domains. Similarly, in 'Towards a new epistemology of the urban?' they identify the 'often-contradictory framing visions, interpretations and cartographies of the urban' (Brenner and Schmid 2015: 164). In contrast, they identify a number of strategies for change that will allow for greater reflexivity as well as unpack 'the underlying conceptual assumptions and cartographic frameworks' (749). Thus, one of the key solutions they propose to this dilemma is stronger collaboration between representation and imagination. Likewise, Madden (2012) refers to the work of the 'global-urban imagination' and draws on Lefebvre as a key foundation for realising 'critical global-urban imaginaries' (774). Or, as Brenner and Schmid maintain, we must 'expand our urban imagination' (2014: 752) and invite more 'heterodox engagements' (749) with our urban questions. Elsewhere it is explained, 'a new cognitive map is urgently needed' (Brenner 2013: 95). Importantly, there is more than a hint of optimism to these formulations that revolve around representation-imagination; 'perhaps the current era of radically intensified urbanisation, even if it now appears as a succession of alienating forms, can repoliticise urbanism and allow for new channels connecting actually-existing cities and the urbanist imagination' (Madden 2015: 301). And this folds into collaborative action grounded in processes of experimentation and imagination: 'The urban is a collective project – it is produced through collective action, negotiation, imagination, experimentation and struggle' (Brenner and Schmid 2015: 178).

Shadowing these fairly enthusiastic declarations of potential, however, are the implications of abstraction. Lefebvre acutely captures how capitalism's appetites and capacity to erase recognitions of suffering and inequality are always a backdrop to concerns around urban research and representation. As a verb, to abstract can mean to extract, to take away from something; in this case, abstraction captures the unbridled resource extractions under urban expansion. But we are also well reminded that an abstract is also a summary of research – something that will frame the detailed analysis enclosed in the research. By implication, this can divert attention away from the initial object, and risks diminishing or weakening its overall stability. In terms of aesthetics, as in painting, it refers to diverse types of non-representational work. Herein rests the underlying concern. Without a critical visuality at work, the representation of planetary urbanisation within critical urbanism verges on its own abstractions that risk detracting from the reach of inequalities these processes represent. Accordingly, representation is another way to describe the tension between the urban as theoretical category and empirical object; 'the urban is thus a theoretical category, not an empirical object: its demarcation as a zone of thought, representation, imagination or action can only occur through a process of theoretical abstraction (Brenner and Schmid 2015: 163). That is, abstraction, when defined both in terms of the translation of research and as aesthetic sensibility can provide, but by no means guarantee, an inroad to a critical spatial aesthetic of the sort that representing expansive urbanisation requires.

To demonstrate, let me turn now to what I view as an example of heterodox engagement (Brenner and Schmid 2014) – a dynamic map that expresses urban displacement and spatial vulnerability as an aesthetic encounter inseparable from global urbanisation. I am suggesting that critical practices in contemporary art can contribute to other practices of experimentation around representation currently being developed in professional practice, and which are also trying to navigate the complexities of abstraction. For example, Harvard University's Urban Theory Lab that 'complicates the task of visualizing urbanization processes' and 'destabilizes inherited assumptions regarding both spatial units and parameters of the urban condition' (Urban Theory Lab 2014: 474).

Contemporary art's fabric of expansive urbanisation

In this section, I develop how the installation work *Woven Chronicle*, 2016 (see Figure 30.1), by Mumbai-based artist Reena Saini Kallat, reinforces a critical spatial aesthetics of planetary

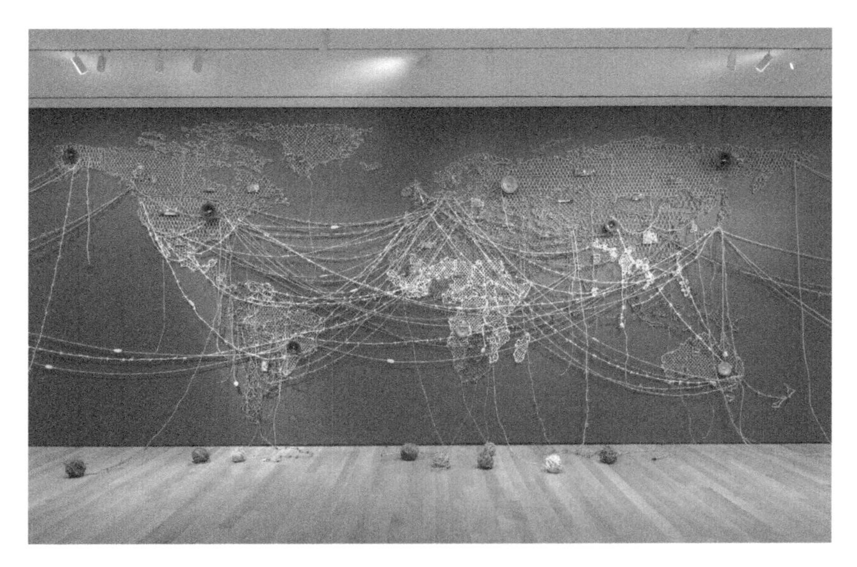

Figure 30.1 Reena Saini Kallat, *Woven Chronicle*, 2016. Circuit boards, speakers, electric wires and fittings, single-channel audio, 10 min. *Source:* author.

urbanisation. This work, which was part of the MoMA exhibition, *Insecurities: Tracing Displacement and Shelter*, 1 October 2016 to 22 January 2017, is an imposing, wall-sized global map of contemporary routes of migration and displacement (MoMA 2016). This map of the world is constructed out of heavy electric cable with multi-coloured strands that draw an outline of the world and demarcate circuits of movement, flight, and prohibition. The wires are ambiguous, both conduits and barriers. The wire-coil pathways intersect, stop, or become suddenly barbed. This map hurts. Piles of wire rest on the floor, unclear if they are part of an unravelling to come, or in wait to make further additions to the routes of displacement. Embedded speakers broadcast a collage of sounds, both sharp and ambient, of high voltage currents, sirens, cell phones, industry: the sounds of urbanisation. Weaving strands of electric wire like wool, the artist describes the work as a mapping of the world from the perspective of the refugee, migrant worker, or indentured labourer. The museum's description of the project stresses how 'the work highlights the inherent contradiction in celebrating an increasingly connected world while stringent immigration laws, closed borders and prejudice face individuals seeking to transgress geographic boundaries' (MoMA 2016). Or, as the artist writes of her process (in reference to an earlier rendition of the installation):

> This work was made with wires that essentially transmit energy and information from one place to another; their linear formations often evoking barbed wires, barriers or different kinds of fencing... By changing the instrument of this quasi-cartographic drawing from a pencil line to a wire, I'm interested in the notion of the map as dynamic, ever changing, streaming and transferring data with the global flows of energies and people.
>
> *(Kallat 2011)*

This is not a cartography of precision (a meticulous charting of exact numbers or routes) but one of aesthetics. This aesthetic perspective captures a dynamic experience of urbanisation processes and their multiple spatial vulnerabilities, an experience that is only activated by the movements of the viewer's inspection. This artwork cuts the potentially abstracting tendencies of techniques of visualising expansive urbanisation by its mixed media dramatisation of a world tied together by displacement. These migrations are yet another feature of planetary urbanisation and its demands, conflicts, and inequalities, but where the aesthetic is an opportunity to reflect on the barbs of these bonds.

The theme of the exhibition in which this work appears is refuge, and this particular project is a representation of the inability to realise the 'right to the city' under an expansive urbanisation that prioritises economic values alongside global hierarchies of power; this artwork is a spatial mapping of precariousness. Schmid (2014) uses Lefebvre to develop a theoretical framework derived from his layered conception of the production of space that stresses the following: networks, borders, and differences. These points, as his ensuing collaborative empirical research project on urbanisation in Switzerland demonstrates, provide a methodological guide to the representation of urban processes. I draw on Schmid's framework here because this artistic representation highlights the inseparability of these three characteristics. However, at this point, I will explore the role of borders in more detail, following Lefebvre's notion of 'incision-suture' (2003: 38) as always ambivalent (Schmid: 78). Borders control, regulate, and structure, but they are also spaces of transitions, of linkages and even potential, where 'new orders, new concepts, new images and new urban configurations emerge from the urban transformation of borders' (78). The migrations in this map are part of an urbanisation-industrialisation pairing that most often forecloses progressive change, but the work of art, in this regard, creates a space of representation that reminds of latent alternatives, alternatives that are embedded in the materiality and name of the work itself, *Woven Chronicle*. These woven cables are an artistic interpretation of our 'urban fabric', another key concept within Lefebvre's theory of complete urbanisation. He employs the term urban fabric to refer to the transformation of rural life through industrialisation and consumption: 'The urban fabric grows, extends its borders, corrodes the residue of agrarian life. This expression, "urban fabric" does not narrowly define the built world of cities but all manifestations of the dominance of the city over the country' (2003: 4). If we associate weaving with a traditional mode of craft, the project presents weaving in the materials of industrialisation, which has clearly changed the nature of the artistic project, but also the art of world-making itself. Yet if we were to remove the barbs, the currents, the volatility of the woven material, we would have a vision/version of urban fabric as solidarity under urbanisation. That is, we would have the remaking of the world in a different image, and through a material, if you will, transformation.

Conclusions

This chapter has touched on key features of Lefebvre's aesthetic orientations. I have described how both academics and artists have adopted Lefebvre's critical spatial aesthetics, and demonstrated in more detail how aesthetic issues play out in distinct ways within recent debates around urban methods and theories under planetary urbanisation. It is easy to miss the centrality of aesthetic insights if we do not combine diverse interpretations of Lefebvre from both the humanities and social sciences. We need dialogue between not only density and periphery within the variegated landscapes of urbanisation but also different domains of aesthetics, aesthetics as not only embedded in everyday life but also as part of questions of urban research and its representation. This chapter underscores that a renewed commitment to a Lefebvrian-inspired triad of research, representation, and imagination can recast the expansiveness of urban processes in critical ways.

References

Balshaw, M. and Kennedy, L. (eds) (2000) *Urban Space and Representation*, Sterling, VA: Pluto Press.
Benjamin, W. (2002) *The Arcades Project*, trans. H. Eiland and K. McLaughlin, Cambridge, MA/London: Belknap.
Brenner, N. (2013) 'Theses on urbanization', *Public Culture*, 25:1 69 85–114.
Brenner, N. (ed.) (2014) *Implosion/Explosions: Towards a Study of Planetary Urbanization*, Berlin: Jovis.
Brenner, N. and Schmid, C. (2014) 'The "Urban Age" in question', *International Journal of Urban and Regional Research*, 38:3 731–55.
Brenner, N. (2015) 'Is "tactical urbanism" an alternative to neoliberal urbanism?' *POST*, 24 March. Online. Available HTTP: <http://post.at.moma.org/content_items/5> (accessed 28 September 2016).
Brenner, N. and Schmid, C. (2015) 'Towards a new epistemology of the urban?', *City*, 19:2–3 151–82.
Castells, M. (1977) *The Urban Question: A Marxist Approach*, trans. A. Sheridan, London: Edward Arnold.
Castells, M. (1997) 'Citizen movements, information and analysis: an interview with Manuel Castells', *City*, 2:7 140–55.
Cohabitation Strategies (2015) Cohabitation Strategies: What We Do. Online. Available HTTP: <http://www.cohstra.org> (accessed 19 January 2017).
Deutsche, R. (1996) *Evictions: Art and Spatial Politics*, Cambridge, MA: MIT Press.
Foster, H. (ed.) (1988) *Vision and Visuality*, Seattle, WA: Bay Press.
Fraser, B. (2015) *Toward an Urban Cultural Studies: Henri Lefebvre and the Humanities*, New York: Palgrave.
Highmore, B. (2005) *Cityscapes: Cultural Readings in the Material and Symbolic City*, New York: Palgrave Macmillan.
Kallat, R.S. (2011) *Untitled (Map Drawing)*. Online. Available HTTP: <http://www.reenakallat.com/map-drawing> (accessed 13 June 2017).
Kittler, F. (1996) 'The city is a medium', *New Literary History*, 27:4 717–29.
Lefebvre, H. (1991) *The Production of Space*, trans. D. Nicholson-Smith, Cambridge, MA/Oxford: Blackwell.

Lefebvre, H. (1996) *Writing on Cities*, trans. E. Kofman and E. Lebas, Cambridge, MA: Blackwell.

Lefebvre, H. (2003) *The Urban Revolution*, trans. R. Bononno, Minneapolis, MN: University of Minnesota Press.

Léger, M.J. (2006) 'Henri Lefebvre and the moment of the aesthetic', in A. Hemingway (ed.) *Marxism and the History of Art*, London: Pluto Press.

Loftus, A. (2012) *Everyday Environmentalism: Creating an Urban Political Ecology*, Minneapolis, MN: University of Minnesota Press.

Madden, D.J. (2012) 'City becoming world: Nancy, Lefebvre, and the global–urban Imagination', *Environment and Planning D: Society and Space*, 30:5 772–87.

Madden, D.J. (2015) 'There is a politics of urban knowledge because urban knowledge is political', *City*, 19:2–3 297–302.

Merrifield, A. (2006) *Henri Lefebvre: A Critical Introduction*, New York: Routledge.

Merrifield, A. (2013) *The Politics of the Encounter: Urban Theory and Protest under Planetary Urbanization*, Geographies of Justice and Social Transformation 19. Athens, GA: University of Georgia Press.

Mitchell, W.J.T. (2004) *What Do Pictures Want? The Lives and Loves of Images*, Chicago, IL: University of Chicago Press.

MoMA (2016) *Insecurities: Tracing Displacement and Shelter*. Online. Available HTTP: <https://www.moma.org/calendar/exhibitions/1653> (accessed 19 January 2017).

Papastergiadis, N. (2010) *Spatial Aesthetics: Art, Place and the Everyday*, Theory on Demand #5, Amsterdam: Institute for Network Cultures.

Pinder, D. (2008) 'Urban interventions: art, politics and pedagogy', *International Journal of Urban and Regional Research*, 32:3 70–6.

Schmid, C. (2014) 'Networks, borders, differences: towards a theory of the urban', in N. Brenner (ed.) *Implosion/Explosions: Towards a study of planetary urbanization*, Berlin: Jovis.

Shields, R. (1996) 'A guide to urban representations and what to do about it', in A.D. King (ed.) *Re-Presenting the City: Ethnicity, Capital and Culture*, New York: New York University Press.

Shields, R. (1999) *Lefebvre, Love, and Struggle: Spatial Dialectics*, London/New York: Routledge.

Urban Theory Lab (2014) 'Visualizing an urbanized planet – materials', in N. Brenner (ed.) *Implosion/Explosions: Towards a Study of Planetary Urbanization*, Berlin: Jovis.

31

Mapping Lefebvre's theory on the production of space to an integrated approach for sustainable urbanism

Florian Wiedmann and Ashraf M. Salama

Introduction

The following analysis on Henri Lefebvre's theories of space production and sustainable urbanism is rooted in his frequently emphasised statement that his theories, observations and interpretations need to be explored and translated in order to serve the overall understanding of social space, which is the precondition for improving and changing our surrounding urban realities. All his main theories were introduced during a time period known for an emerging awareness of rising conflicts within modern urbanisation. The period during the 1960s and 1970s is thus often referred to as the 'crisis of the city' due to the commencing dissolution of cities into fragmented and spreading urban entities (Schmid 2005: 46).

Manuel Castells' *La Question Urbaine* (1977) and David Harvey's *Social Justice and the City* (1973) are only two examples of the growing intellectual movement dealing with the city as a projection of a capitalist society with all its inherited conflicts rooted in social inequity. The previous approach of analysing urban sociology in strict structures, of the Chicago School in the 1920s, was questioned by new dynamics and phenomena requiring new analytical frameworks and new explanations. The dissatisfaction with previous models, which were mainly reflecting developments rather than identifying their roots, was furthermore fuelled by a growing intellectual discourse inspired by the works of Karl Marx and Friedrich Engels questioning capitalism and its effects on society and its space (Schmid 2005: 33).

During the 20th century, urban development became fundamentally dominated by the potential increase of land prices, which defined densities as well as the various land-uses (Alonso 1964). Particularly after World War II, urban planning was often reduced to the calculation and implementation of physical plans that enabled cities to become rationalised entities made of rectangular grids accessible by car, as described by Jane Jacobs in *The Death and Life of American Cities* (1961) or by Kevin Cox in *Urbanization and Conflict in Market Societies* (1978).

The new car-based urbanism led to various conflicts in modern cities. The desire to own properties resulted in a large-scale cultural transformation and the beginning of postmodern consumerism (Dear 2000). Mike Davis added another important reflection on this transformation process and the resulting conflicts by exploring Los Angeles as a major example to illustrate the future path of urbanism in his book *City of Quartz: Excavating the Future in Los Angeles* (2006). In general, it can be stated that during the second half of the 20th century, the urban question was more and more explored with an emphasis on new theories that insist upon the explicit derivation of contemporary urbanisation processes out of the structure of the capitalist mode of production (Dear and Scott 1981).

As a philosopher, Henry Lefebvre approached the urban question from a holistic and elementary point of view. In his two most influential works, *The Urban Revolution* (2003) and *The Production of Space* (1991), Lefebvre introduced his main views on urbanism, its production, its evolution and its key conflicts. In order to tackle the actual roots of any human-made spatial development, Lefebvre followed his own dialectical understanding of how societies shape their surroundings by identifying the physical spatial practice and the mental conception of space as two dialectical poles.

Lefebvre's dialectical thinking is rooted in an in-depth reflection of three major German philosophers of the 19th century. On the one side, Karl Marx's dialectical materialism explains modern urbanisation as the inevitable result of increasing human needs and low productivity, and on the other side, Friedrich Hegel's dialectical idealism, which is rooted in understanding any built reality as a product of thinking and recognition (Schmid 2005: 111). Finally, Friedrich Nietzsche's dialectical philosophical concept of the 'Apollonian and Dionysian' (Nietzsche 2008) is rooted in the understanding of the world from an aesthetic point of view and thus as a product of order and chaos. Lefebvre recognised all three standpoints as valid and developed his dialectical approach by integrating all three perspectives: the physical, the mental and the subjective realms of producing spatial reality.

In this chapter, the authors attempt to link Lefebvre's theories of urbanism to the contemporary conflicts endangering overall sustainability worldwide. While the Global North has entered post-Fordist structures with highly specialised production, the Global South has been locked in labour-intensive mass production serving the advanced markets worldwide. The terms of the ongoing globalisation have created a new dimension of interconnection and dependencies and thus a joint struggle for sustainability. According to Neil Brenner (1997), Lefebvre's theories help to explain the challenges of controlling spatial practices due to the enormous reach of globalised capitalism.

The general theory of a dialectical urban evolution

According to Henri Lefebvre, each urban space has to be seen as a historic product of its society interacting with the surrounding environment. On the basis of his general dialectical understanding, he identified three main dimensions in the process of space production, namely, perceived space, conceived space and lived space (Lefebvre 1991). While perceived space is produced by the spatial practice of all the users of a space due to their daily interaction resulting in complex economies, conceived space is according to his definition the space produced by knowledge and ideologies. In addition to spatial practice and planning (*intellectus*), space is formed by the invisible degree of people's attachment to a certain place. He defined this subjective dimension of space as lived space (*intuitus*) or representational spaces, which comprise complex and often coded symbolisms (Lefebvre 1991: 33).

In parallel to his theory of the three dimensions that produce space, Henri Lefebvre developed his second ontological transformation of space in order to address the evolution of social space through human history. In the past, social space was mainly produced via direct interaction between communities and their environment, which he named absolute space. According to

Lefebvre, the emergence of the open market place in Europe during the 12th century marked the end of the unity between countryside and settlements (Lefebvre 1991: 264). In the 16th century, the newly established mercantile societies developed far-reaching networks and the town overtook the country in terms of its economic and practical weight (Lefebvre 1991: 268).

First, town planning arose and the central control of a few decision-makers led to the end of absolute space, which has always been a direct reflection of a natural symbiotic relationship between settled communities, such as farmers and craftsmen, and nomadic tribes, who both carried and tolerated the special cases of leading social classes as manifested in the Greek and Roman antiquity as well as medieval Europe (Lefebvre 1991: 239). During the Renaissance, the increasing development of trade routes and rapid population growth led to towns becoming hubs and thus places where wealth and subsequently knowledge accumulated within new layers of society dependent on regional trade rather than political or religious leadership (Lefebvre 1991: 262). Subsequently, conceived space, the space of the intellect, began to emerge and formed a new space, known as 'abstract space', named for its characteristic of reducing and dividing spatial aspects to functional and geometrical forms enabling more efficient structures but challenging the integration of diversity (Lefebvre 1991: 361).

Both the new dominance of towns and their spreading economic networks enabled industrialisation in Europe, which heavily relied on the efficient exchange of goods and knowledge, such as technological discoveries. The emerging industrial towns became a spatial manifestation of rationalised planning aiming for increased and highly efficient production but compromising the needs of labourers, as Friedrich Engels discovered in England in 1845 (Engels 1968). The rising conflict between social classes sparked the evolution of modern town planning and its newly discovered importance to secure housing and to avoid civil unrest (Hall 2002: 41).

After World War II, enhanced progress in infrastructural development enabled rapid urbanisation. Thus, the rate of urbanisation between 1950 and 1960 was twice that of the preceding 50 years (Davis 1965: 40). Both technological progress and fast urban growth enabled the rapid introduction of a highly efficient globalisation built on deregulation, capital flows and the subsequent international division of labour (Cohen 1981). The outsourcing of industrial production to the Global North led to the new phenomenon of global service hubs and an unprecedented accumulation of wealth resulting in modern consumerism. The global division of consumers and labourers led to the consolidation of the North-South divide, which is rooted in post-colonial structures and the established dominance of a few world cities (Friedmann 1986; Sassen 1991).

In Lefebvre's view, the end of the 20th century marks the potential beginning of a new kind of space, which he called differential space, that is more dominated by humans' intuition rather than predefined and detached mental frameworks of decision-makers rooted in outdated ideologies. He, however, considers this synthetic evolutionary step as still being in its infancy and an inevitable part of the emerging urban revolution (Lefebvre 1991: 352).

Lefebvre thus thematises the main conflict between a more and more abstract globalised system of capitalism and local regional concerns (Lefebvre 2003: 163). He thus identified the crisis of the cities as 'conflicted space', which is the result of an increasingly challenged abstract space facing its dissolution due to the fact that urbanism has become the result of a 'bureaucratic society of controlled consumption' and thus disconnected from all human needs (Lefebvre 2003: 164).

This globalised 'conflicted space' within the established abstract space is being increasingly questioned with the emergence of public awareness. Collective knowledge and creativity have risen as influential factors within development dynamics by the questioning of existing conditions. Various examples worldwide suggest the potentials of participatory urbanism built on active citizens rather than passive consumers (Merrifield 2002). Thus, Lefebvre expects 'lived space', which he understands as the direct intuitive link between a society and its space, to become an increasingly significant factor within future urbanism (Lefebvre 1991: 399).

In such a scenario, the collective investment of a self-discovering society will overcome the short-term interests of capital movements and the self-management of urban cells will replace top-down and technocratic decision-making. Lefebvre thus argues that the end of this increasingly conflicted urbanism is highly dependent on the proactive participation of an aware society creating a diverse urban space, or as he calls it – a 'differential space' (Lefebvre 1991: 52).

Conflicted space as a key threat to sustainability

The introduction of the car as a preferred mode of transport during the 20th century marked an important milestone in urban developments by making greater distances manageable and thus making the dense cities rooted in the previous form of industrialisation more and more obsolete. More significantly, the new mode of transport enabled a new economy driven by land ownership, which resulted in a complex system of capital accumulation depending on the rising financial debts of societies (Harvey 1985). This new form of urbanism led to fast urban growth and the particular phenomenon of urban sprawl and fragmented settlements due to the division of land uses in order to manage traffic (Hall 2002: 48).

The enhanced consumption and the resulting waste led to complex environmental challenges, whereby cities became the main destroyers of natural habitats worldwide (Newman and Kenworthy 1999: 5). In Lefebvre's opinion, the dominant role of conceived space within abstract space has led to the general conflict between quantity and quality due to the shrinking access of majorities to diversify and differentiate spatial developments. Furthermore, the dependency on continuous growth has become inhibiting for the sustenance and emergence of urban qualities (Lefebvre 1991: 354).

In this context, Lefebvre criticised modern architects, such as Le Corbusier, as followers of the main doctrine to centralise, uniform and dictate urban spaces rather than empowering participation and thus diversification (Lefebvre 2003: 161). The commercialisation of space has created fragmented and segregated urban landscapes, which are described by Lefebvre as agglomerations consisting of either 'spaces of leisure' or 'spaces of labour' (Lefebvre 1991: 383) wherein the role of inhabitants is largely reduced to either the consumption of space or the production of goods and services.

One central argument is that a new pattern of socio-technical organisation, also referred to as the informational mode of development (Castells 1989), initiated new opportunities for the perpetuation and extension of global capitalism. This global realm of capitalism has challenged and weakened the national state as a geographically defined form of governance and led to an ongoing conflict between local concerns and global forces (Sassen 2002: 29).

While Lefebvre's work focused on Europe and its urban history, his theory of conflicted space is applicable to all places. In many parts of the world, there is rather little reflection about contemporary urbanism in relation to its complex contexts in time. The emerging cities of the Global South are often identified as radical examples torn between extensive poverty on one side and unprecedented luxury in small enclaves on the other (Salama and Wiedmann 2013: 235).

The industrialisation and modernisation process, which was described by Lefebvre in the case of European cities, is currently evolving in the Global South in parallel and within less than half a century, instead of a gradual development spread over hundreds of years. But in spite of the different time frames within which modern urbanisation has taken place, most emerging cities in the Global South are exhibitions of Lefebvre's main hypotheses of how conflicted spaces evolve. A few islands of spaces of leisure for higher-income minorities are surrounded by walls and gates for protection and seclusion from the spreading and crowded spaces of labour, often expressed in the form of squatter settlements or monotonous mass housing.

The doctrine of stimulating continuous growth led to enhanced waste and unsustainable structures due to fragmentation and isolation. Modern infrastructure and thus access to global networks can however only be developed and maintained in a few places of the Global South, which have become the gates for international capital flows. The emerging networks of globalisation led to well-established hierarchies, in which cities have to enter a fierce competition in order to attract investment and the headquarters of various types of international cooperation (Sassen 1991: 169).

The established gates to global markets have led to a clear divide between the economic opportunities in urban areas and the emerging struggles in rural areas. The subsequent migration has fuelled the fast growth of urbanisation rates in the developing world. According to Lefebvre the rising accumulation of conflicted space worldwide will eventually lead to the collective reflection of the basic dilemma of today's capitalism, in which a more and more centralised and hierarchical globalisation is facing a strengthened resistance and desire to restructure its basic foundations, which is however endangered by arising protectionism and wars.

The production of urban qualities for sustainability

The commonly used definition of sustainability was made at the Brundtland Commission of the United Nations in 1987, which claimed that sustainable development is development that meets the needs of the present without compromising the ability of future generations to meet their own needs (United Nations 1987). According to this definition, human settlements have to primarily be in balance with the environment by reducing the waste of natural resources and pollution. Thus, any socio-economic development has to be balanced in respect to environmental concerns. This rather simplistic but abstract conception of sustainability, however, is incapable of defining the key factors within the production of spatial structures and their qualities, which either enhance or endanger urban sustainability.

Therefore, core qualities of urban spaces have to be identified in relation to social, economic and environmental concerns. At first, all constructed spaces need to obtain the quality of diversity in order to evolve and adjust to the various environmental challenges so that a community can sustain its livelihood. Rooted in the diverse needs of human beings, any created space has to accommodate the various requirements of a society settled in a specific location, from residential units and private spaces to markets and workshops. Any economic development will lead to an increasingly complex system of dependencies and thus spatial developments (Jacobs 1969). This diversity is the precondition for any future growth and sustenance of a city.

The second quality is rooted in human knowledge, which is accumulating due to interaction and exchange. The empirical findings of how to improve spatial structures are mainly focused on the desire to reduce work and to gain time. The resulting quality of efficiency is usually perceived as key for access to enhanced leisure time, while its key role is the empowerment of communities to use fewer resources and thus reduce their impact on the environment for future generations. Thus, environmentalism is the conscious ideological decision of any community to integrate the needs of future generations.

Last but not least, a third quality, namely identity, needs to be introduced, which is the result of the subjective binding of any community with its surrounding and evolving space. This identification is the result of a long-term process, in which constructed spaces are experienced as nurturing instead of repelling, familiar instead of alienating, and are thus open for individual development. Therefore, a social space can only be healthy if all of its members can identify with the created surrounding conditions. This more general philosophical approach integrates the reality of differing cultural perceptions and lifestyles.

The three urban qualities, introduced as diversity, efficiency and identity, have clear core linkages to both the basic triad of sustainability (balance between economic, environmental and social concerns) and the triadic production of space (perceived, conceived and lived space) as identified by Henri Lefebvre. This triadic relationship is furthermore rooted in human psychology, which differentiates instinctive, mental and emotional factors shaping the personality of each individual. Consequently, any urban space is the dialectical result of the instinctual needs and mental conceptions of a society resulting in emotional attachment or rejection.

This basic theory of how urban sustainability is the product of three core urban qualities can lead to further differentiations and a better understanding of the actual role of urban governance in particular cases (Wiedmann et al. 2014). While spatial practice is diversifying urban environments due to complex needs and networks, conceptions of how spaces can be rationalised will lead to more efficient structures. Finally, the factor of how communities identify and thus attach and invest in certain locations is the direct synthesis of meeting needs and reducing effort. The disconnection of rationalised spaces by social elites and the engineered exclusion of masses can thus be identified as the core conflict of our present form of capitalist urbanism. As with all other urban qualities, urban efficiency is itself based on a dialectical production and thus a triadic relationship. The ideologically influenced visions of how a society and its space should be structured meet the collective knowledge of all available empirical evidence resulting in the actual organisation and governance of a society.

Urban diversity is rooted in the dialectical movement between investment opportunities and actual movements resulting in emerging networks. This abstract concept can be translated into urbanism as follows: investors are taking the risk of developing spaces, which enables inhabitants to follow their needs leading to networks of producers and consumers and thus the economy as a whole. Thus, urban diversity is automatically shrinking if the number of investors and businesses is decreasing.

Finally, urban identity is a dialectical result of a society's intuitive experience of a space as catering and nurturing of basic needs and its familiarity in regard to basic values. These perceptions are, however, merely subjective, and thus belong to the realm of space production, defined as the 'lived space' by Lefebvre. Consequently, any urban identity can be regarded as the collective result of an identification process rooted in evolving perspectives based on images and impressions and, in essence, can never be fully determined due to its fluid and changing nature.

The introduced framework of the production of sustainable urban spaces integrates the three main realms of urban qualities as well as their individual dialectical production leading to a total of three core realms, which are rooted in the basic triadic scheme as introduced by Henri Lefebvre (see Figure 31.1). The framework allows for the identification of the actual dialectical production of each realm in the context of urbanism. Static and moving factors are leading to the three core determinants of urbanism, namely the historically evolved form of governance, current business networks and the future-oriented perspectives of a society. Governance is the direct result of available knowledge and expertise as well as specific decision-making rooted in visions. Physical businesses' economic structures are produced by both working and consuming inhabitants as well as specific investment decisions. And the various perspectives of a society are rooted in the collective access to basic needs and an evolving manifestation of shared values.

Furthermore, the framework emphasises the intersecting and interdependent nature of all realms. While certain cores can be identified, the reality of overlapping and integrated characteristics outweighs any simplified and definite structure due to basic dialectical connections. This

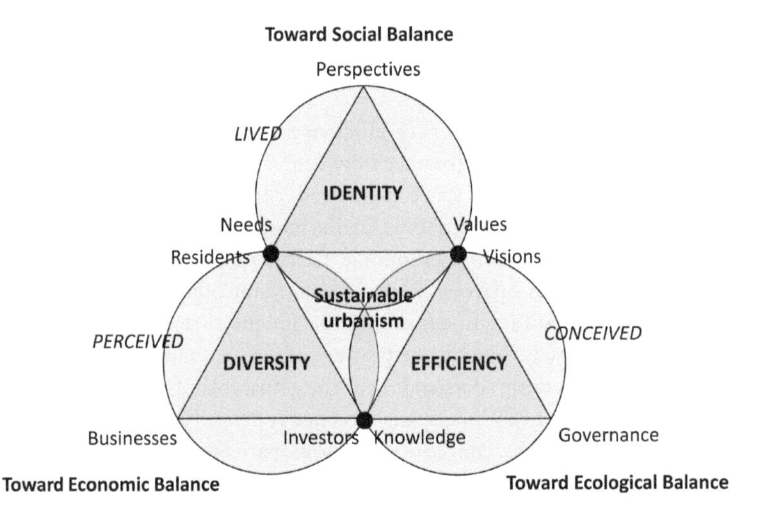

Figure 31.1 The triadic framework of sustainable urbanism. *Source:* authors.

aspect of any framework attempting to integrate all factors within the production of space is often identified as one of the main reasons why Lefebvre prefers an intuitive approach of introducing his hypotheses instead of rigid definitions (Schmid 2005: 111).

The framework of urban qualities rooted in a basic philosophical theory permits the introduction of another layer due to the intersections of certain realms. The first intersection can be found in the case of the basic needs of a human being and his/her spatial practice and thus the driver of diversifying spaces. This spatial practice is therefore connected and related to the individual's experience of how his/her needs have been met in recent times. Accordingly, the first important pillar for sustainable urbanism is an environment, which enables all social groups to take responsibility for their own needs. This instigates the integration of both a collective diversification of spatial structures as well as an identification process.

A second intersection can be found in the case of cultural values and the ideologically influenced vision of how an urban space should be shaped. Both realms are highly complex inner processes usually with historic dimensions. While values are collectively shared and sustained, visions can be isolated and have a clear ideological foundation. Subsequently, the second pillar for sustainable urbanism can be found in a harmonic relationship between newly introduced development visions reflecting the values of a society in order to achieve an overall identity as well as efficiency due to direct implementation of new strategies and policies.

The third intersection is the complex relationship between the available knowledge of a society and its governance, how to develop urban structures and the specific investments in new developments. Any long-term investment heavily relies on a local economic development moving towards consolidation, detached from temporary market dynamics. In this respect, the third pillar of sustainable urbanism is a high level of integration of investment needs on one hand and an in-depth understanding of how urban structures have to develop in order to gain benefits for all inhabitants, on the other. This would strengthen both an increasing efficiency as well as diversity.

The framework and its three pillars of sustainable urbanism offer various explorations of interdependencies within the production of urban environments and help explain why moving towards sustainability has remained a complex challenge. The existing scientific knowledge of new solutions has proven to be a limiting factor for overall change, if the direct link between communities and their inhabited spaces has become compromised. The main reason for this disconnection can be found in elitist structures, which are built on the widespread illusions of continuous consumerism on one hand and the disillusion of ever gaining access to challenge existing structures on the other hand. Henri Lefebvre focused on this core conflict instead of simply proclaiming environmentalism.

Conclusions

Henri Lefebvre's theories on how urban spaces are the direct reflection of social spaces can be an important key to identifying the core problem of any capitalist society beyond the introduction of limited practical solutions. The unequal access to wealth has led to increasingly unequal roles within societies regarding the definition of spaces. Thus, the pure approach to introduce top-down strategies for holistically solving the problem between rapid urbanisation worldwide and the endangered environment needs to be questioned.

The introduced framework of sustainability as a product of three core urban qualities is built on Lefebvre's theories and integrates the various factors defining urban spaces. By analysing these factors, key conflicts can be identified. The three main conflicts endangering any sustainable development are rooted in the divide between top-down decision-making and the values of communities, the divide between masses of labourers and few consumers and finally the divide between widespread speculative tendencies of investments and the implementation of long-term solutions.

These results of the introduced framework are all rooted in one main doctrine of modern capitalist societies, namely the focus on initiating growth. While growth should theoretically be in balance with sufficient resources, the human invention of a debt-driven economy has introduced growth as a necessity to validate the created capital. As a result, cities had to become growth machines attracting capital accumulation and creating complex networks of main control centres and subordinated peripheries (Soja 1989: 184). This divide between global cities and ports in the Global South, which are used for industrial manufacturing and for accessing new resources, has led to a continuous North-South divide despite the officially declared end of colonialism.

Fierce competition between cities to attract growth and thus capital has led to the three mentioned conflicts. Deregulation mechanisms were introduced in order to stimulate investments in emerging cities. The subsequent construction boom fuelled migration, with a large majority of labourers living in informal settlements and a small minority of high-income groups hiding behind gated communities. Subsequently, the majority of communities were obstructed from accessing the main decision-making processes, which resulted in both the danger of social unrest and an increasing disillusion to ever gain a just share in defining the overall urban space.

This circle made of the perception of cities as investment opportunities rather than as home for communities led to a divide between social groups and a shrinking identification between the majority of inhabitants and their surrounding space. The resulting experience of spatial realities as oppressive shells rather than as representations of shared values has led to a fragile foundation for many cities worldwide. While the new form of city-making offers various opportunities for highly qualified knowledge workers, the pressure of future growth is preventing the economic integration of all social groups.

The accelerated globalisation has enabled the transfer of goods and labour worldwide, but its main drivers and shapers cannot be simply identified in the form of free market places. Instead, the main force of enabling complex networks with shrinking boundaries and clear hierarchies has to be seen in the accumulation of capital in few locations and its need for continuous growth in order to reassure its virtual value. Consequently, the overall sustainability debate has to shift from a focus on curing symptoms to the actual core factor producing our conflicted space today: persistent social inequalities.

References

Alonso, W. (1964) *Location and Land Use*. Cambridge, MA: Harvard University Press.

Brenner, N. (1997) 'Global, fragmented, hierarchical: Henri Lefebvre's geographies of globalization', *Public Culture*, 10:1 135–67.

Castells, M. (1977) *The Urban Question*, Cambridge, MA: MIT Press.

Castells, M. (1989) *The Informational City*, Oxford, UK: B. Blackwell.

Cohen, R.B. (1981) 'The new international division of labor, multinational corporations and urban hierarchy', in M. Dear and A. Scott (eds) *Urbanization and Urban Planning in Capitalist Society*, London: Methuen.

Cox, K. (1978) *Urbanization and Conflict in Market Societies*, Chicago: Maaroufa Press.

Davis, K. (1965) 'The urbanization of the human population', *Scientific American*, 213:3 40–53.

Davis, M. (2006) *City of Quartz*, 2nd edn, London: Verso.

Dear, M. (2000) *The Postmodern Urban Condition*, Oxford, England: Blackwell.

Dear, M. and Scott, A. (1981) *Urbanization and Urban Planning in Capitalist Society*, London: Methuen.

Engels, F. (1968/1845) *The Condition of the Working Class in England*, Stanford: Stanford University Press.

Friedmann, J. (1986) 'The world city hypothesis', *Development and Change*, 17:1 69–83.

Hall, P. (2002) *Cities of Tomorrow*, 3rd edn, Oxford, UK: Blackwell.

Harvey, D. (1973) *Social Justice and the City*, Athens: Edward Arnold.

Harvey, D. (1985) *Consciousness and the Urban Experience: Studies in the History and Theory of Capitalist Urbanization*, Baltimore: The Johns Hopkins University Press.

Jacobs, J. (1961) *The Death and Life of Great American Cities*, New York: Random House.

Jacobs, J. (1969) *The Economy of Cities*, New York: Random House.

Lefebvre, H. (1991/1974) *The Production of Space*, Oxford: Blackwell.

Lefebvre, H. (2003/1970) *The Urban Revolution*, Minneapolis: University of Minnesota Press.

Merrifield, A. (2002) *Dialectical Urbanism: Social Struggles in the Capitalist Society*, New York: Monthly Review Press.

Nietzsche, F. (2008) *The Birth of Tragedy: Out of the Spirit of Music*, Oxford: Oxford University Press.

Newman, P. and Kenworthy, J. (1999) *Sustainability and Cities: Overcoming Automobile Dependence*, 2nd edn, Washington: Island Press.

Salama, A. and Wiedmann, F. (2013) *Demystifying Doha: On Architecture and Urbanism in an Emerging City*, London: Routledge.

Sassen, S. (1991) *The Global City: New York, London, Tokyo*, Princeton, NJ: Princeton University Press.

Sassen, S. (2002) 'Locating cities on global circuits', *Environment and Urbanization*, 14 113–30.

Schmid, C. (2005) *Stadt, Raum und Gesellschaft*. Stuttgart: Steiner.

Soja, E. (1989) *Postmodern Geographies: The Reassertion of Space in Critical Social Theory*. London: Verso.

United Nations (1987) Report of the World Commission on environment and development: our common future, accessed online at http://www.un-documents.net/our-common-future.pdf, accessed 10 February 2017.

Wiedmann, F., Salama, A. and Mirincheva, V. (2014) 'Sustainable urban qualities in the emerging city of Doha', *Journal of Urbanism: International Research on Placemaking and Urban Sustainability*, 7:1 62–84.

32

Land use planning, global changes and local responsibilities

Luca P. Marescotti

Introduction

Half a century ago, people claimed that housing ought to represent a service, that everyone had rights to a city. In America and Europe students and workers marched in demonstrations, and not always peacefully. Urban planners, political scientists, philosophers and sociologists alike have all studied the issue and offered their own theories and solutions, as well. Many theories and poorly tested solutions were widely criticised. In the 1970s I was attracted to works by Henri Lefebvre and Manuel Castells, but the former did not believe in urban planning (but see the chapter by Leary-Owhin in Part 6) and the latter supported urban conflict. I believed in urban planning as a science, in political commitment, in Bologna's public building plan and in municipal standards for social services. Things that Lefebvre did not believe in:

> It often happens that talented men believe themselves to be at the centre of knowledge and experience whereas they remain at the centre of systems of writing, projections on paper, visualizations. Architects tending on their part towards a system of significations which they often call 'planning', it is not impossible for analysts of urban reality, grouping together their piecemeal facts, to constitute a somewhat different system of significations that they can also baptize planning while they leave its programming to machines.
>
> *(Lefebvre, in Kofman and Lebas 1996: 117)*

I had no doubts that reformism and urbanistic solidity could limit social imbalances during that period of political turmoil. I searched for practical and theoretical confirmation in the work of Lefebvre and Castells. These were the years of voracious studies that led to my first outline (Marescotti 1979). I never was an architect and at last I became a planner. Much has changed over time.

Another, perhaps fraudulent, message prevailed, saying that property would be a wise savings investment. What we have experienced in recent years is a spreading global crisis, triggered by the housing bubble and consequent subprime mortgages: houses were no longer a service but the financial instruments of the few, who betrayed not only personal investments, but also the state-citizen social pact. The costs are borne by the weaker classes while profits flow up to a few collectors.

Unlimited economic expansion strategies generated an unprecedented financial crisis that has overwhelmed the environment and societies; inequalities and climate change are only secondary effects in a race for profits. Too many voices overlap, producing noise and confusion. Some people hope for technological miracles; someone provides disasters; others wait to replenish their resources with a smile and a show of confidence: do not panic, they say.

The new urban dimension is an 'ecumenopolis' (Doxiadis 1968) that develops in irregular reticulates composed of glittering metropolitan areas, factories as large as cities and amusement towns interspersed with gigantic slums and abandoned areas dominated by illegal occupations until they are renovated and redefined as a luxury. The economic, political and social dynamics were missing from Doxiadis' global conurbation, but I found them in Sassen (1994) and Harvey (1989). I encountered Lefebvre once again.

Cities full of so-called sustainable technologies grow in new areas and new factory cities are established that do not differ in any way from the terrible beginnings of the English industrial revolution. Old disciplines and policies are no longer of any use for shaping a future based on a new urban system, in which suburbs and business centres are no longer localised phenomena, but rather a hotchpotch scattered around the world. The Earth is a complex socio-ecological system and the biosphere is in a metastable equilibrium manipulated by human actions. We possess powerful tools and huge machines and we believe in continuous progress, to what end? Questions with no reassuring answers.

What is to be done? The answer lies in governing politics to govern the land. Although analyses on the productivity of urban areas have improved, urban planning as a discipline is still embedded within administrative procedures. This is the opinion which I have developed over the years while I was looking for potential building blocks to establish urban planning as a science in an environmental sphere. I have never been a Lefebvrian, but the seeds of his reasoning intrigued me and worked on my subconscious, to shape my research.

Similar cities, standardised cities

A technical outlook took hold again in the 1980s and 1990s, but politics fostered a drift in the discipline despite Cassandra's warnings (Harvey 1989; Davis 1994; 2002; 2008): we had to learn, because we only see what we know about. I look at affluent countries: similar cities, standardised cities, which are not the result of errors in the discipline, but rather the success of capitalist planning. We live in a great urban mosaic whose individual tiles are connected but do not fit together: but can it still be called a city, a venue of politics? The historic urban revolution continues all over the world and changes nature: metropolitan areas, urban regions, ecumenopolis, where all forms of integration presuppose a vast literature, supported by social, economic, formal, functional or structural analysis. We must fully understand what is happening on Earth.

I see cities with their uniqueness and identity violated, dominated by simple thoughts and powerful technologies. These cities could be located anywhere in the world, they have no cosmopolitan ambitions; they only exist to provide goods to simplified consumers. At the same time, I see territories designed with wisdom and others ravaged by ignorance. Perhaps we are witnessing something new, based on a single and pervasive way of life: the Anthropocene ecumenopolis, inhabited by more than 4 billion people (2016). So this is the context: we are able to think and to know as never before; despite this, we are unable to reconcile what we have learned and what we are teaching with the changes that are altering the environment and society. These changes give new meanings to the words: the urban space has invaded entire regions; it colonises the subconscious and models our sense of time, rhythms and desires.

Victor Hugo wrote '*Ceci tuera cela. Le livre tuera l'édifice*' (Hugo 1865: 93), meaning that books would replace architecture and become the new popular medium for narration. Unfortunately, what we are encountering now is much worse: technology is killing art, as the new emblem of supremacy in the context of global competitions which are conducted less and less by states and more and more by private enterprise. While competition among kings and princes once depended upon the splendour of public and private spaces with the allure of their artists, nowadays industrial and financial empires are competing with the height of their skyscrapers. *The Professor's Dream*, a Charles Robert Cockerell watercolour (1848), depicting ancient monuments superimposed into a single image, has been transformed into a comparison of the highest and strangest pomposities of technologies prevailing over representation. The seeds of change were planted in Las Vegas: no time to rest and reflect, only entertainment is allowed, for a fee, and the purpose of those exciting shapes, lights, sounds and atmosphere is just to reap money in, to alienate the mind through the imitation of other places, or childhood myths. It mimics urban qualities and it is amazing, but it is nothing more than a 'non-place' in which all the details are calculated to annihilate the individual. Space for the representation of whom?

Then the experiment became a contagious model and spread to other places for the masses, from amusement parks to shopping centres, fake worlds offering false savings; even historical sites have become destinations for fast tourism, where the most terrible aspect is the consumption of time, the alienation from poetry. Lefebvre himself called for poetry to overcome the geometric and timeless approach to producing space (Lefebvre 1991). I now see his calls as trans-disciplinary openings to the complexities of planning.

It really seems that no one wants to envisage wise alternatives. Cities and their historical centres are under siege by business centres, skyscrapers, advertising images and by worldwide fashion brands and ethnic restaurants: all different in their outward appearance, all the same in essence. Global goods which standardise desires, in shopping centres which are monsters devouring territories and personalities, are all tuned in to an absurd race to the ephemeral. Their success lies in the progressive elimination of all other alternatives, in war and in peace. Dreams of democracy and emancipation have been replaced by a non-citizen-consumer reality, supported by a fantasy, manipulated by a repetitive collection of symbols. The same logic creeps in even during wartime: the enemies' cities and places of art and faith are destroyed or dismembered by smuggling. After the war, will the cities be rebuilt according to the latest fashion? The poor do not comply? It does not matter, they are not consumers, but they will not be excluded; something can be taken from them, too.

Both the physical city and its society are unified in 'polis', in a 'political' being. As we move on from words to urban experience, we can observe a double coherence between city and society, because the shape of one moulds the existence of the other, and vice versa. Technological architecture encompasses both a physical shape and international finances: one stands for the other and in doing so they promote systemic developments. While ancient Greek aristocracy put its citizens in first place, the individual is pushed to submission as a consumer in modern cities. Society is divided and there are enormous differences between the haves and the marginalised have-nots, who are either unaware or devoutly integrated. These coherences between urban and social systems are the mirror of reshuffled values through which new identities can arise. Which ones? This is not yet known; the future is open.

The challenge is to reduce economic and financial exploitation and react to pervasive standardisation, indifference to the multiple cultural identities which make cities and regions so fascinating. The priorities of urban planning lie in freedom, emancipation, urban creativity and the transformation of urban reality into protection of the environment. There is a long road ahead full of misconceptions: we must still learn how to inhabit the Earth.

Inhabiting the Earth

Inhabit the Earth, then! Be conscious; look at landscapes with fresh eyes, because everything we need and belongs to us can be found there. Individual or national selfishness, utilitarianism and corruption cause bulimic hoarding and ignore the human condition. That is senseless; to inhabit the Earth is, more than ever and above all, a necessity to politics; 'fraternity, equality and liberty' must overcome all borders. On this urban planet the duality and contrast between cities and countryside, or state and citizens, have no reasons to exist: we are all citizens of the world. The *Universal Declaration of Human Rights* (UDHR) must be voted anew, because it must be applied in every country, because the right to inhabit the Earth in peace (Heidegger 1971: 147), free from poverty and fears, must be given to everyone.

The agricultural revolution took place just ten thousand years ago thanks to a temperate Holocene, and five thousand years ago came the urban revolution in the Fertile Crescent. Urban supremacy consolidated in five centuries through social classes, functional specialisations, writings and thoughts on democracy: no longer phratries, but citizens and societies organised under a Divine King who could govern through power and scribes. It was the beginning of urban and social history.

Worldwide urban revolutions would follow this course without ever being repetitive. The idea of a city always took shape in a singular way, in form and content, and maintained the single living matrix of a village: to be in a safe place, full of water, food and raw materials. These were the principles on which the selection of locations was based, these were the original characteristics which supported the first cities and urban societies. The territories of the city widened beyond the ecological niche, tracing routes through lands, rivers and seas. The borders were interfaces, fringes of contamination or raids. Natural capitals and population, trade and invasions, peace and wars were the main drivers of the growth and death of a city, but also of the transition from a city-state to a territory-state, a fundamental step to overcoming dependence on local resources. Two millennia later, we find proof of the 'crisis' generated by the gap between demographic growth and carrying capacity in ancient Greek colonies, but the limits of their resources appear to have been local episodes that were easily overcome by occupying new lands.

People are increasingly attracted to cities, but they change their habits and lifestyles without ceasing war and carnage. There are no more continents to explore; the frontiers of outer space are easier to explore than those between us and others, influenced by cultural and geopolitical boundaries. Be fruitful and multiply and fill the earth and subdue it: and then? Even if Garrett Hardin acknowledged that freedom to breed is intolerable, his responses and the recognition of necessity are ideological, not scientific (Hardin 1968).

Urban systems are connected together to exploit the poorest members of the population. Industrial manufacturers move where they find a cheaper workforce, because other factors such as raw materials and food can now be found and distributed everywhere, whatever the distance. One just has to pay and be aware that the commodity-money-commodity and the money-commodity-money cycles are inseparable and operate globally at such a speed that material goods turn into virtual goods: a brick is not as solid as it seems. We should have known, because something similar happened in 1873. How can we break away? Although our knowledge has increased, we try to ignore that the world's regions are a single living system and that we are all citizens of the world, with equal rights and dignity.

Proud as always, we have triggered unpredictable, perhaps radical changes with unforeseeable consequences, although some warning signs can be inferred from local and global changes. We pretend that oases, forests and woods are immense, that rivers can be manipulated, and we reject the proof of the absurdity of these claims. An urban planet: this is the last stage of Lefebvre's urban revolution in continuous acceleration. Cities occupy a tiny percentage of land but devour energy and fertile lands to feed their metabolism. Technology enables everything. It is possible to build in the desert, to build on the sea and to fill the world with greenhouses, even vertical ones. One just has to pay. But money does not regenerate nature, it does not compensate for climate changes. We can send people into space, but they can only go there if enormous human and natural resources are extracted from around the world to bring colossal wealth to the few. Money is nothing more than the expression of social relations: imbalances expose different ways of exerting dominion.

Cities will possibly double in extension over the next 20 years and a few affluent decision-makers will dominate and subjugate many proletarians. Or what should we call them, since their offspring will no longer be their wealth? Over the last two centuries innovations have developed faster, the British industrial revolution and advances in hygiene (another revolution) necessitated urbanisation, but administrators were unprepared. The unimaginable and unstoppable urban sprawl of industrial cities quickly became part of the myopic strategies that combined economic power with the size of cities and profits from great differential earnings. The story is well known: there is no break in progressive growth; each step has developed at an increasingly rapid pace from one phase to another, from one country to another. The frenzy has not stopped, nor does it seem possible for anyone to stop it. The word 'planning' means houses, industries, schools and public places: the construction industry is the economic basis for other productive sectors; expansion spreads around the planet, it flows through all countries. Property has become an asset haven, agriculture largely swings between industrialisation and a luxury pastime, the 'natural' value of the soil and environmental protection has been forgotten: natural capital is the source of all capitals.

The symbolic value of architecture dominates everything in diverse forms, but in the late 19th century William Morris sensed a change of an extraordinary scale in architecture: 'A great subject truly, for it embraces the consideration of the whole external surroundings of the life of man; we cannot escape from it if we would so long as we are part of civilisation, for it means the moulding and altering to human needs of the very face of the earth itself, except in the outermost desert' (Morris 1910–15, XXII: 119). It is a radical innovation that transcends his own words, if read in the light of the current knowledge of human environmental impacts, well aware of how difficult it is to identify all their effects: the environment is not a matter of aesthetics! The issue concerns survival, but few strive to build a science to develop theories and then compare, validate, refute them. Do we lack laboratories and experimentation? The biosphere, the environment, territories, landscapes, multiform cities: here is the laboratory, here are our experiments, but we are blind without a science.

Wizards and alchemists

Planning practices pose restrictions on scientific development, but theoretical attempts also appear to be in disarray. We need resources and freedom to build science and theories, to share criteria and methods. Urban analysis and the comparison of environmental observations, as well as philosophical or literary reflections, are all necessary frames of reference for building an ambitious overarching vision that can offer interpretative and operational tools. All the attempts have certainly provided interesting interpretations, they have overcome theoretical critique and offered instruments for land use planning that have been applied all over the world. However, despite the best intentions, scientific ambitions have only rarely been supported by direct analysis of territorial transformations, so they have missed the objective to substantiate theoretical hypotheses with aftermath analysis. Lack of evidence has limited effectiveness and led to conceptual schemes, insinuations and statements which not only are inapplicable in other contexts, but may even lack experimental verification.

The origin of many contributions was confined to the needs of an academic career, or to clearly and concisely describe the contents of the discipline to students through interpretive categories. However, a brief survey will show that the theoretical structure hardly resists the first verification of consistency. If each one of us drew up a list of authors to study to become urban planners, we would reach different results that would probably denote our age groups. Each list would simply be a trace of possible, individually significant pathways, but today's students might find other approaches. Different cultures in time and space are facing each other without common cornerstones: a jumble of ideas like no other.

However, I affirm that logic is not enough to make order between definitions and hypotheses to reach a scientific vision; we need ways to face reality and validate or refute conventional professional practices. This is the point: habits consistently dominate the process in three ways to mould the discipline and tuition: scientific reductionism, localised vision and property assets.

1. The first constant is scientific reductionism. The action and the area of interest are conceptually and territorially forced into watertight compartments. Reductionism operates in empirical destructured ways that are artificially complicated by rules to avoid monitoring transformations. Reductionism consolidates through four, perhaps not strictly consequential, stages:
    - *Planning without programming*: It is not a public authority but rather private interests that decide when and where investments are to be made. Legitimacy does not lie in a response to needs for housing, services, jobs, but only in the market. There is a return to the supremacy of arbitrary use of power in a legally constituted state.
    - *Nullification of the strategic plan*: one talks of process to keep a strong connection between socio-economic transformations and planning. Environmental assessments without measurable objectives have no scientific structure. Ideologies and persuasive discussions prevail over the plan. Only rich regions can afford the cost of strategies, while small municipalities and the least developed countries can only make short-term decisions. The neo-liberalism wave pushes the state out, it diminishes its roles and it merges public law into private law: the era of reformism closes by acceptance of the market, even by the political left. Urban planning is not explicitly mentioned in the Italian constitutional reform of 2001. It is a 'residual matter' (Repubblica Italiana 2001: article 117).
    - *Emphasis on architecture and urban design*: Form is the whole. The metamorphosis is complete. It is worth remembering the Masonic *Plan for Democratic Rebirth* (1982) as the first ever evidence of the neo-liberalism momentum. In that confidential document, socio-economic measures called for anti-urbanisation laws, invested in satellite towns and reduced land use planning to a science of suburban rapid transit (Camera dei Deputati, Senato della Repubblica 1985: 622–3). These few words are not to be underestimated: although it was not a criminal plan, despite the chosen criminal methods, it must not be confused with Aldo Rossi's architecture of the city, or with some urban planning attempts of the 1980s (such as Bologna and Madrid in 1985) which aimed to promote the importance of architecture to speed up the implementation of

the plan. One can possibly find some kinship with circumstances in other countries and universities, where land use planning has been pushed into the corner in favour of other buzzwords like policies, architectural design or urban regeneration. Too many coincidences? It was nothing more than a follow-up to the Chilean repression and the Chicago Boys theories and an anticipation of neo-liberalism, with the arrival of Reagan, Thatcher and Berlusconi.

- *Reaffirmation of land use planning as a local administrative technique*: land use planning is out of control. The predominant characteristics are abandonment of general and shared strategies and, almost consequently, lack of any protocol for analysis and assessment, legislative specificity and weakness, heterogeneity of regional and local applications and exasperated competition to attract investors. No authorities regulate the value and price of land and housing that fluctuate arbitrarily in a state of masked oligopoly. Analyses of shortages of social services and settlements, or strategic environmental assessment and environmental impact assessment, or analyses of carrying capacity, urban metabolism, ecological footprints and environmental impacts, become chapters of stories without a plot. Without quantitative analysis, theory can only be based on examples without a general value.

2. The second constant is the hegemony of local physical planning. Territories and the environment are fragmented by localisms in the same way that urban knowledge is fragmented. Myopic vision tends to reject long-term strategies taken at a higher level and consider them as interferences, thereupon appealing to principles of subsidiarity to approve piecemeal urban regeneration. There is only a hint at governance; everything is decided day by day in a political arena where everyone wants to win. Well-being is confused with increases in house and land prices, although the financial interests in urban transformations are often not local ones. Grassroots participation could express citizens' interests, but it is weakened by consultative assemblies. Without real political will, sustainable development is nothing but a chimera.
3. The third constant is the hegemony of private interests guiding local actions. The territorial economy discards social and environmental costs in favour of profitability and appropriates common and public goods. Ground rents and private accumulation processes become the drivers of development, with the consent of smallholders. The 'differential rent II' described by Marx (a minimum of investment to maximise profits) attracts international investors. The greatest political and technical performance is to build cities in the desert and declare their sustainability.

Land use planning plays a central role in the social and productive system; without it (more or less concealed among the functions of the state) the economic and financial system cannot develop. As in neo-liberalism, it is necessary to conceal its essence, transmute participation, democracy and freedom on the land into icons without substance. It is vital for those who want to understand the sense of the rhythms of life to reflect critically to defend social liberties and to fight for social emancipation: we have to uncover the essence of the discipline. We can only achieve scientific robustness by overcoming these limits with new cognitive models, not with a mixture of architecture and administrative techniques. We have to re-establish the discipline!

Only too rarely do we speak of planning as if it were a normal science; few want to frame it in a broader context. Most people easily forget that cities and their inhabitants are the drivers of global social and environmental changes. They are 'mega-machines' that can produce increasingly more and faster by means of technological acceleration. On the other hand, we must remember that cities are the place of politics and participation, they are catalysts of revolutionary ideas and incubators of emancipation, moulding social identities, giving hospitality, showing civil magnificence in public spaces, hospitals, schools. This contradiction in terms contains the dialectical spring-loaded by command and control, and by acceptance and solidarity. Which soul will prevail? Despite the incredible, belligerent human diversity that unleashes wars to seize resources, only 'cities of meekness' (Bobbio 1995) will dissolve conflicts and find a new alliance with nature and humankind in its wholeness: the construction of a city is primarily an education in citizenship.

Theories cannot be built without science: this is our problem. We cannot understand the changes that cities are causing to the biosphere without science and theories. Both professional and theoretical planners console themselves by saying: 'Artificial land covered by settlements, urban parks and infrastructure only amounts to 0.6 per cent of total land, so it is impossible for it to be responsible for global changes'. Sometimes, they cynically affirm that we should trust in the market, as if we had not already paid enough for the failures accumulated by voracious investors and economists, who are so confident in an invisible hand. The feeling is that neo-liberalist craving has turned water, air and soil into heretical subjects, while I need to focus my teaching on these themes: am I a heretic? Politicians do not comprehend these facts when they transform local power into absolute power with the sole mission to construct great cities, faster and faster.

It is impossible to assess impacts without well-defined quantities, measurements with appropriate protocols, or reliable variables of control. Ecology is necessary to fight disinformation and create awareness in societies that can propose and develop national objectives to give coherence to regional and local planning with a view to environmental protection. What will happen in 2050 when the world's population will probably reach 9 billion, or in 2100 when the global average temperature will rise by 2 centigrade degrees per year? Cities are not mega-machines but living systems which continuously develop new social and environmental characteristics. A mega-machine is controlled mechanically, while a socio-ecological system requires all its members to participate pro-actively to achieve mutual interests: this is not a marginal difference. We are simultaneously the sources and the victims of global changes: how can we create emergent properties in all societies that could reduce our environmental impacts? Can we govern these emergent properties without re-discussing urban planning?

While Colin Clark was convinced that the Earth could feed 47 billion people (Clark 1968: 153), Albert Allen Bartlett lectured on the growth of bacteria confined in the narrow space of a bottle (Bartlett n.d.): the bacterial population doubled every minute, but two minutes before filling the bottle, they only occupied one quarter of it and still believed in an immense space. They were unable to understand. What policies could they think of, poor bacteria? Wars? The serious, current, global question is implicit in the interdependence of social inequalities, demographic growth, consumption of resources, critical factors and metastable equilibrium of the biosphere.

Political will and complexity

Complexity is one of the characteristics of systems. It is revealed when properties emerge that are not present in single elements. We cannot grasp the complexity of the biosphere if we are not willing to see the world as an interaction of biotic and abiotic subsystems that interact in unforeseeable ways. Politicians cannot ignore the warning about the complexity of biological systems:

> Instead of a construct in which the present implies the future, we are going towards a world in which the future is open; and time is a construct in which we can all take part.
>
> *(Nicolis and Prigogine 1989: 3)*

We can and we must take heed!

The increase in land use transformations and consumptions induces cumulative effects that are quite different from their simple addition, because they are new, systemic and unpredictable. Limiting factors for mankind could not only be population growth, food and fuel, but an alteration in the biosphere that would create a different state of metastable equilibrium, a hostile environment for human life equipped with its own resilience. Precisely for this reason, some scientists have begun to investigate the biosphere, to identify the most significant global and local processes that are capable of interacting to the point of forcing the Holocene era out of its equilibrium: this is the known hypothesis of Planetary Boundaries (Steffen et al. 2015). Their main aim is to identify the quantitative limits of human activities. However, I think that the link between land use and other processes is underestimated and that the real limiting factor will be the political capability for cooperation among nations in a wide governance with the manufacturing and trade sectors.

Reductionism has accustomed us to simplify, to avoid the complexities of intertwining social and environmental problems. We must reflect on the territorial dimension of social and environmental issues to give new analytical and operational capabilities to politics and planning. The role of planning on the limits can upset the biotic and abiotic union, between societies and environment. This is what makes it necessary and urgent to reprogramme our minds, so as not to become victims of fascinating but deceptive appearances: *a city is not an architectural problem nor a quasi-artistic creation*; *a landscape is not an aesthetic matter.*

I felt the need to completely revise my text and call for a new reading of history and the present time, for a new role for politics and for establishing an independent, open, interdisciplinary and transdisciplinary science (Marescotti 2008). Looking back, I reread Lefebvre's political project and realised that we have followed a similar itinerary, albeit by different roads (see Leary-Owhin in Part 1 on Lefevbre's project):

> This book has been informed from beginning to end by a project, though this may at times have been discernible only by reading between the lines. I refer to the project of a different society, a different mode of production, where social practice would be governed by different conceptual determinations.
>
> *(Lefebvre 1991: 419)*

and the last sentence:

> I speak of an orientation advisedly. We are concerned with nothing more and nothing less than that. We are concerned with what might be called a 'sense': an organ that perceives, a direction that may be conceived, and a directly lived movement progressing towards the horizon. And we are concerned with nothing that even remotely resembles a system.
>
> *(Lefebvre 1991: 423)*

Procedures should be supported by theoretically robust instruments to strengthen the scientific structure of the discipline, but it is also necessary to link these tools with political choices in a common strategy. We must move away from single cases, however important they may be, to strategies shared by all urban regions. The objective to aim for is the adoption of social and environmental standards: the former to promote a just society with services and infrastructures, the places in which we can build our identities and represent ourselves; the latter to harmonise strategies towards the general objective of caring for the Earth. We must be conscious of biospheric limits and actualise Abraham Lincoln's *Gettysburg Address* (1864). We must convey his message to all human beings, the elderly, adults and children, all over the Earth, because their different ages are emblematic of past and future generations. We must fight to give this planet a new life of freedom, so that the idea of a government of the people, by the people and for the people shall not perish from this Earth. It is both a political and a scientific issue.

Conclusions

I have discussed these subjects: (1) the future of cities in planetary systems based on a standardising model, (2) the need to live on Earth peacefully, (3) the reductionism of urban planning, (4) the limits of the biosphere's metastable equilibrium, (5) the need for scientific independence in urban planning and political commitment.

My interest has moved from the physical city to global urbanisation, to earth sciences and social sciences, from analysis, to the need to share definitions and measuring protocols to verify or refute hypotheses regarding environmental impacts. Above all, I feel the need to strongly argue for political action, however compromised it may seem, because I believe that maturity in an independent discipline equipped with resources and laboratories to study environmental and social dynamics is essentially a political issue. I also feel that schools of all kinds and levels should be involved in this project and foster an education in citizenship. It will only be possible to involve the population in an urban planning which is aware of profound, dramatic ecological interactions if we overcome the bipolar model of the opposition between state and citizens.

Although my path was independent from Lefebvre's studies, many echoes do rebound: analysis of the rhythms, of social imbalances and environmental impacts on a planetary scale. The right to the city expands to the right to live peacefully on Earth. The city of meekness is the city of freedom, equality and solidarity. This is my tribute to Lefebvre's work.

Acknowledgements

I wish to thank Matina Kousidi for translating the first version and Jacqueline C. Ryder for the final English editing.

References

Bartlett, A.A. (n.d.) English transcript of 'Arithmetic, population and energy. A talk by Al Bartlett', accessed online at http://www.albartlett.org, accessed 18 February 2017.

Bobbio, N. (1995) 'In praise of meekness', *Convivium. Journal of Ideas in Italian Studies*, 1:1 21–38.

Camera dei Deputati, Senato della Repubblica (eds) (1985) Piano di Rinascita Democratica. *IX Legislatura, Allegati alla relazione serie II: documentazione raccolta dalla Commissione*. Vol. III, Tomo VII.

Clark, C. (1968) *Population Growth and Land Use*, London: Macmillan St. Martin's Press.

Davis, M. (1994) *Agonia di Los Angeles*, Roma: Datanews.

Davis, M. (2002) *Città morte: storie di inferno metropolitano*, Milano: Feltrinelli.

Davis, M. (2008) *Città di quarzo: indagando sul futuro a Los Angeles*, Roma: Manifestolibri.

Doxiadis, C.A. (1968) 'Ecumenopolis: tomorrow's city', in *Britannica Book of the Year 1968*, 16–38, Chicago: Encyclopaedia Britannica.

Hardin, G. (1968) 'The tragedy of the commons', *Science*, 162:3859 1243–48.

Harvey, D. (1989) *The Condition of Postmodernity. An Enquiry into the Origins of Cultural Change*, Oxford: Blackwell.

Heidegger, M. (1971/1951) 'Building dwelling thinking', in *Poetry, Language, Thought*, New York: Harper Colophon Books.

Hugo, V. (1865) *Notre-Dame de Paris*, Paris: J. Hetzel et A. Lacroix.

Kofman, E. and Lebas, E. (eds) (1996) *Writings on Cities. Henri Lefebvre*, Cambridge, MA: Blackwell Publishers.

Lefebvre, H. (1991/1974) *The Production of Space*, Oxford: Blackwell.

Marescotti, L. (1979) *Urbanistica: Saggio Critico, Testimonianze, Documenti, Bibliografia Ragionata*, Milano: Accademia.

Marescotti, L. (2008) *Urbanistica. Fondamenti e Teoria*, Sant'Arcangelo di Romagna: Maggioli.

Morris, M. (ed.) (1910–15) *The Collected Works of William Morris: With Introductions by His Daughter May Morris*, vol. 24, London: Longmans, Green and Co.

Nicolis, G. and Prigogine I. (1989) *Exploring Complexity: An Introduction*, New York: W.H. Freeman.

Repubblica Italiana (2001) 'Legge Costituzionale 3. Modifiche al Titolo V della Parte Seconda della Costituzione', *Gazzetta Ufficiale*, 248.

Sassen, S. (1994) *Cities in a World Economy. Sociology for a New Century*, Thousand Oaks: Pine Forge Press.

Steffen, W., et al. (2015) 'Planetary boundaries: guiding human development on a changing planet', *Science*, 347:6223 1–10.

Part 5
Rethinking the right to the city

Michael E. Leary-Owhin and John P. McCarthy

Lefebvre was a serious student of the 1871 revolutionary Paris Commune and an instrumental force in the 1968 Paris civil disturbances and subsequent putative 'revolution'. He took an active part in resisting Hitler's fascism during WW2 when the issue of civil rights became paramount. Given this journey we should perhaps not be too surprised that his reaction to Paris 1968 was configured around the issue of rights, more precisely the right to the city (RTC) and urban life. Perhaps even more than Lefebvre's ideas regarding the production of space and the spatial triad, his RTC propositions seem to have captured the imagination worldwide of academics, politicians and activists in the last decade. Lefebvre seemed to delight in providing only indicative sketches of what RTC may mean. We can be sure though that it subsumes the right: to city space, an urban life and to participate in and engage directly with the production of space. The contested understandings of RTC provide ample space for productive debate and a range of strongly expressed views. Advocates seize on RTC as a vital set of governance principles and hopes for urban citizenry, but critics bemoan its vagueness and the usurping of RTC by conservative interests and privileged neoliberal urban elites. Rather than simply a dogmatic reaffirmation of what is after all a highly malleable concept, this part allows a more critical exploration of what RTC might mean, why it continues to fascinate in the early 21st century and if it has outlived its usefulness.

Part 5 commences with a challenge that will unsettle anyone with a naïve devotion to Lefebvre's RTC ideal. Lopes de Souza delivers a wide-ranging critique of Lefebvre's RTC grounded in a complex claim regarding Lefebvre's neglect and underestimation of nature and the relationship of this to Lefebvre's vagueness about a post-capitalist mode of production and post-capitalist politics. Lefebvre is seen in the chapter as a brilliant thinker but one blemished by an intellectual inertia. And although Lopes de Souza sees some value in the RTC, he regards the concept as inspirational but also insufficient and limiting. Emancipatory social-spatial change requires not just an affirmation of the RTC but a more radical rethinking of the spatial, social and environmental ramifications of 21st-century capitalism. Lopes de Souza is adamant that the fundamental basis of Lefebvre's RTC is flawed along ontological, existential and ethical-political lines. Connected to Lefebvre's RTC conceptualisation is his commitment to a utopian future, but his utopian thinking is regarded as timid in several respects. In the chapter Lopes de Souza links RTC with Lefebvre's theorisations regarding the – largely Eurocentric ideas – of the urbanisation of society that lack a radically clear formulation, spatially and politically of a capitalist alternative. Lopes de Souza offers a staunch affirmation that RTC offers fruitful opportunities but has also reached its limits of utility. He concludes it is necessary to discover how to move beyond Lefebvre.

Lopes de Souza's critique is a fitting literary foil against which to test the following seven chapters in Part 5. Lyytinen reveals the results of empirical research in Kampala, the capital city of Uganda. Her chapter examines if and how the RTC concept can be re-thought and applied to understand and support refugees with disabilities in the city. Two things characterise the chapter's approach: the acceptance that a radical meaning must be attached to the RTC and also that its meaning must be allowed to vary across time and space. Lyytinen investigates the issues through a five-faced lens of: access, participation, appropriation, occupation and use of city space. Access includes access to institutions and essential services. She is also sensitive to the fact that refugees with disabilities face additional struggles in respect to everyday life and physical mobility. The interview data in the chapter present first-hand insights into a variety of important problems and responses of refugees in Kampala. Urban inhabitance is seen as including participation in decision-making at all levels regarding the production of space and everyday life. Centrality is posited as a key right (but see Coleman's chapter in Part 6). In rethinking the RTC in the sub-Saharan context of refugees with disabilities, questions of scale, from the individual body to the city, become paramount. The chapter reveals that the granting of formal rights in Ugandan legislation is only part of what the RTC should mean.

Liu and Wong also situate RTC research in a Global South context. Their chapter engages with a variety of Lefebvre's ideas, especially the notion that a dominant discourse on space is dressed in an ideological cloak, making it appear beyond ideology. And this discourse can be deployed as rational aesthetic justification for a state-driven production of space based on the denial of certain rights in the quest for city branding and re-imagining. See the chapter by Haas in Part 1 for a complimentary discussion of state-led suppression of certain rights to the city. Liu and Wong refer to Lefebvre's observation that the violence of power brings forth the violence of subversion, in building the arguments that underpin their research. City growth agendas, the chapter asserts, tend to homogenise space and deny the right to difference. On the ground, this is witnessed in the state's desire to eradicate informal settlements to release the development value of the areas. An unjust two-tier system of rights to city is said to be evidenced in China, whereby the rights of rural migrants to settle securely in cities are denied, enforcing a precarious peripheral existence. The findings of the research point to the need for a new type of urban habitat politics that foregrounds equity-based development rather than state-led growth.

Migration across rather than within nation state borders provides Murrani, in the next chapter, with the chance to develop a Lefebvrian perspective on drifting, displacement and urban creativity. She uses Lefebvre's theorising to bring forth new thinking regarding the inherent complexity of the disrupted spatial mobility of displaced people. She introduces also the concept of, 'a constant state of migrancy'. Right to the city is then viewed through the lens of the lack of rights engendered by perpetual drifting and the emergence of the negotiation of new identities of locatedness and re-rootedness. Centrality and the periphery are also rethought and Murrani seeks to utilise Lefebvre's concepts of plasticity in understanding the production of the RTC at peripheries (see also the chapter by Coleman in Part 6). Displacement and drifting are argued to facilitate urban creativity as migrants continually adapt to the challenges and opportunities they encounter. Such creativity is viewed as a vital contributory factor to the plasticity that is regarded as the essence of the urban. Migration, mobility and drifting, far from being curtailed, are seen as instrumental for the production of urban space. Policymakers and governments are urged to see through the blind field and grasp this alternative insight into the inherent creative capacities of migrant inhabitants.

We stay with the theme of internal movement across borders for Tsavdarolgou's chapter. Here he seeks to assert the newcomers' RTC through the development of the concept of the production of appropriated common spaces in cities. This requires the conceptualisation of: common pool resources, community and commoning. Tsavdarolgou states that refugee common space can be seen as open communities of commons that destabilise discriminatory state policies. It requires a reworking too of the RTC to incorporate a broad range of rights including the right to decent nutrition (see Hilary Shaw in Part 6). The chapter explores emerging migrant practices that, with the support of local solidarity groups in Greek cities, are perceived to produce self-organised collective housing commons spaces. Interview data are used by Tsavdarolgou to build a narrative of resistance to the various inadequacies of state-run refugee camps. In common with the chapters by Murrani, Liinamaa and Lyytinen, migrants are appreciated not as passive but as active agents in the production of urban space. In contrast to the segregated state-run camps, the appropriated common spaces are comprehended as places of diversity where the right to decent housing stimulates the demand for a range of other rights to the city.

In ways that compliment Tsavdarolgou, Bailey illuminates the right to difference as a key factor in the RTC. He also stresses that the RTC must challenge the ideology of participation which Lefebvre argues tends to dull the self-consciousness potential of urban residents to become urban inhabitants. Bailey provides insights into the debates regarding the nature of and potential for RTC to challenge rather than prop up continuing state implicated neoliberal urbanism (see Haas in Part 1). Bailey then explores RTC through three case studies based in England. In common with other critics of the state guarantee of rights to the city, Bailey sees little evidence of a shift in the balance of power in the production of space, away from major property developers and house builders, towards poorly resourced local residents. He asks how far the rights discussed constitute mere procedural concessions, as opposed to genuine contestations of unrestrained neoliberal, profit-oriented urban development. While Bailey sees some affirmation in the case studies for Lefebvre's optimism of an urban society, he wonders if RTC would be more effective if transformed into the co-production of space.

In the subsequent chapter by Hilary Shaw, we remain in the UK but switch the focus to the issue of food poverty. That said, Shaw is mindful that food poverty and the related problems caused by obesity are class-based issues evident also in the Global South. Shaw rethinks the RTC by highlighting the right to food as enshrined in the 1946 Declaration by the United Nations. Shaw identifies access to healthy food and nutrition inequality, i.e. the denial of the right to food, as a significant consequence of neoliberalism. The chapter explicates the spatial aspects of this inequality through an analysis that draws on the spatial triad. Shaw reasons that Lefebvrian space shrinks in an inverse relationship with the expansion of the belly due to obesity. Shaw follows Lefebvre in postulating that those who produce the city through their labour, especially the urban poor, should have the right to reframe urban decision-making. Regarding healthy food and nutrition, Shaw suggests what is needed is a Foodfare State; an element of which would involve encouraging those suffering from poverty to grow and sell food without forfeiting state financial benefits. A culture of autonomous food production would result, that would help secure the right to healthy food and a healthy body.

To close Part 5 we continue the theme of rethinking the RTC in a largely theoretical contribution by Diaz Cordona. He provides a hermeneutic-based discussion of the role of writing in the RTC and two other of Lefebvre's principal texts. He avows that writing and the notion of the text has an insecure place in Lefebvre's work on space. Lefebvre is referring to a type of writing in particular; writing the city or the city as text. At times it seems to affirm things Lefebvre holds dear, such as everyday life, lived space and urban politics. At other times, writing (or its privileging) and the text appear to morph into an enemy of space, becoming signs to be read that may mask and distort concrete reality. Diaz Cordona contends that Lefebvre understands that language is needed to construct a relationship between the real material world and thought and then shows that writing is actually constitutive of the urban form. The challenge is to ensure that mental constructs i.e. signs do not dominate material forms and lived experience, to the detriment of society. Inhabiting runs the risk of being reduced to the status of a reading, evading both history and practice. Diaz Cordona wonders how Lefebvre would react to the explosion of writing the city work distributed by the Internet. And he asks, in the digital age, given the potential of writing to contribute to liberation and the production of differential space, should writing not enter into the rights to the city?

Preliminary conclusions and thoughts

Unsettling or consolidating Lefebvre's ideas

It is clear from this part of the book that a range of opinion still exists regarding Lefebvre's RTC ideas. But a central question would seem to be, not should the concept be ditched but how can it be sharpened and adapted for the present epoch? A variety of researchers are actively deploying Lefebvre's ideas to reveal resistance to the production of abstract space under conditions of neoliberal urbanism, so does it matter that his ideas sometimes constitute sweeping grand narratives that lack precision, or is this part of their liberating strength?

Beyond centrality, the 'city' and the nation state

It is interesting to see how the city itself becomes problematised through the lens of the RTC. Is centrality still crucial to the RTC? How are identities shaped at the nation state and city levels? Has access to services become a vital RTC? When applied to intra- and international mobility and displacement, to what extent does the range of rights as envisaged by Lefebvre need to be rethought and reconfigured? Should and how might migrants and their rights be seen as integral to successful cities?

The distant future and the here and now

RTC requires us to consider temporal dynamics and human agency. When can we expect potential RTC to be fulfilled? Who should we expect to deliver the RTC? Is it to be achieved indirectly through democratic processes or appropriated directly by those most in need and affected by injustice, and who may lack the means and inclination? And if it may be achieved in the longer term, what may be achieved in the short to medium term? We continue to develop some of these themes in the final part of the book.

33

Right to the *city* or to the *planet*?

Why Henri Lefebvre's vision is useful and too narrow at the same time

Marcelo Lopes de Souza

Introduction: a brilliant thinker and his contradictions

The Henri Lefebvre whom so many people today admire emerged only in the second half of the 1960s. Lefebvre was for almost half of his adult life affiliated to a Stalinist party – the French Communist Party (PCF); he only moved away from that party because he was suspended after three decades of belonging to it. Born in 1901, he joined PCF in 1928, at a time when it was already showing a clearly Stalinist line and a tendency to be subservient to Moscow's orientation, both aspects consolidated at the beginning of the following decade. It was only in 1958, when he was finally suspended, that Lefebvre definitely broke with the party. Even the late Lefebvre – the celebrated author of important works such as *The Right to the City*, *Everyday Life in the Modern World*, *The Urban Revolution*, *Marxist Thought and the City* and *The Production of Space* – however, was not free from contradictions and ambiguities.

Henri Lefebvre, it is important to stress, was a philosopher who at the same time as he began to speak (in a rather vague way) of '(generalised) self-management' (*autogestion* or *généralisée*) (Lefebvre 1983; 1998; 2009) from the 1960s onwards, disdained – as a typical heir of the Marxist tradition – Pierre-Joseph Proudhon and the anarchists in general, the first and most notable practitioners and promoters of the principle behind the word *self-management*; moreover, he seems to have spared the so-called 'self-management' in Yugoslavia of Marshal Josip Broz Tito, who for three decades and a half commanded a light version of 'bureaucratic socialism', from any severe objection. Furthermore, Lefebvre is also the thinker who after criticising the working class for the anachronism of their organisations and discourse, found it difficult to value praxis and popular creations without great reservations because of Leninist/vanguardistic remnants which made him believe that only the critical intellectuals would have a vision of totality and '*œuvre*' (as opposed to mere 'product'), which would escape the mass (Lefebvre 1991: 144; see also Lefebvre 1971 and 1983). As we see, from a political-ideological point of view, there is no shortage of problems in his body of work.

Apart from Lefebvre's hesitations and political concessions during the three decades of his membership in the PCF (and to some extent even later), there are a whole set of more subtle and philosophical problems. One of the gaps in Lefebvre's work, and which will serve here as the central axis for discussion, is the relative neglect or underestimation of nature. This relative neglect, however, is neither accidental nor the sign of a pure idiosyncrasy. This kind of gap is closely related to Lefebvre's closeness to Marxist *mainstream* in some sense, in spite of his proverbial (late) heterodoxy in some other senses.

Interestingly, in the 1980s and even in the 1990s, Henri Lefebvre and his work were not much more than objects of academic reflection (and sometimes of worship). And then, at the beginning of the 21st century, there was an explosion of interest: protests and various movements – against gentrification, for more and better public spaces, against obscure business and real estate deals, against mega-sporting events and their consequences for residential segregation and waste of public resources and so on – in Europe and the United States, and soon thereafter in a number of other continents and many other countries – began to have in the right to the city a convenient slogan. As it could have been easily expected, this debate has fed back into academic curiosity and interest. But it also encouraged the willingness of governments and non-governmental organisations (often states' charming appendages within 'civil society') to use the same expression to embellish the discourse of official programs and 'social inclusion' projects. In other words, a wave of co-optation and banalisation has accompanied the worldwide diffusion of the 'right to the city' motto. Increasingly distant from its origins as a radical demand (the right to the city as the *right to full enjoyment of the wealth and culture socially generated and concentrated in the cities*, which presupposes, according to Lefebvre, *another society*), that slogan has been appropriated in a politically debilitating way by disparate agents – from the partnership called 'Cities Alliance', formed jointly by the World Bank and UN-Habitat, to the Brazilian Ministry of Cities – not infrequently for purposes of legitimising the state apparatus and parastatal policies and interventions.

Regardless of the misrepresentations of the idea of the right to the city, my thesis is that it is a valid but *insufficient* one – and if taken as possessing absolute value, also a *limiting* one. It surely deserves to be rescued from attempts to politically emasculate and impoverish it, but on the condition that it is *radically recontextualised and treated as an aspect of a broader and deeper political-social project*. Although it is to a large extent valid and daring, its own daringness also has intrinsic limits within the theoretical and political-philosophical framework of the Lefebvrian body of work. If we look carefully at the body of work in question, we shall see, if we are open to this possibility of interpretation, that the great French philosopher thematised less profoundly or strikingly than it would be desirable a whole range of subjects, such as the problematic nature of capitalist technology, the specific political-organisational forms of the post-revolutionary society of the future and hence the questionable meaningfulness of a 'socialist state' and of political parties – which are formidable challenges and ultimately impasses for any consequent Marxist state. The right to the city is a perspective that brings with it a nucleus of indisputable validity, but that ends up being limited and limiting, and this limitation is expressed in the shape of an underestimation of the scope of the problematic of nature and of the anti-ecological character not only of capitalism but indeed also of typical Marxism itself. This will be manifested in the weaknesses of Lefebvre in relation to his critique of the spatiality and technology of 'bureaucratic socialism', and even regarding his critique of capitalism, not to mention the vagueness in terms of sketching out alternatives.

Emancipatory socio-spatial change can be by no means only a matter of a 'right to the city' – not even within the interpretive framework of the Lefebvrian concept of 'the urban' (*l'urbain*), whose scope goes beyond the geographical entity called 'city'. What is at stake here is the path to a renewed radical theory, as well as to more effective forms of struggle against the capitalist system at the global level. What is at stake, therefore, is the need for a questioning and a practical confrontation of the capitalist (anti)civilisatory model on a planetary scale, which, in my conviction, implies facing up to the challenge of thinking and overcoming: 1) the state apparatus and statism (capitalist or 'socialist') as well as political parties and actually all hierarchical, bureaucratic and vertical modes of collective organisation; 2) the capitalist technological matrix and spatiality; 3) the capitalist ideology of 'economic development' and 'domination of nature', with its economistic/productivistic, Eurocentric, teleological and rationalist premises (somehow shared, albeit in a distinctive and recontextualised way, by typical Marxism). Clearly, at least from a perspective nurtured by the reality from the second half of the 20th century onwards, a critique against the technological matrix and the spatiality inherited from capitalism necessarily includes a radical critique of the anti-ecological aspects of the capitalist mode of production (i.e. its inherent drive for capital accumulation).

It is thus necessary to rethink a whole series of questions regarding spatial organisation (pointing to radical economic-spatial deconcentration and territorial decentralisation, but without slipping to localisms and insularisations), exploitation, alienation and the social division of labour (in which trends of precariousness and 'hyper-precariousness' in the world of labour must be highlighted), ethnocentrism (whose renewed facets in terms of xenophobia, nationalism and racism must be vehemently denounced), the various types of oppression (class, gender, etc.) and heteronomy in general. In order to meet these intellectual and political requirements, a 'right to the city' and an 'urban revolution' are not enough, as inspiring as these ideas undoubtedly can be. Lefebvre's relative neglect of nature is inextricably connected with his ambiguities and vagueness in terms of crucial discussions related to spatial deconcentration and territorial decentralisation, and ultimately to the specific politico-organisational forms of the future – or in other words, to the relationship between a post-revolutionary mode of production and post-revolutionary decision-making processes.

Beyond (but by no means against) the 'right to the city': Lefebvre's limits from a left-libertarian and politico-ecological perspective

Geographer Neil Smith drew attention in a brilliant text to several shortcomings of Lefebvre's formulations, analysing the book *The Production of Space* (Lefebvre 1981 [Smith used the 1991 English edition]). Among the problems detected by Smith is an underestimation and sub-systematisation of what he called the '(social) production of nature'. He argues, after commenting that 'an unresolved contradiction between ontology and history drives much of Lefebvre's vision' (Smith 1998: 51–52) – adding further that, '[t]his is undoubtedly reaffirmed by a sometimes quixotic resort to a philosophism that he himself, in other contexts, would critique' – and that, 'it leads to an uncharacteristically simple dismissal of an unreconstructed nature' (Smith 1998: 52). A few pages later, he mercilessly states:

> Lefebvre's rethinking of nature is poor – far less original than his radical repolitization of space which, in fact, carries along his conceptualization of nature. He leaves nature largely unreconstructed, and with it the relationship between space and nature.
>
> *(Smith 1998: 59)*

According to Neil Smith, '[n]ature for Lefebvre is on the verge of becoming a corpse at the behest of abstract space'. Or, as he immediately adds, for Lefebvre nature, unlike space, 'retains virtually no initiative to and for itself' (Smith 1998: 59). For Smith, such a reading would attribute 'a surprisingly undialectical negativity to the treatment of nature'; actually '[n]ature is reduced to little more than a substratum':

> He [Lefebvre] even adopts the Frankfurt School argument that it is 'domination' of nature and not simply its appropriation, as Marx usually put it, that we witness today. But there is

> little hint of the more active political initiative that the Frankfurt School attributes to nature via the 'revenge of nature', or, in place of that admittedly nostalgic construction, some other sense of a politics of nature. Quite the opposite. Nature is increasingly the raw material of mimetic productions of space in which 'what are produced are the signs of nature or of the natural realm'; 'nature is left, as it were, in a no-man's land'.
>
> *(Smith 1998: 60)*

It seems, however, that Neil Smith's premises and Marxist formation acted as a brake here and there, preventing him from going even further in his criticism – which would beyond a certain point necessarily become reservations *about Marxism itself.*

Smith is quite right in objecting to natural scientists' insistence on a convenient fiction, one that disconnects the study of nature from the cultural-social-historical (and psychological) conditions of the generation of this knowledge. As we know, many natural scientists even commonly perform an 'epistemological and methodological imperialism', by presuming that the study of society should not differ essentially from the study of non-human nature – or if and when it differs, it differs because it is still 'backward' and 'immature' (in the face, of course, of the canons established by sciences such as physics). On the other hand, Marxism (and indeed a large part of the social sciences tradition, albeit in varying ways) tends to convince us that the picture must be completely reversed: 'nature' is itself a kind of chimera, unattainable and not knowable in itself, since everything, all the time, is already mediated by human perception, culture and history. There would be no observed object that would not always bring in its description and explanation the marks of perception by human observers, who are always historically and culturally situated; the very language we use to express any idea connected with nature would already demonstrate it. There would be a 'first nature', yes, but since it is knowable by us only through the mediation of culture and history, it would never appear before us properly as 'first', but always as 'second', even if in some places nature gives us the impression of being very little altered, not significantly artificial or even pristine. Hence this 'first nature' is, for the typical Marxist, of little interest – or even no interest at all. Be that as it may, Neil Smith criticises Lefebvre, and rightly so; but he would have been much more precise if he had realised that in this regard (as in several others too) the 'heretic' Lefebvre was simply being a coherent Marxist.

For Lefebvre, 'first nature' is little more than a residue, as Neil Smith acknowledged. For the former, the production of space is, in the end, ceaseless *reproduction* of space – incessant reproduction and retransformation of the 'second nature'. 'First nature' would be linear and uninteresting in itself. (Lefebvre here, incidentally, follows especially the tradition of 'Western Marxism'.) This neglect and contemptuous treatment of nature resembles a kind of 'revenge' against the arrogance and prepotency of the natural sciences; but how productive and balanced could such an epistemological and ontological retaliation be? Despite all dialectics, this view is simplistic; it restores a dichotomy 'society/nature' whose overcoming was already glimpsed over a century ago by anarchist geographer Élisée Reclus under the sign of the maxim '*l'Homme est la nature prenant conscience d'elle-même*' (Reclus 1905–1908).

Lefebvre's view – which is intertwined with his difficulty to move beyond a right to the city towards a broader 'planetary' approach – is flawed for three main reasons:

1) *Ontological*: 'first nature' is ineliminable and can never be reduced to a kind of 'residue'; what is necessary is to verify *how* it is constantly historically recontextualised, politically reassessed, materially reworked/impacted and culturally-symbolically re-signified, often to the point of posing threats to the well-being and even health of entire population groups within the framework of its instrumentalisation to heteronomous purposes, or on the

basis of heteronomous social relations (in the sense systematically explored by Bookchin (2005)).

2) *Existential*: global climate change shows how the ecological problem (or rather the ecological-metabolic dimension of the social crisis generated by capitalism) should not be underestimated by anti-systemic forces. Marxist productivism, as tributary of the capitalist imaginary, needs to be challenged – and this requires more than a simple criticism of Lefebvre, something Neil Smith certainly was not willing to admit. It is necessary to see that taking this challenge seriously has nothing to do with reviving Malthusianism or throwing oneself into the arms of a socially regressive ecocentrism compatible with conservative and even reactionary ideologies.
3) *Ethical-political*: ignoring the socio-environmental struggles of social groups and sometimes whole peoples (especially in the 'Global South', but not only) reveals a lack of political and ethical-cultural sensitivity. It is, strictly speaking, a Eurocentric stance, it matters little whether through ignorance or choice.

Having as place of enunciation invariably the big city, many simplifications have already been committed by many Marxist urban researchers, with biases not only exaggeratedly 'rural-phobic' but also Eurocentric. The process of economic, political and cultural domination of the countryside by the (large) city under industrial capitalism, brilliantly described and controversially exalted by Marx and Engels in the *Communist Manifesto* (Marx and Engels 1982), reflects both the 'colonisation' of the countryside by the city in the wake of the rise of capitalism and the colonisation imposed by the Christian and urban-industrial West, even before industrialisation, (from the 16th century) on other peoples and cultural matrices – and with this the coloniality of hegemonic knowledge itself. It is worth recalling how, in the *Communist Manifesto*, the most emblematic passages about the domination of the countryside by the city and the domination of nature by the triumphant bourgeoisie are separated from each other by only one paragraph, showing, when they are uttered in one breath, that they are two aspects of the same process (Marx and Engels 1982: 111).

At this juncture, we have reached the point where it is necessary to focus, although succinctly, on one of the most controversial questions for current critical theorisation, especially for political ecology: was Marx insensitive to environmental degradation or, on the contrary, was he an ecologist *avant la lettre*? Authors such as Massimo Quaini (1982) and Alfred Schmidt (2016) tried to convince us that there was an 'ecological Marx' that has been underestimated (Schmidt even makes a self-criticism in this regard), while superficial critiques on the contrary seek to demonstrate that Marx ignored the theme of 'environmental protection' altogether. Both approaches to the matter are severely biased and therefore wrong. Authors such as Quaini and Schmidt, in fact, only demonstrate that Marx did have some conservationist concerns, although Schmidt himself admits that these appear only sparsely in his work. As a shrewd observer of 'man's and nature's metabolism' (*Stoffwechsel von Mensch und Natur*), Marx revealed 'ecological' concerns at times when he observed phenomena of devastation caused by irrational uses of the soil or by the sheer logic of capital. However, this does not invalidate or neutralise what sets the tone for his work: productivism-economism, 'domination of nature', rationalism and finally relatively uncritical urban-philia and a somewhat exaggerated rural-phobic (on these problems, Cornelius Castoriadis provided abundant material, full of evidence and backed by solid argument: (see, for example, Castoriadis 1975; 1978a; 1978b and Souza 2006; 2012). As for this discussion, the relevant question is not 'did Marx pay attention to specific environmental problems?' but rather 'what place do environmental problems generated by capitalism occupy in Marx's work?' Considering that Marx, as a 19th-century author, was a son of the European

Enlightenment and of the industrial revolution, the fact that he did not extensively deal with environmental problems or cultivate 'ecological' concerns is somewhat understandable, just as it is not difficult to understand the ambiguous ideological relationship he maintained with the productive forces inherited from capitalism in the midst of a teleological scheme based on a supposed 'dialectic' between productive forces and production relations. Far less understandable, however, is Lefebvre being unclear or reticent about these points.

The classical 'city-countryside opposition' should not be oversimplified, since the interdependence between activities and the interpenetration of types of space have always been real and complex. It cannot be denied, however, that at the dawn of capitalism the differences between countryside and city were evident (to begin with, as regards the typical social classes: in the countryside, traditional landowners and peasantry; in the large cities, the nascent proletariat and industrial bourgeoisie, with the latter eclipsing the merely commercial bourgeoisie and the former very often replacing the artisans).

On the basis of Marxian historical materialism, Henri Lefebvre sought to reflect on the late tendencies of the process initially described by Marx and Engels at the dawn of modern capitalism, showing the contours that it presented in the second half of the 20th century: according to him, a process of becoming an *urban* and no longer only an industrial society. As he says in *Space and Politics*, '[t]he whole society becomes urban' (Lefebvre 1976: 67). It is the thesis of a complete urbanisation of society as a tendency, in which the countryside becomes a kind of less dense version of the city, but completely urbanised from the point of view of culture, social relations (in terms of production relations and social classes) and technology. The contrast between the city and the non-city (the 'countryside') would be much more quantitative than qualitative.

Nevertheless, although the differences between the countryside and the (big) city suffer a leap in quality, first with the advent of industrial capitalism, and in the 20th century with the gradual emergence of what Lefebvre calls 'urban society', many nuances and complexities remain. Yes, it is true that, on the one hand, a growing percentage of humanity lives in cities – according to UN data, more than half (54% in 2014: see United Nations 2015). But it is also true, on the other hand, that a probably small but not insignificant portion of that population lives – especially at the (semi)periphery of global capitalism – in cities whose strict and strongly urban character is debatable. This seems to suggest that data like those of the UN are probably exaggerated. In Brazil, for example, where official data show more than 80% of urban population, several scholars began to draw attention to the underestimation of the presence of rurality in the Brazilian socio-spatial fabric (see, for example, Bitoun et al. 2015).

This brings us to an interesting review of the Lefebvrian analysis: if on one hand, in general, and especially in the Global North socio-spatial contexts, the countryside has undoubtedly increasingly urbanised. On the other hand, there are also 'cities', especially in the 'Global South', which still bring strong brands of rurality and pre-capitalist traditions. It is not a question of proposing a disqualification of the analyses (by Lefebvre and others) that indicate an 'urbanisation of humankind'. It is a question of relativising the Lefebvrian thesis of an 'urban society' and of a 'complete urbanisation of society', We see the persistence and reproduction of phenomena which, despite the presence of globalised capitalism in all corners of the world (penetrating and transforming the countryside and the villages, as well as small and medium-sized cities), continue to exist and challenge the complete economic and cultural pasteurisation of humankind and planetary space. Lefebvre himself had already conceded that in the countries of the periphery and semi-periphery of capitalism there would be a very special complexity, due to the coexistence of distinct spaces-times (the urban, industrial and agrarian eras) (Lefebvre 1983). But the Lefebvrian reading itself induces some over-simplification; it invites a Eurocentric look, which has influenced academia even in the countries of the Global South.

Neither urban-philic nor urban-phobic

Lefebvre's thought, though committed to a utopian vision (utopia in a positive sense, namely of seeking a tomorrow that is radically different and better from today, but without underestimating existing constraints), is timid in several respects, however. It is characteristically urban-philic, and this urban-philia is none other than an expression of how even the Lefebvre that emerged in the 1960s, although detached from the French Communist Party, remained close to various aspects of the Marxist mainstream. Lefebvre continued to belong to the Marxist tradition, with all its merits but also with its limitations and vices. And it is to this tradition, of which Lefebvre was a representative, that we must address our reservations in the first place if we are to have greater political and intellectual confidence in the objections we raise. Actually, one of the factors that makes Lefebvre interesting in this context is precisely the fact that he was a brilliant thinker and that he was seen by other Marxists as a remarkable heterodox.

Prisoners of the dichotomy 'city/countryside' for a long time (a dichotomy that is the sister of 'society/nature'), we do not realise that the thesis of the 'urbanisation of society', although largely valid, is somewhat Eurocentric; moreover, it does not solve the problem of cultivating as a politico-social goal a spatially radically different society from the capitalist one. Marxists – beginning with Marx himself, in the spirit so clearly announced by the *Communist Manifesto* – have accepted and admired the material and technological achievements embedded in capitalist urbanisation rather than systematically reflected on socio-spatial formats that promise their surpassing. Sure, already Marx's closest friend and collaborator, Friedrich Engels, addressed this problem in a clear way, as in the *Anti-Dühring* (Engels 1990), but his message was almost forgotten among subsequent Marxists. Symptomatically, Engels was sharply criticised by Lefebvre himself in *Marxist Thought and the City* (Lefebvre 1972). Taking heed of problems – such as the petty-bourgeois moralism of young Engels in *The Condition of the Working Class in England*, where large cities are seen (sometimes in a simplistic way) as places of a supposed 'moral degradation' of the working class (prostitution, alcoholism), and also considering Engels' tendency (sometimes poorly conducted, but also at times productive) to value the 'philosophy of nature' and to insist on the challenge of reflecting on an spatial organisation alternative to capitalism – Lefebvre violently attacked the *Anti-Dühring* as well as *The Condition of the Working Class in England* and even *The Housing Question*.

Among the left-libertarians, the approaches were not homogeneous either, but there were at least no traces of that kind of self-censorship and lack of interest in thinking about the themes of 'environmental protection' and overcoming the capitalist spatial organisation that until at least the mid-20th century were still so characteristic of Marxist thought. Anarchist geographer Piotr Kropotkin, for example, offered us precious insights about economic-spatial deconcentration and territorial decentralisation (see, for instance, Kropotkin 2002), which constitute a formidable source of inspiration. If Kropotkin took a slightly 'urban-phobic' stance at times, the same cannot be said about his friend and comrade Élisée Reclus, who examined and deplored the miseries of an unjust society, spatially condensed in large cities (e.g. segregation, poverty and insalubrity) (Reclus 1905–1908), and pioneered a kind of dialectical and critical conservationist approach (Reclus 1864; 1898); at the same time, however, he was by no means simplistically 'urban-phobic': without turning to 'urban-philia' in a Marxist style, he nevertheless was unimpressed by the formation of metropolises and sought to open our eyes to pragmatically (even if critically) dealing with this circumstance (Reclus 1895; 1905–1908). These classical contributions by Kropotkin and Reclus were systematically enriched in the second half of the 20th century by neo-anarchist Murray Bookchin (see, for instance, Bookchin 1974; 1992; 2004; 2005; 2007), among others.

In spite of all the classical and contemporary contributions, much remains to be done – and in this regard Lefebvre's contribution is ultimately more a curb than a stimulus. It is contradictory and insufficient: it opposes capitalism, but its somewhat ambiguous rejection of bureaucratic 'socialism' leaves us without a clear and operational view on alternative technology, spatial deconcentration/territorial decentralisation and the need to reduce eco-stress ('entropy') – themes on which the left-libertarians have made a significant contribution. Without intending to provide any blueprint or recipe, but rather indications and orientations, Bookchin's contributions on social ecology in general (Bookchin 2004; 2005; 2007) and on the topic of 'urbanisation without cities' in particular (Bookchin 1992; see also Bookchin 1974) offer a background that seems especially rich and stimulating, against which alternatives can be thought and tested. He also made advances in the task of (re)thinking the problem of technology (Bookchin 2004; see also Castoriadis 1978a) and also reflected on the possibilities of emancipatory action and organisation in the contemporary world. This is important considering the challenging room of manoeuvre provided by the spatiality of today's large cities, such as the strategy of 'libertarian municipalism' (Bookchin 1992; 2007), which despite controversial aspects, should not be underestimated, much less ignored.

When we compare this rich debate, developed over a century and a half, with certain discussions of the last decades, it is quite embarrassing to see how we have been able to ignore information, arguments and warnings already well-known. What current 'sustainable development' enthusiasts typically aim at is something similar to squaring the circle: ecologically and socially sustainable capitalism. But capitalism is by its own premises (imperative of growth/accumulation, exploitation of labour, heteronomy) ecologically and socially *unsustainable*. Or rather, it is only made sustainable on the basis of huge doses of alienation and repression, in varying combinations according to country and moment. As far as 'real socialism' is concerned, in which heterodox Marxists insist on seeing only a 'misrepresentation' of the 'true Marx' (and not the realisation, albeit caricatured, of certain problems embedded from the beginning in Marxian thought), it has never been a real alternative, and academic Marxism (Lefebvre included) has always had great difficulties in reaching the ultimate consequences of rejecting the authoritarian and productivist aspects of the Marxian legacy.

A new socio-spatial order (which includes, of course, the political-ecological dimension) requires much more than a right to the city: it demands a *right to the planet*. This right to the planet, it is worth adding, must be based on *individual and collective autonomy* (Castoriadis 1975; 1983; 1990; 1996; see also Souza 2006) as a principle, rather on premises such as 'democratic centralism', 'dictatorship of the proletariat' and the building of a 'socialist state' which resulted in Stalinist totalitarianism but were already present in Bolshevism and at least in the 'authoritarian' facets of Marx's thought itself (see, on the 'authoritarian'-heteronomous component of Marxian thought, Castoriadis 1983, among other works; also the objections raised by Bakunin amidst his struggles with Marx in the context of the First International, as well as Proudhon's premonitory critiques). This right to the planet, on the other hand, presupposes more than an 'urban revolution', it presupposes a *socio-spatial revolution* founded on the rejection of Ethnocentrism, exploitation and all types of oppression (in short, of heteronomy). Several of the main components of the socio-political project of a 'right to the planet' were explicitly explored in Souza (2015), but they are also implicitly present on the previous pages: a deep and balanced concern with what we could call ecological prudence, with social rights and with socio-spatial inequalities, all this implying a radically anti-capitalist reorganisation of space. A much more visceral questioning, therefore, of the model of socio-spatial organisation inherited from capitalism than the one promised by the project of the right to the city. The details, it goes without saying, cannot be discussed in a brief chapter such as this; furthermore, it is necessary to emphasise that such an effort of imagining a radically alternative scenario will be essentially collective (also at the intellectual level, not to mention the praxical one) – that is – it cannot be a purely theoretical and individual exercise.

Conclusions: an overestimated legacy?

It was only several years after the death of Henri Lefebvre and a few decades after the publication of books such as *The Right to the City* and *The Urban Revolution* that Lefebvre's ideas (or rather interpretations supposedly inspired by Lefebvre's ideas, starting with the 'right to the city') began to be massively publicised at the beginning of the 21st century. This occurred many years after they had already been intensely discussed not only in the Francophone intellectual environment but also in a semi-peripheral country such as Brazil, where some circles have focused on Lefebvre's work since the early 1980s. Typically, for the last four decades, something only happens (or exists) truly if it happens (or exists) in English, the *lingua franca* of the *Pax Americana*. And so, after the translation of Henri Lefebvre's major books into English, the English-speaking academic world sanctioned and opened space for the diffusion of certain debates to finally take place. At the same time, the diffusion of Lefebvrian ideas has been accompanied by an annoying and often embarrassing dilution of the original critical message. From institutions of the state apparatus to non-government organisations to multilateral organisations, everyone feels able to invoke, in an authoritative manner, a motto like the 'right to the city' – usually to reduce it, in the best of all cases, to a reformist perspective based on obtaining specific material improvements and some 'popular participation'. In this context, actions and even discourse show only a pale resemblance with the radicality of Lefebvre's critique. In contrast to that, Lefebvre left no doubt in questioning capitalist society itself, and not only some of its aspects and derivations (neoliberalism and lack of genuine popular participation).

However, Lefebvre was to some extent co-responsible for the seeming ease with which his formulations have been co-opted by the status quo, or transformed into an object of uncritical reproduction by a large part of the progressive academic world. After all, to a considerable degree his analyses remained vague, allowing therefore, various interpretations according to the 'customer's' taste. Above all, demands such as 'generalised self-management', the critique of the working class and the critique of the state remained superficial and very incomplete at an operational (and more explicitly political) level. This vagueness serves to foster both fruitful debates and illusions and self-deception about what the 'master' really meant. In this sense, Lefebvre's work, both for its virtues and for its defects, is potentially inexhaustible, since it can continue to feed layers and more layers of reinterpretations, glosses and re-readings (as it is also the case with Marx's body of work).

Although almost inexhaustible in the above sense, I venture to say that, in a way, the Lefebvrian message, on the other hand, could also be regarded as having reached its limits. As I have tried to show in this chapter, this is so not because it has become totally useless or because it no longer has anything relevant to communicate or inspire, but because it falls short of the needs of the present moment. I even venture to say that when it was formulated, between the 1960s and 1980s, this message already bore the germ of a certain anachronism, though mixed with a keen awareness of new trends. But it is now, when Lefebvre's contributions have proved highly influential and to a large extent fertile, even though they are excessively unilateral and somewhat Eurocentric, that it is necessary to know how to advance *beyond* Lefebvre. Concretely, that means moving beyond the assumptions and arguments proposed by him – or in some cases beyond the mental schemes of which he never knew how (or wanted to) get rid of completely. This advancement must benefit from the help of authors that the hegemony of Marxist thought and the 'Lefebvre cult' has left in the shadows, such as Murray Bookchin and his social ecology. Similarly useful are the left-libertarian contributions, which long before and after Lefebvre represent deeper and more consequential criticisms and indicate alternatives to capitalism and heteronomy in general in a more coherent and systematic way than the great French thinker did.

References

Bitoun, J., et al. (2015) 'As ruralidades Brasileiras e os desafios para o planejamento urbano e regional'. *Anais do XVI ENANPUR (Sessão Temática 3). Belo Horizonte.*
Bookchin, M. (1974) *The Limits of the City*, New York: Harper Colophon Books.
Bookchin, M. (1992) *Urbanization Without Cities. The Rise and the Decline of Citizenship*, Montreal: Black Rose Books.
Bookchin, M. (2004/1971) *Post-Scarcity Anarchism*, 3rd edn, Oakland, CA: AK Press.
Bookchin, M. (2005/1982) *The Ecology of Freedom: The Emergence and Dissolution of Hierarchy*, Oakland, CA: AK Press.
Bookchin, M. (2007) *Social Ecology and Communalism*, Oakland, CA: AK Press.
Castoriadis, C. (1975) *L'institution imaginaire de la société*, Paris: Seuil.
Castoriadis, C. (1978a/1973) 'Technique', in *Les carrefours du Labyrinthe*, Paris: Seuil, pp. 221–48.
Castoriadis, C. (1978b/1975) 'Valeur, égalité, justice, politique: de Marx à Aristote et d'Aristote à nous', in *Les carrefours du labyrinthe*, Paris: Seuil, pp. 249–316.
Castoriadis, C. (1983/1979) 'Introdução: socialismo e sociedade autônoma' in *Socialismo ou barbárie: O conteúdo do socialismo*, São Paulo: Brasiliense.
Castoriadis, C. (1990/1988) 'Pouvoir, politique, autonomie', in *Le monde morcelé – Les carrefours du labyrinthe III*, Paris: Seuil, pp. 113–39.
Castoriadis, C. (1996/1994) 'La démocratie comme procédure et comme régime', in *La montée de l'insignifiance – Les carrefours du labyrinthe IV*, Paris: Seuil, pp. 221–41.
Engels, F. (1990/1878) 'Herrn Eugen Dührings Umwälzung der Wissenschaft' ['Anti-Dühring'], in *Marx-Engels Werke (MEW), Band 20*, Berlin: Dietz Verlag.
Kropotkin, P. (2002/1899) *Fields, Factories and Workshops*. Online (12/01/2002). Available HTTP: <http://dwardmac.pitzer.edu/Anarchist_Archives/kropotkin/fields.html>.
Lefebvre, H. (1971) 'La classe ouvrière est-elle révolutionnaire?', *L'Homme et la société*, 21 149–56.
Lefebvre, H. (1972) *La pensée marxiste et la ville*, Paris: Casterman.
Lefebvre, H. (1976/1972) *Espacio y política: El derecho a la ciudad, II*, Barcelona: Península.
Lefebvre, H. (1981/1974) *La production de l'espace*, Paris: Anthropos.
Lefebvre, H. (1983/1970) *La revolución urbana*, 4th edn, Madrid: Alianza Editorial.
Lefebvre, H. (1991/1968) *O direito à cidade*, São Paulo: Moraes.
Lefebvre, H. (1998/1968) *L'irruption: de Nanterre au sommet*, Paris: Syllepse.
Lefebvre, H. (2009/1966) 'Theoretical problems of *autogestion*', in N. Brenner and S. Elden (eds) *State, Space, World*, Minneapolis: University of Minnesota Press, pp. 138–52.
Marx, K. and Engels, F. (1982/1848) 'Manifesto do Partido Comunista', in Marx/Engels *Obras escolhidas*, Moscow and Lisbon: Edições Progresso and Edições "Avante!".
Quaini, M. (1982/1974) *Geography and Marxism*, Oxford: Blackwell.
Reclus, E. (1864) 'L'Homme et la Nature: De l'action humaine sur la géographie physique', *Revue des Deux Mondes*, 54 762–71.
Reclus, E. (1895) 'The evolution of cities', *The Contemporary Review*, 67:2 246–64. Online (Facsimile Reproduction): Librairie Nationale Française. Available HTTP: <http://gallica.bnf.fr/ark:/12148/bpt6k660448.r=reclus.langPT>.
Reclus, E. (1898) 'La grande famille', *Le Magazine International*, pp. 8–12, January. Online (Facsimile Reproduction): Librairie Nationale Française. Available HTTP: <http://gallica.bnf.fr/>.
Reclus, E. (1905–1908) *L'Homme et la Terre*, vol. 6, Paris: Librairie Universelle. Online (Facsimile Reproduction): Librairie Nationale Française. Available HTTP: <http://gallica.bnf.fr>.
Schmidt, A. (2016/1962) *Der Begriff der Natur in der Lehre von Marx*, 5th edn, Hamburg: CEP Europäische Verlagsanstalt.
Smith, N. (1998) 'Antinomies of space and nature in Henri Lefebvre's *The Production of Space*', in A. Light and J.M. Smith (eds) *The Production of Public Space*, Lanham, MD: Rowman & Littlefield.

de Souza, M.L. (2006) 'A prisão e a ágora: Reflexões em torno da democratização do planejamento e da gestão das cidades', *Rio de Janeiro: Bertrand Brasil.*
de Souza, M.L. (2012) 'The city in libertarian thought: from Élisée Reclus to Murray Bookchin – and beyond', *City*, 16:1–2 4–33.
de Souza, M.L. (2015) 'From the "right to the city" to the right to the *planet*: reinterpreting our contemporary challenges for socio-spatial development', *City*, 19:4 408–43.
United Nations (2015) *World Urbanization Prospects – The 2014 Revision*, New York: United Nations.

34

'In a group you feel OK, but outside there you are ready to die'

The role of a support group in disabled refugees' struggles for their 'right to the city' in Kampala, Uganda

Eveliina Lyytinen

Introduction

In this chapter, I examine urban refugees' everyday life, and the associated feelings of insecurity and protection, in the capital city of Uganda, Kampala. My overall aim is to demonstrate the applicability of the idea of 'right to the city' (RTC) regarding the investigation of refugee protection in a group setting – a theoretical approach so far lacking in-depth investigation. I commence this chapter by deliberating on the original idea of the RTC and presenting my application of it. This conceptual part is partially based on my article published in the *Journal of Eastern African Studies* (Lyytinen 2015a). Yet I further develop my conceptual thinking of the RTC by focusing on the scale of the group, and subsequently base my analysis on new empirical data from RWDs.

The concept of the RTC, originally developed by Lefebvre (1996), has been a focus of much academic scholarship (Purcell 2003, Marcuse 2009, Attoh 2011, Butler 2012), and it has been increasingly used in advocacy by different organisations (Kuymulu 2013). Often the attempts to apply it have, however, lacked the radical meaning that was its original characteristic (Purcell 2002, De Souza 2010, Butler 2012, Kuymulu 2013). The RTC is interpreted in this chapter in an open manner as suggested by Iveson (2011: 258), who argues that 'what "RTC" means simply cannot, indeed should not, be answered in the same way in different times and places'. Rather, the RTC includes multiple, interrelated rights that form a collectivity of rights. These rights are interpreted in this chapter through three overlapping spatial domains: participation, access and appropriation (Lyytinen 2015a).

First, urban inhabitance, for Lefebvre (1996), is characterised by participation in decision-making over the production of urban space at all levels. Effective participation is based on mechanisms that are controlled by the urban inhabitants; not imposed on them. Participation therefore requires self-management, autogestion. In this chapter, disabled refugees' participation in decision-making regarding their protection is analysed as a collective right argued for in the support group and pursued through various methods of action to counteract individual and collective exclusion.

Second, for Lefebvre, the RTC in its most basic form is a right not to be expelled from the city centre (Mitchell 2003). The RTC, this 'transformed and renewed right to urban life' which is 'like a cry and a demand' (Lefebvre 1996: 158), refers to the freedom to physically access, occupy and use urban space. Moreover, the issue of exclusion is central to Lefebvre's writings, as he refers to the RTC as the right to information, the rights to use of multiple services and the right of users to make known their ideas on the space and time of their activities in urban areas (Lefebvre 1991, cited in Marcuse 2009: 189). In this chapter, I analyse disabled refugees' struggles to access, occupy and use urban space – both at the city level and at the level of protection institutions and services.

Third, appropriation is presented by Lefebvre as a means of counteracting oppression by complete usage of space (Butler 2012). For him (1996: 147), appropriation refers to the collective attempts to produce city space as a creativity, or 'oeuvre'. In this chapter, I largely interpret appropriation of space as the collective attempts by the disabled refugees to occupy and transform the insecure spaces of exile into places of protection. Subsequently, I argue, along Lefebvrian lines, that it is necessary to conceive refugees' right to use space simultaneously as physical, imagined and social, creating the spatial triad (Lefebvre 1974: 1991).

In his later writings, Lefebvre perceives the RTC more as a 'right to space', thus opening up the discussion regarding the scale of these rights (Brenner 2000, cited in Butler 2012: 139). In this regard, Parnell and Pieterse (2010: 149) argue for conceptualising the RTC through different scales ranging from the individual, household, neighbourhood or city to the freedom from external risks. Pushing this scalar thinking further, in this chapter I focus on the scale of the group. This is important as in urban contexts refugee protection is increasingly conducted in group or community settings (UNHCR 2009, Lyytinen 2013).

The analytical focus on a support group is associated with my answer to the important question of 'whose right'. I subscribe to Marcuse's (2009) interpretation where he distinguishes between the demand of the RTC by those deprived of basic material and legal rights, and the aspiration for the future by those discontented with life and the limitations it imposes for growth and creativity. The RTC is a right based on ethics, justice and morality, and it belongs to:

> those deprived of basic material and existing legal rights,... those discontented with life... and those directly in want, directly oppressed, those for whom even their immediate needs are not fulfilled: the homeless, the hungry, the imprisoned and the *persecuted* on gender, religious, racial grounds.
>
> *(Marcuse 2009: 190, emphasis added)*

Consequently, my empirical focus is on those refugees who are deprived of and/or discontented/alienated with their exilic life given their particular character not only as the persecuted but also as the disabled (or caretakers of the disabled). The analysis that follows the description of the data and field context focuses on disabled refugees' RTC as communal rights to participate in the activities of this group, to access and occupy urban space, and to appropriate insecure spaces into spaces of protection.

Data and field context

Uganda is a party to the 1951 Convention Relating to the Status of Refugees and its 1967 Protocol and is also a signatory to the 1969 OAU Convention Governing the Specific Aspects of the Refugee Problem in Africa. There have been some positive changes in refugee law and policy in Uganda, namely the implementation of the 2006 Refugee Act and the adoption of the 2009 UNHCR urban refugee policy, both of which recognised cities as legitimate places for refugees to reside. Despite these changes, 'what exists today – and for the foreseeable future – is a policy that focuses assistance and protection on refugees living in settlements, and not those refugees who chose, for various reasons, to live outside such restrictive spaces' (Bernstein and Okello 2007: 47). Thus, it can be argued that the protection of urban refugees in Uganda has not improved significantly despite the newly adopted legal frameworks and policies (RLP undated). Yet the number of refugees in Kampala had expanded rapidly in recent years. In July 2011, of the total 150,000 recognised refugees and asylum seekers in Uganda, more than 26 per cent were living in Kampala. Out of the nearly 40,000 registered refugees and asylum seekers in Kampala, 45 per cent were Congolese. Other major nationalities included Somali, Rwandan, Burundian, Ethiopian and Eritrean refugees (UNHCR 2011).

This chapter is limited to a distinct refugee support group, that of the refugees with disabilities (RWDs), which exists in relation to Refugee Law Project's (RLP) community-based work in Kampala. RLP is a centre for justice and forced migration established in 1999 under the School of Law, Makerere University in Kampala. Over the years, RLP has helped to establish 15 support groups that are directly and indirectly affiliated with it. These include, among others, groups for torture survivors, women, youth, elderly, disabled and male refugees who have experienced sexual violence. The experiences of members and of the RLP staff working with different groups are utilised to contextualise the analysis focused on the RWDs group. Here I use the shorthand of RWDs to refer to this support group for refugees with disabilities. In the official documents this group is, however, referred to as people with disabilities (PWDs), but I wish to emphasise the characteristic of the members as both refugees and as PWDs. It should also be noted that even though some people have been disabled since birth, others were disabled during the violent conflicts in their home countries, or at the time of living in exile. Thus, sometimes people's refugee background and disability were very much connected.

The qualitative primary data collected in 2011 with the RWDs group include semi-structured interviews ($N = 6$) and focus group discussions ($N = 1$) with the leaders and members of this groups. RLP staff, mostly Legal officers and Psychosocial officers, who work closely with the RLP support groups, were also interviewed ($N = 6$). Additionally, group meetings were observed and secondary data, such as the constitutions, were analysed. Most of the interviews and focus groups were audio recorded and transcribed. Transcriptions were coded in NVivo 9 software, and analysed with discourse analysis.

The RTC as participation

In various UN Refugee Agency (UNHCR) policies there is an explicit link between refugee protection and community, particularly in the urban settings of the Global South (UNHCR 2009). Yet, UNHCR has realised that providing protection through a community-based approach may be challenging as territorially defined community structures are often weakly developed in cities. Therefore, creating and recognising new types of communities is crucial in order to reinforce refugees' participation in their protection. One of these new community structures is the support groups – the focus of this chapter.

In the context of Kampala, the RLP has also realised the importance of 'empowering the refugees and giving the refugees an opportunity to participate in the design and implementation of RLP services to their communities' (RLP 2011a: 5). Furthermore, after recognising that most refugee populations have originated as a result of tribal conflicts, RLP was committed to prevent these conflicts from reoccurring in Uganda. One of the attempts to prevent this was the creation of support groups which were open to refugees from all nationalities and ethnicities.

As these groups were formed around the idea of shared experiences and challenges of the deprived and the discontented (Marcuse 2009), participation in them provided means to fight against ethnic conflicts. Refugee groups' protective nature should not, however, be taken for granted nor should the issues of competition and mistrust in and between these communities be dismissed (Lyytinen 2013, 2015b). Yet, at their best, these groups can provide their members essential spaces of protection and platforms for claiming their RTC as is demonstrated in this chapter.

In Uganda, refugees have a legal right to form associations based on the 2006 Refugee Act. Initiating a refugee support group in association with the RLP typically commenced by forming a core group that mobilised a larger interest group. Often the initiative of forming a new group came from the RLP officers who worked with the clients with similar needs and backgrounds. This was also the case with the RWDs group. Thus, it can be asked to what extent was this participation in the creation of the group imposed from above on the refugees. As according to Lefebvre (1996) participation requires self-management, autogestion, it was clear that in order to function well, the RWDs had to take ownership of the group. Self-management seems to have succeeded if we look into the growth of this group. The group was established in 2010 with just eight core members. In 2011 the RWDs had around 70 members, their own constitution and they had officially been registered as an association with the Kampala Capital City Authorities (KCCA). In 2015 this group already had 200 members. Thus, participation in terms of numbers had increased significantly over time.

Regarding participation, it is also important to discuss the question of 'participation in what?' For many RWDs the mere fact that they took part in the monthly meetings held at the RLP was important. Yet, it was participation in fulfilling the core objectives of the group that motivated the members. The objectives were not only agreed collectively, but the actions to achieve them were also conducted jointly. The main objectives of the RWDs included:

> provision of psychosocial support to refugee PWDs [i.e. RWDs] and their families; fighting stigma and discrimination against PWDs; providing a collective voice to advocate for PWDs' rights; advocacy for mainstreaming of PWD issues in organizational plans; promotion of self-reliance through income generating activities.
>
> *(RLP 2015, no page)*

Thus, participation included tangible activities that the members conducted together to achieve not only basic material and legal rights but also growth and creativity (Marcuse 2009). The strengths of this group regarding participation were twofold. On the one hand, they had been able to shift from the initial participation in meetings into participation in tangible joint activities. On the other hand, they had succeeded in combining the 'cry and the demand' (Lefebvre 1996) of both the deprived and the discontented/alienated (Marcuse 2009).

Despite the clear strengths of participation, the RWDs group also faced a number of challenges to function well and to attract new members. Even though the number of members kept growing, the group struggled with finding the most deprived RWDs. There were a number of reasons for this. First, since the majority of the leadership and members were men, their concern was not being able to reach the women, particularly single women with disabled children that often were living in the most precarious situation. Second, the overwhelming majority of the members were Congolese. This was partly due to the fact that the president of the RWDs was a Congolese man, and his personal role in forming the group was paramount. Activating the most deprived and alienated RWDs (Marcuse 2009) from other nationalities was, however, identified as a challenge to be addressed. Third, the group struggled with the issue of identifying 'real members' as their criteria for membership was not necessarily something to be seen – not all disabilities are visible or physical. Given this, self-identification became an important determinant of participation even though it was sometimes questioned by other members.

The RTC as access to city space

The RWDs support group faced challenges regarding their rights to access, occupy and use space (Lefebvre 1996) in Kampala. Thus, their RTC in its most fundamental manner was limited. In this section, I focus on two spatial difficulties expressed by the RWDs: to live and to move in their city of exile, and to access official protection institutions and public services. This analytical take is based on Lefebvre's writings where he refers to the RTC as the right to information, the right to use services and the right to use the centre (Lefebvre 1991, cited in Marcuse 2009: 189).

First, RWDs expressed challenges in living and moving in their city of exile. According to the RWDs, they were often encouraged by the officials working on their cases to live in the rural refugee settlements. Therefore, their mere existence in the city was questioned, and the officials struggled to know what to do with these 'undesirables' in the centre (Mitchell 2003). Yet, shifting spatially to the rural camps would have meant major obstacles for the RWDs, as explained a female group member:

> When you are talking about the settlement, in settlement you are disabled, and the life in the settlement is digging. So how are you going to dig when you are disabled? As disabled, how are you going to survive there apart just from dying?
>
> *(Interview 5 August 2011)*

Spatial relocation over longer distances was not seen either as a realistic option by many of the Kampala-bound RWDs. They argued that even though they did not receive sufficient medical care in Uganda, they had minimal prospects for resettlement to a third country in the Global North. Some of the RWDs had been accepted as the official 'case load' of urban refugees who were financially and materially assisted for up to six months, but this assistance was not perceived by them as a permanent solution to their struggle for 'transformed and renewed right to urban life' (Lefebvre 1996: 158). They had hardly any options but to stay put in Kampala – or in other words to live 'between a rock and a hard place' (RLP 2011b).

The everyday life for many RWDs in Kampala meant struggles over their ability to spatially navigate the city. One of the objectives of the support group was to reach out to RWDs who were unable to move from their homes. The leadership and the members of the group visited different neighbourhoods of the city to assist and to encourage those who were living in spatial limbo. This meant challenges for refugees with physical disabilities, as moving by foot was sometimes the only option due to lack of money for transportation. RWDs' mobility could have been enhanced with special equipment, such as crutches or supportive belts, but getting them was difficult. At times RWDs had established arrangements whereby *boda boda* [motorbike] drivers known to them would provide free or discounted transportation. Without these arrangements or financial means, RWDs could 'spend even two months without reaching any place' as suggested by a male group member (interview 20 July 2011). Parents of the disabled children also struggled with the fact that they had become 'disabled' because they were so attached to their children: 'For us who are having children with disabilities, it is very difficult to leave him at home. They have become like a mobile phone; you have to move with him everywhere you go' (interview 4 August 2011). Subsequently, everyday life and running the group activities that required physical movement in the city were restraining.

Second, as the RTC can be interpreted as collective rights 'to information and the rights to use of multiple services' (Lefebvre 1991, cited in Marcuse 2009: 189), I examine how the members of the RWDs support group managed to access official protection institutions, such as UNHCR, its implementing partner InterAidUganda (IAU) and public services. As I have demonstrated elsewhere (Lyytinen 2013, 2015b), many of the urban refugees, in particular the Congolese, struggled to gain access to the UNHCR services that were implemented in the city by IAU. In addition, RWDs suggest that they were not respected by some of the officers working on refugee protection. RWDs also argued that they had a worse chance of getting an appointment to see a counsellor or a protection officer than able-bodied refugees. This was because they were not capable of going through the physically and mentally tough system of acquiring an appointment: 'Even to talk to the offices, those different organisations, you have to compete with the people who are physically fit. The staff of those different organisations do not care about people with disabilities' (interview 5 August 2016). In 2011, getting access to the IAU and UNHCR officers was restricted, and many refugees, particularly the Congolese whose appointment day was on Mondays, had to sleep outside the IAU office and/or bribe people to get an appointment. This was explained by a refugee mother of a disabled child:

> I used to go and sleep outside IAU. I go and book on Sunday morning, and in the evening I come and sleep. In the [Monday] morning if I get the chance, I get the appointment now. I go there around 8PM [on Sunday]. I sleep there until the morning. I do not feel safe, I do not feel OK, because I am sick and I do not feel free. I sleep there alone. They bring him [her disabled son] in the Monday morning.
>
> *(Interview 4 September 2011)*

The RWDs also struggled to access public services, such as schools and hospitals. By the 2006 Refugee Act, refugees in Uganda are entitled to all basic services, such as medical care and primary education. In Kampala there are no parallel services for refugees as they are expected to access the mainstream services meant for all of the urban inhabitants. Like in the case of protection institutions, these difficulties were not only about accessing the physical, material space of these services, but they also had to do with acquiring access to the imagined or social space (Lefebvre 1991) of these services – the attitudes, relations and emotions. Regarding schooling for refugee children with disabilities, their parents struggled to keep their children in school even though schooling was often prescribed as 'treatment' by doctors. Challenges had to do with lack of money, requirement to be present in the school to assist the child, and the general discrimination against disabled children in public schools. These challenges were articulated, for instance, by a disabled refugee mother who was unable to educate her children: 'Because you are disabled, you feel guilty for making other people to suffer... The child was supposed to enjoy the status of the child, but he is now occupying the position of the parent' (FGD 27 August 2011).

Like with schooling, RWDs faced challenges when trying to access medical services in hospitals. The most common issues had to do with not getting the needed treatment or not having the financial means to access health services. At times, refugees suggested that Ugandan doctors had to be bribed, and a common story was that when refugees needed a surgery, the doctors provided them with painkillers. Subsequently, some of the RWDs argued that they had been physically disabled due to lack of treatment in Kampala. Accessing hospitals often required persistence and time. A mother of a disabled child explained how, during her everyday life in

Kampala, she often spent significant amounts of time in hospitals: 'Most of the time I spend going up and down looking for livelihood. But since my kid is sick, I spend most of my time at hospitals' (interview 5 August 2011). Another RWD, deprived of basic material and financial means, explained how her treatment was not adequate due to protection institutions arguing who to pay her medical bills and which hospitals to access:

> I am unable to move, walk, sit... I went with the [medical] bill to IAU.... They told that the bill cannot be paid by IAU, but it must be send to UNHCR, so I had to wait until they got that money from UNHCR.... It was OK, I could not resist; do whatever you want to do, because I am the one suffering. They refused to pay that bill, and I continue taking tablets.
> *(Interview 4 August 2011)*

As demonstrated by these examples, RWDs faced challenges when trying to promote their RTC as rights to access particular offices, spaces and services (Marcuse 2009). Due to these problems of accessing and occupying not only physical spaces, but also imagined and social, or lived, spaces (Lefebvre 1991), they were often not only deprived of their basic material and social rights but of their rights to use urban space – the fundamentals of the RTC (Lefebvre 1996).

The RTC as appropriation

The RTC does not only manifest in the deprived and discontent (Marcuse 2009) refugees' participation in the support groups and their access to city space and services, but it also involves an element of appropriation of space. Together participation in the spatial production of the city and appropriation of urban space create a 'twin element' of the RTC (Butler 2012). Appropriation refers to the collective attempts to produce city space as a creativity, or 'oeuvre' (Lefebvre 1996). Here, I largely interpret appropriation of space as the collective attempts by the RWDs to transform the insecure spaces of exile into places of protection and belonging. In this analysis Lefebvre's (1974: 1991) idea of lived space is emphasised. Lived or social space for him refers to 'representational space' that is produced in everyday social interactions, i.e. it is the 'lived space of sensations, the imaginations, emotions and meanings incorporated into our everyday lives and practices' (Harvey 2004: 8).

Refugees who were members of the RWDs support group felt that through their active and meaningful membership in this group they were supported individually and collectively in numerous ways – often the tangible support not being the most important issue. Overall, their aim was to transform the timid and lonely spaces of exile into spaces of protection and collective support. These transformative attempts are interpreted here as appropriation of space (Lefebvre 1996; Butler 2012).

Members of the RWDs expressed value in being able to share their experiences with other refugees with similar backgrounds: 'When we try to put ourselves together, it brings again strength to us. It will give you more strength' (FGD 27 August 2011). A female member of the RWDs group explained her sense of belonging in the following words:

> The group that I belong here is just PWDs [i.e. RWDs]. When we started our group, it really helped me. This group really helped me, because when I was alone at home, I was thinking that I was the only one suffering with these issues. But when I came to this group, I saw that I was not alone. I saw that there were some who were suffering more than me.
> *(Interview 4 August 2011)*

By transforming their everyday urban life that used to be characterised by alienation and distress into meaningful lived spaces (Lefebvre 1974: 1991) of encounters and support with other RWDs, the refugees were enacting their right to appropriate space.

The members also highlighted the inclusiveness of the RWDs group, and how they had found a safe space within this group even if they were discriminated against by others, including different national or ethnic groups. The RWDs perceived that they were, in general, discriminated against because of their condition, and some of them were struggling with feelings of anger and shame. They saw that being able to express their frustration within the group was helpful in order to overcome the feelings of exclusion and rejection:

> In the group, when we share experiences, it helps us to move when we are angry... and you get some help in the group, some counseling. It even moves something which could make you angry in life. I have no other communities.
>
> *(FGD 27 August 2011)*

Moreover, by coming together, the members of the RWDs group had been able to increase their agency and potential to act on their own behalf. Members of this group, for instance, suggested that because they were united and organised, they had been able to change the RLP system of seeing clients in their favour. Through their communal action they were able to access the RWDs officer every Wednesday without having to go thought the normal tombola system of getting the appointments. Overall, they were more recognised once they started to advocate for their rights as a collective. This was also acknowledged by the RLP officer:

> It is very, very important [for the RWDs to get together]. It has actually proven to be very important. First of all, they kind of feel that they are recognised; that they exist. They feel that they can speak out now that they are part of the group. So, they feel like they can easily approach you and get access to the services. It is kind of empowering to them.
>
> *(Interview 26 August 2011)*

In addition, the RWDs support group had established advocacy campaigns together with the RLP to advocate for their RTC – as access, participation and appropriation of space (Lyytinen 2015a). The RWDs, for instance, celebrated the International Day of Persons with Disabilities the 3rd of December and cooperated on this event with the National Umbrella for Disabled Persons in Uganda (NUDIPU). Through this cooperation, the RWDs advocated for their rights in a broader setting of PWDs in Uganda. Additionally, the group had taken part in the production of short documentaries produced by the RLP's video advocacy unit. These documents demonstrated visually the RWDs' everyday struggles for their RTC in Kampala, their city of exile.

Conclusions

In this chapter, the concept of the 'right to the city' (RTC) was rethought in relation to urban refugee protection in the context of Kampala, Uganda. Analytically the focus was on the scale of the group. This group-based analysis of refugees' cry and demand for their RTC was noteworthy, for the examination of the RTC has been lacking sophistication and critical approach regarding the scale of the group or community. Yet, the scalar analysis of the RTC is very much needed, and this article has contributed to the group-based rethinking of the RTC. Analysis focused on the scale of the group was also important in order to enhance our understanding of urban refugee protection. In urban contexts, particularly in the Global South, refugee protection is increasingly conducted in group or community settings. Thus, new conceptual frameworks, for instance based on rethinking the RTC, are needed.

Empirically the applicability of the RTC was analysed regarding the disabled refugees living in Kampala, Uganda. This chapter is one of the few pieces of research focusing on reapplying the RTC in the sub-Saharan context, particularly in relation to refugees living in urban areas. The focus on RWDs also enabled answering the question of 'whose right'. The examination paid attention to the most deprived and discontent or alienated refugees with various mental and physical disabilities, and refugees who were caretakers of these RWDs. The RTC was, therefore, analysed with regard to people not only persecuted but also disabled.

The notion of the RTC was applied in this chapter in a threefold manner: as participation, access and appropriation. First, refugees' participation was analysed as a collective right to be organised and advocate for their rights in the group setting. In the case of the RWDs group, participation was functioning well both in terms of increasing number of members, but also as recognition and meaningful communal action. Second, RWDs' right to access, occupy and use space – be it the city, the protection institutions or other services – was limited and constantly pushed forward by the refugees. Access to space was analysed through Lefebvre's spatial triad which sees space simultaneously as physical, imagined and social/lived. In urban refugee studies, perceiving space in this manner has proven helpful. Third, appropriation of space was featured in refugees' discourses on how to transform their insecure everyday life in urban exile into protective spaces of collective support and belonging. This was achieved through support and sharing that led to an enhanced sense of security. To conclude, the analysis demonstrated the usefulness of applying the RTC, as rethought here, not only regarding the scale of the group, but also in the context of urban refugee protection.

References

Attoh, K.A. (2011) 'What kind of right is the right to the city?', *Progress in Human Geography*, 35:5 669–85.

Bernstein, J. and Okello, M.C. (2007) 'To be or not to be: urban refugees in Kampala', *Refuge*, 24:1 46–56.

Butler, C. (2012) *Henri Lefebvre: Spatial Politics, Everyday Life and the Right to the City*, Abington: Routledge.

De Souza, M.L. (2010) 'Which right to which city? In defence of political-strategic clarity', *Interface*, 2:1 315–33.

Harvey, D. (2004) 'Space as a key word', paper for Marx and Philosophy Conference, Institute of Education, London, 29 May 2004.

Iveson, K. (2011) 'Social or spatial justice? Marcuse and Soja on the right to the city', *City*, 15:2 250–59.

Kuymulu, M.B. (2013) 'The vortex of rights: "right to the city" at a crossroads', *International Journal of Urban and Regional Research*, 37:3 923–40.

Lefebvre, H. (1991/1974). *The Production of Space*, Oxford: Blackwell.

Lefebvre, H. (1996) *Writings on Cities*, trans. Kofman, E. and Lebas, E. (eds). Oxford: Blackwell.

Lyytinen, E. (2013) 'Spaces of trust and mistrust: Congolese refugees, institutions and protection in Kampala, Uganda', unpublished thesis, University of Oxford.

Lyytinen, E. (2015a) 'Congolese refugees' "right to the city" and urban (in)security in Kampala, Uganda', *Journal of Eastern African Studies*, 9:4 593–611.

Lyytinen, E. (2015b) 'The politics of mistrust amongst and between Congolese refugees and the institutions providing refugee protection in Kampala, Uganda', in K. Koizumi and G. Hoffstaedter (eds) *Urban Refugees Challenges in Protection, Service and Policy*, London: Routledge.

Marcuse, P. (2009) 'From critical urban theory to the right to the city', *City*, 13:2–3 185–97.

Mitchell, D. (2003) *The Right to the City. Social Justice and the Fight for Public Space*, London: The Guilford Press.

Parnell, S. and Pieterse, E. (2010) 'The "right to the city": institutional imperatives of a developmental state', *International Journal of Urban and Regional Research*, 34:1 146–62.

Purcell, M. (2002) 'Excavating Lefebvre: the right to the city and its urban politics of the inhabitant', *GeoJournal*, 58 99–108.

Purcell, M. (2003) 'Citizenship and the right to the global city: reimagining the capitalist world order', *International Journal of Urban and Regional Research*, 27:3 564–90.
RLP (2011a) *Legal and Psychosocial Department End of Year Report 2010*, Kampala: Refugee Law Project. Online. Available HTTP: <http://www.refugeelawproject.org/> (accessed 14 September 2012).
RLP (2011b) *Between a Rock and a Hard Place, An Advocacy Video for PWDs*, Kampala: Refugee Law Project. Online. Available HTTP: <https://www.youtube.com/> (accessed 2 March 2017).
RLP (2015) *Profile of Support Groups at Refugee Law Project*, Kampala: Refugee Law Project. Online. Available HTTP: <http://www.refugeelawproject.org/groups_profile.pdf> (accessed 28 February 2017).
RLP (undated) *Critique of the Refugee Act 2006*, Kampala: Refugee Law Project.
UNHCR (2009) *UNHCR Policy on Refugee Protection and Solutions in Urban Areas*, Geneva: UNHCR. Online. Available HTTP: <http://www.unhcr.org/protection/hcdialogue%20/4ab356ab6/unhcr-policy-refugee-protection-solutions-urban-areas.html> (accessed 28 February 2017).
UNHCR (2011) *Uganda Statistics as of July 01*, Kampala: UNHCR.

35

'Right to the city' *versus* neoliberal urbanism in globalising cities in China

Ran Liu and Tai-Chee Wong

Introduction

Almost half a century ago, Henri Lefebvre in his book entitled *The Production of Space* discussed the appropriation and reappropriation of space, in which he defined the city as a place of revaluation or market value production. As examples for the elaboration of his point, Lefebvre cited the renewal of *Les Halles Centrales*, Paris, and the reworking of the selected leisure-oriented Mediterranean sites; highlighting that:

> all this seemingly non-productive expense is planned with the greatest care: centralized, organized, hierarchized, symbolized and programmed to the *n*th degree.
>
> *(Lefebvre 1991a: 59)*

Comparing cities as conceptual places of wealth accumulation and conflicts, the Lefebvrian approach to socio-spatial theories and politics has enlightened ideas about the 'right to the city' and the 'right to the oeuvre', which are respectively associated with restricted access to created wealth and use value change (Lefebvre 1996: 34, 66, 158). Lefebvre's notion of the 'right to be different' (1991a: 64, 205) was inclined, however, to reflect acts of neoliberal urbanism and their destructive consequences, producing 'alienated' urbanites and 'alienated' everyday life in different historical periods, urban contexts, and spatial scales. In citing Marx's interpretation of the role of the state in the transformation of cities, Lefebvre's analysis of the capitalist growth and crisis has implied a revolutionary concept of 'citizenship', or citizens' rights when their rights could be restricted by productivist ideologies and developmentalism, with destructive consequences for everyday life (Lefebvre 1982; 2009).

The Lefebvrian critiques of modern capitalism fetishism and the state-implicated mode of production, together with his advocacy of: human-faced development instead of merely economic growth; the right to the city instead of alienation; the right to difference instead of state-imposed homogeneity or consumerism; and the ideal of *autogestion* (self-management of grassroots) instead of state power, have all become part of the theoretical basis for studies of urban politics (Harvey 2008: 2012).

What, then, are the implications of the Lefebvrian notion of the production of space, with reference to the right to the city in China's modernist urbanism today? There is, indeed, a strong relevance. Drawing on the neoliberal approach adopted by the Chinese authorities in attempts to eliminate slum-like dwellings and upgrade property market values in China's globalising cities, our analysis employs the Lefebvrian critique, which closely demonstrates the conflicts between the state-controlled spatial order and the sprawling and informal settlements of low-wage migrant dwellers in urban China. The analysis investigates the state's attempts to reconcile its planning for growth with the peasant workers' right to habitat and to inhabit. This area of concern resonates well with the Lefebvrian utopia of a new habitat policy, which addresses the interplay of:

> spatial (territorial) *autogestion*, direct democracy and democratic control, affirmation of the differences produced in and through that struggle taking place in the cities.
>
> *(Lefebvre 2009: 251)*

In citing Lefebvrian urban theories, this chapter focuses on the right to the city in contemporary China which, under state and market-led urbanisation with a strong Chinese character, has rapidly developed into the world's second largest economy. In this process of transformation, a clash between neoliberal urbanism and slum urbanism in urban China is reviewed, with Beijing offered as an example. It highlights the chronic socio-political injustice and spatial inequality, in terms of the hierarchisation of dwellings, and the opposition of formal and informal habitats. The differentiation in access to formal habitat, by locals and non-locals, has occurred within the paradigm of city image-making, developmental urbanism favoured by Chinese state authorities. Our analysis concludes with a visionary hope for a state-initiated transformation that will shift away from the less popular 'GDPism' towards a more people-oriented mode of the production of urban space. Ultimately, the welfare of the entire workforce, including peasant workers, will be considered on the basis of equal citizenship in the cities, in which their contribution will create collective wealth for the urban world. At this stage, nevertheless, peasant workers still face high levels of discrimination in spatial politics.

Spatial politics: developmentalism *versus* right to habitat

From 1949 to 1978, the Chinese communist government adopted an anti-urban approach by restricting the size of the population of cities to be engaged in industrial production, public administration, and other services. The residency system (*hukou*) introduced in the mid-1950s was aimed at controlling the rural population's ability to migrate to the cities, which were organised into *danweis* or work units, from the communes established in the rural sector. Free choice of workplace or residency, and of rural-urban or inter-city migration, was virtually forbidden during the period 1949 to 1978, until the rigid socialist conformity was dismantled by the reformist leader, Deng Xiaoping. Since this opening up and pro-market reforms, a new city-centred, export-oriented, and state-dominated urbanism mode has been introduced to China, which has witnessed an influx of surplus farm labour to the coastal provinces and primary cities as migrant workers offer their cheap labour to help build the urban economy. These people are considered sojourners rather than full citizens, as they are only given provisional residency permits (Chan 1994; Solinger 1999).

In terms of residence, the low occupational rewards of unskilled or low-skilled migrant workers have restricted them from moving into strategic and prime areas which fetch high land values and rentals. Instead, they are typically found in illegally built housing clusters in urban villages within the cities, and in the suburbs. China's pro-growth developmental urbanism has sought to exploit the high-profit returns from urban land and other resources, which are seen as key national assets in wealth creation, as well as instruments to consolidate state authority and legitimise the reformist ruling communist party (Wallerstein 2005; Lin 2007). Lefebvre, in his *Critique of Everyday Life* (1991b), extended the Marxian notion of the alienation of labour to a cognitive form of every aspect of life associated with spectacular commodities in the modern world. This has created a false consciousness ('mystification') in the minds of workers, which deprives them of an objective vision in relation to their roles in the reproduction of labour. Accordingly, the ideological tenet of developmentalism has become subservient and dehumanised (1991b: 91), like the cogs in the great clockwork wheel of the state, which have been set in motion in the interests of rapid urban growth and national wealth accumulation.

Thus, the alienation of ordinary workers is not only economic, social, political, ideological, and philosophical; but also spatial, as space can be defined in terms of a reifying alienation which itself has become a commodity to be sold, in both wholesale and retail markets (Lefebvre 1991a: 50; 1991b: 249). When developmentalism and urbanism are combined, developmental urbanism is a novel form of urban development. Such development is characterised by a pro-growth orientation by state- and urban-led entrepreneurialism, with:

> a strong relationship with modernist place and image making, and supported by sustainable spatial planning and design leading to a highly compact, highly vibrant and livable city environment.
>
> *(Wong and Liu 2016: 3)*

Since the advent of tax sharing and land conveyance reforms between the Chinese central government and municipal governments in 1994, China's emerging real estate market has employed city imaging practices aimed at enhancing national competition in the international marketplace, including in the area of financial services. In this respect, Fulong Wu (2005: 7–8) raised an interesting question: does the making of new urbanism in China, enriched by Western-style architectural motifs and transplanted cityscapes, such as gated communities, enclosed villas, golf communities, nightscapes, and skyscrapers, reflect real globalisation? Wu suggested that this is more a sort of imagined globalisation, using global motifs to sell off local products, and an opportunity to give the pro-growth elite 'the credibility that they are the builders of a vision of the good life'.

In elaborating the mirage effects (mystifications) which are comparable to those in the production of the unique space of Venice, Lefebvre (1991a: 76, 189) pointed out the illusion of a new life by including cityscapes in an aesthetic scenario to suit the tastes of people who have dominated the grand designs and projects. Whether city imaging is really as rational as the urbanists or planners claim, Lefebvre (1991a: 317) argued that:

> The dominant discourse on space – describing what is seen by eyes affected by far more serious congenital defects than myopia or astigmatism – robs reality of meaning by dressing it in an ideological garb that does not appear as such, but instead gives the impression of being non-ideological (or else 'beyond ideology'). These vestments, to be more specific, are those of aesthetics and aestheticism, of rationality and rationalism.

City imaging and eviction of socially weak groups

Legitimised within a developmentalist paradigm, city imaging has acted as a growth engine to attract increasingly mobile global investment, talent and tourism revenues, enhance the city's liveability and beautification, and further strengthen civic cohesiveness. China's major cities have undergone development with megaprojects and property-led regeneration in prime locations. Such developments have included skyscrapers, magnificent squares, Economic Development Zones, high-tech parks, five-star hotels, villas, golf courses, and other entertainment centres. Looking at these areas of improvement and at the spectacular mega events in the cities, the negative aspects of commercial gentrification have been concealed in order for the developments to appear positive and constructive (He and Wu 2009).

This seemingly 'rational' spatial practice, however, has led to massive land grabs and the dispossession of many low-income residents in the inner cities and suburbs. Rhetorically disguising such activities as environmental improvements, Chinese local governments have promoted the re-imaging of their cities by commercial gentrification, land acquisition for renewal. There is denial of the claims of low-income residents for affordable housing and territorial rights, despite their protests against eviction (Hsing 2010; Wu 2016). Citing such acts that occurred in central London, New York, Delhi, Seoul, Mumbai, and Beijing, Harvey (2008: 34) referred to them as an 'accumulation by dispossession', characterised by 'the mirror-image of capital absorption through urban redevelopment'. These acts have given rise to numerous conflicts between profit-making ventures via land revaluation and displacement of underprivileged low-income migrant tenants.

From 'city as spectacle' to 'city as oeuvre'

In Lefebvre's view, there were state-led hegemonic forces that had segregated urban space and alienated urbanites into different worlds where grassroots transgressed the state hegemony and fought to build a more just city. In this regard, Lefebvre (1991a: 23) wrote:

> In this same space there are, however, other forces on the boil, because the rationality of the state, of its techniques, plans and programmes, provokes opposition. The violence of power is answered by the violence of subversion… State-imposed normality makes permanent transgression inevitable… These seething forces are still capable of rattling the lid of the cauldron of the state and its space, for differences can never be totally quieted. Though defeated, they live on, and from time to time they begin fighting ferociously to reassert themselves and transform themselves through struggle.

A similar spatial opposition and struggle is observable in China's metropolises today. In the modernist and image-building Chinese state, new urbanism is deemed to be a good way of satisfying the needs of capable consumers or property stakeholders (as gentrifiers); while slum-like urban villages (*chengzhongcun*) are treated as transitional landscapes with impending elimination, where the peri-urban peasants have built substandard informal houses for low-income earners. When the demolition of this informal housing begins, sharp land use conflicts arise between the state-dominated city-branding movement and the great need for low-rental areas. While new private estates have edged out low-income tenants, the informal habitats are replicated in ever more peripheral sites (Wu et al. 2013; Liu 2015). In this sense, the dispersion, segregation, and dispossession of low-income migrants can be decoded as the spatial consequences of ceaseless social contradictions and inequalities, dominated by the top-down strategic aims of the highest order, which is pro-growth oriented. These current state-initiated spatial practices in China are classified as 'planning for growth' (Wu 2015: 192), and demonstrate the limitations of planning in achieving a more sustainable and just urban future within the present context of state- and urban-led entrepreneurialism.

Arguably, this tremendous change and development in urban space has proven to be a very critical area for political control and grassroots struggle. Accordingly, Lefebvre (1991a: 416) again argued that the whole issue lies:

> in space, on a worldwide scale, that each idea of 'value' acquires or loses its distinctiveness through confrontation with the other values and ideas that it encounters there. Moreover – and more importantly – groups, classes or fractions of classes cannot constitute themselves, or recognize one another, as 'subjects' unless they generate (or produce) a space.

Therefore, the notion of the appropriation of space is far more exigent than in his previous thesis, taking space as a mirror (Lefebvre 1991b). These oppositions and antagonisms were conceptualised in the famous Lefebvrian perceived-conceived-lived triad (in spatial terms: spatial practice, representations of space, and representational spaces), wherein the:

> 'lived experience' (i.e., spaces of 'inhabitants' and 'users') is crushed, vanquished by what is 'conceived of' (i.e., dominant spaces and modes of production by planners, urbanists, bureaucracy.
> *(Lefebvre 1991a: 38–39: 51)*

In his spatial triad, Lefebvre elaborated the two oppositional spaces – abstract space, which is produced and manipulated by the authorities as a logical system of power in the world of commodities – and the new differential space, as the 'right to be different', and the struggles to establish differences within the existing abstract space (Lefebvre 1991a: 53, 64). The developmental urbanism of Chinese character falls into the domain of the Lefebvrian abstract space, the most dominant force, which Lefebvre criticises for being: violent, illusory, political, and institutional. But why did Lefebvre call this kind of institutionalised space and its instrumental homogeneity an illusion? Lefebvre (1991a: 285) explained thus:

> Its practical character vanishes and it is transformed in philosophical fashion into a kind of absolute. In face of this fetishized abstraction, 'users' spontaneously turn themselves, their presence, their 'lived experience' and their bodies into abstractions too. Fetishized abstract space thus gives rise to two practical abstractions: 'users' who cannot recognize themselves within it, and a thought which cannot conceive of adopting a critical stance towards it.

But how have the 'contradictions of space' compelled people into different lives and to be engaged in different modes of production? Lefebvre (1991a: 55) advocated that it was the outcomes of class struggle that have generated such differences, which are not intrinsic to the nature of economic growth itself. In the case of migrant tenants renting illegally built premises, such a 'difference' is rooted in the distinction between formality and informality. Slums, or slum-like housing without legal title, are typical informalities to be dealt with (Davis 2006), based on which urban dwellers (*citadin*) are deprived of the right to habitat as citizens and as users of multiple services, inclusive of those available in the city centre as a privileged place (Lefebvre 1996: 34).

Harvey (2008) shared the Lefebvrian notion of the right to the city beyond the range of city space, touching more on the citizen's right to participate in the process of city building, and sharing a greater democratic control over the production and utilisation of the surplus value created. In summary, such struggles would bring a revolutionary new mode of production, whether gradual or sudden, moving away from 'the city as a spectacle' to the 'city as an oeuvre', to respond to people's claim to 'totality' instead of being alienated or mystified by illusionary fetishism (Lefebvre 1995: 86).

The next section turns to the contention in Chinese metropolises between the state-led pro-market forces and the poor-image habitats of the low-income residents that are in conflict with the state's city branding ambitions. We examine how developmental urbanism, as a typically top-down abstract space, has justified the exclusion of migrant workers in the urbanisation process in China recently.

Right to the city in China's urban space

Unlike the spontaneous squatting behaviour in Latin America and South Asia, which is largely free from state control, urban space formation in China is characterised by the state's artificial control of admission, which prevents rural non-local migrants from permanently and automatically settling in the cities where they work. Local peasants who live in the rural zones within the municipal boundary can only convert their residency status from rural to urban once their farms are formally urbanised and integrated into the municipality. In this rapid urbanising process, a broad range of dispossessed local residents, who were living in low-quality apartment blocks in the inner-city areas and the dilapidated work-unit compounds, have been displaced or relocated as their areas of residency are redeveloped to modernist norms.

In the present-day growth-driven city imaging and spatial-sectoral upgrading, a 'permanent' right to the city is conditional and selective. Welcomed social groups include: investors, those who possess a talent, property owners, and the traditional work-unit workers and their families, who are expected to best adapt and contribute to the neoliberal market forces and newly alienated but modernised city spaces (see Huang and Jiang 2009). Together with these groups, peri-urban peasants who are apparently disadvantaged but rightful rural property owners and holders of local residency are accepted and provided with government assistance and other benefits. Their right to ownership of private property is deemed by Lefebvre to be the ability to enjoy the same real rights as private individuals (Lefebvre 1982: 130). Such rights, which are supported by the Property Law of the People's Republic of China and new regulations over property expropriation and compensation, made a decade ago, have marked a remarkable milestone in China's legislative progress in rights recognition (Hess 2010). The rights of citizens are for those having an imaginary sovereignty (Lefebvre 1982: 130); those who are of rural origin, however, are denied.

Low-wage migrants' access to shelter is predominantly dependent on illegitimate lease deals, whereby local landlords are their 'patrons' and migrant entrepreneurs are their agents. In the book *Strangers in the City* (Zhang 2001), a story is told about *Zhejiangcun* in the 1990s, which was once Beijing's most famous migrant enclave, in which Wenzhou migrant entrepreneurs struggled to develop their own community and control over the key social spaces (residential areas, working sites, and market places). The Beijing government, however, took prompt action to eliminate the presence of the enclave. It was seen as an emerging informal and illegitimate social space. It was perceived by political authorities as a potential threat to state control, and a violation of the modernist norms of aestheticism and rationalism (Lefebvre 1991a: 317).

Urban space is produced as a result of the relationship between (re)production and domination, and its physical outcome falls under the Lefebvrian schema of 'homogeneity – fragmentation – hierarchization' (Lefebvre 2009: 212). By analogy, urban politics in transitional China are built on the opposing factors of centre and periphery. Here, the centre:

> exercises its control at all (organizational, administrative, juridical, fiscal, police, etc.) points of view over peripheries that are both dominated and broken apart… [and] link up the peripheries, coordinate them, and submit them to the global strategy of the State'
>
> *(Lefebvre 2009: 215)*

As they are hierarchised at the 'periphery', unskilled and low-wage migrants are neither accepted nor given a chance to establish themselves, unless they enter the 'centre' as capable consumers or homeowners. In a material sense, city rights may include such basic necessities as decent housing, legalised residency status, and education services. However, in terms of the Lefebvrian notion of the right to the city, associated with urban space formation, city rights are considered to be people's ability to exercise their 'collective power over the process of urbanization' in order to be included as an equal power, and to freely make and remake their cities for themselves (Harvey 2012: 4).

At the periphery of the Chinese metropolises, space is highly fragmented by a diversity of occupiers. It is common to see a division in villages similar to the situation in peri-urban Beijing, where half the villages have been expropriated and redeveloped into villas or gated communities, and the other half into substandard informal tenements for low-income earners. These low use values and informalities at the periphery, however, are expected to disappear as an inevitable result of developmental urbanism, which exerts its authority to homogenise the fragments scattered through space and finally establish an absolute primacy of exchange value (Lefebvre 1991a: 339). The elimination of urban villages therefore becomes the crux of the contradictions between local governments (dominant space with its repressive and assimilative capacity) and low-wage migrants (periphery as dominated space). How to grant the new centrality to the periphery (including everyday life and the lived space of inhabitants) is central to the grassroots practices to reclaim the residents' rights to the city.

Transitional space for sojourners instead of full citizens

China's reform and integration into the world market are characterised by large-scale mobilisation of its surplus farm labour and a massive release of land assets into the mainstream of capital accumulation (Harvey 2003: 149; Zhang 2004). This mobilisation and release has created a situation of 'man' being 'in transition', moving towards being a 'total man' crossing through alienation; in other words, this transitional man is 'disalienating' himself (Lefebvre 1991b: 66).

At the present post-reform historical juncture, the debate on the Chinese right to the city focuses on two major issues: the problematic lack of a proper residency status of migrants in the city; and the absorption of local urban and rural residents in the municipal areas when cities undergo renewal and expansion, respectively (Wang 2005; Hsing 2010; Liu et al. 2012). For the first issue, the *hukou* system, a residency card system, was innovatively reinvented in 2016 to be more inclusive and aims to better serve municipal interests (*Xinhua News* 2015). Residency control has continued to be harsh in China's first-tier cities, however, where migrants are required to satisfy a five-year residency period, and to have made social security contributions and paid taxes in order to qualify for subsidised public housing or property purchase in the host cities. Additionally, migrants' children are admissible only to inferior local schools and are barred from participating in local college entrance examinations. This implies that large numbers of migrant workers will be forced to continue to seek cheap informal habitats. The second issue impacts local citizens who are affected by renewals and the urbanisation of rural zones. These people have witnessed a similar conflict-laden 'transition' of their right to the city, which lies in closely-negotiated power relations with the state authority. Such relations are being dealt with by the state on two fronts: with inner-city residents over their resistance to demolition and eviction; and with local peasants over their claims to land value gains.

Conclusions

The Lefebvrian catchphrases of the right to the city and the urban revolution (1991: 147–159; 2003) have endured as visionary democratic ideals. They have even evoked participatory and collaborative approaches towards city planning in the Global South (Esther 2010). In the eyes of the modernising state and developmental cadres of China today, however, slum urbanism is seen as a form of 'subversion' and an 'uncivilised' part of the city (Lefebvre 1991a). Although pro-growth measures since the 1980s have attracted over 200 million rural migrant workers into Chinese cities, the largescale spread of informal settlements has been avoided as a result of absolutely state-controlled land ownership, residency control, and the state and local authorities' firm belief in city image building. In particular, developmental urbanism is a novel form of growth-led pathway which places a strong emphasis on modernism and image making, supported by sustainable spatial planning and design concepts aimed at eventually building a long-term liveable and slum-free city environment (Wong and Liu 2016). In this urban-led pro-growth paradigm, developmentalism has created two differentiated populations administered under local *hukou* status and non-*hukou* status. They exist in differently regulated spaces of tradeable urban land and untradeable rural land, as well as in the space situated in between and referred to as urban villages (*chengzhongcun*). This space is a 'transitional' space for the 'transitional man', who is 'alienated' in the urbanisation process, but trying to 'disalienate' himself towards a 'total man' (Lefebvre 1991b: 66, 90).

Backed by an improved economy and as a developmental state, there is now an awareness in China of the social conflicts arising from the widening gaps. As a result, China is becoming more responsive to migrants' misfortune by granting them access to more public services and welfare, albeit in a progressive manner. By virtue of the *hukou* reform, issued by the State Council since 2014, a more people-oriented development path under a 'new urbanisation model' is being implemented during China's 13th Five Year Plan (2016–2020). During this period, up to 100 million rural migrants, and regular residents from other regions, will be issued local residency permits in their host cities, which will provide the basic welfare and services according to their financial capacity (State Council 2014). These increased rights for migrants are deemed the 'Six Services and Seven Advantages' in an innovative residency card system which has, since 2016, begun to replace the present segregative agricultural and non-agricultural *hukou* system (*Xinhua News* 2015). Access to affordable housing, however, has not been included as a requisite right of these migrants. Easily removed by a developmental regime, the migrant workers are largely too disorganised to act through their collective claims and lack recognition as formal and localised actors in the use of city space (Wong and Liu 2016).

Until recently, providing migrants with a decent home has been a challenge for local governments, especially for the first-tier 'paradigm cities', characterised by their world-class image buildings and high-value property developments (Huang and Tao 2015). The fear of the rising collective strength of migrant enclaves is another driving factor towards the elimination of the spatial grouping of low-end migrants. What has previously occurred in the low-income concentrations of the French suburban towns or the inner cities of American major cities, which are characterised by a vicious circle of poverty and high rates of crime and riot, has provided a further justification for a 'clear up' instead of a 'beef up' (Merrifield and Swyngedouw 1996). If the spatial formation of migrants is not allowed, a full usage of places for a renewed urban life, inclusive of a self-management scheme, would not be permitted (Lefebvre 1991a: 416).

China's reformist model aims to ultimately remove discrimination and injustice towards the identity of migrant workers and extend full rights and equal welfare to all citizens. Preference will be given to those who have resided in the city for a longer period, and to those people who possess a higher calibre of employability and job adaptability. The grading system converting local *hukou* to migrants has been introduced in Shanghai and Beijing, wherein higher points are given to investors, overseas trained professionals, highly educated workers, and those hired by high-tech or large-scale enterprises. Understandably, by virtue of their lowest ranking in the queue, unskilled and poorly educated migrant workers may have to wait a long time for full rights to decent housing and higher quality public services. Apart from the conversion of more qualified migrants to full citizens, a new type of urban habitat politics should be constructed to entail equity-based development instead of growth; the right to the city instead of alienation or dispossession; and the 'right to the difference' instead of state-imposed homogeneity or consumerism.

Funding statement

This research is supported by the Humanity and Social Science Youth Foundation of the Ministry of Education of China (Grant Number: 16YJCZH060).

References

Chan, K.W. (1994) *Cities with Invisible Walls: Reinterpreting Urbanization in Post-1949 China*, Hong Kong: Oxford University Press.
Davis, M. (2006) *Planet of Slums*, London: Verso.
Esther, H.M. (2010) 'Social inclusion through participation: the case of the participatory budget in São Paulo', *International Journal of Urban and Regional Research*, 34:3 512–32.
Harvey, D. (2003) *The New Imperialism*, New York: Oxford University Press.
Harvey, D. (2008) 'The right to the city', *New Left Review*, 53: 23–40.
Harvey, D. (2012) *Rebel Cities: From the Right to the City to the Urban Revolution*, London: Verso.
He, S. and Wu, F.L. (2009) 'China's emerging neoliberal urbanism: perspectives from urban redevelopment', *Antipode*, 41:2 282–304.
Hess, S. (2010) 'Nail houses, lands rights and frames of injustice on China's protest landscape', *Asian Survey*, 50:5 908–26.
Hsing, Y. (2010) *The Great Urban Transformation: Politics of Land and Property in China*, New York: Oxford University Press.
Huang, Y.Q. and Jiang, L.W. (2009) 'Housing inequality in transitional Beijing', *International Journal of Urban and Regional Research*, 33:4 936–56.
Huang, Y.Q. and Tao, R. (2015) 'Housing migrants in Chinese cities: current status and policy design', *Environment and Planning C: Government and Policy*, 33:3 640–60.
Lefebvre, H. (1982) *The Sociology of Marx*, New York: Columbia University Press.
Lefebvre, H. (1991a) *The Production of Space*, Oxford: Basil Blackwell.
Lefebvre, H. (1991b) *Critique of Everyday Life Volume I: Introduction*, London: Verso.
Lefebvre, H. (1995) *Introduction to Modernity: Twelve Preludes September*, London: Verso.
Lefebvre, H. (1996/1968) 'Right to the city', in E. Kofman and E. Lebas (eds) *Writings on Cities*, Cambridge, MA: Blackwell Publishers.
Lefebvre, H. (2003/1970) *The Urban Revolution*, Minneapolis, MN: University of Minnesota Press.
Lefebvre, H. (2009/1970) *State, Space, World: Selected Essays*, Minneapolis, MN: University of Minnesota Press.
Lin, G. (2007) 'Chinese urbanism in question: state, society, and the reproduction of urban spaces', *Urban Geography*, 28:1 7–29.
Liu, R. (2015) *Spatial Mobility of Migrant Workers in Beijing, China*, London: Springer.
Liu, R., Wong, T.C. and Liu, S.H. (2012) 'Peasants' counterplots against the state monopoly of the rural urbanization process: urban villages and 'small property housing' in Beijing, China', *Environment and Planning A*, 44 1219–40.
Merrifield, A. and Swyngedouw, E. (1996) 'Social justice and the urban experience', in A. Merrifield and E. Swyngedouw (eds) *The Urbanization of Injustice*, London: Lawrence & Wishart.

Solinger, J. (1999) *Contesting Citizenship in Urban China: Peasant Migrants, the State, and the Logic of the Market*, Berkeley: University of California Press.

State Council (2014) A full account of the *Hukou* System Reform, issued July 2014 and gazetted August 2014, No. 25, accessed online at http://www.gov.cn/zhuanti/2014hjzdgg/, accessed January 2017.

Wallerstein, I. (2005) 'After developmentalism and globalization, what?', *Social Forces*, 83:3 1263–78.

Wang, F.L. (2005) *Organizing Through Division and Exclusion: China's Hukou System*, Stanford: Stanford University Press.

Wong, T.C. and Liu, R. (2016) 'Developmental urbanism, city image branding and the 'Right to the City' in transitional China', *Urban Policy and Research*.

Wu, F.L. (2005) 'Beyond gradualism: China's urban revolution and emerging cities', in F.L. Wu (ed.) *China's Emerging Cities: The Making of New Urbanism*, New York: Routledge.

Wu, F.L. (2015) *Planning for Growth: Urban and Regional Planning in China*, New York: Routledge.

Wu, F.L. (2016) 'State dominance in urban redevelopment: beyond gentrification in urban China', *Urban Affairs Review*, 52:5 631–58.

Wu, F.L. Zhang, F. and Webster, C. (2013) 'Informality and the development and demolition of urban villages in the Chinese peri-urban area', *Urban Studies*, 50:10 1919–34.

Xinhua News (2015) Chi juzhuzheng ke xiang liu da fuwu he qi xiang bianli [Residency Card grants migrants six services and seven advantages], accessed online at http://news.sina.com.cn/, accessed January 2017.

Zhang, L. (2001) *Strangers in the City: Reconfiguration of Space, Power, and Social Networks within China's Floating Population*, Stanford, CA: Stanford University Press.

Zhang, L (2004) *China's Limited Urbanization under Socialism and Beyond*, New York: Nova Science Publishers.

36

Urban creativity through displacement and spatial disruption

Sana Murrani

Introduction: urban displacement

This chapter takes Henri Lefebvre's 1968 concept, the 'right to the city' (RTC), as a theoretical guide in its critique of the current understanding of the meaning of the term 'right', particularly in relation to the 'right to drift'. It questions and contests the relevance of the territorial and spatial dimensions of the notions of 'rights' and the 'city' in the context of mass displacement, which poses an immanent challenge to the meanings of these important concepts. Alongside this, it extends Lefebvre's (2003) concept of the 'blind field' to encompass a more holistic view of the meaning of 'urban', with particular reference to the sites that emerge out of displacement and spatial disruption. The incessant process of drifting and displacement prompts the question: can the constant frustrations provoked by disruptions to the mobility of the displaced result in a spatial blind field.

The increase in armed conflict, natural disasters, famine and poor economic prospects at the cusp of the 21st century triggered a resurgent academic interest in migration, displacement and refugee studies in recent years. This is due to the fact that these conditions have collectively contributed to the soaring figures of mass displacement and had a profound impact on the spatial locatedness, sense of belonging and the right to space of many millions of displaced people. The UNHCR has pronounced 2015 and 2016 as the years of 'refugee and migrant crises', with a staggering 12,400,000 people internally or externally displaced in 2015 alone, culminating in a total of 65,300,000 forcibly displaced by the end of that year (UNHCR 2015).

Displacement, as a condition, arises from the forced movement of people from their locality and carries negative connotations both socially and spatially. As a product of social change, urban displacement also refers to the morphological and collective social impact of human displacement on a given context. It manifests itself in such conditions of urban change as gentrification, mass urbanisation and urban sprawl – the direct results of globalisation and the associated politics of neoliberalism. These conditions engender a slow process of displacement, where both the causes and the effects develop over a long period of time.

The work of Ipsita Chatterjee (2014) on the urban exploitation and territorialisation of an almost-dry river bed in Ahmedabad, Gujarat, by displaced communities provides an insight into entrenched attitudes to urban gentrification within the Global South. Meanwhile, *Cities for People, Not for Profit*, edited by Neil Brenner, Peter Marcuse and Margit Mayer (2012), has helped to position urban displacement as a condition that sits at the heart of current debates in urban theory, human geography, urban planning and anthropology. In this volume, scholars from these fields address the necessity to reposition what they refer to as 'capitalist cities', prioritising the construction of cities that respond to social needs rather than the spatial imperatives of the capitalist pursuit of profit (Brenner et al. 2012).

The type of urban displacement this chapter focuses on, however, is of the sudden (yet protracted), violent and forced kind. While acknowledging the inevitably negative nature of this condition, I would argue that it is essential to also harness constructive and positive interpretations of this highly disruptive social and spatial predicament. Hence, I particularly focus on the creativity displaced people display in navigating their way to places of safety. The majority of the public debate (as communicated by various media channels) and research on policymaking concerning the refugee crisis has understandably focused on the disturbing facts of devastation and death, and the loss of livelihoods, heritage and culture. By contrast, there has been scant research into the effect of their obstructed journeys (including the negotiation of intermediate destinations) on the creative spatial and navigational abilities of the displaced.

There is an inherent complexity in the disrupted spatial mobility of displaced people. A recent study of the protracted and fragmented journeys taken by refugees arriving in Italy, Greece, Turkey and Malta uses a series of maps to reveal the intricate and decentralised networks of mobility and routes (Crawley et al. 2016). These maps offer a glimpse into the complexity of the nonlinear drifts of protracted mobility: they do not follow a single flow of movement from places of conflict and human rights abuses to places of safety (in this case, Europe); rather, they include a number of 'sub-flows' involving different countries and different timescales (Crawley et al. 2016).

This drifting to and fro in a constant state of migrancy has become one of the main characteristics of the lives of the displaced as they struggle with what this represents in terms of becoming uprooted from place and identity. Stephen Cairns (2004), however, explores the potential synergies emerging from an unorthodox reading of two seemingly diametrically opposed terms: 'architecture', which delineates rootedness and the groundedness of place, and 'migrancy', which represents displacement in both the spatial and social dimensions. Cairns (2004: 42) establishes that the architecture of migrancy overlaps with the traditional rootedness and locatedness of architecture and place through the act of 'drifting', 'a particular kind of movement that carries ongoing, multiple, intermittent and intensified investments in place'.

Despite a recognition of the challenges the current refugee and migrant crisis pose in terms of the human rights of those seeking refuge at the borders of Europe (and elsewhere), their freedom to cross these borders, to leave or enter different states, does not exist (Miller 2016). It is in the context of this lack of rights and the notion of perpetual drifting, in social and spatial terms, that we can begin to understand 'displacement' and 'emplacement' as ideas that permanently negotiate their locatedness, rootedness and positioning within space, place and time. A number of studies have considered the initiatives of marginalised and displaced people to attempt to situate and re-root themselves within their host countries (Coleman et al. 2012). Maroussia Hajdukowski-Ahmed (2012) has identified, through her studies of projects with refugee women who have suffered trauma, a connection between creativity and building resilience, using the theories of 'being as event' and 'dialectic imagination' developed by Mikhail Bakhtin. Bakhtin portrays creativity as a projection and manifestation of our agency, past and present, cultural and social, in space and time (Hajdukowski-Ahmed 2012: 220–1).

We can thus see two distinct processes emerging from the literature: one which relates to drifting, mobility and the right, or the lack of it, to traverse borders, and another that asserts the ambiguity and perpetual negotiation of that right. The Lefebvrian concepts of 'the right to the city' and the 'blind field' are vital ingredients in the emergence of the creative agency of urban society.

The right to drift

The word 'drift' replaces 'city' in the conceptual term, 'the right to the city', as it offers a flexible dimension that crosses and meanders between multiple fields, including the three identified by Lefebvre (the rural, the industrial and the urban). This meandering enables a much more malleable reading of the relationship between these fields. Lefebvre first introduced the idea of RTC in his work, *Le Droit à la Ville*, during the revolutionary protests of 1968 in Paris. RTC, therefore, has both theoretical and practical dimensions: Lefebvre (1996: 63) asserts its conceptual relevance as an ideology and a practice that opens up existing social, economic, political and spatial systems and structures to new creative possibilities and an 'urban society'. Fundamentally, Lefebvre's original concern, as expressed in RTC, is with two types of rights to urban space: the right to 'appropriate' urban space (that is, the right to its complete use) and the right to 'participate', not just peripherally but centrally, in its 'production'. He calls for cities to be creative centres for the production of the everyday lives of their inhabitants – cities as '*oeuvres*', rather than simply sites of commerce or consumption (Lefebvre 1996: 149).

At this point, it is important to digress in order to consider an earlier idea Lefebvre (2003: 32) established in which he defines the aforementioned three types of fields (or layers of theories and social practices) through which RTC is articulated: the rural or agrarian (encompassing nature – a place of limited production); the industrial (places of fetishised productivity); and the urban (the field of enjoyment, where history meets society and the production of urban society emerges). These fields are not morphologically or spatially distinct, they can enclose or morph into each other. Accordingly, the 'right' referred to in the 'right to the city' meanders and drifts through and between the three fields, not in a nostalgic and traditional sense (as in the desire to return to what is natural and rural), but rather in a newly formulated way, or as Lefebvre (1996: 158) postulates, 'as a transformed and renewed *right to urban life*'.

Marcelo Lopes de Souza (2010) argues that Lefebvre's idea of RTC does not only mean the right to a better life on the basis of an improved and/or reformed democracy, it also means the existence of new possibilities and new worlds, which are constantly negotiated, reinvented and articulated. For example, strategies such as mutual aid, free association and the formation of networks and confederations are seen as tools with which to overcome class exploitation and oppression – local actions with a potentially global impact.

Emphasising the urgent need to interpret the concept of RTC as it was put forward by Lefebvre, Marcuse (2009: 190) eloquently sums up its collective, multiple nature: 'Lefebvre's right is both a cry and a demand, a cry out of necessity and a demand for something more'. It does not merely signify a single individual's right to multiple services, neither is it simply a right to information or to the use of city centres, but rather it encompasses a much more collective meaning: 'the right to a totality, a complexity, in which each of the parts is part of a single whole to which the right is demanded' (Marcuse 2009: 193). David Harvey (2008) echoes Marcuse's interpretation, stating that Lefebvre was not referring to the individual's right to urban existence, but to the collective right to urbanisation in relation to the production and use of space, which Harvey identifies as the true urban revolution.

It is perhaps apposite here to give a brief historical account of the current state of displacement, since it has been fuelled predominantly by attempts to claim these collective urban rights. This is particularly the case across the Middle East, the region from which the vast majority of refugees and migrants currently emerge. The political unrest that erupted in the West during the 1950s and 1960s is well documented, but the contemporaneous revolutions in the Middle East are less widely acknowledged. The attraction of socialism as a political ideal and a way of life spread across the region during this time. However, these movements were not without pitfalls; they often led to brutal oppression and coup d'états (in 1952 in Egypt and 1963 in Iraq), civil war (in 1970 in Jordan/Palestine and 1982 in Lebanon), decades of wars and invasions (between Israel and Palestine in 1967, Iraq and Iran in 1980, followed by the two Gulf Wars in Iraq in 1990/1 and 2003), and finally in 2011, the eruption of the Arab Spring and its suppression, the rise of the so-called Islamic State in the region, and the current war in Syria. Socialism has failed the Middle East; capitalism is now manifest across the region in many different ways, but is mainly experienced through oppression (Al-Ali 2014). Displacement has become a choice, as well as a force in the pursuit of urban social rights (Murrani 2016: 197).

Concurrently, urban social movements in Latin and North America and across Europe are involved in practical interpretations of the RTC concept. These social movements have emerged in response to the global rise of neoliberal urban development in the current era: in some areas they seek to secure and protect the right of all to participate in the city, while elsewhere they attempt to create the right to more open and democratic cities through social and political agency (Mayer 2012: 64). Besides these urban social movements, Lefebvre's RTC concept increasingly features on the agendas of governments, non-governmental organisations (NGOs) and international organisations across the world – in 2016 it appeared at the heart of the United Nations' HABITAT III report on the *New Urban Agenda* (2016: 2), in which it shares its vision of 'cities for all'. However, this ideal of the right to participate in, and openly and democratically appropriate, urban social life can only hold true for those inhabitants who are situated and rooted in a given city; how can RTC and its related concepts reach out to the marginalised, the displaced and the uprooted?

In order to address this question it is necessary to unpack a very important yet convoluted concept in Lefebvre's thought: his reference to the third field, the urban, which implicitly explains the concept of the city. Each of these notions has equivocal characteristics. In certain contexts, this ambiguous idiosyncrasy could be seen as a challenge to the appropriateness of Lefebvre's RTC, yet the concept's very ambiguity simultaneously allows new interpretations to emerge.

Prior to 2015, over 60 per cent of the world's refugees and internally displaced persons were living in urban environments (UNHCR 2001). However, this percentage has now been superseded by external displacement, with people residing in refugee camps and on the borders of states. These spaces are mainly peripheral in a spatial and morphological sense, located on the edges and borders of nations, in deserts or across the sea. Can such spatial elements be regarded as features of the urban? What, in fact, is the urban?

Lefebvre (2003: 116) asserts that the plasticity of the term 'urban' is embedded in its characteristic of being located at the conjuncture of networks of production of space and society, thus inferring that any point in this network can become the centre of urban space-time. He emphasises this plasticity in several definitions of the term: '[T]he urban is a highly complex field of tensions, a virtuality, a possible-impossible that attracts the accomplished, an ever-renewed and always demanding presence-absence' (Lefebvre 2003: 40).

Lefebvre puts forward the idea of the city in association with the concept of the urban as a dialectic relationship, in which the city is the creative hub of activity, consumption and production of urban society. This is predominantly why cities are always associated with the concept and application of the urban. Nonetheless, the idiosyncratic plasticity of the urban as a concept allows for other centres to emerge from the fissures and cracks generated by the misdistribution of wealth, consumption and power, not only within cities but also on their margins and elsewhere (Lefebvre 2003: 118). This confirms that the production of the urban as a concept is bound, in a spatial and morphological sense, not only to cities but also to new societal centres and networks existing on the peripheries – at borders, in camps and deserts, across the sea or in the virtual world. The urban is everywhere.

Lefebvre's notions of RTC and the urban are made manifest in Edward Soja's (1996) 'thirdspace' ('a space of radical openness') and bell hooks's (1990) 'margin' (real and imagined spaces on the margins), thus revealing themselves to be malleable concepts. This plasticity disrupts urban and social centres in pursuit of a 'radical creative space', where consciousness is rooted in identity and the experience of everyday life (Soja 1996: 99), and which, as hooks (1990: 153) eloquently asserts, 'gives us a new location from which to articulate our sense of the world'. The conditions of protracted drifting and displacement impose an element of openness and continuous disruption on space and social existence, yielding the potential for new creative possibilities.

Urban creativity through the concept of the 'blind field'

According to Lefebvre, 'blindness' as an ideology is situated between the 'presentation' of facts as they appear to us and their 're-presentation' (the interpretation of the facts). These moments do not follow each other smoothly; there are ruptures and conflicts between the presented and the re-presented. The blindness or the blind field occurs during these moments of misinterpretation and misunderstanding (leading to a failure of recognition, and mental and social false consciousness) as our consciousness oscillates between the presentation and the re-presentation (Lefebvre 2003: 30). Rob Shields (2014) explains that, for Lefebvre, the blind field is a manifestation within the urban of 'enigmatic and excessive' forces of knowledge that cannot be fully perceived or comprehended from any one viewpoint. He further asserts that the blindness or dark moments are attempts at actualising the virtuality of the urban (Shields 2014: 53).

Since its introduction in 1970, the Lefebvrian notion of the blind field has featured in academic literature concerned with spatial practice, place-making, mapping and urban studies. However, while acknowledging Lefebvre's influence on spatial theories, Les Roberts (2012: 16) critiques the limitations and rigidity of the bonds between the Lefebvrian concepts and their spatial grounding – in particular, between manifestations of the concept of the blind field and the urban. Lefebvre himself insists that the urban field is new and still unknown; hence, its association with the blind field is due to its embrace and signification of 'difference'. He asserts that the urban field is a manifestation of urban thought; it is a 'reflection of urban society', not urbanism (Lefebvre 2003: 36–7).

This challenge to earlier perspectives on urban thought has contributed to the eruption within the field of urban theory of what Neil Brenner (2013: 92) describes as the critique of 'every imaginable issue – from the conceptualization of *what* urbanists are (or should be) trying to study to the justification for *why* they are (or should be) doing so and the elaboration of *how* best to pursue their agendas'. Brenner (2013: 109) elaborates on nine different conceptualisations for the mapping of the planetary urban condition, favouring the identification of a 'unit-like' urban character, which this chapter infers to comprise the creative agency of urban society.

Philosophers, psychologists, sociologists and artists have all been captivated by the idea of creativity and its association with individual and collective consciousness. However, the study of creativity as a phenomenon did not encompass the spatial dimension until geographers in the late 20th century became more interested in the impact of spatial situations and the environment on the emergence of creativity, as opposed to the prevailing theories (Meusburger et al. 2009: 2). Until that point, creativity had been defined as the by-product of a mind lacking a pre-ordered plan or formulated goal (Bohm 2004: 32), or alternatively, as a network of organisation and interaction between the consciousness of a number of individuals and their environment (Csikszentmihalyi 1998).

The mid-20th century brought with it the first theoretical insight into the origins of creativity, manifested in the psychodynamic approach. At its centre was the earlier notion propounded by Sigmund Freud (1908) that the unfolding of creativity is rooted in the tensions between conscious reality and unconscious drivers in the environment. Ernst Kris (1952) subsequently put forward a theory of psychoanalysis in which he connected creativity to frustration, or what he termed the 'adaptive regression' to unmodulated thoughts that can occur during problem-solving activities. This theory was later abandoned due to its over-reliance on evidence from case studies of highly creative individuals and the uncontrolled nature of the environments in which these case studies were conducted – the field of psychology at the time placed greater value on replicable, measured and strictly controlled scientific experiments (Sternberg 1998: 6).

Later research and experiments, however, further developed the link between creativity and frustration, showing that sudden and unexpected disruptions to everyday routines could lead to creative outcomes. When a strike recently affected London's underground system, for example, commuters quickly responded by collectively finding alternative ways to reach their workplaces or homes. Economists examined the data generated by the use of travel cards during and after the strike and realised that not only had commuters found alternative ways to navigate their city, but also one in 20 went on using their new commuting routes for various reasons after the dispute was resolved (Harford 2016: 15).

Tim Harford (2016) writes of the intrinsic nature of human beings to seek to improve a particular condition, especially in situations where daily routines are subject to constant disruptions. He suggests that creativity cannot exist in isolation from external stimuli, both social and spatial, but is fundamentally triggered by the process of problem-solving, in which failed attempts in one area add to the collective accumulation of fresh insights and the generation of new tools that might work elsewhere (Harford 2016: 25). Harford was not referring to the spatial disruption experienced by refugees attempting to cross the Mediterranean or walk across the borders of mainland Europe, but if the concept is transferable, then these displaced people arriving in Europe potentially have developed a heightened sense of spatial creativity. This hypothesis, which I term 'urban creativity', is not to be confused with street art (Neves 2016); also, it is 'urban', not in the sense of inhabiting a city, but rather in the social and spatial, open-network sense of the concept established in the previous section.

It is important to note that creativity, which was originally thought to be an innate characteristic of a single individual's consciousness (Boden 1994), arises from complex and lengthy dialogues and dynamic interactions between creators/actors and their environment (Simonton 2000). Peter Meusburger (2009: 111) emphasises that spatiality and locatedness are both equally crucial in triggering creativity. This idea has led to ethnographic explorations of creativity in makeshift refugee camps (Agier 2002), as well as collaborative research with refugees and displaced people in their host countries. Recent initiatives such as Counterpoints Arts, a London-based cultural arts programme, and Chatterbox, a social enterprise based at School of African and Oriental Studies, University of London, provide a platform whereby migrants and refugees can connect with local communities through practising the arts and participating in cultural enrichment programmes.

Such initiatives require systematic development, but once they gain momentum, they can be instrumental in helping the integration of refugees within local communities. However, although there have been similar studies of displacement and creativity within situated spatial environments such as camps and cities (Betts et al. 2015), research on displaced persons' creativity while they are mobile – that is, during their journeys – has been scarce. However, one such piece of research conducted by the Open University in the UK, in collaboration with France Médias Monde, has focused on mapping refugees' media journeys to Europe through an analysis of their creative use of smartphones and social media (Gillespie et al. 2016). *En route*, refugees rely on mobile technologies and social media to navigate their way across Europe and to stay abreast of the latest news on border controls and route diversions (Gillespie et al. 2016: 11).

At this point, I return to the concept of drifting in relation to mobility. Cairns (2004: 40) believes that drifting 'signifies a discomfort that arises when the bonds between proper being and place are denaturalized'. This precise discomfort, and the associated frustration of dis-locatedness, situates migrants and refugees in a space of contingent mobility, riddled with disruptions. Studies have shown that disruptions to everyday spatial mobility have been proven to increase and stimulate new patterns of creative thinking (Meusburger 2009: 140), yet the intensity of any form of disruption to everyday life, which affects emotion and memory, has a profound nonlinear influence on the levels of creativity generated (Byron et al. 2010). It is, however, suggested that in focused situations of spatial problem-solving, where the individuals are subjected to high levels of negative emotions and stress, they are likely to experience an increase in the production of creative thought that is persistently directed towards finding solutions (De Dreu et al. 2012). This finding confirms the validity of exploring creativity in the context outlined in this chapter.

The continuous implementation of increasingly rigorous border controls, such as fencing, policing and camps, overlooks the needs of people in favour of 'border politics' (Andersson 2014) and exacerbates the problems facing migrants by making their paths to safety even more treacherous. Simultaneously, the altered state of disruption to their spatial existence contributes to a heightened creative sense of the social navigation of space. This protracted oscillation between the camp and the journey, the node and the network, the point and the field, is precisely how Lefebvre's concept of the blind field manifests itself within urban displacement.

Policymakers and nations have been blind to the wider impact of knee-jerk border policies on protracted displacement, while aid agencies, depending on their focus, have assisted some and alienated others, and by doing so, have been blind to the larger network of support existing in the countries of origin and those *en route*, and in the nations of arrival. This is in addition to the deliberate moral blindness of people traffickers who routinely place vulnerable lives in danger. As the displaced are distracted by this blindness, they search for other, more illuminated paths to explore, for new possibilities where such blindness, as Lefebvre (2003: 31) asserts, represents 'an enclosure to break out of, a consecration to transgress'. The creativity of the displaced lies within the slippages and gaps between such blind fields.

Conclusions

The perpetual drifting of the displaced, in spatial and social terms, drives a contingent negotiation between notions of displacement and emplacement through the acts of positioning, rootedness and locatedness in place and time. While the right to drift is a contested concept for many who are uprooted and displaced, drifting and meandering as forms of mobility allow a flexible and malleable conjuncture that is ambiguous, both spatially and socially. At the same time, protracted displacement as a condition impacts the spatial locality and social rootedness of the displaced and disturbs their ecologies, their networks and their way of life. On the one hand, then, this disturbance clearly obstructs their sense of locatedness and belonging; yet on the other, it creates places of renewal, experimentation and opportunity, where new creative directions emerge out of the plasticity of this condition.

This plasticity is the essence of the urban, a conceptual thought that this chapter has argued also applies to the contested spaces of the marginalised. Similarly, the notion of rights, as in the right to the city, shares with the urban those flexible characteristics that enable it to encompass the total collection of rights of representation, as well as the creative production of space and society through the right to drift between the camp and the journey, between the point and the field.

To drift is to flow to and fro, and to drift amid displacement is to continuously flow in space and time. The challenges faced by the displaced through protracted drifting and the spatial disruption to their daily lives stimulate the emergence of new patterns of creative thinking that are specific to spatial problem-solving. Thus, drifting becomes the essence of the renewal and re-emergence of their creative agency, an urban creativity amongst the blind fields. By focusing on this specific creative agency, an argument emerges in which spatial disruptions to the mobility of the displaced generate new urban centres through the process of protracted drifting. Therefore, migration, mobility and drifting are processes that are instrumental for urban renewal, future planning and urban policy; these disciplines could be facilitated and enhanced by a greater understanding of the creative drifting of refugees and migrants. Likewise, policymakers, governments and host communities would benefit greatly from this alternative insight into the creative agency of the displaced.

References

Agier, M. (2002) 'Between war and city: towards an urban anthropology of refugee camps', *Ethnography*, 3:3 317–41.

Al-Ali, Z. (2014) *The Struggle for Iraq's Future: How Corruption, Incompetence and Sectarianism Have Undermined Democracy*, New Haven: Yale University Press.

Andersson, R. (2014) *Illegality, Inc.: Clandestine Migration and the Business of Bordering Europe*, Berkeley, CA, United States: University of California Press.

Betts, A., Bloom, L. and Weaver, N. (2015) *Refugee Innovation: Humanitarian Innovation that Starts with Communities*. Available HTTP: <https://www.rsc.ox.ac.uk/> (accessed February 2017).

Boden, M.A. (ed.) (1994) *Dimensions of Creativity*, Cambridge, MA, United States: The MIT Press.

Bohm, D. (2004) *On Creativity*, 2nd edn, New York: Routledge.

Brenner, N. (2013) 'Theses on urbanization', *Public Culture*, 25:1 85–114.

Brenner, N., Marcuse, P. and Mayer, M. (eds) (2012) *Cities for People, Not for Profit: Critical Urban Theory and the Right to the City*, London: Routledge.

Byron, K., Khazanchi, S. and Nazarian, D. (2010) 'The relationship between stressors and creativity: a meta-analysis examining competing theoretical models', *Journal of Applied Psychology*, 95:1 201–12.

Cairns, S. (2004) *Drifting Architecture and Migrancy*, London: Routledge.

Chatterjee, I. (2014) *Displacement, Revolution, and the New Urban Condition: Theories and Case Studies*, London: Sage.

Coleman, D., Glanville, E.G., Hasan, W. and Kramer-Hamstra, A. (eds) (2012) *Countering Displacements: The Creativity and Resilience of Indigenous and Refugee-ed Peoples*, Alberta: University of Alberta Press.

Crawley, H., Düvell, F. and Sigona, N. (2016) *No Direct Flight: New Maps Show the Fragmented Journeys of Migrants and Refugees to Europe*. Available HTTP: <http://theconversation.com/> (accessed January 2017).

Csikszentmihalyi, M. (1998) 'Implications of a systems perspective for the study of creativity', in R.J. Sternberg (ed.) *Handbook of Creativity*, Cambridge: Cambridge University Press.

de Dreu, C.K., Nijstad, B.A., Baas, M., Wolsink, I. and Roskes, M. (2012) 'Working memory benefits creative insight, musical improvisation, and original ideation through maintained task-focused attention', *Personality and Social Psychology Bulletin*, 38:5 656–69.

Freud, S. (1908) 'The relation of the poet to day-dreaming', *Collated Papers*, 4 173–83.

Gillespie, M., Ampofo, L., Cheesman, M., Faith, B., Iliadou, E., Issa, A., Osseiran, S. and Skleparis, D. (2016) *Mapping Refugee Media Journeys: Smartphones and Social Media Networks Research Report*. Available HTTP: <http://www.open.ac.uk/> (accessed February 2017).

Hajdukowski-Ahmed, M. (2012) 'Creativity as a form of resilience in forced migration', in D. Coleman, E.G. Glanville, W. Hasan and A. Kramer-Hamstra (eds) *Countering Displacements: The Creativity and Resilience of Indigenous and Refugee-ed Peoples*, Alberta: University of Alberta Press.
Harford, T. (2016) *Messy: How to Be Creative and Resilient in a Tidy-Minded World*, London: Little, Brown Book Group.
Harvey, D. (2008) 'The right to the city', *New Left Review*, 53 23–40.
hooks, b. (1990) *Yearning: Race, Gender, and Cultural Politics*, Toronto: Between-the-Lines.
Kris, E. (1952) *Psychoanalytic Exploration in Art*, New York: International Universities Press Inc.
Lefebvre, H. (1996) 'Right to the city', in E. Kofman and E. Lebas (eds) *Writings on Cities*, Cambridge, MA: Blackwell Publishers.
Lefebvre, H. (2003/1970) *The Urban Revolution*, Minneapolis, MN: University of Minnesota Press.
Marcuse, P. (2009) 'From critical urban theory to the right to the city', *City*, 13:2–3 185–96.
Mayer, M. (2012) '"The right to the city" in urban social movements', in N. Brenner, P. Marcuse and M. Mayer (eds) *Cities for People, Not for Profit: Critical Urban Theory and the Right to the City*, Oxford: Routledge.
Meusburger, P. (2009) 'Milieus of creativity: the role of places, environments, and spatial contexts', in P. Meusburger, J. Funke and E. Wunder (eds) *Milieus of Creativity: An Interdisciplinary Approach to Spatiality of Creativity*, Dordrecht: Springer-Verlag.
Meusburger, P., Funke, J. and Wunder, E. (2009) *Milieus of Creativity: An Interdisciplinary Approach to Spatiality of Creativity*, Dordrecht: Springer-Verlag.
Miller, D. (2016) 'Is there a human right to immigrate?', in S. Fine and L. Ypi (eds) *Migration in Political Theory: The Ethics of Movement and Membership*, Oxford: Oxford University Press.
Murrani, S. (2016) 'Baghdad's thirdspace: between liminality, anti-structures and territorial mappings', *Cultural Dynamics*, 28:2 189–210.
Neves, P.S. (2016) *Street Art & Urban Creativity Scientific Journal: Centre, Periphery: Theory*, 2:2. Available HTTP: <http://www.urbancreativity.org/> (accessed February 2017).
Roberts, L. (2012) 'A spatial anthropology', in L. Roberts (ed.) *Mapping Cultures: Place, Practice, Performance*, New York: Palgrave Macmillan.
Simonton, D.K. (2000) 'Creativity: cognitive, personal, developmental, and social aspects', *American Psychologist*, 55:1 151–58.
Shields, R. (2014) 'The virtuality of urban culture: blanks, dark moments, and blind fields', in M. Darroch and J. Marchessault (eds) *Cartographies of Place: Navigating the Urban*, Montreal: McGill-Queen's University Press.
Soja, E.W. (1996) *Thirdspace: Journeys to Los Angeles and Other Real-and-Imagined Places*, Malden, MA: Blackwell Publishers.
de Souza, M.L. (2010) 'Which right to which city? In defence of political-strategic clarity', *Interface: A Journal For and About Social Movements*, 2:1 315–33.
Sternberg, R.J. (ed.) (1998) *Handbook of Creativity*, Cambridge: Cambridge University Press.
UN HABITAT III (2016) *New Urban Agenda*. Available HTTP: <https://www2.habitat3.org/> (accessed November 2016).
UNHCR (2001) *Urban Refugees*. Available HTTP: <http://www.unhcr.org/uk/> (accessed February 2017).
UNHCR (2015) *Global Trends: Forced Displacement in 2015*. Available HTTP: <http://www.unhcr.org/> (accessed September 2016).

37

The 'newcomers'' right to the city

Producing common spaces in Athens and Thessaloniki

Charalampos Tsavdarolgou

Introduction

The so-called migration crisis in Greece has been a major issue during 2015–16. According to the United Nations (UN 2016), in one year 851,319 people have entered and crossed the country. On 8 March 2016, following a gradual restriction of access to the Balkan route based on ethnic origin criteria, the border between Greece and Former Yugoslav Republic Of Macedonia (F.Y.R.O.M.) was closed for all third-country migrants. In the aftermath of this closure, and following the implementation of the EU-Turkey deal (European Commission 2016), over 60,000 refugees are suddenly trapped in Greece, half of them in Athens and Thessaloniki (Coordination Centre for the Management of Refugee Crisis in Greece 2016). More than 10,000 refugees are settled in 14 state-run camps in the outskirts of Athens, 20,000 in 11 state-run camps in the outskirts of Thessaloniki and about 2,000 in self-organised occupied buildings in the urban core of both cities. Focused on this context, this paper examines the newcomers' right to the city as it is expressed by Greek state policies and the solidarity practices of newly arrived refugees.

Specifically, in this paper I aim to examine the emerging spatial commoning practices of migrants and refugees. Although there is a vast literature (Gabiam 2012; Mountz et al. 2013; Ihlen et al. 2015) on social philanthropy, humanitarianism, Non-Governmental Organisations' (NGOs') activities and state immigration policies, there have been few attempts to research the ongoing refugees' self-organised actions that produce seemingly anonymous, however highly personal and collective housing common spaces.

In the above context my basic argument is that despite the vivid and increasingly popular discussion on commons (De Angelis 2007; Hardt and Negri 2009), there have been few attempts to make the connection with the ongoing migrant crisis. In recent years, the discussion on urban commons has revolved mainly around critical geographers' approaches that focus on accumulation by dispossession (Glassman 2006; Harvey 2012; Hodkinson 2012) and conceptualise commons as a new version of the right to the city (Brenner et al. 2009; Mayer 2009; Kuymulu 2013). At the same time, during the current migrant crisis, the newcomers are settled in inadequate housing facilities on the outskirts of cities, which gradually become ghettoised, and face discriminatory access to facilities essential for health, security, comfort and nutrition. However, the previously described migrant urban policies do not stay uncontested. In the case of Athens and

Thessaloniki, the newcomers claim spatial justice and visibility as well as the right to city and to adequate housing; and in collaboration with solidarity groups they occupy abandoned buildings in the urban core and tend to transform them into housing common spaces. Moreover, in their effort to survive, migrants do not only challenge the state-run camps, but seek to negotiate and go beyond cultural, class, gender, religious and political identities.

For the purposes of the paper the social data is collected from both qualitative and quantitative processes and a methodological tool, which is applied for the determination of these dynamic characteristics approved by participatory action research, ethnographic analysis, semi-structured interviews and the collection of articles of local presses and Web pages. It is clear that refugee research participants are a relatively vulnerable research population due to their legal status. During my research in 2016–2017, some participants felt uncomfortable discussing and reflecting on the conditions of their shelter. The anonymisation of data ensured that any potential uneasiness which may arise is addressed and no physical, psychological or social adversities affect the participants as a result of taking part in the study. Thus, it is necessary to mention that the names of most interviewed individuals have been changed, using culturally appropriate names, to protect their identity.

The paper is structured as follows. The next section engages with the theoretical discussion on open dialectics and a post-colonial approach to the production of the common space. The following section explores the features of the refugees' right to the city and to adequate housing vis–à–vis state-run camps in Athens and Thessaloniki. The next one explores the socio-spatial features of the refugees' common spaces in Athens and Thessaloniki. The final section draws some concluding remarks on the social relations and modes of communication, through which the communities of the refugee common space are formed.

Open dialectics and autonomy of migration on the production of common space

Several critical scholarly analyses (De Angelis 2007; Hardt and Negri 2009; Caffentzis 2010) conceptualise the commons using three things at the same time: common pool resource, community and commoning. The people who, through commoning, constitute communities that self-organise, sharing the common pool resources in non-commercial ways, are called 'commoners'. According to Harvey (2012: 73), the commons are constructed as an unstable and malleable social relation between 'a particular self-defined social group and those aspects of its actually existing or yet-to-be-created social and/or physical environment'. Furthermore, De Angelis (2010: 955) makes the point that 'there are no commons without incessant activities of commoning', it is through (re)production in common that 'communities... decide for themselves the norms, values and measures of things' (ibid.: 955). Moreover, several scholars (Caffentzis 2010; Mattei 2011) make the point that the commons have to be separated from the dipole of private or state management. In this brief review on the commons a point worth mentioning is Blomley's (2008: 320) proposal that 'the commons... is [sic] not so much found as produced... the commons is [sic]a form of place-making'. Finally, Stavrides (2014: 548) suggests that the spaces of common emerge as 'thresholds', which are 'open to usage, open to newcomers'.

In order to contribute to a fuller understanding and accurate connection between the concepts of 'commons' and the 'space', I draw on Lefebvre's open dialectic spatial approach. Lefebvre (1991) argues that space is not an empty container filled with actions, images, relationships and ideologies, but it constitutes a social product or a complex social construction based on social values and the social production of meanings, which affects spatial practices and perceptions.

Lefebvre's main method is based on trialectic analysis, i.e. space is diversified into the spatial triad, i.e. perceived-conceived-lived space (ibid.: 11; 38; 39; 40; 50; 53; 68; 73). The perceived is materialised socially produced space, which can be determined empirically. It can be measured and described. Perceived space with spatial practice embraces production and reproduction, and the particular locations and spatial sets characteristic of each social formation (ibid.: 38). Conceived space is mentally constructed, it is tied to the relations of production and to the 'order' which those relations impose, and hence to knowledge, to signs and to codes (ibid.: 33). Lived space is directly lived space; it is alive; it speaks. It is practically and directly experienced social space. It embraces the affective, bodily lived experience, the sense of passion, of action and lived situations; it is formatted from everyday life; this is the space of the everyday activities of inhabitants (ibid.: 39; 40).

In this theoretical framework, I propose to connect the Lefebvrian approach with the aforementioned analysis on commons, in order to conceptualise the concept of the 'common space'. In the common space the physical-perceived space is the spatial practice of collective sharing of the means of (re)production and existence. The physical space of common pool resources is constituted, generated or reclaimed each time by social-commoning spatial practices. Finally, commoners, the users of common space, through commoning practices, establish their communities.

Moreover, in order to conceptualise the power relations in the production of the refugee common space, I build on the 'autonomy of migration' (Papadopoulos and Tsianos 2013: 184), which refers to a rapidly developing series of ideas that reflect a kind of 'Copernican turn in migration studies' (Casas-Cortes et al. 2015: 895). According to the autonomy of migration idea, the focus has to shift from the apparatuses of control to the multiple and diverse ways in which migration responds to, operates independently from, and in turn shapes those apparatuses and their corresponding institutions and practices.

From this point of view, contemporary refugee housing common spaces could be seen as open communities of commoners, which through their spatial practices of commoning destabilise the state-led policies and seek to (re)claim both the physical and the social space of the city producing unique collective common spaces. Such a framework seems helpful in explaining the spatialities of recent refugee common spaces.

The refugees' right to the city and to adequate housing vis-à-vis state-run camps in Athens and Thessaloniki

In order to explore the refugees' right to the city I draw attention to Lefebvre's work *Right to the City* (1996). One of the basic theses and point of departure for Lefebvre was that:

> the city [is] a projection of society on the ground that is, not only on the actual site, but at a specific level, perceived and conceived by thought… the city [is] the place of confrontations and of (conflictual) relations(…), the city [is] the 'site of desire'… and site of revolutions.
>
> *(Ibid.: 109)*

For Lefebvre the right to the city embodies and goes beyond:

> the rights of ages and sexes (the woman, the child and the elderly), rights of conditions (the proletarian, the peasant), rights to training and education, to work, to culture, to rest, to health, to housing.
>
> *(Ibid.: 157)*

Finally, Lefebvre's concept of the right to the city challenges the notion of citizen. In his thought, citizenship is not defined by membership in the nation-state but is based on membership in inhabitance. As Purcell (2003: 577) notes:

> everyday life is the central pivot of the right to the city: those who go about their daily routines in the city, both living in and creating space, are those who possess a legitimate right to the city.

After World War II the refugees' right to the city and to adequate housing was recognised as part of the right to an adequate standard of living in the 1948 Universal Declaration of Human Rights (UN 1948) and in the 1966 International Covenant on Economic, Social and Cultural Rights (UN 1966). Moreover, the United Nations Committee on Economic, Social and Cultural Rights has underlined that the right to adequate housing should not be interpreted narrowly (UN 2009). Rather, it should be seen as the right to live somewhere in security, peace and dignity. Furthermore, the European Council (ECRE 2007) recognises that the living environment and conditions in terms of housing are key to the integration of refugees and migrants. Since 2007 Greece has adapted the Council Directive for the minimum standards for the reception of refugees (Presidential Decree 2007). The characteristics of the right to the city and to adequate housing are clarified mainly in the Committee's general comments No. 4 (UN 1991) and must meet the following criteria: security of tenure, availability of services, materials, facilities and infrastructure, affordability, habitability, accessibility and cultural adequacy. Finally, it is emphasised that housing is not adequate if it is cut off from employment opportunities, health-care services, schools, childcare centres and other social facilities, or if it is located in polluted or dangerous areas. In contrast to the above criteria, the state-run refugee camps in the case of Athens and Thessaloniki are overcrowded dilapidated former factories and old military bases, where there is a dire lack of basic necessities, such as running water. Such facilities include derelict warehouses in filthy conditions that appear unfit for habitation.

According to the Syrian refugee Ahmed, who is living in Oreokastro camp in Thessaloniki:

> The whole situation is disastrous. Immigrants' rights have been totally destroyed. Camps are full of germs and diseases. There's unbearable heat in the summer, unbelievable cold in the winter. The camps are all situated outside the city, none of them is anywhere near other people.
>
> *(Research interview, 4 November 2016)*

Moreover, according to several NGOs' reports (Amnesty International 2016; International Rescue Committee 2016; Médecins Sans Frontières 2016) and the report of the Parliamentary Assembly of the Council of Europe (2016), the camps do not meet international standards. They are located in extremely polluted and dangerous environments, close to or inside industrial zones, oil refineries and pesticides facilities. Infrastructures, schools, supermarkets and social life are remote, and most of the camps are not connected with public transportation. The reports reveal dirt-strewn warehouses lined with tents pitched on filthy concrete floors. The tents have been placed too tightly together, the air circulation is poor, and supplies of food, water, toilets, showers and electricity are insufficient. Consequently, the refugees have to survive in appalling and precarious housing conditions, struggling against: cold or hot weather, illnesses, psychosocial distress, lack of food, energy and water supplies.

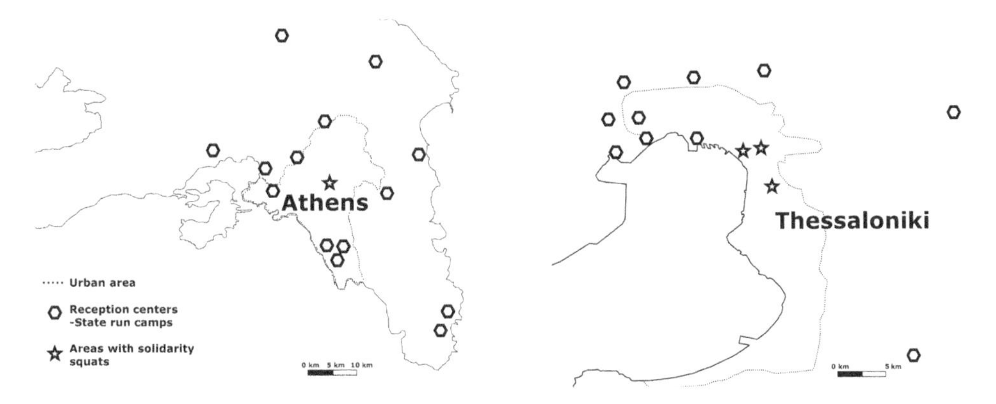

Figure 37.1 State-run refugees' camps in Athens and Thessaloniki. *Source*: the author.

Refugee common spaces

In recent years Athens and Thessaloniki have been hit by an unprecedented turmoil that is expressed socially, economically and spatially (Hadjimichalis 2011; Koutrolikou 2016). One of the main consequences of the socio-spatial crisis was that several public (schools, hospitals) and private buildings (houses, hotels) were abandoned in the centre of these cities (Ministry of Environment and Energy 2014). From autumn 2015 to the summer of 2016 refugees' solidarity groups occupied several of these empty buildings and turned them into housing projects for hundreds of newcomers. According to Moving Europe (2016), about 2,000 refugees are hosted in squats, which are run by both refugees and solidarity groups.

Each squat has a different level of political influence and a distinct character. For instance, in Thessaloniki, housing squat for immigrants Orfanotrofio is both a housing project and a social centre for refugees; Hürriya community squat provides solidarity-based housing for refugee families; and Nikis squat was an anti-authoritarian housing project that is transformed to a refugee shelter. In Athens, Notara 26 is a self-organised housing structure that accommodated approximately 3,500 people until summer 2016; School Squat 2 evokes boisterous, family-style living; School-5th Likio is currently housing 400 people; Strephi Squat is only for women and their children; Dervenion 56 functions as a hub for various activities, such as: a kitchen, food, clothes, hygiene and medicine supplies. City Plaza, the most publicised, is billed as the Best Hotel in Europe.

According to the Housing Squat for Refugees and Immigrants Notara 26, (2016: 2):

> We are squatting an empty public building in Athens, 26 Notara Str, in order to territorialize our solidarity towards refugees/immigrants to cover their immediate needs. This project doesn't stand for philanthropy, state or private, but rather for a self-organized solidarity project, wherein locals and refugees-immigrants decide together. The decisive body is the squat's open assembly where everyone is welcome to participate.

As outlined by the Solidarity Initiative to Economic and Political Refugees (2016), refugee families from different nationalities, together with people of solidarity, are working collectively for the cleaning, repairing and organisation of several occupied spaces. They can be seen therefore as projects of self-organisation and solidarity, as centres of struggle against racism and exclusion, for the right to the city, decent living conditions and equal rights. Collective kitchens, kindergartens, medicine and clothes stores are set up in the self-managed and self-financed structures. Furthermore, according to Theodorou, lawyer and member of the occupied City Plaza Hotel:

> It was a gesture to reclaim the right of the visibility of refugees because we feel that [the Greek government] is trying to hide them on the outskirts of the city.
>
> *(Strickland 2016: 3)*

According to several reports, scholars and interviews (Haddad 2016; Kantor 2016; Karyotis 2016), the occupied refugee shelters are managed as commons through participatory processes. Locals and refugees cook together and eat around the same table; they take decisions together in horizontal assemblies; they recognise each other's culture and customs and overcome preconceptions and stereotypes. In the words of two informants, Hassan and Gamal, two Palestinians in Notara squat, 'Here, we are free. We decide on common matters together. It's better than being locked up in military camps' (research interview, 15 June 2016). Each squat is run by its own assembly, which usually takes decisions by consensus. According to the informant Murad:

> Squats are run without government or NGOs' influence and rely on donations and manpower from independent volunteers. Responsibility is divided among the residents.
>
> *(Research interview, 17 June 2016)*

Moreover, in the words of the informant Alfarawan:

> We manage to create a solid unity with the trustworthy people. Migrants and locals operate as a collective force. Every week we decide, together, about the operation of the house. Every resident is responsible for certain things, for the protection of the building, for the preparation of meals, for cleaning parts of the building, for maintenance. Of course, those who are more specialized in certain things got to work more on those things. But I must admit most people are trying to learn from [each] other so that they could best participate and act responsibly in the collective endeavor.
>
> *(Research interview, 16 December 2016)*

Against the enforced segregation, solidarity initiatives create a common language and common spaces of action for locals and refugees. In contrast with the charitable and sometimes victim-centric ethos of many organisations working in the state-run camps, the aim is to build a culture of mutual respect. Hassan worked in information technology in Syria and now he is working to set up the Wi-Fi network in School-5th Likio squat in Athens; Fatima was an Arabic teacher in Syria and now she teaches class every day from 5–7 p.m. in Micropolis in Thessaloniki. Ahmed from Afghanistan started giving language classes to the other residents, and he says:

> I like so much giving classes and it is very good to have something to do that makes sense. I would like to do more than this. Until now I was just surviving in Greece. Now I can say I am living.
>
> *(Welcome to Europe 2016)*

Raha from Syria says about her experience in the occupied buildings:

> Our everyday life is full of vitality, it is not about mere survival or about sticking to certain habits and routines, it is much more interesting and joyful than the usual everyday life of an ordinary person. First of all, we do not have a boss, we do not have to wake up every

> morning and obey the orders of some superior. Nobody order [sic] us what to do or not to do, we make decisions about everything together. Every person has to develop a conscience and then to act according to their conscience. This is our life, it is not a life revolving around money and work, it is a life of friendship and sharing.
>
> *(Research interview, 27 December 2016)*

Thus it can be argued that in the emerging common spaces, the refugees shape the sense of belonging, security and personal well-being, and along with the support of volunteers, they have access to food, healthcare, education and employment. For this to occur, the mode of communication, the characteristics and identities of the participants, both locals and refugees, are confronted with their limits, modified and troubled. The process of setting up the housing common spaces is based on collective practices, mutual aid and respect, horizontal organisation and emotional, communicative and aesthetic interactions.

The transformation of the physical space of the occupied buildings into a common space took place as the buildings started to acquire characteristics of the 'threshold'. According to Stavrides (2012: 589):

> Common spaces emerge as threshold spaces, spaces not demarcated by a defining perimeter. Whereas public space bears the mark of a prevailing authority that defines it, common space is opened space, space in a process of opening toward newcomers.

Thus common spaces are 'porous, spaces in movement, space passages' (ibid.). The mode of communication and the social relations of the participants, both locals and refugees, give the occupied buildings their porous, threshold character.

Until the day of the occupation, the aforementioned buildings had the typical characteristics of enclosed spaces, with clear borders between private and public space. Specifically, the majority of the buildings are state public spaces (i.e. schools or hospitals) or private hotels, where the government and municipal authorities authorized the permitted uses and functions. Conversely, following the occupation, the squatted buildings acquired the features of common space. The social relations and the commoning practices of the participants have destabilised and altered the boundaries between private and public, personal and political. The occupied buildings combine elements of collective space and personal space. For this reason, the multitude of participants who take the buildings in their hands passionately and consistently take care and defend them, as if they are their personal space, and simultaneously in collective ways protect them both from state power and the varied and constantly reproducing systems of domination.

Exemplary, *inter alia*, are the practices of improvisations and experimental modes of communication expressed by several groups. In each squat there is a: reception group, kitchen-cooking group, cleaning group, technical support group, education and childcare group, multimedia-communication-radio group, legal group, medical care group, guard-security group and translation group. Moreover, art groups, library groups and 'composure' groups have been established. In addition, lectures are organised as well as poetry, music, and theatre events. At the same time, the participants' action repertoires, mode of communication and commoning practices included, among others, dance festivals, vegetable gardens, collective sleeping places and many more components of a self-sufficient commune life. Finally, the paintings, the photographs, the handmade t-shirts, the makeshift placards, the anti-government, anti-racist and anti-fascist slogans and the soundscape of the occupied buildings with the voices of the people, the percussion instruments, form an unpredictable and subversive common space.

According to the report of Kantor (2016: 4) in the School Squat 2 in Athens:

> There is familiarity and freedom. As Mohammad prepares dinner, a cluster of Syrian and Lebanese guys in their 20s debate which music to play on the loudspeaker, finally deciding on an Arabic remix of Adele. A Syrian Kurdish woman peels through a large milk carton of onions, and inside the tiny toolshed, cucumbers are being cut lengthwise twice then sliced [for] Arabic-salad style. The smell of boiling eggplant carries out of the open door and past children playing obstacle games with Spanish volunteers.

Furthermore, it can be argued that the participants' commoning, the various modes of communication and their social relations, have developed a culture of coexistence, in which the multiple identities are troubled and questioned, as the multitude of people is constantly confronted with their political, cultural, class, racial and gender identities. Rima, a young Syrian woman, says:

> For the first time since I am in Greece we are living in an atmosphere that makes it possible to speak with our neighbors no matter where they come from. It is the first time that I found friends from Afghanistan and the first time that we start to understand that only united we can be strong.
>
> *(Welcome to Europe 2016)*

Moreover, it is worth noting that in the occupied solidarity common spaces, volunteers and activists work to protect the basic dignity of vulnerable groups like women, children and disabled people, creating safe places. These spaces represent for many of the refugees the only opportunity to express their cultural practices and gender identities openly. It also allows for people of different faiths, socio-economic backgrounds, ages, abilities, ethnicities, ages and skin colours to converge.

When I asked Amena, who fled with her child from the threats of her violent ex-husband, what is most important for her in the occupied hotel City Plaza, she did not need to think one second about it:

> For me the most important is that I found safety. I have for the first time since long ago a room with a door that I can lock if I need this. There is always someone at the entrance of the hotel [City Plaza], checking who is coming and going. And there are a lot of people here I can go to when I get afraid.
>
> *(Research interview, 2016)*

Conclusions

In the previous sections, I have sought to show that in the case of Athens and Thessaloniki, the occupied buildings can be recognised as physical-perceived space. It becomes the common pool resource of the commoners', both locals and refugees' – their community. In fact, it has emerged not only as a perceived but also a lived space, as it has formed a fluid and open community with no boundaries, concerning its members, but with specific forms of commoning and communication practices between them; hence a nexus of micro-communities or mini-societies has emerged inside the urban core. The process of setting up the common space follows the Lefebvrian triad on perceived-conceived-lived space, as it is based on the multitude of solidarity gestures, the emotional, communicative, cultural and aesthetic interactions, which seek to overcome the bipolar contrasts of native-immigrant, young-old, worker-unemployed, male-female, Greek speakers-Arabic speakers, Syrian-Afghan, Iraqi-Moroccan, Alevi-Sunni, Sunni-Shiite etc. In doing so, it constitutes intermediate commoning social relations and modes of communication, through which the communities of the common space are formed. Consequently, I argue that the study of the refugee common spaces enriches the Lefebvrian concept of production of space.

Moreover, the refugee housing commons enrich the concept of the Lefebvrian right to the city with the plethora of human rights represented. Against the segregation of state-led refugee policies the occupied structures reveal that the multiple refugee rights to the city are interdependent, indivisible and interrelated. In other words, the violation of the refugees' right to the city may affect the enjoyment of a wide range of other human rights, including the rights to work, health, social security, privacy, transportation, sexual orientation or education. The right to the city does not just mean that the physical structure of the house itself must be adequate. There must also be sustainable and non-discriminatory access to facilities essential for health, security, comfort and nutrition. My research for the case studies in Athens and Thessaloniki reveals that the self-organised and occupied refugee common spaces could much better fulfil refugees' needs rather than state-run camps.

Finally, I have to emphasise my argument that the refugees' commoning practice of squatting is not necessarily related only to housing needs; it is also associated with the (re)claiming of right to the city, that is the right to the multiple aspects of the everyday life: like the public and political sphere, social and cultural relations or even the space of imagination and representation. Hence the idea behind the squatting common spaces is not just to provide shelter but also to provide tools for the refugees to help manage their own lives. The overarching aim is to help the newcomers regain their humanity by escaping social marginalisation and creating new social bonds. By actively participating in decision-making at the domestic and political level as well as fulfilling the everyday commoning tasks regarding the place where they live, migrants and refugees and locals develop avenues to take part in the social and political life of the city.

References

Amnesty International (2016) 'Trapped in Greece: refugees stranded in dire conditions as Europe drags its heels', accessed online at www.amnesty.org/, accessed 5 March 2017.

Blomley, N. (2008) 'Enclosure, common right and the property of the poor', *Social & Legal Studies*, 17:3 311–31.

Brenner, N., Marcuse, P. and Mayer, M. (2009) 'Cities for people, not for profit', *City*, 13:2–3 176–84.

Caffentzis, G. (2010) 'The future of "The Commons": neoliberalism's "Plan B" or the original disaccumulation of capital?', *New Formations*, 69:1 23–41.

Casas-Cortes, M., Cobarrubias, S. and Pickles, J. (2015) 'Riding routes and itinerant borders: autonomy of migration and border externalization', *Antipode*, 47:4 894–914.

Coordination Centre for the Management of Refugee Crisis in Greece (2016) *Summary Statement of Refugee Flows at 12.08.2016*. Online. Available HTTP: <http://www.media.gov.gr/> (accessed 5 March 2017).

De Angelis, M. (2007) *The Beginning of History*, London & Ann Arbor, MI: Pluto Press.

De Angelis, M. (2010) 'The production of commons and the "Explosion" of the middle class', *Antipode*, 42:4 954–77.

ECRE (2007) *ECRE Policy Briefing on Housing for Refugees and Migrants in Europe*. Online. Available HTTP: <http://www.epim.info/> (accessed 5 March 2017).

Gabiam, N. (2012) 'When "humanitarianism" becomes "development": the politics of international aid in Syria's Palestinian refugee camps', *American Anthropologist*, 114:1 95–107.

Glassman, J. (2006) 'Primitive accumulation, accumulation by dispossession, accumulation by "extra-economic" means', *Progress in Human Geography*, 30:5, 608–25.

Haddad, E. (2016) Solidarity, squats and self-management assisting migrants in Greece, *Equal Times*, May 2. Online. Available HTTP:<http://www.equaltimes.org/> (accessed 5 March 2017).

Hadjimichalis, C. (2011) 'Uneven geographical development and socio-spatial justice and solidarity: European regions after the 2009 financial crisis', *European Urban and Regional Studies*, 18:3 254–74.

Hardt, M. and Negri A. (2009) *Commonwealth*, Cambridge, Massachusetts: Harvard University Press.

Harvey, D. (2012) *Rebel Cities: From the Right to the City to the Urban Revolution*, London:Verso.

Hodkinson, St. (2012) 'The new urban enclosures', *City*, 16:5 500–18.

Housing Squat for Refugees and Immigrants Notara 26 (2016) *Let's Make the Refugees' Odyssey of Survival a Journey of Humanity Towards Freedom, Athens: Author.*

Ihlen, Ø., Figenschou T.U. and Larsen A.G. (2015) 'Behind the framing scenes: challenges and opportunities for NGOs and authorities framing irregular immigration', *American Behavioral Scientist*, 59:7 822–38.

International Rescue Committee (2016) *IRC Deeply Concerned Over Poor Humanitarian Standards at Refugee Transit Sites in Greece*. Online. Available HTTP: <http://www.rescue.org/> (accessed 5 March 2017).

Kantor, J. (2016) 'Welcome to Greece's refugee squats', *The Washington Post*, August 5. Online. Available HTTP: <https://www.washingtonpost.com/> (accessed 5 March 2017).

Karyotis, T. (2016) Criminalizing solidarity: Syriza's war on the movements, *Roar Magazine*, July 31. Online. Available HTTP: <https://roarmag.org/> (accessed 5 March 2017).

Koutrolikou, P. (2016) 'Governmentalities of urban crises in inner-city Athens, Greece', *Antipode*, 48:1 172–92.

Kuymulu, M.B. (2013) 'Reclaiming the right to the city: reflections on the urban uprisings in Turkey', *City*, 17:3 274–78.

Lefebvre, H. (1991/1974) *The Production of Space*, Oxford: Blackwell.

Lefebvre, H. (1996/1968) *Writings on Cities*, Oxford: Blackwell.

Mattei, U. (2011) *The State, the Market, and Some Preliminary Question About the Commons*. Online. Available HTTP: <http://works.bepress.com/ugo_mattei/40/> (accessed 5 March 2017).

Mayer, M. (2009) 'The "Right to the City" in the context of shifting mottos of urban social movements', *City*, 13:2–3 362–74.

Médecins Sans Frontières (2016) *Greece: Involuntary Eviction from Idomeni Creates Further Hardship for Refugees*. Online. Available HTTP: <http://www.msf.org/> (accessed 5 March 2017).

Ministry of Environment and Energy (2014) *Draft Law: Settings for Abandoned, Empty and Unidentified Buildings, Intervention Procedures in Selectable Areas*. Online. Available HTTP: <http://www.tovima.gr/> (accessed 5 March 2017).

Mountz, A., Coddington, K., Catania, T.tR. and Loyd, J.M. (2013) 'Conceptualizing detention: mobility, containment, bordering, and exclusion', *Progress in Human Geography*, 37:4 522–41.

Moving Europe (2016) *Refugee-Squats in Athens*. Online. Available HTTP: <http://moving-europe.org/> (accessed 5 March 2017).

Papadopoulos, D. and Tsianos, V. (2013) 'After citizenship: autonomy of migration, organisational ontology, and mobile commons', *Citizenship Studies*, 17:2 178–96.

Parliamentary Assembly of the Council of Europe (2016) *Refugees at Risk in Greece*. Online. Available HTTP: <http://assembly.coe.int/> (accessed 5 March 2017).

Presidential Decree (2007) 'Transposing Council Directive 2003/9/EC from January 2003 laying down minimum standards for the reception of asylum seekers', *Government Gazette Issue GG/251/A/13.11.2007*, Athens: National Printing House.

Purcell, M. (2003) 'Citizenship and the right to the global city: reimagining the capitalist world order', *International Journal of Urban and Regional Research*, 27:3 564–90.

Solidarity Initiative to Economic and Political Refugees (2016) *Refugee Accommodation Center City Plaza*. Online. Available HTTP: <http://solidarity2refugees.gr/> (accessed 5 March 2017).

Stavrides, S. (2012) 'Square in movement', *The South Atlantic Quarterly*, 111:3 585–96.

Stavrides, S. (2014) 'Emerging common spaces as a challenge to the city of crisis', *City*, 18:4–5 546–50.

Strickland, P. (2016) 'Greek leftists turn deserted hotel into refugee homes', *Aljazeera*, July 3. Online. Available HTTP: <http://www.aljazeera.com/> (accessed 5 March 2017).

UN (1948). The universal declaration of human rights. *United Nations General Assembly in Paris on 10 December 1948*. Online. Available HTTP: <http://www.un.org/> (accessed 5 March 2017).

UN (1966) *International Covenant on Economic, Social and Cultural Rights (ICESCR)*. Online. Available HTTP: <http://www.ohchr.org/> (accessed 5 March 2017).

UN (1991) *General comment no. 4: the right to adequate housing, Committee on Economic, Social and Cultural Rights (CESCR)*. Online. Available HTTP: <http://www.refworld.org/> (accessed 5 March 2017).

UN (2009) The right to adequate housing. Fact sheet no 21, *Office of the United Nations High Commissioner for Human Rights & UN Habitat*. Online. Available HTTP: <http://www.ohchr.org/> (accessed 5 March 2017).

UN (2016) *Refugee/Migrants Emergency Response – Mediterranean*. Online. Available HTTP: <http://data.unhcr.org/> (accessed 5 March 2017).

Welcome to Europe (2016) *City Plaza Hotel Athens (Greece)*, May 29. Online. Available HTTP: <http://infomobile.w2eu.net/> (23 August 2016).

38

The right to the city

Evaluating the changing role of community participation in urban planning in England

Nick Bailey

Introduction

The writing of Henri Lefebvre on the Right to the City (RTC) has given rise to a whole new field of academic discussion and speculation about his precise meaning, particularly in the light of recent developments of neoliberal societies of the Global North after the financial crisis of 2007–8 (see for example Lefebvre 1996; Purcell 2002; 2013; Merrifield 2011; Butler 2012). The debate continues about how far Lefebvre's writing on the RTC represents an impossible utopian aspiration which can only be realised after a complete social revolution or an aspiration and inspiration which can empower citizens to argue for new forms of local democracy which increase their influence over decisions that affect their lives in the city.

Differences of perspective and interpretation vary over time. Lebas, for example, reminds us that Lefebvre was writing at a time of political turmoil in France and questions whether his writings on the city are appropriate now. She argues: 'Henri Lefebvre's The Right to the City isn't about compromise; "planning gain", "participation", "putting right" or "best practice", but about revolution' (Lebas 1996: 184). On the other hand, Purcell argues:

> The right to the city is one vital element of this movement towards the urban. That movement is set in motion when inhabitants decide to rise up and reclaim space in the city, when they assert use value over exchange value, encounter over consumption, interaction over segregation, free activity and play over work. As they appropriate space, as they develop the ability to manage the city for themselves, they give shape to the urban.
>
> *(Purcell 2013: 151)*

A third interpretation acknowledges that Purcell's writing represents 'The most sophisticated account of how Lefebvre's right to the city can be incorporated practically into a new form of spatial citizenship' (Butler 2012: 149), but Butler goes on to assert that the RTC must be seen as only one element of Lefebvre's writing on the 'urban' (158).

This debate about how far the RTC is relevant and applicable today raises interesting questions for those engaged in the study of social movements, public participation and community engagement. How far are legal and administrative provisions which promote citizen engagement a genuine shift towards recognising that citizens have a right to contribute to decision-making and deliberation about the structuring of urban spaces, or do they merely represent a cynical concession to placate and at best inform the public (Arnstein 1969)? What changes have occurred in the local context that have given rise to new opportunities for citizen engagement, and how far do these represent a genuine shift in power relations? The RTC has been adopted by both global social media-based organisations such as the Occupy Movement as well as more traditional organisations such as UNESCO which has produced a *World Charter for the Right to the City* (UNESCO 2005). A similar declaration was made for the Habitat III conference in Quito in September 2016 (UN 2016). There are many other similar statements produced by global assemblies and forums (Mayer 2009), but it remains uncertain how far these have been adopted and implemented by national governments at central or local levels.

This chapter aims to review the relevance of Lefebvre's work to current practice in England. The UK has a long history of experimenting with mainly local consultative and participatory organisational arrangements in relation to planning and urban development. The chapter begins by discussing some of the key concepts linked to the RTC and then explores in more detail three examples of community engagement in the field of urban planning and development. The conclusions attempt an evaluation of how far the examples identified might best be defined as procedural concessions and how far they challenge the dominant neoliberal consensus which privileges unconstrained profit-orientated urban development. Is the RTC merely a utopian rallying cry or does it offer insights which are relevant to current practice today?

Interpreting the right to the city

Lefebvre's key works on the city and urban space set out both an analysis of the current urban predicament and suggestions about how a fully liberated society might be created. First, he distinguishes between the 'city' and the 'urban'. The right to the city is a vital element towards creating the 'urban'. The *urban* is 'a possible world, a society yet to come. It is what Lefebvre calls an urgent utopia, which involves a style of thinking turned towards the possible in all areas' (Lefebvre 2009: 288; Purcell 2013: 151).

Closely linked to the RTC is a second spatial demand, the *right to difference*. This right emerges 'from the essential qualities of the urban as a place of encounter and simultaneity, and as an ensemble of differences' (Butler 2012: 152). The right to difference has been applied successfully to the urban scene whereby the needs and aspirations of particular sub-groups are asserted – those defined by age, gender, race and sexual orientation. But it can also be applied spatially in that under the neoliberal capitalist system certain groups – often the poor and ill-housed – can become excluded from the centre of urban areas and marginalised socially, politically and geographically. Fundamentally, the right to difference enables citizens to fully engage with others of all kinds, to go beyond the RTC in building the fully 'urban' society.

A further important element of the RTC is participation. Lefebvre argues that citizens rarely have more than a nominal role in making decisions. He calls this the ideology of participation, which:

> Lefebvre calls for 'real and active participation' whereby residents 'come into consciousness of themselves as inhabitants, as embedded in a web of social connections, as dependent on and stewards of the 'urban'... allows those in power to obtain, at a small price, the

> acquiescence of concerned citizens. After a show trial more or less devoid of information and social activity, citizens sink back into their tranquil passivity.
>
> *(Purcell 2013: 150–1)*

This process of the active engagement of all citizens leads Lefebvre to assert the principle of *autogestion* or self-management. Rather than advocating the Marxist assumption of the withering away of the state after the revolution, Lefebvre assumes that citizens will increasingly challenge centralised forms of the state which will be replaced by more transparent and collaborative modes of state activity. As Brenner argues, the question is less one of:

> The erosion of state power as such than the possibility of its qualitative transformation into a non-productivistic, decentralised and participatory institutional framework that not only permits social struggles and contradictions, but actively provokes them.
>
> *(Brenner 2008: 240)*

In Lefebvre's view, to be fully effective and meaningful, participation has to be linked to the concept of self-management and under the direct control of residents:

> Without self-management, 'participation' has no meaning; it becomes an ideology, and makes manipulation possible. Self-management is the only thing that can make participation real, by inserting it in a process that tends towards the global.
>
> *(Lefebvre 1976: 120)*

Thus Lefebvre appears to be arguing that the RTC can best be understood as a 'transformed and renewed right to urban life, which links to the essential characteristics of the urban as both a creative work and a space of 'centrality', 'gathering' and 'convergence' (Butler 2012: 144). Through initially engaging with participation and celebrating the right to difference, residents can take on responsibilities of self-management and assume democratic control of the city. The dominance of exchange value over use value is reversed, the right to difference prevents social and political exclusion and power is more widely dispersed through self-managing organisations. Lefebvre's vision is based on a combination of Marxist, anarchist and utopian philosophies born out of the milieu in which he was writing. He takes us beyond simply the RTC and asserts that the 'urban' is only achieved when the much larger national and global commanding heights of capitalism have been secured. However, he offers relatively little guidance on how this can be achieved, particularly when real dangers exist of falling into the 'local trap' of privileging democratic change at the easiest, most local level first without addressing higher-tier powers (Purcell 2006).

Applying Lefebvre's conceptual framework to planning and urban development in England

In order to explore the implications of Lefebvre's conceptual framework more fully, the following examples of community participation and engagement in England have been selected for more detailed evaluation:

- Public participation which is limited and partial and enables residents and interested parties to object to planning applications and to comment on the preparation of development plans.

- Provisions where neighbourhoods and parishes can prepare neighbourhood development plans based on extensive public engagement and consultation.
- Opportunities where local residents and other interests form autonomous organisations to acquire assets and carry out their own development to meet locally determined needs.

The key questions arising here are: how far does each option enable citizens to challenge the neoliberal status quo in property development and house building? How far are citizens in control of the participation process and in the production of urban plans which affect them? And, to what extent is there evidence of participation leading to self-management (*autogestion*), the right to difference and what might be called the right to the city?

Public participation in development management and statutory plans

In the year ending September 2016, 483,600 planning applications were submitted to local authorities in England. Of these, 434,600 were determined, and 88 per cent were approved, although this proportion varied significantly between local authorities (DCLG 2016). Of these applications, almost all could be subject to public consultation, and major developments could have been determined by an Inspector or even the Secretary of State of the Department for Communities and Local Government through a complex appeals procedure.

At least since Town and Country Planning Act 1968 in England and Wales, the public and a variety of statutory agencies have had a right in law to be consulted about both individual planning applications and a right to be involved in the preparation of plans (Baker et al. 2007). The same year the Skeffington Committee was set up 'to consider and report on the best methods, including publicity, of securing the participation of the public at the formative stage in the making of development plans in their area' (Skeffington 1969: 1). A number of potentially far-reaching recommendations were made in the Skeffington report *People and Planning*, although many were not pursued with any vigour or had no resources attached. Local residents, amenity societies and parish councils receive regular notification of planning applications and draft plans affecting their area, but it was often the case that few were motivated to respond. As one commentator noted:

> The propensity of citizens to engage with planning is typically in direct proportion to the distance from their home, and often manifests only when a development becomes visible, i.e. when a bulldozer or planning application notice appears on site.
>
> *(Cullingworth et al. 2015: 509)*

The Planning and Compulsory Purchase Act 2004 added a further refinement whereby local authorities in England and Wales were required to publish a Statement of Community Involvement. This sets out for each local authority the procedures to be used in consulting residents and other interests about all aspects of planning. When it became recognised that the planning system had become overloaded with guidance and policy documents, a shorter, simplified statement was introduced called the National Planning Policy Framework (DCLG 2012; revised 2018) for England. This continued to assert the often repeated rhetoric that:

> Early and meaningful engagement and collaboration with neighbourhoods, local organisations and businesses is essential. A wide section of the community should be proactively engaged, so that Local Plans, as far as possible, reflect a collective vision and a set of agreed

> priorities for the sustainable development of the area, including those contained in any neighbourhood plans that have been made.
>
> *(DCLG 2012: 37)*

Public and community participation in the English planning system remained for at least 40 years from 1968 excessively bureaucratic and rule-bound. Moreover, practitioners and administrators often lacked adequate academic training in this important aspect of their work (Shipley and Utz 2012). There was no attempt to shift the balance of power away from the major developers and house builders and towards the much less well resourced local interests, despite occasional support from central government to fund voluntary 'planning aid' services to residents. Moreover, it required both a high level of motivation and professional skills to launch a successful campaign against public sector and commercial interests seeking to carry out development in the face of strong opposition from local opinion. As a result, cases where major developments were successfully resisted were few and far between. In fact many groups found greater success through attempting to influence elected members or committee structures rather than engaging with highly paid barristers and consultants at, for example, planning inquiries. As successive British governments deregulated planning controls over development, the influence residents and amenity societies could bring to bear was at best very marginal and at worst ineffective.

Neighbourhood planning

The UK Coalition Government of 2010–15 introduced a number of policy and legislative changes in order to both liberalise urban development while also devolving decision-making to lower levels (Clarke and Cochrane 2013). The Localism Act 2011 (England and Wales) enabled local communities – either existing civil parish councils or newly formed neighbourhood forums – to prepare and seek approval for neighbourhood development plans. These required local authority approval at several stages: in defining the boundaries of the 'neighbourhood', in forming a suitable organisation and in ensuring that the draft plan was in general conformity with the strategic policies of the local authority. In addition, the neighbourhood plan could only become part of the statutory planning framework after an examination by a suitably qualified person (an Examiner) and a referendum of all residents living in the designated area. By March 2017, 2,052 neighbourhoods had been designated in England, 316 had held referenda and 293 plans had been 'made' or achieved statutory status (PlanningResource 2017).

Neighbourhood plans differ from previous initiatives in that their preparation is entirely voluntary, and only very limited resources are made available from central government to assist in collecting the evidence base and preparing suitable policies. The expectation is that much of the preparatory work is done by volunteers. Civil parish councils in more affluent rural areas (Brookfield 2016) tended to respond first to this opportunity, and a number had the resources to employ consultants to prepare the draft plan and undertake public consultation. This is an important element of the process in that a 'consultation statement' is part of the documentation to be submitted to the local authority and formed part of the examination. After this stage the Examiner could authorise the referendum to proceed if it was determined that the submission meets the 'basic conditions' defined in the legislation.

It has been argued that neighbourhood planning is part of a wider trend towards neoliberalism and the withdrawal of the state from its strategic planning functions in an era of growing austerity (Davoudi and Madanipour 2015). On the other hand, there is growing evidence that it has opened up new opportunities for public engagement in the planning process. Indeed, a number of neighbourhoods have pursued innovative approaches to consulting, engaging and involving groups and interests in their neighbourhoods who have not previously participated (Bailey 2016). This may be because designated neighbourhoods often have relatively small populations, making intensive participation processes more manageable, and because of the need to ensure a good turnout when the referendum takes place. There is limited research to date into the impact of neighbourhood planning on levels of engagement and on the planning process itself. A survey of neighbourhood planning groups carried out by Parker et al. (2014) found that they often found the process over-complex and bewildering and wished to see more guidance on processes and procedures. Yet the ability to influence planning policy for their local area proved a strong motivation to exert as much influence as possible within the legislative constraints.

A particularly contentious issue is the designation of sites for housing development in a neighbourhood plan because it is often assumed that local communities, particularly in urban fringe or rural areas, resist new house building which may reduce their perceived level of amenity. Two surveys of the recent neighbourhood plan suggest that some groups see the neighbourhood plan as an opportunity to challenge the monopoly of the big house builders who often prefer to develop large greenfield sites or urban extensions (Bailey 2017; Bradley and Sparling 2017). Instead, local communities tend to favour more housing provision so long as it is well designed, affordable and preferably developed on small-scale infill sites. There is also growing evidence that government ministers and the Planning Inspectorate tend to uphold neighbourhood plans when challenged by major developers whose sites are not designated for housing (Bailey 2017: 7).

Just how far there has been a shift in power relations towards neighbourhoods and away from the impersonal market forces which are often reinforced by the neoliberal planning system remains hard to determine. Clearly the procedures for adopting neighbourhood plans are heavily prescribed by the legislation, regulations and guidance where this is provided and not least because of the local authority approvals and examination which are needed. However, they do legitimise in many cases more intensive and systematic processes of participation which generate positive responses from residents who feel they can exert some influence over planning and housing processes which have a real impact on their locality and quality of life. As Bradley and Sparling conclude:

> Neighbourhood planning endorsed the spatial practices of lived or representational space in contrast to the exchange values promoted in the current market model. It directed citizens to an awareness of housing needs not the market needs of the volume house-builders. The balance between community engagement and spatial liberalism appeared to have undergone a similar qualitative shift with empowerment and sense of place emerging as new arbiters of development planning.
>
> *(Bradley and Sparling 2017: 117)*

In some cases local communities form organisations which can go beyond influencing the planning process and actually exert more direct control over the provision of services and the use of land and buildings by acquiring assets for development themselves. These organisations will be discussed in the third example in the next section.

Community-based developments

The United Kingdom has a long tradition of social enterprise going back to Robert Owen's construction of a new industrial settlement at New Lanark in Scotland and the formation of the co-operative movement in Rochdale in 1847. More recently, and again possibly a response to changing economic relations and neoliberalism, there has been the growth of social enterprises which have operated as 'hybrid' organisations between the public and private sectors. Pearce (2003: 25) identifies six defining characteristics fundamental to social enterprise:

- Having a social purpose or purposes.
- Achieving the social purpose by, at least in part, engaging in trade in the marketplace.
- Not distributing profits to individuals.
- Holding assets and wealth in trust for community benefit.
- Democratically involving members of its constituency in the governance of the organisation.
- Being independent organisations accountable to a defined constituency and to the wider community.

There are many forms of organisation which meet this definition, including co-operatives, housing associations, credit unions and other mutual organisations (Mayo and Moore 2001). A subset of this sector are community-based enterprises or community businesses. These organisations meet the criteria above but also are established in order to benefit a defined locality or section of the population. They can take a number of different legal forms, such as by registering as a company limited by guarantee, which protects the trustees from any debts arising from trading activities. Most also become charities, which provides certain tax advantages. Many are located in areas of high levels of deprivation and may contribute towards local regeneration strategies (Bailey 2012; Healey 2015).

Community-based enterprises have emerged in both urban and rural areas largely as a result of bottom-up community aspirations rather than top-down government policy. They are supported by a national organisation called *Locality* with a membership of about 600. In a number of cases they were a community response to urban interventions such as road building or 'slum' clearance, while in others funding opportunities or sponsorship provided a catalyst. For example, the Westway Trust was one of the first in London and was formed in 1971 in order to acquire a lease on 9.3 hectares of vacant land under an elevated motorway. The Londonderry Inner City Trust was established in 1981 as a non-sectarian organisation to provide training and employment at the height of the Troubles in Northern Ireland. In Central London the Coin Street Community Builders was set up in 1984 when 5.3 hectares of vacant land was transferred to the Trust by the soon to be abolished Greater London Council. All three organisations represent areas with long histories of community activism, and this was a characteristic of many more recent examples. Each has developed land and other assets to provide a mix of commercial space to generate an income in order to cross-subsidise community and sports facilities and in some cases affordable housing.

The relevance of these organisations to a discussion of the RTC is that they represent opportunities for local residents to play a direct role in contributing to and managing these enterprises not for personal gain but in order to generate new opportunities for employment, social and leisure activities, arts and entertainment which would not otherwise be available. The ability to control the type and speed of development is much greater with full legal ownership rather than through the planning system, and in areas where property values and development pressures are low there are considerable opportunities for community enterprise. Additional provisions in the Localism Act 2011 enable community organisations to register an interest in 'assets of community value' for which there is a 'right to bid' should they come up for sale. Many buildings which are surplus to requirements, such as schools, community centres and libraries, have been transferred from the public sector to community enterprises.

Healey (2015) argues that there are two main reasons for the growth of 'small-scale civil society initiatives' in the UK. First, there is the realisation that the state can no longer provide and there may be previous experience of public services poorly co-ordinated or delivered in the past. Second, there has been a shift in public attitudes whereby the dominance and authority of certain professions has been challenged and those representing or working for public agencies have become more willing to facilitate independent, civil society initiatives. There can also be symbiotic relationships established with local authorities whereby trust is developed in order to promote mutual benefits. For example, the local authority can transfer land and buildings at below market value whereas the community enterprise can access funding from other sources and draw on goodwill in the community, for example through volunteering and crowdsourcing.

On the other hand, community enterprise is open to questions about representation and accountability. There is always a danger that these organisations become inward looking and exclusionary and do not fully reflect the needs and aspirations of the communities they serve. In many ways they enter a local political arena where the only real defence is the need to take extraordinary steps to ensure they remain in touch through regular consultation and by encouraging a regular turnover of key representatives.

Thus although community enterprises can acquire greater control over resources in their area, and a stronger sense of empowerment, they must still operate in the wider commercial context of property development and a pro-commercial development planning system. They must also ensure that they remain viable and sustainable, which means operating on a commercial and sometime profit-making basis. This can cause resentment and opposition from those residents who only favour community-orientated facilities.

Of the three examples outlined here, none fully meets Lefebvre's definition of the 'urban' but might just trigger the revolutionary imagination which could constitute urgent utopia: 'For Lefebvre the urban constitutes a revolution, but one that requires millions of everyday acts of resistance and creation' (Purcell 2013: 151). What the examples do illustrate is that in many different arenas, and with varying levels of commitment and engagement, citizens are confronting arbitrary decisions that impact on how they wish to live and the quality of life in their locality. Rather than attempting to overthrow agencies of power in the public and private sectors with Lefebvre's concept of the urban, the examples above tend to suggest an alternative approach, which is about enabling change through deliberation, collaboration and what Gaventa calls 'deepening democracy' (Gaventa 2006). This has also been described as 'progressive localism', whereby struggles are not merely defensive but can 'reconfigure existing communities around emergent agendas for social justice, participation and tolerance' (Featherstone et al. 2012: 179).

Closer collaboration with powerful agencies of either public or private sectors brings real dangers of co-option and legitimisation through weak and ineffective processes of consultation. Those seeking improvements through democratic change and collaboration need to walk a fine line between adopting the interests and perspectives of those in authority and knowing when to oppose and argue for more equitable and inclusive strategies. This also relates to the concept of co-production (Bovaird and Loeffler 2012), which has been defined as delivering public services in an equal and reciprocal relationship between professionals and people using services (NESTA 2011). Innovative approaches to developing governance systems based on co-production have been particularly associated with the third sector throughout Europe and other developed economies (Pestoff et al. 2012).

Conclusions

This chapter has discussed the meaning of Lefebvre's key concepts and then sets out three case studies of different types of community engagement in the urban context in England. It has reviewed the powers available for residents to comment on planning applications; to form neighbourhood forums to prepare neighbourhood plans; and to establish community enterprises able to operate as developers in the community interest. The relevance of Lefebvre's work to residents' involvement in the urban realm is that it provides a glimpse of what might be, and a rallying cry for those seeking change. For example, he stresses that the urban is already here and operating in the present city. 'Every space in the city… bears within it the seeds of the urban, a not-yet-realised potential for urban life' (quoted in Purcell 2013: 151). This is a common right, argues Harvey (2008), since 'this transformation inevitably depends upon the exercise of a collective power to reshape the processes of urbanisation' (2008: 23). What is not so clear in Lefebvre's writing is the road map to achieving the urban, given the trends towards neoliberalism and the growing power of multi-national corporations, which increasingly determine the quality of everyday life and where wealth and life chances become increasingly polarised.

References

Arnstein, S. (1969) 'A ladder of citizen participation in the USA', *Journal of the American Institute of Planners*, 35:4 216–24.

Bailey, N. (2012) 'The role, organization and contribution of community enterprise to urban regeneration policy in the UK', *Progress in Planning*, 77:1 1–35.

Bailey, N. (2016) 'Localism in the UK: how will it impact on urban governance and local democracy in a period of financial austerity?', paper presented to the European Urban Regeneration Association Conference, Turin, June 2016 (unpublished).

Bailey, N. (2017) 'Housing at the neighbourhood level: a review of the initial approaches to neighbourhood development plans under the Localism Act 2011 in England', *Journal of Urbanism: International Research on Placemaking and Urban Sustainability*, 10:1 1–14.

Baker, M., Coaffee, J. and Sherriff, G. (2007) 'Achieving successful participation in the new UK spatial planning system', *Planning Practice & Research*, 22:1 79–93.

Bovaird, T. and Loeffler, E. (2012) 'From engagement to co-production: the contribution of users and communities to outcomes and public value', *Voluntas*, 23 1119–38.

Bradley, Q. and Sparling, W. (2017) 'The impact of neighbourhood planning and localism on house-building in England', *Housing, Theory and Society*, 34:1 106–18.

Brenner, N. (2008) 'Henri Lefebvre's critique of state productivism', in K. Goonewardena, S. Kipfer, R. Milgrom and C. Schmid (eds) *Space, Difference, Everyday Life: Reading Henri Lefebvre*, New York: Routledge.

Brookfield, K. (2016) 'Getting involved in plan-making: participation in neighbourhood planning in England', *Environment and Planning C: Government and Policy*, published online, pp. 1–20.

Butler, C. (2012) *Henri Lefebvre: Spatial Politics, Everyday Life and the Right to the City*, Abingdon: Routledge.

Clarke, N. and Cochrane, A. (2013) 'Geographies and policies of localism: the localism of the United Kingdom's coalition government', *Political Geography*, 34 10–23.

Cullingworth, B., Nadin, V., Hart, T., Davoudi, S., Pendlebury, J., Vigar, G., Webb, D. and Townshend, T. (2015) *Town and Country Planning in the UK*, Abingdon: Routledge.

Davoudi, S. and Madanipour, A. (eds) (2015) *Reconsidering Localism*, London: Routledge.

Department for Communities & Local Government (DCLG) (2012) *National Planning Policy Framework*, London: DCLG.

Department for Communities & Local Government (DCLG) (2016) *Planning Applications in England: July to September 2016*, London: DCLG.

Featherstone, D., Ince, A., Mackinnon, D., Strauss, K. and Cumbers, A. (2012) 'Progressive localism and the construction of political alternatives', *Transactions of the Institute of British Geographers*, 37 177–82.

Gaventa, J. (2006) *Triumph, Deficit or Contestation? Deepening the 'Deepening Democracy' Debate*, working paper 264, Brighton: Institute of Development Studies.

Harvey, D. (2008) 'The right to the city', *New Left Review*, 53, September/October, pp. 23–40.

Healey, P. (2015) 'Citizen-generated local development initiative: recent English experience', *International Journal of Urban Sciences*, 19:2 109–18.

Lebas, E. (1996) 'The right to the city'. *City*, 1:3–4 184–85.

Lefebvre, H. (1976) *The Survival of Capitalism: Reproductions of the Relations of Production*, London: Allison & Busby.
Lefebvre, H. (1996) 'The right to the city', in E. Kofman and E. Lebas (eds) *Writings on Cities*, Cambridge, Massachusetts: Wiley-Blackwell.
Lefebvre, H. (2009) 'The worldwide experience', in N. Brenner and S. Elden (eds) *State, Space, World: Selected Essays*, Minneapolis: University of Minnesota Press.
Mayer, M. (2009) 'The "Right to the City" in the context of shifting mottos of urban social movements', *City*, 13:2–3 362–74.
Mayo, E. and Moore, H. (2001) *The Mutual State: How Local Communities Can Run Local Services*, London: New Economics Foundation.
Merrifield, A. (2011) 'The right to the city and beyond: notes on a Lefebvrian re-conceptualisation', *City*, 15:3–4 173–481.
NESTA (2011) *Co-Production Phase 2: Taking Co-Production to Scale in Services for Patients with Long Term Health Conditions. Strategic Partners – Call for Proposals*, London: NESTA.
Parker, G. with Lynn, T., Wargent, M. and Locality (2014) *User Experience of Neighbourhood Plans in England Research*, London: Locality.
Pearce, J. (2003) *Social Enterprise in Anytown*, London: Calouste Gulbenkian Foundation.
Pestoff, V., Brandsen, T. and Verschuere, B. (eds) (2012) *New Public Governance, the Third Sector and Co-production*, London: Routledge.
PlanningResource (2017) Map: neighbourhood plan applications, accessed online at http://www.planningresource.co.uk/article/1212813/map-neighbourhood-plan-applications, accessed 11 April 2017.
Purcell, M. (2002) 'Excavating Lefebvre: the right to the city and its urban politics of the inhabitant', *GeoJournal*, 58 99–108.
Purcell, M. (2006) 'Urban democracy and the local trap', *Urban Studies*, 43:11 1921–41.
Purcell, M. (2013) 'Possible worlds: Henri Lefebvre and the right to the city', *Journal of Urban Affairs*, 36:1 141–54.
Shipley, R. and Utz, S. (2012) 'Making it count: a review of the value and technique for public participation', *Journal of Planning Literature*, 27:1 22–42.
Skeffington, A.M. (1969) *People and Planning: Report of the Committee on Public Participation in Planning*, London: HMSO.
UNESCO (2005) *World Charter for the Right to the City*, Paris: UNESCO.
United Nations (UN) (2016) *Habitat III: New Urban Agenda*, New York: UN.

39

Lefebvre and the inequity of obesity

Slim chance of food justice for the urban poor

Hillary J. Shaw

Introduction: the right to food in the city

A city is an organised and differentiated agglomeration whose citizens, removed in a Marxian sense from their means of livelihood, must travel to access resources. Mere physical access of course is insufficient; the Marxian working class must also commute to obtain the money to access these necessary resources. As Sir James Mathew, 19th-century Irish judge, cynically observed, in England, justice is open to all, like the Ritz Hotel. Increasingly, however, cities are perceived as denying rather than providing access to life's necessities, for many of their citizens. This is a serious accusation given that the majority of mankind, Global South and North, now live in urban areas. Back in 1968 Lefebvre asserted a need for a right to the city, which is far more than the individual liberty to access urban resources: it is a right to change ourselves by changing the city (Harvey 2008). This right to the city has been expanded to cover a panoply of rights and injustices across the world, from alleviating poverty to providing affordable housing to environmental sustainability; there has even been proposed a digital right to the city (Shaw and Graham 2017). However, a fundamental human right must be the right to food. After air and water it is the third most essential requirement of life itself. The United Nations recognised this right in their 1948 Declaration of Human Rights. This right to food goes beyond merely enough calories to survive, with Article 25 of the 1948 Declaration stating, 'Everyone has the right to a standard of living adequate for the health and well-being of himself and his family, including food'. As Lefebvre notes, the urban economy is fed by people, and in producing the city, a large workforce is employed, in the maintenance and feeding of machines (dos Santos 2014); however, cities are not good so good at feeding their people.

The city of Glasgow, Scotland, was one of the earliest locales where the issue of a poorly-fed population became apparent. In a 1996 survey of deprivation in that city, the Low Income Project Team encountered a respondent who described the area as a 'food desert', meaning there was a lack of shops selling good-quality, reasonably priced food in the area (Cummins and Macintyre 2002). The term 'food desert' rapidly entered both academic research and the media, as researchers found them in other cities. *The Guardian* on 17 March 1999 (cited in Shaw 2014: 105) wrote of the food deserts phenomenon in Britain:

> On the poorer estates of Coventry, low cost, good quality, food is not available to the poorest. These people either have to shop at expensive local stores or pay for transport and lug small children for miles and back with shopping.

Coventry, a former industrial city in the Midlands area of England is, like most cities, highly spatially segregated by class, and the least affluent districts have the poorest facilities, the lowest performing schools and doctors, and the most impoverished selection of fresh produce in local retailers; the inverse care law proposed by Hart (1971) as a pun on the inverse square law in physics is evident here. Social class intersects with urban space to continually deepen economic divisions.

The class wars of food

The space produced in cities is increasingly class-riven and divided; as Lefebvre (1991: 55) wrote, 'today, more than ever, the class struggle is inscribed in space'. Superficially, however, urban food spaces appear to have no role, or even an inverse role, in this class struggle. The more deprived urban areas are often awash with convenience food shops, ethnic-minority-oriented stores, local supermarket branches, not to mention fast food outlets and other sources of ready-to-eat food. Conversely, affluent suburbs have fewer food stores, just the local town centre and a peripheral hypermarket. The poor do not consume fewer calories than the wealthy; in fact, they have swapped places. In Brazil, for example, in 1975 women in the lowest income quartile were half as likely to be obese as those in the top quartile, but in 1997 these lower-quartile women were ten per cent more likely to be obese than their wealthier compatriots (Power and Schulkin 2009: 129–30). The same trend has occurred at all spatial scales, as poorer nations have become more obese, lower-GDP per capita regions within countries, and poorer social groups within societies all become more overweight.

So, the poor consume more calories than the wealthy. Are they choosing to do this? Unless someone is force-feeding them, how is this inequitable? The answer is that economics is forcing them to do this, through the mechanism of the 'health premium'. Global food chain developments from vast container ships and industrial-scale farms to the consumer convenience of supermarkets and microwave meals have greatly reduced the proportion of income spent on food. However, it is the processed, sugar preserved, and cheaply flavoured foods that have reduced in price. The health premium, the extra cost of a healthy diet, with fresh fruit and vegetables over an unhealthy one, has increased across the UK from 18 per cent in 1988 to 48 per cent in 2007. In affluent areas the difference has widened from 16 per cent to 39 per cent; in deprived areas, from 20 per cent to 58 per cent (Shaw 2014: 119). Healthy food affordability and the cost of living are linked to wages, especially at the bottom end of the jobs market, when both capital and labour are highly mobile. The neoliberal necessity to maintain a low-paid reserve army of labour but the unwillingness of taxpayers to overly subsidise this army produces a basic state income of just enough to exist on; in turn the competitiveness of capital produces a state minimum wage only just above this level. One can exist, albeit on the borders of health, on 'ridiculously cheap' (Anonymous 2012) burgers and pizzas. A UK worker on the minimum wage, eating 2,000 calories a day and spending 20 per cent of income on food, can afford potatoes, meat and tinned tomatoes, but very likely not fresh broccoli or lettuce (Shaw 2014: 120). Increasingly the existence of a flexible zero-hours precariat is eroding even that slight financial advantage of employment over unemployment.

The shops in deprived urban areas may offer a small selection of fresh produce, but at higher prices than a distant supermarket, accessible in theory by all, but in practice a car is needed to bring fresh produce home, and at higher total cost than most local residents can afford. Transporting food is a major issue for the poor without cars because not only is fresh produce more expensive per calorie than processed pizzas, fresh food is also heavier in terms of weight per calorie. For those who shop by bus or on foot, there is a disposable carrying capacity, just as there is a disposable income. Poorer households will only spend on food once essentials like rent and electricity have been paid, because being evicted or having power disconnected is worse than filling a hungry stomach with cheap biscuits, and 2,000 calories of biscuits can be purchased for 30 pence, so they are always affordable; likewise, there is only so much food one can carry once the shopping bag is filled with toilet paper and washing up liquid. Perhaps the local burger outlet, always closer than the supermarket, obviates the need to carry much food at all, especially when one has children in tow, as *The Guardian* quote above (as cited in Shaw 2014: 105) pointed out.

Casualties of war

Wars produce casualties; injuries, deaths, also psychological trauma, and the food class wars are no exception. The primary effect of a diet high in cheap processed foods and low in expensive fresh produce is obesity. Physically, excess weight leads to diabetes, sleep apnoea, arthritis, cancer, and cardiovascular disease. In turn, these conditions predispose to osteoporosis, blindness, amputations, gallstones, and chronic fatigue. These are chronic conditions that do not kill straightaway, so the burden falling on both public and private healthcare costs can be considerable. Such costs are extremely hard to quantify, as some of these ailments would have occurred without obesity, and extra patients add to hospital costs in a non-linear way. However, in 2011 the UK Government estimated that obesity was costing the National Health Service around £4 billion a year, and other costs such as private medical care, informal care, and lost work days took the total annual cost to the UK to between £15 and £20 billion a year. Each individual may cost the UK economy between £1,500 and £2,000 extra for each year they are obese (Shaw 2014: 28). Being obese shortens the lifespan by around 11 years, reducing state pension entitlement by around £70,000, possibly outweighing the excess medical costs of obesity.

Less visibly, obesity has profound psychological effects, and these effects can begin much sooner than the physical complications, even in childhood. Obese children suffer lower self-esteem at school, reduced learning ability, and lower qualifications. Poorer educational performance perpetuates the very poverty that causes obesity down the generations. These children also are less able to participate in school games, which always seem to be competitive team sports, where they may be teased for 'letting their own side down'; they might prefer non-competitive individual sports such as cross-country jogging. Obese adults lose out at job interviews and have lower earnings, as well as fewer social opportunities and networking chances (Härkönen et al. 2011; Reichert 2013), further perpetuating generational poverty. Their Lefebvrian social space shrinks, and they may find their main societal role is to be the anti-example, the bogeyman, in Global Northern societies that exalt svelte slimness whilst simultaneously making it economically and environmentally ever harder to reach this ideal.

The shrinkage of Lefebvrian space, money, and time

As the belly gets bigger, Lefebvrian space shrinks around you. The Lefebvrian triad of produced space comprises *l'espace percu*, the space perceived, experienced; *l'espace concu*, the space conceived; and *l'espace vecu*, lived, social space. As noted earlier, the poor have to rely on public transport or walking to access food retail, whereas the wealthy drive. With childcare being relatively expensive, *l'espace percu* shrinks further, and with heavy shopping to carry or lug on and off buses, distances are harder to cover. Less affluent areas are less attractive to walk in, suffer more pollution from industry, traffic fumes and noise, contain fewer trees and parks, and the crime risk is higher and *l'espace percu* is reduced. For some, living in very deprived areas, even the places they can imagine going to, *l'espace concu*, are reduced. In 2016 the extent of deprivation, obesity, and barriers to healthy eating was investigated in the Welsh Valleys region of south Wales, UK, funded by the Centre for Urban Research on Austerity (CURA 2016). Life for some in this part of Wales illustrates all too well the restrictions imposed by poverty upon Lefebvrian space. In the Welsh Valleys, a former coal mining and now deindustrialised region immediately north of Cardiff, 'there are some who have never left their home village' (CURA 2016). On the Gurnos Estate, Merthyr Tydfil, a local authority housing area noted for its ill-health, short life expectancy, and poverty even compared to the rest of the Valleys, there is actually a free bus twice a week to the Asda supermarket two kilometres away. However, many less affluent shoppers, especially those with children, avoid large hypermarkets because they fear the temptation, or the nagging from their offspring, to buy more than they can afford. More insidiously, poverty, especially food poverty, reduces both the motivation and ability to entertain, to keep friendships going. A poor diet may lead to ill-health and ongoing fatigue, and then there is the shame of not being able to provide the same food treats as your neighbours. Cheap sugary foods may make children more hyperactive and tiring for the parents. Ultimately the stresses of poverty, of being unable to afford enough living space, can result in family break-up and divorce, so *l'espace vecu*, social space, shrinks.

If space is a socially constructed concept then surely money is even more of an artificial construct, and the financial situation of the poor can be analysed in Lefebvrian terms of *l'argent percu*, *l'argent concu*, and *l'argent vecu*. As noted earlier, the concept of food deserts relies heavily on the spending power of households and the price of fresh produce relative to household income, as well as their distance, perceived and actual, from fresh food retailing. Otherwise, it would make little sense to regard a better-off household living in a deprived area as facing barriers to a healthy diet; likewise, very prosperous UK suburbs such as London's Bishops Avenue or the affluent Four Oaks private estate in Birmingham, where there are houses over two kilometres from any shops, would count as food deserts. Purchasing food may require additional expenditure for travel as well as the actual purchases at the shop; even virtual travel, shopping on the Internet, carries costs in the form of buying a computer, subscribing to the Internet, and of course possession of a credit or debit card. Many less affluent households do not have the first two items, and the very poor may not even have a bank card, being restricted to basic bank accounts; any credit they need must come from expensive payday loans companies or pawnshops. Just as the United Nations has recognised food as an essential right, Baroness Neville-Rolfe in the UK's House of Lords has defined the Internet as a crucial part of modern life, essential for full participation in society (Shaw and Shaw 2015: 236). *L'argent percu* restricts shopping to the essentials, even to where some must choose between eating or heating, or must forego a decent meal to pay the rent. Then there is *l'argent concu*, what one can imagine doing with the funds one has to spend. The difference between *l'argent percu* and *l'argent concu* is related to the famous researcher into poverty in Victorian Britain, Charles Booth. Booth investigated the finances of unskilled labouring households in York and distinguished between primary poverty, what Lefebvre might have called *l'argent percu*, and secondary poverty, *l'argent concu*. Primary poverty was when household income was simply insufficient for necessities such as food, rent, fuel, clothes. However, Booth found that the unskilled labourer households of York were paid 20

Table 39.1 A second triad: space, money, and time. Lefebvrian space, money, time: food purchasing and the constraints on the poor

	percu	*concu*	*vecu*
l'espace (space)	Travel arduous, carry shopping by hand	Lack of motivation, fear of overspending	Shame, social stress, family stress
l'argent (money)	Fresh produce unaffordable, no car or Internet	Vice/gratification, spending over necessities	Unaffordability of conspicuous gift giving
le temps (time)	Low wages	Short time horizons, so healthy food less important	Instability, future uncertainty

shillings (one UK pound) a week, just enough for the bare necessities, except that many of them spent up to a third of that income on alcohol, creating what Booth termed secondary poverty (Scott 1994: 38). Fast-forward a century and 'vice-spending', on things like alcohol, smoking, and gambling, is still highest amongst the lower income deciles. Lack of financial resources then restricts social life and entertaining (*l'argent vecu*) in ways discussed above, especially in an age of conspicuous consumerism when social events become a competition as, for example, to who can bring the most expensive party gift.

Deprivation also brings short time horizons, a desire for immediate gratification. Almost everybody suffers from time myopia; ask people to choose between a gift of £100 now, or £120 in 12 months' time, and most will pick the first option; in economics terms, they have just turned down an investment with a 20 per cent annual return. But poverty can bring desperation for immediate relief, which is why Booth found households where the man would prefer drunkenness now even if that meant hunger or eviction later. In some districts things have not changed much. With food, the case for taking a long-term view is even less clear; choose between sugary salty fatty tasty but unhealthy food now, or spend more and eat, perhaps less palatably, and you may live another few years. As George Orwell wrote in *The Road to Wigan Pier*, 'A millionaire may enjoy breakfasting off orange juice and Ryvita biscuits, an unemployed man doesn't' (2001: 108). Fresh vegetables and other healthy food also carry a 'time penalty' in terms of cooking; it is much easier to microwave and do something fun whilst awaiting the ping; the same perceived time penalty now, for future uncertain benefits, applies to learning to cook (when one cannot afford these healthy foods, or even access them locally). Low wages, and lack of money to buy time-saving devices such as a car, mean the poor suffer reduced *le temps percu*, and their *le temps concu* is shortened by the attractiveness of scarce gratification now, versus an uncertain future. Future certainty is something else that money tends to buy, when one is not dependent on shifting rented accommodation, the whims of the State Benefits Office, or an employer who sees employees as casual disposable labour. This restricts the formation of stable social contacts and relationships, reducing *le temps vecu* of the poor. In terms of lifestyle factors, many of which impinge on the propensity to eat healthily, one can sum up the triad of Lefebvrian space and extend space into a further intersecting triad of space, money, and time. Table 39.1 summarises some of the factors bearing down on the disadvantaged in society.

Dantean Space: from precariats to cold boxes

In Dante's *Inferno* there were concentric Circles of Hell, each one worse than the last. In extremely deprived areas, such as the Welsh Valleys, reported to be poorer than some regions of Eastern Europe, one can find successive layers of poverty, descending to circles probably unimagined by the citizens of Cardiff, spatially close with its shiny St David's Shopping Centre, ranked sixth in the UK by Experian. One respondent said (CURA 2016): 'It's expensive to be poor'; and those five words sum up the devastating synergies of flexible neoliberal capitalism, austerity government on the lives of the poorest in society. The poor pay more for transport; bus fares in the Valleys of £4 for a journey that would cost £1 by car are frequently necessary to access: work, interviews to claim state benefits, or even food banks. Because the poor are digitally excluded they must also travel to banks to pay bills, and if they fall into debt they pay the highest interest rates, and when the power bills are not paid on time they are forced onto the most expensive tariffs, charged by pre-pay meters.

'Work remains the best route out of poverty', said Iain Duncan Smith, Work and Pensions Secretary for the UK's Conservative Government (Smith 2014). However, statistics say otherwise. In 2014, 53 per cent of Britons 'living in absolute poverty after housing costs' were in work, as opposed to 47 per cent of them having no work (*The Economist* 2016). The core of the 'working yet poor' problem is the zero-hours contract, where income falls short of the level needed to escape the Welfare Benefits system, so working-claimants of the precariat must report each week just how many hours they worked. Because the state benefits system responds to changing weekly income with the agility of a rhinoceros, weekly benefits are frequently under- or over-paid, and overpayments may be clawed back weeks later when work is scarce. Other events can tip someone into destitution through state benefits sanctions, basically the withdrawal of even a minimum survival income by the government; appointments missed because: the bus was late, an under-maintained car broke down, or the children caused the claimant to oversleep. An axiom amongst charities for the homeless is, we are all just two bad decisions way from the pavement. That is when the precariat needs the food bank for sheer survival, but it is a £4 bus fare away. Sanctions 'are applied as a blunt instrument but should be applied as a last resort' (Blackman 2017); the problem is, these two principles can co-exist. Sanctions are supposed to be a last resort, but are bluntly applied to all regardless of personal circumstances. Meanwhile food is one of the first expenditures to be curtailed; and some are deterred from food banks by the bus fare, others by the stigma of being seen to need charity: the *l'espace vecu* of hunger. Health is bound to suffer in the long run. Furthermore, the Trussell Trust (a UK food bank operator) does not want to become a free supermarket (CURA 2016) and limits any individual to a maximum five weeks' usage of their local food bank in any 12-month period. However, some food banks will give out food in excess of this limit if, after checks with agencies, it is evident that the recipient will have no choice but to borrow, beg, or steal food or attempt to source it from churches. Churches are also often very limited in what they can do, offering perhaps just one hot meal a week. Food banks can only offer tinned or dried foods, not fresh produce, for logistical reasons, yet there are circles of poverty even below this.

The penultimate level is the kettle box. The kettle box is designed specifically for the homeless poor, who may exist in hostels but have no access to a kitchen, just an electric kettle. The box contains items such as powdered soup, pot noodles, and pasta sauces that can be 'cooked', with no more facilities than hot water. They are nutritionally poor but provide some sustenance and avoid people going to bed hungry. Then, the ultimate level below kettle boxes is the cold box. As the Trussell Trust stated in 2014:

> For even more destitute clients, a cold box food parcel has been created, containing three days' worth of mainly tinned groceries that can be prepared without the need for heating or hot water.
>
> *(In Butler 2014)*

This is the rock-bottom of poverty, the end result when the benefits of flexible capitalism all go to capital but the flexibility is suffered by the poor workers. Flexibility for the working poor becomes contortion almost to breaking point when the wage-depressing effect of the (near-destitute) reserve army of labour is added in.

Minimum wages, bare survival, the precariat, workers with just enough income for food as fuel, paid just enough to function until discarded at retirement; this is, from capital's point of view, just what capital is supposed to do. Capitalism is always on the edge, exploiting the last tiny niche of profit, extracting the last sliver of value wherever it can. But workers are not robots; they have human value and dignity. As Harvey (2012: 129–30) argues:

> Urbanization is itself produced. Thousands of workers are engaged in its production, and their work is productive of value and of surplus value. Why not focus, therefore, on the city rather than the factory as the prime site of surplus value production?

Lefebvre argues that the city must be reclaimed, for those who really produce it, its citizens; he calls this reclamation *autogestion* (Purcell 2002), where urban citizens manage collective decisions themselves rather than surrendering those decisions to a cadre of state officials (Lefebvre 2003: 187–8). Can urban food systems be reclaimed, even to the point where food banks and cold boxes become a relic of the past?

Reclaiming the city: urban agriculture and the Foodfare state

The socio-spatial infrastructure for reclaiming the urban food chain in fact already exists, in fragmented form, but is continually at risk of being dismantled or co-opted by capital interests. This infrastructure has three main elements: allotments, urban farms, and more radically, 'Dig for Victory' (see Granzow and Shields in Part 4 for another view of urban agriculture). Allotments still exist, but the UK has lost over 80 per cent of them since their peak of 1.5 million during World War Two (WW2). In the Welsh Valleys, waiting lists exist for allotments in some towns whilst elsewhere they lie derelict. With pressure for building land (why not utilise brownfield sites, often plentiful in deprived urban areas) and the squeeze on local authority finances, it is much more lucrative for local authorities to allow building on allotment land and then reap the local government property tax, the council tax. Most large cities now have urban farms, but these are essentially tourist attractions, places to take middle class urban kids to stroke the lambs after the Sunday roast has been consumed. 'Dig for Victory', originally a British campaign during WW2, to use every available scrap of land to produce food, and replicated in Cuba in the 1990s after support from Russia collapsed, appears today to lack an essential element of social solidarity and cohesion. Where the urban lawns of social housing projects have been converted to vegetable growing, problems of vandalism and theft have often emerged. In Britain, local authority housing of the 1930s through to the 60s often had large gardens, to facilitate the tenants growing some fruit and vegetables, but leisure opportunities, cheap unhealthy takeaway food, and lack of both gardening and cooking knowledge mean this ideal is seldom achieved.

Britain also has a history of preventing the poor from being too self-sufficient in food. Thomas Rudge, in his 1813 work *General View of the County of Gloucester*, encouraged landowners not to allow edible fruits to grow in hedgerows, as this would discourage the idle poor from seeking work. The 19th-century owners of Stiperstones mine, Shropshire, England, resented the fact that mine workers' cottages had vegetable plots, 'reducing their dependence on earnings at the mine' (Francis et al. 2000: 59). However, when political food restrictions such as the 19th-century Corn Laws begin to work against the industrial economy, they were soon repealed. The early 19th-century Corn Laws benefitted the rural gentry, preventing imports from lowering bread prices, but by the second half of the 19th century they had been repealed, as they increased the necessary wages that urban factory owners had to pay. Furthermore, the Corn Laws were precipitating urban discontent, a famous example being the Peterloo Massacre in Manchester, England, in 1819, where 15 protestors were killed and hundreds injured by the army during a protest partly caused by high bread prices. In the 20th century, state and capital were willing to finance Beveridge's Welfare State, set up from 1942, because slaying the Five Giants of Want, Disease, Squalor, Ignorance, and Idleness was of mutual benefit to the nation, society, and the economy, improving the health and productivity of the workforce during WW2 and the post-war recovery period.

A similar list of Five Giants could apply to today's food situation amongst the urban poor. We have Want (poverty, unaffordability of healthy food), Disease (obesity, diabetes), Sugar (excess in cheap food), Ignorance (of cooking, nutrition), and Idleness (not cooking, not growing food where possible). To fulfil the United Nations Right to Food, our cities need a 21st-century Foodfare State, which would be a sort of partnership between the citizen and state. An important consideration would be to allow benefits claimants to sell surplus fresh produce they might grow in gardens or allotments, without their benefits being reduced pound for pound as at present. That would encourage a culture of food production and healthy eating, besides all the environmental benefits. Concomitantly, the government could provide food preparation and cooking advice, such as leaflets and even household food budgeting tips, given out at Jobcentres and other benefits offices. These educational initiatives could also be usefully extended to other state service points such as schools, doctors' surgeries, and hospitals. If one has both motivation and knowledge, it is possible to eat both healthily and cheaply now, by sourcing fresh produce at street markets, by ignoring the arcanities of fancy TV chef programmes and discovering that many fresh vegetables can be cooked quite simply, and that halfway houses to healthy eating do exist, for example adding some fresh vegetables to microwave-ready meals. The level of benefits and the minimum wage should be revised to allow for the health premium cost of eating healthily, not just eating enough calories. In practice the health premium should reduce over time if the other measures kick-started demand for fresh produce and local independent retailers then discovered they could gain economies of scale through wholesale purchasing. On the fiscal side, local government could be given incentives to retain allotments, and central government could apply variable Value Added Tax rates to food and even garden centres; lower rates for fresh produce and gardening equipment, higher rates for non-productive garden ornaments and unhealthy food. Local business rates could be raised or lowered according to the contribution of the business to health and diet, and land taxes could be utilised to encourage food production and discourage disuse of brownfield sites. Tax incentives could be given to employers to promote healthy eating amongst their employees, also moderate exercise such as using the office stairs and not the lift. Further information dissemination of these initiatives could be done through household leafleting and television channels, both direct advertising and via programmes that encourage gardening and healthy simple cooking, perhaps utilising popular soap operas for this as well as children's programmes. A programme for reconnecting the urban poor, and indeed all urban citizens, with healthy food should be no less radical or comprehensive than was Beveridge's Welfare State in 1942.

Conclusions

The degree of poverty and malnourishment in British cities has reached levels that are unacceptable. The Welsh Valleys are an extreme, but sadly by no means unique, example of this poverty.

Food banks should not exist in UK cities, let alone food banks that also provide kettle boxes and cold boxes for the ultra-poor. History has shown that, left to the powers of state and capitalism alone, human suffering is insufficient to trigger change; rather, there is strong, perhaps violent, state resistance to change, for example at Peterloo as noted above. Lefebvre calls for a better system, a reframing of the urban system so that those who produce the city also have a right to frame the decision-making of the city: *autogestion* rather than surrendering those decisions to the state. Unfortunately, that surrender has already taken place, perhaps at the ballot boxes of 1979 when the vanguard of flexible neoliberal capitalism was ushered in by British Prime Minister Margaret Thatcher (and US President Ronald Reagan) (Shaw and Shaw 2016). It is always harder to regain than to not surrender, but with rising poverty and discontent, even the current wave of populism that threatens globalisation itself, there may be the political will for the re-empowerment of *autogestion*. Perhaps rising diet-related health costs will compel the institution of something like the Foodfare State proposed above. Hopefully we will not have to wait until 2042 to see this happen.

References

Anonymous (2012) Research interview with a Senior Dietician, Birmingham, UK.

Blackman, R. (2017) 'Conservative MP for East Harrow, London', *BBC News Interview*, 21 February.

Butler, P. (2014) 'Food bank issues parcels for those too poor to heat dinner', *The Guardian*. Online. Available HTTP: <https://www.theguardian.com> (accessed 5 February 2017).

Cummins, S. and Macintyre, S. (2002) 'A systematic study of an urban foodscape: the price and availability of food in greater Glasgow', *Urban Studies*, 39:11 2115–30.

CURA (2016) *Initial Summary*. Online. Available HTTP: <http://fooddeserts.org/> (accessed 1 February 2017).

The Economist (2016) 'When a job is not enough', 25 June.

Francis, P., Price, J. and Yapp, K. (2000) *Never on a Sunday*, Shropshire, UK: Scenesetters.

Härkönen, J., Räsänen, P. and Näsi, M. (2011) 'Obesity, unemployment, and earnings', *Nordic Journal of Working Life Studies*, 1:2 23–38.

Hart, J. (1971) 'The inverse care law', *The Lancet*, 297 405–12.

Harvey, D. (2008) 'The right to the city', *New Left Review*, II, 53 23–40.

Harvey, D. (2012) *Rebel Cities: From the Right to the City to the Urban Revolution*, London: Verso.

Lefebvre, H. (1991/1974) *The Production of Space*, Oxford: Blackwell.

Lefebvre, H. (2003/1970) *The Urban Revolution*, Minneapolis: University of Minnesota Press.

Orwell, G. (2001/1937) *The Road to Wigan Pier*, London: Penguin Classics.

Power, M.L. and Schulkin, J. (2009) *The Evolution of Obesity*, Baltimore, MA: John Hopkins University Press.

Purcell, M. (2002) 'Excavating Lefebvre: the right to the city and its urban politics of the Inhabitant', *GeoJournal,* 58 99–108.

Reichert, A. (2013) *Obesity, Weight Loss, and Employment Prospects: Evidence from a Randomized Trial*, Working Paper, University of York.

dos Santos, O. (2014) 'Urban common space, heterotopia and the right to the city: reflections on the ideas of Henri Lefebvre and David Harvey', *Brazilian Journal of Urban Management,* 6:2 146–55.

Scott, J. (1994) *Poverty and Wealth: Citizenship, Deprivation and Privilege*, Harlow, UK: Longman.

Shaw, H. J. (2014) *The Consuming Geographies of Food*, Abingdon, UK: Routledge.

Shaw, J. and Graham, M. (2017) *Our Digital Rights to the City*. Meatspace Press. Online. Available HTTP: <http://meatspacepress.org/> (accessed 5 February 2017).

Shaw, J. J. A. and Shaw, H. J. (2015) 'The politics and poetics of spaces and places: mapping the multiple geographies of identity in a cultural posthuman era', *Journal of Organisational Transformation & Social Change*, 12:3 234–56.

Shaw, J.J.A. and Shaw, H.J. (2016) 'Mapping the technologies of spatial (in)justice in the Anthropocene', *Information and Communications Technology Law*, 2:1 32–49.

Smith, I. D. (2014) *Child Poverty Strategy Launched*. Press Release Online. Available HTTP: <https://www.gov.uk/> (accessed 10 July 2016). <https://www.gov.uk/government/news/child-poverty-strategy-launched>

40

The urban and the written in Lefebvre's urban texts

Rebio Diaz Cardona

Introduction

Writing has an unstable place in Lefebvre's work on space. Lefebvre acknowledges the importance of writing and symbolic representation in the urban process but is militantly cautious against the linguistic imperialism that had resulted from the French structuralism of the 1950s and 60s, of which the 'city-as-text' and 'writing-the-city' metaphors are frequently a proxy. Statements in his three main urban volumes, *The Right to the City* (RTC), *The Urban Revolution* (UR), and *The Production of Space* (POS), portray writing as relevant to how the city came into being, historically, but as never the whole story, and the semiotics of the city as interesting but ultimately misguided and lacking explanatory power. This is not surprising as the appeal of Lefebvre's work comes partly from his way of directly engaging the intellectual landscape of his time, especially when he pushes against the vigorous reductionisms (economic, linguistic, psychoanalytic, urbanistic, informational) that mined it. In the case of the relation between city and writing, surveying his many references to writing in the three books reveals how the writing metaphor may have been more at the heart of Lefebvre's view of the urban than Lefebvre explicitly states, and despite the frequently negative way in which writing is cast in his texts. Lefebvre's own theoretical arsenal, however, specifically the notions of production of space and right to the city, can be used to explore the relationship between social space and writing and the status of writing in contemporary social space. In particular, one can ask, can the notion of production of space be extended to writing, perhaps by recasting written language as directly producing space? And, would it make sense to think of written language as an urban material implicitly included in the notion of the right to the city, texts as portions of space to which we have spatial rights?

Writing in *The Right to the City*

In 'On urban form', Section 12 of RTC, Lefebvre, seeking to clarify the meaning of urban, and having stated that the urban is a form, sets out to offer a theory of forms. To elucidate 'the meaning of form', 'one will have to refer to a very general, very abstract theory, the theory of forms', one which is 'close to a philosophical theory of knowledge' but is also very different, since 'it designates its own historical and cultural conditions' as well as 'rests upon difficult logico-mathematical considerations' (1996: 133). The brief schematic exposition that then follows is rather abstract and hard to connect with other parts of his work. I am interested in it because of where it locates 'written form'. Lefebvre had made his first frontal attempt at defining the urban a few pages before, in Section 11, titled 'Around the critical point':

> The urban cannot be defined either as attached to a material morphology (on the ground, in the practico-material), or as being able to detach itself from it. It is not an intemporal essence, nor a system among other systems or above other systems. It is a mental and social form, that of simultaneity, of gathering, of convergence, of encounter (or rather, encounters). It is a *quality* born from quantities (space, objects, products). It is a *difference*, or rather, an ensemble of differences.
>
> *(Lefebvre 1996: 131)*

And in the next paragraph:

> Urban society, a collection of acts taking place in time, privileging a space (site, place) and privileged by it, in turn signifiers and signified, has a logic different from that of merchandise. It is another world. The *urban* is based on use value.
>
> *(Lefebvre 1996: 131)*

The urban, it seems, is the utopian dimension of the city, the overcoming of fragmentation, homogenisation, and abstraction. But what type of status should it have? What, or *where*, is it? Reflection, he states:

> can elaborate a *scheme of forms*. It is a sort of analytical grid to decipher the relations between the real and thought. This (provisional and modifiable) grid moves from the most abstract to the most concrete, and therefore from the least to the most immediate. Each form presents itself in its double existence as mental and social.
>
> *(1996: 135–136)*

Lefebvre's 'grid' lists and describes schematically eight forms, each one defined in two parts, 'mentally' and 'socially'. The order (with keywords from Lefebvre's descriptions) is as follows: 1) logical form (centred around the 'principle of identity'); 2) mathematical form ('identity and difference, equality in difference'); 3) form of language ('coherence', 'cohesion' of signification, meaning, messages); 4) form of exchange (centred around 'equivalence'; 5) contractual form (centred around 'reciprocity'); 6) 'form of the practico-material object' (involving 'equilibrium', 'symmetry'); 7)'written form' (involving 'recurrence', 'fixation'); and finally, 8) 'urban form' ('simultaneity', 'encounter', 'concentration'). All eight forms are defined and described in the space of two pages.

The order itself is intriguing and suggests at least three interesting things. First, it suggests that Lefebvre thought of writing (form seven) as different enough from language (form three) to assign them different places in his scheme. He could have reasonably grouped writing and language together as belonging in the same 'form', or list the forms consecutively, contiguous to each other. While the translation of 'language' is enigmatic, as it may refer to the French *langue* (a specific language) or *langage* (language in general), the distance between language and writing in the grid suggests that Lefebvre, at least at the time of writing RTC, saw writing as something relatively independent from language (or a language) and, given what he states about the sequencing of the forms, as more 'concrete' and 'immediate' than it.

Second, Lefebvre locates 'written form' (form seven) the closest to, and right before, urban form (form eight). This is intriguing especially because, as mentioned, Lefebvre has told us that the grid moves 'from the most abstract to the most concrete, and therefore from the least to the most immediate' (1996: 135), which would suggest that 'written form' is more concrete and immediate than 'the form of the practico material object' (form six) which precedes it.

Third, similarly, locating 'written form' in between 'the form of the practico material object' and 'urban form' could be taken to suggest that it is considered by Lefebvre as perhaps a means or a step in the urban's 'ascent to the concrete'. It is not entirely clear if Lefebvre is using the term 'written form' here to refer to written language in its restricted sense or writing in the metaphorical, *writing-the-city*, *city-as-text* sense that he attributes to others (and uses himself) throughout his urban texts. However, since Lefebvre uses the term 'written form' throughout the section, while he uses the terms 'writing' and 'the written word' mostly everywhere else, it seems safe to assume that he is referring to writing in the limited sense (again, as opposed to the metaphorical, 'inscription in general' sense that is still widely used in textualised portrayals of the city).

Lefebvre does not say it, but the placement suggests that the written is intimately linked to the urban; it could even be taken to imply that 'written form' offers itself as a kind of material support for 'urban form', as a form that announces the urban, and is at its service, literally laying the ground for the urban. But are we reading too much into this? We could, of course, simply take written form's placement in the scheme, and the whole scheme for that matter, as the result of yet another of those creative eruptions that Neil Smith referred to in his heartening evocation of Lefebvre's profuse output, in his foreword to UR (2003: xxii). After all, most references to writing in later work, and even in RTC, seem to take a more dismissive, or ambivalent tone, when not hostile. We could also, as we do with other passages, take his scheme of forms as highly suggestive, first, of tensions in Lefebvre's own thinking, and second, of much of what the future of social space would bring as far as writing is concerned, and as it unfolds today.

Specifically, the scheme defines 'written form' in the following terms:

> Mentally: recurrence, synchronic fixation of what has occurred over time going backwards and returning along a fixed becoming.
>
> Socially: the accumulation in time on the basis of fixation and the conversation of what is acquired, the constraint of writing and writings, terror before the written and struggle of the spirit against the letter, the power of speech against the inscribed and the prescribed, the becoming against the immutable and the reified.
>
> *(1996: 137)*

As is clear from this quote, it is not the case that Lefebvre portrays writing entirely in positive terms. For that matter his description seems more one-sidedly positive in the brief entries on 'form of exchange' (broadly focused on 'equivalence') and 'contractual form' (broadly focused on 'reciprocity'), both of which, given their relation to the commodity form and to the law, one could have reasonably expected Lefebvre to portray a more two-sided, at least partly negative way. But is Lefebvre simply making a proto-historical point about writing as being a precondition for the emergence of cities and the urban? Is the placement in the sequence simply having to do with the notion that writing was invented after exchange value and contracts and practico-material objects like 'houses, buildings, utensils, instruments' were? Or, is writing the negation of negation, the negative moment of the dialectic against which the urban, as *aufhebung*, will rise? As mentioned, the 'urban form' section is very brief and schematic, and Lefebvre will not spend much time explaining his grid or, for that matter, using it. His main purpose for including it, one could speculate, would have been to have a framework in which to encumber the urban at the top, as the most 'concrete' form (in the 'synthesis of multiple determinations' sense that this word has in Marxian thought), while at the same time providing the ensemble with a utopian, integrative dimension or orientation. Still, at the very least, this ordering suggests that Lefebvre: 1) considered writing important, 2) perhaps saw it as closer to urban form than any other form listed, and 3) considered it as more 'concrete' than any other form except for urban form itself. I take the 'urban form' section as strong evidence that Lefebvre, at least at the time of writing RTC, considered writing to be a fundamental ingredient in the road to the 'urban', one which performed crucial work for the urban, while at the same time he struggled with what the term writing had come to stand for in the French intellectual context of the time. In that context, his commitments to a form of dualism, in particular, his commitment to presenting the urban as aligned with lived experience, practice, and use value, led him to lump writing together with the bureaucratic, technocratic, formalistic, urbanistic, and ultimately statist ideologies he saw as enemies of the urban. In many places, mainly in POS, playing on old 'phonocentric' (Derrida 1998) clichés, Lefebvre casts writing as more dead than alive, as entangled with power, as a tool of domination, fragmentation, and abstraction, thus making it an easy opposite to lived experience, practice, use value, simultaneity, the encounter, the *fete*, the *oeuvre*, and everything that breathes life in the picture that his urban (as well as other) works trace. This made it harder to articulate a view of writing as one would expect from reading the 'urban form' section discussed above, writing understood in constructive terms, as a crucial ingredient in the production of urban simultaneity. In fact, when Lefebvre talks (briefly) about the role of writing, in the limited sense, in creating the city he is almost exclusively talking about the past, in particular what he calls the 'political city', which he broadly identifies with antiquity (Lefebvre 2003: 8–15). The political city precedes the 'merchant city' and the 'industrial city' in a sort of urban spatial version of the periodisation by mode of production. About it he states '[t]he political city was populated primarily by priests, warriors, princes, "nobles", and military leaders, but administrators and scribes were also present. The political city is inconceivable without writing: documents, laws, inventories, tax collection' (2003: 8).

Lefebvre alludes to writing multiple times in RTC before the section on urban form, in particular in the three short sections between pages 101–117. In these, writing comes up mostly in the context of critical commentary on the promise and perils of the project of an 'urban semiology', something he will engage in frequently in subsequent work. The basic points are laid out in RTC and will not change much in UR or POS. In the section entitled 'Continuities and discontinuities', 'destructurations and restructurations' (1996: 107) and 'general relations' (1996: 108) are said to be 'inscribed in the practico material, written in the urban text' but coming from elsewhere: from history and becoming', 'not from the supersensible, but from another level'. However, he cautions, one must not reduce city to text:

> Yes the city can be read because it writes, because it was writing. However, it is not enough to examine this without recourse to context. To write on this writing or language, to elaborate a *metalanguage of the city* is not to know the city and the urban.
>
> *(1996:108)*

Much is 'below' ('daily life, immediate relations') and 'above' ('institutions, ideologies') the text. 'The city cannot therefore be considered *a* signifying system, determined and closed as a system'.... 'Nonetheless the city has this singular capacity of appropriating all significations for saying them, for writing them (to stipulate and to "signify" them)' (1996: 108). And so he states:

'Therefore the semiology of the city is of the greatest theoretical and practical interest. The city receives and emits messages'. Nevertheless, he continues,

> it is not without the greatest reservation or without precautions that one can consider the city as a *system*, as a unique system of significations and meanings and therefore of values. Moreover, semiology does not exhaust the practical and ideological reality of the city. The theory of the city as a system of signification tends towards an ideology; it separates the urban from its morphological basis and from social practice, by reducing it to a 'signifier-signified' relation and by extrapolating from actually perceived significations. 'This is not without a great naiveté.'
>
> *(1996: 114)*

This mix of mild excitement and irritation or even hostility seems to persist throughout Lefebvre's urban output. It seems clear that Lefebvre is using the term writing in two different senses: 1) in the more restrictive sense of the activities and objects associated with letters, books, documents, and scribes alluded to in the passage about the political city above, and 2) in the semiological, extended sense in which a city is viewed as a readable text, a surface on which history leaves its marks. We can reasonably infer that the writing that these references to urban semiology refer to, writing in the second, metaphorical, city-as-text sense, is not the same as the 'written form' that he refers to in the 'On urban form' section a few pages later or in the references to the political city from UR.

Writing in *The Urban Revolution*

In the *UR*, writing makes its entrance early on, with the reference, cited above, to the effect that the political city 'is inconceivable without writing': 'It is completely given over to orders and decrees to power' (2003: 6). And later... 'The city had writing; it had secrets and powers, and clarified the opposition between urbanity (cultured) and rusticity (naïve and brutal)' (2003: 12). Chapter 6, which has the same title as the section from RTC cited above, 'On urban form', does not feature a 'scheme of forms' or anything along its lines, and the term writing is used, as is more typical in Lefebvre, in its extended sense: 'The city writes itself on its walls and in its streets. But that writing is never completed. The book never ends and contains many blank or torn pages. It is nothing but a draft, more a collection of scratches than writing' (2003: 121). Ultimately, Lefebvre reiterates his view of the semiology of the city as a case of 'fragmentary science', not necessarily false, but ultimately limited, and ideological insofar as it promotes a view of the social as a closed system, thus unintentionally promoting a passive, consumption-based view of the city and society, 'a bureaucratic society of controlled consumption' (2003: 163). As far as theorising the relationship between writing and the urban explicitly, UR does not add much to what we have seen in RTC. But some passages lend themselves to some theoretical exploration of that relationship. For example, a passage on the genesis of *centrality*, that toward which the urban (not a system and not an object, but a form) tends. Lefebvre states, 'With the first gathering and collection of objects existing separately in nature, from the first cairn or pile of fruit, centrality came into being, and with it its virtual realization' (2003: 123). Of course, one could raise the point that many species of nonhuman animals 'gather' as well, and yet we do not attribute any 'urban' implications to their gathering. Considered from the perspective of the relationship between writing and the urban, however, one can argue that writing, in the strict sense, can be seen precisely as fundamentally a *gathering* operation, and that the specific kind of gathering that writing enables (gathering of 'speech', meaning, ideas, stories, accounting practices, captured, stored, and made retrievable by written texts) became a crucial infrastructure for the urban right from the start. This is what Lefebvre's highlighting of writing (scribes, administrators, documents, inventories...) as an important ingredient of the 'political city', the birthplace of the urban, seems to imply.

Writing in *The Production of Space*

In *The Production of Space* (1991), writing (which unfortunately but perhaps understandably, Lefebvre usually conflates with semiology) also comes into play early on. In page 7, Lefebvre states that when codes worked up from literary texts are applied to space, we 'reduce that space itself to the status of a *message*, and the inhabiting of it to the status of a *reading*', and thus we 'evade both history and practice'. The fact remains, however, that 'an already produced space can be decoded, can be *read*. Such a space implies a process of signification' (1991:17).

Discussing the 'illusion of transparency' that characterises those treatments of space that conflate mental space with social space, he alludes to the widespread ideology that sees language and communication as merely transpositions of space; space as becoming transparent when written and talked about.

> Closely bound up with Western 'culture', this ideology stresses speech, and overemphasises the written word to the detriment of a social practice which it is indeed designated to conceal. The fetishism of the spoken word, or ideology of speech is reinforced by the fetishism and ideology of writing.
>
> *(1991:28)*

Later, discussing the passivity into which technocratic society lulls its citizens, he talks about how perhaps

> the place of social space as a whole has been usurped by a part of that space endowed with an illusory special status – namely, the part which is concerned with writing and imagery, underpinned by the written text (journalism, literature), and broadcast by the media; a part, in short, that amounts to abstraction wielding awesome reductionistic force vis-à-vis 'lived' experience.
>
> *(1991:52)*

Here writing falls squarely on the opposite side of lived, concrete, practical experience. But if any doubts remain of the negative role that writing takes up in this account, one may consider this quote, which seems to obliquely make writing the opposite, even the enemy of space: 'To underestimate, ignore and diminish space amounts to the overestimation of texts, written matter, and writing systems, along with the readable and the visible, to the point of assigning to these a monopoly on intelligibility' (1991:62). Writing (or, rather, its privileging) is the reductionistic default from which *space* can save us, and one must assume, by reversal, an obstacle to understanding it. Later, in sections on the genesis of space the tone turns surprisingly more positive. But as is clear in his text, Lefebvre is now using writing as a guiding metaphor to refer broadly to material inscription in space, as he imagines distant historical or constituting processes, far from his belligerent present:

> In the history of space as such, on the other hand, the historical and diachronic realms and the generative past are forever leaving their inscriptions upon the writing-tablet,

> so to speak, of space. The uncertain traces left by events are not the only marks on (or in) space: society in its actuality also deposits its script, the result and product of social activities.
>
> *(1991: 110)*

A few pages later is an evocative passage in which Lefebvre describes the emergence of social space through first inscriptions of human activity on the rhythms of nature:

> Traversed now by pathways and patterned by networks, natural space changes: one might say that practical activity writes upon nature, albeit in a scrawling hand, and that this writing implies a particular representation of space. Places are marked, noted, named. Between them, within the 'holes in the net', are blank or marginal spaces.
>
> *(Lefebvre 1991: 117–118)*

The writing metaphor, it seems, is just too convenient, or too tempting to resist, and becomes readily available as soon as one seeks to describe constituting processes of social space. An unavoidable metaphor, or at least a benign and forgivable one, so long as Lefebvre himself is the one using it. But do not get confused, a recomposed Lefebvre seemingly warns the reader, his use of the city-as-text metaphor does not make him a simple semiologist. Though used by semiologists, notions such as those of marks, marking, and traces do not actually originate with them. Anthropologists, among others, used them earlier:

> It is reasonable to ask, however, whether one may properly speak of a production of space so long as marking and symbolization of this kind are the only way of relating to space. And the answer to this question has to be: not as yet... If and to the extent that production occurs, it will be restricted for a long time to marks, signs and symbols, and these will not significantly affect the material reality upon which they are imprinted.
>
> *(Lefebvre 1991: 141)*

This can have two meanings: 1) marks or symbols are not more than a mere overlay, they do not fundamentally change what things are, or; 2) markings understood in the more material sense are so minute and superficial (physically, materially) that they don't change material reality much, and so in order to have something we can call 'the production of space', more has to be created than mere markings and symbols. In either case, writing does not amount to the production of space. On the other hand, does it make sense to speak of a 'reading' of space? Lefebvre asks and replies, yes and no, and after some elaboration he concludes, 'That space signifies is incontestable. But what it signifies is dos and don'ts – and this brings us back to power' (1991: 142). And dismissively, in the following page:

> The reading of space is thus merely a secondary and practically irrelevant upshot, a rather superfluous reward for the individual for blind, spontaneous and *lived* obedience.... This space was *produced* before being *read*; nor was it produced in order to be read and grasped, but rather in order to be *lived* by people with bodies and lives in their own particular urban context.
>
> *(1991: 143)*

This sense in which (active) production is to *lived* what (passive) consumption is to *read*, permeates much of what Lefebvre has to say about semiology; its focus on signs and systems, as opposed to people and practices, entails 'the reduction of "reality" to the semiosphere' (1991: 296) and ends up promoting abstraction and fragmentation, in knowledge and ultimately, when informing the practices of technocratic urbanism, in life. His discussion of abstract space, unsurprisingly, locates writing squarely on the side of the abstract:

> This immense process starts out from physical truth (the presence of the body) and imposes the primacy of the written word, of plans, of the visual realm, and of a flattening tendency even within that realm itself. Abstract space thus simultaneously embraces the hypertrophied analytic intellect; the state and bureaucratic *raison d'etat*; 'pure' knowledge; and the discourse of power.
>
> *(1991: 308)*

Implications

Overall, even though the number of places in which Lefebvre refers to writing in RTC, UR, and POS are plenty, there are five main ways in which writing is presented: 1) 'written form' conceived as a surprisingly 'concrete' form, occupying the seventh spot in an enigmatic 'scheme of forms' that goes 'from the most abstract to the most concrete' and comprises eight forms, with 'urban form' in the eighth spot (as discussed in the section 'On urban form' in RTC); 2) written language or the written word briefly portrayed as a necessary ingredient in the historical emergence of the city (as alluded to in UR); 3) references to 'reading the city', 'writing the city', and 'city as text' metaphors; something that semiologists engage in and which Lefebvre occasionally lauds but mostly criticises (RTC, UR, and POS); 4) the 'writing the city' metaphor in a more positive light, as an inscriptive metaphor that Lefebvre himself uses to theorise the historical genesis, production, and transformation of social space (RTC, UR, and strongly in POS), and finally; 5) writing portrayed as being in the company of, intimately linked, or simply lumped together with, vision, power, patriarchic society, and as a tool or a process that promotes abstraction, domination, and fragmentation, thus running counter to speech, experience, history, life, time, and even space (POS).

There is also a certain amount of throwing around of the concept in broad, evocative fashion, casually lumping it together with other terms in ways that mobilise the concrete/abstract, use/exchange, lived/dead, spontaneous/constrained dualisms referred to above. Whenever Lefebvre is directly engaging in the critique of urbanistic ideology, technocratic power, passive consumer society, structuralist ideology, or information ideology, all seen by Lefebvre as forms of privileging or fetishising the abstract, writing is likely to receive a hit, either as a tool of domination and abstraction or as the wrong approach. It is frequently in those not uncommon 'how should we go about this?' passages, in which Lefebvre seems to make his search for models explicit, that recourse to writing occurs. Reluctantly at times, with no apologies at other times, he comes back to the writing metaphor, as if it were just too hard to avoid. Then reminds himself and us of what the main issue with all forms of 'reading the city', 'writing-the-city', 'city-as-text' tropes are: they cannot move beyond the descriptive level on to the explanatory, theoretical level. And insofar as they assume closed systems of signification, they share in the ideological, passive acceptance, celebration even, of things as they are, thus failing to make room for the possible and the utopian dimension. Lefebvre is at war with the idea of closed systems when applied to the social, so any attempt at viewing the city in semiological terms, as a closed system of signification, puts him on high alert against reification, neutralisation, abstraction, and the erasure of social practices. So important is this to Lefebvre that the very last thing he says in POS is that social space is anything *but* a system (1991: 423). The fire, however, is there from the start, hardly concealed in those otherwise not very combative few pages of the 'On urban form' section of RTC discussed above:

> [T]he constraint of writing and writings, terror before the written and struggle of the spirit against the letter, the power of speech against the inscribed and the prescribed, the becoming against the immutable and the reified.
>
> *(1996: 137)*

What Lefebvre has to say about writing is better understood by reference to the context in which the urban works were written. Text-centred and writing-centred discourse had broad appeal at the time when Lefebvre started publishing his urban works. The author of *Le langage et la societé* (1966) must have had a hard time garnering attention for his book, published the same year as Lacan's *Ecrits* and Foucault's *Les mots et les choses*, just to be followed by three major works by Derrida the following year. Kofman and Lebas (1996) remind us that Lefebvre was 'profoundly hostile to structuralist analyses' (1996: 17). It is a lasting hostility, as the 1975 reissue of his 1971 book against structuralism, '*L'ideologie structuraliste*' shows. In the preface Lefebvre states that as he saw structuralism become broadly rejected in 1975. His only regret was not having pressed his case harder ('*le seule regret de l'auteur de ce recueil, c'est de n'avoir pas mené plus loin et plus fortement encore la polemique*' 1975: 11). The year before the publication of RTC, Barthes (who Lefebvre praised at times), one of the major figures associated with structuralism and semiology, published a brief text, *Semiology and the Urban* (1967). In the short text, Barthes famously states that 'the city is a discourse and this discourse is truly a language: the city speaks to its inhabitants, we speak our city, the city where we are, simply by living in it, by wandering through it, by looking at it'. Rarely cited, however, Barthes already poses the problem of the 'language of the city' metaphor in the lines that follow:

> Still the problem is to bring an expression like 'the language of the city' out of the purely metaphorical stage. It is very easy metaphorically to speak of a language of the city as we speak of the language of the cinema or the language of flowers. The real scientific leap will be realized when we speak of a language of the city without metaphor.
>
> *(Barthes 1997: 168)*

I am not about to advocate for a 'scientific leap' regarding the relationship between writing and the urban, but I think one can argue that the reading and writing the city metaphor may well operate as an 'epistemological obstacle' (in the sense of Bachelard, whom Lefebvre viewed favourably, and of Althusser, whom he did not) when it comes to studying the relationship between writing and the urban. In particular, without denying its value in guiding students of the city to raise interesting questions, the 'city as text' metaphor has at least one disadvantage, in that its use tends to obscure the question about the specific ways in which writing, in the strict sense, contributes to the production of space. This is something that Lefebvre did not explore in any sustained way. In fact, Lefebvre's engagement with writing as a topic remains always very general. Lefebvre is clearly very interested in the structuralists and above all in countering their influence, from Levi-Strauss to Althusser.

But beyond that, a work like that of Annales School founder Lucien Febvre and H.J. Martin's *The Coming of the Book* (1958), admired as a supreme example of historical scholarly work, and focusing on the spread and impact of printing and print culture in Europe between the 15th and 18th centuries, is not even mentioned in Lefebvre's urban works. McLuhan, who popularised some of the ideas about the impact of writing and literacy on society that others like Innis

(2007) and Goody and Watts (1963) had explored (but had Lefebvre read them?), is mentioned only in passing in POS (1991: 261, 286), and mostly in connection to the rise of a visual logic. Because he is not engaging with it, Lefebvre is unable, for example, to attribute to written language and print culture the constructive roles played in the creation of the public sphere, public space, and modern nationhood that work by others like Habermas (1991), Anderson (1991), and Warner (1990), will ascribe to it. Or in a darker light, he cannot clarify the specific complicities of writing and literacy with power, domination, dispossession, marginalisation, extermination, and colonialism that de Certeau (1984) directly theorises and that Rama (1996), Mignolo (1995), and others will carefully explore. Yet others, like Petrucci (1993) and Henkin (1998), have focused on written texts with a historical focus on urban material culture, in ways that align nicely, at least in principle, with the study and conceptualisation of the production of space that Lefebvre inaugurated.

It would be anachronistic to blame Lefebvre for not having engaged with work published after his death. But we ought to find it somewhat puzzling that Lefebvre all but ignores work on writing, literacy, print, and print culture (including the reach of print matter into daily life and the life of social institutions) that in retrospect would seem directly relevant to the study of the production of space and could have reasonably raised Lefebvre's interest and sparked his theoretical imagination. This sort of, I dare say, 'motivated forgetting' fits well with, and may simply be a by-product of Lefebvre's unrelenting rejection of 'the primacy of the written word' and related theoretical positions, as described above. But it led Lefebvre to leave unexplored the ways in which people use reading and writing, and text-supported practices more broadly, to do much of what they do in the urban context, as well as the ways in which all kinds of institutions in societies with writing are profoundly shaped by and imbricated with writing practices and written matter. To use Lefebvre's spatial triad, while Lefebvre seemed to have considered writing in terms of 'spaces of representation' (as he variously tried to answer the question, are space, the city, society adequately conceived as text, written and read?), writing is also directly related to urban life in terms of 'spatial practices' and 'representational space', so far as it has become a material scaffold for all sorts of text-supported lived social practices. In this sense writing is intimately implicated in all three dimensions of the triad, mental, social, and physical, as it supports the ontological, socio-spatial richness of society itself, 'lived' and 'perceived', and not only 'conceived'. If, as Lefebvre states, '(social) space is a (social) product' (1991: 26) and '[e]very society produces space, its own space' (1991:31), then writing ought to be understood as one of the means whereby society directly produces social space; literally, in other words, writing could be understood as a form of direct spatialisation, and not only by virtue of its referring to space or spatial processes (Diaz Cardona 2016).

In the years after Lefebvre's death, the 'writing the city' metaphor and its variations have continued to be widely in use, and have even been extended into the digital realm, for example in Thrift and French's treatment of coding as writing in their discussion of the 'automatic production of space' (Thrift and French 2002). But also, as suggested above, significant work has been produced that can be seen as exploring the ways in which writing, in the more limited sense, becomes one of the ways in which, to borrow Merrifield's apt phrase 'space gets actively produced' (Merrifield 2006: 105). One could speculate on how Lefebvre himself would have reacted to the socio-spatial life of writing in contemporary society, in particular, the ways in which this more than five-thousand-year-old technology has been put to use, or made itself useful, in the context of the so-called 'digital age'. It is probably not completely unreasonable to imagine that the enduring aptness of writing, and in particular its key role in enabling information and communication technologies, to the extent that they are text-based, to operate widely, would not have escaped his attention. And neither would the emergence of new kinds of text-based action from below, or the way these support new forms of politically effervescent simultaneity. Perhaps it is not impossible that he would have changed his mind, or simply made up his mind, so far as writing is concerned, this old but ever new socio-spatial technology, a veritable people-space relations game changer, with stabilising *and* destabilising powers, actively and directly producing space, potentially contributing to abstract but also differential space, put to use by forces of domination *and* liberation, and perhaps deserving of inclusion in any meaningful list of rights to the urban and the city that may be compiled. Maybe he would even remember why, otherwise unexplainably, he decided to place writing high up there, underneath the urban, closest to the concrete, farthest from the abstract, in a 'scheme of forms' contained in a book on our right to the city some 50 years ago.

Conclusions

In conclusion, there are five principal ways in which references to writing work in Lefebvre's main urban texts, briefly:

1) 'Written form' conceived (surprisingly, I have argued) as a highly concrete form in the enigmatic 'scheme of forms' (presented by Lefebvre in the 'On urban form' in RTC).
2) Writing briefly referred to as a necessary ingredient in the historical emergence of the city (as alluded to in UR).
3) The 'reading the city', 'writing the city', 'city as text' metaphors characteristic of the urban semiologies, of which Lefebvre sometimes approves and disapproves (in RTC, UR, and POS).
4) The 'writing space' metaphor as used by Lefebvre to theorise the historical genesis, production, and transformation of social space (RTC, UR, and more strongly in POS).
5) And finally, writing as loosely associated with vision, power, patriarchic society, and with the furthering of processes of reification, abstraction, domination, and fragmentation (in POS).

Lefebvre's references to writing seem to fall primarily within one of the elements of the spatial triad, spaces of representation, or mental space, insofar as Lefebvre's focus is most often on whether it makes sense to conceive of space as a written text. Alternatively, relations between writing and representations of space (physical, material, perceived space), and relations between writing and spatial practices or lived space are left largely unexplored. Although there was published and widely known work on writing that could have led to a consideration of writing from the point of view of spatial practices or of physical and material space, Lefebvre seems to have either not been aware of it or considered it not relevant to a theory of the production of space. I argued that his view of writing is significantly constrained by the polemical context of his writing, in which writing becomes a proxy for structuralism, to which Lefebvre is a declared enemy. This programmatic commitment has a simplifying effect on Lefebvre's view of writing, pulling him, with few exceptions, to a negative portrayal of writing in relation to space.

Finally, there remains the possibility, however, of using Lefebvrian concepts to explore the life of writing and text-supported practices in the urban context and the ways in which these contribute to the production of urban space, adding to the socio-spatial and material richness of the urban. In particular, the persistent role of writing as a stable socio-technical support on which digital information and communication technologies seem to depend for their functioning, insofar as they continue to be heavily text-based, may prompt students of spatial theory to consider upgrading writing to a critical socio-spatial infrastructure, a crucial ingredient in the production of contemporary urban space.

References

Anderson, B. (1991) *Imagined Communities: Reflections on the Origin and Spread of Nationalism*, London, New York: Verso.
Barthes, R. (1997/1967) 'Semiology and the urban', in N. Leach (ed.) (1997) *Rethinking Architecture: A Reader in Cultural Theory*, New York: Routledge.
de Certeau, M. (1984/1980) *The Practice of Everyday Life*, Berkeley: University of California Press.
Derrida, J. (1998/1967) *Of Grammatology*, Baltimore: Johns Hopkins University Press.
Diaz Cardona, R. (2016) 'Ambient text and the becoming space of writing', *Environment and Planning D: Society and Space*, 34:4 637–54.
Febvre, L. and Martin, J. (1997/1958) *The Coming of the Book: The Impact of Printing 1450–1800*, trans. David Gerard, London, New York: Verso.
Foucault, M. (1994/1966) *The Order of Things: An Archeology of the Human Sciences*, New York: Vintage Books.
Goody, J. and Watt, I. (1963) 'The consequences of literacy', *Comparative Studies in Society and History*, 5:3 304–45.
Habermas, J. (1991/1962) *The Structural Transformation of the Public Sphere: An Inquiry into a Category of Bourgeois Society*, Cambridge, MA: MIT Press.
Henkin, D. (1998) *City Reading: Written Words and Public Space in Antebellum New York*, New York: Columbia University Press.
Innis, H. (2007/1950) *Empire and Communications*, Lanham, MD: Rowan & Littlefield.
Kofman, E. and Lebas, E. (1996) 'Lost in transposition', in E. Kofman and E. Lebas (eds) (1996) *Henri Lefebvre: Writings on Cities*, Oxford: Blackwell.
Lefebvre, H. (1966) *Le langage et la societé*, Saint-Amand: Gallimard.
Lefebvre, H. (1991/1974) *The Production of Space*, Oxford: Blackwell.
Lefebvre, H. (1996/1967) 'The right to the city', in E. Kofman and E. Lebas (eds) (1996) *Henri Lefebvre: Writings on Cities*, Oxford: Blackwell.
Lefebvre, H. (2003/1970) *The Urban Revolution*, Minneapolis: University of Minnesota Press.
Merrifield, A. (2006) *Henri Lefebvre: A Critical Introduction*, New York: Routledge.
Mignolo, W. (1995) *The Darker Side of the Renaissance: Literacy, Territoriality, and Colonization*, Ann Arbor: The University of Michigan Press.
Petrucci, A. (1993/1980) *Public Lettering: Script, Power, and Culture*, Chicago: The University of Chicago Press.
Rama, A. (1996) *The Lettered City*, London: Duke University Press.
Smith, N. (2003) *Foreword to Henri Lefebvre, The Urban Revolution*, Minneapolis: University of Minnesota Press.
Thrift, N. and French, S. (2002) 'The automatic production of space', *Transactions of the Institute of British Geographers*, 27:3 309–35.
Warner, M. (1990) *The Letters of the Republic: Publication and the Public Sphere in Eighteenth-Century America*, Cambridge, MA: Harvard University Press.

Part 6

Right to the city, differential space and urban utopias

Michael E. Leary-Owhin and John P. McCarthy

In this part of the book the focus remains on right to the city (RTC) but also broadens out to embrace two other key concepts central to Lefebvre's theorising regarding cities and urban society. Lefebvre was an optimist; many say a Utopian, when it came to thinking about the long-term future for cities and urban society. In theorising utopia, Lefebvre was continuing a powerful Marxian tradition (Gardener 2006). His utopianism does not necessarily rely on a particular closed architectural arrangement of the material city, but rather he allows us, through his ideas about the production of space, everyday life and RTC, to dare to imagine alternatives to neo-capitalist abstract space (Harvey 2000). Nowhere is this more apparent than in his ideas about differential space and concrete utopia. If his RTC ideas germinated in the Paris disturbances of 1968, his thoughts on differential space may well have been developed in the context of South American informal settlements. These multi-faceted ideas have been springboards for much research and street-level activism. A key feature of Part 6 is the manner in which differential space and the RTC are regarded as being intimately entwined. Rather like the spatial triad itself, this part argues that the RTC and differential space are metaphorical but also concern the exercise of real rights in the concrete public spaces of cities worldwide through contingent processes (Sliwinski 2016).

The United Nations deployment of RTC ideas is viewed cynically by some observers, who see it as a tactic to neuter radical political change. Given that Lefebvre was explicit that RTC represents power from below, not rights bestowed by the state, he would probably be surprised by its adoption by governments, but he may well not reject it if it brings benefits for inhabitants. In contrast, it is also a rallying cry for many city-based 'Occupy' movements and a variety of urban activist campaigns in the Global North and South. Global human rights discourses add greater salience to RTC and differential space ideas. For Lefebvre, differential space initiated by the political right to the city action has the potential to subvert abstract space in the long term in ways sympathetic to everyday use value. In the short term there are myriad opportunities to produce locally important inclusive democratic urban space, especially in cities. Part 6 offers the chance to re-consider three of Lefebvre's most radical but contested ideas.

Part 6 commences with a chapter by Butler that seeks to contrast abstract space with concrete utopia in ways that foreground the salience of the RTC. This concrete utopia has a spatial form which Lefebvre calls differential space. Butler explains how the achievement of a differential space post-capitalist society is a long-term project and one that, therefore, would not be bestowed by the neoliberal state. Rather Butler argues, it is through the appropriation of space and a series of rights that a better society, an urban society, may be achieved. The chapter charts some of the attempts by governments and multilateral agencies to adopt RTC policies and enact legislation. For Butler, although rights gained through legal means, such as in Brazil, may be important and valuable, the challenge Lefebvre throws down is that the RTC, differential space and an urban society must be brought to fruition by processes of *autogestion*. Butler shows how Lefebvre encourages us to think beyond existing structures and institutions and imagine a concrete utopia, one that is already anticipated in everyday materialities and social relationships. It implies not just a different arrangement of physical space but also of social and power relations and the mode of production. Butler accepts that complex questions remain concerning how this transformation could and should be achieved and what a concrete utopia may look like. That said, RTC understood as appropriation through the collective efforts of inhabitants can facilitate resistance to abstract space.

One of the examples Lefebvre provides of differential space is in Brazil, and it is interesting that Brazil is one of the few countries to draw on Lefebvre's ideas to legislate for certain RTC. Huchzermeyer in the following chapter starts with Lefebvre's critique of traditional profit-driven urban development, which she contrasts with the development of what she calls 'shanty towns' (favelas, barrios, ranchos); Huchzermeyer prefers the term informal settlements. The chapter then presents a detailed engagement with Lefebvre's analysis of informal settlements, drawing on her experiences during visits to Peru and Brazil in 1972. In complimentary fashion, the chapter is based partly on conversations with Brazilian scholars in Rio de Janeiro. Although official planning and urban development policies tend to seek the eradication of informal settlements, Huchzermeyer notes that Lefebvre sees them as the result of the contradictions stemming from the capitalist mode of production, ergo an inherent element of it that cannot be eradicated. The chapter draws attention to Lefebvre's claim that informal settlements have high degrees of appropriation, one of the key attributes of the RTC. Informal settlements are regarded in the chapter as the opposite of dominated or abstract space, providing certain rights to the city. Huchzermeyer shows that Lefebvre considers informal settlements provide pointers to the urban society, and she has some suggestions for politicians and policymakers.

Lefebvre's visit to South America came after he spent a decade observing the impacts of government-led new town planning in France. No doubt the regimented order of the new town and the spontaneity, fluidity and intense social life of the informal settlement provided an informative contrast for his analysis. Leary-Owhin's chapter elucidates the emergence of Lefebvre's urban theorising. He does this through an analysis of Lefebvre's direct engagement with the new town of Mourenx and his involvement and subsequent analysis of the events of Paris (1968). He asserts that this dual investigation assists in understanding of Lefebvre's re-evaluation of the city and the urban, leading to his conceptualisation of urban society. This chapter warns us not to see Lefebvre as only critical of modernist new town planning, Lefebvre is fair-minded enough to praise where appropriate. In the chapter, Leary-Owhin stresses the importance of differential space as a desired outcome of the production of space in the transition to an urban society. Two kinds of differential space are proffered concerning the here and now and a distant utopia. Leary-Owhin closes by suggesting we re-think our use of the term differential space.

Resistance by inhabitants themselves is a key feature of Lefebvre's RTC. Erdi in the next chapter documents local resistance to state-led urban redevelopment in two Turkish cities. Her chapter investigates how inhabitants perceive and appropriate city space and in so doing utilise the RTC concept alongside the concept of spatial justice. A major contention is that no one actor, however powerful, can control totally the production of space in their own interests. Resistance, especially by civil society groups, forces the various spatial actors to adapt their tactics to maintain their influence. The chapter considers rural to urban migration and the RTC potential of informal neighbourhoods (see Huchzermeyer in this part). Interview data and analyses are presented to enrich the narrative power of the chapter. They also serve to highlight but normalise the everyday struggles and resistance of inhabitants. Erdi finds that the appropriation of space inspired by the RTC is a more complex process than usually realised; it also requires decades rather than years.

A set of different experiences unfold in Ng's chapter that provides an assessment of two case study examples of RTC struggles in Hong Kong. Ng argues that a dystopia of growth machine urban redevelopment emerged in the current neoliberal context. Ng applies ideas of dystopian utopia and another of Lefebvre's controversial concepts, enlightened elites. The research asks to what extent they can assist disempowered communities in achieving their RTC and differential space. A critical role is proposed whereby enlightened elites or transformative intellectuals provide long-term support in resisting the encroachment of exchange value urban development. The chapter shows the complexity of applying a RTC analysis, where in the case studies, long term, settled inhabitants in the city have different legal status and rights compared with incoming rural migrants. Conflicts occur not just between the state and local inhabitants but between them and rural migrants. The struggle for the RTC in Hong Kong space is seen as protracted and one that requires stoic determination to cope with setbacks.

Lefebvre was critical of philosophers who operated solely at an abstract level. Alasdair Jones, like Ng, seeks to ground his research in the context of an everyday (im)possible concrete utopia. In this chapter he takes Lefebvre's intense and prolonged engagement with architects and planners and with the materialities of real places as a key entry point into an analysis of urban public space in London. Drawing on Lefebvre's ideas about play and concrete utopia in the production of space, Jones first presents a discussion of how these concepts have been interpreted, then presents the findings of his own ethnographic research regarding London's Southbank Centre. Of great importance for the chapter is the association of 'Disneyfication', with claims for the end of public space. Jones is interested in the emancipatory possibilities of urban public space and also the possibilities that emerge when Lefebvre's notion of experimental utopia is taken seriously. In addition, the optimistic edge of Lefebvre's thinking is seen not to involve a predetermined or planned utopia but an everyday utopianism, grounded in seeking the impossible to achieve the possible. In the continual reconfiguration of public space of the Southbank Centre, Jones sees grounds for optimism.

In the final substantive chapter, Coleman provides a theoretical investigation of the role of centrality in the RTC; his line of reasoning is then illustrated by a study of Nanni Moretti's (1993) film *On My Vespa*. Lefebvre's complex understandings of urban centrality and cities and the process of *embourgeoisement* are mobilised by Coleman in seeking to explore the issues generated by a series of questions. First, does the RTC necessarily encompass the right to enjoy an urban life at the city centre? Coleman is conscious that this question challenges Lefebvre's association of the RTC with geographical centrality. Second, should centrality as a key RTC be rejected as an act of resistance to neoliberalism? Third, what is the role of the RTC in stimulating an appreciation of the possibility of a different mode of urban existence: a Lefebvrian utopia? Following Lefebvre, the realisation of the RTC is understood as a practical necessity rather than a far-off dream. Coleman shares with Butler (in this part of the book) a wariness about the appropriation of the RTC by state institutions in ways that neutralise its revolutionary-utopian potential. Coleman accepts Lefebvre's criticism of the visual in the spectacularisation of the city but believes he would see merit in the application of his RTC notion to cinematic analysis.

Moretti's film is shown by Coleman to reveal an alternative Rome with an unexpected rejection of the over-consumed centre and an embrace of reimagined peripheries.

Preliminary conclusions and thoughts

Lefebvre's ideas and utopia

Lefebvre contrasted his concrete utopian ideas, of a fairer post-capitalist society arising partly out of direct action by citizens appropriating space in claiming various rights to the city, with the abstract utopianism of some architects and planners. This theme, along with that of state intervention for urban improvement (and utopia), runs through several chapters in this part of the book. Lefebvre's RTC is anchored partly in the city centre and his concept of centrality. Would the rejection of geographical centrality make the achievement of some of the other aspects of RTC more or less feasible?

Lefebvre's ideas and differential space

Related to the RTC is Lefebvrian differential space that manifests geographically in material space and a variety of alternative social and economic configurations. The global tendency towards homogenisation through the production of abstract space is even stronger now than when Lefebvre was writing. A theme of this part that also appears elsewhere in the book is the confirmation of Lefebvre's claims regarding the inability of abstract space to resolve its inherent contradictions; hence its fragilities are unable to suppress entirely the production of differential space or spaces of difference. Are the benefits we see resulting from the production of differential space worth having, or are they minor successes in state attempts to reconcile neoliberal spatial contradictions?

References

Gardiner, M.E. (2006) 'Marxism and the convergence of utopia and the everyday', *History of the Human Sciences*, 19:3 1–32.

Harvey, D. (2000) *Spaces of Hope*. Oxford: Blackwell.

Sliwinski, A. (2016) 'The value of promising spaces: hope and everyday utopia in a Salvadoran town', *History and Anthropology*, 27:4 430–446.

41

Exploring the contours of the right to the city

Abstraction, appropriation and utopia

Chris Butler

Introduction

One of the most compelling and influential elements of Lefebvre's theory of the production of space is his portrait of abstract space as the mode of spatial production generated by contemporary capitalism. Despite the apparent ubiquity of a constellation of abstract spatial tendencies towards homogeneity, fragmentation and hierarchical ordering, Lefebvre is adamant that abstract space never achieves absolute dominance. As much as it 'relates negatively' towards the alternative forms of spatial production that it harbours, it can never totally suppress or resolve the contradictions that engender them (Lefebvre 1991: 50). Lefebvre describes this new potential mode of spatialisation as 'differential space' – a space oriented towards appropriation rather than domination. The central means that Lefebvre envisages for the political assertion of differential space is the 'right to the city' (Lefebvre 1996; Leary-Owhin 2016). There is now an extensive secondary literature which has considered both these aspects of Lefebvre's thought, but in fields such as socio-legal studies and the interdisciplinary field of law and geography, his conceptualisation of the right to the city has largely been subsumed within a positivist and bureaucratically reformist agenda, which inevitably obscures and undermines its most radical implications. In this chapter I will suggest that an alternative to such anodyne formulations lies in a rethinking of the relationships between abstraction, appropriation and utopia. It is only through viewing the right to the city as a demand for the political and aesthetic appropriation of space, which rejects both the violent abstractions of juridification and the moralistic repudiation of abstraction per se, that we may glimpse how this concept can be harnessed in struggles which challenge the artificial boundaries between the possible and the impossible.

I will begin with Lefebvre's account of abstract space as the dominant mode of spatial production within contemporary capitalism and highlight an alternative form of differential space, which he envisages as emerging from within the contradictions of abstract space. While abstract space is the product of technologies and practices which are focused on the domination of space and the extraction of exchange value, differential space is oriented towards the appropriation of spatial uses and points towards the possibility of restoring the relationships between the body and its space. Lefebvre understands the transition to differential space as a political struggle which depends on the right to the city as a radical demand for the creative appropriation of space. But in charting this transition, it is important to avoid establishing a rigid dichotomy between appropriation and abstraction. The right to the city is best understood as a means of navigating through these two poles of social practice – linking emancipatory forms of praxis to alternative models of social organisation. In this way, differential space can be understood, not as an abstract utopia, disconnected from everyday life, but as a concrete utopia which transforms previously fixed social relations and networks of power.

The emergence of differential space

A widely discussed element of Lefebvre's classic work *The Production of Space* is a history of spatial modes of production, in which he charts both the gradual development of the abstract space that defines contemporary capitalism, and the contradictions which this form of space harbours. In Lefebvre's account, a crucial moment in the passage to abstract space was a shift from the 'metaphysical closure' of the Greek civil order to the novel forms of secular difference and relativity that characterised the Roman city-state (Lefebvre 1991: 239). But this breakdown of the 'internal unity' of form, function and structure in ancient Greece was accompanied by new manifestations of authority, such as the legal codification of private land ownership and the rise of patriarchal social relations, which subordinated space 'to the unifying but abstract principle of property' (Lefebvre 1991: 252) and effectively 'promoted abstraction to the rank of a law of thought' (Lefebvre 1991: 243). In this sense, while the organisation of space in Rome opened up greater possibilities for diversity, Lefebvre is quite explicit in his characterisation of this city as a 'space of power' (Lefebvre 1991: 244–245), which provided a crucial juridico-political framework for the development and consolidation of an abstract mode of spatial production from the late Middle Ages through to the turn of the 20th century. While processes of primitive accumulation are obviously of critical importance here, Lefebvre is also insistent that the introduction of the juridical relations of private property over the entirety of physical space and the law's imposition of the contractual form over customary relationships were essential elements in the rise of abstract space (Lefebvre 1991: 263; Cunningham 2008: 455–456). The Roman world:

> introduced a conception of space the characteristics of which would continue to manifest themselves in later times: the dissociation of component elements... subordination to the unifying, but abstract principle of property; and the incorporation into space of this same principle, which is in itself impossible to live... because it is juridical in nature, and hence external... to 'lived experience'.
>
> *(Lefebvre 1991: 252)*

In Lefebvre's account, abstract space has reached its apogee with the development of contemporary capitalist social relations and is characterised by tendencies towards the fragmentation, homogenisation and hierarchical ordering of space (Lefebvre 2003a: 210; Lefebvre 1976; Martins 1982: 177–178). The fragmentation of space is manifest in its physical division into privatised lots and the submission of these parcels of space to both commodification and the intellectual divisions imposed by the specialist technologies of intellectual disciplines such as economics, law, geography, sociology, planning and architecture. A second tendency that Lefebvre attributes to abstract space is the drive towards homogenisation, which is facilitated by both the flattening out of the diversity of uses through the commodification of space and the expansion of technical capacities for quantification and measurement (Martins 1982: 177). While abstract space is certainly not homogeneous in absolute terms, Lefebvre regards it as having 'homogeneity as its goal, its orientation, its "lens"' (Lefebvre 1991: 287). These tendencies are not opposed, but mutually reinforce each other and are supplemented by a third orientation – towards the hierarchical ordering of space by dominant economic, technological and administrative powers. This stratified ordering of space is a strategic endeavour pursued by the combined forces of juridical and political institutions and processes of capital accumulation. One of Lefebvre's primary examples here is the spatial contradiction between the urban centre and its peripheral spaces, through which the centre organises that which is 'around it, arranging and hierarchising the peripheries' (Lefebvre 1976: 17).

Lefebvre argues that it is through abstract space's tendencies towards fragmentation, homogeneity and hierarchy that capitalist social relations have been able to entrench and reproduce themselves within everyday life (Lefebvre 1976: 21; Wilson 2013). While liberal capitalist societies are usually presented as embodying normative attachments to civil order, political consensus and a social compact based on a commitment to non-violence, the production of abstract space also depends on economic, social and juridico-political processes which are inherently coercive. This is particularly so in relation to the forms of violence which are inflicted on the body, and are characteristic of what Lefebvre, perhaps polemically, depicts as 'a violence intrinsic to abstraction' (Lefebvre 1991: 289).

While the brutal and violent tendencies that are inherent to abstract space are a central focus of Lefebvre's narrative in *The Production of Space*, he also makes clear that there are cracks in the façade. Abstract space is of course a project that is never entirely accomplished, but is constantly confronted by contradictions which undermine naïve presumptions about its totalising capacity to eliminate alternative spatial possibilities. In this context, Lefebvre identifies the potential of a post-capitalist mode of spatial production, which he characterises as 'differential space' (Lefebvre 1991: 50). This form of space encompasses physical, mental and social uses which nurture differences and particularities, and it holds out the prospect of a restoration of unity to the 'functions, elements and moments of social practice' that are fractured by abstract space (Lefebvre 1991: 52). In this sense, differential space emerges from the contradictions that cannot be suppressed by abstract space's tendencies towards homogeneity, fragmentation and hierarchy.

A crucial aspect of Lefebvre's argument is that, whereas abstract space is founded on the domination of nature and of the body, differential space is oriented towards 'appropriation'. This distinction between the 'appropriation' and 'domination' of space is one that Lefebvre suggests was not sufficiently clarified by Marx and led many of his followers throughout the 20th century to endorse a Promethean view of humanity's productivist mastery over material nature (Lefebvre 1991: 165; but see Burkett 1999; Foster 2000). For Lefebvre, this focus on the role of appropriation in defining differential space is not just a rhetorical matter but reveals the necessity of the appropriation of the body – its rhythms, practices and symbols – in any challenge to the dominance (and the dominating logics) of abstract space. Accordingly, the production of differential space requires the restoration of the relationship between the body (as a deployment of energy) and its own space (Lefebvre 1991: 170). This of course leads us to the fact that the production of differential space is a political project, which relies upon the active agency of the inhabitants of space. Lefebvre depicts this project as a massive intervention of personal and collective uses of space which points:

> towards the space of a different (social) life and of a different mode of production... [and] straddles the breach between science and utopia, reality and ideality, conceived and lived. It aspires to surmount these oppositions by exploring the dialectical relationship between 'possible' and 'impossible'.
>
> *(Lefebvre 1991: 60)*

The contours of the right to the city

The most prominent social and political form through which Lefebvre envisages that the appropriation of differential space may be pursued is what he refers to as the right to the city or the right to urban space (Lefebvre 1996). This concept is one that Lefebvre introduced in the context of the revolutionary events of May '68 and was further developed through his later collaborative work with the Group de Navarrenx (Lefebvre 1990; 2003b). In recent years the right to the city has become his most well-known contribution to transdisciplinary debates within politics, law and urban studies (Attoh 2011; Butler 2009, 2012; Dikeç and Gilbert 2002; Fernandes 2007; Gilbert and Dikeç 2008; Harvey 2008; Lebas 1996; Marcuse 2009; Mayer 2009; Merrifield 2011; Mitchell 2003; Purcell 2002, 2008, 2013, 2014; Souza 2010; Wall 2012). Indeed, there has been a somewhat surprising recognition and adoption of the terminology of the right to the city within mainstream policy development, and it has become the subject of a voluminous literature which includes both advocacy for the broadening of access to urban services and proposals for expanding mechanisms of participation in urban decision-making (Brown and Kristiansen 2009; Cities Alliance 2010; Coggin and Pieterse 2012; HABITAT III 2016; UNESCO 2006; UN-HABITAT 2010).

In Lefebvre's original formulation, the right to the city is put forward as an assertion of the entitlement of all urban inhabitants to collaborate in the collective and creative appropriation of space and its use values, over its domination by capital, exchange value and the technocratic expertise of state bureaucracies (Lefebvre 1996; 2003b). This draws out Lefebvre's longstanding interest in the necessity of emancipatory political tendencies being generated by the self-management or *autogestion* of space, rather than relying on the institutions and systems of the liberal capitalist state to act as a guardian of the public interest (Lefebvre 1991: 383; Lefebvre 2009; Rose 1978). Indeed, autogestion is understood by Lefebvre as a direct confrontation with the state's roles in the organisation, control and policing of space. As he makes clear in an essay written prior to his mature writings on both the politics of space and the character of contemporary state power:

> The principal contradiction that *autogestion* introduces and stimulates is its own contradiction with the State. In essence, *autogestion* calls the State into question as a constraining force erected above society as a whole, capturing and demanding the rationality that is inherent to social relations (to social practice)... it cannot escape this brutal obligation: to constitute itself as a power which is not that of the State.
>
> *(Lefebvre 2009: 147, emphasis in original)*

Given the philosophical and political origins of this concept, it seems curious at first glance to recognise how widely the right to the city has been adopted within an increasing range of contexts, including scholarship and policy-making around urban governance and service provision which operate at some distance from the radical vision of this right in Lefebvre's writings. While in one sense this explosion of interest in the right to the city is encouraging, it is also indicative of a widespread tendency to co-opt, institutionalise and, in some cases, to legally codify the right to the city, in order to position it as yet one more addition to a long list of orthodox liberal human rights claims (Brown and Kristiansen 2009; Coggin and Pieterse 2012). While these associations of the right to the city with a broadly social liberal, distributive agenda have gained strength within mainstream scholarship and policy discourse, they have largely sidestepped or obscured the full philosophical implications of Lefebvre's original idea.

A prominent instance of this can be observed in Don Mitchell's widely read (and otherwise laudable) attempt to invoke the right to the city in the service of a radical democratisation of public space and resistance to the coercive spatial ordering that has been imposed by neoliberal urbanisation (Mitchell 2003). Despite an explicit commitment to a transformative political agenda, Mitchell tends to cast the right to the city narrowly, as a symbolic category for the gradual enhancement of a set of orthodox social and economic rights (such as the right to housing and the right to free speech) that are contested in an urban setting (Wall 2012: 140). Another more recent example can be seen in Edward Soja's *Seeking Spatial Justice*, where he laments the degree to which Lefebvre's explicitly spatial orientation and his radical political objectives have been absorbed into a weakly defined, policy-oriented, liberal version of municipal rights (Soja 2010: 107). But as David Cunningham points out, Soja's own attempt to generate a 'justice politics' is linked to this increasingly popular usage of the right to the city, which 'all-too-easily oscillates between… reformist or even essentially conservative/nostalgic modalities, and… future oriented, "progressive" ones', and leaves Soja's project constantly in danger of lapsing into a nebulous, left-liberal reformism (Cunningham 2010: 604–605).

In addition to these scholarly attempts to utilise the right to the city as a ground for reinvigorating an urban human rights agenda, there have also been a number of attempts to institutionalise it as a positive, juridically enforceable right at both domestic and international levels. The most celebrated example is Brazil's enactment of the Federal City Statute in 2001, which was the product of activism amongst both the urban poor and the recognition by federal legislators of the juridico-political importance of municipalities in urban planning and management (Fernandes 2007; 2010). Based on Articles 182 and 183 of Brazil's 1988 Federal Constitution, the City Statute (Law No. 10.257) is a legislative scheme for the implementation of a comprehensive urban policy throughout Brazil through the following aims:

1) Regularising informal settlements (favelas) and integrating them into formal state and economic systems.
2) Prioritising the social functions of urban land over its commercial values and hence limiting incentives for speculation.
3) Providing mechanisms for the democratic involvement of urban inhabitants in urban planning and governance. (Fernandes 2007; Friendly 2013)

The last of these reforms builds on an already strong tradition of 'participatory budgeting' processes in a number of Brazilian cities, which allows for widespread public consultation and debate about annual public budgeting priorities for local authorities (Santos 1998; Wampler 2007). Edesio Fernandes acknowledges that the success of such legislative reforms will always depend upon the continued socio-political mobilisation of the most marginalised urban inhabitants, and he argues that, in this regard, Lefebvre's work has provided a significant inspiration to urban reform movements in Latin America during past decades (Fernandes 2007: 208). While there have been a range of predictable obstacles to the full implementation of the aims of the City Statute (Friendly 2013), Fernandes characterises this piece of legislation as a major development, which has effectively 'laid the foundations of the "right to the city"' in Brazil (Fernandes 2007: 215). It is on this basis that he argues for the need to embrace an immanent critique of the limitations of the urban legal order, and for critical scholars to present a more sustained defence of the right to the city as an enforceable legal right (Fernandes 2007: 208). Such an argument feeds into the international campaign, largely driven by non-governmental organisations, for the recognition of a 'World Charter on the Right to the City' under the auspices of UNESCO and UN-HABITAT (Brown and Kristiansen 2009; UNESCO 2006; UN-HABITAT 2010). The most recent development in this project is the limited official recognition of institutionalised versions of the right to the city by the United Nations General Assembly resolution endorsing the 'New Urban Agenda' developed by the United Nations Conference on Housing and Sustainable Urban Development (United Nations 2017; HABITAT III 2016).

There are certainly powerful reasons why campaigns for the development of legal safeguards for reforms in urban governance and the extension of democratic involvement in decision-making are worth defending. Apart from anything else, they may ensure that successful struggles for the recognition of urban rights do not have to be continually re-fought on the same terms. But it also needs to be recognised that there are dangers in mandating a positivist road to the implementation of the right to the city, which necessarily places so much reliance on the support of sympathetic state actors in order to achieve domestic or international legislative reforms in this area. In circumstances where activists are confronted with the implacable opposition of neoliberal and revanchist authoritarian administrations, the codification of the right to the city may be counter-productive to the overall aims of social movements which mobilise to challenge existing forms of spatial domination and exclusion. In addition, there is a deep contradiction between the incorporation of urban struggles within state-controlled institutions and the radical contestation with state power envisaged by Lefebvre's version of autogestion. While not rejecting the strategic use of legal mechanisms to further political demands, he explicitly rejects the idea that the right to the city can be reduced to either a 'natural' human right or a positivist legal right (Lefebvre 1996: 194).

A recognition of Lefebvre's reluctance to limit the right to the city to such orthodox categories is one of the reasons why Mark Purcell emphasises the connection between Lefebvre's formulation of the right to the city and forms of political utopianism. Far from undermining its plausibility or efficacy as a strategic tool for urban politics, Purcell argues that its utopian orientation is precisely its strength, in that it provides a means through which inhabitants may appropriate their urban environments and challenge the limitations of political reality:

> Lefebvre sees 'the urban' not merely as urbanization, but as a society beyond capitalism, one characterized by meaningful engagement among inhabitants embedded in a web of social connections. The urban is thus what we might call a possible world, a society yet to come.
>
> *(Purcell 2014: 151)*

Concrete utopia: between appropriation and abstraction

This emphasis on the need to extend the parameters of what is considered possible – in effect to 'demand the impossible' – is central to the way in which Lefebvre understands the right to the city as providing an opening for the emergence of differential space (Pinder 2015: 34). It also reflects what Nathaniel Coleman has recently referred to as a 'utopian prospect' that runs through many of Lefebvre's writings, from his multi-volume *Critique of Everyday Life*, through his contributions on the production of space and his critique of state power in the late 1970s (Coleman 2013: 361). But it is also important to recognise that Lefebvre is particularly interested in a specific form of utopianism – a striving towards 'concrete utopia', which he distinguishes from 'abstract' types of utopianism (Lefebvre 1965; 1969). Interestingly, this is a categorisation which is also made by the philosopher Ernst Bloch in his magisterial *The Principle of Hope* (Bloch 1995). Because abstract utopia remains mired within the 'possibilities' of what is considered achievable and is unable to politically confront the impasses presented by the status quo, it must rely on flights of fantasy or technological dreams to engender faith in the idea of progress.

This can be contrasted with the way in which concrete utopia is grounded in the material possibilities of everyday life and is directed towards the transformation of social relations and perhaps 'even the rule of impossibility' (Lefebvre 2002: 347). As Lefebvre states in the recently rediscovered manuscript *Toward an Architecture of Enjoyment*:

> An opposition is continuously at work between abstract and concrete utopias... Today... abstract utopia relies on technocrats; they are the ones who want to build the perfect city. They concern themselves with the "real": needs, services, transport, the various subsystems of urban reality, and the urban itself as a system. They want to arrange the pieces of the puzzle to create an ideal. Contrast this with concrete utopia, which is negative. It takes as a strategic hypothesis the negation of the everyday, of work, of the exchange economy.
>
> *(Lefebvre 2014: 148)*

This distinction between concrete and abstract utopias is important to our understanding of the ways in which the right to the city might be harnessed towards the production of differential space. To some extent, simply emphasising the utopian character of the right to the city in isolation does little more than restate the theoretical dichotomy between appropriation and abstraction, and thereby reinforces the categorical divide between differential and abstract space. This neatly fits in with a widespread tendency in much of the secondary literature on Lefebvre to associate his work with a polemical attitude towards all forms of abstraction, quantification and representation – treating them as inherently repressive – while regarding qualitative phenomena and manifestations of difference and appropriation as potentially emancipatory. Such interpretations are supported by numerous passages in *The Production of Space*, but when treated apart from Lefebvre's broader work, this perspective tends to simplify his approach to abstraction and risks lapsing into a nihilistic rejection of rationality in general (Butler 2016; Dimendberg 1998: 37).

Indeed, as David Cunningham so perceptively points out, an attachment to pure notions of appropriation or social difference, 'which somehow lie beyond abstraction per se', prevents us from understanding the ways in which differential space could ever emerge in a concrete form (Cunningham 2005: 23). Cunningham identifies the limitations of resorting to a critique of the violence of abstraction in general, and instead argues for the importance of locating 'the historically specific forms of violence as domination that it generates under certain conditions' (Cunningham 2008: 467). Such a perspective allows us to acknowledge the necessity of some forms of abstraction in even imagining the pursuit of potentially emancipatory political strategies or the generation of alternative modes of social organisation. In making this argument, Cunningham highlights the inadequacy of those interpretations of Lefebvre which seek to invoke his critique of abstract space in the service of an absolutist rejection of abstraction as a philosophical category. A more helpful reading of this critique would recognise how it is equally concerned with the necessity of techniques of abstraction for comprehending 'the processual materiality of lived space', as it is directed towards the negativity of specific forms of abstraction (McCormack 2012: 719). This reflects Lefebvre's sentiments in a lecture from the early 1960s, that one of the defining markers of modernity is the inescapable need to navigate a path through the continual production of abstractions:

> Modernity is doomed to explore and to live through abstraction. Abstraction is a bitter chalice, but modernity must drain it to the dregs and, reeling in simulated inebriation, proclaim it the ambrosia of the gods. Abstraction perceived as something concrete, antinature and a growing nostalgia for nature which has somehow been mislaid – such is the conflict lived out by 'modern' man [sic].
>
> *(Lefebvre 1995: 193)*

This acknowledgement of the ubiquity of abstraction in both its intellectual and practical manifestations allows us to go beyond the polemical aspects of Lefebvre's account of abstract space in recognising the role of abstractions in the reproduction of social relations. In this context, Lefebvre's introduction of the right to the city is instructive, as it operates as both a strategic and utopian demand, which reconceptualises the connections between appropriation and abstraction. It does so as a political and aesthetic appropriation of space, which rejects both the violent abstraction of positivist formulations of juridical rights and the moralistic repudiation of all forms of political organisation and institutional order. In this sense it presents a potential example of how it might be possible to struggle within and through abstraction by means of what Peter Osborne tentatively identifies as a form of 'appropriation *within* abstraction' (Osborne 2004: 27).

Conclusions

While abstract space is the dominant mode of spatial production within contemporary capitalism, it carries within it the seeds of an alternative form of space, which is based on appropriation and the defence of use values. It has been suggested in this chapter that the right to the city can be understood as a crucial political and social form through which struggles for the production of differential space may be waged. While the right to the city has been widely recognised within a range of disciplines and urban policy contexts, its formulation in much recent scholarship has been constrained within a juridical model of rights. But even acknowledging that the production of differential space cannot be pursued solely through the legislative and juridical codification of the right to the city, this does not eliminate complex questions regarding how the active and creative transformation of urban life may take place, and what forms of social ordering this might entail. It is in this context that it has been argued here that the most effective way of conceptualising the right to the city is by linking it to Lefebvre's specific understanding of concrete utopianism, which arises from the material possibilities of the everyday, but points towards the transformation of social relations. In this way the right to the city provides a means of mediating between the violence of legal and economic abstractions and an idealist faith in the power of pure appropriation. It is only through understanding the right to the city as a demand for the political and aesthetic appropriation of space, which rejects both the limitations and potential violence of juridification, and the moral rejection of abstraction as such, that it might be possible to deploy this concept as a ground for resistance to dominant forms of spatial production.

References

Attoh, K.A. (2011) 'What kind of right is the right to the city?', *Progress in Human Geography*, 35:5 669–85.

Bloch, E. (1995) *The Principle of Hope: Volume 1*, Cambridge, MA: MIT Press.

Brown, A. and Kristiansen, A. (2009) *Urban Policies and the Right to the City: Rights, Responsibilities and Citizenship*, Paris: UNESCO.

Burkett, P. (1999) *Marx and Nature: A Red and Green Perspective*, New York: St. Martin's Press.

Butler, C. (2009) 'Critical legal studies and the politics of space', *Social and Legal Studies*, 18:3 313–32.

Butler, C. (2012) *Henri Lefebvre: Spatial Politics, Everyday Life and the Right to the City*, Abingdon: Routledge.

Butler, C. (2016) 'Abstraction beyond a "law of thought": on space, appropriation and concrete abstraction', *Law and Critique*, 27:3 247–68.

Cities Alliance (2010) *The City Statute of Brazil: A Commentary*, São Paulo: Cities Alliance and Ministry of Cities, Brazil.

Coggin, T. and Pieterse, M. (2012) 'Rights and the city: an exploration of the interaction between socio-economic rights and the city', *Urban Forum*, 23:3 257–78.

Coleman, N. (2013) 'Utopian prospect of Henri Lefebvre', *Space and Culture*, 16:3 349–63.
Cunningham, D. (2005) 'The concept of metropolis: philosophy and urban form', *Radical Philosophy*, 133 13–25.
Cunningham, D. (2008) 'Spacing abstraction: capitalism, law and the metropolis', *Griffith Law Review*, 17:2 454–69.
Cunningham, D. (2010) 'Rights, politics and strategy: a response to *Seeking Spatial Justice*', *City*, 14:6 604–6.
Dikeç, M. and Gilbert, L. (2002) 'Right to the city: homage or a new societal ethics', *Capitalism, Nature, Socialism*, 13:2 58–74.
Dimendberg, E. (1998) 'Henri Lefebvre on abstract space', in A. Light and J. Smith (eds) *Philosophy and Geography II: The Production of Public Space*, Lanham: Rowman & Littlefield.
Fernandes, E. (2007) 'Constructing the "right to the city" in Brazil', *Social and Legal Studies*, 16:2 210–19.
Fernandes, E. (2010) 'The City Statute and the legal-urban order', in Cities Alliance, *The City Statute of Brazil: A Commentary*, São Paulo: Cities Alliance and Ministry of Cities, Brazil.
Foster, J.B. (2000) *Marx's Ecology: Materialism and Nature*, New York, Monthly Review Press.
Friendly, A. (2013) 'The right to the city: theory and practice in Brazil', *Planning Theory & Practice*, 14:2 158–79.
Gilbert, L. and Dikeç, M. (2008) 'Right to the city: politics of citizenship', in K. Goonewardena, S. Kipfer, R. Milgrom and C. Schmid (eds) *Space, Difference, Everyday Life: Reading Henri Lefebvre*, New York: Routledge.
HABITAT III (2016) *Quito Declaration on Sustainable Cities and Human Settlements for All*. Available HTTP: <https://habitat3.org/the-new-urban-agenda> (accessed 16 May 2017).
Harvey, D. (2008) 'The right to the city', *New Left Review*, 53 23–40.
Leary-Owhin, M.E. (2016) *Exploring the Production of Urban Space: Differential Space in Three Post-Industrial Cities*, Bristol: Policy Press.
Lebas, E. (1996) 'The right to the city', *City*, 1:3–4 184–85.
Lefebvre, H. (1965) *La Proclamation de la Commune*, Paris: Gallimard.
Lefebvre, H. (1969/1968) *The Explosion: Marxism and the French Revolution*, New York: Monthly Review Press.
Lefebvre, H. (1976/1973) *The Survival of Capitalism: Reproduction of the Relations of Production*, London: Allison & Busby.
Lefebvre, H. (1990) 'Du pacte social au contrat de citoyenneté', in Group de Navarrenx (ed.) *Du Contrat de Citoyenneté*, Paris: Editions Syllepse et Editions Périscope.
Lefebvre, H. (1991/1974) *The Production of Space*, Oxford: Blackwell.
Lefebvre, H. (1995/1962) *Introduction to Modernity*, London: Verso.
Lefebvre, H. (1996/1968) 'Right to the city', in E. Kofman and E. Lebas (eds) *Writings on Cities*, Oxford: Blackwell.
Lefebvre, H. (2002/1961) *Critique of Everyday Life II: Foundations for a Sociology of the Everyday*, London: Verso.
Lefebvre, H. (2003a/1986) 'Preface to the new edition: the production of space', in S. Elden, E. Lebas and E. Kofman (eds) *Key Writings*, London: Continuum.
Lefebvre, H. (2003b/1990) 'From the social pact to the contract of citizenship', in S. Elden, E. Lebas and E. Kofman (eds) *Key Writings*, London: Continuum.
Lefebvre, H. (2009/1966) 'Theoretical problems of *autogestion*', in N. Brenner and S. Elden (eds) *State, Space, World: Selected Essays*, Minneapolis: University of Minnesota Press.
Lefebvre, H. (2014/1973) *Toward an Architecture of Enjoyment*, in Ł. Stanek (ed.), Minneapolis: University of Minnesota Press.
Marcuse, P. (2009) 'From critical urban theory to the right to the city', *City*, 13:2–3 185–97.
Martins, M.R. (1982) 'The theory of social space in the work of Henri Lefebvre', in R. Forrest, et al. (eds) *Urban Political Economy and Social Theory*, Aldershot: Gower.
Mayer, M. (2009) 'The "right to the city" in the context of shifting mottos of urban social movements', *City*, 13:2–3 362–74.
McCormack, D. (2012) 'Geography and abstraction: towards an affirmative critique', *Progress in Human Geography*, 36:6 715–34.
Merrifield, A. (2011) 'The right to the city and beyond: notes on a Lefebvrian reconceptualization', *City*, 15:3–4 473–81.
Mitchell, D. (2003) *The Right to the City: Social Justice and the Fight for Public Space*, New York: Guilford Press.
Osborne, P. (2004) 'The reproach of abstraction', *Radical Philosophy*, 127 21–8.

Pinder, D. (2015) 'Reconstituting the possible: Lefebvre, utopia and the urban question', *International Journal of Urban and Regional Research*, 39:1 28–45.
Purcell, M. (2002) 'Excavating Lefebvre: the right to the city and its urban politics of the inhabitant', *GeoJournal*, 58:2 99–108.
Purcell, M. (2008) *Recapturing Democracy: Neoliberalization and the Struggle for Alternative Urban Futures*, New York: Routledge.
Purcell, M. (2013) 'To inhabit well: counterhegemonic movements and the right to the city', *Urban Geography*, 34:4 560–74.
Purcell, M. (2014) 'Possible worlds: Henri Lefebvre and the right to the city'. *Journal of Urban Affairs*, 36:1 141–54.
Rose, E. (1978) 'Generalized self-management: the position of Henri Lefebvre', *Human Relations*, 31:7 617–30.
Santos, B. (1998) 'Participatory budgeting in Porto Alegre: toward a redistributive democracy', *Politics and Society*, 26:4 461–510.
Soja, E. (2010) *Seeking Spatial Justice*, Minneapolis: University of Minnesota Press.
de Souza, M.L. (2010) 'Which right to which city? In defence of political-strategic clarity', *Interface*, 2:1 315–33.
UNESCO (2006) *Urban Policies and the Right to the City: International Public Debates*, Paris: UNESCO.
UN-HABITAT (2010) *The Right to the City: Bridging the Urban Divide*, Rio de Janeiro: World Urban Forum, United Nations.
United Nations (2017) *General Assembly Resolution 71/256: New Urban Agenda*. Online. Available HTTP: <https://habitat3.org/> (accessed 16 May 2017).
Wall, I. (2012) *Human Rights and Constituent Power: Without Model or Warranty*, Abingdon: Routledge.
Wampler, B. (2007) *Participatory Budgeting in Brazil: Contestation, Cooperation, and Accountability*, University Park, PA: Pennsylvania State University Press.
Wilson, J. (2013) 'The devastating conquest of the lived by the conceived': the concept of abstract space in the work of Henri Lefebvre, *Space and Culture*, 16:3 364–80.

42

Informal settlements and shantytowns as differential space

Marie Huchzermeyer

Introduction

Informal settlements, or in Lefebvre's translators' words, 'shantytowns', form a continuity of as well as a contradiction to the planned urban order. In Lefebvre's (1970; 2003) dialectic analysis, this planned urban order aides a combination of state control and neoliberal agenda. In certain instances, informal settlements or 'shantytowns' serve this same purpose. The idea of informal settlements forming part of the planned order, thus being willed by the state, resonates with Roy's (2009) concept, based on work in India, of informality as a 'strategy' or an 'idiom' of planning, and her earlier observation that states may deliberately suspend order (Roy 2005). Roy (2009) also writes about insurgence and in that sense touches on aspects of a highly nuanced lens that Lefebvre developed over time and which he offers for an understanding of informal settlements.

In his critique of French urban development, Lefebvre sees informal settlements arising from displacements and intentions to segregate by those determining the planned order. This applies to the repressive use of spatial strategies in colonialism and apartheid South Africa (Lemon 1991) and their incomplete dismantling in the post-colonial era. But informal settlements also wilfully transgress. They contravene building regulations, minimum space standards and layout requirements. In South Africa today, from where this chapter is written, informal settlements may insert themselves, if only temporarily, in areas legally designated for non-residential uses or for which protracted planning is underway. Some occupy areas which are designated for middle-class residence, contravening and frustrating spatial regulations governing subdivision in anticipation of economic growth through a property market (Huchzermeyer 2011). This tension between abstract space and space that of necessity is inserted and produced from below is included in what Lefebvre calls 'contradictions of space', which are 'added to and superimposed on the entrenched contradictions of the capitalist mode of production' (Lefebvre 2009a: 235). Lefebvre begins to acknowledge this directly after visiting Latin America in 1972.

The dominant urban discourse internationally frames informal settlements as a problem, to be solved, undone or eradicated. The United Nations (UN) (1997) defines informal settlements as unplanned and illegal occupation of land, not complying with building and planning regulations. Since 2000, the term 'slum' has taken precedence within the UN. 'Slum' is likewise defined officially through inadequacies, be they in safety, access to basic services or tenure security (UN-Habitat 2003) and carries the unintended burden of legitimising 'slum' clearance (Huchzermeyer 2014).

Notwithstanding the limitations of the term 'informal settlement', I use it throughout this chapter due to its more neutral meaning than that of 'slum', and given the derogatory connotation that the apartheid state in South Africa lent alternative terms such as shantytown and squatter settlement. With this choice, I do not align with technocratic discourse and expertise which sets out merely to correct the in-formality through relocation or at best adjustment and formalisation, most recently captured in the United Nation's *New Urban Agenda* (NUA). Adopted in Quito in 2016, NUA commits signatories to retrofit, integrate and upgrade informal settlements to ideals of accessibility, safety, quality of space, inclusivity and affordability through conventional approaches (Habitat III 2016: S77, 97, 109).

Lefebvre would term this the imposition of 'quantitative space', simultaneously homogenous (in its parcelling into equivalent portions for the purpose of exchange or speculation) and fractured (through the individualisation) (Lefebvre 2009a: 233, 234). Given a persistent state tendency, not only in South Africa, to aim for eradication or complete removal of informal settlements, necessary advocacy has sought to promote less disruptive and more poverty-responsive approaches to upgrading. In South Africa, this requires demands for a redefinition of what is technically considered possible in terms of in situ interventions (Huchzermeyer 2011). In limited ways reference to global consensus captured in the *New Urban Agenda* strengthens this endeavour. But this does not go beyond identifying a different way of solving the 'problem' posed by informal settlements, and is often accompanied by a superficial though not unimportant slogan inspired by a simplification of the work of John F.C. Turner and his contemporaries (Mangin 1967; Turner 1968) that informal settlements should be seen not as a problem but as part of the solution.

Lefebvre is concerned with the way urban phenomena are understood, and the effects of these understandings. He notes that 'the state and political power seek to become, and indeed succeed in becoming, reducers of contradictions' (Lefebvre 1991: 106). This is possible only 'via the mediation of knowledge, and this means a strategy based on an admixture of science and ideology' (ibid.). The approach that his work offers requires 'a reversal of the conventional way of looking at things' (ibid.: 139). It critically analyses the thought that informs formal spatial decision-making. It understands informal settlements as a contrast to what emerges from this thought. It affords informal settlements attributes of what he deems urban, imbued with the political potential, though constrained, for achieving a different future.

This chapter examines Lefebvre's critique of urbanism, which encompasses the approach of 'state-bound specialists', namely architects, planners and developers, and his critique of the dominant space which urbanism plays a role in realising. It examines the contrasting approach that Lefebvre puts forward for the understanding of and strategising for urban space, and within this the way he treats spatial difference. His reference to informal settlements or 'shantytowns' in this analysis shifts from a treatment of these spaces as part of the dominant segregation, to a recognition of their potential for political opening. This latter approach is brought into focus in a passage in *The Production of Space*, following his mission to Latin America. Here he employs concepts that are attributes of the urban, but qualified through the repressive consequence of dominant space and spatial doctrine.

Lefebvre's conceptual work is extensive, interlinked and circular. Any attempt at addressing or applying it to a particular topic or phenomenon necessitates clarification. As the urban phenomenon of informal settlements has not received attention by Anglophone Lefebvrian scholars, this clarification is to a large extent not mediated by 'critical commentaries' or the 'secondary route', instead it draws largely on Lefebvre's original writing.

From the critique of urbanism to the affirmation of difference through a political strategy

Lefebvre's theory of the urban informs his recommendations on urban strategy. Lefebvre calls for a 'radical critique' to 'define a *strategy*' – both strategy of knowledge and political strategy, without these being separate (182003: 141). Consequently he elaborates an extensive critique of urbanism which he refers to as a 'body of doctrine' (ibid.: 165–166). Urbanism is 'an activity that claims to control the process of urbanization and urban practice and to subject it to its order' (ibid.: 151). It takes the form of humanist urbanism which 'proposes abstract utopias', developer urbanism which seeks to sell 'happiness, a lifestyle [and] a certain social standing' and state urbanism which separates into two aspects, institutions and ideology; despite claiming order, this results in incoherence and 'chaos' (ibid.). Lefebvre understands chaos also as inherent in the spatial contradiction between the 'goals of individual property developers', who produce space and the 'general (strategic) goals of the State' (Lefebvre 2009a: 238, 239).

Seen through Lefebvre's critical lens, mainstream urbanism as a doctrine lacks 'dialectical thought', ignoring the internal contradictions of urbanism as manifested in space (Lefebvre 2003: 171) and mirroring the contradictions inherent in the state (Lefebvre 2009a). Indeed, the urban order and the segregation that urbanism imposes are closely related to state power (2003: 92). This involves a 'sophisticated exploitation and carefully controlled passivity' (ibid.: 140) and the organisation of inhabited space, controlling 'consumption of space and the habitat' (ibid.: 164).

Lefebvre (2003: 163–164) defines urbanism as a superstructure distinct from practice, from social relations and from society, of neo-capitalist society, 'a *bureaucratic society of controlled consumption*'. It 'establishes a repressive space that is represented as objective, scientific, and neutral' (ibid.: 181). Urbanism produces dominant space, a concept that 'attains its full meaning only when it is contrasted with the opposite and inseparable concept of *appropriation*' (Lefebvre 1991: 165). As I show below, Lefebvre deems appropriation key to understanding informal settlements. Dominant space in turn 'forbids the transgressions that tend to produce a different space (whatever that may be)' (Lefebvre 2009a: 240). Lefebvre makes this point in a discussion about the necessary decline of the state, which cannot overcome its contradictions (ibid.). This decline can only come about through 'control by the base', territorial self-management, subduing 'state logic through a spatial dialectic', direct democracy and democratic control and 'affirmation of the differences produced in and through that struggle' (Lefebvre 2009a: 250, 251). Lefebvre builds this into a 'political strategy' in which self-management does not stand on its own. Alongside it, he includes the foregrounding of the 'urban problematic' in 'political life' or the 'politicization of urban issues' and the introduction of a 'right to the city' into 'the enlarged, transformed, concretized contractual system' (Lefebvre 2003: 148,150). Thus the right to the city, as concretised through a particular type of policy, programme and legislation, must be seen as tied to a political discourse as well as a recognition of self-management practices, including those in informal settlements.

Towards the political possibility inherent in produced difference

Methodologically, Lefebvre promotes an integration or political economy approach which examines an element in its wider context, allowing understanding and reasoning, and moving between scales (Lefebvre 2009b: 114–115, 165). He requires that as a 'starting point philosophy, ideology and institutional discourse' be abandoned, as they make up the 'customary scientificity that limits thought to an existing framework and prevents it from exploring possibilities through form'

(Lefebvre 2003: 122). Lefebvre's interest in possibilities or the possible, often expressed as the 'impossible-possible future' (ibid.: 105) or how the impossible may become possible, noting the dialectic between the two (Lefebvre 1991: 60). This drives his concern for 'premature closure' (ibid.: 84) in urban space, and through this in the growth of the individual person inhabiting urban space. Lefebvre expresses 'a sense of loss for an urban reality that is slipping away' (ibid.: 166). With 'ideology, consumption, and the predominance of the rational' and with everything needing to 'be part of an order', Lefebvre points to the exception of 'a residue of disorder and freedom, which is sometimes tolerated, sometimes hunted down with overwhelming repressive force' (ibid.: 36). This would be a residue of 'the urban', which Lefebvre defines (throughout time) as 'place of desire, permanent disequilibrium, seat of the dissolution of normalities and constraints, the moment of play and of the unpredictable' (Lefebvre 1996: 129). He challenges those wishing to 'strengthen the kernel, the *urban*, which survives in the fissures of planned and programmed order' to go beyond liberal humanism (one could add humanist urbanism) and beyond Marx's 'incomplete teachings' on the urban, to understand how the 'socialization of society… has urbanization as its essence' but has turned the 'signs… of urban life… over to consumption' (ibid.: 129, 130).

Lefebvre extends this to the political: 'the social holds the secret of the political' (Lefebvre 2009c: 60). He places the social at a level between the economic and the political. The choices 'between political possibilities' therefore depend 'on the social forces that are mobilized or mobilizing, and the intensity of their action'. Thus 'if social forces stagnate' or 'neutralize each other', the state 'remains a bloc' with 'no fissures… no crack in the State apparatus' (ibid.). In such periods, the only possibilities are actions from within the state, 'through the acceptance of its structures' (ibid.). However, these structures serve also 'to structure social forces and economic life, practical life, and society as a whole' (ibid.):

> once social forces begin moving, everything happens, as if under this house, under this edifice that seemed solid and balanced, the earth begins to move. And there promptly appear fissures where we once saw a vertical rock face. And what appeared to be a simple crack in the walls promptly becomes a crevasse and deepens. Everything immediately shifts in this gigantic edifice, in this State apparatus, and the bloc begins to move. And so there appears the possibility of changing something in this bloc, though not without difficulty, not without danger.
>
> *(Lefebvre 2009c: 60–61)*

Lefebvre (2009c: 61) underlines the constant need, within democracy, for a struggle against the state. Democracy is never an end state, and therefore democracy at various depths is revolution at those same depths (ibid.). Lefebvre adds urban revolution to peasant and workers' revolutions (Lefebvre 2009d: 291) and defines it as 'the transformations that affect contemporary society', sudden or gradual, nonviolent or possibly violent (Lefebvre 1970/2003: 5). Pinder (2015: 42) refers to Lefebvre's 'attempts to engage with the possible-impossible through understanding urban revolution as a mutual transformation of space and social life, combined with efforts to explore possibilities through interventions in the present'. This captures Lefebvre's 'dialectic utopianism' (ibid.). Pinder explains Lefebvre's thinking on utopia mainly in relation to the 'proliferation of urban utopias within France and Western Europe' at the time of Lefebvre's writing (ibid.). Pinder hints only at 'the visions of urban worlds embodied in current… landscapes' which Lefebvre mentions in his work (ibid.), and which could include landscapes of constrained informality.

Emancipatory futures, whether European utopian experiments or informal settlements, represent difference in urban space. This interests Lefebvre precisely because of the potential of certain difference for wider change and indeed for revolution. He contrasts actual difference with the seeming difference of segregation and separation induced by urbanism in the service of the state:

> Difference is incompatible with segregation, which caricatures it. When we speak of difference, we speak of relationships, and therefore proximity relations that are conceived and perceived, and inserted in a twofold space-time order: near and distant. Separation and segregation break this relationship. They constitute a totalitarian order, whose strategic goal is to break down concrete totality, to break the urban. Segregation complicates and destroys complexity.
>
> *(Lefebvre 2003: 133)*

Lefebvre calls for a methodological approach that understands 'the urban as a differential field' (Lefebvre 2003: 53), breaking with a reductive representation that 'is based on a logistics of restricted rationality and motivates a strategy that destroys the differential space of the urban and "habiting" by reducing them' (ibid.: 48). Lefebvre (1991: 371) draws on 'still incomplete theory of difference' to distinguish between differences in two ways. First, through the extent of the difference, a simple logic with two categories: 'minimal' difference and 'maximal' difference (ibid.: 372). Second, through 'the theory of dialectical movement', which distinguishes whether the difference is within or beyond 'a set or system generated according to a particular law' (ibid.). Thus 'induced' difference remains within, whereas 'produced difference presupposes the shattering of a system; it is born of an explosion; it emerges from the chasm opened up when a closed universe ruptures' (ibid.). The distinction between induced and produced difference is distinct from Lefebvre's concepts of isotopy and heterotopy. Isotopy (sameness) and heterotopy (otherness) are always relative to one another. Heterotopy can be difference in the form of 'a highly marked contrast' (this would be maximal difference) 'all the way to conflict' (this could be produced difference) (Lefebvre 2003: 38).

Critical theory texts on Lefebvre's work on difference, e.g. Kipfer (2008), conflate the distinction between minimal/maximal and induced/produced, using them instead as synonyms. For a reading from South Africa and the Global South more broadly, where informal settlements mark the urban landscape in constant, if tolerated, tension with the normative spatial aspirations and manifestations of the state, the question as to whether the difference is merely large or small (minimal or maximal) or in fact induced or produced and therefore whether it holds a revolutionary-transformative potential is critical.

A shift in Lefebvre's understanding of informal settlements: from induced to produced difference

In the late 1960s, Lefebvre sees the 'shantytowns', whether in France or beyond, as forming part of the process of segregation and peripheralisation, involving no more than induced difference. In *Right to the City*, Lefebvre (1996: 70–71) frames shantytowns as an implication of industrialisation, which is 'unable to employ and fix available labour', resulting in cities, for instance in Latin America, being 'encircled by shanty towns' (ibid.). In the same book, he likens shantytowns in 'poor countries' with peripheralised suburbs in 'highly industrialised countries' (ibid.: 125), shantytowns and peripheral suburbs, both resulting from segregationist tendencies (ibid.: 140), thus representing induced difference. Of interest to Lefebvre at this stage of his work is when 'workers, chased from the centre of the city to its outskirts, returned to the centre occupied by the bourgeoisie' during the Paris Commune of 1871, 'with its myth and ideology, its utopia'

(Lefebvre 2003: 110). In the *Urban Revolution*, Lefebvre goes further to acknowledge 'guerrilla activity' in shantytowns in South America (1970/2003: 146).

Lefebvre develops his theory on difference in a publication in 1970 and on induced and produced difference in 1971 (Lefebvre 1974/1991: 373). Thus he approaches his first-hand experience of life in Latin American shantytowns in the following year with different theoretical insight. According to Shields (1999: 183), he stayed in Brazilian *favelas*. Information on Lefebvre's Latin American travels is scant in published biographies, but a footnote in Hess (1988: 274–275) mentions an official mission to Peru and Brazil in November and December of 1972. This experience evidently leads Lefebvre to treat shantytowns explicitly as struggles against the state. Shields (1999: 183), though providing no detail of Lefebvre's engagements in Brazil, notes that Lefebvre 'detected' a 'reawakening of a "politics of difference"'. Brazilian scholars at Rio de Janeiro's Federal University, with whom I enquired about the Brazilian recollection if any of Lefebvre's visit to that country, observed that having visited shantytowns in Latin America, Lefebvre 'had the intuition that slums would be like an embryo of a new society, a new form of sociability, thus breaking with the "destruction of urbanity"' (Fridman 2015 personal communication). In *The Production of Space* (Lefebvre 1991), published two years after his visit to Brazil, Lefebvre provides his longest and perhaps most impassioned passage about informal settlements. He first emphasises the threat of this produced difference being 'absorbed' into the dominant homogenised space as induced difference (Lefebvre 1991: 373). He then demonstrates in more detail how the theory of difference applies, and this is worth quoting at length:

> The vast shantytowns of Latin America (*favelas, barrios, ranchos*) manifest a social life far more intense than the bourgeois districts of the cities. This social life is transposed [or transferred] onto the level of urban morphology but it only survives inasmuch as it fights in self-defence and goes on the attack in the course of class struggle in its modern forms. Their poverty notwithstanding, these districts sometimes so effectively order their space – houses, walls, public spaces – as to elicit a nervous admiration. *Appropriation* of a remarkably high order is to be found here. The spontaneous architecture and planning ('wild' forms, according to a would-be elegant terminology) prove greatly superior to the organization of space by specialists who effectively translate the social order into territorial reality with or without direct orders from economic and political authorities. The result – on the ground – is an extraordinary *spatial duality*. And the duality in space itself creates the strong impression that there exists a duality of political power: an equilibrium so threatened that an explosion is inevitable – and in short order. This impression is nonetheless mistaken – a measure, precisely, of the repressive and assimilative capacity of the dominant space. The duality will persist, certainly; and, failing any reversal of the situation, dominated space will simply be weakened. 'Duality' means contradiction and conflict; a conflict of this kind eventuates either in the emergence of unforeseen differences or in its own absorption, in which case only induced differences arise (i.e. differences internal to the dominant form of space). A conflictual duality, which is a transitional state between opposition (induced difference) and contradiction/ transcendence (produced difference), cannot last forever; it can sustain itself, however, around an 'equilibrium' deemed optimal by a particular ideology.
>
> *(Lefebvre 1991: 373–374)*

While relatively short in a book of over 400 pages, this text may be read as a prism which focuses Lefebvre's thought and in turn refracts. Whereas Pinder (2015: 42) warns that at times, Lefebvre's writing has been 'prone to being decontextualized and romanticized', Lefebvre's passage on the shantytowns of Latin America provides a bridge between his urban theory and the pervasive condition of informal settlements in the Global South. It legitimises the transfer of his urban theory and strategy to unevenly developing cities marked by informal settlements.The concepts he employs in this compact text connect to his wider work. Indeed, an attempt to grasp its meaning I draw in the next section on texts spanning from 1966 to 1976.

Lefebvre's understanding of informal settlements, and its political meaning

In the passage quoted above, Lefebvre employs attributes of the urban alongside induced and produced difference: social life, urban morphology, the effective ordering of space, appropriation and spontaneity. Lefebvre shows their tension with dominated space. But he also points to a tension internal to informal settlements between produced and induced space, and the ideology that keeps these two in balance, holding back the real possibility for a different future.

Lefebvre refers to the intensity of 'social life' in these settlements.While providing no straightforward definition for social life, the concept features throughout his urban texts, pointing to its significance in the passage about informal settlements. He associates 'social life' with 'relationships', distinct from 'individual and private life' (Lefebvre 1991: 154). Social life is where people are brought together (Lefebvre 2003: 21). This can be achieved in a controlled way, by monuments, which merely project 'a conception of the world' or 'a sense of being *elsewhere*', whereas cities project 'social life' (ibid.: 22). Lefebvre likens this with 'globality', meaning the entire or whole (ibid.). Participation in social life, similar to 'community', is closely linked to the concept of 'to inhabit' (Lefebvre 1968/1996: 76), that is 'the plasticity of space, its modelling and the appropriation by groups and individuals of the conditions of their existence' (ibid.: 79). Social life is situated in 'social space' and also has a 'clandestine or underground side' (Lefebvre 1991: 33, 35). This relates to informal settlements – the benign escape from an exclusionary regime, but also the non-benign exploitation of informality through criminal networks, drug trafficking in the case of contemporary Brazilian *favelas*.The *New Urban Agenda* document, perhaps excessively and in response to dominant urban paranoia in the Global North, goes as far as to link 'slums and informal settlements' to the potential harbouring of terrorism (Habitat III 2016: S103), by implication excluding any positive embrace of the clandestine.

Lefebvre (2003: 175) contrasts social life with 'material' life, noting that the disaggregation between social and material is produced by segregation in order to 'resolve conflicts'. Social life is poorly represented by urbanists (ibid.: 153, 188). Urbanist doctrine tends to assume that architects or urbanists create 'social life and social relations', but Lefebvre points out that this is not the case (ibid.: 156), 'although under certain conditions' architects and sociologists 'can help trends to be formulated (to take shape). Only social life (praxis) in its global capacity possesses such powers' (Lefebvre 1991: 151).

Lefebvre's passage on Latin American shantytowns refers to social life being 'transposed to the level of urban morphology' but without explanation. However, in *The Urban Revolution*, he associates 'morphology' with the 'urban practice of groups and classes – that is, their way of life' (Lefebvre 2003: 137). Here, as in his concept of social life, the collective comes to the fore. His emphasis on 'urban' morphology relates to urban life and, associated to this, inhabiting.

Linked to morphology is Lefebvre's observation about the effective ordering of space in informal settlements. This relates to his interest in self-management or *autogestion*, which has direct relevance for political opening, or 'opening towards the possible' (Lefebvre 2009e: 150; see also Elden 2004: 229). Lefebvre (2009e: 147) notes that '[o]nce aimed at ground level, in a fissure, this humble plant [namely *autogestion*] comes to threaten the huge state edifice'. Lefebvre (2003: 150) places hope in self-management of industry and of urban life, though not in isolation of one another. He takes particular interest in the self-organisation of space. In an interview with *Autogestion et Socialisme* two years after the publication of the text on informal settlements, Lefebvre alludes to 'some extraordinary examples' of the organisation of space (rather than enterprise), 'like in a Mexican shantytown, where two hundred thousand inhabitants are under complete *autogestion*' (Lefebvre 1976/2009f: 160). He uses this example to illustrate that self-management had possibly become more significant in the ordering of space than the ordering of enterprise (ibid.). This adds political significance to informal settlements, but also explains severe repression of self-management in informal settlements, for instance in contemporary South Africa (see Pithouse 2014).

While Lefebvre applauds effective ordering of space in informal settlements, he also highlights the spontaneity. He treats these two concepts as distinct. Lefebvre takes interest in the 'spontaneous city', and the decline thereof through urbanism (2003: 160). He acknowledges in particular that the theory and methodology of 'transduction', which involves an 'incessant feed back between the conceptual framework used and empirical observations', can achieve 'certain spontaneous mental operations of the planner, the architect' and others (1996: 151). But he argues that 'it is by no means a simple matter of return to spontaneism'; rather, self-management holds the key to taking 'over development, to orient growth (recognised and controlled as such) towards social needs' (1976: 40).

Lefebvre refers to a 'remarkably high order of appropriation' in informal settlements. Kofman and Lebas (1996: 20) explain that Lefebvre associates appropriation closely with use value, in contrast to property and exchange value. Lefebvre (1974/1991: 166) expands the meaning of appropriation in a further dimension that has relevance for understanding informal settlements: 'appropriation cannot be understood apart from the rhythms of time and of life'. Lefebvre (e.g. 1968/1996: 131) frequently uses the term 'space-time' or 'time-space' to emphasise the importance of past and future and of rhythms and moments, for a dynamic and spatialised understanding of the urban present, thus understanding space and time in relation to one another. 'Urban time-space' is distinct from the cyclical and locally specific agrarian time-space, and the homogenous, rational, planned and constrained industrial time-space (Lefebvre 2003: 37). Lefebvre calls for urban time-space not to be defined by 'industrial rationality' (ibid.), which would be linear and homogenised, but rather as '*differential*', with 'contrasts and oppositions' connecting to 'a whole', thus at the same time 'dualistic' and 'unitary' (Lefebvre 2003: 37). Lefebvre notes that during societal transformation, whether industrialisation or urbanisation, the 'true nature' of space is revealed as:

> (1) a political space, the site and object of various strategies, and (2) a projection of time, reacting against and enabling us to dominate time, and consequently to exploit it to death, as it does today – which presages [or forebodes] the liberation of time-space.
>
> *(Lefebvre 2003: 37)*

We can therefore read into Lefebvre's understanding of informal settlements a particular, liberated time-space, which has political significance which should inform strategy.

Appropriation is explained also by its opposite, dominated space, 'closed, sterilized, emptied out' (Lefebvre 1991: 165). But these two, though opposites, are 'inseparable'; with direct relevance for strategy for informal settlements, Lefebvre adds 'ideally…, they ought to be combined' (ibid.: 166). However, with the rise of military, state and political power, 'dominated space has subjugated appropriation' (ibid.). Can state intervention in informal settlements represent the ideal of a combination of dominant space and appropriation, or does it lead to what Lefebvre (1991: 319) terms 'negative appropriation', prohibitions ('the negative basis, so to speak, of the social order') (Lefebvre 1991: 319) inscribed in space? 'Prohibition is the reverse side and the carapace [or hard shell] of property, of the negative appropriation of space under the reign of private property' (ibid.). Prohibitions induce the transgressions that result in informal settlements in the first place. Transgressions 'reveal tendencies', and '*tensions* and the direction taken by these tensions' (Lefebvre 1976: 35). Lefebvre places transgressions in 'the realm of desire' – 'transgressions can point towards... a project that expressly proposes a radically different way of living... but they cannot realise it' (ibid.: 34). Yet transgressions 'disclose... the *possible* and the *impossible*... in order to extend the possible, it is necessary to proclaim and desire the impossible. Action and strategy consist in making possible tomorrow what is impossible today' (ibid.: 36). The use of the term 'informal' in dominant urban discourse implies a focus on transgression but ignores the political significance and possibility that Lefebvre associates with transgression.

Transgressions toward making the impossible possible occur in everyday practices. Of direct relevance to an understanding of informal settlements, Lefebvre emphasises the importance of everyday life as 'the inevitable starting point for the realization of the possible' (Lefebvre 1968 in Pinder 2015: 42). Lefebvre notes that 'above all it is urban life and the everyday where the project takes the form of practical elaborations and attempts at a radical change', giving 'priority to *social needs*, not individual needs' (Lefebvre 1976: 36). Social needs are 'above all urban needs... producing and managing a space that will correspond to the possibilities and technology and knowledge, and also to the demands made on social life by and for the "masses"' (ibid.: 37).

Conclusions

Informal settlements or shantytowns represent a varied condition. Lefebvre provides the conceptual vocabulary with which to recognise this as different intensities and forms of social life, appropriation, spontaneity, time-space, transgression, ordering of space and self-management. In their varied manifestations, these exist in tension with dominated space, resisting spatial domination. In that sense, they form contradictions in space. Far from recommending a reduction in this contradiction, Lefebvre places hope in the liberated time-space that such urban formations represent. Thus Lefebvre does not only offer his concepts for a nuanced understanding of informal settlements. By framing informal settlements or shantytowns as the possible-impossible and as potential for political opening, he directs engagement with strategy towards overcoming the constraints to such opening, particularly in professional doctrine, codes and legislation and in political ideology that shuns self-management and ignores the importance of the urban for social life. Any endeavour to address informal settlements through the right to the city must follow these leads.

References

Elden, S. (2004) *Understanding Henri Lefebvre: Theory and the Possible*, London: Continuum.

Fridman, F. (2015) Questions from South Africa about the Influence of Henri Lefebvre in Brazil in the 1970s, email to author (30 October).

Hess, R. (1988) *Henri Lefebvre et l'Aventure du Siècle*, Paris: AM Métailié.

Habitat III (2016) *New Urban Agenda*, Quito: United Nations.

Huchzermeyer, M. (2011) *Cities with 'Slums': from Informal Settlement Eradication to a Right to the City in Africa*, Cape Town: University of Cape Town Press.

Huchzermeyer, M. (2014) 'Troubling continuities: use and utilities of the term "slum"', in S. Parnell and S. Oldfield (eds) *The Routledge Handbook on Cities of the Global South*, London: Routledge.

Kipfer, S. (2008) 'How Lefebvre urbanized Gramsci: hegemony, everyday life, and difference', in K. Goonewardena, S. Kipfer, R. Milgrom and C. Schmid (eds) *Space Difference and Everyday Life: Reading Henri Lefebvre*, London: Taylor Francis.

Kofman, E. and Lebas, E. (1996) 'Lost in transposition – time, space and the city', in E. Kofman and E. Lebas (eds) *Writings on Cities*, Oxford: Blackwell Publishing.
Lefebvre, H. (1969/1968) *Explosion: Marxism and the French Upheaval*, New York: Monthly Review Press.
Lefebvre, H. (1976/1973) *The Survival of Capitalism*, New York: St. Martin's Press.
Lefebvre, H. (1991/1974) *The Production of Space*, Oxford: Blackwell Publishing.
Lefebvre, H. (1996/1968) 'Right to the city', in E. Kofman and E. Lebas (eds) (1996) *Writings on Cities*, Oxford: Blackwell Publishing.
Lefebvre, H. (2003/1970) *The Urban Revolution*, London: University of Minnesota Press.
Lefebvre, H. (2009a/1978) 'Space and the state', in N. Brenner and S. Elden (eds) *State, Space, World: Selected Essays/Henri Lefebvre*, London: University of Minnesota Press.
Lefebvre, H. (2009b/1940) 'Preface to the new edition', in H. Lefebvre (ed.) (2009) *Dialectical Materialism*, London: University of Minnesota Press.
Lefebvre, H. (2009c/1964) 'State and society', in N. Brenner and S. Elden (eds) *State, Space, World: Selected Essays/Henri Lefebvre*, London: University of Minnesota Press, pp. 51–68.
Lefebvre, H. (2009d/1986) 'Revolutions', in N. Brenner and S. Elden (eds) *State, Space, World: Selected Essays/Henri Lefebvre*, London: University of Minnesota Press, pp. 290–306.
Lefebvre, H. (2009e/1966) 'The theoretical problems of *autogestion*', in N. Brenner and S. Elden (eds) *State, Space, World: Selected Essays/Henri Lefebvre*, London: University of Minnesota Press.
Lefebvre, H. (2009f/1976) 'It is the world that has changed. Interview with *Autogestion et Socialisme*', in N. Brenner and S. Elden (eds) *State, Space, World: Selected Essays/Henri Lefebvre*, London: University of Minnesota Press.
Lemon, A. (ed.) (1991) *South Africa's Segregated Cities*, Bloomington: Indiana University Press.
Mangin, W. (1967) 'Latin American squatter settlements: a problem and a solution', *Latin American Research Review*, 2 65–98.
Pinder, D. (2015) 'Reconstituting the possible: Henri Lefebvre, utopia and the urban question', *International Journal of Urban and Regional Research*, 39:1 28–45.
Pithouse, R. (2014) 'Undoing the silencing of the present: the imperative to recognise the shack settlement as a site of politics', in C. Haferburg and M. Huchzermeyer (eds) *Urban Governance in Post-apartheid Cities: Modes of Engagement in South Africa's Metropoles*, Pietermaritzburg: University to KwaZulu-Natal Press.
Roy, A. (2005) 'Urban informality: towards an epistemology of planning', *Journal of the American Planning Association*, 71:2 147–58.
Roy, A. (2009) 'Why India cannot plan its cities: informality, insurgence and the idiom of urbanization', *Planning Theory*, 8:1 7–11.
Shields, R. (1999) *Lefebvre, Love and Struggle: Spatial Dialectics*, London: Routledge.
Turner, J.F.C. (1968) 'Uncontrolled urban settlements: problems and policies', *International Social Development Review*, 1 107–30.
UN (1997) *Glossary of Environmental Statistics, Studies in Methods*, Series F, No. 67, New York: United Nations.
UN-Habitat (2003) *The Challenge of Slums, Global Report on Human Settlements 2003, Nairobi: United Nations Human Settlements Programme*, London: Earthscan.

43

From Mourenx to spaces of difference

Michael E. Leary-Owhin

Suddenly I saw a town being built with amazing rapidity, the decision for it being taken at a high level, the bulldozers arrived, leaving the peasants traumatised. It was a period of high drama in the Mourenx area. It was then that I began to study the urban phenomenon.

(Lefebvre in Burgel et al. 1987: 32)

From a less pessimistic standpoint, it can be shown that abstract space harbours specific contradictions... Thus, despite – or rather because of – its negativity, abstract space carries within itself the seeds of a new kind of space. I shall call that new space 'differential space', because, inasmuch as abstract space tends towards homogeneity, towards the elimination of existing differences or peculiarities, a new space cannot be born (produced) unless it accentuates differences.

(Lefebvre 1991: 52)

Introduction

Two places stand out in the intellectual and research development of Henri Lefebvre: Navarrenx in the Pyrenees where he was raised and Mourenx where he witnessed the state imposition of a modernist new town in the late 1950s. I travelled to both towns in April 2017 to see and feel the places that so influenced Lefebvre and by proxy generations of scholars, researchers and activists thereafter. Driving from Toulouse and exploring Lefebvre's terrain was part spring road trip, part research reconnoitre. Time is the major difference between the two places. Navarrenx evolved slowly over two millennia, while Mourenx erupted across the rural landscape at bewildering speed. Lefebvre's intellectual and research trajectory is full of interesting shifts and patterns:

We can easily divide Lefebvre's work into periods and map out zones in his wide ranging intellectual geography. There is an abiding interest in emancipation and the condition of the human. If early on his interest is focused on the self-liberation of the individual, this later shifts to his commitment to Communism and socialist forms of autonomous management. He develops an interest first in his mother's birthplace in the Pyrenees. From this

> 'motherland' he moves to a focus on rural and peasant life, later to become best known for his work on urban space.
>
> *(Shields 1999: 2)*

The year 1968 is replete with contested revolutionary meaning in France and is often flagged as the temporal marker for the rural-urban swing in Lefebvre's work. Lefebvre confirms this shift was triggered by the Mourenx new town as the first epigraph at the head of this chapter indicates. It is evident that Lefebvre's Mourenx moment and his involvement in the May 1968 'revolution' were crucial points in the evolution of his ideas about cities, the urban, the production of space and everyday life. I argue here that we need to understand Lefebvre's utopianism in the context of his Marxist approach to the world and the development of key ideas principally: planetary urbanisation, urban society, right to the city and differential space.

Lefebvre: reader and researcher

Lefebvre is most associated with his prodigious written output, but he was also a prolific reader and researcher. While based in the Pyrenees during World War Two (WW2) Lefebvre used his copious spare time to carry out empirical rural sociological research. He engaged in ethnographic participant observation and scoured local archives for relevant data concerning the Occitanian peasant way of life, asking what persists and what is mutable and why? This body of research eventually became his doctoral thesis, *Docteur d'État*, awarded by the University of Paris in 1954. His doctorate documented the decline of the Pyrenean peasant way of life brought about by the increasing modernisation, accentuated by the industrialisation and urbanisation of the Béarn countryside. In 1965 he was appointed professor of sociology at the new modernist extension to the Sorbonne at Nanterre to the west of Paris and retained this position until 'retirement' in 1973. Lefebvre the worker continued to labour in various roles including doctoral supervision until his untimely death in 1991.

Lefebvre was an 'unstintingly industrious' writer, covering a 'belief-beggaring range of subjects' (Alvarez 2007: 51). He authored hundreds of publications including: about 70 books (depending on what counts as a book and a 'new' book), hundreds of academic and popular articles and research reports. Relatively few are translated into English and other languages. His whole canon is now dwarfed, in quantity rather than status, by the thousands of texts that constitute the secondary literature. He founded with Anatole Kopp in 1970 the prestigious academic journal *Espace et Societes*. It is still in print and remains true to its original objective, to synthesise diverse knowledge concerning the relationships of societies with their spaces. Much of his colossal output depended an equally massive amount of reading, and his grasp of ideas and detail across many disciplines is evident in his writing. His work is full of references to those who influenced and inspired him. And holding the strong views that he did, it is also replete with criticisms notably of Louis Althusser, Ferdinand de Saussure, Charles-Édouard Jeanneret (Le Corbusier) and Jean-Paul Sartre, but there is also mutual respect.

He published his most significant and influential works on cities and the urban after his 60th year. Notably, his prodigious written output tended to arrive in bursts. One such burst, critical for his city and urban literature, was the period 1968 to 1974: a literary flurry book-ended by *The Right to the City* (1968) and *The Production of Space* (TPOS) (1974). In between, he penned his account of the May 1968 French 'revolution' in the book *The Explosion. Marxism and the French Upheaval* (1969) and *The Urban Revolution* (1970). It is generally agreed that TPOS is his most comprehensive and influential intervention (translated into English) in the broad field of urban theory: the pinnacle of his theoretical and intellectual achievements regarding the city and the urban, and its importance is rightly recognised in this volume. In a less well-known book, *Introduction to Modernity* (1962), Lefebvre charts the analysis of a new town that would become a key moment in the life of his research and theorising.

The Mourenx moment

We cannot understand Lefebvre's reaction to Mourenx without thinking about its comparison with Navarrenx. Lefebvre lived his childhood and formative years in this historic medieval town. It evolved slowly, receiving renaissance, 18th- and 19th-century additions and adaptations. Lefebvre likens its unhurried development to that of a seashell. In subtle ways Lefebvre clearly understands Navarrenx as an open urban constellation of: houses, streets, courtyards, gardens and businesses that blur smoothly into the countryside in ways which developed organically over the centuries. This openness pertains despite the town's medieval ramparts. Ironically though, Lefebvre is not afraid to admit that Navarrenx actually became a fully planned new town, albeit a pre-modern and pre-capitalist one.

In the 20th century, only a few necessary holes were punched through Navarrenx's medieval ramparts to allow vehicular access, despite the disdain of post -WW2 modernist architecture and town planning for the historic environment. This example of historic conservation is due mainly to the work of the French planning system. In France the protection of the historic built environment and heritage, *patrimoine*, is taken seriously, with strict regimes operating in conservation areas overseen by the formidable *Architectes des Bâtiments de France* (ABF), specialist planning officials based in the *prefecture*. Since WW2 the town has benefitted from historic area protection that prevents unsympathetic demolition and redevelopment. It is now designated a conservation area, *Zone de Protection du Patrimoine Architectural, Urbain et Paysager* (ZPPAUP), including the whole of the ramparts. My research visit to Navarrenx was positive and enjoyable, people were helpful and friendly and the town is full of pleasant intimate public spaces, historic gems and surprises. In contrast, Merrifield (2006: xxvi) found little more than a swarm of hungry mosquitos and 'the odd melancholy café'. What is noticeable though and something ignored by Merrifield and Lefebvre, is that Navarrenx in the 1960s and still today has predominantly an ethnically homogenous, White French population and I comment on matters of ethnicity below.

A keen observer of the everyday, Lefebvre perceived the rhythms of small town and rural life as he grew into manhood. His thoughts and feelings about the rural and urban were to be impacted dramatically by events millions of years in the making through geological time and by the creation of a French state planning organisation. Geological accident led to the formation of petroleum deposits in southern France. A huge natural gas field was discovered in 1951 in the Lacq commune near the historic hamlet of Mourenx, a few miles east of Navarrenx, by the French petroleum company Total. A large-scale industrial complex was planned to include petro-chemicals and an alumina refinery for the production of aluminium powered by the natural gas. To accommodate the workforce of about 10,000, French government planners entered into partnership with Total to build a new town to be called Mourenx. Modernist architecture and planning principles underpinned the masterplan, by the architects and urban planners René-André Coulon, Philippe Douillet and Jean-Benjamin Maneval. A national public financial institution and provider of public housing, *Société civile immobilière de la caisse de dépôts et consignations* (SCIC), was the lead developer, and indeed most of the original housing was in today's terms affordable, secure, social rent tenure. Mourenx was constructed between the late 1950s and mid-60s.

For several years Lefebvre studied Mourenx in great detail, writing a series of accounts, most notably, *Seventh Prelude: Notes on the New Town* (Lefebvre 1995) – an essay mostly ignored in the literature (but see Stanek 2011; Wilson 2011). One of the hills overlooking Mourenx was adopted by Lefebvre as a favourite vantage point. He remarks that as an intellectual of the Left, he would sit on this hill to meditate on the destiny of Mourenx, knowing this may be regarded as 'ridiculous' by some people (Lefebvre 1995: 122). Sitting on that same hill looking down on Mourenx, I could appreciate the impact the urbanisation of Béarn must have had on Lefebvre (see Figure 43.1). Modernism produced a uniformity of form and function across Europe and North America. When I visited Mourenx in April 2017, it reminded me of many public housing estates in England such as the Aylesbury Estate in London and parts of new towns in England such as: Corby, Crawley and Hatfield. It reminded me also of modernist civic centres of Wythenshawe, Manchester and Gaborone, Botswana, designed in the 1950s by British architects and planners.

Mourenx's modernist planning was a major impetus for Lefebvre's refocus on the urban in the early 1960s. He critiqued the rationalist approach of the *Délégation à l'Aménagement du Territoire et à l'Action Régionale* (Delegation for Regional Planning and Regional Action, DATAR), the state agency that among other things designated the first modernist French new towns, an initiative Lefebvre:

> followed very closely and which was not solely a descriptive scientific project but one that involved accurate prediction. Something new was happening; an idea of spatial planning and practice was born... However, my initiation was neither from the point of view of philosophy, nor sociology, though these were present implicitly, nor was it historical or geographical. Rather it was the emergence of a new social and political practice. DATAR aimed to reorganise France from questionable, sometimes catastrophic, perspectives.
>
> *(Lefebvre in Burgel et al. 1987: 28)*

He compared Mourenx unfavourably with historic Navarrenx and was hypercritical, saying, 'Whenever I step foot in Mourenx I am filled with dread' (Lefebvre 1995: 118). The key trope

Figure 43.1 Foreground, Mourenx New Town; Background Lacq industrial complex; the view from 'Lefebvre's hill'. *Source:* © Michael Leary-Owhin (2017).

of abstract space is alluded to, as are: segregation of land uses, disregard of everyday life and the masking of the possible; themes to which he would return continuously in the following decades:

> Yet every time I see these 'machines for living in' I feel terrified... In Mourenx, modernity opens its pages to me. It is just like a 'novel of objects' (no, I must ask contemporary novelists to excuse me, that is not right – I mean just like a propaganda leaflet). Here I cannot read the centuries, not time, nor the past, nor what is possible. Instead I read the fears modernity can arouse: the abstraction which rides roughshod over everyday life – the debilitating analysis which divides, cuts up, separates – the illusory synthesis which has lost all ability to reconstruct anything active – the fossilized structures, powerless to produce or reproduce anything living, but still capable of suppressing it.
>
> *(Lefebvre 1995: 119–120)*

The 'machines for living' phrase is an oblique reference to Le Corbusier, of whom Lefebvre was critical, but he was more critical of the Bauhaus School of architecture and city planning (Maycroft 2002). It is also a reference to the emerging approach of the French state-led new town movement of the 1950s, influenced greatly by Le Corbusier's modernist ideas. Rather like in the UK and elsewhere, modernist planning ideals were applied in France to public housing. The French result was the building of collective housing, *grands ensembles* or *banlieues* at the peripheries of many cities.

In Mourenx the urban form reproduced the social hierarchy of industrial production: workers lived in large tower blocks (see Figure 43.2), their managers in houses (*pavillons*) This concretisation in urban form of socio-productive relations was a feature of previous new towns such as Saltaire in Yorkshire, England – a company new town built by the 'philanthropic' textile magnate, Titus Salt, in the 1850s. Lefebvre's assessment highlights the fact, perhaps following Engels, that 'he was there' and was able to draw on first-hand knowledge of his research subject.

Lefebvre's Marxism made him suspicious of state involvement, in direct provision of social housing, due to its complicity in the production of abstract space but also in the reproduction

Figure 43.2 Workers' residential tower blocks, Mourenx, French New Town. *Source:* © Michael Leary-Owhin (2017).

of labour. His direct experience of the speed of planning and construction of Mourenx and the devastating impact on rural communities and environment had a decisive impact on his thinking about the urban. Lefebvre's subsequent analysis of Mourenx was a 'pivotal moment' in the development of his urban and production of space theories (Moore in Lefebvre 1995: 391). It may well be the first time he engaged in an exposition of the production of urban space. Providing workers with decent housing, at a reasonable cost, close to their workplace and at a time when many citizens were living in terrible housing conditions, are things Lefebvre would not criticise; on the contrary he offers praise for the modernist planners:

> Yet the new town has a lot going for it. The overall plan (the master blueprint) has a certain attractiveness: the lines of the tower blocks alternate horizontals and verticals. The break between the landscape – wooded hills, moorland, vineyards – and the city may be rather abrupt, but it is bearable; it is relatively easy on the eye. The blocks of flats look well planned and properly built; we know that they are very inexpensive, and offer their residents bathrooms or showers, drying rooms, well-lit accommodation where they can sit with their radios and television sets and contemplate the world from the comfort of their own homes... Over here, state capitalism does things rather well. Our technicists and technocrats have their hearts in the right place, even if it is what they have in their minds which is given priority. It is difficult to see where or how state socialism could do any differently and any better.
>
> *(Lefebvre 1995: 118)*

At times there is a subtlety and balance in the analysis, and Lefebvre was prepared to acknowledge the potential for positive outcomes through the French new towns initiative. He does though weight the essay toward damning critique:

> [In Mourenx] Everything is trivial. Everything is closure and materialized system. The text of the town is totally legible, as impoverished as it is clear, despite the architects' efforts to vary the lines. Surprise? Possibilities? From this place, which should have been the home of all that is possible, they have vanished without trace.
>
> *(Lefebvre 1995: 119)*

Lefebvre's research approach saw him generating original data through interviews, which strangely he mentions only in passing, not letting us hear the voices of local residents. We know that some of them were unhappy and thought that Mourenx was less new town and more public housing estate or *grand ensemble* (quoted in Stanek 2011: 116).

Nowadays, faced with rampant neoliberal hollowing out of state social provision, it would be easy to lambast Lefebvre for this critique of high-quality social housing, but context is crucial and the context for Lefebvre was the transition from industrial society to urban society. The 1950s was a time in France, the UK and many parts of the Global North when many working-class people still lived in appalling housing conditions in (often high-rent) properties dating from the 19th century. State housing intervention did in the 1960s achieve improvements, but it brought problems in Mourenx to which Lefebvre draws attention:

> functionalist ensembles [public housing development] were expressing the paternalism of the state and they were anachronistic, since they did not account for the society moving beyond Fordism, in which the urban space was about to replace the factory as the place of socialization, exploitation, and struggle.
>
> *(Stanek 2014: xxvii–xxviii)*

Lefebvre goes further, after he 'poured unending scorn' on Mourenx (Wakeman 2016: 296) he muses that paradoxically, the town's boredom is pregnant with unrealised possibilities and freedoms and, 'a magnificent life is waiting', nearby yet 'far far away' (Lefebvre 1995: 124). Another example of Lefebvre the Utopian? Definitely. So did Lefebvre answer the housing question? No and yes. He did not offer any retro-design solutions, but he did focus attention on how to make the best of what Mourenx had to offer. The task (then and now) is 'to construct everyday life, to produce it, consciously to create it' (ibid.).

Lefebvre's theoretically grounded and empirically informed critique of Mourenx grew out of his everyday life experiences and heterodox Marxism. It is an example of how his research methodology, ethnographic observations and powerful intellect combined to produce new insights. He argues often that his work is his life. Of course his life was much more than his work, 'but in his case, perhaps more legitimately than with many other writers, the work and the life were closely interrelated' (Elden 2004: 2). He published other research about Mourenx documented in Stanek (2011). Most notable is his analysis of the rent strike of 1962, which saw local residents organise themselves to oppose various housing management requirements of the SCIC. They appropriated Mourenx public space for demonstrations and protests. They achieved a measure of victory. So the new town abstract space of state and big capital was not totally dominant.

Lefebvre was researching Mourenx at the start of the long French modernist new town programme. And while he railed against French town planning, ironically it was the historic conservation arm of the system, the ABF, using the ZPPAUP that protected the character of Navarrenx, the town he cherished. The issue of when to evaluate the outcomes of the production of space is always a moot point. Those outcomes are in ceaseless dialectical flux. At the time of his Mourenx research, construction work was still in progress. He accepts pragmatically that he cannot give a definitive verdict on the new town. He ponders whether we are on the threshold of socialism 'or supercapitalism?' Will it be the 'city of joy or the world of unredeemable boredom?' (Lefebvre 1995: 119). One wonders what Lefebvre would make of the more diverse Mourenx population today, now it is home to various minority ethnic and religious communities, especially those originating in North Africa? Putting that aside, it is important to consider his critique of Mourenx, which offered no cogent suggestions for improvement, with his observations of May 1968, which resulted in his explication of heterogeneity, concrete utopianism and differential space.

From Nanterre to urban society

Lefebvre apparently despised ivory tower philosophers who ruminate on social issues but do not act politically. His Marxism required him to try to transform, not just think about the world. For Lefebvre history is punctuated by key events that have the power to bring change, but they must be analysed and understood:

> Events are always original, but they become reabsorbed into the general situation; and their particularities in no way exclude analyses, references, repetitions, and fresh starts... Although they are belittled during stagnant periods for the benefit of those who preserve stagnation – those who show contempt for history and are preoccupied with stability – events reactivate the movement of both thought and practice. They pull thinkers out of their comfortable seats and plunge them headlong into a wave of contradictions. Those who are obsessed with stability lose their smiling confidence and good humour. Good and bad conscience, ideological labels, and scraps of obsolete practices are swept

> up like refuse. Under the impact of events, people and ideas are revealed for what they are.
>
> *(Lefebvre 1969: 7–8)*

The event he has in mind is the 'urban revolution', signified by Paris, May 1968. This event provoked in Lefebvre the realisation that a different kind of society was possible, an urban society – different from that produced under conditions of state-regulated and supported neo-capitalism, 'state capitalism' – different from Mourenx.

Lefebvre witnessed first-hand the events of May 1968 which started at Nanterre University. Rather incongruously, the campus was parachuted into a peripheral area of industry, working-class housing and (mainly North African immigrant) shanty towns. It is hard to appreciate now that such areas of extreme poverty existed in Paris in the 1960s. Lefebvre witnessed the brutal removal of these housing areas to make way for the new university campus: another example of modernist state planning. According to Lefebvre, French state modernist planning produced an abstract regimented space of social and functional segregation. Tucked away from the bustling street life and excitement of central Paris, the mainly middle-class students suffered intense boredom. Their campus was an implanted student 'ghetto', replete with alienation and hopelessness that sowed the seeds of urban rebellion.

It was on this problematic campus enclave in late March that the civil disturbances now labelled May 1968 began, prompted partly by Professor Lefebvre's provocative Marxist sociology lectures. Lefebvre documents his interpretation of these momentous events in the book *The Explosion. Marxism and the French Upheaval* (1969). Nanterre university campus occupied an area that 'contains misery, shantytowns and excavations for an express subway line', cheek by jowl with 'low-income housing projects for workers, and industrial enterprises: it is a 'desolate and strange landscape' (Lefebvre 1969: 104). With the modernist campus came multiple segregation, 'functional and social, and industrial and urban' (Lefebvre 1969: 105). Lefebvre sees ghettos. Ghettos of abandoned working-class and marginalised immigrant populations. In the midst of these are ghettos of students and teachers. An absurd modernist planning ideology created the sterile campus, 'utterly devoid of character' (ibid.). Students reject the slightest university prohibition or regulation, which become intolerable: they are rejected because they *symbolise* repression. Malaise, boredom and hostility to authority coalesce into political rebellion with students in the vanguard (Lefebvre 1969: 106).

In late March student unrest boiled over and, joined by a hotchpotch of political activists, they occupied a Nanterre University administrative building. Their grievances targeted: bourgeois university management, capitalism, class discrimination, rampant consumerism, urban poverty and American imperialism – all the subject of Lefebvre's lectures. Anger was expressed in emotive anti-authority slogans and graffiti. Following increasing confrontation, Nanterre University was shut down by the authorities. Protests then spread to the Sorbonne in central Paris; many workers left their factories and joined students in dissent on the streets. Dialectical struggle saw the periphery challenge the centre. A large demonstration on 6 May in the university Latin Quarter of Paris was broken up viciously by the notorious riot police, the CRS (*Compagnies Républicaines de Sécurité*). Further mass demonstrations followed in Paris and other cities. Later the protestors were joined spontaneously by French workers, eventually 10 million, from many sectors, and by French North Africans. Lefebvre wonders if there was 'Mixing of classes? Fusion?' but decides that 'Interaction would be a better word' (Lefebvre 1969: 106). There was some harmony between diverse groups, but relationships were complex and at times strained (Bracke 2009). On 13 May there was an unofficial general strike, nationwide occupations, widespread civil disorder and further strikes that lasted for several weeks.

A diverse political coalition of protest emerged. Lefebvre calls it 'the movement', it was 'profoundly political from the outset' (Lefebvre 1969: 112). It sought to challenge the state and all manifestations of officialdom. Workers occupied factories and declared industrial self-management (*autogestion*) would replace capitalist production. Streets were blockaded; cars and the Stock Exchange, a controversial symbol of capitalism, were torched. Students and workers occupied Les Halles, the site of Paris' fresh produce market, and Lefebvre saw in this appropriation the production of differential space (discussed below). But with conflagration and conflict came parties, festivals and celebrations. Workers and students revelled in the freedom from routine work. Paris exploded with violence but also with short-lived laughter, unfettered speech, humour, art and song. Previously hidden contradictions were revealed. Organising committees, or communes, were established by protestors to negotiate with the authorities. Economic and educational demands were voiced. Concessions were made by employers, universities and the government regarding wages, working conditions and educational management. Was a Lefebvrian urban society glimpsed? Momentarily perhaps, but it brought its own problems and requirements. By mid-June, however, support for the strikes and protests had peaked. The trade unions and French Communist Party helped the police and government restore order. Striking workers returned to work in July.

We can divine elements of the regressive-progressive method (see Leary-Owhin in Part 1) in Lefebvre's ensuing analysis of May '68. After description, he draws historical comparisons with the 'French Revolution' of March 1871. He then provides evaluation and explanation of current events. In Paris (1968) and other cities, art met politics through the creation of street murals. Formerly, dead streets became alive with effervescences, sparking animation and playfulness but always for a political purpose (Lefebvre 1995: 116). Lefebvre is careful though to set his ethnographic and observational details of May '68 and the modernist planning that preceded it in a broad Marxist political, geographical, colonial and historical context that infers explanation:

> A country which oppresses other countries cannot be free. The ignorant or passive accomplices of oppression are themselves bound by chains with which they bind the oppressed... The population in the metropolis is regrouped into ghettos (suburbs, foreigners, factories, students), and the new cities are to some extent reminiscent of colonial cities.
>
> *(Lefebvre 1969: 92–93)*

For a flickering hopeful moment, cracks in homogenised abstract space are exploited to reveal unexpected possibilities. Out of dialectical interaction between the marginality and heterotopia of the Nanterre campus and centrality of Paris, a temporary concrete utopia emerged (Lefebvre 1969: 118). It was a heterotopia of non-work and cultural diversity, a coming together of social difference for political dissent. Heterotopia is mentioned briefly in the analysis of May 1968. For Smith (in Lefebvre 2003: xii), heterotopia is rooted 'in a sense of political and historical deviance from social norms'. But Lefebvre's interest in heterotopia seems to fade, and ideas of utopia and differential space take precedence and continue to be developed in *The Production of Space* (Smith in Lefebvre 2003: xiii). When it emerges, temporary concrete utopia is precious and needs to be celebrated, protected and documented. Lefebvre's critiques related to ghettos and alienation appear today more prescient, given the continuing social crisis of the *banlieues* (Smith 2005; Chrisafis 2015). On the 50th anniversary, French historian Éric Alary agreed with Lefebvre that possibilities materialised, 'May 68 is seen as a period when audacious moves seemed possible' and 'society profoundly changed' (in Smith 2018). Possible futures occupied much of Lefebvre's thinking after May 1968. Differential space is rather neglected in Lefebvrian literature but is nevertheless of crucial importance. Lefebvre associates it with the production of space, and the concept is developed through the related ideas of heterotopia, utopia and urban society.

Differential space or spaces of difference?

One of the reasons for the confusion regarding Lefebvre's ideas about space is that he is ambivalent about the outcome(s) of the production of space. Kipfer et al. (2008: 9) criticise Soja and Harvey's interpretation of the spatial triad, asking, 'How did Soja get so lost looking for Lefebvre in the prison-house of spatial ontology?' Their answer is that Soja tried too hard to see Lefebvre as postmodern. I argue that none of the three spatial moments of the triad can be *the* outcome of the production of space because *together* in dialectical tension they produce space. So what is the outcome? Lefebvre is insistent over many decades that social struggle against unpleasant aspects of neo-capitalism can produce a fairer more inclusive space: an urban society brought about through the production of differential space. Lefebvre imbues differential space with at least two temporal meanings. Firstly, it is a distant utopian alternative to oppressive state-supported neo-capitalism:

> On the horizon, then, at the furthest edge of the possible... a planet-wide space as the social foundation of a transformed everyday life open to myriad possibilities – such is the dawn now beginning to break on the far horizon.
>
> *(Lefebvre 1991: 422)*

Secondly, it is also the 'here-and-now' utopian space of the favela or Paris '68. Lefebvre regards both these utopias as types of differential space. Like urban society, differential space results from contradiction, it is transition and is both process and outcome. Lefebvre provides sufficient insights as to what it is, might be and how it may be produced, but he was not precise or dogmatic about it. This has led to a variety of legitimate interpretations of differential space.

Shields perceives duality: post-capitalist society and transformed everyday space (1999: 183). Smith posits that differential space was Lefebvre's code for socialism (in Lefebvre 2003: xiv). Kofman and Lebas claim 'the production of differential space and plural times, have direct resonances in Nietzschean thought' (1996: 5). Only half-jokingly, Merrifield (2006: 120) postulates that the project of differential space can 'begin this afternoon' through academics 'reclaiming our own workspace', by giving a nod to disruption rather than co-optation, a nod 'to real difference rather than cowering conformity'. It is the possibility of unorthodox, experimental urban politics, despite the apparent failures of multiculturalism (Goonewardena and Kipfer 2005). Lefebvre's differential space is a fundamental concept: it is the hard-won space of social use value, it contradicts neo-capitalist exchange value and homogeneity:

> Our present approach is also based on an analysis of the overall process and its negative aspects, on an analysis that is tied to practice. The transition here considered is characterized first of all by its contradictions: contradictions between (economic) growth and (social) development, between the social and the political, between power and knowledge (*connaissance*), and between abstract and differential space.
>
> *(Lefebvre 1991: 408)*

Differential space is therefore one of the desired outcomes of the production of space for its immediate societal benefits and its propulsive role in the transition to an urban society. For Lefebvre, leisure and non-work time were crucial for combatting the alienation of capitalist production, allowing for the emergence of differential space and a concrete utopia.

One of the most significant challenges Lefebvre posed in his urban research and writing was to identify, analyse and produce this alternative possibility. He called this alternative 'urban society' and associated it with differential space. Throughout his metaphorical life journey from the rural to the urban, he never actually left the rural behind. This is most noticeable in his ideas about the production of space, urbanisation and urban society where the rural constitutes an integral part of the story and the 'city is simultaneously ruralized' (Lefebvre 2016: 149). Lefebvre is careful not to posit a smooth linear historicist transition from one era to the next, especially capitalism to urban society. He is careful also to stress the need for struggle and the uncertainty of immediate outcome, emphasising rhythms and cycles. However, the teleological aspects of Lefebvre's historical dialectic in which an inevitable transition unfolds, from the absolute space of nature to capitalist abstract space, finally reaching utopian differential space and urban society, is imputed (e.g. Keith and Pile 1993: 24–25). I agree with Elden (2004: 7), who detects no such teleological naivety in Lefebvre.

Lefebvre uses the term 'differential' in a variety of contexts including architecture and semiotics, but I contend it is a surprising way for a philosopher critical of Cartesian mathematics to describe space. Lefebvre is less than helpful; since in the TPOS chapter entitled 'From the contradictions of space to differential space', the latter term does not appear. In English the term differential space (or *espace différentiel*) has strong mathematical overtones. I contend that the term differential space has something of the weakness of the term representational space – its meaning is rather opaque. Some translation background is useful here. Remarkably perhaps, although the term 'spaces of representation' is used widely, it does not appear in TPOS. Nicholson-Smith translated '*les spaces de représentation*' as 'representational spaces'. The term spaces of representation is used instead and first appeared in English in Frank Bryant's translation of Lefebvre's *The Survival of Capitalism* (Lefebvre 1976: 26). The term spaces of representation is preferable because the Nicholson-Smith translation makes the triad 'more difficult to comprehend' (Shields 1999: 161).

Lefebvre is not definitive regarding how differential space is produced but leaves a variety of clues. Principally, it can be brought about by spatial coalitions or social movements through counter-projects. So an appreciation of social and spatial diversity is critical for the production of space and urban society. A defining feature of Lefebvre's urban theories is the importance of power relationships and the dialectical tensions between civil society, the private sector and the state. And over recent decades the importance of civil society has become increasingly apparent. For Lefebvre the production of a new space, a counter-space, can never be brought about by any one particular social group and must necessarily result from relationships between diverse groups which may include: 'reactionaries', 'liberals', 'democrats' and 'radicals' so:

> There should therefore be no cause for surprise when a space-related issue spurs collaboration... between very different kinds of people... Such coalitions around some particular counter-project or counter-plan, promoting a counter-space in opposition to the one embodied in the strategies of power, occur all over the world.
>
> *(Lefebvre 1991: 380–381)*

Here Lefebvre seems to be drawing on his Nanterre analysis. It is strange that his thoughts on the importance of coalitions in the production of space are largely overlooked in the literature. Despite this, spatial coalitions are not unlike the well-known idea of counter-publics and the impact of coalitions' pursuit of counter-projects is documented in Leary-Owhin (2016) and McFarlane (2018).

Spaces of difference of the utopian kind herald the urban society, the possible-impossible on the 'far horizon' (Lefebvre 1991: 422). Lefebvre sees this potential as contested everyday present reality 'at the margins of the homogenized realm' (Lefebvre 1991: 373) in the self-managed

Brazilian favelas (ibid., and see Huchzermeyer in this part of the book). Albeit that favela self-management today includes organised crime drug cartels and the direction is towards the production of abstract space (Lacerda 2018). Lefebvre sees this potential in Les Halles in May 1968, which became for a while an alternative space: a space of use value, 'a scene of permanent festival' (Lefebvre 1991: 167). Harvey helps bring Lefebvre up to date but prefers the related term heterotopia and relates it to the 'transgressive social practices' and 'oppositional movements' (2012: 110). He announces heterotopia to be simultaneously a utopian post-revolutionary space and a prosaic space of everyday life. It can be produced through what people feel, sense and do in the lived space of everyday life. Heterotopic spaces can and do pop up across various realms of urban space (Harvey 2012: xvii).

Lefebvre sees the potential to contradict abstract space when land and property is abandoned by capital interests *or* the state. This withdrawal from space occurs continually in urban areas even in city centres. From his Marxist perspective, Lefebvre highlights the potential for ordinary inhabitants to seize new rights and produce differential space from abandoned abstract space:

> An existing space may outlive its original purpose and the *raison d'être* which determines its forms, functions, and structures; it may thus in a sense become vacant, and susceptible of being diverted, reappropriated and put to a purpose quite different from its initial use.
>
> *(Lefebvre 1991: 167)*

Diverted purposes may well, I argue, produce differential space. In spite of the definitional issues, differential space has proved useful conceptually and as a springboard for research in differing contexts (e.g. Groth and Corijn 2005; Andres 2013; McFarlane 2018). Even though I have deployed the term in recent research (Leary-Owhin 2016), I suggest moving beyond Lefebvre – the term spaces of difference seems more compatible with his explanation of differing types of differential space. I show in Figure 43.3 how we might venture beyond Lefebvre and suggest

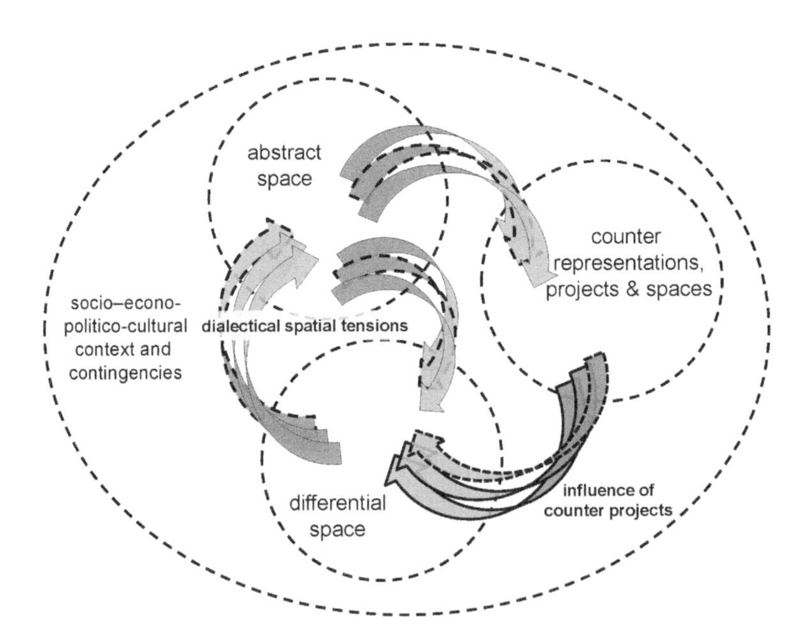

Figure 43.3 Suggestions for a 'spaces of difference triad'. *Source:* © Michael Leary-Owhin.

how spaces of difference could be produced through the dialectical production of space (based on Leary-Owhin 2016: 306). Figure 43.3 tries to convey the production of spaces of difference as being: non-hierarchical, non-linear, non-chaotic, uncertain, unpredictable with the 'traditional' spatial triad symbolised by the arrow triplets.

The right to difference

Lefebvre encourages us to see differential space or spaces of difference and associated social relations as critiques *and* examples of everyday life. Differential space is related intimately to his concept of the right to the city. He identifies a range of rights which are not necessarily legal rights (Lefebvre 1996). There is a large amount of literature affirming and critiquing the right to the city (see Aalbers and Gibb 2014; Purcell 2014). Merrifield wonders if its co-optation by elites and conservative interests means it should be abandoned as a radical idea (2011). I focus here on its relationship with differential space. I argue the three fundamental aspects of the right to the city are: 1) the right to an urban life (presumably within the emerging urban society), 2) the right to geographical and other forms of centrality and 3) the right to difference. Lefebvre distinguishes this last right to the city from the capitalist right to property (presumably real estate):

> The 'right to difference' is a formal designation for something that may be achieved through practical action, through effective struggle – namely, concrete differences. The right to difference implies no entitlements that do not have to be bitterly fought for. This is a 'right' whose only justification lies in its content; it is thus diametrically opposed to the right of property.
>
> *(Lefebvre 1991: 396)*

Patently, difference and the claiming of rights to the city can be a source of solidarity and/or conflict in any society. Lefebvre stressed that protest based on *particularities* should not be confused with opposition to repression or exploitation or with an *awareness* and acknowledgement of difference. Urban difference can include ethnic, linguistic, local and regional particularities. Despite the potential for conflict, the urban and urban society 'can be defined as a place where differences know one another and, through their mutual recognition, test one another, and in this way are strengthened or weakened' (Lefebvre 2003: 96).

Right to difference should be understood in the context of strong French universalism since the First Republic and its testing relation to the particular, which resonates today (Samuels 2016). This collection of Lefebvrian rights inherent in the right to difference needs to be grabbed through the appropriation of urban space. We see it in the appropriation by spatial coalitions of abandoned and underused space for its use value. We see it in the appropriation and transformation of space for the purposes of overt political expression from the small-scale spontaneous to the large-scale organised demonstration (Leary-Owhin 2016). We see it in the concept of *autogestion*. Literally this translates as self-management and was used to describe workers taking control of factory production. Lefebvre extended this concept and linked it theoretically and politically to Marx and Lenin's idea of the withering away of the state. Workers and communities take control of their everyday lives and decide the organisation of work and urban space. (Lefebvre 2009; Purcell 2014: 147–148). It may at some time in the future provoke a seismic shift in neo-capitalism, the possible-impossible utopian horizon may be reached. For now though, total transformation is not necessary for genuine everyday spaces of difference to be produced; they are here and now.

Conclusions

This chapter has highlighted the importance of: Lefebvre's life experiences, his research during WW2 and his analysis of Mourenx and May '68 for the development of key theoretical ideas. His assessment of Mourenx is mostly critical, but he also praised the French modernist new town planners. Overall there is ambivalence. He was not so dogmatic as to reject French modernist new town planning entirely. Lefebvre is shown in this chapter to have used his analysis of May 1968 to develop ideas about post-capitalist urban society. Spaces of representation and the right to the city dominate Lefebvrian research attention, but everyday political action can produce, and research can reveal, the production of differential space. Once created, differential space is simultaneously robust and fragile. It shares those characteristics with abstract space. A host of processes are involved in the production of what I call spaces of difference: centrality, marginality, celebration, urban guerrilla warfare, everyday existence and the counter-projects of coalitions. Concrete utopia, or urban society, as a core transition within the abstract space of neo-capitalism, is revealed in this chapter as a major dialectical contradiction in Lefebvre's urban theorising stimulated by May 1968. But post-capitalist urban society will emerge through struggle rather than through a teleological process. I conclude that several kinds of differential space or spaces of difference seem possible with different temporal characteristics, from the enduring to the transient. There are always opportunities for the production of spaces of difference and urban society. They inhere, especially in the fraught functioning of neoliberal state neo-capitalism.

Acknowledgements

I am most grateful to Catherine and Léa Régulier-Lefebvre for their generous hospitality and for their assistance with this research, when I visited Navarrenx during Easter 2017. Thanks also to: Vincent Berdoulay, Stuart Elden, Andy Merrifield and Rob Shields for their advice regarding this trip.

References

Aalbers, M.B. and Gibb, K. (2014) 'Housing and the right to the city: introduction to the special issue', *International Journal of Housing Policy*, 14:3 207–13.

Alvarez, D. (2007) 'The return of Henri Lefebvre', *Grand Valley Review*, 31:1 51–74.

Andres, L. (2013) 'Differential spaces, power hierarchy and collaborative planning: a critique of the role of temporary uses in shaping and making places', *Urban Studies*, 50:4 759–75.

Bracke, M.A. (2009) 'May 1968 and Algerian immigrants in France: trajectories of mobilization and encounter', in G.K. Bhambra and I. Demir (eds) *1968 In Retrospect: History, Theory, Alterity*, London: Palgrave Macmillan.

Burgel, G., Burgel, G. and Dezes, M.G. (1987) 'An interview with Henri Lefebvre', *Environment and Planning D: Society and Space*, 5:1 27–38.

Chrisafis, A. (2015) 'Nothing's changed': 10 years after French riots, banlieues remain in crisis, *The Guardian* (22 October).

Elden, S. (ed.) (2004) *Understanding Henri Lefebvre: theory and the possible*, London: Continuum.

Goonewardena, K. and Kipfer, S. (2005) 'Spaces of difference: reflections from Toronto on multiculturalism, bourgeois urbanism and the possibility of radical urban politics', *International Journal of Urban and Regional Research*, 29:3 670–78.

Groth, J. and Corijn, E. (2005) 'Reclaiming urbanity: indeterminate spaces, informal actors and urban agenda setting', *Urban Studies*, 42:3 503–26.

Harvey, D. (2012) *Rebel Cities: From the Right to the City to the Urban Revolution*, London: Verso.

Keith, M. and Pile, S. (eds) (1993) *Place and the Politics of Identity*, London: Routledge.

Kipfer, S., Goonewardena, K., Schmid, C. and Milgrom, R. (2008) 'On the production of Henri Lefebvre', in K. Goonewardena, S. Kipfer, R. Milgrom and C. Schmid (eds) *Space Difference and Everyday Life: Reading Henri Lefebvre*, London: Taylor Francis.

Lacerda, D.S. (2018) 'The "visible hand" of the state: urbanization of favelas as a violent abstraction of space', in K. Dale, S.F. Kingma and V. Wasserman (eds) *Organisational Space and Beyond: The Significance of Henri Lefebvre for Organisation Studies*, Abingdon: Routledge.

Leary-Owhin, M.E. (2016) *Exploring the Production of Urban Space: Differential Space in Three Post-Industrial Cities*, Bristol: Policy Press.

Lefebvre, H. (1969/1968) *The Explosion. Marxism and the French Upheaval*, New York: Monthly Review Press.

Lefebvre, H. (1976/1973) *The Survival of Capitalism: Reproduction of the Relations of Production*, London: Allison & Busby.

Lefebvre, H. (1991/1974) *The Production of Space*, Oxford: Blackwell.

Lefebvre, H. (1995/1960) *Introduction to Modernity: Twelve Preludes*, New York: Verso.

Lefebvre, H. (1996/1968) 'Right to the city', in E. Kofman and E. Lebas (eds) *Writings on Cities*, Cambridge, MA: Blackwell.

Lefebvre, H. (2003) *The Urban Revolution*, Minneapolis MN: University of Minnesota Press.

Lefebvre, H. (2009/1966) 'Theoretical problems of autogestion', in N. Brenner and S. Elden (eds) *State, Space, World*, Minneapolis MN: University of Minnesota Press.

Lefebvre, H. (2014/1973) *Toward an Architecture of Enjoyment*, Minneapolis, MN: University of Minnesota Press.

Lefebvre, H. (2016/1972) *Marxist Thought and the City*, Minneapolis: University of Minnesota Press.

Maycroft, N. (2002) 'Repetition and difference: Lefebvre, Le Corbusier and modernity's (im)moral landscape: a commentary', *Ethics, Place and Environment*, 5:2 135–44.

McFarlane, C. (2018) 'Fragment urbanism: politics at the margins of the city', *Environment and Planning D: Society and Space*, 36:6 1007–25.

Merrifield, A. (2006) *Henri Lefebvre: A Critical Introduction*, London: Taylor & Francis Ltd.

Merrifield, A. (2009) 'Review essay. The whole and the rest: Remi Hess and les Lefebvriens Francais', *Environment and Planning D: Society and Space*, 27:5 936–49.

Merrifield, A. (2011) 'The right to the city and beyond: notes on a Lefebvrian re-conceptualization', *City*, 15:3–4 473–81.

Moore, J. (1995/1962) 'Footnote', in H. Lefebvre (ed.) (1995) *Introduction to Modernity: Twelve Preludes*, London: Verso.

Purcell, M. (2014) 'Possible worlds: Henri Lefebvre and the right to the city', *Journal of Urban Affairs*, 31:1 141–54.

Samuels, M. (2016) *The Right to Difference: French Universalism and the Jews*, Chicago: University of Chicago Press.

Shields, R. (1999) *Lefebvre, Love, and Struggle: Spatial Dialectics*, London: Taylor Francis.

Smith, A.W. (2018) 'The Paris riots of May 1968: how the frustrations of youth brought France to the brink of revolution', *The Independent* (6 May).

Smith, C. (2005) 'Inside French housing project, feelings of being the outsiders', *New York Times* (9 November).

Soja, E. (1996) *Thirdspace: Journeys to Los Angeles and Other Real-and-Imagined Places*, Oxford: Blackwell.

Stanek, L. (2011) *Henri Lefebvre on Space: Architecture, Urban Research, and the Production of Theory*, Minneapolis: University of Minnesota Press.

Stanek, L. (2014) 'Introduction', in H. Lefebvre (ed.) (2014) *Toward an Architecture of Enjoyment*, Minneapolis, MN: University of Minnesota Press.

Stanek, L., Schmid, C. and Moravánszky, A. (eds) (2014) *Urban Revolution Now: Henri Lefebvre in Social Research and Architecture*, London: Routledge.

Wakeman, R. (2016) *Practicing Utopia: An Intellectual History of the New Town Movement*, Chicago: University of Chicago Press.

Wilson, J. (2011) 'Notes on the rural city: Henri Lefebvre and the transformation of everyday life in Chiapas, Mexico', *Environment and Planning D: Society and Space*, 29:6 993–1009.

44

Right to the city and urban resistance in Turkey

A comparative perspective

Gülçin Erdi

Introduction

In the book *The Production of Space*, Henri Lefebvre contended that:

> the space serves as an instrument of thought and action; it is at the same time a means of production, a means of control, hence of domination and power, but it partially escapes, as such, from those whom it serves.
>
> *(1974: 35)*

This statement shows how it is complicated for a specific actor to control entirely the production of space. Even if public and private actors want to design the space according to their own objectives – whether political or for profit – there will always be other kinds of power relations and attempts at resistance from inhabitants and/or civil society in order to participate in their design. The interaction of all these actors often transforms the urban space, and this transformed space also constrains these actors so they constantly adapt their behaviour, tactics and strategies in order to maintain their place and influence.

This chapter aims to show how this interaction between different actors for the appropriation of space takes place in Turkey, especially in the context of neoliberal urban planning under the rule of the Party of Justice and Development (AKP) since 2003. I propose to analyse this interaction from the perspective of the concept of right to the city and spatial justice. Drawing on Lefebvre's theory, I think that both concepts are closely related and complementary. Edward Soja, who largely reinstated Lefebvre's thought in urban studies by his theory of the spatial turn, emphasises this relationship:

> Paris in the 1960s and especially the still understudied co-presence of Henri Lefebvre and Michel Foucault, becomes the most generative site for the creation of a radically new conceptualization of space and spatiality, and for a specifically urban and spatial concept of justice, encapsulated most insightfully in Lefebvre's call for taking back control over the right to the city and the right to difference.
>
> *(Soja 2009: no page)*

The purpose of this chapter is to study different urban resistances in Turkey in the light of the right to the city and spatial justice and to analyse how these concepts are perceived and mobilised by inhabitants. My field study is based on two neighbourhoods, Dikmen Valley in Ankara and May Day neighbourhood in Istanbul, both composed mostly of informal settlements. Their creation and development were similar. Both neighbourhoods emerged as a result of the housing need of rural migrants coming from different parts of Anatolia in the 1970s. In the beginning the first settlements were founded by the initiative of radical left revolutionary groups that were relatively powerful in metropolitan cities during this period. Both neighbourhoods therefore have an important heritage of political and collective action. Another common point is that they now have a high estate value because of their central location and are threatened by large-scale urban transformation projects. Unsurprisingly, some inhabitants in these neighbourhoods have for several years been organising resistance and mobilisation in order to protect their everyday life spaces and houses and to participate in the decision process of these projects.

First I will present in detail these neighbourhoods by locating them in the context of neoliberal urban planning in Turkey. Then, I will focus on the emergence of different forms of resistance despite their similarities. The right to the city and the attempt to appropriate space will be central to my purpose.

Informal neighbourhoods in the newly restructured urban system

Urbanisation in Turkey can be explained through three historical phases: the period before 1980, the period between 1980 and 2001 and the period after 2001. The first phase corresponds to the beginning of urbanisation in the 1950s and is clearly related to the high rate of rural migration (Öncü 1988). This period is marked by the absence of public policy on housing. The housing question was never entirely addressed in Turkey as a policy in the political agenda of changing governments, and housing needs were long managed by individual initiatives by constructing informal settlements generally called *gecekondus* (Türkün 2011). After the 1980s, these informal settlements have been progressively legalised by different governments who saw these places as a potential source of political support for the elections.

During the phase after 2001, metropolitan cities in Turkey have become central to the reproduction and continual development of neoliberalism itself, constituting increasingly important geographical targets and laboratories for a variety of neoliberal policy experiments, all aimed at increasing the value of the land. This process, observed in many cities in the world, was particularly accelerated in Turkey during the AKP rule and became the main tool for economic growth. The AKP decided to restructure the governance of Turkish real-estate markets and urban planning through the implementation of a number of legal and institutional reforms, with significant consequences for the socio-economic geography of cities and the rural environment. Social policies in the city and the search for spatial justice fostering social diversity and support for disadvantaged populations have been gradually downgraded. It also combined neoliberal strategies with the conservative-Islamist restructuring of urban space.

Neoliberal urban regeneration policies have three major characteristics in order to legitimise this process and to reduce potential channels of resistance. First, they are supported by a wide-ranging legal framework as indicated above, that government adapts according to needs and conditions. Second, urban security discourses are used for these policies in the public arena in order to legitimise human consequences such as forced displacements and the destruction of homes. In this perspective, many informal and degraded neighbourhoods have been presented as areas of crime, terrorism and social alienation. Third, this neoliberal restructuring of cities in Turkey has an authoritarian character since it ignores the demands and desires of the majority of residents, namely lower-middle and poor classes, and privileges the market priorities in order to integrate cities like Istanbul and Ankara in global economic, financial and cultural flows.

To ensure sustainability for the construction sector considered as the catalyst of economic growth, all *gecekondu* areas and old 'unhealthy' neighbourhoods have been opened up to regeneration and gentrification (Türkün 2011). The neighbourhoods inhabited by the city's poorest, which at the same time carry the highest potential in terms of the rising value of urban land, are refashioned by local municipalities and the Mass Housing Administration of Turkey (TOKI) in order to launch urban transformation projects. In this sense, the neoliberal restructuring of the urban space raises some questions about the meaning, the production and the appropriation of the space by those who structure it and those who live in it. Lefebvre argues that the production of space not only manifests various forms of injustice but also produces and reproduces them, thereby maintaining established relations of domination and oppression (Lefebvre 1974: 41).

In this perspective, social groups with low incomes occupying old and unhealthy neighbourhoods and the inner-city, *gecekondus* are now considered undesirable. Current urban renewal projects displace these communities in order to confine them to new resettlement areas far away from their previous neighbourhoods. This is observable in many urban projects in Istanbul where the inhabitants are the last ones to know about the details of such projects. This process reflects what Lefebvre calls difference of perception between conceived and perceived space, which is often adversarial in capitalist urbanization (Lefebvre 1974).

In an urban space conceived under a neoliberal logic based on market value of place and without a participative process taking into account the needs and desires of inhabitants, neighbourhood becomes the place where many social groups (including minorities, and political and/or religious groups) create enclaves wherein their identity is recognised without repression, and these environments enhance the development of a relatively shared identity, connected to the neighbourhood, inside the community. Many inhabitants, especially in informal neighbourhoods concerned with several planning projects, try to organise resistance even if it is sometimes weak and not a general reaction. This is the case for the Dikmen Valley and May Day neighbourhood. These communities have sometimes organised themselves in independent structures and developed their own local protests. Their protest campaigns and their daily, unspectacular survival strategies challenge the connection between urbanisation and civilisation as claimed in neoliberal 'development' concepts (Mayer 2012: 79). In other words, the state's desire to renew these areas constitutes a direct threat to the community's shared identity, thereby triggering resistance in order to protect it. Resistance, here, is understood as active but also passive and sometimes invisible actions, strategies and tactics used by inhabitants individually or collectively in everyday life in order to protect their way of life and the social and cultural specificities of their community which are widely shaped inside their neighbourhood. In this article, this type of resistance is analysed in connection with the right to the city that Lefebvre developed in order to explain the right of all people living in the city to participate in decisions concerning their everyday life and to exist with their own life-styles, identity and cultural habits in the city. The right to the city is considered then as a resistance to the standardisation of city life. Michel de Certeau (1980) analyses this as a tactic used by the 'weak' through which everyday spatial hegemonies are covertly transgressed, and identity is projected through claims to a particular space of power which is neighbourhood.

Two cases from Turkey are selected in order to discuss this relationship between the neoliberal city and right to the city due to the intensive neoliberal urbanisation undertaken by the AKP government, and because this process has dramatically affected the living environments of these different communities. The first case, the May Day neighbourhood (the name was unofficially given by inhabitants to refer to the Labour Day), is a poor working-class neighbourhood situated on a hill in the Ümraniye district of Istanbul The urban transformation of the Ümraniye district, whose inhabitants live mainly in *gecekondus*, is one of the priorities of the Metropolitan Town Hall of Istanbul, largely due to this area's geological resistance to earthquakes. Attempts have been made to destroy the shantytowns on several occasions in the past, but each met with a high-level resistance from the May Day neighbourhood. This location, close to two of the city's main motorways and the new location of the National Bank of Turkey, is only 15 minutes away from the Bosporus by car, making it extremely desirable to the private construction industry (Gülhan 2011) (see Figure 44.1).

This neighbourhood, located on the Asian side of the city, grew in stages in the 1970s via the *gecekondus* built by economic migrants from rural Anatolia. Initially the building plots were sold by the land mafia, but from 1976, left-wing socialist organisations started to settle in the neighbourhood (Aslan 2004). During the second half of the 1970s, control of the neighbourhood passed over to these political organisations, resulting in land distribution which favoured dissenting families, particularly the Alevis from the Kurdish cities of Tunceli, Kahramanmaraş and Sivas. The Alevis are one of the main religious minorities in Turkey and they interpret Islam in a secular way, have religious rituals dramatically different from Sunni interpretations and they are often politically close to Kemalist left in Turkey (Massicard 2012). In May Day, various services and institutions were gradually established across the *gecekondus*. Community centres, schools and co-operatives for purchasing food, fuel and construction materials were designed and managed by the residents via popular committees. The district earned a reputation for political dissidence by repelling the *gecekondu* demolition teams three times in the1970s, making it a site of resistance in the public imagination. However, after the military coup of 1980, in 1983 the neighbourhood was 'legalised' by the military regime which took over its control. A *muhtar* (legal administrator democratically elected) was chosen, and the neighbourhood was renamed Mustafa Kemal, which remains its official name today. In 2008 the neighbourhood was divided and each portion attached to a different district, so that the historic May Day neighbourhood now composes parts of the Aşık Veysel and Mustafa Kemal neighbourhoods attached to the Ataşehir district. Even so, kinship networks between people coming from the same Anatolian city are still important, with many hometown associations still operating in the area (Toumarkine and Hersant 2005).

Figure 44.1 The May Day neighbourhood, Istanbul. *Source:* author.

The second neighbourhood is the Dikmen Valley in Ankara. Initially, what is now called the Dikmen Valley was a small village with some vineyard gardens, relatively far from the city centre of Ankara (the new capital of the young Republic of Turkey), designed in 1928 according to the urbanisation plans of German planner Herman Jansen. These plans were based on a city of 300,000 inhabitants by the 1980s and planned to preserve the village of Dikmen and its surroundings in order to maintain its agricultural activities, and to preserve green areas

Subsequently, strong demographic pressure, unexpected and therefore not planned for, rendered Jansen's master plans obsolete. The urgent need for housing for rural migrants, expecting to find jobs in the new capital city, has gradually widened the city's borders, and the Dikmen Valley is one of the areas where *gecekondus* have multiplied since the 1960s. The construction of these dwellings was accelerated in the 1970s under the impulse of the left-wing radical groups who were looking for unoccupied lands in various *Ankariot* districts in order to distribute them – sometimes by using force – to working-class immigrants. This tendency was observable in many other neighbourhoods in metropolitan cities in the country such as Mamak Tuzluçayır in Ankara, May Day or Gazi neighbourhoods in Istanbul and Gültepe in Izmir. In the case of Dikmen, the neighbourhood was composed of people coming from different cities of Central Anatolia like Çankırı, Çorum or Sivas. Some of them were Alevis, as in May Day, but the majority of inhabitants were Sunnis. Before the launch of the project, the ethnic, religious and political diversity of the neighbourhood had as a corollary the absence of organised political actions or political consciousness, which was able to generate collective solidarity. Compared to May Day, the different parts of the neighbourhood were not interacting daily. Some respondents on this point highlight, for example, the tendency of families from the same Anatolian city to gather together by constructing their houses next to each other. Daily exchanges related to the use of the grocer, the bus stop or the neighbourhood cafe existed, but in the absence of a common risk of destruction of their homes or life spaces, they were limited. Everyone preferred to socialise in their street, with their neighbours, the people they knew through family or town of origin.

At its creation, the Dikmen Valley was hardly connected to the city centre of Ankara and difficult to access. It had no basic infrastructure or services such as roads, electricity networks or

Figure 44.2 Dikmen Valley in Ankara. *Source:* author.

water. Moreover, the connections to these latter two services were carried out clandestinely. Due to the rapid increase in population and *gecekondus*, policymakers were finally forced to administratively recognise the neighbourhood during the 1980s and provide basic public services. Some of the inhabitants of *gecekondus* even managed to obtain certificates of ownership from the municipality during this period. If this official acknowledgement was necessary in view of the size of the neighbourhood and the population, it also largely met electoral considerations: with the legalisation of the building and the recognition of the district, successive governments hoped to obtain votes.

In the early 1990s, the Ankara Metropolitan Municipality and the Çankaya Municipality, both led by the People's Republican Party (CHP), decided to carry out an urban renovation and included the Dikmen Valley project in the master plans of Ankara in 1989. At the beginning, this renovation project was fairly participative and tried to respect the right of owners. However, the project evolved in the 2000s under the AKP rule of Ankara and became entirely a 'for-profit' housing project, and the new mayor refused to cooperate and to negotiate with inhabitants.

Compared to the May Day neighbourhood, known for its historical political involvement in the 1970s and for the presence of activists affiliated to various extreme left political organisations, politicisation of a large part of the inhabitants of Dikmen seems to be more closely linked to the spatial changes and stigmatisation. The neighbourhood constitutes a life space in which social ties, networks and social solidarities are forged and contributes, to a large extent, to the socialisation of individuals. It thus plays an important role in the processes of building individual and community identity, particularly in the sense attributed to it during collective mobilisation and resistance. As a result, any attempt to transform and destroy it is seen by some inhabitants as a threat to their existence within the city.

In Dikmen Valley, as in May Day, ordinary people are mobilised in a context characterised by permanent social uncertainty and urgency, like the deficiencies of basic equipment and services. These two districts constitute zones of economic, social and communitarian segregation and could be considered also as differentiated spaces in a Lefebvrian sense as they have a historical political heritage and a collective identity construction related to belonging to their neighbourhood, which is different from the other parts of the city. In both cases the municipal authorities have sought to change the composition of these neighbourhoods by destroying the *gecekondus*. The police forces have confronted the inhabitants several times either to enable the demolition of houses or to repress political mobilisation related to some commemorative events like the 2 September 1977 resistance in the May Day neighbourhood against the destruction of *gecekondus*, even causing the deaths of some protestors. In both instances, public organisations have been unable to transform these districts either culturally or politically.

These confrontations have also come at a cost for the inhabitants. The precariousness of the housing, due to the constant risk of demolition, and the resultant difficulty in accessing safe and legal housing, remain common problems in both districts. In the following section, I will explore the place of the right to the city concept in the development of neighbourhood resistance and especially the way inhabitants represent this concept in both case study neighbourhoods.

Right to the city in neighbourhood resistance

Lefebvre perceived the right to the city as a way of legitimating 'the refusal to allow oneself to be removed from urban reality by a discriminatory and segregative organization' (1996: 197). For Lefebvre, the urban is not simply limited to the boundaries of a city, but includes its social system of production. Hence the right to the city is a claim for the recognition of the urban as the (re)producer of social relations of power, and the right to participation in it (Gilbert and Dikeç

2008). In that sense, the right to the city could be described as a right to the appropriation and the participation of the inhabitants. As Marcuse explains, it is at the same time:

> a right to produce the city as well as to enjoy it and the two are integrally linked. It is not only the right to a choice of what is produced after it is produced, but a right to determine what is produced and how it is produced and to participate in its production.
>
> *(2012: 36)*

Lefebvre's conception of the right to the city is for inhabitants to retain the ability to produce their spaces without conforming to the dominant modes of spatial production, and to participate in reshaping the existing norms and forces in which space is being produced within the neoliberal order, rather than being themselves engulfed in those modes (Fawaz 2009). The concept of right to the city in this sense has been introduced in Turkey by some urban activists and their organizations and transmitted to inhabitants in their collaboration to resist the urban projects. These activists are often urban planners and architects with Marxist ideas, defending an alternative vision of city-making against that of public and private actors involved in urbanisation. They are often members of the *Union of Chambers of Architects and Engineers* or associations like *Solidarity Studio* or *One Hope*. In the case of May Day and Dikmen Valley, these organisations were present at the beginning of the mobilisation but progressively left the place to inhabitants themselves but also to leftist political groups. In any case, the inhabitants adopted the term, combining it with spatial justice. During the mobilisation they built neighbourhood associations, and these associations, thanks to their interaction with professional activists, developed their own discourse of justice and right to the city. In fact, the sense of the term was present from the beginning of the resistance as the mobilisation was never limited to housing rights but broadly to the right to exist in the city with the choice of life space, the access to all city facilities and the right to enjoy their own cultural and political tools and social relations. For one of the urban activists:

> The right to the city is a revolutionary right, not a mere right of dwelling; right of access to centrality or to the urban services. It must be elaborated as related to the use-value of space with a democratic urban imagination. Urban struggle must aim for the right to the city, targeting a more democratic, just city based on the use value.
>
> *(Ergin and Rittesberger 2014: 51)*

Concretely, both neighbourhoods I have studied started to resist the urban transformation projects from 2007. Unlike Dikmen Valley, May Day was not directly threatened but with the construction of a large gated community opposite the neighbourhood, rumours started to be spread about the future of this informal settlement, which was the target of the government for several years because of political and cultural reasons, and the opportunity presented by its high estate value. Because of the iconic case of Sulukule, a Roma neighbourhood in Istanbul which was destroyed and gentrified entirely by a state-led urban project (Erdi-Lelandais 2013), inhabitants in May Day were determined to stay active and to be ready to defend their neighbourhood against threats of destruction. The importance of resistance is underlined by one of the founders who still remains in May Day:

> I think we should keep our neighbourhood alive because we earned it by the blood of our comrades. There are resistance memories in every corner of it. This neighbourhood marked our social memory. It gives always a place for revolutionary activities and leads the

> production of progressive ideas. Despite all problems and political measures targeting the neighbourhood, it is still a revolutionary and red neighbourhood.
>
> *(Hüseyin, age 60, interview)*

Neighbourhood cements a shared identity related to the place among residents, but not just in relation to the specific codes and practices associated with ethnicity. This identification is also enriched by: traditional customs, social networks, rituals, symbols, collective memories and mechanisms of mutual aid that exist only within the physical living environment of that community. According to an inhabitant:

> May Day is a place where there is no oppression and where everybody can find their place and live. It's a left-wing neighbourhood. The human being is central to the vision of people here. I understand the value of May Day when I'm outside. Other neighbourhoods are stand-offish, conservative. Individual rights are limited. When I come back here, I relax. In May Day, there are opportunities for people to develop themselves. They [state institutions] tried to destroy it but they didn't succeed.
>
> *(Kamil, age 41, interview)*

In May Day, the inhabitants argued that state agencies were seeking to pervert the revolutionary character of their neighbourhood by encouraging criminal activity. For them, this was a government strategy stigmatising May Day in order to legitimise an urban transformation project, so their action was focused on this issue. A neighbourhood website was created with the title 'Against degeneration, defend your neighbourhood' (this site does not exist anymore). Regular meetings are held by the two main political organisations, the Socialist Party of Oppressed People and the Revolutionary Popular Front. In addition, the Association of 2 September organises cultural events, dramas, study programmes and an annual festival in order to maintain the neighbourhood's collective identity. In this case, collective action appears to be in keeping with a highly charged and constantly reiterated local history, allowing the district to remain in a state of a permanent memorial mobilisation. Resistance is framed and organised by one or two political organisations that control access to the neighbourhood, both physically and in relation to land and property ownership, so the protest is not generated externally. There are often confrontations between the police and young activists, mainly members of the Revolutionary Popular Front, some of whose members are currently serving custodial sentences.

In the case of Dikmen Valley, this leftist character is rather in the background of the mobilisation even if the role of leftist activists and their historical capital of activism is undeniable in the organisation of resistance. The specificity of Dikmen Valley was the primary role of inhabitants and especially women in the acts of resistance. Strong ties were established during the resistance thanks to a socialist organization, *Halkevleri*, whose activists, having lived in the neighbourhood for several years, knew how to promote political consciousness around the right to the city amongst its inhabitants. This connection was partial in May Day, as some inhabitants were tired of street battles between leftist groups and the police. In Dikmen Valley, street battle was only engaged in to prevent the destruction of the neighbourhood, and the resistance was organised entirely by one unique organisation, the *Office for Housing Rights*, which was constituted by inhabitants themselves and whose representatives were chosen by the inhabitants of each street.

In that sense, the neighbourhood and the city give rise to a strong sense of belonging, and the resistance aims above all to preserve the place and the social and cultural practices of everyday life that are related to it. In the present case, this resistance is expressed by a proclamation of the inhabitants, which reminds us that this territory is theirs and is constituted and supported by a collective identity. They claim their right to occupy this area, to own it and to preserve it:

> Before this struggle, I did not know many people in the neighbourhood. The struggle has brought us together. Without making religious, language or ethnic distinctions, we have become like sisters to share our misfortunes, our happiness, our hopes. We women have learned that life is not only in our homes but also in the barricades. While we hardly ever go out and end the day between four walls, today we are everywhere. We have learned to claim the life we desire and not the one we are forced to do. The valley has become a great learning school for women.
>
> *(Sultan, age 39, interview)*

In the *Office for Housing Rights*, for example, it was decided that every 'owner' of *gecekondu* should ensure that his garden is beautiful, tidy and welcoming, that everyone should participate in the cleansing of the neighbourhood and commit to not dumping rubbish. The installation of Syrian refugees and waste collectors in abandoned houses was also discussed. The inhabitants believe that both were incited by the Mayor of Ankara with the aim of creating a climate of conflict in the neighbourhood. However, in order to reinforce their reputation for tolerance and friendliness and to defeat the mayor's strategy, it was decided to accept these newcomers and invite them to meetings in order to establish collective rules of 'living together'. In this strategy, the will of the appropriation of space by inhabitants is clearly present.

Conclusions

In the examples of May Day and Dikmen Valley, resistance opposes both the discrimination apparent in the spatial organisation of the city and the rules of institutional citizenship which disciplines space by fixed power relations. The inhabitants of these threatened neighbourhoods organise resistance through what Michel de Certeau (1990) defines as the 'tactics' of making do. In both May Day and Dikmen Valley, the inhabitants who were interviewed emphasised the particular character of their neighbourhood, and the strong solidarity of its residents. Through this discourse, their living-space was used to construct and reflect another vision of the neighbourhood, far away from its common image of being an undesirable area. Through their everyday rituals, by maintaining their customary habits and by creating alternative lifestyles, they assert their right to exist in the city.

The example of two neighbourhoods studied in this chapter proposes alternative ways of thinking about the conception and the use of urban space. As Lefebvre points out in *The Production of Space*, space becomes a place of struggle for its appropriation and conception between public actors and their opponents. In this struggle, the right to the city is chosen as a tool by urban dwellers in order to legitimise their right to 'be' in the city. Neoliberal hegemony tends to absorb alternative logics and shape them to its ends.

The case studies outlined in this research also show the importance of space and its appropriation through the organisation of daily resistance. This means that even the neoliberalisation of public planning is radically reorganising the supra-urban scalar hierarchies in which cities are embedded; cities thus remain strategic arenas for socio-political struggles (Purcell 2008). The recent protests organised for the protection of Gezi Park in Taksim Square in Istanbul during May–June 2013 showed the importance of symbolic spaces (neighbourhoods, squares, parks) in the emergence of resistance, as the *citadins* would not allow top-down public decisions concerning their life space, which is considered part of their identity and a mark of their everyday life

(Erdi-Lelandais 2016). The acts of resistance described here also show that the appropriation of space is a more complex process than is usually acknowledged. Despite all the tools of physical and symbolic domination wielded by state institutions, social activism and resistance within the city is striving to transform the socio-territorial organisation of capitalism itself on multiple geographical scales (Purcell 2008). If the right to the city shows the will of mobilised inhabitants to remain in the city and to participate in its everyday construction, this will is also closely connected to the claim of spatial justice. Inhabitants refuse also to be stuck in specific areas of cities and want access to all amenities and public services as urban dwellers regardless of their social and economic status.

References

Aslan, Ş. (2004) *1 Mayis Mahallesi. 1980 Öncesi Toplumsal Muhalefet ve Kent*, Istanbul: İletişim.

De Certeau, M. (1980) 'On the oppositional practices of everyday life', *Social Text*, 1:3 3–43.

De Certeau, M. (1990) *L'invention du quotidien. 1'arts de faire*, Paris: Folio.

Erdi Lelandais, G. (2013) 'Citizenship, minorities and the struggle for a right to the city in Istanbul', *Citizenship Studies*, 17:6–7 817–36.

Erdi Lelandais, G. (2016) 'Gezi protests and beyond: urban resistance under neoliberal urbanism in Turkey', in M. Mayer, C. Thörn and H. Thörn (eds) *Urban Uprisings. Challenging Neoliberal Urbanism in Europe*, London: Palgrave.

Ergin N.B. and Rittesberger H. (2014) 'The right to the city: right(s) to "possible-impossible" versus a mere slogan in practice?' in G. Erdi-Lelandais (ed.) *Understanding the City. Henri Lefebvre and Urban Studies*, Newcastle-upon-Tyne: Cambridge Scholars Publishing.

Fawaz, M. (2009) 'Neoliberal urbanity and the right to the city: a view from Beirut's periphery', *Development and Change*, 40:5 827–52.

Gilbert L. and Dikeç M. (2008) 'The right to the city? Politics of citizenship', in K. Goonewardena, S. Kipfer, R. Milgrom and C. Schmid (eds) *Space, Difference, Everyday Life. Reading Henri Lefebvre*, London: Routledge.

Gülhan, S.T. (2011) 'Devlet Müteahhitlerinden Gayrimenkul Geliştiricilerine, Türkiye'de Kentsel Rant ve Bir Meta Olarak Konut Üreticiliği: Konuta Hücum', *Birikim*, 270, October.

Lefebvre H. (1996/1968) 'Right to the city', in E. Kofman and E. Lebas (eds) (1996) *Writings on Cities*, Oxford: Blackwell Publishing.

Lefebvre, H. (1974) *La Production de l'Espace*, Paris: Anthropos.

Marcuse, P. (2012) 'Whose right(s) to what city?', in N. Brenner, P. Marcuse and M. Mayer (eds) *Cities for People not for Profit. Critical Urban Theory and the Right to the City*, London: Routledge.

Massicard, E. (2012) *The Alevis in Turkey and Europe: Identity and Managing Territorial Diversity*, London: Routledge.

Mayer, M. (2012) 'The right to the city in urban social movements', in N. Brenner, P. Marcuse and M. Mayer (eds) *Cities for People Not for Profit. Critical Urban Theory and the Right to the City*, London: Routledge.

Öncü, A. (1988) 'The politics of urban land market in Turkey: 1950–1980', *International Journal of Urban and Regional Research*, 12:1 38–64.

Purcell, M. (2008) *Recapturing Democracy: Neoliberalization and the Struggle for Alternative Urban Futures*, New York: Routledge.

Soja, E.W. (2009) 'The city and spatial justice', *Justice Spatiale-Spatial Justice*, vol. 1, accessed online at www.jssj.org, accessed May 2017.

Toumarkine, A. and Hersant J. (eds) (2005) 'Hometown organizations in Turkey' (special issue), *European Journal of Turkish Studies*, 2.

Türkün, A. (2011) 'Urban regeneration and hegemonic power relationships', *International Planning Studies*, 16:1 61–72.

45

Dystopian utopia? Utopian dystopia?

A tale of two struggles for the right to the city

Mee Kam Ng

Introduction

'Every ideology contains a utopia' (Sargent 2006: 12). As epitomised in the 2017 Policy Address of the Chief Executive in Hong Kong: 'We believe that only through economic development can we improve people's livelihood and promote social harmony and inclusion' (Government of Hong Kong 2017: 1). The Policy Address continues to boast of Hong Kong being the world's freest economy for 22 consecutive years (Heritage Foundation 2017); the city's dual advantages as the 'super-connector' between China and the world under China's renewed Belt and Road Initiative (to expand global trade and cultural exchanges); and the one country, two systems arrangements for capitalist Hong Kong to be ruled by socialist China. However, this 'utopian' economic picture is perhaps true only for certain sectors in Hong Kong – the self-crowned 'Asian World City' is very much a polarised society. According to the UNDP, Hong Kong ranked as the most polarised society (Cagape 2009), and in 2011 Hong Kong's Gini coefficient (to measure wealth disparity) was 0.54 (Hong Kong Council of Social Service (HKCSS 2017) (a score of 0.5 or more is considered high i.e. showing high wealth inequality). In 2014 almost 20 per cent of the population were considered poor (*South China Morning Post* [SCMP 2015]), and depending on who you ask, Hong Kong is both heaven and hell (Cagape 2009). This chapter tries to interrogate this heaven and hell reality through the lens of utopia and dystopia, highlighting how a dystopian context can trigger people's utopia-inspired acts to defend their lived spaces. The chapter first examines the dialectical relationships of utopia and dystopia. Then the story of two communities threatened by a pro-growth ideology, defending their right to lived spaces is discussed. Their experiences reveal that the choice to produce differential spaces for nurturing relationship-rich communities has to be made on a daily basis – to enact relentlessly a utopia amidst every dystopian situation.

The dialectics of utopia and dystopia

Gordin et al. (2010: 1) argue that 'utopias and dystopias are histories of the present'. To them, dystopia is a utopia gone wrong or one that functions only for a particular segment of society. A dystopia emerges when the dominant groups apply their (utopian) visions and hegemonic solutions to resolve society's 'perceived deficiencies' (Gunder and Hillier 2007: 467). The hegemonic rule of the dominant groups in this age of planetary urbanisation (their utopia) unfortunately is often serviced by 'scientists, technicians, administrators, physicians and soldiers' (Mumford 1965: 289) whose knowledge becomes an ideological tool to 'prove' the impossibility of alternative urbanisms. Hence, everyday life as well as the institutional set-up of such a faulty or dystopian utopia is permeated with ideological qualities, twisting the dreams of many, limiting their lives, leaving them with few options (Stillman 2003: 15). Spaces created in a dystopian utopia can for Lefebvre be spaces of: catastrophe, violently produced by exploitation, inequality, class domination, environmental degradation, stultification and oppression, homogenising, universalising and totalising. The pursuit of economic competitiveness and profit is seen as 'progress' even though the erosion of use values in space has threatened the well-being of every territory-bounded urbanite. If this is the case, how can people realise that they are indeed living in a faulty or dystopian utopia? As argued by Lefebvre, 'the silence of the "users" [of space] is indeed a problem' (1991: 364). Will these silent users be awakened and start to voice the conditions of their dire reality?

Wilson and Bayon provide an answer using 'the black hole capitalism metaphor' (2016: 352). While planetary urbanisation razes the planet, turning it into 'an infernal machine' for 'endless valorisation of value' (Jameson 2011 in Wilson and Bayon 2016: 354–5), at the 'border of oblivion' where the annihilation of space is swallowing up all possible hope, people directly affected may then perceive for the first time their 'nightmarish reality', coercing them to make a desperate decision to re-assert their equal rights to planet Earth, thus forging the birth of a real utopia of differentiated spaces (Wilson and Bayon 2016: 357, 362). In other words, when people recognise the real face of a hegemonic utopia as dystopia, they will begin to search for their own utopia in the midst of the dystopian contexts that hopefully will lead them to real alternative possibilities (Bloch 1986: 223).

Interestingly, these 'awakening' moments, if they ever take place, are often preceded by utopian work done by 'enlightened elites' who, as argued by Lefebvre (1991: 51), always lead the protests even though they should be the privileged ones in planetary urbanisation. Ng (2014: 3) calls these 'enlightened elites' 'system-transforming intellectuals' (these could be 'sympathetic insiders' within the government or 'critical experts' in civil society) who help 're-problematise or re-conceptualise existing situations to provide directions for changing cultural and habitual practices'. Ng (2016: 282–3) further suggests that in the Chinese culture, intellectuals usually are expected to be loyal to a functioning system. However, if the system is corrupt, it is the calling of intellectuals to criticise it boldly and educate the 'common people', encouraging them to undertake utopian acts of righteousness and benevolence.

Whether by system-transforming intellectuals or crisis-ridden awakening communities, the enactment of a utopian dystopia can be a powerful tool, inspiring 'hopes in the darkness itself' (Bloch 2000: 201), excavating possibilities for alternative and more just urbanisation processes (Gordin et al. 2010: 2; Macleod and Ward 2002: 164; Pinder 2002: 239). The tactic is to appropriate and activate fully the possibilities that lie hidden within the everyday 'ordinary' setting as active sites for the continuous removal of the dystopian reality (Brown 2011; Pinder 2010). It is to acquire once again the ability to see places as a product of 'various interactions of people with each other and with that place' (Brown 2011: 14), to rekindle the dream of a desirable urbanism that is open-ended, creative, shared and inherently unique (Madden 2012: 775) where 'everyday life can flourish' convivially (Coleman 2013: 354). These utopian acts of critiquing the present and developing ideas of the future can only happen when people come together to deliberate on what it means to create and inhabit places that are valuable to people's lives to meet their everyday social needs (city-bound collective 2012; Sliwinski 2016: 439).

Indeed, concrete utopia (Bloch 1986: 223) – utopian places of encounters, play, work, creativity and exchange that do not pass into exchange value, commerce and profit (Lefebvre 1996: 148) but give people dignity (Bloch 1986) – sometimes can be found in urban spaces described by the hegemonic discourse as 'dystopia', which in fact can turn out to be 'transgressive lived spaces of escape, refuge, employment and entertainment' (MacLeod and Ward 2002: 164). However, it is not easy to map these utopic spaces as there is no paperwork, no law and no formalised plans but complicated 'social property relations' (city-bound collective 2012: 604–5). Discerning these places would require denaturalising 'customary ways of framing and resolving social problems' through sharing values to rebuild 'commons' (Sliwinski 2016: 440), commons that satisfy 'the legitimate needs, desires and capacities of human beings' (Gardiner 1992: 28). And this demands dialogical interaction of various stakeholders, co-endeavouring to develop mutual recognition, support and trust (Gardiner 1992: 40; Martin 2003: 732) for the building of a dedicated 'beloved community' (Royce 2001: 9). Lefebvre (1996: 151) calls these experimental utopias, where there is 'incessant feed-back between the conceptual framework [utopian thoughts] used and empirical observations'.

The building of a utopian dystopia, the enactment of progressive convivial spaces in a hegemonic dystopian utopia, is to transform an exchange value-dominated urban realm to one that allows all its stakeholders a right to use urban spaces and turn them into places that nourish people's dignity. Value is indeed a social relation (Wilson and Bayon 2016: 354). It is because through the making and remaking of these experimental utopias that 'we make a place and we construct ourselves through this making', finding ourselves and the true meanings of our lives (Brown 2011: 15).

Government plans (conceived space): it is all about growth...

The pro-growth ideology as a hegemonic utopian goal intensified when Hong Kong returned to Chinese rule in 1997 because China was eager to prevent capital flight (Chu and So 2013). However, paralleling a culture that stresses the importance of economic stability and prosperity is a long history of political activism in the city (Lam 2004). For instance, from 2005 to 2014, there were 49,508 public meetings and processions, averaging 14 events per day (Ng 2016: 284). The post-colonial city has indeed faced many planning controversies in its two decades of existence (Ng 2013). The sections examine two episodes of a close encounter of dystopia and utopia in urban and rural Hong Kong.

Wanchai, an old urban area east of the Central Business District, has been facing huge (re) development pressure. Over the years, it has witnessed what Wilson and Bayon (2016) called 'black hole capitalism', where rows after rows of four to five-storey tenement buildings were razed to make way for high-rise luxury apartments. In the past decade many redevelopment projects, including four large-scale ones led by the Urban Renewal Authority (URA), were conducted in Wanchai, with the remaining community being pushed to the border of oblivion. As argued by Wilson and Bayon (2016: 352), a real utopia is like a black hole environment where 'a space of wild creative energies' meets 'a vortex of destruction'.

In 2006, when a community-based battle against the URA redevelopment of 'Wedding Card' street (a street with low-rise tenement buildings and diversified design and printing shops at the ground level) was hardly over, the URA announced another HK$100 million (US$12.8 million) project to convert a grade one-listed heritage building, the Blue House, into a tourist attraction. However, the Blue House has a long history of being valued highly by the local community, serving once as: a temple, then a hospital and school. So the local community objected and put forward a counterproposal called *Living in the Living Museum*

(Lai 2006), as in 2005 they had applied to set up a 'Wanchai Livelihood Museum' in the Blue House.

The destructive power of black hole capitalism is also evident in the New Territories. Since China's Open Door Policy and Hong Kong's economic integration with the mainland, the once rural agricultural fields have become 'container territory' (i.e. turned into storage sites for such things as containers and construction materials). It was not until after 1990 that the urban planning rules and regulations were extended to control land use development in the New Territories. But by then, the damage had been done.

In 2008–9, because of the global financial tsunami, China tried to stimulate its economy through the development of a high-speed railway network. In April 2008, the government of Hong Kong announced the construction of a HK$66.9 billion (US$8.6 billion) 26-km Express Rail Link (ERL). In November 2008, without prior notification or consultation, the government announced that Choi Yuen Village, a non-indigenous rural village (villages not eligible for reconstruction when resumed (acquired through compulsory purchase or eminent domain) for public purposes because their ancestors came to Hong Kong after the arrival of the British colonisers) in the New Territories, had to be removed in 2010 to make way for an emergency rescue station along the ERL. The Mass Transit Railway Corporation (MTRC) responsible for building the ERL then claimed that it would connect Hong Kong with the 16,000-km National High Speed Railway network, create 11,000 jobs, save approximately 42 million hours of travel time and gain the benefits of a much bigger economic circle (MTRC 2017). This conceived plan is full of 'violence' (Lefebvre 1991: 358), as it aims to annihilate the Choi Yuen Village. Not only was the ERL project met with strong societal objection, but the villagers were adamantly against the removal of their village, and very quickly the two groups joined forces to resist the implementation of the project. Before we continue with these stories, let us have a deeper understanding of the two affected communities when they faced the official plan that would exterminate their lived space.

Representational space at the 'vortex of destruction'

With the completion of the URA's redevelopment projects, local stakeholders started to recognise the 'horrible' face of the redevelopment machine. They were shocked when they learned that the redevelopment projects did not provide a continued right for them to live or earn a living in the district. Collective memories, histories and culture would be destroyed. In the name of redeveloping for the 'public interest', long-term residents and social networks cultivated over the years were removed (Wong 2006):

> I grew up in Wanchai… I feel very safe walking on the streets… just like going home… We have invested all our money and efforts in our shop and now we are deprived of a harmonious community… this cannot be compensated by money… our sacrifice is too big.
>
> *(16–17)*

> When the economy was bad in the 1990s, the landlord automatically lowered our rents… The Urban Renewal Authority required us to show them tenant proofs in the past two to three decades.
>
> *(16–18)*

When the government announced the project to revitalise the Blue House, many residents were retired elderly people who were extremely worried about the prospect of being displaced.

Cheap rent in the Blue House, old markets, street corner shops, convenient transportation networks and strong social capital meant that they could lead an affordable life (Lee 2009: 94). Given the threat of displacement, they had no choice but to work with other social activists to fight for their continued right to live in the Blue House:

> After the introduction of the official revitalization project, we have spent more time for the meeting with other locals and outsiders. In the past, we would not spend so much time chatting with neighbours.
>
> *(Lee 2009: 156)*

There are two classes of citizens in rural Hong Kong. The indigenous landowners (those who came before the British colonisers) are entitled to build their own houses and the reconstruction of their villages when their land is resumed for public purposes. Many indigenous villagers had turned their unused farmland into open storage or car parks. Hence, indigenous landowners usually are very happy when infrastructure projects demand the resumption of their land. These arrangements do not apply to the non-indigenous villagers, who are usually farmers, tilling land they either rent or bought from indigenous villagers, and who do not have the entitlement to a reconstructed village following the expropriation of their land by the state. As remarked by a journalist turned social activist: 'all the fields [in non-indigenous Choi Yuen Village] have been carefully tilled... this is the only place in Shek Kong not ruined by car dumps or open warehouses' (Chu 2009).

Choi Yuen Villagers were very angry at the imminent demise of their homes and their preferred and dignified lifestyle:

> After learning the news, my mom could not sleep... We have strong relationships in the village... Why do we need compassionate rehousing? We can move if we want to!... The kind of development embedded in the ERL is too irrelevant to me... It is development for 'you' [the government] at the expense of us minorities.
>
> *(Ip 2009a)*
>
> I am 80 years old and have lived here for 50 to 60 years... I cannot live in apartment buildings... When I visit my daughter in the city, I get lost... Here I can continue to grow flowers, vegetables... I need to do something... I cannot just stay at home.
>
> *(Ip 2009b)*

> I realised how free it is to be a farmer... When you work for others... you are seldom recognised for your contribution... much more rewarding to farm... its value is not measured only with money... It stimulates my creativity.
>
> *(Lam 2009a)*

The critical roles of 'enlightened elites'

The two struggles might be a totally different story if not for the help of 'enlightened elites' or 'transformative intellectuals' who rendered their long-term or timely expertise to resist the encroachment of the hegemonic dystopian utopia onto the lived spaces of the two communities. In the case of Wanchai, the help has been long term. Back in 1949, Bishop Hall of the Hong Kong Anglican Church, modelling on the Settlement Movement in Oxford (Britain), set up the St. James Settlement (SJS) in Wanchai to provide poor local children with education (Lam 2009b: 5). Since then, social workers in SJS have been an indispensable voice for the local community. A few months before the announcement of the Blue House revitalisation project in 2006, SJS had already applied to the Lands Department to transform the Blue House into a 'Wanchai Livelihood Museum', exhibiting items donated by local residents as a living record of the history of the place. The preparatory work for the launching of the Museum pulled SJS and local enthusiasts together, forming a 'beloved community' (Royce 2001). They organised local tours, public exhibitions and workshops to gather information and opinions of local residents.

When the revitalisation project was announced, SJS, including members of Community Cultural Concern who offer advice on community development in SJS, was instrumental in organising community members and waged a multi-front battle to save their rights to lived spaces. After consulting local stakeholders, social workers in SJS put forward the 'Living in the Living Museum' counterproposal, presenting the local community's utopian vision of transforming the Blue House into their community place with spaces for living, housing resident artists, cultural workshops, community uses and social enterprises. At the same time, a 'Blue House Conservation Group' comprising local residents, social workers, university professors, architects and urban planners (many have been comrades in the fight against the redevelopment of Wedding Card Street) worked through the Town Planning Ordinance and managed to convince the Town Planning Board to change the land use zoning to include the residential use as flats, paving the way for residents to fight for their right to stay in the Blue House. The successful move prompted the relevant authorities to reconsider the feasibility of the revitalisation options.

The timely involvement of Heritage Hong Kong, founded in 2008 by a wealthy and socially active expatriate couple in the surveying business, was a pivotal point in reshaping the fate of the Blue House. Not only was the group willing to financially support the conservation of the Blue House, it had also exercised its network and social capital to persuade the government to pursue an alternative way of revitalising the building complex (Leung 2007).

The anti-ERL drive and the resistance movement by the Choi Yuan Villagers were fuelled also by enlightened elites in the city, and interestingly many have been comrades in various social movements in the city. According to an in-depth interview with a local leader (Ko 2014), the villagers were at first rather helpless as local councillors were not particularly helpful. The situation changed when a journalist turned social activist (he was later elected to the Legislative Council in 2016) started listening to individual villagers and writing their oral histories, which moved the villagers to accept him. Paralleling the villagers' resistance movement was a drive to stop the construction of the expensive ERL. Many protestors were students and young adults who were called the 'post-80s' (born after 1980). To them, the ERL symbolised everything they opposed: expensive infrastructure for the rich, absence of genuine public engagement and unequal treatment of rural indigenous and non-indigenous villages (Lai 2010). To the post-80s, 'neighbourhoods, heritage, countryside, traditional communities and small businesses' have priority over economic growth and planetary urbanisation (Chan 2011).

Throughout the Finance Committee of the Legislative Council debates on funding support for the ERL, there were demonstrations outside the council building. One group even performed a 'prostrating walk' like the Tibetan pilgrims (walking 26 steps – symbolising the length of ERL – before kneeling down and touching the ground) in five districts and around the Legislative Council building for two days before the budget for the project was approved. On the night when the budget was finally approved, the protestors stormed and blocked all the exits of the Legislative Council building, attracting wide media attention.

One reason for the post-80s' proactive involvement probably had to do with their encounter with rural village life (Ko 2014). For many of them, their visit to Choi Yuen Village during the campaign was eye-opening and life-transforming. As mostly urbanites, they were somehow captivated by the simple lives and warm human relationships in the village (Ko 2014). Many enjoyed staying with the 'extended families' in the village, listening to stories told by the elderly and enjoying the village food they missed in the city. Very quickly the village became their 'beloved community' (Royce 2001) and their actions proved their devotion to their newly found 'utopia'. The approval of the ERL budget meant that the village had to move, but with the help of different enlightened elites the villagers eventually decided to fight for their right to rebuild a new village to create a utopia out of a dystopia.

The fight for differential space

Ravaged by four large-scale redevelopment projects, residents in Wanchai, assisted by SJS and other social activists, gathered momentum when the saga of the Blue House revitalisation project was announced. They worked together on a counter-proposal based on not just preserving the building typology but also the environment encompassing the existing living culture and economy to promote sustainable development (SJS 2007: 3). Through bottom-up participatory planning, they aimed at conserving the original architectural character of the Blue House as a testimony to traditional developments in Hong Kong, hence: conserving community history and culture, improving liveability (for those who wanted to move out or stay), building social capital and a sense of community, developing social enterprises and preventing gentrification (SJS 2007: 4–5).

With the help of Heritage Hong Kong and other social activists, the government was persuaded to include the Blue House in the 'Revitalising Historic Buildings Through Partnership Scheme' that aims at balancing conservation and development through adaptive re-use of government-owned historic buildings. In August 2009, the government invited proposals for the revitalisation of the Blue House with the condition that existing residents can continue living there. In September 2010, SJS together with Community Cultural Concern and Heritage Hong Kong won the bid. The building complex will regain its past multi-functional role as a community space for different stakeholders: existing tenants, good neighbours (tenants who are required to contribute to community building), social enterprises and spaces for promoting local community culture such as workshop spaces for learning local arts and crafts, the House of Stories, Community Currency Shop (shop to exchange services or second-hand goods) as well as an open space for local markets and community activities. It will be an oasis for rebuilding the local economy and community through the working together of local residents, SJS, social activists as well as built environment-related professionals in the rapidly gentrifying Wanchai.

The fight for their lived spaces and the quest for differential rural and farming spaces were much more eventful for Choi Yuan Village. The indigenous villagers supporting land expropriation by the government argued that they understood 'the spirit of sacrificing oneself for the greater good of the community' (Lam 2009c). However, when indigenous villagers 'sacrificed', meaning selling their land to the government, their village would be rebuilt, and they would be generously compensated. Yet, for the non-indigenous villagers, when their villages are expropriated for public purposes, they are not entitled to rebuild their homes. At the beginning, Choi Yuan Villagers adopted the 'no move, no demolition' strategy. However, after the funding was approved by the Legislative Council in January 2010, the Village Concern Group announced in late February 2010 that they had decided to put forward a counter-proposal to the government for 'Rebuilding Choi Yuan Village'. Ninety households announced that they would, with the help of professionals, find land to rebuild a model eco-village so that their rural lifestyles and social networks could be continued.

Although the government does not have a policy to rebuild non-indigenous villages after land expropriation, it has a rather tokenistic 'Agricultural Land Rehabilitation Scheme' of allowing the re-instatement of housing and farmland when farmers are affected by infrastructure development. The then chairman of the Rural Council, controlled by indigenous villagers, also helped search for appropriate land for the rebuilding of Choi Yuan Village. Community Cultural Concern, the group that facilitated the Blue House project, and other social activists established the 'Choi Yuan Village Community Building Office' in March 2010 and recruited volunteers from different professions such as: planners, architects, landscape designers, engineers, ecologists and farmers to engage in participatory planning. Eventually a plan for 'New Choi Yuan Eco-village' was formulated with the principles of a car-free environment; saving 40 per cent of land for collective farming; conserving the existing cultural landscape including orchards and fish ponds; and practising water recycling. The fight for their differential lived spaces was not smooth. As of February 2017, the new Village is still not fully developed.

Conclusions: everyday utopianism in face of a dystopian utopia

Lefebvre (1991: 356) argues that 'it is the political use of space... that does the most to reinstate use value'. Politics exist when those who have no right exert their rights to be counted, symbolising 'a return of the repressed' (Žižek 1989: 55). As of February 2017, the Blue House building complex is still under renovation and its utopian vision is threatened as the district continues to gentrify with the growth of bars, high-end health shops and expensive and exclusive shopping malls for the new and affluent residents. As indicated by a local resident-volunteer (SJS 2014), they have succeeded in conserving the building cluster but Wanchai has changed and gentrified – they have fought to stay but the community has changed. After all, the pro-growth ideology has an inherent bias towards privileging the production of spaces for exchange values.

Choi Yuan Villagers' quest for their lived spaces has been an uphill struggle in the face of the hegemonic power of the government and the indigenous villagers. The Agricultural Land Rehabilitation Scheme provides no solid policy or substantive support for infrastructure and utility development. Even though they managed to purchase land from indigenous villagers, neighbouring indigenous villagers refused to grant them road access unless they made extra payment. Because of all these difficulties, when construction costs escalated, the project was delayed, forcing them to compromise on the original ecological design (Ngo 2012). To co-build an eco-village is difficult enough, and the engagement of a less than professional contractor to save costs did not help. As of February 2017, although the New Village has not been fully developed, villagers managed to celebrate the Chinese New Year in their new homes.

The struggles to fight for their lived and differential spaces have continued to unfold. These two amazing dramas of unsung heroes and heroines point to the importance of hope and the courage to dream of utopia in the midst of a dystopian situation. These two stories testify the importance of 'enlightened elites' or 'system-transforming intellectuals' whose 'will to truth' (Lefebvre, 1991: 399) will not only awaken the silent space-users but also inspire them to fight with their beloved communities for their lived spaces.

Acknowledgement

This research is supported by the Research Grants Council of the Hong Kong, Special Administrative Region, China (project numbers: CUHK447713 and CUHK14408314).

References

Bloch, E. (1986/1954) *The Principle of Hope Volume One*, Cambridge MA: MIT Press.
Bloch, E. (2000/1923) *The Spirit of Utopia*, Standard, CA: Stanford University Press.

Brown, R. (2011) 'You can't go home again: the place of tradition in *Firefly*'s dystopian utopia and utopian dystopia', *Traditional Dwellings and Settlements Review*, 22:2 7–18.
Cagape, E.W. (2009) 'Hong Kong tops world rich-poor gap', *Asian Correspondent*. Online. Available HTTP: <https://asiancorrespondent.com/> (accessed 22 February).
Chan, B. (2011) 'Today's young "radicals", tomorrow's visionaries', *South China Morning Post*, 25 February.
Chu, Y.W. and So, A.Y. (2013) 'Can Hong Kong design a new growth engine?' in J. Cheng (ed.) *HKSAR in its First Decade*, Hong Kong: City University Press.
Chu Hoi-dick, E. (2009) 'Choi Yuen village's self-help movement in the past month', *InMedia*, 16 May, Online. Available HTTP: <http://www.inmediahk.net/> (in Chinese) (accessed 22 February 2017).
City-Bound Collective (2012) 'Notes on NEOutopia', *City*, 16:5 595–606.
Coleman, N. (2013) 'Utopian prospect of Henri Lefebvre', *Space and Culture*, 16:3 349–63.
Gardiner, M. (1992) 'Bakhtin's carnival: utopia as critique', *Utopian Studies*, 3:2 21–49.
Gordin, M.D., Tilley, H. and Prakash, G. (eds) (2010) *Utopia/Dystopia: Conditions of Historical Possibility*, New Jersey: Princeton University Press.
Government of Hong Kong (2017) *The 2017 Policy Address*, Hong Kong: Government Printer. Online. Available HTTP: <http://www.policyaddress.gov.hk/> (accessed 22 February).
Gunder, M. and Hillier, J. (2007) 'Planning as urban therapeutic', *Environment and Planning A*, 39 467–86.
Heritage Foundation (2017) *2017 Index of Economic Freedom*. Online. Available HTTP: <http://www.heritage.org/> (accessed 22 February).
HKCSS (2017) *Social indicators of Hong Kong: Gini Coefficient*. Online. Available HTTP: <http://www.socialindicators.org.hk/> (accessed 22 February 2017).
Ip, P.L. (2009a) 'I need no compassionate removal from you!', *InMedia*, 12 May 2009. Online. Available HTTP: <http://www.inmediahk.net/> (in Chinese) (accessed 22 February 2017).
Ip, P.L. (2009b) 'Eighty year old Grandma Ko in Choi Yuen village: "I will not move"', *InMedia*, 18 May. Online. Available HTTP: <http://www.inmediahk.net/> (in Chinese) (accessed 22 February 2017).
Ko, C.H. (2014) Research interview, 2 May, it lasted three hours.
Lai, C. (2006) 'Community at odds over future of historic buildings', *South China Morning Post*, 30 July, Online. Available HTTP: <http://www.scmp.com> (accessed 22 February).
Lai, C. (2010) 'Railroaded into the front line', *South China Morning Post*, 7 March 2010.
Lam, A. (2009a) 'Growing, growing, gone', *South China Morning Post*, 29 July.
Lam, A. (2009c) 'More cash for villagers affected by high-speed rail link', *South China Morning Post*, 10 October, EDT3.
Lam, K.W. (2009b) 'Community development of St. James Settlement: retrospect and prospect', in Y.M. Yu (ed.) *Starting from the Stone Nullah Lane: 20 Years of Community Services by the St. James Settlement*, Hong Kong: Masterpress Ltd (in Chinese).
Lam, W.M. (2004) *Understanding the Political Culture of Hong Kong: The Paradox of Activism and Depoliticization*, Armonk, New York: M.E. Sharpe.
Lee, C.M. (2009) 'Whose heritage? A study of a conservation movement in Hong Kong urban building complex', Unpublished MPhil, The Chinese University of Hong Kong, Hong Kong.
Lefebvre, H. (1991/1974) *The Production of Space*, Oxford: Blackwell Publishing.
Lefebvre, H. (1996/1968) 'Right to city', in E. Kofman and E. Lebas (eds) *Writings on Cities*, Oxford: Blackwell.
Leung, S.Y. (2007) 'Local Foundation willing to sponsor the conservation of the Blue House', *Hong Kong Daily News*, 8 October.
MacLeod, G. and Ward, K. (2002) 'Spaces of utopia and dystopia: landscaping the contemporary city', *Geografiska Annaler. Series B, Human Geography*, 84:3/4 153–70.
Madden, D.J. (2012) 'City becoming world: Nancy, Lefebvre, and the global-urban imagination', *Environment and Planning D: Society and Space*, 30 772–87.
Martin, D.G. (2003) '"Place-framing" as place-making: constituting a neighbourhood for organizing and activism', *Annals of the Association of American Geographers*, 93:3 730–50.
MTRC (2017) *Express Rail Link*. Online. Available HTTP: <http://www.expressraillink.hk/> (accessed 22 February 2017).
Mumford, L. (1965) 'Utopia, the city and the machine', *Daedalus*, 94:2 271–92.
Ngo, J. (2012) 'New town going down the tubes', *South China Morning Post*, 20 January 2012, CITY4.
Ng, M.K. (2013) 'Got the controversial urban planning job done? An institutional perspective', in J. Cheng (ed.) *Evaluating the Tsang Years 2005–2012*, Hong Kong: City University of Hong Kong Press.
Ng, M.K. (2014) 'Intellectuals and the production of space in the urban renewal process in Hong Kong and Taipei', *Planning Theory and Practice*, 15:1 77–92.

Ng, M.K. (2016) 'From Xinhai Revolution (1911) to the Umbrella Movement (2014): insurgent citizenship, radical planning and Chinese culture in the Hong Kong SAR', in H. Rangan, M.K. Ng, J. Chase and L. Porter (2016) *Insurgencies and Revolutions: Reflections on John Friedmann's Contributions to Planning Theory and Practice*, New York: Routledge.
Pinder, D. (2002) 'In defence of utopian urbanism: imagining cities after the "end of utopia"', *Geografiska Annaler: Series B, Human Geography*, 84(3–4) 229–41.
Pinder, D. (2010) 'The breath of the possible: everyday utopianism and the street in modernist urbanism', in M.D. Gordin, H. Tilley and G. Prakash (eds) (2010) *Utopia/Dystopia: Conditions of Historical Possibility*, New Jersey: Princeton University Press.
Royce, J. (2001/1913) *The Problem of Christianity*, Washington, DC: Catholic University of America Press.
Sargent, L.T. (2006) 'In defense of utopia', *Diogenes* 209 11–7.
SCMP (2015) *Editorial*, 21 October. Online. Available HTTP: <http://www.scmp.com/> (accessed 22 February 2017).
SJS (2007) 'New thoughts on community conservation: starting from "Blue House", proceedings of the 6th Conference of the Pacific Rim Community Design Network, Quanzhou, Fujian, China, 18–21 June 2007. Online. Available HTTP: <http://courses.washington.edu/> (accessed 22 February 2017).
SJS (2014) Research interview, 30 May, it lasted about two hours.
Sliwinski, A. (2016) 'The value of promising spaces: hope and everyday utopia in a Salvadoran Town', *History and Anthropology*, 27:4 430–46.
Stillman, P.G. (2003) 'Dystopian critiques, utopian possibilities, and human purposes in Octavia Butler's Parables', *Utopian Studies*, 14:1 15–35.
Wilson, J. and Bayon, M. (2016) 'Black hole capitalism: utopian dimensions of planetary urbanization', *City*, 20:3 350–67.
Wong, A. (2006) *Wanchai Street Paper: Final Report of the Urban Renewal Strategy Review*, Hong Kong: SEE Network (in Chinese).
Žižek, S. (1989) *The Sublime Object of Ideology*, London: Verso.

46

'Something more, something better, something else, is needed'

A renewed 'fête' on London's South Bank

Alasdair J.H. Jones

Introduction

> there is today... no theory without utopia. The architects, like the planners, know this perfectly well.
>
> *(Lefebvre 2009: 179)*

Although Lefebvre's writing on rural sociology has been the subject of recent revisits (e.g. Elden and Morton 2015), his work has left an indelible mark on urban sociology and in particular on scholars interested in the nexus of urban design, architecture and social theory (especially Stanek 2011; Stanek et al. 2014). Within this body of work, a number of ideas developed in Lefebvre's writings have been increasingly influential in scholarship concerned with the urban public realm, in particular, the material focus of this chapter. Of the many conceptual approaches articulated by Lefebvre, three feature most prominently in the urban space literature.

First, Lefebvre's (1991) spatial triad or trialectic has been theoretically advocated (see especially Harvey 1990; Soja 1996) and empirically deployed as a means to explore processes of urban spatial (re-)production (Low 2017: 40–42). For instance, describing the approach taken in her comparative study of the sensuous dimensions of public space in Barcelona and Manchester, Degen (2008: 10) states how:

> In order to link transformations of space with an analysis of sensuous experience this study follows the tradition of thought initiated by Lefebvre (1991), who began to look beyond space as a 'container' for social action and instead started to interpret space as a product and producer of multiple forms of spatial practice.

Such foundational invocations of Lefebvre's work on the dialectical relationship between urban morphology and social relations are routine in scholarship on the production of urban public space (e.g. Low 2000; Jones 2014; Moravánsky 2014; Leary-Ohwin 2016), and in particular on resistance as a constituent element of spatial (re-)production in such settings (e.g. McCann 1999; Dhaliwal 2012).

Linked to this interest in how urban public space *is produced*, and in particular how it is produced through practice, various scholars have turned to rhythmanalysis (Lefebvre 2004) as a means to explore the productive qualities of everyday life in cities (e.g. Borden 2001; Williamson 2016). Third, Lefebvre's (1996) writings on the right to the city are at the heart of another body of work, in particular in social geography, concerned with arguments about public space and social justice (especially Mitchell 2003; also Low and Smith 2006; Butler 2013). In addition to these deployments of some of Lefebvre's theoretical work, Lefebvrian concepts (including those developed in *The Production of Space*) have also featured routinely in scholarship concerned with users or 'practitioners' of public space, including, but not limited to, skateboarders (Borden 2001; and see the chapter in this volume), immigrants (Ugolotti and Moyer 2016), traceurs (Ameel and Tani 2012; Kidder 2012; Daskalaki, Stara and Imas 2008), cyclists (Spinney 2010), homeless people (Speer 2016), 'street-involved youth' (Kennelly and Watt 2011) and walkers (Fenton 2005; Williamson 2016).

Reviewing the social scientific urban public space literature, then, it is hard to find works that fail to draw heavily on Lefebvre's scholarship, let alone merely reference it. What is striking about the deployment of Lefebvre's work in this literature, however, is the routine use of his conceptual treatises as the basis of critiques of the urban condition (of the pernicious effects of top-down urban visions (de Certeau 1984) and of the dilution of rights to the city in particular). In practical terms, there has been an emphasis in considerations of Lefebvre's work vis-à-vis public space on problematising contemporary urban (public) space production and on situating tactical uses of space at the interstices of an increasingly unforgiving and revanchist urban fabric (especially Mitchell 2003). In this respect, Lefebvre's work is used not only to reveal the multifaceted nature of processes of spatial production but also to theoretically account for struggle for radically different city forms and the social and spatial orders that structure them (e.g. Brenner et al. 2012; Harvey 2012).

But what of the more optimistic, utopian strands of Lefebvre's writings on cities and urban space? As Stanek et al. (2014: 3) put it, 'Lefebvre not only critically analysed the phenomenon of urbanization and its implications, but at the same time explored and revealed its potentials'. Lefebvre's own interest in the transgressive potential of urban space – and in particular in the importance of 'play, festival and creativity in the remaking of public space' (Butler 2013: 137) – has been relatively underplayed in the subsequent literature (but see Pinder 2015). In the following discussion I draw on the findings of my own work on the (re-)production of public space on London's South Bank (Jones 2013; 2014) to sketch a more optimistic reading of the production of public space inflected by the theoretical work of Lefebvre. In doing so I hope to encourage other analyses of *realised* urban form that have an eye for the utopian as well as the dystopian.

Fête-less space: be wary of Disneyfication

For recent accounts of urban public space, its transformations and practices therein, Michael Sorkin's (1992) edited volume *Variations on a Theme Park: The New American City and the end of Public Space* is a highly influential work. This volume arguably prefigured a range of more and less empirical engagements with urbanisation processes linked to the titular 'end of public space'. Among other things, authors have written about the role of: revanchism (Rogers and Coaffee 2005), securitisation (Atkinson 2003; Raco 2003; Dixon et al. 2006; Ruppert 2006; Herbert and Beckett 2009; Minton 2009; Blomley 2010), fortification (Davis 1992), commodification (Loukaitou-Sideris and Banerjee 1998), urban entrepreneurialism (Boyle and Hughes 1994) and gentrification (Paton 2014) in the demise of urban public space.

Most important for this chapter, however, is the *association* Sorkin makes between the (re)production of urban spaces that have been increasingly endowed with the aesthetic and regulatory characteristics of 'theme park' design and the end of public space. This association, described elsewhere as the 'Disneyfication' of public space (Zukin 1995; Bridge and Watson 2000; Webb 2014: 193–205), has arguably underpinned many critical accounts of the contemporary urban public realm (see Carmona 2010), such that planning efforts to enliven and regenerate urban public space are interpreted inversely; as signalling the loss and effacement of fragments of public realm that may provide room for the 'pleasure principle' and *jouissance* (Lefebvre 2003: 32, 85).

It is towards this paradox that this chapter is oriented, and that Lefebvre's (1996) discussions of the role of the fête are deployed. It is *La fête*, 'a celebration which consumes unproductively' that must be revitalised (Lefebvre 1996: 66). Via recourse to my fieldwork on London's South Bank I want to make the case for more optimistic readings of urban interventions that could be characterised as displaying theme park tropes. Here I am interested not only in the 'emancipatory possibilities' (Lees 2004) of urban public space, but also material *realisations* of urban public space that embody the qualities of Lefebvre's notion of the fête and that play with, rather than foreclose, 'the Dionysian dimensions of life' (Tonkiss 2005: 136). Against a social scientific literature in which the role utopia can 'play in contemporary critical urban studies... has often been treated warily, side-lined or dismissed' (Pinder 2015: 28), this article seeks to make an empirical contribution to a burgeoning interest in exploring 'urban worlds that are different and better' (Ibid.).

Extending the possible

In his recent critical engagement with Lefebvre's work on utopia, David Pinder (2015: 28) 'explores Lefebvre's emphasis on the possible, and in particular the importance he attached to extending and realising the possible through struggling for what seems impossible' (see Lefebvre 1976: 36). While Lefebvre's work has been used to make claims about the possibilities of and for urban life (Harvey 2000; Lees 2004) and public space (Stevens 2007), and for how we might more broadly reimagine the urban (Amin and Thrift 2002), it is worth noting that Lefebvre himself was interested in the *realisation* of the possible; in moving beyond the rhetoric of imminence that characterises much work on (potentially) virtuous qualities of the urban condition. This interest of Lefebvre's speaks to a crucial distinction made a number of times in his work between 'utopist' and 'utopian'. As Pinder (2015: 32) puts it:

> In contrast to abstract and transcendental ideal plans for living associated with the former, he [Lefebvre] favoured more 'concrete' explorations of what was possible that were rooted in everyday life and space.

In this chapter, I will use my ethnographic work on the ongoing transformation of London's South Bank to explore an example of the (at least partial) realisation of a *utopian* urban vision. In the analysis that follows I will draw, in particular, on Lefebvre's (1996) work on the fête in relation to the urban, using his conceptual musings on this topic to ground my analysis of the transformation of the public realm of the Southbank Centre and of how this transformation was envisioned and described by those responsible for it.

The 'possible-impossible' on the South Bank

The Southbank Centre is Europe's largest arts complex. Comprising a number of arts institutions (including the Royal Festival Hall, the Queen Elizabeth Hall and the Hayward Gallery) as well as substantial areas of constituent public realm, the Centre occupies a 27-acre estate at the heart of the wider 'South Bank' district on the south embankment of the River Thames in central London.

This estate, which the Southbank Centre (as a charitable organisation) manages 'on a long lease from the Arts Council, who hold the freehold on behalf of the government' (House of Commons Select Committee on Culture, Media and Sport 2002), has been the subject of a sequence of architectural and urban design 'masterplans' that have envisioned its redevelopment. The ongoing 'transformation' of the Southbank Centre is guided by one of these (Rick Mather Architects 2017).

However, the first such visioning exercise is of particular importance, namely Cedric Price's (1983) proposals for renovating the site. Cedric Price has been dubbed 'the most influential architect you've never heard of' (Milmo 2014), well known for radical architectural proposals that were not built at the time of their design but have subsequently been constructed (Hardingham 1983). Among these proposals that Price is famous for not building is a giant Ferris wheel by the Thames, which was a central component of his architectural vision for the Southbank Centre and is a structure that has since been realised, the 'London Eye'.

In a very real way, urban design interventions dismissed as fanciful and impossible at the time of their conception have, and are being, realised on the South Bank – radical architectural motifs in the work not only of Cedric Price, but also the avant-garde architectural group Archigram. Notably, this modernist and avant-garde architectural group contemporaneous to Lefebvre developed projects that were very much of the kind to which he was drawn (i.e. those working across theory and practice. In fact, there are clear links between Lefebvre, Cedric Price and Archigram. For instance, Jean-Paul Jungman (the 1967–70 editor of the journal of the French architectural group Utopie – a group that included Lefebvre's former assistants Jean Baudrillard and Hubert Tonka [Pinder 2015: 41]) – recalls in an interview how all three men attended a conference by Utopie (Buckley 2010: 350).

Among the members of Archigram were Ron Herron and Warren Chalk, who were part of a team of architects at London County Council responsible for the design of the Queen Elizabeth Hall and the Hayward Gallery complex of buildings on the South Bank (Borden 2014). Significantly, some of the ways that this complex was envisioned as being used have recently been reinvigorated. This includes the addition of 'clip-on' design elements that were a central design concept for the group (e.g. 'A Room for London' [David Kohn Architects] – a boat-themed architectural installation constructed on top of the Queen Elizabeth Hall) and the creation of a rooftop garden on the Queen Elizabeth Hall [Figure 46.1] that, after decades of delay, starts to take seriously Archigram's utopian vision for the site (Kelly 2013).

Reversing a longstanding policy to preclude the use of some of the raised terraces characteristic of the buildings of the South Bank (Figure 46.1) and so to effectively abandon these spaces, this resuscitation of much of the local public realm has been achieved primarily through the leadership of the Southbank Centre's committed pursuit of a transformation of the area premised on the idea of 'melting the walls' of the Centre's constituent institutions and bringing the arts outside (Jones 2014: 215).

Notably, another architectural feature of the Hayward Gallery/Queen Elizabeth Hall complex of buildings – namely the undercroft of the Queen Elizabeth Hall that has since the 1970s been appropriated as a skate spot (and, latterly, as a graffiti canvas as well) – has recently been preserved precisely through the popular 'struggle' to which Lefebvre (1976: 36) refers. A *de facto* skate park has been *produced*, through campaigns organised by the 'Long Live South Bank' group (Blayney et al. 2014), in a central London setting where critiques of the redevelopment of public space would have us expect commodified spaces characterised by revanchist management policies designed to exclude non-consumers, minority groups and un-aesthetic public space users (Rogers and Coaffee 2005: 321–2); policies that have been argued to affect young people (given their relative lack of wealth and often noisy presence) in particular (e.g. Valentine 1996). Against such representations, through institutional will, and popular struggle (and an at times highly

Figure 46.1 A rooftop garden on the upper terraces of the Queen Elizabeth Hall (photo credit: Alasdair Jones, 2013).

antagonistic relationship between these forces), 'something else' (Lefebvre 1996: 173) has been realised on the South Bank.

Notably, this utopian fragment, or spatiotemporal 'moment' in the language of Lefebvre (Pinder 2015: 36), is not a purpose-built 'skateplaza' (Vivoni 2009) but rather the result of a characteristically dialectical spatial production process. When they designed the material space now understood to be the South Bank skate park (by users and passers-by alike), the architects 'really – and quite deliberately – had no exact idea about... who would use the Undercroft and in what ways' (Borden 2014: 67). The space was subsequently appropriated, and often physically adapted, by skateboarders and finally 'preserved' (and effectively produced) as skate park through ultimately successful social activism that culminated in a decision by the Southbank Centre to shelve their plans to redevelop the space (Blayney et al. 2014).

A renewed 'fête': putting art at the service of the urban

However, the Queen Elizabeth Hall undercroft does not demarcate the extent of a Lefebvre-inspired utopian reading of the South Bank. Rather, the inscription of a definitively playful use value in that space can be seen to reflect a wider disposition towards play, spectacle and encounter that underpins the production and performance of public space around the South Bank (Jones 2013; Spinney 2010). At the heart of this is an institutional desire to revisit and revive some of the tenets, including what might be understood as more utopian ideas, that (whether or not they were brought into being) informed the various planning proposals that have been drawn up for the Southbank Centre since the site's germination as part of the 1951 Festival of Britain (Mullins 2007).

Thus, as well as actively pursuing Archigram's ideas about how the Hayward Gallery/Queen Elizabeth Hall complex might be used (Kelly 2013), the transformation of the site has also been influenced by the ideals of the earlier Festival of Britain. This was described as a gigantic toyshop for adults in *Brief City*, a short 1952 film about the Festival of Britain by Jacques Brunius and

Maurice Harvey. As a senior Southbank Centre executive responsible for shaping the overarching vision of the Centre put it to me in interview:

> the most profound influence on me about the site was its original purpose... this phrase, that they [the Festival of Britain organizers] used, 'landscape of the imagination', it seems to me to be a unique heritage.
>
> *(2013)*

As part of this re-visioning, a key implication for urban public realm in-and-around the Southbank Centre is that it would no longer be treated, or left, as merely the space between buildings. Rather, it was better understood as an extension of the internal space of the constituent arts institutions of the Centre. Just as Archigram had intended (albeit in an ideologically loose way), the external surfaces of the Southbank Centre buildings, and the public realm in-between, would be reconceptualised as canvas or stage. This process – of revisiting former creative visions in the appropriation of space, and of rejecting the sanctity of built architectural form – strongly resonates with Lefebvre's writing. As he puts it:

> To put art at the service of the urban does not mean to prettify urban space with works of art... Rather, this means that time-spaces become works of art and that former art reconsiders itself as source and model of *appropriation* of space and time... Coming back to style and to the *oeuvre*, that is, to the meaning of the monument and the space appropriated in the *fête*, art can create 'structures of enchantment'. Architecture taken separately and on its own, could neither restrict nor create possibilities. Something more, something better, something else, is needed.
>
> *(Lefebvre 1996: 173 emphasis in original)*

Lefebvre's invocation of 'something else' as the basis for 'space appropriated in the fête' can be read into both the cause and effect of the utopian features of space being realised on the South Bank. Thus, not only do the spaces realised there have heterotopian qualities – qualities that enable these spaces to 'stand outside a conventional order of space' (Tonkiss 2005: 131) – but the conditions of their emergence are also characteristically 'other'. Rather than being driven by commercial considerations, a 'creative vision' for the Southbank Centre (albeit one that includes commercial interests) underpins its transformation (Jones 2014: 232–233); a vision that revisits the heritage of the site embodied in the profoundly utopian and social democratic remit of the 'Festival of Britain' (Mullins 2007).

Lefebvre's interest in the fête is a relatively underexplored feature of his work on the 'right to the city' (Lefebvre 1996). While, as Dijkstra (2000: 7) laments, 'Lefebvre did not suggest a way of operationalising that right [to the city] or, for that matter, even a way of measuring a lack of it', he does provide clues as to how it might be conceived. Among these, of particular importance to this chapter is the 'playful' character of 'the *urban*, which survives in the fissures of planned and programmed order' (Lefebvre 1996: 129). The importance of play to Lefebvre is clear at the start of his essay *The Right to the City*, where he posits that the 'commercial and cultural infrastructures' with which planners are typically concerned do not satisfy 'the need for creative activity, for the *oeuvre*..., symbolism, the imaginary and play' (Lefebvre 1996: 147 [emphasis in original]). Lefebvre (1996: 147) goes on:

> Through these specified needs lives and survives a fundamental desire of which play... sport, creative activity, art and knowledge are particular expressions and *moments*, which can more or less overcome the fragmentary division of tasks.
>
> *[emphasis in original]*

In Lefebvre's reading, play is not superfluous, but rather expressive of a decidedly human need (Jones 2013). Furthermore, playful 'moments' allow urban dwellers to transcend the rational fragmentation of the city and rediscover the urban. As Lefebvre (1996: 158) puts it, 'the *right to the city*... can only be formulated as a transformed and renewed *right to urban life*' (emphasis in original). Play is of 'supreme value' in this formulation for Lefebvre (1996: 172), and is identified as the source of the contents of the principle of assembly or urban public life. As he puts it:

> should [old places of assembly] find a meaning again [this] does not preclude the creation of places appropriate to a renewed *fête* fundamentally linked to play.
>
> *(Lefebvre 1996: 171)*

In Lefebvre's conceptual work, therefore, the realisation of meaning in the urban should take seriously the renewal of the fête and the contribution of play.

As well as places of assembly, play can be seen as constitutive of social spaces for Lefebvre. These are spaces he describes as follows:

> social spaces are related to social times and rhythms which are prioritized... To *inhabit* finds again its place over habitat. The quality which is promoted presents and represents as *playful*.... Already, to city people the urban centre is movement, the unpredictable, the possible and encounters. For them, it is either 'spontaneous theatre' or nothing.
>
> *(Lefebvre 1996: 172 emphasis in original)*

As a productive force, Lefebvre argues that play has been marginalised in urban development. It has only survived, as he puts it, 'in the holes of a serious society which perceives itself as structured and systematical and which claims to be technical' (Lefebvre 1996: 171). So, how to recover play in the production of space? Lefebvre's work, while typically not conclusive on this count, points to a few principles for action.

Among these, Lefebvre (1996: 166) makes a purposive distinction between strategic and tactical variables when it comes to relating urban research 'to the concrete of urban drama', the former comprising 'the transformation of everyday life', the latter being weak interventions in society. As an example of the former, Lefebvre (1996: 166) proposes the 'constitution of a very simple apparatus of social pedagogy' – the purposive provision of spaces *for* youth – as a transformative gesture in cities.

Elsewhere, and with specific reference to his discussion of '*fête*', Lefebvre (1996: 171) iterates that '[t]he proposition of this project is to gather together *by subordinating to play* rather than to subordinate play to the "seriousness" of culturalism and scientificism' [emphasis added]. Here, Lefebvre (1996: 173) draws our attention to the 'structures of enchantment' that meaningfully 'putting art at the service of the urban' can achieve. This notion of enchantment can be seen to characterise not only architectural interventions at the Southbank Centre [Figure 46.1], but also ways that being at the Southbank Centre was experienced by visitors (as recorded in my fieldnotes [Jones 2014: 250–1]).

Finally, and of methodological interest, Lefebvre also makes the case for 'an experimental approach towards utopia' (Pinder 2015: 37). Reflecting his concerted, but often overlooked, engagement with a range of architects, planners and other urban design professionals (Pinder 2015: 37), here Lefebvre was interested in possibilities for evaluating the implications and consequences of utopia on the ground. In very practical terms, as Pinder (2015: 37) argues, this included 'considering the criteria by which places are judged "successful"', and 'the times and rhythms of daily life which are inscribed and prescribed in these "successful" spaces favourable to happiness' (Lefebvre 1996: 151).

Conclusions

This chapter has explored, via recourse to ethnographic fieldwork on the transformation of London's Southbank Centre, the tension between pejorative accounts of the 'Disneyfication' of the urban public realm and Lefebvre's (1996) assertion in *The Right to the City* that *la fête* must be revitalised. At the Southbank Centre, with its functional roots in the Festival of Britain and architectural heritage in modernist utopianism, the case can be made that Lefebvre's demand is, at least partially, being met. Beneath the surficial prettification of space with works of art, a more meaningful desire to allow former art to be recognised 'as source and model of appropriation of space and time' (Lefebvre 1996: 173) is evident.

Importantly, Lefebvre's 'everyday utopianism' (Gardiner 2013) needs to be understood as a distinctive conceptualisation because 'it did not involve prescribing an already fully formed ideal as the term utopia is often construed and, indeed, derided' (Pinder 2015: 32). Rather, it allowed for an engagement with the real and the possible – a position no doubt informed by Lefebvre's active engagement with urban design and planning practice (and practitioners). Intriguingly, Lefebvre's own emphasis on such material engagement is often sorely missing in critical urban studies deployments of his work that operate at a more abstract level. This is certainly the case in discussions of urban public space, and in this chapter I have therefore thought through Lefebvre's discussion of a renewed fête, in particular, via recourse to an in-depth empirical study of a particular set of urban public spaces.

The optimistic edge to Lefebvre's work, and his willingness to consider the value of transgressive spatiotemporal moments in urban development, invites us to think again critically about urban spaces characterised by tropes of theme park design. In its current morphology, much of the public realm of the Southbank Centre can be characterised in this way, as those responsible for the site seek to revive the ideological impulses of designs on the space since the Festival of Britain. Through Lefebvre, however, rather than uncritically characterise such spaces as Disneyfied, we can not only seek out and celebrate moments of unproductive consumption – of fête – among them, but also think through the implications of such space for urban practice. At the South Bank, I would argue, surrounded by hard, worn concrete, the 'possible-impossible', and the utopian possibilities of putting art at the service of the urban, can be glimpsed.

References

Ameel, L. and Tani, S. (2012) 'Parkour: creating loose spaces?', *Geografiska Annaler: Series B, Human Geography*, 94:1 17–30.
Amin, A. and Thrift, N. (2002) *Cities: Reimagining the Urban*, Cambridge: Polity.
Atkinson, R. (2003) 'Domestication by Cappuccino or a revenge on urban space? Control and empowerment in the management of public spaces', *Urban Studies*, 40:9 1829–43.
Blayney, S., et al. (2014) *Long Live Southbank*, London: Heni Publishing.
Blomley, N. (2010) 'The right to pass freely: circulation, begging, and the bounded self', *Social & Legal Studies*, 19 331–50.
Borden, I. (2001) *Skateboarding, Space and the City: Architecture and the Body*, Oxford: Berg.
Borden, I. (2014) 'The architecture', in S. Blayney, et al. (eds) (2014) *Long Live Southbank*, London: Heni Publishing.
Boyle, M. and Hughes, G. (1994) 'The politics of urban entrepreneurialism in Glasgow', *Geoforum*, 25:4 453–70.

Brenner, N., Mayer, M. and Marcuse, P. (eds) (2012) *Cities for People, Not for Profit: Critical Urban Theory and the Right to the City*, Abingdon: Routledge.
Bridge, G. and Watson, S. (2000) 'City publics', in G. Bridge and S. Watson (eds) *A Companion to the City*, Oxford: Blackwell.
Buckley, C. (2010) 'Interview with Jean-Paul Jungman', in B. Colomina and C. Buckley (eds) *Clip, Stamp, Fold: The Radical Architecture of Little Magazines 196X to 197X*, New York: Actar.
Butler, C. (2013) *Henri Lefebvre: Spatial Politics, Everyday Life and the Right to the City*, Abingdon: Routledge.
Carmona, M. (2010) 'Contemporary public space: critique and classification, part one: critique', *Journal of Urban Design*, 15:1 123–48.
Daskalaki, M., Stara, A. and Imas, M. (2008) 'The "Parkour Organisation": inhabitation of corporate spaces', *Culture and Organization*, 14:1 49–64.
Davis, M. (1992) *City of Quartz: Excavating the Future in Los Angeles*, New York: Vintage.
de Certeau, M. (1984) *The Practice of Everyday Life*, Berkeley, CA: University of California Press.
Degen, M. (2008) *Sensing Cities: Regenerating Public Life in Barcelona and Manchester*, Abingdon: Routledge.
Dhaliwal, P. (2012) 'Public squares and resistance: the politics of space and the Indignados movement', *Interface*, 4:1 251–73.
Dijkstra, L. (2000) 'Public spaces: a comparative discussion of the criteria for public space', in R. Hutchison (ed.) *Research in Urban Sociology Volume 5: Constructions of Urban Space*, Greenwich, CT: JAI Press.
Dixon, J., Levine, M. and McAuley, R. (2006) 'Locating impropriety: street drinking, moral order, and the ideological dilemma of public space', *Political Psychology*, 27:2 187–206.
Elden, S. and Morton, A.D. (2015) 'Thinking past Henri Lefebvre: introducing "The theory of ground rent and rural sociology"', *Antipode*, 48:1 57–66.
Fenton, J. (2005) 'Space, chance, time: walking backwards through the hours on the left and right banks of Paris', *Cultural Geographies*, 12:4 412–28.
Gardiner, M. (2013) *Weak Messianism: Essays on Utopia and Everyday Life*, Oxford: Peter Lang.
Hardingham, S. (ed.) (1983) *Cedric Price: Opera Architectural Monographs*, Chichester: Wiley-Academy.
Harvey, D. (1990) *The Condition of Postmodernity*, Oxford: Blackwell.
Harvey, D. (2000) *Spaces of Hope*, Edinburgh: Edinburgh University Press.
Harvey, D. (2012) *Rebel Cities: From the Right to the City to the Urban Revolution*, London: Verso.
Herbert, S. and Beckett, K. (2009) 'Zoning out disorder: assessing contemporary practices of urban social control', *Studies in Law, Politics, and Society*, 47 1–25.
House of Commons Select Committee on Culture, Media and Sport (2002) *Third Report: Arts Development. Memorandum Submitted by the Southbank Centre*, London: HMSO.
Jones, A.J.H. (2013) 'A tripartite conceptualisation of urban public space as a site for play: evidence from South Bank, London', *Urban Geography*, 34:8 1144–70.
Jones, A.J.H. (2014) *On South Bank: The Production of Public Space*, Abingdon: Routledge.
Kelly, J. (2013) 'Why the original dream of the South Bank is still worth striving for', *The Guardian*, 11 December.
Kennelly, J. and Watt, P. (2011) 'Sanitizing public space in Olympic host cities: the spatial experiences of marginalized youth in 2010 Vancouver and 2012 London', *Sociology*, 45:5 765–81.
Kidder, J.L. (2012) 'Parkour, the affective appropriation of urban space, and the real/virtual dialectic', *City & Community*, 11:3 229–53.
Leary-Owhin, M. (2016) *Exploring the Production of Urban Space: Differential Space in Three Post-Industrial Cities*, Bristol: Policy Press.
Lees, L. (2004) *The Emancipatory City? Paradoxes and Possibilities*, London: Sage.
Lefebvre, H. (1976/1973) *The Survival of Capitalism*, London: Allison and Busby.
Lefebvre, H. (1991/1974) *The Production of Space*, Oxford: Blackwell.
Lefebvre, H. (1996/1968) 'Right to the city', in E. Kofman and E. Lebas (eds) (1996) *Writings on Cities*, Cambridge, MA: Blackwell Publishers.
Lefebvre, H. (2003/1970) *The Urban Revolution*, Minneapolis, MN: University of Minnesota Press.
Lefebvre, H. (2004/1992) *Rhythmanalysis: Space, Time and Everyday Life*, London: Continuum.
Lefebvre, H. (2009/1970) 'Reflections on the politics of space', in N. Brenner and S. Elden (eds) *State, Space, World: Selected Essays*, Minneapolis, MN: University of Minnesota Press.
Loukaitou-Sideris, A. and Banerjee, T. (1998) *Urban Design Downtown: Poetics and Politics of Form*, Berkeley, CA: University of California Press.
Low, S. (2000) *On the Plaza: The Politics of Public Space and Culture*, Austin, TX: University of Texas Press.
Low, S. (2017) *Spatializing Culture: The Ethnography of Space and Place*, Abingdon: Routledge.

Low, S. and Smith, N. (eds) (2006) *The Politics of Public Space*, Abingdon: Routledge.
McCann, E.J. (1999) 'Race, protest, and public space: contextualizing Lefebvre in the U.S. city', *Antipode*, 31:2 163–84.
Milmo, C. (2014) 'Cedric Price: the most influential architect you've never heard of', *The Independent*, 10 November.
Minton, A. (2009) *Ground Control: Fear and Happiness in the Twenty-First-Century City*, London: Penguin.
Mitchell, D. (2003) *The Right to the City: Social Justice and the Fight for Public Space*, New York: The Guildford Press.
Moravánsky, A. (2014) 'The space of the square: a Lefebvrean archaeology of Budapest', in L. Stanek, C. Schmid and A. Moravánsky (eds) *Urban Revolution Now: Henri Lefebvre in Social Research and Architecture*, Abingdon: Routledge.
Mullins, C. (2007) *A Festival on the River: The Story of the Southbank Centre*, London: Penguin.
Paton, K. (2014) *Gentrification: A Working-Class Perspective*, Abingdon: Routledge.
Pinder, D. (2015) 'Reconstituting the possible: Lefebvre, utopia and the urban question', *International Journal of Urban and Regional Research*, 39:1 28–45.
Raco, M. (2003) 'Remaking place and securitising space: urban regeneration and the strategies, tactics and practices of policing in the UK', *Urban Studies*, 40:9 1869–87.
Rick Mather Architects (2017) *Rick Mather Architects: Southbank Centre Masterplan*. Online (accessed: 21 March 2017).
Rogers, P. and Coaffee, J. (2005) 'Moral panics and urban renaissance: policy, tactics and youth in public space', *City*, 9:3 321–40.
Ruppert, E.S. (2006) 'Rights to public space: regulatory reconfigurations of liberty', *Urban Geography*, 27:3 271–92.
Soja, E.W. (1996) *Thirdspace: Journeys to Los Angeles and Other Real-and-Imagined Places*, Oxford: Blackwell.
Sorkin, M. (ed.) (1992) *Variations on a Theme Park: The New American City and the End of Public Space*, New York: Hill & Wang.
Speer, J. (2016) '"It's not like your home": homeless encampments, housing projects, and the struggle over domestic space', *Antipode*, 49:2 517–35.
Spinney, J. (2010) 'Performing resistance? Re-reading practices of urban cycling on London's South Bank', *Environment and Planning A*, 42:12 2914–37.
Stanek, Ł. (2011) *Henri Lefebvre on Space: Architecture, Urban Research, and the Production of Theory*, Minneapolis, MN: University of Minneapolis Press.
Stanek, Ł., Schmid, C. and Moravánsky, A. (eds) (2014) *Urban Revolution Now: Henri Lefebvre in Social Research and Architecture*, Abingdon: Routledge.
Stevens, Q. (2007) *The Ludic City: Exploring the Potential of Public Spaces*, Abingdon: Routledge.
Tonkiss, F. (2005) *Space, the City and Social Theory*, Cambridge: Polity.
Ugolotti, N. and Moyer, E. M. (2016) '"If I can find a way to climb a wall of ten meters": migrant youth using Capoeira and Parkour to negotiate integration and social change in Italian public spaces', *Patterns of Prejudice*, 50:2 188–206.
Valentine, G. (1996) 'Children should be seen and not heard: the production and transgression of adult's public space', *Urban Geography*, 17:3 205–20.
Vivoni, F. (2009) 'Spots of spatial desire: skateparks, skateplazas, and urban politics', *Journal of Sport and Social Issues*, 33:2 130–49.
Webb, P. (2014) *Homeless Lives in American Cities: Interrogating Myth and Locating Community*, London: Palgrave Macmillan.
Williamson, K.A. (2016) 'From shipping to shopping: Providence's Capital Center, nervous landscapes, and a phenomenological analysis of walkability', in E. Brown and T. Shortell (eds) *Walking in Cities: Quotidian Mobility as Urban Theory, Method, and Practice*, Philadelphia, PA: Temple University Press.
Zukin, S. (1995) *The Cultures of Cities*, Oxford: Blackwell.

47

The right to the city

Centre or periphery?

Nathaniel Coleman

Introduction

> The right to the city implies nothing less than a revolutionary concept of citizenship.
>
> *(Lefebvre 2014: 205)*

> 'The city may be dead', Lefebvre seems to say, but 'long live the city!' So is pursuit of the right to the city the pursuit of a chimera? In purely physical terms this is certainly so. But political struggles are animated by visions as much as by practicalities.
>
> *(Harvey 2012: xvi)*

In this chapter, one of Lefebvre's best-known concepts, the right to the city, is considered to determine whether it always presumes the right to the centre (Lefebvre 1991; 1996). Although Lefebvre may have construed the right to the city as extending primarily to the centre, during the 1960s and 70s, city centres have diminished in significance as they became the domain of smaller segments of urban populations. Consider London, New York and Rome, for example. Nevertheless, identifying the right to the city with a right to the centre spatialises the concept in a concrete manner by identifying it with a particular sort of place generally recognisable.

'Centre' refers primarily to the city centre, to the core (or heart) of any city; that part of it traditionally most intensely lived; alluding also to older quarters of cities. The benefit of identifying the right to the city with any given city's historic, symbolic, and cultural core is that it rescues the concept from slipping into abstraction as a generalised 'natural', 'universal', or 'inalienable right', which, under conditions of alienation quickly lose their force, easily becoming guilt offerings to disenfranchised populations, or banal marketing slogans attached to already commodified cities.

As David Harvey has observed, 'The right to the city is far more than the individual liberty to access urban resources: it is a right to change ourselves by changing the city' (Harvey 2008: 23). Crucially, the right to the city is a right of transformation, of self and city. But how might we actually practice such a right? Does Lefebvre offer any clues? Joseph Rykwert asserts that practising rights to the city is performative: only through 'constant community participation and involvement' can we lay claim to the city (Rykwert 2000: 246). Rykwert's proposition, that we can transform reality (incrementally) through activist engagement with it, clearly echoes Lefebvre. However, Andy Merrifield asks: 'what would the right to the city actually look like?' (Merrifield 2011: 471). Is it 'the right to centrality... a right to participate in life at the core' (Merrifield 2011: 471). Or, he wonders, is it ultimately just that 'the right to the city quite simply isn't the right right that needs articulating' (Merrifield 2011: 478)?

While Merrifield's doubt is reasonable, especially considering the questionable continuing relevance of the city as a meaningful figure in the elaboration of 'a new revolutionary conception of citizenship' (which Lefebvre believed would be cause and effect of the right to the city), Merrifield's reading is perhaps too literal (Lefebvre in Merrifield 2011: 470). The key is participation in decision-making, conceptualised as a new kind of citizenship entailing co-production of city and self, alongside power. Nevertheless, Merrifield identifies some significant tensions in Lefebvre's assertion of the right to the city:

> If urbanization is planetary, if the urban – or urban society – is everywhere, is this right to the city the right to the metropolitan region, right to the whole urban agglomeration? Or does it just mean the right to a certain neighborhood, to the city's downtown, the right to centrality?
>
> *(Merrifield 2011: 470)*

Here, Merrifield outlines problems of scale – from neighbourhood to world – but if the right to the city is primarily a right to the centre, then gentrification has all but exhausted the concept. Missing from Merrifield's attempt to define exactly what city and the right to it might be at the scale of planetary urbanism is the market's seemingly inexhaustible capacity for producing space in its own image, which neutralises the concept of the city as centre and citizens as participants in its co-production, while anaesthetising claims to rights over either. However, Lefebvre's conviction that any apparently closed system, including global capitalism, is 'a totality that is in fact decidedly open – so open, indeed, that it must rely on violence to endure', suggests openings onto other possibilities (Lefebvre 1991: 11).

Moving away from the centre – as too consumed, and out of reach, to matter – makes it possible to simultaneously retain the right to the city concept while extending. Abandoning the centre constitutes a form of utopian resistance to gentrification, neoliberal spatial practices, and the spectacular city they produce. Shifting emphasis to the periphery arguably enhances the concept of the right to the city, without doing violence to it, and without neglecting Lefebvre's conviction that the seeds of radical transformation reside in everyday life, coalescing in urban settings.

Consideration of the empowering potential of retreat from the centre is supported by reflection on *In Vespa* (*On My Vespa*), the first chapter of Italian director Nanni Moretti's film *Caro Diario* (*Dear Diary*) (1993), in which the absence of central, touristic, and monumental Rome is crucial. *On My Vespa* (discussed in detail later), I argue, can be read as challenging judgements of a retreat from the centre as inevitably pessimistic or acquiescent, rather than as constituting a radically utopian gesture.

Legal rights to the city?

Documents like *The World Charter on the Right to the City* (Unnamed 2004) following the Social Forum of the Americas and the World Urban Forum, operate negatively on Lefebvre's right to the city concept: no matter how ambitious or well-meaning the precepts of the charter may be, they colonise the right to the city, and, in so doing, neutralise it. Although Lefebvre engaged with the present as the concrete setting of real transformation, he believed that a different social and economic order could only emerge out of alternative modes of production. Accordingly, the conditions under which the right to the city would be operative constituted a utopia for Lefebvre; neither state capitalism nor state socialism are qualified to produce the necessary conditions for fulfilling the right to the city as a moral imperative, rather than as a putatively legal one.

According to Lefebvre, 'Today, the right to the city, fully understood appears as *utopia* (not to say pejoratively, utopist)' (Lefebvre 1996: 196). Notwithstanding its utopian status, Lefebvre argues that achieving the right to the city ought to be a central aim of all plans, projects and programmes, no matter how restrictive the cost might appear (Lefebvre 1996: 196). This is key: the right to the city only seems impossible under the current order but is quickly revealed as possible when horizons shift beyond the limits of what is conceivable. Putting 'the right to the city and some other rights of man and of the citizen [into practice]' would require 'a great increase of social wealth' in tandem with 'profound alterations in social relations themselves (the mode of production)' (Lefebvre 1996: 196). In a utopian statement of the preconditions for making the right to the city operative, Lefebvre continues:

> Such a development supposes an orientation of economic growth which would no longer carry within it its 'finality', and no longer aim at (exponential) accumulation for itself, but would instead serve superior 'ends'.
>
> *(Lefebvre 1996: 196)*

Lefebvre did not require that everything must change before anything could. For example, shifts in perspective, toward serving superior ends, are enough to initiate a process of fundamental transformation. However, increasing social wealth depends on alterations to modes of production and a recalibration of values, which can only take shape outside of the logic of rapacious capitalist accumulation. Despite the homoeostasis of capital accumulation making real change seem little more than a faraway fantasy, Lefebvre asserts that realising the utopian conditions under which the right to the city could be practised is a practical necessity:

> While waiting for something better, one can suppose that the social costs of negation of the right to the city (and of a few others), accepting that we could price them, would be much higher than those of their realization. To estimate the proclamation of the right to the city as more 'realistic' than its abandonment is not a paradox.
>
> *(Lefebvre 1996: 196–97)*

The wait for something better hints at the virtues of seeking salvation away from the corrosive aspects of the centre that conspire against citizenship and social life. On the periphery, one might even have more mental space for thinking about how to subvert apparently intransient conditions, in part because cracks in the system are often more visible where its reach is less decisive. At the very least, systemic contradictions will be in sharper relief away from the centre, if for no other reason than that individual and collective rhythms there are somewhat less dominated by overly rationalised clockwork time.

The conceptual attractiveness of the right to city risks it becoming just another empty slogan. Its appropriation by governments, or even non-governmental organisations (NGOs), carries with it permanent threats to neutralise the revolutionary-utopian import of Lefebvre's phrase.

Systematisation of the right to the city may bring some benefits but alienates it from the spirit of Lefebvre's thought. In this regard, legal scholar Chris Butler (see Butler's chapter in Part 6) observes that:

> there are dangers in mandating the positivist road to the implementation of the right to the city… legislative reforms… require the support of sympathetic governments; in the absence of such support, the codification of the right to the city is likely to be counterproductive to the overall aims of social movements mobilising to challenge existing forms of spatial domination and exclusion.
>
> *(Butler 2012: 148)*

Butler continues by observing that for Lefebvre:

> autogestion [self-organization] is an active political process that is not satisfied by either abstract models of transparency and public interest, or a retreat to pure dialogue between profoundly unequal parties.
>
> *(Butler 2012: 146)*

As such, discourses on rights of the sort governments (including opposition parties) and even NGOs engage in is not what Lefebvre had in mind in asserting the right to the city:

> If… the words and concepts 'city', 'urban', 'space', correspond to a global reality… the right to the city refers to the globality thus aimed at. Certainly, it is not a natural right, nor a contractual one. In the most 'positive' of terms it signifies the right of citizens and city dwellers, and of groups they… constitute, to appear on all the networks and circuits of communication, information and exchange.
>
> *(Lefebvre 1996: 194–95)*

Involvement is key: citizens must be able to contribute to co-producing the urban, on at least as equal a footing as all other participants in the networks that shape social life and its settings. As a global reality, the right to the city encompasses spatial forms and social processes of the urban at all scales, from village to world, and from centre to periphery.

Lefebvre's attention to scale and reach is significant. With the collapse of cities as centres of everyday life, production, and social relations, partly because of expansion beyond comprehensibility but also obscured by encroaching – homogenising – planetary urbanism and the flows of global capital, the social and productive significance of historic city centres diminishes:

> There was a time when city centers were active and productive, and thus belonged to the workers (populaire). In this epoch, moreover, the City (cité) operated primarily through its center. The dislocation of this urban form began in the late 19th century, resulting in the deportation of all that the population considered active and productive into suburbs (banlieues), which were being located ever further away.
>
> *(Lefebvre 2014: 203)*

Although expansion beyond historic centres may have been inevitable – lest cities be choked off by polluting industries, according to Lefebvre – moving outward served other interests as well: cheaper land outside of cities made industrial expansion easier, along with facilitating transport (Lefebvre 2014: 203). Equally, labour is easier for the dominant class to control outside of city centres (where social interaction within and across classes was less restricted, facilitating both the organisation of labour and greater visibility of activism). Population displacement may initially empty city centres but ultimately creates opportunities for gentrification and transformation of the city into a space of exchange and consumption and a centre of decision-making and services (Tabb and Sawers 1978). Such zoning of diversified urban life outwards from the centre has little to do with the usability of the city, which is why Lefebvre asserts that the right to the city:

> depends neither upon an urbanistic ideology, nor upon an architectural intervention, but upon an essential quality or property of urban space: centrality... there is no urban reality without a centre.
>
> *(Lefebvre 1996: 195)*

Even when a vaguely geometric (or historic) centre remains identifiable, cities remade in the image of neoliberal global capitalist spatial practices lack traditional centrality. Moreover, throughout the global periphery (from cities in the hinterlands to inner and outer suburbs), there is very limited space for 'encounter, actual or possible' (Lefebvre 1996: 195). Even where spaces for gathering together exist, many have been colonised by commercial activities or are under constant threat of this.

Lefebvre's preoccupation with the city, and centrality, derives from his identification of traditional cities with pre-modern, pre-capitalist, spatial organisation, which makes them models for a possible future urbanism. Hence his special affection for Florence, as a thriving historic city, with new factories on its edges. Obviously, coherent urban arrangements are more immediately comprehensible to the body and mind, than ever-expanding cities with multiple centres. As such, traditional urbanism – especially Mediterranean – provided Lefebvre with a powerful foundation for his critique of capitalist spatial practices.

In Lefebvre's terms, city branding and 'regeneration' reveal today's apparently successful cities as products (akin to mass-produced consumer goods) rather than as works (akin to works of fine art): largely uniform in character, and reproducible, rather than unique. For Lefebvre, the return of affluence and population brought to the centre by gentrification are symptomatic of the unmaking of cities into spectacular places of programmed consumption; the antithesis of dynamic urban settings of the unexpected.

The tension between Lefebvre's observation of the necessity of centrality for practising the right to the city and the loss of city centres as gathering places outside the domain of spectacle and exchange suggests that 'there is no [longer any] urban reality [because cities now largely exist] without a centre'. He continues, 'It is true, of course, that the city endures, but only as museum and as spectacle', as does urban life (away from the centre), in some form (Lefebvre 2014: 204). According to Lefebvre, cities may be finished, and with them any persisting diversified and productive vitality.

Urban regeneration and the expansion of cities, including the establishment of multiple centres, may seem to put the lie to the assertion that *cities are finished*, but close readings of Lefebvre reveal that the apparent success of urban expansion, and supposed rebirth of city centres, would not have impressed him. While this might confirm Lefebvre's diminishing relevance, arguably, his thinking still challenges urbanists (architects, planners, urban designers, and geographers, amongst others) to make deeper analyses of the intensifying inequalities of cities, mirrored by their homogenisation and transformation into branded products, which may serve the many but largely benefit only the few.

In Lefebvre's view:

> the political benefit for the dominant classes is clear: the gentrification (embourgeoisement) of city centers, the replacement of the earlier productive centrality with a center for decision making and services.
>
> *(Lefebvre 2014: 203)*

Homeostatic aspirations of capital for an eternal hold on power (economic and political) take shape in tandem with the seemingly contradictory absorption of surplus through urban transformation, which entailed repeated bouts of urban restructuring through 'creative destruction' (Harvey 2012: 16). The violence capitalist urbanisation requires to construct 'the new urban world on the wreckage of the old, not only absorbs capital surpluses, but transforms the existing city physically and socially, through displacement, usually of the poor, the underprivileged, and those marginalised from political power (Harvey 2012: 87, 16, 22).

For Lefebvre, '[t]he urban center is not only transformed into a site of consumption; it also becomes an object of consumption, and is valued as such' (Lefebvre 2014: 203). Paradoxically, 'the producers, who had earlier been exported – or more accurately deported – to the suburbs, now return as tourists to the center from which they had been dispossessed and expropriated' (Lefebvre 2014: 203–4). Touristic return to the centre does not, however, amount to recuperating rights to the city or the lost centre. Rather, in revisiting the centre, 'Peripheral populations are today reclaiming urban centers as places of leisure, of empty and unscheduled time' (Lefebvre 2014: 204). Arguably, these new spatial arrangements are manifestations of the 'society of the spectacle' (Debord 1977). As Lefebvre observes: 'In this way the urban phenomenon is profoundly transformed. The historic center has disappeared as such. All that remains are, on the one hand, centers for power and decision making and, on the other, fake and artificial spaces' (Lefebvre 2014: 204).

Ruined (or consumed) centres and bereft (or alienated) peripheries make the emergence of a new form of citizenship and a new urbanism that could render the right to the city operative matters of urgency. Until a new city arises, as a product of citizen self-determination, within which more just and equitable claims can be made and satisfied, the periphery will have to suffice as an outpost of possibility, on the edges of world cities, and in the hinterlands of nations. Opting out of the centre may entail sacrifices but is a most radical gesture that makes it possible to outmanoeuvre the apparent trappings of cosmopolitanism, including participation in the insatiable forces of gentrification. But even at the edges – particularly in provincial cities – capital lays claim to the remnants of civic space and community, in the form of city branding and culture-led regeneration. Although more visible in the periphery, the pace of destruction is slower there. Halting the capture of every space within the 'world interior of capital' depends on the remnants of community joining together in rejecting 'sameness' in favour of 'difference'; a form of resistance so far made possible by the logic of unequal development, even if it seems as though there can be 'no salvation away from the centre' (Lefebvre 1996: 205–8; Sloterdijk 2013). Yet, so long as centralised political economies dominate, few other options remain.

Because we can adapt to even the most unpropitious of social and environmental conditions, demands for transformation arise only when prevailing conditions are perceived as so bad that change is necessary as a matter of survival. Shifts in consciousness provoked in this way enable social dreams of difference. Utopias catalyse this process through the 'education of desire', which cultivates the 'desire for a better way of being' (Levitas 2011: 140–2; 2013: 113–6). Confronted with an obstructing consciousness in paradoxical relation to the utopian vocation of educating desire for a better way of being, description alone is insufficient. Rather, models are required to show how a radical utopian subject might take flesh, to act upon self and city to transform both through action. Moretti's *On My Vespa*, introduced earlier and discussed shortly, is a cinematic mediation on this; outlining different ways of experiencing the city, suggestive of alternative modes of appropriating the urban, while engaging in the sorts of encounters that occur there.

The problem of the visual

Clearly, using a film to support the reading of the right to the city developed here is fraught with problems, even if it proposes how cities could be used differently (see Jones in Part 3 of the book for a different Lefebvrian approach). Foremost is film's complicity in spectacularising life and cities. Because constituted exclusively out of messages to the eye, whatever carnality film depicts will be a simulacrum of actual fleshy existence, or experience. Indeed, for Lefebvre, ocular dominance makes the 'arts of image' handmaidens of alienation:

> by assimilation, or perhaps by simulation, all of social life becomes the mere decipherment of messages by the eyes, the mere reading of texts. Any non-optical impression – a tactile one, for example, or a muscular (rhythmic) one – is no longer anything more than a symbolic form of, or a transitional step towards, the visual. An object felt, tested by the hands, serves merely as an 'analogon' for the object perceived by sight.
>
> *(Lefebvre 1991: 286)*

Lefebvre's insights into capitalist spatial practices as based on exchange (including of images), on spectacularity, and on the consumerist entertainment miracle of city resurgence as primarily simulacra, are particularly revealing:

> That which is merely *seen* is reduced to an image – and to an icy coldness... Inasmuch as the act of seeing and what is seen are confused, both become impotent. By the time this process is complete, space has no social existence independently of an intense, aggressive and repressive visualization. It is thus – not symbolically but in fact – a purely visual space. The rise of the visual realm entails a series of substitutions and displacements by means of which it overwhelms the whole body and usurps its role.
>
> *(Lefebvre 1991: 286)*

In short, the dominance of the eye reproduces the logic of capitalist spatial practices, coldly reducing the lived realm to purely visual spaces that usurp the body, serving up everything as readily exchangeable products. Consequently, Lefebvre was not convinced that film (amongst other visual communication) could 'expose errors concerning space' because 'the image is more likely to secrete it [errors and illusions of space] and reinforce it than to reveal it[:] images fragment; they are themselves fragments of space' (Lefebvre 1991: 286).

Despite his reservations, surely Lefebvre could tolerate the use of a film developed here. Moreover, he was involved in the making of at least one film; '*La Droit à la Ville*', a 26-minute documentary, completed in 1975, for which he provided the text – 'The Other Parises' – that formed its basis (Lefebvre 2003 [1974/1975]: 151–9). In his unique manner, the text oscillates between apparent romantic nostalgia and the most radical of utopian prospects for a socially resurgent city (Coleman 2015). He searches for the city beyond the 'easily available' 'trite Paris', where it might be possible '[t]o wander through a modern city pursuing the 'reveries of a solitary stroller (Lefebvre 2003: 151, 152). Thus, while the critique of regeneration developed in this chapter depends on Lefebvre, so does the use of *On My Vespa* as a model in the following section.

Caro Diario and different centres

In the modern, spectacular city of consumption, wandering constitutes a radical act of resistance, free of purpose in any quantifiable sense. It is in this spirit that Moretti's first chapter of *Dear Diary*, (*On My Vespa*), is considered. *On My Vespa* is the antithesis of most films in which the city of Rome is protagonist, including Federico Fellini's *Roma* (1972), Peter Greenaway's *The Belly of an Architect* (1987), and Paolo Sorentino's later *La Grande Bellezza* (2013). If *Roma* is overheated and chaotic, *The Belly of an Architect* is at times overwhelming in its pretentiousness, and in *La Grande Bellezza*, Rome is a decoration for nocturnal debaucheries, *On My Vespa* is so ordinary as to be radical by comparison.

In gentle and subtle ways, Moretti models an alternative way of being; for the self in the city, with others. His mostly solitary Vespa journey through Rome reveals less about any possible misanthropy than it does about how social change begins with the individual person. As Harvey observes:

> Through changing our world we change ourselves. How, then, can any of us talk about social change without at the same time being prepared, both mentally and physically, to change ourselves? Conversely, how can we change ourselves without changing our world?
>
> *(Harvey 2000: 234–35)*

The unique rhythms of *On My Vespa* recall the spirit of Lefebvre's posthumously published final book, *Rhythmanalysis* (Lefebvre 2004 [1992]). The rhythms observable in the everyday were of great interest to Lefebvre, not least because they constitute a reservoir of resistance to institutionalised time. According to Lefebvre, the rhythmanalytical method and its interdisciplinary practitioners are intertwined, both beginning with the body, its senses and its rhythms. The rhythmanalyst's own body is his or her primary tool (Coleman 2015: 91–123; Lefebvre 2004). Arguably, Moretti enacts Lefebvrian practices of rhythmanalytical method leading to reimaginings of the city, including the periphery.

In no small way, in *On My Vespa* Moretti asks a question similar to one posed earlier (in a Lefebvrian register) by another Italian filmmaker, Pier Paolo Pasolini:

> What is Rome? Which is Rome? Where does Rome start and where does it end...? For the foreigner and visitor, Rome is the city contained within the old Renaissance walls: the rest is vague and anonymous periphery that is not worth seeing.
>
> *(Pasolini 1995: 119)*

Moretti's project is to expand the conventional boundaries of Rome (as conceived by tourists, expats, and privileged Italians), even beyond the extensions of the neo realists. In contrast to them, for Moretti, Rome is one city, or ought to be (not socially fragmented, or impossibly divided by class). That said, Moretti invites us into alternate Romes, in which its native spirit persists in the periphery, far outliving its displacement from the already consumed *Centro*. While Fellini hints at this in *Roma*, his film is animated by unsustainable optimism about the centre's capacity to incorporate all manner of diversity and contradictions. For Moretti, the *Centro Storico* (Rome's historic centre) is relatively unimportant. In *On My Vespa*, intramural locations figure in just a few shots. Ruins are relatively unimportant for Moretti, since the promise of a more civil society lies in the future, even if its roots are in the past and it arises out of the present.

For the self-identified alternatives of Moretti's generation, different Romes needed to be discovered or invented, away from the stomping grounds of the characters in *ROMA*, *The Belly of an Architect*, or *La Grande Bellezza*. Those Romes may still be beautiful, but the average Roman citizen's right to them – to reside in the centre as an active participant – has long been conceded. If Fellini's motorcycle barbarians that roar through Rome at the end of *Roma* supercharges the touristic scooter ride; in William Wyler's aptly titled *Roman Holiday* (1953); Moretti's Vespa journey acknowledges what is lost, that Fellini imagined could endure.

Whereas Greenaway's *The Belly of an Architect* oscillates between detachment and misanthropy and the more intimate portrait of the city, Fellini's *Roma* profits from an artistic pose of intense curiosity, proclaiming the right to observe; to represent but not intervene. Moretti portrays himself as a more fully engaged citizen and political being, who co-produces the city by living it beyond conventional boundaries. Although Moretti is surely an observer of the city, appreciative of its spirit and historical inheritance, he is not primarily an aesthete (Greenaway) or film industry insider (Fellini). His art is surely entertaining, but his self-deprecating presentation discloses a desire to share his curiosity for, and enjoyment of, life in his city, as a coalescence of lived rhythms. Ultimately, Moretti is decidedly active rather than passive in his critical project. He is far more *operative* than *objective* in his criticism of early 1990s complacency and blind ambition, and feels no need to apologise for this, or to supress it, all of which is largely played out in a Rome away from the historic centre, and from the corrosive forces then taking shape in the city, nationally and globally.

Moretti's vision of Rome is the most dialectical of the films introduced above. His radical geography, and conviction that for a right to the city to persist, it must be claimed and enacted, even invented against unpropitious backdrops, places him close to Lefebvre. They also share a clear sense that space and time are inseparable. Spaces surely unfold through time, but for Moretti, the interplay between determined spaces, already given, and the social processes unfolding within them, that they facilitate, might, by way of considered effort, encourage a more civil society to take shape, through transformations suited to the needs of individuals and communities. Spaces of difference can be new or existing, 'perceived', 'conceived', or 'lived' differently (Lefebvre 1991). Moretti is too self-effacing to shout this in his films; his overarching interest in houses and neighbourhoods – in Rome and elsewhere – pronounces it, as does his self-acknowledged envy of those who can do things he cannot, all the while content to watch them excel at it. A charming example of which is his fanboy excitement over Jennifer Beals, star of the eighties film *Flashdance* (1983), who he encounters outside of Rome's city walls.

The subject of Moretti's art is the self, because real change begins with the individual person's (re)shaping of space by moving through it and identifying cracks; other spaces, different centres, and alternative reference points in society and the city, suited to the emergence of different selves and alternative communities. Moretti considers himself an *autarchico*, in the sense of valuing self-sufficiency in the form of as much independence from the state as is possible. He is thus an individualist, but not in isolation, and is not a narcissist. Rather, for him, the state's incompetence makes self-sufficiency the only viable option, but not primarily as an expression of anti-state sentiment. Situating socially engaged individuals as instruments of possible concrete utopias is another explicit point of convergence with Lefebvre.

The urban: place of encounter

Although Lefebvre's conceptualisation of the right to the city is inevitably identified with the centre of Paris (more specifically the Marais), it is worth recollecting that his favourite city was Florence, and that he extolled the virtues of Mediterranean cities generally, particularly in his final book, precisely because he identified their morphology with potentialities for 'being free in the city outside the state' (Lefebvre 1996: 208; Lefebvre and Régulier 2004: 96). For him,

> The *right to the city* cannot be conceived of as a simple visiting right or as a return to traditional cities. it can only be formulated as a transformed and renewed *right to urban life*. It does not matter whether the urban fabric encloses the countryside and what survives of peasant life, as long as the 'urban', place of encounter, priority of use value, inscription in space of a time promoted to the rank of a supreme resource amongst all resources, finds its morphological base and its practico-material realization.
>
> *(Lefebvre 1996: 158)*

While it might seem a small point, the dominant assumptions that Lefebvre's assertion of the right to the city refers to the existing city (whichever one prefers), that it is a legal right, rather than a moral one (see Butler in this part of the book), and that it encompasses only city centres, ultimately obscures what is most radical in his urban thinking: only utopian visions of urban difference – at the scale of traditional cities and exploding urbanism – could redeem the human habitat by emphasising use over exchange, and the concrete over the abstract, in the production of heterogeneous, rather than homogeneous spaces of difference.

Conclusions

In this chapter, the problems of conflating Lefebvre's key idea of the right to the city with a presumed right to the centre (or to 'centrality') have been outlined (Lefebvre 1991; 1996). The process of gentrification makes construing the right to the city with city centres irrelevant. Nevertheless, continuing to associate a right to the city with the centre spatialises the right to the city, at least initially, which secures it from becoming an abstract generalised inalienable right, even if only temporarily.

The unexpected assertion developed in this chapter is that before practising any rights to the city becomes possible, fantasies of the centre must be relinquished, because no matter how resurgent contemporary regenerated spectacular cities may seem, they are largely spaces of exclusion that hinder, rather than facilitate, the dynamics of everyday life. As consumed spaces, and as space of consumption, gentrified cities constitute a distortion of vital social life and project illusions of their centres as loci of participation in the co-production of the city (as setting and process). Even so, Lefebvre alerts us that the apparent totality of the loss of city centres and of our rights is equally illusory, masking just how fragile the dominant system is at any given moment. Because of its vocation for demonstrating how the apparently impossible can become possible, with just a gentle shift of horizons, utopia is superior to all other imaginaries for unmasking illusions of totality, including apparent permanent exclusion from the centre.

A key reason for moving conceptualisations of the right to the city beyond the centre is that expansion of cities far beyond their original boundaries has rendered traditional centres less meaningful, partly because of changing proximities to them in relation to peripheries but also because expanded conurbations with multiple centres stretch the capacity to cognitively map them. This raises a paradox: as cities expand and more residents come to live away from the core, centres lose significance; through transformation into development opportunities for different, more affluent classes, cities are remade into domains of spectacle and exchange (of services and consumer products), and into touristic destinations, including for those who have been excluded. Ultimately, retreat from the Centre is proposed as empowering; utopian, rather than simply acquiescence to prevailing conditions.

References

Butler, C. (2012) *Spatial Politics, Everyday Life and the Right to the City*, Abingdon: Routledge.

Coleman, N. (2015) *Lefebvre for Architects*, Abingdon, Routledge.

Debord G. (1977/1967) *Society of the Spectacle*, Detroit, Michigan: Black and Red.

Fellini, F. (Director) (1972) *Roma* [Film], Ultra Film, Les Productions Artistes Associés.

Greenaway, P. (Director) (1987) *The Belly of an Architect* [Film], Callender Company, Mondial, Tangram Film, Film Four International (in association with), British Screen Productions.

Harvey, D. (2000) *Spaces of Hope*, Edinburgh: Edinburgh University Press.

Harvey, D. (2008) 'The right to the city', *The New Left Review*, 53 23–40.

Harvey, D. (2012) *Rebel Cities: From the Right to the City to the Urban Revolution*, London: Verso.

Lefebvre, H. (1991/1974) *The Production of Space*, Oxford: Wiley-Blackwell.

Lefebvre, H. (1996/1968) 'Right to the city', in E. Kofman and E. Lebas (eds) *Writings on Cities*, Oxford: Blackwell.

Lefebvre, H. (2003/1974–75) 'The other Parises', in S. Elden, E. Kofman and E. Lebas (eds) *Henri Lefebvre: Key Writings*, London: Continuum.

Lefebvre, H. and Régulier, C. (2004/1986) 'Attempt at the rhythmanalysis of Mediterranean cities', in *Rhythmanalysis: Space, Time and Everyday Life*, London: Continuum.

Lefebvre, H. (2004/1992) 'Elements of rhythmanalysis: an introduction to the understanding of rhythms', in Lefebvre, H. (2004) *Rhythmanalysis: Space, Time and Everyday Life*, London: Continuum.

Lefebvre, H. (2014/1989) 'Dissolving city, planetary metamorphosis', *Environment and Planning D: Society and Space*, 32:2 203–5.

Levitas, R. (2011/1990) *The Concept of Utopia*, Oxford: Peter Lang.

Levitas, R. (2013) *Utopia as Method: The Imaginary Reconstitution of Society*, Basingstoke, Hampshire: Palgrave Macmillan.

Lyne, A. (Director) (1983) *Flashdance* [Film], Paramount Pictures, USA, PolyGram Filmed Entertainment.

Merrifield, A. (2011) 'The right to the city and beyond', *City*, 15:3–4 469–76.

Moretti, N. (Director) (1993) *Caro Diario* [Film], 'Capitolo Uno: In Vespa (On My Vespa)', Italy, France: Sacher Film, Banfilm, La Sept Cinéma.

Pasolini P.P. (1995) *Storie della città di dio Racconti e cronache romane 1950–1966*, Torino: Einaudi.

Rykwert, J. (2000) *The Seduction of Place: The City in the Twenty-First Century*, New York: Random House.

Sloterdijk, P. (2013/2005) *In the World Interior of Capital: Towards a Philosophical Theory of Globalization*, Cambridge: Polity.

Sorentino, P. (Director) (2013) *La Grande Bellezza (The Great Beauty)* [Film] Indigo Film, Medusa Film, Babe Film, Pathé, France 2 Cinéma. Mediaset.

Tabb, W.K. and Sawers, L. (eds) (1978) *Marxism and the Metropolis: New Perspectives in Urban Political Economy*, New York: Oxford University Press.

Unnamed (2004) *World Charter on the Right to the City*. Available HTTP: <http://webcache.googleusercontent.com> (accessed 5 March 2017).

Wyler, W. (1953) *Roman Holiday*, USA: Paramount Pictures.

Conclusions
The future-possible

John P. McCarthy and Michael E. Leary-Owhin

> What many people look upon as the conclusion of a well-defined period, as the end of this or that (capitalism, poverty, history, art, etc.), or else as the institution of something new and definitive (an equilibrium, a system, etc.), should really be conceived of solely as a transition.
>
> *(Lefebvre 1991: 408)*

> It may be a possibility which is impossible to realise. But change implies the pursuit of possibilities. What is impossible today becomes possible tomorrow.
>
> *(Lefebvre 1969: 141)*

Introduction

Our understanding of the research outlined in this book leads us to four overriding conclusions encapsulated in the two epigraphs above. These conclusions are as follows. First, in seeking what seems impossible, the worthwhile but difficult-possible will be achieved, though not as a conclusion but rather as a process of transition. Second, in achieving the difficult-possible, the future-impossible seems less daunting. Third, this closing section of the book is not an end point but rather a transition, as indicated by our suggestion for a research agenda. Fourth, the book promotes Lefebvre's city and urban society ideas while necessarily problematising them in terms of the current context and issues. Our conclusions are intended to assist the reader by highlighting the key themes that have become evident throughout the book, emphasising how theoretical interpretations have evolved and been applied. They also underline the specific implications for policy and practice in relation to cities, the urban and land use planning, so that more sensitive and effective ways of engaging with and applying the ideas of Henri Lefebvre may be developed. The conclusions culminate in a new research agenda.

Certainly, there seems no better time in recent decades to reengage with Lefebvre on the global scale. As we started writing these conclusions (April 2018), French police were attempting to clear 'anti-capitalist squatters' from the site of an international airport project at Notre-Dame-des-Landes (Nantes). The site had been occupied for a decade in a bid to stop the state-led construction of the airport. After decades of opposition and prevarication, the government abandoned the plans early in 2018 (Willsher 2018). But the squatters remained, indicating they wished to follow an alternative lifestyle, as a utopian experiment in autonomous living. In fact, the original tents and makeshift homes on the site were replaced by permanent homes, shacks or cabins, and some protestors occupied abandoned farms, making them habitable, and planted abandoned farmland. A de facto informal settlement was created. They also established a *boulangerie*, brewery and vegetable market, as well as an online newspaper and pirate radio station, and they indicated the wish to follow a simpler form of existence based on satisfying basic needs. This occupation is taking place at the same time as the wider resistance in France to the modernisation process initiated by President Macron, with echoes of 1968 arising from the broad coalition including workers and students that has arisen and made its presence felt in a range of protests. Indeed, the struggle at Notre-Dame-des-Landes has become a symbol of resistance against the state in general and the Macron administration in particular. Such protests are also reflected in many other contexts globally.

So the time is right for a review of the ideas of Henri Lefebvre regarding the city, urbanisation and urban society. The struggle faced by the community at Notre-Dame-des-Landes as set out above foregrounds a range of issues examined throughout the chapters of this book. They include for instance the key issue of how initiatives such as this can potentially be co-opted by the state via a process of formalisation and regulation of land use and economic activity. Indeed, the French 'squatters' were offered the chance to stay if they agreed to operate their farming enterprise in a regulated and controlled way according to legal requirements of the state – but most refused since they wished to run the site as a collective and also pursue non-agricultural projects, reflecting a Lefebvrian vision. These activists, like many others appearing in the preceding pages, signal the importance of imagining a different future and making an apparently impossible future-possible. In part, this book facilitates this process.

Themes

The themes that have emerged in this book are clearly complex and intertwined, but the following broad categories may be distinguished in order to allow us to move towards conclusions in relation to implications for research, policy, practice and political action. While navigation through the morass of interlinked and cross-cutting issues is difficult, some cross-cutting themes are set out below.

'Operationalisation 101': challenges to creating a 'concrete utopia'

Many authors in the book clearly set out their view of the rather abstract way in which Lefebvre's ideas are sometimes expressed. This is precisely articulated by Filion's chapter, which shows how Lefebvre's ideas are interpretable in many different ways; he highlights how this leads to a degree of flexibility which has proved to be a positive advantage, since it allows researchers and practitioners to weave their own needs and interpretations into the fabric of Lefebvrian theory. However, this means in turn that his ideas are difficult to pin down in terms of a specific research methodology or a set of ideas for how urban development practitioners can promote better cities. This conundrum is reflected in Liinamaa's chapter, which suggests that Lefebvre's ideas are rather removed from empirical research, and also by de Souza in his chapter which shows how some of Lefebvre's ideas are rather vague and open to contrasting interpretation. That said, Leary-Owhin's chapter (in Part 1) demonstrates the powerful methodology that underpins Lefebvre's research and which could be deployed in the future. Even if the view is taken, as argued repeatedly in the book, that significant political change is needed in order to push forward a Lefebvrian agenda, it is not clear how best local activists and others might actually progress political action and work in a more effective manner, given the limited evident success of circumscribed notions of 'occupation'. This is pursued in the next theme.

So Filion and others show how the interpretability of Lefebvre is both a serious distraction *and* attraction and indeed might be the key source of his popularity. Nevertheless, the lack of any obvious linkage to empiricism, Filion suggests, is problematic, though several chapters show how we can test Lefebvre's ideas. Linked to the empirical challenge is the abstract nature of Lefebvre's concepts, such as rhythmanalysis, discussed below, which leads to problems for those interested in amassing a comprehensive and systematic body of evidence related to his concepts. Equally, it presents problems for those who seek to translate his concepts into policy and practice. These issues are examined below in relation to urban planning policy and practice.

For policy, one way forward of course might be to articulate a discourse for the production of space that is participative and inclusive in the spirit of empowerment. These are concepts familiar to land use planners as articulated for instance in Bailey's chapter, which shows how many have furthered notions of progressive localism in order to show how local democracy can align to aims for a truly inclusive approach. But Bailey also shows how such approaches have been rather tokenistic and lacking in real empowerment, with many activists and actors co-opted by those interests that align directly with the state and/or the private development industry.

Hence, several chapters in the book highlight real difficulties in attempting to apply Lefebvre's theories in any straightforward way. He does not offer a simple 'recipe book' for solving the problems inherent in neoliberal capitalism. Lefebvre's ideas pose several conundrums especially regarding how inhabitants might participate in the production of space, achieve empowerment and *autogestion*. Bailey and others clarify how true empowerment would seem to necessitate the control of the commanding heights of the economic power nexus or a dismantling of those commanding structures, but they also argue that Lefebvre gives us no road map for either route. More problematic still, while many considered notions have developed in recent decades in relation to new modes of participation such as those in relation to 'collaborative planning', Bailey's chapter shows how this has essentially failed to achieve or even point a way to achieving true empowerment. To be specific: how should we use Lefebvre to advise those preparing a neighbourhood plan in England or the equivalent elsewhere? This of course relates directly to what we mean by space and place. It relates also to Lefebvre's notion of differential space and how inhabitants' and communities' resistance can achieve this by appropriation of land and real property in their own interests.

Working with this idea, there is an even greater problem in empowering those who currently have little or no 'voice' in the context of urban development. Mukhopadhyay's chapter identifies how this relates to artists in her case study in India. She explains how her research reveals inhabitants who have no effective voice or means of resistance against the processes of state-led gentrification which produce their environment. This is echoed also in the context of India by Spacek's analysis in his chapter of processes of dispossession and displacement following the Indian state's promotion of infrastructure such as airports. Equally and from a more extreme perspective, the experience of transient and migratory populations features in the chapters for instance by Tsavdaroglou, Murrani and Cassegård. Such populations exhibit a degree of marginalisation and indeed invisibility which renders their plight particularly precarious and compelling, underlined of course by the extreme nature of their suffering to which neoliberal governments sadly contribute in their attempts to assuage the right to the city claims of the privileged and powerful.

'Operationalisation 102': enabling real resistance

Developing the above theme, we can see that the state's attempt to facilitate participation in land use decisions is often inherently trivial, tokenistic and irrelevant. We can also see from Hesketh's chapter that the class struggle is inherent in Lefebvrian philosophy, which implies a direct contradiction with the current neoliberal consensus (or what may be seen as such). Moreover, several chapters such as those by Liu and Wong, Peng and Lyytinen show clearly the pernicious nature of such processes and their effect on cities via elements such as displacement, linked to the privileging of exchange value over use value – issues developed further under the theme of gentrification below.

This is also linked directly to the problematisation and operationalisation of the notion of the 'right to the city' which, as Butler's chapter illustrates, is often interpreted in a limited and circumscribed way as the legal rights to land and property. While of course the contemporary scourge of 'gated communities' and 'control' of public space to eradicate difference and 'the other' is only too evident, Filion and Huchzermeyer show in their chapter how the right to the city implies a fundamental right not only of access but also of the right to produce space itself – linking back to Bailey's assertion of inherently tokenistic 'participation'. This is echoed by Heinickel and Dallach's chapter, which explains how the right to the city is about more than access and residential affordability. De Souza's chapter goes further, claiming that the concept of the right to the city has been banalised and co-opted by the interests of those with power, legitimising the role of the neoliberal state and justifying limited forms of participation. Equally, other interpretations are possible, including the notion in Murrani's chapter of the right to drift, from the viewpoint of migrants to the city. Similarly, Lyytinen's chapter argues for the promotion of rights for groups such as international migrants with disabilities. There is a link here also to the processes of neo-colonisation highlighted in the chapter by Haas which shows how state planning may serve to support this banalisation. In complementary fashion, Torres Garcia's chapter explores the regressive effects of neo-colonialist modernist planning in Spain. He demonstrates how it has clear concrete effects, for instance in lack of space for casual encounter or collective social interaction.

Several authors in the book argue, like Batuman in his chapter, that the 'occupy' strategy has often proved ineffective, with the implication of the need to progress 'non-occupy' protest activism. This is in relation of course to political activity, and many such as Batuman but also for instance Hanson call for real political work to follow readings of Lefebvre. Specifically, Batuman calls for overt political solutions rather than more modest or circumscribed forms of applying protest. Also, in a positive context, Olivier-Didier's chapter shows how negative perceptions of the Bronx in New York gave rise to effective resistance and protest, illustrating the importance of language in motivating and facilitating change and resistance.

Problematising gentrification: a complex pervasive urban process

The cross-cutting, pervasive issue of gentrification or *embourgeoisement* is manifest in the book in, for instance, chapters by Jorge, Ng, Liu and Wong, and Haas. Lefebvre was prescient in his warnings of the potential harm that could result from a middle-class takeover of the central city, and he advocates the encouragement of difference to counter and subvert the homogenising tendencies of abstract space. Ironically, increased difference, albeit superficial, would seem to be produced in the early stages of state-led regeneration via mixed communities. From a Lefebvrian perspective, the main questions are: does this infusion of difference based on a middle-class influx happen at the expense of the right to the city of established residents? And to what extent is there a positive role for enlightened elites in the equitable production of space? At the same time, several chapters document genuinely productive resistance by inhabitants that can and does undermine the production of neoliberal abstract space.

The centrality question: Lefebvre in the suburbs (and beyond)

As Filion's chapter indicates, Lefebvre's ideas focus on the city as centralised by nature, with particular needs for diversity for instance. It also shows how a focus on urban centrality can produce exclusivity, pointing to notions of culture and city image as drivers of urban growth. Rather than benefitting everybody, Ng and Liu and Wong argue in their chapters that it is the privileged elite and urban creative professionals who benefit. In contrast, Marinic's chapter takes Lefebvre's ideas and applies them directly to suburban 'dead malls'. More fundamentally, Granzow and Shields' chapter points to the implications of Lefebvrian-inspired urban agriculture for broader urbanisation debate; this is reflected also in de Souza's chapter, which focuses on Lefebvre-neglected concerns in relation to nature and the broader 'right to the planet'. Liu and Wong in their chapter ask how and if centrality can be instilled in the periphery. Given that the peripheralisation of working-class inhabitants, as observed by Lefebvre, is now well advanced, Coleman's provocation in his chapter regarding the rejection of centrality leads us to conclude that issues of centrality produce no consensus but do suggest that bringing Lefebvre up-to-date is no simple matter.

The sustainable and just city: recognising inequity and resilience

Wiedmann and Salama's chapter highlights Lefebvre's implications for sustainability, linked to diversity, efficiency and identity. This leads in turn to considerations of urban inequalities brought forth by several authors in the book, including for instance Coleman. Certainly, the issue of spatial inequality is one of the most serious threats to the security and stability, in a positive sense, of our cities. Julia Shaw in her chapter stresses the need for notions of spatial justice within our conception of the city. In addition, Erdi's chapter throws light on the issue of spatial justice, and Hilary Shaw's chapter points to the way in which the precariat are treated in terms of the urban context and food health inequality. Ford demonstrates how inequality and spatial injustice extend to the production of air, and Shaw and Ford indicate how Lefebvrian perspectives can be taken in new directions to address issues of contemporary concern. Of course, the broader context of the right to the city is also of critical importance here, as emphasised in the chapters by Butler and Leary-Owhin (Part 1). We conclude therefore that sustainability and spatial inequality debates have been enlivened and focused by the Lefebvrian-inspired interventions.

The 'touristified' city: focusing the gaze

The issue of touristification is linked to gentrification, as shown for instance by Mukhopadhyay's chapter, which illustrates how redevelopment of an artists' colony in India was in part motivated by the need to attract tourists (rather than, as suggested by state interests, to improve amenity for artists themselves). In addition, Coleman's chapter illustrates the implications of Lefebvre's focus in his writings on the experience of cities that have a clear appeal to tourists, linking again to the broader right to the city, in terms of rights for whom? Other chapters also link to issues of touristification, including for instance those ofYung – in relation to heritage and tourism – and Coleman – in terms of the cinematic city. So while the book reveals some of the benefits that a touristic influx can bring, it also points to the potential threats to certain rights to the city linked to such an influx.

Autogestion as self-build

The notion of individual and collective empowerment over space production for instance via self-build is a key element in Lefebvre's relevance for contemporary cities and urbanisation. This is referred to in several chapters; for instance, Viegas' chapter shows how the state has the potential to subsidise and encourage self-build, though in practice this may fail to ensure adequate services. It nevertheless shows the potential for co-production involving the state, which can include self-build settlements, not just one-off houses. In addition, Cassegård's chapter points to the need for empowerment of the homeless by allowing them to make use of 'counter-space' via the production of differential space which, while transgressive, meets their needs for shelter. Thompson's chapter is instructive in its illumination of the potential for cooperative design for housing, which can stimulate counter-intuitive design solutions such as cul-de-sacs, and this argument can be extended to cover for instance co-construction in a broader sense, including self-build. Perhaps the ultimate expression of this globally is the experience of informal settlements such as in Brazilian and South African cities as illustrated in the chapter by Huchzermeyer, illustrating what she calls effective urbanism.

Visualising rhythmanalysis in the city

The notion of rhythmanalysis is considered in several chapters in this book. For instance, Paiva's chapter highlights the need to take into account the stability of natural rhythms. Many authors also assert that the concept can offer specific pathways for researchers and practitioners in relation to urban development and use. Hence Lee's chapter asserts – from an urban anthropological viewpoint – that this concept might offer a future-possible way of applying anthropological ideas in the urban realm by identifying the lived practice of everyday rhythms. In addition, Revol's chapter points to the linkage of this concept to notions of the architecture of enjoyment and an experimental utopia, though it acknowledges the lack of clear explanation or articulation of this concept. Nevertheless, she identifies the methods of psychogeography – 'walking without purpose' – allowing for casual and chance encounter as practised by the 'flaneur', as a means by which a form of rhythmanalysis might be considered in practice. She also identifies the role of lived experience as an artistic activity which might be linked to this concept, perhaps via music and dance. While each practitioner must develop their own method of application, these will of course be subject to the trials of implementation and to critique from practice and from the academy.

Heinickel and Dallach's chapter offers hope that Lefebvre can indeed be operationalised, for instance in terms of encouraging creativity, innovation and encounter in the use of public space. They also focus on rhythmanalysis, asserting for example that this has been taken up by many practitioner urbanists for instance via space-time design approaches. They set out a proposal for a methodology model in this context, but temper this with a series of acknowledgements relating to the limits to application in practice of Lefebvre's ideas by planners. For instance, they question the privileging of hypermobility, virtualisation and efficiency in the use of space-time, asserting that this erodes the authenticity of place, with the corresponding need for personal and individual experiences and autonomy to be prioritised in the use of space and mobility.

The digitised city: democracy and demonisation

The widespread use of social media and digitisation of the city has clear implications for urban futures. In this context, Julia Shaw's chapter draws attention to networked space and spatial-ethical values, asking if a consensus on these might be possible. This links to the ideas set out in Diaz Cardona's chapter in its consideration of the role of writing in Lefebvre, linking to the use of data digitisation and writing or discourse as socio-spatial infrastructure. In addition, Heinickel and Dallach's chapter considers the negative implications of hypermobility and efficiency as enabled by technology more generally, which has implications in this context linking in particular to transport and housing as well as other aspects of urban land use. Of course, this may also have implications for participatory methods with increasing attempts in many contexts to digitise democracy.

The festival city

The issue of city as festival is a vital concept in Lefebvre's understanding of the city under capitalism. It emerges in Filion's chapter; this shows how Lefebvre prioritised the nature of the city as enabling creativity and fulfilment by means of a festival function which also enables a high degree of social interaction linked to a high degree of diversification of uses, and access for inhabitants to a range of amenities. Non-work time and space are also powerful antidotes to alienation. Borden's and Alasdair Jones' chapters for instance highlight the need for city space for leisure, and Ellison's chapter refers to the need for transient uses which offer elements of carnival and forms of protest. The issue of play more broadly is a particular concern in Alasdair Jones' chapter, with significant implications for cities that offer the possibility for transcendence via exploitation of fissures in the planned order of abstract space, but the chapter also grounds his suggestions in the assumption of the need to embed such considerations in everyday life. The narrower issue of public art and its emancipatory potential are referred to in several chapters including that by Queirós, and the potential linkage to participation via use of culture in a more general sense is also made evident. We conclude more broadly that there are also concerns here of the dangers of the city as theme park, linked to processes of 'Disneyfication', which also links to the right to the city, again in terms of rights for whom?

What's my identity? City branding and the local

Mukhopadhyay's chapter sets out clearly how issues of identity underpinned the resistance of artists to the refurbishment of their colony since new buildings lacked linkage with the histories, customs and practice of the artists, for instance by preventing them from following their previous practices of drying out the artworks in the sun. More broadly, it shows how the new urban space in this case also failed to link to the historical and cultural identity of the artists based on their ancestral roots in the area, leading to a sense of detachment stemming from evident failure to reflect their perceived narratives of space, place and identity. Identity formation and instability figures in different ways in the lives of migrants in Murrani's chapter and is important by implication in the chapter by Tsavdaroglou. These issues are also reflected in Yung's chapter, which considers the lack of appreciation of the importance of culture as a tool to address the needs of residents in view of redevelopment possibilities, linked in this case to heritage. We see a role here for urban anthropology as suggested by Lee's chapter. The issue of identity is also taken up in Batuman's chapter in the context of the symbolic importance for local communities of Gezi

Park, Istanbul and the Presidential Compound in Ankara. Identity and the production of space is thus identified as an emerging theme in the book.

The planned city: scourge or utopian vision?

Lefebvre often contrasts concrete everyday utopia with planned utopia. Following from the issue of how Lefebvre's ideas are and can be operationalised, discussed above, an important theme arises from consideration of the extent to which Lefebvre's ideas can be translated into policy/practice specifically by urban planners, architects and other urbanists who seek to improve cities. This is difficult conceptually given Lefebvre's critiques of planning in particular. Certainly, many authors in the book set out a vision of what such urbanists might seek to achieve in terms of urban outcomes. Filion's chapter, for instance, points to Levebre's vision for a multi-functional city (for land use) which is diversified, with good quality amenities, though he, like other authors, highlights Lefebvre's distrust of regulation and control that leads to alienation. The idea of use value is helpful here since it implies priority for land uses related to need and everyday life, not demand and the requirements of investors. The outcome of diversity is also highlighted in Spacek's chapter, which points to Lefebvre's notion of cities as becoming homogenised to the detriment of the experience of their residents. Moreover, Marinic's chapter highlights the need for a 'heterotopia', again prioritising diversity of use, and, in the context of mobility, Scott's chapter illustrates the anti-car role for planning.

But this of course leads to the question: how to achieve such outcomes? Indeed, there are hints here in a range of chapters. Marinic for instance offers the specific case of a kind of *autogestion* 'regeneration' of 'dead malls' which have encompassed uses such as pop-up shops, flea markets, carnivals and independent stores. We consider that planners and others have a role in enabling such diversified uses by appropriate (possibly relaxed) policies and even active interventions. There is indeed much history and practice of 'planning for informality' by positively encouraging street markets, pop-ups and other such uses. Indeed, Borden and Alasdair Jones show in their chapters how cities such as Montreal and London respectively encourage skateboarding and informality – normally seen as transgressive. Several chapters discuss temporary and informal uses in cities, showing how inhabitants often make use of liminal spaces hitherto ignored by planners and architects, as highlighted for instance in Huchzermeyer's chapter. In addition, Julia Shaw's chapter shows how planners might encourage positive elements of public space by appropriate consideration of design and materials. But issues remain of the equitable spatial distribution of such interventions.

Certainly, the book illustrates how well-meaning planning goes wrong. Torres-Garcia's chapter, for instance, illustrates the inherent shortcomings of the modernist-style planning of colonial towns in Spain, based upon top-down mechanistic notions privileging the family unit and largely ignoring the need for collective space and space for casual encounter. Leary-Owhin's chapter (Part 6) underlines Lefebvre's concerns about the harmful impacts on everyday life of the new town concept and its implementation. Equally, however, it is clear that formal planning can encourage urban innovation and creativity, as shown for instance in Thompson's chapter focusing on the important potential role for citizen participation in an organically evolving city, including elements such as guerrilla gardening. This is also shown in Borden's chapter, which highlights the role for planners in enabling and encouraging skateboarding, and in Ellison's chapter via consideration of (flexi-)space creation to facilitate encounter and cross-fertilisation of ideas. Overall, these examples demonstrate the need for strategic and tactical approaches to encouraging transformative spaces of empowerment and enchantment. This is far from the centralised control and regulation of uniformity and homogeneity that Wiedmann and Salama's chapter associates with the modernist approaches to urban design of Le Corbusier, for instance. We agree with Granzow and Shields and others on the dangers of prioritising tidiness in the urban realm, drawn from overly neat representations of space, with implications of the need for a more innovative visioning approach by planners and urbanists, as opposed to what Marescotti's chapter associates with an essentially deadening administrative role.

A new research agenda

Having considered the themes evident across the book, we can now turn to consider a research agenda, linking back also to the preliminary concluding thoughts posed in the introductions to the book's constitutive parts. One issue that arises from the discussion above is the need espoused by many authors to operationalise the ideas of Lefebvre in terms of enabling the testing of his ideas via a methodology. This in turn may enable operationalisation of his ideas via refinement of policy and practice implications (returned to below). Again, this is difficult in view of (as Filion, de Souza and Liinamaa's chapters highlight) the rather philosophically infused nature of some of Lefebvre's writings, and his resistance to plotting an explicit route for empirical investigation. Nevertheless, authors in this book consistently demonstrate the contemporary relevance of his ideas, their motivational potential as a means to engender interest in cities and the urban and their capacity to be adapted and shaped into research questions and methods. Hence several authors link his ideas to such disparate issues as: the use of public space by migrants (Murrani, Tsavdaroglou); use by homeless people (Cassegård); use by people with disabilities (Lyytinen); use by skateboarders (Borden); implications for interior space (Ellison); implications for sub-urban malls (Marinic); implications for informal settlements (Huchzermeyer); implications for urban mobility including cycling (Scott); and implications for temporary and transgressive uses (many chapters). Linked to such implications, we suggest the use of Lefebvre's ideas in relation to wider issues, for instance in terms of surveillance and control of public space in the light of enhanced security issues in many cities.

A point of entry into thinking about future research potential is the critical importance of the right to the city concept and its implications for research. This emerged as the key cross-cutting notion in the book, notwithstanding Parts 5 and 6, as interpreted for instance in chapters by Butler, Coleman, Erdi; Heinickel and Wallach; and Granzow and Shields, as well as others. This has particular relevance for a progressive interpretation and application of participative (as opposed to coercive) approaches to urban development and use (as set out in Butler's chapter). Rather than abandon the right to the city concept, we see it as a critical starting point for a new research agenda. Research that encourages a contemporary theoretical reworking of right to the city is needed. This should be backed up by research focused on how the idea has been interpreted and adapted in different global contexts and by different stakeholders. It should link to questions such as: whose interests are served by the different interpretations that percolate through myriad research publications?

Further, of course, the idea of the right to the city also has clear implications for the way in which urban public space is produced, used and managed. For instance, further research is needed to understand the state's willingness to respond to claims for certain rights to the city from middle-class groups, but its unwillingness to respond to claims from other more marginal groups. Thus business improvement districts present a contemporary phenomenon in many cities which many argue is serving to substitute private subsidy and control for formerly public funding and control (often in central office districts but increasingly encompassing other areas and uses). The stakeholders promoting such districts would seem to represent the relatively wealthy and powerful, but the need for representation of others is set out for instance in the chapters by Haas, Lyytinen and Tsavdaroglou.

Security and surveillance are not issues with which Lefebvre was too concerned. These are nevertheless strong themes of neoliberal urbanism and there is certainly a need for research to engage directly with security issues in the contemporary context of cities globally. What methods might we adapt via a Lefebvrian lens to address the increasing perceived need for overt and covert security in cities such as London, with media reporting constantly highlighting street crime often related (so it is suggested) to so-called youth gangs? Equally, research might usefully build upon Borden's consideration of transgressive uses such as skateboarding and the implications for the privatisation and management of public space. This might also consider mobility within cities such as via cycling, as highlighted by Scott. This might be broadened out to consider potential new modes of transport in cities such as self-drive vehicles, drones and electric scooters. In addition, the broader and global issues of displacement and unequal rights – addressed by many authors in the book including Jorge and Ng – could prove useful foci for Lefebvrian analysis.

So there are many issues to be explored. But this leads in turn to the question of what specific methods might be adapted from Lefebvre's ideas, for research? Following Leary-Owhin's chapter in Part 1, the time is right for research that reclaims Lefebvre as an empirical researcher and takes seriously his regressive-progressive method. Related to this, we recommend more research derived from quantitative approaches to match the qualitative research, mainly cross-sectional research, important though it is, as represented in this book. Quantitative research, for example that which provides robust measurement and analysis of the concrete and theoretical impacts of the right to the city, is necessary. Longitudinal, comprehensive, systematic case study research is needed to complement and bolster the small-sample case study approach of the recent past. This would re-balance Lefebvrian scholarship to reflect his acknowledgement of the power of quantitative and historical research. An obvious arena for this should centre on the urgent need for city/urban planning, the discipline, policy and practice which most directly contributes to the production of space. We suggest that this implies the need for research that addresses the vexed question of whether and how formal planning can contribute to the transition towards a more just urban society or whether this can be achieved despite planners. As Marescotti makes clear in his chapter, there is a need for research that is able to analyse the effects of urban planning to inform practice via evidence-based empirical research. While his advocacy centres on quantitative approaches, there are nevertheless palpable implications of the need for clarity in and development of all relevant techniques and methods, something advocated by Lefebvre. Some authors in the book, such as Heinickel and Dallach, propose methods and models in this context, related for instance to rhythmanalysis, and Lee's chapter points to the need to develop Lefebvrian-inspired research methods specifically for urban anthropology.

So future research should seek to enhance and develop such methods in relation to the contemporary themes set out above. This is hampered somewhat by the perception by some observers of Lefebvre's ideas, at least in terms of empirical research, as rather obscure and lacking in detailed application. But this is not the whole reality, as shown conclusively by many authors in this book. What is needed then is research and research dissemination of the relevance of Lefebvre for those interested in cities and the urban. Future research across a range of disciplines should allow a dialogue on how to apply his ideas to cross-cutting research questions. The potential is clear: what is needed is research that encourages wider interest in Lefebvre in terms of empirical analysis to match the huge interest in his right to the city ideas of inhabitants' access to – and genuine participation in – the production of city space and services.

Furthermore, research is required that considers how Lefebvre's ideas should be applied to newly emerging fields of inquiry, particularly with a view to explanation and advocacy in relation to the more inclusive use of space. In this context, the chapters by Nick Jones and Coleman offer clues that point to the potential for cinematic research from a Lefebvrian perspective. This might focus on the progressive and transgressive use of liminal space, and it should be extended to include other modes of artistic expression including television, literature, social media and computer gaming. This should in turn provide clear pointers for practice in view of the current focus of many on techniques for public involvement in land-use decisions involving art and culture in various forms. This would allow for instance techniques derived from rhythmanalysis to explore how inhabitants use space through time, and also how conflictual uses of space, for instance via potentially transgressive activities, might be managed, take on board Lefebvre's advocacy for space as encounter, facilitated in ways that produce positive outcomes which enhance everyday lived space.

The future potential for rhythmanalysis research should be seen as important. For example, Lehtovuori, Tartia and Cerrone's chapter demonstrates the need for research of the everyday lived experience of city inhabitants, particularly in terms of their interaction with ephemeral spaces, often ignored by planners, other urban professionals and decision-makers. In addition, the centrality of notions of spatial justice should be springboards for Lefebvrian-inspired research. This should be furthered via consideration of the visualisation of urban futures in genres such as science fiction for instance, through the common trope of the central city as a prison and a consequent 'turning-over' of spaces of privilege and desire. This links to broader issues of spatial justice since it is the logical extension of current trends in many cities (related for instance to gated communities as well as increasing spatial inequalities). Also of relevance here is the use of art in the public realm, as demonstrated in Queirós' chapter in the context of Lisbon, where public art has a privileged and powerful position in relation to the city's image and identity. This is also reflected by Liinamaa's chapter, which considers aesthetics and the emancipatory potential of art in the public realm. These are relatively new fields for Lefebvrian research that should be explored further.

Of course, all these ideas are linked to education and potential pedagogical research within urbanist and other relevant disciplines. But – linked to the exploration of practice implications below – this is more problematic within professional disciplines which inevitably centre on the specific empirical elements that have proved elusive within the Lefebvrian canon. Hence pedagogical interdisciplinary research in such fields as urban planning, architecture, geography, anthropology and sociology could usefully assess the extent of application of Lefebvre's ideas, potentially in different global contexts. It is instructive that a recent review of urban planning teaching (Alterman 2017) suggests that of the 300 programmes in this context in China, none teach planning theory, albeit circumscribed in the context of conceptual analysis based on procedure (see for instance Healey 1997) and thereby distinct from broader urban theory within which the writings of Lefebvre would seem to be located more appropriately. Fundamental research into the contemporary applicability of Lefebvre's ideas for interdisciplinary approaches in the main disciplines, that helps deliver the production of space, is essential. That said, robust inductive, empirical research should lead to implications for theoretical development. This, we would stress, is a pivotal element of Lefebvre's regressive-progressive model.

But this leads to the question: to what extent do courses in the key professional disciplines address Lefebvre at all? Research is needed to answer this important question. This again links to questions of how his ideas might be more effectively communicated, perhaps via media-related research and scholarship centring on how Lefebvre's ideas are filtered in mainstream media and the implications of this for research and practice. Equally, of course, such research might examine the ways in which Lefebvrian ideas might be taught and integrated within professional disciplines, for instance via studio-based project work in architecture and design programmes, and via reflective analysis and writing. Just as important is the development of research that interrogates how and the extent to which university and other researchers should take seriously an active and radical engagement with local communities seeking spatial justice along Lefebvrian lines. There is a long history of public engagement practice within many teaching institutions, but this is often based on rather traditional ideas of public participation; a Lefebvrian dynamic would therefore add power and robustness.

Lefebvre's intellectual journey was made possible by his engagement with his contemporaries and with historical philosophers and theorists. So too in this book we reveal how Lefebvrian scholars draw regularly on other theorists to inform their work, though recognising that this can pose significant challenges. In the spirit of Soja's (1996 and 2010) (not uncontested) multi-theoretical approaches, new research is necessary that attempts to weave Lefebvre's ideas into current related theoretical approaches that we see in the book regarding, for example, automobility (Scott) and creativity (Murrani). We feel that research is also needed that compares Lefebvrian approaches to the city and the urban with his (near) contemporaries such as: Pierre Bourdieu, Christine Boyer, Gordon Cullen, Michel de Certeau, Michel Foucault, Dolores Hayden, Kevin Lynch, Gayatri Spivak and of course Le Corbusier. It is salient that space matters not just to geographers and planners but to a host of other related disciplines, and future research that explores the Lefebvrian implications for spatial justice would provide fitting continuity to some of Lefebvre's seminal work.

Policy, practice and activism implications

The strong philosophical and Marxian grounding of Lefebvre's city and urban society ideas, and the lack of focus in much of the literature on his empirical research, poses problems for urban professionals who tend to seek straightforward, immediate solutions to urban problems. The book provides ample evidence that a Lefebvrian perspective has and should prove beneficial when applied to real-world problems, issues and opportunities. Certainly, as Heinickel and Dallach's chapter shows, concepts such as rhythmanalysis have been applied by architects and designers in terms of space-time design. And there is increasing interest in the notion of place management, as opposed to place-making or even place-shaping. So the way in which public space is imagined, conceived, used and maintained in cities is now an important issue for many policymakers, including those concerned with initial design of space and organisation of land use. There are policy and practice implications also for specific ludic uses such as play, interpreted broadly, as set out for instance in the chapters by Alasdair Jones and Ng.

In this context, however, we identify a potential awareness gap: how many practitioners and policymakers are even aware of Lefebvre? This is an uncomfortable question for those embedded in Lefebvrian scholarship, no doubt resulting in part from professional education as considered above. Ellison's chapter for instance illustrates the lack of engagement with Lefebvre in any area of the facilities management field, though he illustrates the potential for this via the use of interior knowledge workspace, which is increasingly flexible and agile in orientation with implications for users' lived experience and the appropriation of interior space. This is corroborated by much contemporary research which highlights the critical importance of user participation and control over individual workspace, often contrary to the exhortation for, and practice of, 'top-down' approaches. Indeed, such research illustrates that even enlightened 'top-down' approaches are unlikely to be effective in view of the central importance of the process of user control over space design (Harford 2016). So perhaps what is needed is consciousness-raising on the applicability of Lefebvre's ideas for urban professionals and urbanists. Even if this falls short, perhaps appropriately, of advocating all-embracing solutions, it will serve nevertheless as a motivating force for researchers and policymakers and their advisors to explore Lefebvre's ideas and consider their implications. We recommend research that discovers how this might be facilitated by social media.

Building on this, we distinguish particular elements of participative (and inclusive) practice which could usefully be served by a Lefebvrian lens. This is of particular note since it is a cross-cutting concern in all urbanist professions, especially architecture, housing planning and tourism. Moreover, the need for practitioners with knowledge and expertise in this area would seem extremely valuable now. Innovative research exploring the use of art and culture in public engagement using Lefebvre's concepts should be applied here as indicated above. The book reveals specific techniques already applied in this context, including for instance *charrettes*, as reflected in the arguments in Thompson's chapter. Research can inform practice and policy as to how residents' everyday lived experiences produce desires for features such as communal facilities, which appear counter-intuitive in the context of calls for privatised, securitised space and individual autonomy and control.

The value of autonomy over space production is alluded to also in Huchzermeyer's chapter, which stresses the value of spontaneous development for instance via informal housing as in many Brazilian cities, sometimes pejoratively perceived as illegal and unplanned but increasingly seen as reflecting dynamic and effective urbanism. We should also not forget the importance of even minor or small-scale elements in the public realm since these are important particularly for the disadvantaged, as shown by Queirós in her chapter. Negative perceptions, we argue, need not prevail, if urban professionals applied a truly inclusive approach, based on existing research findings, taking on board the needs of not just urban elites but also more marginalised inhabitants of the urban realm.

The book illustrates how practices of participation in the production of space involve several components such as the possibility of self-build, not just of individual houses but also settlements, and this is an area of concern for urbanists across the Global North and South. It is an area of current policy relevance not only in terms of planning procedures but also in terms of the provision of genuinely affordable housing. This is most evident in the Global South where many areas of informal housing have been developed, with obvious resonance with Lefebvrian ideas, but with – as shown by several authors such as Erdi – limited understanding on the part of local and central government authorities, who remain wedded to the eradication of informal settlements using a rationale based on an insanitary and unsafe discourse. Research is needed which explores the range of values that self-build and informal settlements, linked to wider notions of identity, can confer. Another important area for future research is the use value of fragmentary, informal, ephemeral or liminal space, which can be characterised, in Lefebvrian terms, as counter-space. Awareness of the importance of counter-space, indeed since the ideas of Jane Jacobs in the 1960s, has not overcome planners' presumptions in favour of functional zoning (Harford 2016). Explanatory research which acknowledges the importance of use value, the informal and ephemeral is needed to unpack the apparent conflicts with 'tidy' notions of space as being the desired outcome.

Thompson, Granzow and Shields and Hilary Shaw's considerations (in their respective chapters) of 'guerrilla gardening' are also instructive here, and link to broader notions of tactical urbanism (as outlined in Marescotti's chapter). These research findings have evident implications for practice, with positive and negative aspects, in cities throughout the Global North and South. There are implications too for policy and practice to appreciate more the potential for 'pop-up' and temporary uses which can cater to diverse and unmet needs. Research is required that considers linkages between spatial justice, gentrification and the changing nature of urban neighbourhoods which in some cases is inhabitant-led. A better understanding of the conflicts and unintended consequences that arise from well-intentioned policy interventions and how to resolve or mitigate them is needed. Scott's research is a useful fillip to this research direction.

Concerns for the production of heterogeneous use value space also have implications for notions of the city that emphasise play and performance, which can further the aims for diversity of uses of urban land and space. We conclude that there are policy and practice implications for how cities might usefully maximise the value of fissures or gaps within abstract space and therefore within mainstream functions and uses. Certainly, the everyday social use of space is a central theme in the book, with obvious connotations for potential future research and practice. This is illustrated directly for instance in Hilary Shaw's chapter in the context of urban agriculture, which shows the importance of space that allows encounter, movement, spontaneity and unpredictability.

In addition, as Wiedmann's chapter suggests, Lefebvre's ideas have the potential to address sustainability as broadly defined, with implications for more general elements of policy and practice for cities. He identifies here core urban qualities incorporating diversity, efficiency and identity. Diversity relates most directly to land use with the implication of a need for a shift from modernist approaches to those that might encourage what Marinic in his chapter calls 'heterotopia', involving a mistrust of uniformity and homogeneity. Efficiency relates to mobility with the (erroneous) assumption that more efficient space-time design is better, including in relation to leisure; in this respect Marinic's arguments are endorsed by Heinickel and Wallach. Identity relates to the need to nurture and respect individual conceptions of the self and their implications for cities, as endorsed by Yung for instance in her chapter. Together, these arguments would seem to point to the relevance of Lefebvre's ideas on a broader policy canvas, potentially linking to the UN Sustainable Development Goals and consequently to even broader issues of addressing climate change. Certainly, Wiedmann's interpretation highlights the critical importance of urban inequalities (again as key within Lefebvre's notions, and highlighted in Coleman's chapter) as a key element of broader notions of sustainability and the need for urban resilience as opposed to fragility.

In addition, of course, the increasing currency in the UK and elsewhere for arguments in relation to a reinterpretation of land value would seem to relate directly to Lefebvre's notions of exchange value as opposed to use value (as set out in many chapters, including for instance those by Erdi and Coleman). While many call for reform of compulsory purchase and similar legal mechanisms which require the public sector to acquire land at its development value, Lefebvrian-inspired research can provide a rationale for more progressive systems which reflect how cities are actually used by all inhabitants, including public or private sector tenants, children, the elderly and the homeless, rather than prioritising the narrow financial interests of land and property owners.

And moving outside the realm of urbanist policy and practice, it is clear from the chapters for instance by Yung, Mukhopadhyay and Coleman that activities such as tourism management and heritage protection can also benefit from the application of Lefebvrian ideas. This links to the management of conflict over city use, as illustrated for instance by the unprecedented actions being taken by many city authorities to limit tourist numbers, linked to chapters such as those by Coleman and Yung. The cases of Venice and Barcelona are clear examples here, but other cities such as Paris and Amsterdam are progressing policies to address the proliferation of short-term letting via online platforms in view of the frequent loss of amenity and gentrification effects as outlined above in the context of research implications. We might also look to the changing form of cities and how this has implications for practice – for instance Ford highlights the importance of the spaces above buildings in relation to 'air rights' for development. This is particularly pertinent with the contemporary focus in London on high-rise build, and an increased focus globally on 'transit-oriented development' often implying high densities above public transport nodes. So issues of transport and mobility are also relevant here, and sustainable modes of transport and relevant policy to encourage these might arguably be addressed via the prism of Lefebvrian research.

The good news for activists and inhabitants is that neo-liberal capitalism is not an impregnable fortress. Its cracks and inherent contradictions allow for and indeed require successful interventions that challenge the formalised production of space, which goes on without and despite them, however success is measured. A salutary lesson is that the road to successful appropriation of space is neither short or straight nor smooth. Activists and their supporters should understand that spatial struggle, especially agonism, is a long-term endeavour. Difficult-possible gains do not come easily, and the urban society is still a long way off.

Successful activism and tactical urbanism would benefit from research which can demonstrate its potential capacity beyond local resistance to formal planning and urban development proposals, though these do bring tangible benefits. While activism appears to deny the role of formal professional procedural urban planning, it should be bolstered by research that helps planners and policymakers appreciate its value, rather than (paradoxically) serving to reinforce neoliberal logic. This is because at present tactical urbanism tends to be perceived as limited to the local, with little prospect of scaling-up. An important implication of the book suggests, counter-intuitively, that urban planning has a valuable role in promoting an essentially 'anti-[traditional] planning' or 'anti-urban design' approach, as distinct perhaps from 'anti-development' or 'anti-regeneration'. While concerns about planning's role in activism might seem peripheral in the present global context of capitalist urbanisation, professional resistance to the hegemonic global rhetoric of planning's role being predominantly about promoting and achieving economic growth, city image branding and marketing, is important here.

Consequently, many would argue that Lefebvre's ideas might assist mobilisation in favour of a 'pro-planning' agenda which serves the needs of the underprivileged and dispossessed. This may well be a kind of contemporary Lefebvrian advocacy planning. The chapters in this book give insights into how this might be possible, if only we can ensure that urban professionals such as planners avoid outdated and insensitive technocratic 'fixes' and over-tidy or functional approaches, in historical terms associated with modernism and a focus on administrative control and pragmatism. This might of course involve an acceptance of the need to relinquish elements of regulation and control perhaps in conjunction with a broader visioning approach concentrating on potential outcomes rather than processes, and applying truly democratic, bottom-up and empowering methods. This might facilitate rather than preventing appropriate activities and uses in relation for instance to cycling, skateboarding and public art.

Consequently, this book offers a degree of cautious utopian optimism in contrast to an unhelpful nihilistic rejection of all rationality or a naïve modernist utopianism. But Lefebvrian notions might be harnessed further to promote such an enlightened urbanist approach. There are implications here for how tactical interventions can prioritise for instance spaces which cater for young people, with a transformative potential that has implications for other dimensions of the urban problematic such as the co-optation of Lefebvre's ideas by privileged elites. Plainly, therefore, as many authors in the book indicate, urban policy and planning practice need not continue to privilege uniformity, homogenisation and fragmentation but can instead promote diversity, flexibility and inclusiveness. We do not claim the book provides neat answers to such intractable problems. But we feel it has achieved its major objectives, in showcasing exciting research, enlivening the debate, providing some surprises and challenges and giving directions and support for future practice and research.

References

Alterman, R. (2017) 'From a minor to major profession: can planning and planning theory meet the challenge of globalisation?', *Transactions of the Association of European Schools of Planning*, 1, 1–17.
Harford, T. (2016) *Messy: How to Be Creative and Resilient in a Tidy-Minded World*, London: Little, Brown.
Healey, P. (1997) *Collaborative Planning: Shaping Places in Fragmented Societies*, Vancouver: UBC Press.
Soja, E. (1996) *Thirdspace: Journeys to Los Angeles and Other Real-and-Imagined Places*, Oxford: Blackwell.
Soja, E. (2010) *Seeking Spatial Justice*, Minneapolis, MN: University of Minnesota Press.
Willsher, K. (2018) 'France abandons plan for €580m airport and orders squatters off site', *The Guardian*, 17 January.

Index